THE HUMAN BODY
Concepts of Anatomy and Physiology

BRUCE D. WINGERD

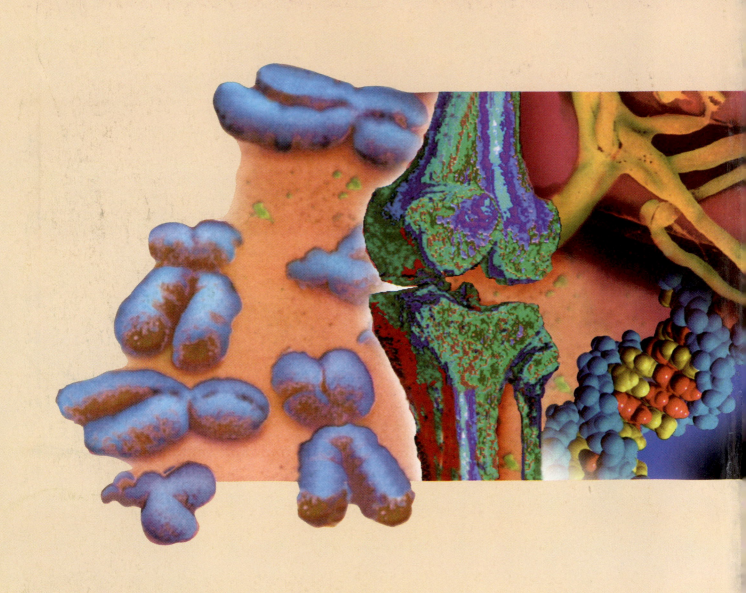

THE HUMAN BODY
Concepts of Anatomy and Physiology

BRUCE D. WINGERD

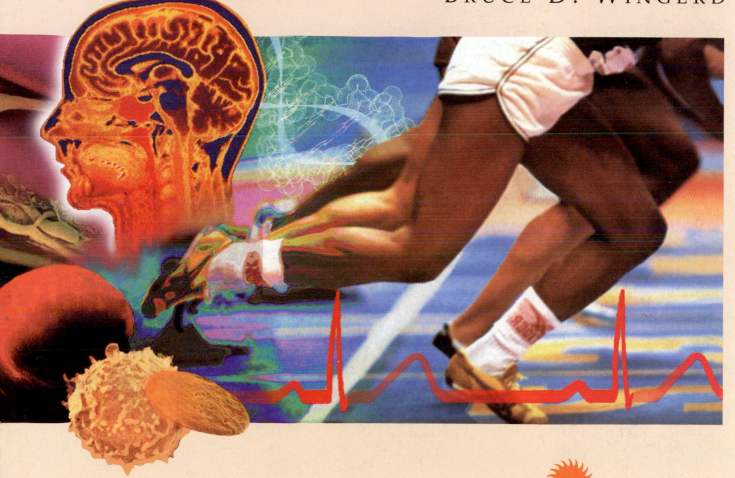

SAUNDERS COLLEGE PUBLISHING
Harcourt Brace College Publishers

Fort Worth • Philadelphia • San Diego • New York • Orlando
San Antonio • Toronto • Montreal • London • Sydney • Tokyo

Text Typeface: Palatino
Compositor: York Graphic Services, Inc.
Acquisitions Editor: Julie Levin Alexander
Development Editor: Cathleen Petree
Art Development: Jane Whiteley
Managing Editor: Carol Field
Project Editor: Nancy Lubars
Manager of Art and Design: Carol Bleistine
Art Director: Christine Schueler
Text Designer: Nanci Kappel
Layout Artist: Tracy Baldwin
Text Artwork: Rolin Graphics
Cover Designer: Lawrence R. Didona
Photo Permissions: Amy Ellis
Marketing Manager: Sue Westmoreland
Director of EDP: Tim Frelick
Production Manager: Joanne Cassetti

Cover Credit: Montage designed by Lawrence R. Didona
Photographs: *Runners:* © Dave Kingdon, Index Stock Photography. *Computer generated DNA:* © Will and Deni McIntyre, Science Source / Photo Researchers. *MRI of mid-sagittal section through the head:* © Mehau Kulyk, Science Source / Photo Researchers. *SEM of human blood cells:* © NIBSC, Science Source / Photo Researchers. *SEM of synapses:* © Don Fawcett, Science Source / Photo Researchers. *Molecular computer graphic of B-DNA:* © NIH, Science Source / Photo Researchers. *Computer image of human knee joint:* © Thomas Porett, Science Source / Photo Researchers. *Human chromosomes:* © Biophoto Associates, Science Source / Photo Researchers.

Printed in the United States of America

THE HUMAN BODY: Concepts of Anatomy and Physiology

ISBN 0-03-055507-8

Library of Congress Catalog Card Number: 93-085932

4 5 6 7   032   9 8 7 6 5 4 3 2 1

PREFACE

The Human Body: Concepts of Anatomy and Physiology presents the essential information for understanding structure and function of the human body. My motivation for writing this book originated from a need to provide introductory level students with a text that they can rely on for instructional guidance, visual assistance, and accuracy, without the feeling of intimidation that is often associated with an abundance of new information. It is my hope that the student will not only gain important insights into the human body by using this text, but also experience a sense of marvel and wonder at the body's amazing design and intricate workings.

■ AUDIENCE

The Human Body: Concepts of Anatomy and Physiology is primarily designed for use in single-term courses in human anatomy and physiology and in human biology. These courses are often preparatory for students pursuing careers in the allied health fields, and, in some cases, for students majoring in physical education, art, psychology, or anthropology. Additionally, some institutions offer these courses as electives for nonscience majors, in order to fulfill a science course requirement. In each of these cases, students are introduced for the first time to the structure and function of the human body, often without the benefit of prior courses in general biology and chemistry. This introductory level textbook was specifically written with the educational needs of these students in mind.

■ APPROACH AND MAIN THEME

This is a unique text whose main goal is to depict the integration of human body structure and function for students who are not biological sciences majors. Its style is based on the notion that learning new technical information becomes more attainable and complete if a conceptual framework is provided. A manageable number of concepts are set forth at the beginning of each of the six units by means of a brief overview paragraph that serves to summarize and to integrate key ideas. Individual concepts are further underscored in the chapter outline and learning objectives that open each chapter. Key ideas continue to be highlighted within section-opening, single-concept boxes, and by means of section-ending checkpoint questions.

Throughout this text, the conceptual framework is reinforced by clear, easy-to-understand language, logical explanations that tie new information with previously learned material, correlations with every-day experiences, accurate and current information, and many high quality figures and photographs that correspond directly to the text material. A great effort was made to limit the number of terms to those that are regarded as essential for understanding the concepts of structure and function. Instead of increasing student stress levels by listing banks of abstract terms, this classroom-tested method of teaching provides valuable assistance to the student who wants to master the material.

The theme of the text is the remarkable feature of body function that maintains stability, **homeostasis.** How the body achieves homeostasis despite changing conditions, the body components involved in its maintenance, and the consequences of the body's failure to maintain it effectively are topics that are interwoven throughout the text. In many chapters, the role of body tissues, organs, and systems in maintaining and supporting homeostasis is discussed along with their other functions. To stress the importance of homeostasis in the daily maintenance of health, this vital body function is also presented as a separate topic at the end of most chapters. This feature serves to connect homeostasis with the main functions of the body system discussed in the chapter.

■ ORGANIZATION

The book studies human structure and function using the time-tested systemic approach, in which the body systems are explored individually following the introductory chapters. The text is divided into twenty chap-

v

ters, which are grouped by common themes into six units.

Unit 1 provides the fundamentals for study. It includes the first five chapters of the text, all of which introduce the student to the organization of the body. The progression of chapters in **Unit 1** uses a "building block" approach, beginning with the most basic information and building with each subsequent chapter into more complex information.

Chapter 1, "Introduction to the Human Body," introduces the concepts of structure and function, homeostasis, and body organization from building blocks of chemicals, cells, tissues, organs, and systems. There is a unique discussion in this chapter on health and disease, which is accompanied by a boxed presentation of infectious agents (viruses, bacteria, fungi, and protozoans).

In Chapter 2, "The Chemical Basis of the Body," the basic concepts of chemistry are provided. It includes discussions on basic atomic theory, inorganic chemistry, and the organic chemical building blocks of the body (carbohydrates, lipids, proteins, and nucleic acids).

Chapter 3, "Cells: The Basis of Life," focuses on the most basic living subunit of the body, the cell. The chapter discusses the structural components of a cell and the current theories regarding their primary functions. It also includes discussions on cell division and protein synthesis.

The next level of structural organization, tissues, is the theme in Chapter 4, "Tissues." The four main types of tissues, along with each of their various subtypes, are discussed in this chapter. Membranes are also introduced.

Chapter 5, "Organs and Systems: Overview of the Human Body," concludes Unit 1. This chapter clearly distinguishes between organs and systems, and presents each of the eleven systems by introducing the organs each contains and their main functions. The purpose of this chapter is to expose the student to this general information in order to "see the big picture" before moving on to a more detailed study of individual systems.

Unit 2, "Systems that Cover, Support, and Move the Body," begins the more detailed coverage of the eleven systems. It present the integumentary system, skeletal system, and muscular system by emphasizing how their building-block components are organized to accomplish their important functions.

Chapter 6 is "The Integumentary System." In this chapter, the skin is introduced as the primary organ, and the receptors, glands, hair, and nails as secondary organs. The role of the integumentary system in maintaining homeostasis by its effect on body temperature is also discussed.

"The Skeletal System" is Chapter 7. The discussion includes topics on bone growth, bone remodeling, and skeletal system functions, although the main focus in this chapter is on bone structure. The structure of joints and the various types of movement they provide are also included. The importance of the skeletal system's role in homeostasis concludes the chapter.

Chapter 8, "The Muscular System," contains a basic description of muscle histology, as well as explanations of the sliding filament mechanism and other aspects of muscle physiology. It also contains information on the major muscles of the body, including their name, location, origin, insertion, and primary action. The chapter concludes with an explanation of the muscular system's role in homeostasis.

The main theme of **Unit 3, "Systems that Control by Communication,"** is the control or regulation of body activities. It includes three chapters, which cover the nervous and endocrine systems.

Chapter 9, "Organization of the Nervous System," introduces the structure and function of the nervous system. It includes explanations on nerve tissue physiology; the structure and organization of nerve tissue which forms the brain, spinal cord, and peripheral nerves; and the organization of these organs to form the nervous system. The autonomic nervous system is also included. Chapter 9 concludes with an important summary of the nervous system's role in maintaining homeostasis.

Chapter 10 continues coverage of the nervous system. Called "The Special Senses and Functional Aspects of the Nervous System," this chapter explains the sensory, integrative, and motor functions of the nervous system. Its emphasis is on sensory functions, which includes complete descriptions of each of the four special senses.

Chapter 11 is "The Endocrine System." It is organized into discussions of hormonal mechanisms, anatomy of the primary endocrine glands and their roles in body regulation, and the importance of endocrine gland function in maintaining homeostasis.

Unit 4 is concerned with the systems of the body that transport fluids and protect the body from infection. Called **"Systems that Transport and Protect,"** it includes three chapters that cover the blood, the cardiovascular system, and the lymphatic system.

The first chapter of this unit, Chapter 12, is called "The Blood." It introduces the functional significance of blood, and describes its structural composition. It also includes discussions about hemostasis and blood grouping. The chapter concludes with an explanation of blood's role in maintaining homeostasis.

Chapter 13 is called "The Cardiovascular System." This important chapter contains discussions of heart structure, heart physiology, the structure of vessels, blood pressure, and the circulatory pathways. It also describes the role of the cardiovascular system in maintaining homeostasis.

Chapter 14, the "Lymphatic System," contains two parts. The first part is a description of lymphatic system structural organization, which includes lymphatic circulation and lymphatic organ structure and function. The second part is an explanation of the defense mechanisms of the body, which includes descriptions of cell-mediated immunity and humoral immunity. The chapter contains a very current discussion of immune suppression and AIDS, and concludes with a discussion of homeostasis and immunity.

Unit 5 is called **"Metabolic Processing Systems."** It unifies four chapters on the common theme of metabolism. The systems included in this unit are the respiratory, digestive, and urinary systems.

Chapter 15 is "The Respiratory System." It includes the structure of the respiratory system, breathing mechanics, and physiology of gas exchange. The importance of respiratory system function in maintaining homeostasis concludes the chapter.

Chapter 16, "The Digestive System," deals with the structure and function of the digestive system. The organs of the alimentary canal (GI tract) are presented with the accessory organs. The role of the digestive system in maintaining homeostasis is also discussed.

In Chapter 17, "Nutrition and Metabolism," students are introduced to the basic concepts of nutrition and metabolism. Nutrition describes the various nutrients, including their importance in maintaining health. Nutrient transport throughout the body is also discussed. This is followed by a description of metabolism, including cellular respiration and the concepts of metabolic rate and heat production.

Chapter 18, "The Urinary System," focuses on the structure and function of the kidneys. Emphasis is placed on the functions of filtration, reabsorption, and secretion, although other functions of the kidneys are described as well. The remaining urinary organs are also presented. The importance of urinary functions in maintaining homeostasis is emphasized, and is summarized at the end of the chapter.

The final unit, **Unit 6,** is called **"The Cycle of Life."** Its main topics are the reproductive system, human development, and genetics.

Chapter 19 is "The Reproductive System." It includes discussions about the structure of the male reproductive system, male reproductive physiology, the structure of the female reproductive system, and female reproductive physiology.

Chapter 20, "Human Development and Inheritance," emphasizes prenatal development. It also includes discussions about parturition, postnatal development, and patterns of genetic inheritance.

■ LEARNING AIDS

In courses that survey the human body, a large volume of new information must be retained by the student. To provide for this challenging task, learning aids are included throughout the book. Each learning aid has been carefully developed in an effort to make learning as enjoyable as possible.

UNIT AND CHAPTER OPENERS

The opening page of each unit contains an **Introductory Statement** that summarizes and integrates the main points within the unit, as well as introduces the chapters that are included. The opening of each chapter includes a **Chapter Outline,** which provides the student with an overview of the material, and a list of **Learning Objectives,** which identify the goals that must be accomplished in order to master the chapter material. An introductory statement of the chapter material is also provided.

CONCEPT-STATEMENT HEADINGS AND CONCEPTS CHECK QUESTIONS

Within the body of each chapter, **Concept-Statement Headings** are provided at various intervals, and identify a key idea that will be discussed within subsequent paragraphs. At the end of each section following the concept-statement heading, a list of questions is provided, called **Concepts Check,** which promotes active learning.

TABLES AND FIGURES

Each chapter contains numerous boxed summaries of information, or **tables,** in order to provide a quick review of material within the text. Also, an abundant number of **figures** are provided, each of which has been carefully rendered or reproduced to effectively support concepts explained in the text. Accompanying most figures is a figure legend that ends with a question. The figure legend question challenges the student to take an active part in the learning process; the answer to each question is provided at the end of the chapter to reinforce this learning technique.

HEALTH CLINIC AND SPORTS CLINIC

Selected topics in biological research and medicine are provided in special boxes called **Health Clinic** in every chapter. These topics are accurate and current, since they are based on recent information obtained from research publications including *The New England Journal of Medicine, Scientific American, Nature,* and *Science.* Also, selected topics that relate text material to sports activities are provided in special boxes called **Sports Clinic.** This information is also based on current publications, and is thereby accurate and current. The careful selection and placement of "Health Clinic" and "Sports Clinic" topics in the text serve to spark further interest in the learning material.

PHONETIC PRONUNCIATIONS AND WORD ROOT ORIGINS

Learning new terminology is a difficult task, but it helps if the student is shown how to pronounce the term correctly. This is provided by **phonetic pronunciation guides,** which follow the new term to be learned in parentheses. It also helps the learning process if the student is informed of the actual meaning of the term, based on its Latin and Greek stems. These **word root origins** are provided at the bottom of the page in colored boxes.

END-OF-CHAPTER MATERIAL

At the end of each chapter, a list of **Clinical Terms** that are associated with the chapter material is provided. This list, which includes a brief description of each term, is mainly concerned with pathological disturbances in homeostasis. After the clinical terms list is a **Chapter Summary,** presented in outline form to provide a quick review of chapter material. A list of **Key Terms** follows the summary, each of which is keyed to the page number on which it was first defined. This is followed by **Questions for Review,** in which objective questions and essay questions pertaining to the chapter material are provided. Each chapter concludes with a list of answers to short-answer questions that are associated with each chapter figure, called **Answers to Art Legend Questions.**

SELECTED READINGS AND GLOSSARY

The end of the book offers a list of **Selected Readings,** which provide an avenue of supplemental reading for the motivated learner. These references were selected on the basis of their popular availability in most libraries, and on their readability to students who may not be focused in biology. Following this list of references is a **Glossary,** which contains most terms indicated in boldface type within the text, accompanied by their definitions.

■ SUPPLEMENTS

The following supplemental material is available to further facilitate learning and teaching.

INSTRUCTOR'S MANUAL, *by Bruce D. Wingerd*

The Instructor's Manual is a complimentary publication that provides a guide to course preparation. It includes Chapter Overviews, Learning Objectives, Key Concepts, and Chapter Outlines. It also contains a com-

prehensive list of audio-visual films, videos, and computer software.

TEST BANK, *by Annalisa Berta, San Diego State University*

This complimentary booklet contains a series of test questions for each of the twenty chapters. There are more than 1,300 questions arranged in multiple-choice, true/false, essay, and fill-in formats to test student comprehension. The Test Bank is also available on disk for the IBM and Macintosh computers.

OVERHEAD TRANSPARENCY ACETATES

A set of 200 full-color acetate transparencies taken directly from figures in the book is included in this complimentary packet.

LABORATORY MANUAL, *by Craig W. Clifford, Northeastern University*

A fully illustrated, comprehensive, problem-solving guide to laboratories appropriate for single-term courses in human anatomy and physiology or human biology is presented in this manual, which is for sale to students.

COLORING ATLAS OF HUMAN ANATOMY, SECOND EDITION, *by Edwin Chin, Jr., Ph.D., and Joanne Thorner Kerr, Ph.D., both of San José State University*

Approximately 100 carefully drawn plates offer students a creative visual opportunity to enhance and reinforce learning through coloring and labeling structures of the human body.

STUDENT STUDY GUIDE, *by Robert Bauman, Jr., Amarillo College*

Organized to follow the textbook, this manual reviews each chapter with effective study techniques such as multiple-choice, matching, and true/false questions; sentence, list, and chart completion exercises; figure labeling; word dissections; problem solving; and study suggestions.

HUMAN ANATOMY AND PHYSIOLOGY FLASHCARDS, *by Jane Schneider, Westchester Community College*

A box of 225 black/white cards provides additional study tools for students. The cards are designed with medical terminology exercises, blank diagrams to label, and short-answer questions on one side, with the correct responses on the other side.

MULTI MEDIA ANCILLARIES

The Saunders Anatomy and Physiology Videodisc contains an Image Bank of still photographs from the textbook, live action video, and animation. In coordination with the videodisc, Lecture Active Software (available in IBM and Macintosh formats) lists all video clip and still frame data from the videodisc, enabling professors to easily create custom lectures.

VIDEOS

The Infinite Voyage Videos, "The Champion Within" and "A Taste of Success," are available to adopters. This video series is also available in videodisc format.

BODY ART

A selection of 100 pieces of line art from the text are reproduced in black and white in this booklet for additional work in anatomical structural relationships. Students may use BODY ART for notetaking, labeling, or coloring.

■ ACKNOWLEDGMENTS

Although I am the only author indicated on the cover of this book, this is by no means an independent effort. Rather, it represents a body of information that has been drawn from many sources, in many different ways. To begin with, I wish to extend appreciation to my family, Mala, Josh, and Ryan, for their support, encouragement, and constant source of fresh ideas. Mala's years of teaching experience in all levels of education, including Junior High School, High School, and College, became particularly helpful in developing the appropriate language level. Thanks to Cathleen Petree, of SciQ Publishing on freelance assignment to Saunders College Publishing, who saw promise in a preliminary manuscript years ago and, as Developmental Editor, played a prominent role in developing it into the book it is today.

I appreciate the support of my Publisher, Elizabeth Widdicombe, and my Acquisitions Editor, Julie Levin Alexander. Julie played a key role in determining the philosophical approach and theme of the book, in the book's design, and in keeping production on target with her expert management skills.

I wish to thank the book's freelance Art Developmental Editor, Jane Whiteley, for her outstanding contributions to the art program. Her background in medical illustration and understanding of clinical medicine proved to be valuable assets for our goal of presenting the art in an accurately clear, crisp manner. I appreciate the special efforts of Amy Ellis Research of Orlando, Florida, for collecting and, in some cases, shooting the very appealing photographic program for this text. I also wish to thank the book's Project Editor, Nancy Lubars, who guided the project through the complex process of production, Christine Schueler, who expertly directed the art program, and the other dedicated professionals at Saunders who provided the skill and attention required to produce *The Human Body*.

During the review process of the book, many professional eyes scrutinized every chapter for accuracy, clarity, balance, and organizational acceptance. Their helpful suggestions proved to be valuable contributions to the final product. The reviewers are:

Robert Bauman, Jr., Amarillo College
Emily H. Boegli, Savannah Technical Institute
Craig Clifford, Northeastern State University
Vincent Coffey, Odessa College
James V.A. Conkey, Truckee Meadows Community College
Judy Donaldson, North Seattle Community College
Ralph E. Ferges, Palomar College
George P. Ferrari, Broome Community College
Greg Garman, Centralia College
Luanne Gogolin, Ferris State University
Ann Harmer, Orange Coast College
Karen Kucharski, Cornell University
Lewis Milner, North Central Technical College
Betsy Ott, Tyler Junior College
William Perrotti, Mohawk Valley Community College
Richard Pflanzer, Indiana University/Indianapolis
Kurt E. Redborg, Coe College
Jane Schneider, Westchester Community College
John Sheard, Eastern Michigan University
Michael E. Smith, Valdosta College
Ian Tizard, Texas A & M University

As a final note, I wish to invite your reactions, comments, and suggestions to be sent to me so that subsequent editions may reflect your actual educational needs.

Bruce D. Wingerd
Biology Department
San Diego State University
San Diego, CA 92182

A NOTE TO THE STUDENT

The Human Body: Concepts of Anatomy and Physiology contains a number of features that are designed to help you learn the material. However, to benefit from these features you must be able to recognize them and use them properly. This introduction serves to identify these features so that when you begin to study, you will be ready to maximize your learning.

When you begin your study of a particular chapter, read through the chapter outline and learning objectives first. The chapter outline gives you a bird's eye view of the topics in that chapter, and the learning objectives tell you what you need to learn in order to master the chapter material.

Within the body of each chapter are other helpful features: Beginning most major sections is a boxed summary of information. This summary informs you of the concepts you will be learning. Following the major section is another type of box with short-answer questions, called **Concepts Check.** It is important for you to attempt to answer these questions immediately after reading the section. If you cannot, go back through the section to find the information you need for the correct answer. In doing so, you will reinforce the concepts, or important ideas, in your mind.

At the end of each chapter is a **Chapter Summary,** which summarizes the important information in the chapter without listing all of the terms. A list of **Key Terms,** which identifies the page number where the definition of each term appears, follows the summary. A number of **Review Questions,** which provide you an opportunity to quiz yourself after reading through the chapter, are also at the end of each chapter. Following the Review Questions is a list of answers to questions that are associated with the figures. As you read the text and look at a referenced figure to think about what you have just read, it is helpful to try to answer the question associated with the figure in order to visualize what you have learned. As mentioned, the answers to these questions can be found at the end of the chapter.

Most medical and scientific terms that you may be required to learn are provided with a word root, which informs you of the Latin and/or Greek origins of the word. With new terms, it is often helpful to learn how the word is constructed in order to remember what it means. Once you have mastered a few key word roots, you will be able to recognize and to understand many, many scientific and medical words. Words that relate to the same body organ or to the same condition often share the same word stems. For example, carditis (card = heart + itis = inflammation) means simply "inflammation of the heart," and cardiology (card = heart + ology = study of) means "study of the heart." Dermatitis (derma = skin + itis) means "skin inflammation" while epidermis (epi = outer + dermis = skin) is the outer layer of skin. If you enjoy puzzles, you will have fun even while you master a new vocabulary.

Word stems are provided at the bottom of the page on which they first appear. Also, when you come upon a new term that you do not know how to pronounce, use the pronunciation guide that is usually provided in the text next to the term (in parentheses). If a pronunciation cannot be found, look up the term in the Glossary at the back of the book. Pronunciation guides are based on a pronunciation key, which is provided here and in the Glossary.

■ PRONUNCIATION KEY

1. The syllable with the strongest accent appears in capital letters. For example, science (SI-ens) and learning (LER-ning).

2. In words with a secondary accent, it is identified by a single quote mark ('). For example, homeostasis (ho'-me-o-STA-sis).

3. Vowels with a line above the letter indicate the hard sound. For example, feet (fēt), base (bās), and site (sīt).

4. Vowels with no line above the letter indicate the soft sound. For example, atom (A-tum), duct (dukt), and retina (RE-ti-na).

5. Other phonetic clues to sounds may be found in the following examples:
   oo as in blue
   ar as in fair
   oy as in oil
   ah as in father (FAH-ther)

FEATURES OF THE BOOK

SPORTS CLINIC

Muscle Hypertrophy and Disuse Atrophy

Heavy exercise and good nutrition can lead to increased muscle mass. When muscles enlarge in size, they are said to *hypertrophy*. At the present time, most researchers agree that actual enlargement of a muscle is mainly due to an increase in the size of individual muscle fibers. In each fiber, the mitochondria increase in number and new filaments of actin and myosin are added to the sarcomeres. Eventually, the addition of new protein may lead to the de-

velopment of additional myofibrils. Hypertrophy of a muscle provides for an increased force of contraction, as the strength of muscle contraction is directly related to the diameter of the muscle fibers.

If muscles are not kept active but become immobilized for an extended period of time, such as during enforced bed rest or loss of nervous stimulation, muscles begin to lose their protein mass. As a result, muscle fibers decrease in size.

This condition is called *disuse atrophy*, and can result in a decrease of muscle strength at the surprising rate of 5% each day. As atrophy progresses in long-term disuse, most of the muscle tissue is replaced by fibrous connective tissue, making muscle rehabilitation nearly impossible. In short-term disuse (less than about two years of immobilization), exercise is usually successful in returning muscle fibers to their original size.

The Sports Clinic discusses specific health conditions that may result from athletics.

thick and thin filaments that are arranged into sarcomeres. The presence of alternating light and dark bands in the sarcomeres produces striations, as we have seen in skeletal muscle cells. As a result of the regular organization of proteins, cardiac muscle contraction is forceful, although slightly less than skeletal muscle contraction. Also, cardiac muscle tissue does not develop an oxygen debt and does not fatigue. Its contraction is autorhythmic, which means it does not require an external stimulus to begin the contraction cycle.

Smooth muscle cells are relatively small and are spindle-shaped, with a single nucleus in each cell. Internally, they contain a protein distribution different from that found in skeletal muscle cells. There are no troponin and fewer actin fibers in the thin filaments, and the filaments are not organized into sarcomeres. Consequently, smooth muscle cells are not striated. Also, no T tubules or sarcoplasmic reticula are present. Smooth muscle cells contract more slowly than skeletal muscle and cardiac muscle cells, and with less force. However, they have the greatest ability to sustain a contraction relative to skeletal and cardiac cells. Like cardiac cells, they do not develop an oxygen debt. Under most circumstances, smooth muscle cells require an external stimulus to contract, which can be in the form of nerve stimulation or hormone stimulation.

MUSCULAR RES...

A muscle fiber respon... by contracting. The na... a whole may vary acco... responding, the frequen... sion is applied.

We have learned in th... a motor neuron may c... However, our nervous... the strength and freque... this affect the muscle's... question and others in...

All-or-None Respon...

The weakest stimulus th... called a **threshold stim**... weaker will not cause ev... is called a **subthreshold s**... receives a stimulus of thr... will contract to its compl... a muscle fiber does not pa... it all the way, or not at all.... is called the **all-or-none r**...

HEALTH CLINIC

Tissues, Tumors, and Cancer

A **tumor**, or **neoplasm**, is an overgrowth of cells to form a tissue that has no useful purpose to the body. It becomes a threat to health when it physically replaces healthy tissue and competes with surrounding tissues (see the photomicrographs). A tumor originates in a single cell that, as a result of damage to the DNA, has lost its ability to regulate growth and division. An alteration of chromosomes that affects the DNA is called a **mutation**. Most mutations are caused by exposure to environmental agents, called **carcinogens**. A carcinogen may be a source of ionizing radiation (either ultraviolet light from the sun or X-rays) or a chemical. One of the more well-known groups of carcinogens is found in cigarette tobacco, which is the

documented cause of 90% of all *lung cancers* and may influence the development of other cancers as well.

A mutation may also be caused by a virus. Experiments performed on animals and on human cells grown in culture have shown that certain viruses are able to enter a cell and change the genetic code by incorporating their own genetic material into that of the cell. The result is a mutant cell that manufactures proteins that are different from those of normal cells. If this mutant cell evades detection by white blood cells of the immune system and is stimulated by the viral genes to divide more rapidly than normal, a tumor will develop. An example of a virus-induced tumor occurs in the disease *adult T-cell leukemia-*

A tumor arises from normal tissue. As the tumor grows, it competes with healthy tissue by overcrowding it and reducing its ability to perform normal functions. (a) A healthy bronchial passageway to the lungs is lined with ciliated cells. The cilia sweep dust and other foreign particles away from the lungs. Magnification 4000x. (b) Tumor cells (*green*) invade the bronchial wall and crowd out normal cells lining the bronchus. Magnification 3000x.

(a)

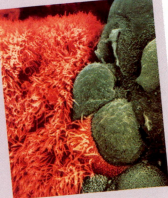
(b)

The Health Clinic applies structural and functional concepts to common pathophysiological conditions in order to demonstrate what happens when homeostasis fails.

The exceptional line art emphasizes physiological processes wherever possible to reinforce text explanations.

ARTWORK

Photomicrographs are paired with line art wherever possible to offer a dual perspective for learning

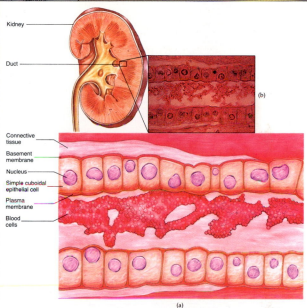

Acromioclavicular joint
Coracoid process
Clavicle
Acromion process
Head of humerus
Glenohumeral joint
Scapula
Rib
Costal cartilage
Humerus
Sternum
Ulna
Radius

sternum directly by means of the costal cartilages. The remaining five pairs, called **false ribs,** have an indirect connection or no connection at all. The last two (or sometimes three) pairs of false ribs are also known as **floating ribs,** since they lack the costal cartilage completely.

A typical rib has several distinguishing features that set it apart from other bones (Fig. 7-16). It contains a long, flattened **shaft** that curves around in a semicircle. The **head** of a rib is an enlargement at its posterior, or proximal, end that articulates with the body of one or two thoracic vertebrae. Near the head is a small projection called a **tubercle,** which articulates with the transverse process

Bones of the Appendicu...

Pectoral Girdles

The **pectoral girdles** (should... nection between the axial ske... (Fig. 7-18). Each girdle conta... or collarbone; and a scapul... gether these bones support t... cles that move the arms.

Clavicles

The **clavicles** are slender, r... the letter S in shape. Each o... tween the sternum and the... neck (Fig. 7-18). At its un... of two joints that form the...

Figure 7–18
The pectoral girdle. This drawing of the right pectoral girdle shows an anterior view and includes the sternum and the humerus that connect to it.
What two bones articulate with the clavicle?

Figure 7–19
The scapula. (a) Anterior, (b) lateral, and (c) posterior views of the left scapula.
What bone articulates with the scapula at the glenoid cavity?

Coracoid process
Acromion process
Coracoid process
Acromion process
Glenoid cavity
Glenoi... cavity
Subscapular fossa
Sp...
Infras... foss...
Lateral border
Medial border
(a)
(b)

Kidney
Duct
(b)
Connective tissue
Basement membrane
Nucleus
Simple cuboidal epithelial cell
Plasma membrane
Blood cells
(a)

Figure 4–2
Simple cuboidal epithelium. (a) The tissue may be found lining tubules and ducts in the kidney, and its cells are cube-shaped and are arranged in a single layer. The blood cells in the duct space are an abnormal finding; their presence usually indicates kidney disease. (b) A photomicrograph of a sectioned duct from a kidney that is lined with simple cuboidal epithelium. Magnification 250x.
Where in the body would you expect to find simple cuboidal epithelium?

Simple Cuboidal

Composed of a single layer of cube-shaped cells, **simple cuboidal epithelium** resembles closely fitted polygons (Fig. 4–2). Its cells have a centrally located nucleus and often contain cilia and/or microvilli along their free border. This tissue commonly forms the walls of small tubes, or *ducts,* that carry secretions from one region of the body to another. It is found in the kidneys, in the liver, and in many glands.

Simple Columnar

Simple columnar epithelium consists of a single layer of elongated, cylindrical cells whose nuclei typically lie near the basement membrane (Fig. 4–3). Its cells fre-

Another major goal has been to clearly relate macroscopic subjects to their microscopic components. Wherever appropriate, reference locator insets are provided with line art and photomicrographs to offer students proper orientation and perspective, with respect to the whole organ or anatomical structure.

Most figure legends end with a question to stimulate critical thinking and learning.

PEDAGOGICAL HIGHLIGHTS

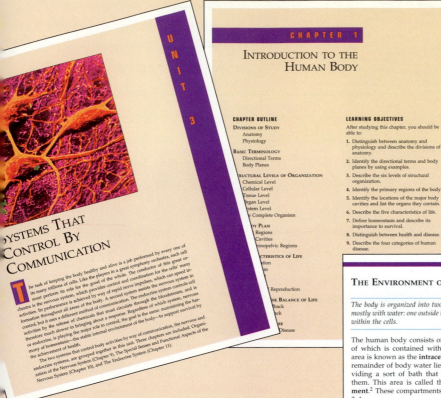

A conceptual framework for learning includes a brief overview paragraph in the unit opener, a chapter-opening outline, and learning objectives.

UNIT 3

SYSTEMS THAT CONTROL BY COMMUNICATION

CHAPTER 1

INTRODUCTION TO THE HUMAN BODY

CHAPTER OUTLINE

DIVISIONS OF STUDY
  Anatomy
  Physiology

BASIC TERMINOLOGY
  Directional Terms
  Body Planes

STRUCTURAL LEVELS OF ORGANIZATION
  Chemical Level
  Cellular Level
  Tissue Level
  Organ Level
  System Level
  The Complete Organism

BODY PLAN
  Regions
  Cavities
  Thoracinopelvic Regions

CHARACTERISTICS OF LIFE

LEARNING OBJECTIVES

After studying this chapter, you should be able to:

1. Distinguish between anatomy and physiology and describe the divisions of anatomy.

2. Identify the directional terms and body planes by using examples.

3. Describe the six levels of structural organization.

4. Identify the primary regions of the body.

5. Identify the locations of the major body cavities and list the organs they contain.

6. Describe the five characteristics of life.

7. Define homeostasis and describe its importance to survival.

8. Distinguish between health and disease.

9. Describe the four categories of human disease.

Section-opening essentials boxes, each highlighting a single concept, are reinforced by a section-ending box of checkpoint questions.

THE ENVIRONMENT OF THE CELL

The body is organized into two major compartments filled mostly with water: one outside the cells of the body, and one within the cells.

The human body consists of about 65% water, much of which is contained within cells. This fluid-filled area is known as the **intracellular environment**.[1] The remainder of body water lies outside cells, often providing a sort of bath that surrounds and supports them. This area is called the **extracellular environment**.[2] These compartments are illustrated in Figure 3–1.

Extracellular Environment

The area outside a cell contains various substances in addition to water. They include a mixture of dissolved gases, salts, food particles, and cellular **products**. A product is any material that is manufactured, or *synthesized*, by mechanisms inside the cell and released to the extracellular environment by a process called **secretion**. Some of the more common products are proteins, hormones, and vitamins. The mixture of water, products, and other substances in the extracellular environment forms a slightly thickened, syruplike liquid known as **extracellular fluid (ECF)**.

There are two types of ECF in the body. One type is located within blood vessels and the chambers of the heart and is called **plasma**. Plasma provides a liquid medium for the transport of substances in the blood. The other type of ECF is located between the cells of the body. It is called **interstitial fluid**[3] and provides a pathway for substances that are en route to and from neighboring cells. Plasma and interstitial fluid make the passage of substances possible, by providing a liquid "freeway" through which materials can travel. In some areas of the body, however, the extracellular environment contains abundant strands of protein that form a dense matting, called **matrix**, with little interstitial fluid leaking through it. This type of arrangement provides structural support for body parts but restricts the free passage of materials.

The compartment of the extracellular environment that lies between adjacent cells is called the **intercellular environment**. Where cells are arranged close together, this space often contains molecular bridges that connect adjacent cells. The bridges may also provide a means of direct communication between the cells.

CONCEPTS CHECK

1. What is ECF composed of?

2. What is the area outside a cell called?

3. What important role is performed by interstitial fluid?

Intracellular Environment

The substance of a cell is called **protoplasm**.[4] It is composed of chemicals that include water, proteins, carbohydrates, fats, nucleic acids, and electrolytes. When these basic building blocks are organized in such a way as to carry out activities characteristic of living things—organization, metabolism, excitability, movement, growth, and reproduction—protoplasm is said to be *alive* (see p. 13 in Chapter 1 for a discussion of these activities if you don't recall them).

The portion of the protoplasm that separates extracellular space from the space inside the cell is the **plasma membrane** (cell membrane), which is shown in Figure 3–1. The plasma membrane surrounds and contains the rest of the cell like a sack, creating a space inside. This internal space is known as the **intracellular environment**. The fluid mixture of water, proteins, and other chemicals within the intracellular environment is called **intracellular fluid**, or **ICF**.

In addition to the plasma membrane, there is a second major component of protoplasm: a thickened, gel-like fluid called **cytoplasm**,[5] which usually occupies most of the intracellular environment. The cytoplasm contains many small, highly organized structures that play important functions in the cell, known as **organelles**.

A third major component of protoplasm is a large, oval-shaped structure known as the **nucleus**. Like cytoplasm, the nucleus also lies within the intracellular environment. It regulates the activities of the cell.

The intracellular environment is the site where most internal cell functions occur. Activities such as energy production, energy storage, and synthesis of new products for growth or secretion are examples of these important functions.

---

[1]Intracellular: *intra-* = within + *cellular* = of the cell.

[2]Extracellular: *extra-* = outer + *cellular*.

[3]Interstitial: Latin *inter-* = between + *-stitial* = standing (referring to cells).

[4]Protoplasm: Greek *proto-* = first + *plasma* = formable material.

[5]Cytoplasm: Greek *cyto-* = cell + *plasma*.

Significant medical and scientific terms are broken into their word root forms and appear at the bottom of the page on which they are used to help students build their medical terminology.

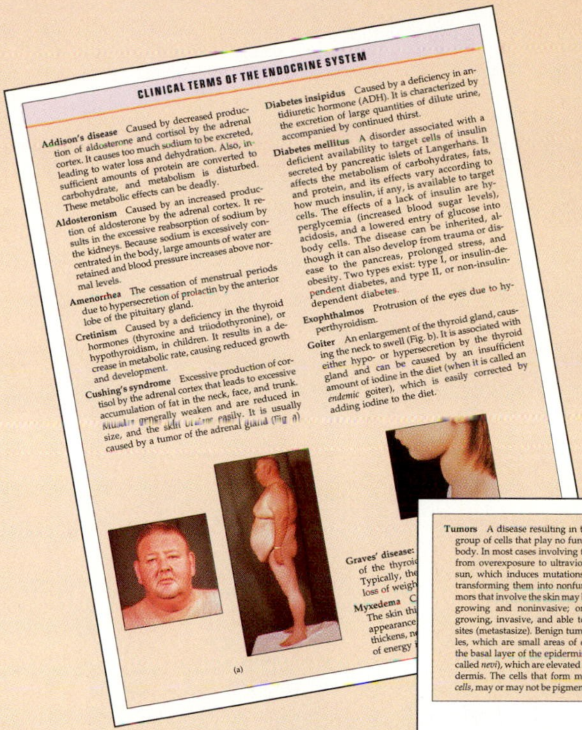

End-of-chapter pedagogy includes a list of common Clinical Terms and their descriptions, a Chapter Summary in outline form, Key Terms with page references, Questions for Review, and Answers to the Art Legend Questions.

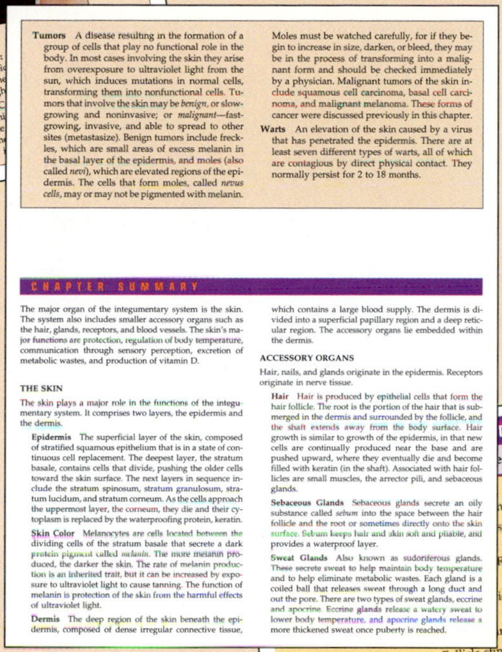

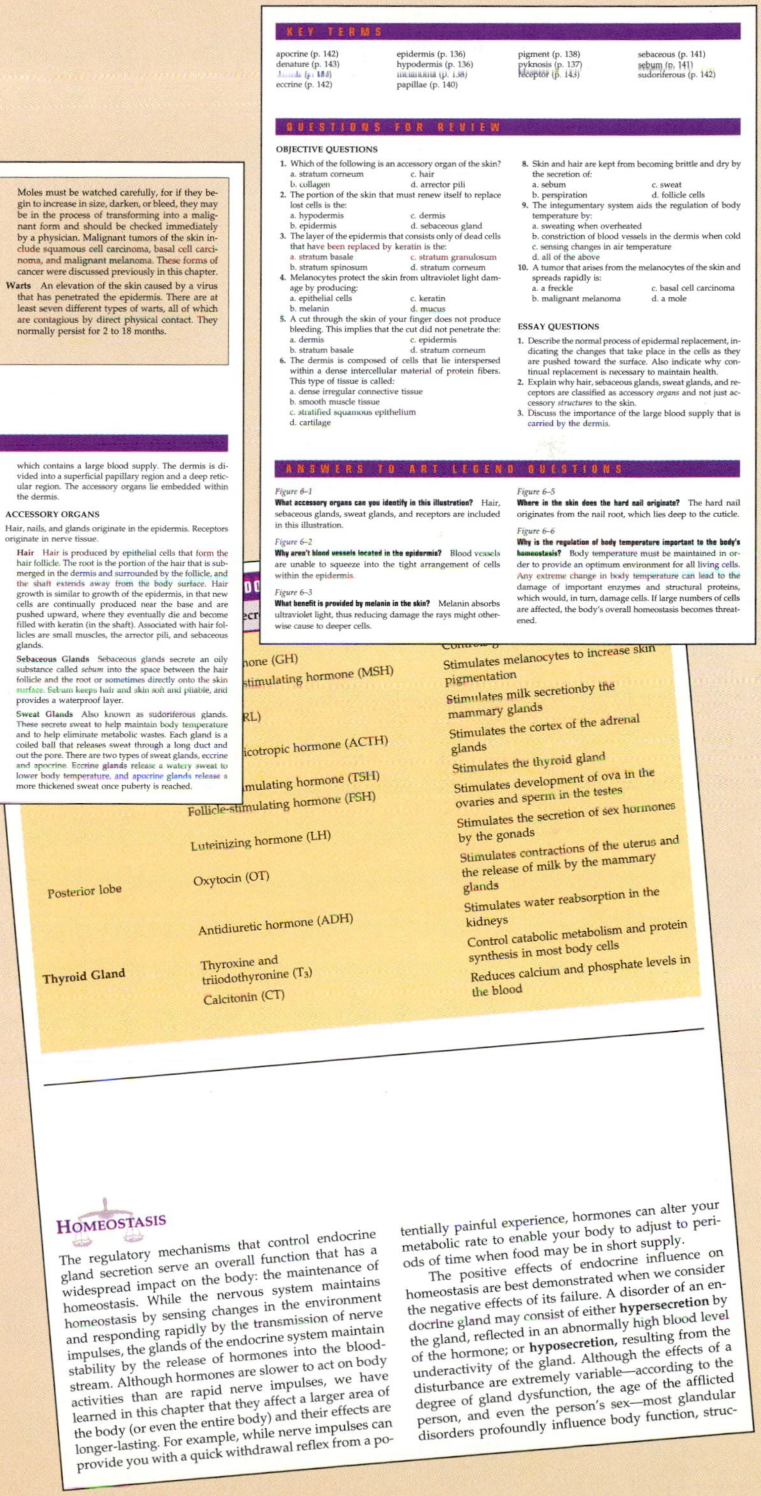

CLINICAL TERMS OF THE ENDOCRINE SYSTEM

Addison's disease Caused by decreased production of aldosterone and cortisol by the adrenal cortex. It causes too much sodium to be excreted, leading to water loss and dehydration. Also, insufficient amounts of protein are converted to carbohydrate, and metabolism is disturbed. These metabolic effects can be deadly.

Aldosteronism Caused by an increased production of aldosterone by the adrenal cortex. It results in the excessive reabsorption of sodium by the kidneys. Because sodium is excessively concentrated in the body, large amounts of water are retained and blood pressure increases above normal levels.

Amenorrhea The cessation of menstrual periods due to hypersecretion of prolactin by the anterior lobe of the pituitary gland.

Cretinism Caused by a deficiency in the thyroid hormones (thyroxine and triiodothyronine), or hypothyroidism, in children. It results in a decrease in metabolic rate, causing reduced growth and development.

Cushing's syndrome Excessive production of cortisol by the adrenal cortex that leads to excessive accumulation of fat in the neck, face, and trunk. Muscles generally weaken and are reduced in size, and the skin bruises easily. It is usually caused by a tumor of the adrenal gland (Fig. a).

Diabetes insipidus Caused by a deficiency in antidiuretic hormone (ADH). It is characterized by the excretion of large quantities of dilute urine, accompanied by continued thirst.

Diabetes mellitus A disorder associated with a deficient availability to target cells of insulin secreted by pancreatic islets of Langerhans. It affects the metabolism of carbohydrates, fats, and protein, and its effects vary according to how much insulin, if any, is available to target cells. The effects of a lack of insulin are hyperglycemia (increased blood sugar levels), acidosis, and a lowered entry of glucose into body cells. The disease can be inherited, although it can also develop from trauma or disease to the pancreas, prolonged stress, and obesity. Two types exist: type I, or insulin-dependent diabetes, and type II, or non-insulin-dependent diabetes.

Exophthalmos Protrusion of the eyes due to hyperthyroidism.

Goiter An enlargement of the thyroid gland, causing the neck to swell (Fig. b). It is associated with either hypo- or hypersecretion by the thyroid gland and can be caused by an insufficient amount of iodine in the diet (when it is called an *endemic* goiter), which is easily corrected by adding iodine to the diet.

(a)

Graves' disease: ... of the thyroid ... Typically, the ... loss of weight ...

Myxedema C... The skin th... appearance ... thickens, n... of energy ...

CHAPTER SUMMARY

The major organ of the integumentary system is the skin. The system also includes smaller accessory organs such as the hair, glands, receptors, and blood vessels. The skin's major functions are protection, regulation of body temperature, communication through sensory perception, excretion of metabolic wastes, and production of vitamin D.

THE SKIN

The skin plays a major role in the functions of the integumentary system. It comprises two layers, the epidermis and the dermis.

Epidermis The superficial layer of the skin, composed of stratified squamous epithelium that is in a state of continuous cell replacement. The deepest layer, the stratum basale, contains cells that divide, pushing the older cells toward the skin surface. The next layers in sequence include the stratum spinosum, stratum granulosum, stratum lucidum, and stratum corneum. As the cells approach the uppermost layer, the corneum, they die and their cytoplasm is replaced by the waterproofing protein, keratin.

Skin Color Melanocytes are cells located between the dividing cells of the stratum basale that secrete a dark protein pigment called *melanin*. The more melanin produced, the darker the skin. The rate of melanin production is an inherited trait, but it can be increased by exposure to ultraviolet light to cause tanning. The function of melanin is protection of the skin from the harmful effects of ultraviolet light.

Dermis The deep region of the skin beneath the epidermis, composed of dense irregular connective tissue,

which contains a large blood supply. The dermis is divided into a superficial papillary region and a deep reticular region. The accessory organs lie embedded within the dermis.

ACCESSORY ORGANS

Hair, nails, and glands originate in the epidermis. Receptors originate in nerve tissue.

Hair Hair is produced by epithelial cells that form the hair follicle. The root is the portion of the hair that is submerged in the dermis and surrounded by the follicle, and the shaft extends away from the body surface. Hair growth is similar to growth of the epidermis, in that new cells are continually produced near the base and are pushed upward, where they eventually die and become filled with keratin (in the shaft). Associated with hair follicles are small muscles, the arrector pili, and sebaceous glands.

Sebaceous Glands Sebaceous glands secrete an oily substance called *sebum* into the space between the hair follicle and the root or sometimes directly onto the skin surface. Sebum keeps hair and skin soft and pliable, and provides a waterproof layer.

Sweat Glands Also known as sudoriferous glands. These secrete sweat to help maintain body temperature and to help eliminate metabolic wastes. Each gland is a coiled ball that releases sweat through a long duct and out the pore. There are two types of sweat glands, eccrine and apocrine. Eccrine glands release a watery sweat to lower body temperature, and apocrine glands release a more thickened sweat once puberty is reached.

Tumors A disease resulting in the formation of a group of cells that play no functional role in the body. In most cases involving the skin they arise from overexposure to ultraviolet light from the sun, which induces mutations in normal cells, transforming them into nonfunctional cells. Tumors that involve the skin may be *benign*, or slow-growing and noninvasive; or *malignant*—fast-growing, invasive, and able to spread to other sites (metastasize). Benign tumors include freckles, which are small areas of excess melanin in the basal layer of the epidermis, and moles (also called *nevi*), which are elevated regions of the epidermis. The cells that form moles, called *nevus cells*, may or may not be pigmented with melanin.

Moles must be watched carefully, for if they begin to increase in size, darken, or bleed, they may be in the process of transforming into a malignant form and should be checked immediately by a physician. Malignant tumors of the skin include squamous cell carcinoma, basal cell carcinoma, and malignant melanoma. These forms of cancer were discussed previously in this chapter.

Warts An elevation of the skin caused by a virus that has penetrated the epidermis. There are at least seven different types of warts, all of which are contagious by direct physical contact. They normally persist for 2 to 18 months.

KEY TERMS

apocrine (p. 142)	epidermis (p. 136)	pigment (p. 138)	sebaceous (p. 141)
denature (p. 143)	hypodermis (p. 136)	pyknosis (p. 137)	sebum (p. 141)
dermis (p. 134)	melanin (p. 138)	receptor (p. 143)	sudoriferous (p. 142)
eccrine (p. 142)	papillae (p. 140)		

QUESTIONS FOR REVIEW

OBJECTIVE QUESTIONS

1. Which of the following is an accessory organ of the skin?
   a. stratum corneum
   b. collagen
   c. hair
   d. arrector pili
2. The portion of the skin that must renew itself to replace lost cells is the:
   a. hypodermis
   b. epidermis
   c. dermis
   d. sebaceous gland
3. The layer of the epidermis that consists only of dead cells that have been replaced by keratin is the:
   a. stratum basale
   b. stratum spinosum
   c. stratum granulosum
   d. stratum corneum
4. Melanocytes protect the skin from ultraviolet light damage by producing:
   a. epithelial cells
   b. melanin
   c. keratin
   d. mucus
5. A cut through the skin of your finger does not produce bleeding. This implies that the cut did not penetrate the:
   a. dermis
   b. stratum basale
   c. epidermis
   d. stratum corneum
6. The dermis is composed of cells that lie interspersed within a dense intercellular material of protein fibers. This type of tissue is called:
   a. dense irregular connective tissue
   b. smooth muscle tissue
   c. stratified squamous epithelium
   d. cartilage

7. (item partially obscured)
8. Skin and hair are kept from becoming brittle and dry by the secretion of:
   a. sebum
   b. perspiration
   c. sweat
   d. follicle cells
9. The integumentary system aids the regulation of body temperature by:
   a. sweating when overheated
   b. constriction of blood vessels in the dermis when cold
   c. sensing changes in air temperature
   d. all of the above
10. A tumor that arises from the melanocytes of the skin and spreads rapidly is:
    a. a freckle
    b. malignant melanoma
    c. basal cell carcinoma
    d. a mole

ESSAY QUESTIONS

1. Describe the normal process of epidermal replacement, indicating the changes that take place in the cells as they are pushed toward the surface. Also indicate why continual replacement is necessary to maintain health.
2. Explain why hair, sebaceous glands, sweat glands, and receptors are classified as accessory *organs* and not just accessory *structures* to the skin.
3. Discuss the importance of the large blood supply that is carried by the dermis.

ANSWERS TO ART LEGEND QUESTIONS

Figure 6–1
What accessory organs can you identify in this illustration? Hair, sebaceous glands, sweat glands, and receptors are included in this illustration.

Figure 6–2
Why aren't blood vessels located in the epidermis? Blood vessels are unable to squeeze into the tight arrangement of cells within the epidermis.

Figure 6–3
What benefit is provided by melanin in the skin? Melanin absorbs ultraviolet light, thus reducing damage the rays might otherwise cause to deeper cells.

Figure 6–5
Where in the skin does the hard nail originate? The hard nail originates from the nail root, which lies deep to the cuticle.

Figure 6–6
Why is the regulation of body temperature important to the body's homeostasis? Body temperature must be maintained in order to provide an optimum environment for all living cells. Any extreme change in body temperature can lead to the damage of important enzymes and structural proteins, which would, in turn, damage cells. If large numbers of cells are affected, the body's overall homeostasis becomes threatened.

	...one (GH)	Control ...
	...stimulating hormone (MSH)	Stimulates melanocytes to increase skin pigmentation
	...RL)	Stimulates milk secretion by the mammary glands
	...icotropic hormone (ACTH)	Stimulates the cortex of the adrenal glands
	...mulating hormone (TSH)	Stimulates the thyroid gland
	Follicle-stimulating hormone (FSH)	Stimulates development of ova in the ovaries and sperm in the testes
	Luteinizing hormone (LH)	Stimulates the secretion of sex hormones by the gonads
Posterior lobe	Oxytocin (OT)	Stimulates contractions of the uterus and the release of milk by the mammary glands
	Antidiuretic hormone (ADH)	Stimulates water reabsorption in the kidneys
Thyroid Gland	Thyroxine and triiodothyronine (T_3)	Control catabolic metabolism and protein synthesis in most body cells
	Calcitonin (CT)	Reduces calcium and phosphate levels in the blood

---

HOMEOSTASIS

The regulatory mechanisms that control endocrine gland secretion serve an overall function that has a widespread impact on the body: the maintenance of homeostasis. While the nervous system maintains homeostasis by sensing changes in the environment and responding rapidly by the transmission of nerve impulses, the glands of the endocrine system maintain stability by the release of hormones into the bloodstream. Although hormones are slower to act on body activities than are rapid nerve impulses, we have learned in this chapter that they affect a larger area of the body (or even the entire body) and their effects are longer-lasting. For example, while nerve impulses can provide you with a quick withdrawal reflex from a po-

tentially painful experience, hormones can alter your metabolic rate to enable your body to adjust to periods of time when food may be in short supply.

The positive effects of endocrine influence on homeostasis are best demonstrated when we consider the negative effects of its failure. A disorder of an endocrine gland may consist of either **hypersecretion** by the gland, reflected in an abnormally high blood level of the hormone; or **hyposecretion**, resulting from the underactivity of the gland. Although the effects of a disturbance are extremely variable—according to the degree of gland dysfunction, the age of the afflicted person, and even the person's sex—most glandular disorders profoundly influence body function, struc-

---

Most systems chapters conclude with a homeostasis section that reviews the homeostatic mechanisms in each system in order to highlight this important theme for students.

BRIEF TABLE OF CONTENTS

CONTENTS

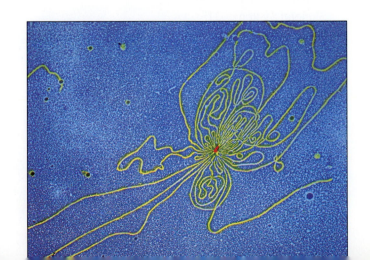

■ UNIT 2:
SYSTEMS THAT COVER, SUPPORT, OR MOVE THE BODY 133

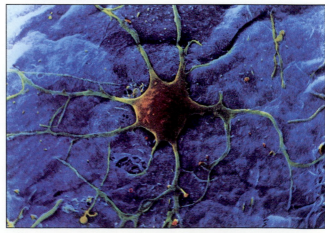

■ UNIT 4:
SYSTEMS THAT TRANSPORT AND PROTECT 343

■ UNIT 5:
METABOLIC PROCESSING SYSTEMS 441

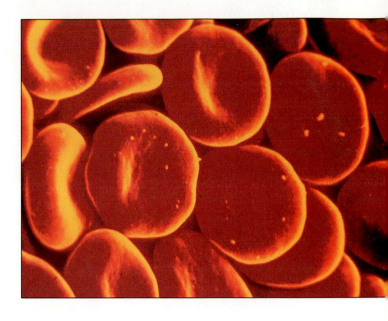

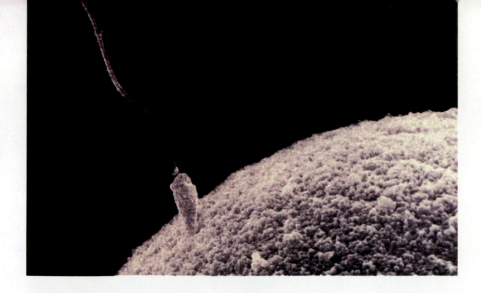

ORGANIZATION OF THE BODY

The human body is a beautiful showcase of nature's design and organization. Its many parts operate in a carefully orchestrated manner with the primary goal of maintaining the body's health in an oftentimes unhealthy world. As we are about to discover in this book, health and, ultimately, survival depend upon the body's ability to maintain its proper structure and functions even under adverse conditions.

In order to understand the importance of body maintenance and the meaning of health, we must first learn how the body is organized and how some of its components operate. This is the goal of our first unit, which introduces the basic design and organization of the body. It begins in Chapter 1 by introducing some basic definitions. The next several chapters examine the fundamental components that provide the basis for body structure and function. Finally, Chapter 5 presents an overview of body organization that provides a framework for future chapters, which examine structure and function in more detail.

Introduction to the Human Body

LEARNING OBJECTIVES

After studying this chapter, you should be able to:

1. Distinguish between anatomy and physiology and describe the divisions of anatomy.

2. Identify the directional terms and body planes by using examples.

3. Describe the six levels of structural organization.

4. Identify the primary regions of the body.

5. Identify the locations of the major body cavities and list the organs they contain.

6. Describe the five characteristics of life.

7. Define homeostasis and describe its importance to survival.

8. Distinguish between health and disease.

9. Describe the four categories of human disease.

In past years when medical science was in its infancy, the intricately complex design of the human body seemed to many like a mysterious miracle of nature. Now that body structure and many of its functions have been explained by the efforts of scientific investigation, some of the mystery may be gone, but the seemingly miraculous complexity of structure and function still remains. At first thought, it may seem impossible to learn about and understand this complexity for the first time—but not so, if you begin your study by establishing a strong foundation of basic understanding first.

This chapter begins your study of human body structure and function. It serves as a starting point by establishing a learning foundation of basic information, upon which new information will be gradually built in the chapters that follow. You will begin to build your foundation of understanding by learning some basic terms, how these and other terms are actually formed, and the general organization of body parts. You will also study the factors that distinguish between living systems and those that are not, the delicate balance in your body that is maintained every day between health and disease, and the very nature of human disease itself.

DIVISIONS OF STUDY

The study of the human body is an interdisciplinary science. It consists of fields that focus on structure or function.

The study of the human body is divided into two primary areas of discipline: anatomy and physiology. Anatomy is the study of body structure, and physiology is the study of body function.

Anatomy

Anatomy[1] is the field of study that describes the locations, appearances, and relationships of body parts. Its goal is to answer the basic questions: Where is it located? What does it look like? How does it relate to other body parts? As knowledge of body structure has grown over the years, it has become necessary to divide anatomy into more specific areas of study. For example, **gross anatomy** is the study of body structures that are visible without the aid of a microscope, whereas **microanatomy** examines structures on a microscopic level only. Microanatomy that focuses on the study of tissues is known as **histology.** Also, **systematic anatomy** is an approach that studies body structure within a given organ *system*, such as the skeletal system or muscular system. **Regional anatomy,** by contrast, examines all structures within a given *region* of the body, such as the head or leg. This book is organized using the systematic approach to anatomy and focuses on applications of gross anatomy.

Physiology

Physiology[2] is a discipline explaining the mechanisms that operate body activities. Its goal is to answer the simple question, How does it work? The answer to this question often focuses on the ways in which the body attempts to maintain a steady, stable state. As you will discover, the body is continually making internal adjustments to changes that occur in the external world around us, such as changes in temperature, heat, light, and food availability. These adjustments require a continual source of energy. However, energy is obtained from the food we eat and is, consequently, always in limited supply. Therefore, the use of energy must be carefully monitored and controlled. The various ways in which the body obtains energy and manages its use in order to maintain stability are the focus of physiology.

CONCEPTS CHECK

1. What is the focus of study in anatomy?

2. What question does the physiologist seek to answer?

BASIC TERMINOLOGY

The language used to describe the human body is universal, with an established set of terms.

Language is a tool that we use to communicate ideas. The language of science—and specifically the language describing the human body—expresses certain ideas in a precise way so that people around the world can understand. We call this language a *universal* language. It is based on Greek and Latin word parts that are arranged in a certain order to form a new word. The word, or term, that is formed is often descriptive of the body part or function with which it is associated. For example, the muscle that forms the walls of the heart is called the *myocardium.* This term literally means the "muscle (Greek *myo-*) of the heart (*kardia*)." Another example is the term for red blood cell, *erythrocyte* (Greek *erythro-* = red + *-cyte* = cell). Throughout this book the major terms of the human body are broken into their Greek and Latin word parts, in the hope that this will help you in your task of learning.

Directional Terms

The language that is used to describe the location of a body structure relative to another is **directional terminology.** It is a valuable tool because it abbreviates otherwise lengthy descriptions, and removes the question of what position the body is in as a consideration. For example, let's say that you wish to describe the location of the heart to someone who doesn't know where it is but does know where the head is. Using directional terminology, you would simply say, "the heart is inferior to the head." Without this language tool you would have to explain that the heart is below

---

[1]Anatomy: Greek *ana* = up + *tome* = cutting.

[2]Physiology: Greek *physio-* = nature + *-logia* = study.

Table 1–1 DESCRIPTIVE TERMS

Term	Definition	Example
Superior (cranial)	Toward the head end or upper part of the body	The head is superior to the neck.
Inferior (caudal)	Away from the head end or toward the lower part of the body	The neck is inferior to the head.
Anterior (ventral)	Toward the front or belly side	The eyes are on the anterior side of the head.
Posterior (dorsal)	Toward the back	The vertebral column (backbone) extends down the posterior side of the body.
Medial	Toward the midline, which is an imaginary line that extends vertically down the middle of the front outside surface of the body.	The nose is medial to the ears.
Lateral	Away from the midline	The ears are lateral to the nose.
Superficial (external)	Toward the surface of the body	The location of the skin is superficial to the muscles.
Deep (internal)	Inward from the surface of the body	The heart lies deep to the rib cage.
Proximal	Toward a structure's origin or point of attachment to the trunk	The upper arm is proximal to the wrist
Distal	Away from a structure's origin or point of attachment to the trunk	The knee is distal to the thigh.

the head when the body is standing upright, or above the head when the body is hanging upside down, or to one side of the head if the body is lying horizontally—and possibly all in a foreign language!

The directional terms that are in common application are presented with their definitions in Table 1–1. The most frequently used terms are illustrated in Figure 1–1.

Directional terminology is based on a universally accepted position of the body. This is the **anatomical position,** and it is defined as the position of a person standing erect (upright) facing the observer, with the arms at the sides and the toes and palms turned forward (Fig. 1–2). This position provides a point of orientation, like a direction key on a map pointing north, south, east, and west; it gives you your bearings when you are studying a "map" of the human body.

Body Planes

The task of describing the locations of body structures is aided by the use of **planes.** A plane is an imaginary flat surface. It may be oriented in space in any direction so that it passes through the body at a particular angle. When a drawing of the body reveals internal structures, the artist has, in his or her imagination, sliced through the specimen along one of these planes. There are three basic types of body planes: **sagittal** (SAJ-i-tal), **frontal** (**coronal**)**,** and **horizontal.** They are shown in Figure 1–3.

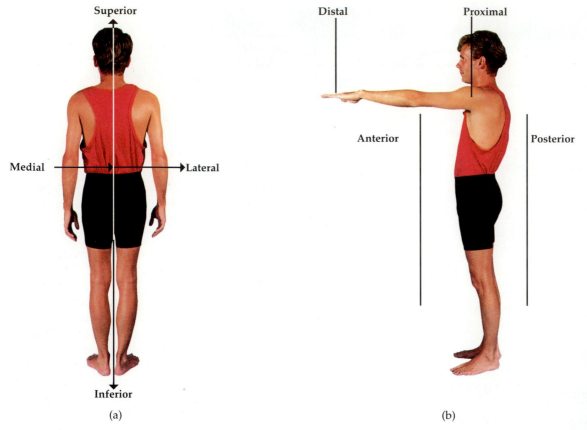

Figure 1–1
(a) Posterior view and (b) lateral view of the body
illustrating the most frequently used directional terms.
These terms help describe the locations of body parts.

Can you identify a body part that is proximal to the hand?

Figure 1–2
The human figure in the anatomical position. This position
provides a point of reference for studying the body.

How are the hands oriented in the anatomical position?

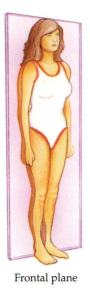

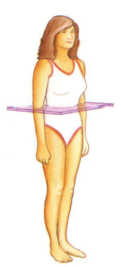

Frontal plane

Sagittal plane

Horizontal plane

Figure 1–3
Body planes. The body may be sectioned along any of these planes to observe internal parts.

A sagittal plane divides the body into what parts?

A sagittal plane extends parallel to the long axis of the body (along the body's length). When the subject is in the anatomical position, a sagittal plane extends in the vertical direction. It divides the body into right and left portions. A sagittal plane that divides the body into *equal* right and left halves is called **midsagittal,** and one that divides unequally is termed **parasagittal.** A frontal plane also extends along the body's long axis, but it divides the body into anterior (front) and posterior (back) portions. A horizontal plane is a plane that extends in a direction perpendicular to the sagittal and frontal planes, dividing the body into superior (upper) and inferior (lower) portions. A horizontal plane is also called a **transverse plane,** and the section it makes through the body is often referred to as a *cross section.*

CONCEPTS CHECK

1. What is the purpose of constructing scientific terms from Latin and Greek word parts?

2. Why should you use directional terms when describing the location of body parts instead of more common descriptions such as *on top of, below,* or *to the side?*

---

3. What is a plane?

4. How may planes be used to view body structures?

STRUCTURAL LEVELS OF ORGANIZATION

Body structure is organized by a series of building-block components, going from simple to complex. Health depends upon every level functioning properly.

A building that is in the early stages of construction looks quite different from the way it will look when completed: it lacks structural organization. The materials that are used to provide its organization begin as individual particles of sand and clay or as shapeless masses of steel and wood. These materials are formed into larger structures: concrete and plaster from sand and clay, boards from wood, and supports from steel. Eventually, these more useful materials are organized to form the basic foundation and frame of the building. When finishing materials are added, the building is structurally complete and the activities it was intended to support may begin.

Although the human body is not composed of sand, clay, steel, and wood, it is in a similar way composed of a series of building blocks that increase in size and complexity. Their proper organization is vital to the health of the whole individual. The building blocks of the body are organized into six distinct levels, called **structural levels of organization,** that proceed from simple to complex as shown in Figure 1–4.

Chemical Level

Chemicals are substances that cannot be simplified further under natural conditions. They are composed of **atoms,** which may react together to form **ions** and **molecules.**[3] The large molecules, or *macromolecules,* that provide the structural foundation for the body are proteins, fats, carbohydrates, and nucleic acids. Examples of an atom and a molecule are shown in Figure 1–4.

Cellular Level

Molecules combine in an organized way to form the **cell.** Cells are the basic structural and functional units of life; they are the smallest living parts of the body.

---

[3]Molecule: Latin *moles* = mass + *-cule* = tiny.

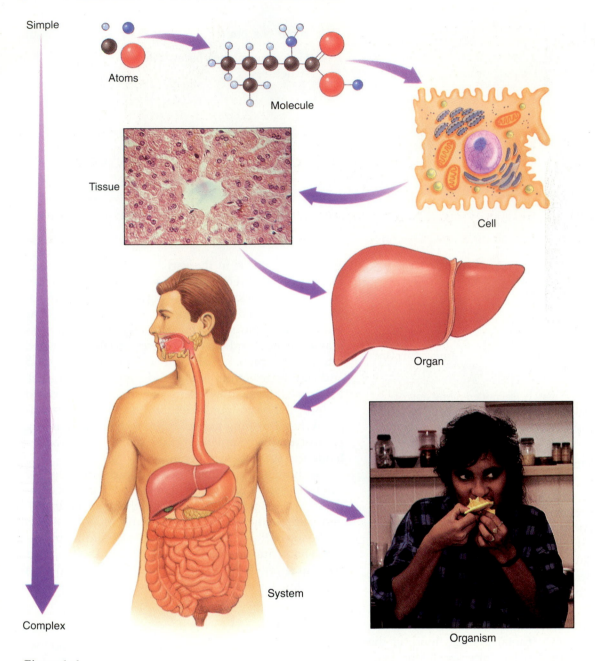

Figure 1–4

Structural levels of organization. The human organism is made of small parts that combine to form more complex larger parts.

What is the organized combination of two or more different types of tissue into a single structure called?

As a living unit, each cell performs functions that are necessary to sustain life. These will be discussed later in this chapter. There are many types of cells in the body, each with its own particular role to play for the benefit of the body as a whole. Figure 1–4 shows an example of an individual cell.

Tissue Level

A **tissue** is a group of similar cells that combine to perform a common function. The common function may be to protect other body structures, provide movement to a body part, or provide a means for communication between body parts. There are only four major types of tissues in the body: **epithelial, connective, muscle,** and **nervous.** Figure 1–4 shows a group of cells organized to form a tissue. This particular example is a type of *epithelial tissue.*

Organ Level

An **organ** consists of two or more different types of tissues that, when combined, perform a general function. Organs can usually be distinguished from other structures because they have a distinct shape. The organ in Figure 1–4 is the *liver,* which performs the functions of blood detoxification, energy storage, and interconversion of nutrients.

System Level

A **system** is an organization of two or more organs and their associated structures. When combined as a unit, they perform a more general function. The liver is one of several organs in the *digestive system,* shown in Figure 1–4. The digestive system performs the general function of preparing food for its incorporation into body cells. A list of the 11 body systems with their functions is provided in Table 1–2.

The Complete Organism

The **organism** is composed of many systems, which depend on one another to perform their tasks. When

Table 1–2 THE SYSTEMS OF THE BODY		
System	**Major Organs**	**General Function**
Integumentary	Skin	Protection of underlying structures from damage and from loss of body fluid
Muscular	Muscles that are attached to bones	Movement of the body
Skeletal	Bones	Support and protection of softer body parts
Nervous	Brain, spinal cord, nerves	Control homeostasis by stimulating muscles to contract and glands to secrete
Endocrine	Pituitary gland, thyroid gland, adrenal glands, pancreas, gonads	Control homeostasis by releasing hormones that alter body processes
Cardiovascular	Heart, arteries, veins	Transportation of materials to and from body cells
Lymphatic	Spleen, thymus, tonsils, lymphatic vessels	Removal of dead cells and foreign bodies from body fluids
Respiratory	Larynx, trachea, lungs	Exchange of gases between the bloodstream and the external environment
Digestive	Esophagus, stomach, small intestine, liver, pancreas, large intestine	Break apart food into small particles for their absorption into the bloodstream
Urinary	Kidneys, ureters, urinary bladder, urethra	Maintain homeostasis by controlling water and salt balance in the bloodstream and by removing metabolic waste materials
Reproductive	Testes, urethra, penis; ovaries, uterus, vagina	Provide for production of new individuals

all of the systems of the body are functioning properly, the organism, or whole individual, is capable of surviving. Survival and reproduction are the ultimate goals underlying all of the body's internal activities.

> ### CONCEPTS CHECK
>
> 1. How is the structural organization of the body similiar to that of a building?
>
> 2. What is the basic structural and functional unit of life?
>
> 3. What are the definitions of a tissue, an organ, a system, and an organism?

THE BODY PLAN

The human body is divided into regions. Some regions contain spaces called cavities that house organs.

If you were asked by a visitor who had no knowledge of the human body to describe the structural design of human beings, how would you respond? Perhaps you would begin by describing the general body areas, such as the head, neck, trunk, arms, and legs. Once this was understood, you might perhaps proceed internally by describing how major structures such as organs are arranged. This would provide the visitor with a general picture of the human body. This is the format that we shall follow in our discussion of the structural plan of the body.

Body Regions

The major areas of the body that are structurally distinguishable are called **regions.** They include the **head,** the **neck,** the **trunk,** the **upper appendages,** and the **lower appendages.** Each major region is divided further into smaller regions. For example, the anterior side of the trunk is divided into an upper portion, the **thorax,** a middle portion, the **abdomen,** and a lower portion, the **pelvis.** The posterior side of the trunk is simply called the **back** region. The major regions of the body and their useful divisions are shown in Figure 1–5. Surface features that are routinely used by physicians during physical examinations are also included in Figure 1–5. Knowledge of the regions of the body and their surface features aids the anatomist in describing the relative locations of parts, and the physi-

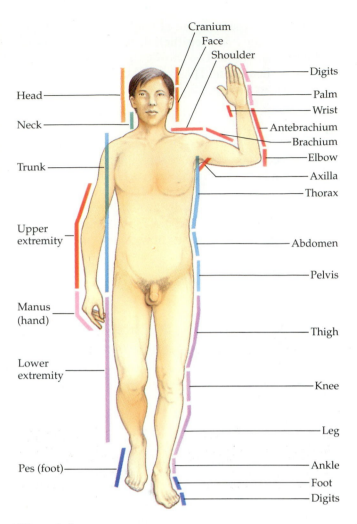

Figure 1–5
Regions of the body. The regions shown in this anterior view represent many of the important body regions.

What is the location of the axilla relative to the thigh (in directional terms)?

cian in identifying internal causes of surface pain. The major body regions, their divisions, and important surface features are summarized in Table 1–3.

Body Cavities

The body is internally divided into several spaces, or **cavities,** that contain many of the organs. These are shown in Figure 1–6. The two major cavities are the **dorsal cavity** and the **ventral cavity.**

The dorsal cavity is located on the posterior (dorsal) side of the body. It contains a **cranial cavity** within the skull and a **vertebral canal** that extends through the center of the vertebral column (backbone). The cranial cavity is a well-protected space formed by the bony plates of the skull, which contains the brain. The

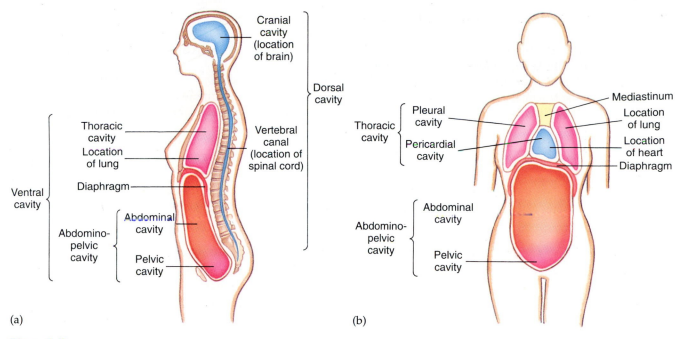

(a)   (b)

Figure 1–6
Body cavities. (a) The two main cavities are the dorsal and ventral cavities. (b) The
ventral cavity contains the thoracic cavity in the chest and the abdominopelvic
cavity in the abdomen and pelvis. They are separated by the diaphragm.

What major organs occupy the thoracic cavity?

Table 1–3	REGIONS OF THE BODY
Major Body Regions	**Subdivisions**
Head	Face
	Cranium
Neck	Anterior neck
	Posterior neck
Trunk	Thorax
	Abdomen
	Pelvis
	Back
Upper appendages	Shoulder
	Axilla (armpit)
	Brachium (upper arm)
	Elbow
	Antebrachium (forearm)
	Carpus (wrist)
	Manus (hand)
	Palm
	Digits
Lower appendages	Gluteus (buttock)
	Femorus (thigh)
	Knee
	Crus (leg)
	Tarsus (ankle)
	Pes (foot)
	Sole
	Digits

vertebral canal is protected by the vertebrae and houses the spinal cord.

The ventral cavity is on the anterior (ventral) side of the body. It is divided into an upper portion called the **thoracic**[4] (thō-RAS-ik) **cavity** and a lower portion called the **abdominopelvic**[5] (ab-do'-mi-nō-PEL-vik) **cavity** by a thin sheet of muscle known as the **diaphragm**[6] (DĪ-a-fram).

The thoracic cavity is protected in part by the rib cage anteriorly and the vertebral column posteriorly. It contains three smaller cavities: two **pleural**[7] **cavities,** which are small spaces between two membranes surrounding each lung; and the **pericardial**[8] (par-i-KAR-dē-al) **cavity,** which is a small space between two membranes that surround the heart. With the exception of the two lungs, all remaining structures within the thoracic cavity form a partition, or *septum,*

[4]Thoracic: Greek *thorac-* = chest + *-ic* = pertaining to.

[5]Abdominopelvic: Latin *abdomino-* = trunk midsection + *pelvis* = bowl.

[6]Diaphragm: Greek *diaphragma* = partition or barrier.

[7]Pleural: Greek *pleura* = lung.

[8]Pericardial: Greek *peri* = around + *kardia* = heart.

11

between the two pleural cavities in the center of the chest. They are collectively referred to as the **mediastinum** (mē′-dē-as-TĪ-num) and include the heart, the thymus, part of the trachea, part of the esophagus, and the major vessels of the heart.

In the region inferior to the diaphragm, the abdominopelvic cavity is divided into two cavities that are separated by an imaginary line extending between the upper tips of the hipbones (called the *iliac crests*). The upper area is called the **abdominal cavity** and contains many organs including the stomach, small intestine, liver, pancreas, spleen, and most of the large intestine. The smaller **pelvic cavity** lies below the iliac crests and is in the shape of a bowl that is formed by the hipbones. It contains the urinary bladder, the final segment of the large intestine, and internal reproductive organs.

Abdominopelvic Regions

The abdominopelvic cavity is the single largest cavity in the body. Consequently, it contains a large number of organs. To aid the physician in locating internal body parts and in identifying sources of surface pain that might arise from this large area, the abdominopelvic cavity is divided further into smaller regions. The regions are separated by invisible lines that are similar in application to the latitudinal and longitudinal lines on a map. In one method of division, there are two horizontal and two vertical lines that divide the cavity into nine regions. These are shown in Figure 1–7a. An alternative method that is also used in hospitals includes just two lines, one horizontal and one vertical, to divide the cavity into four regions, or *quadrants*. These are shown in Figure 1–7b.

Figure 1–7
The regions within the abdominopelvic cavity. It may be divided (a) into nine regions or (b) into four regions. Note how the terms are descriptive of their locations.

Can you identify the two organs that occupy much of the epigastric region?

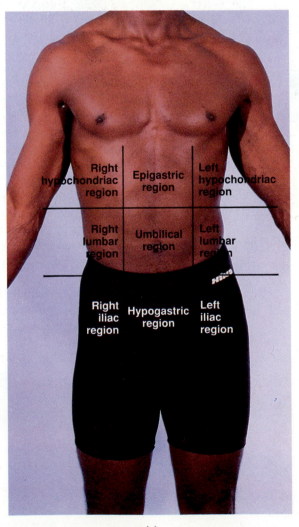

(a)   (b)

CONCEPTS CHECK

1. What are the major regions of the body?

2. What is a body cavity?

3. What are the two major cavities in the body, and what cavities does each contain in turn?

4. What are the ways in which the abdominopelvic cavity may be further divided into smaller regions?

THE CHARACTERISTICS OF LIFE

All living organisms share a set of activities that are characteristic of being alive. These activities are focused at the level of the cell.

All living cells in the human body share a number of characteristic activities that are considered vital to their continued existence. Their collective goal is to maintain the structural and functional organization of the cell. Should one or more of these activities fail, the life of the cell becomes threatened. The vital functions of living cells include organization, metabolism, movement, excitability, and growth and reproduction.

Organization

The molecules that make up our world are governed by a set of physical laws. These laws control the properties of the molecules—that is, how heavy they are, how they react to each other, and how fast they move. Because all the molecules that make up the cell are governed by the same physical laws, they are able to provide the cell with a structural basis that is relatively stable. The stable structure, or **organization,** of the cell makes it possible for the cell to perform its various functions. The stable organization of the cell, in turn, provides a structural foundation for the organization of the body as well.

Metabolism

The process by which the body obtains and uses energy is called **metabolism**[9] (me-TAB-ō-lizm). It requires the exchange of materials with the external environment, for we are not capable of producing our own energy as plants do. When food is consumed, it is brought into the body to be broken down into smaller particles. The particles that are useful to the body as fuel find their way into cells, where they are broken down further to release energy. The energy is either used immediately by the cell or stored within molecules for later use. Energy is used to power all of life's activities, including the synthesis of new materials, the movement of cells and their components, the transport of materials, and the generation of heat.

There are two categories of metabolic processes: **anabolism**[10] (a-NAB-ō-lizm) and **catabolism**[11] (ka-TAB-ō-lizm). Anabolic processes are the ways in which the body uses energy to build large molecules, cells, and tissues from simple molecules; they are the processes of growth and repair. Catabolic processes break apart large molecules, reducing them to simple molecules for the purpose of releasing energy that is immediately available to power body functions.

Movement

The constant movement of molecules within and around a cell is an important feature of a cell's dynamic nature. It is necessary for the transport of vital materials into and out of a cell, the transport of materials through different regions of a cell, and the transport of waste products out of a cell. The vital materials of a cell, which include oxygen and nutrients that are necessary ingredients for anabolic and catabolic processes, must be provided continuously if the cell is to survive. Without these materials the cell would be unable to manufacture important molecules or produce energy to power its functions, and would consequently perish. Oxygen and nutrients must be obtained from the environment, and thus must be transported into the cell. Also, waste products resulting from catabolic processes, such as carbon dioxide and urea, must be transported out of the cell to prevent their poisoning effects.

The cell itself may also move about its environment. For example, many white blood cells wander throughout the body actively searching for invading microorganisms. Muscle cells that are attached to bone also move about, for they change their length by contracting and relaxing to produce the movement of body parts such as arms, legs, and fingers.

---

[9]Metabolism: Greek *metabole* = change + *-ism* = a condition.

[10]Anabolism: Greek *anaballein* = to throw upward + *-ism.*

[11]Catabolism: Greek *kataballein* = to throw downward + *-ism.*

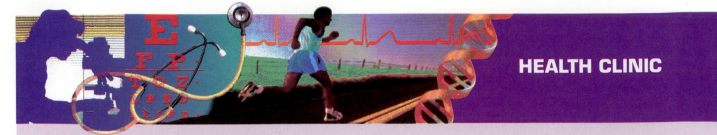

Seeing Through the Body: Diagnostic Imaging

A common problem facing members of the medical community whenever they must help a suffering patient is in determining the source of the afflicting disease. In past years, physicians had to rely on a knowledge of anatomy and physiology that was obtained through studying, through prior experience with other patients, and through dangerously invasive procedures such as surgery. In recent years, however, this knowledge has been given a tremendous boost with technologically advanced instruments that allow one to actually see into the body without having to cut through tissues. The procedures involving the use of these instruments are collectively known as *diagnostic imaging,* and include **computed axial tomography (CAT), positron-emission tomography (PET), ultrasound imaging,** and **magnetic-resonance imaging (MRI).**

CAT scans are produced by beams of energized particles (called *X-rays*) that are focused on a specific plane of the body. The beam is projected from different positions to permit scanning from multiple angles while the patient remains stationary, and the information from the scan is relayed to a computer. The computer processes the information to produce cross-sectional images, or "slices," of body regions. With the aid of additional computer enhancement, three-dimensional images may be generated. CAT scans are useful when cross-sectional images of organs in the chest or abdomen, muscles, and bone are needed. Their speed and relatively low cost make CAT scans the standard for evaluation of trauma to most areas of the body.

PET scans are provided by a scanner that tracks the path and concentration of a radioactive chemical tracer as it moves through the bloodstream and into a body part being imaged. By way of powerful computers, the information can be processed into full-color video maps that distinguish between areas of varying metabolic activity. PET scans are thereby useful in analyzing body function (more so than are still pictures of anatomical structure). For example, they can identify areas where blood flow to the brain is blocked, show how different parts of the brain are using energy, and determine where epileptic seizures originate. PET scans have also been instrumental in providing information on specific brain functions.

Ultrasound imaging, or sonography, involves the pulsation of harmless sound waves through a body region. As the waves travel through tissues of varying density, they produce echoes that can be collected by the instrument. A computer analyzes the echoes and constructs a sectional image that outlines internal body structures. Because of its harmless nature, ultrasound imaging has proven useful in prenatal care by allowing an early view of a developing fetus (a child before birth). However, it is not effective for viewing structures surrounded by bone (the brain and spinal cord), since sound waves cannot penetrate dense objects.

Among the diagnostic imaging techniques available, MRI has generated the most excitement in the medical and research communities. It offers the clearest, most complete images of soft tissues that are currently possible. Although MRI is continually being improved and upgraded, in its present form it utilizes a powerful magnetic field generated within a chamber in which the patient lies. The magnetic field traces the element hydrogen, which is a component of water, in the patient's body. Since bones contain very little water compared with soft tissues, MRI can peer directly through them. In fact, MRI distinguishes between internal structures on the basis of their differences in water content. Once the hydrogen atoms have been detected, the information is analyzed by a computer, which creates a three-dimensional image. Multiple colors may be added by computer enhancement to make the image appear more realistic, or to distinguish more clearly between structures of varying water content. In a recent development, scientists have been successful in tracing blood flow through various organs by MRI—an advance that may open new doors in the study of organ functions. Although the MRI procedure is considered safe, there is no information yet on the long-term effects of strong magnetic fields on the human body.

Opposite

Diagnostic imaging. (a) A CAT scan of a normal brain looking down onto its upper surface (called a *top view*) revealing the brain's anatomical features. (b) A PET scan of a normal brain, top view, which shows (*in color*) the areas of greatest brain activity. (c) A PET scan of a brain affected by Alzheimer's disease. (d) Ultrasound imaging through a pregnant abdomen, revealing a lateral view of a fetus lying on its back. (e) MRI digitized photograph of the brain, top view. Notice the improved clarity and definition in the MRI photograph. (f) A collage of MRI photographs of the brain, all lateral views. This collection was provided courtesy of the Dr. Erik Courcesne research group at San Diego Children's Hospital and UCSD, and demonstrates the variations of MRI currently availiable.

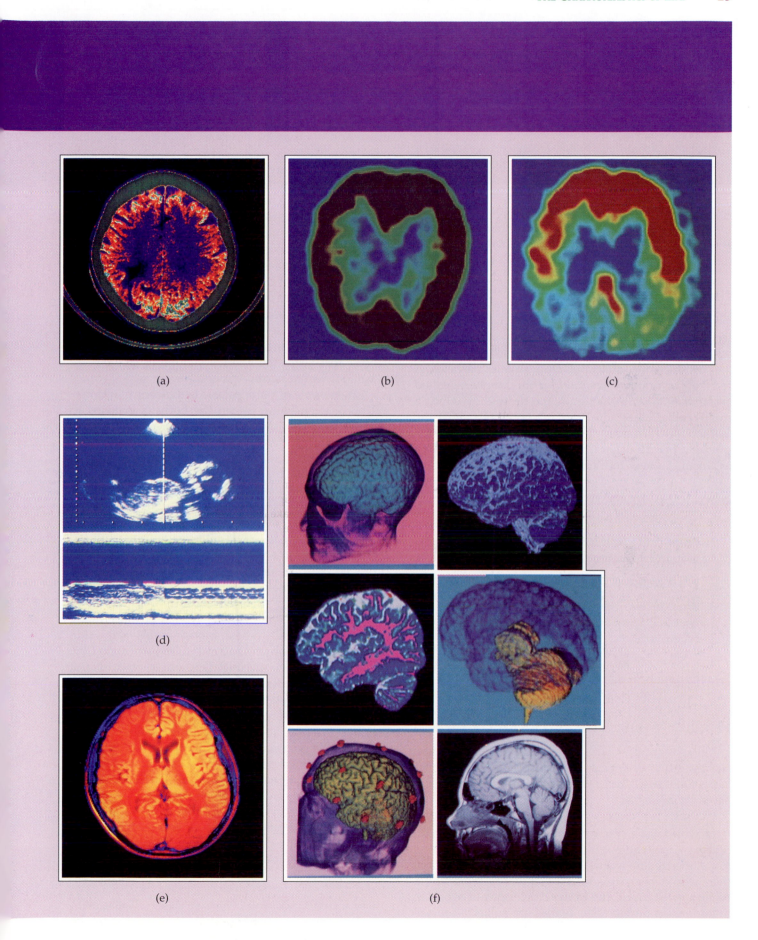

(a)

(b)

(c)

(d)

(e)

(f)

Excitability

The capability of a cell to respond to changes in its environment is called **excitability**, or irritability. An environmental change may be a change in temperature, a change in pressure, an invasion by a foreign substance, exposure to a form of radiation, or exposure to a chemical substance. Environmental changes that influence cells are called *stimuli*. Every cell is capable of receiving stimuli and responding to them. However, different types of cells respond in different ways. For example, certain cells of the nervous and endocrine systems are highly specialized to respond quickly to stimuli and pass their response to other cells. Their goal in this regard is to maintain the body in a stable state despite changes that occur in the environment.

Growth and Reproduction

All cells are capable of growth and reproduction at some stage in their life history. Growth occurs when a cell increases in size as a result of anabolic activities that produce new molecules from smaller particles. A cell that has undergone growth in this manner is called *hypertrophied*. Reproduction is the process by which a single cell divides into two or more cells. It is the method in which dead cells are replaced, growth of tissues and organs occurs, and—in the complete human organism—a new individual originates.

CONCEPTS CHECK

1. What is the collective goal of the functions that are characteristic of life?

2. What is metabolism?

3. What is the difference between the two metabolic processes, anabolism and catabolism?

4. How is the movement of molecules important for life?

HOMEOSTASIS: THE BALANCE OF LIFE

Homeostasis is the process by which the internal environment of the body is kept relatively stable despite changes in the world around us.

If the world were a perfect place to live, survival would be very easy. There would be plenty of healthy food to eat, the climate would remain comfortably constant, there would be no disease; in other words, there would be little *stress* upon the body. However, our world is by no means perfect. Our bodies are subjected to un-

predictable changes in environmental temperature, pressure, water and salt availability, and—in many areas of the world—food availability; microorganisms that are well suited for invading our cells and destroying them abound; and we are being exposed to increasing levels of radiation and chemicals that can alter and destroy cells. How are we able to survive in this world of changing environmental conditions and other hazards?

The human body has the remarkable ability to sense a change in the environment, such as a change in temperature or the invasion of a population of microorganisms, and to respond by making changes in body functions. As a result of the changes in functions, the body's internal environment is kept relatively stable. The process by which the body maintains a stable internal environment despite changes in the external environment is called **homeostasis**[12] (hō′-mē-ō-STĀ-sis). Although the literal translation of the term *homeostasis* is "sameness" or "a state without change," it is not a static, or unchanging, process. Rather, it is a dynamic process that keeps the internal conditions of the body in balance within narrow ranges.

An example of homeostasis is the control of internal body temperature when the body is exposed to cold weather (Fig. 1–8). When you are cold, sensory receptors in your skin that can detect temperature changes relay this information to the brain. The region of the brain that receives this information, the *hypothalamus*, functions as a thermostat for the body. It operates in much the same way as the thermostat in your house: when the temperature is perceived as being too cold, it "turns the heat on" and keeps it on until the temperature returns to the desired level. Body heat is provided by the contraction of small groups of muscles that are stimulated involuntarily by the hypothalamus. Contraction of these muscles causes you to "shiver" from the cold. While groups of muscles are contracting to produce heat, blood vessels in the skin are directed by the hypothalamus to reduce blood flow by closing up, or *constricting*. This reduces the amount of heat that is normally lost through the surface of the skin. The overall effect of making heat by muscle contraction and reducing heat loss through the skin is to stabilize internal body temperature, enabling other body functions to proceed normally despite the drop in the external temperature.

The process of homeostasis is also active when the body becomes overheated. When the hypothalamus receives the information of increasing temperature from nerve cells, it directs the body to make changes that will keep it from becoming too hot. They include perspiring, which cools the skin surface as the water

---

[12]Homeostasis: Greek *homeo-* = similar + *stasis* = standing still.

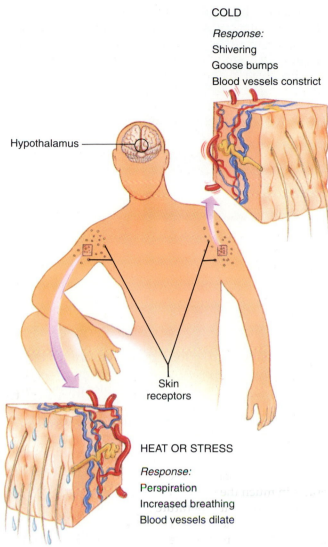

COLD

Response:
Shivering
Goose bumps
Blood vessels constrict

Hypothalamus

Skin receptors

HEAT OR STRESS

Response:
Perspiration
Increased breathing
Blood vessels dilate

Figure 1–8
An example of homeostasis: the regulation of body temperature by the hypothalamus. The hypothalamus responds to changes in body temperature by activating changes in blood-vessel diameter in the skin, and muscle contraction. The result is to maintain body temperature within a narrow margin, which thereby helps maintain body homeostasis.

What are the responses to heat that your body uses to maintain homeostasis?

evaporates; opening the blood vessels in the skin, called *dilation,* to allow an increased volume of blood to carry heat from the deeper regions of the body for its loss through the skin surface; and an increased breathing rate, which moves a larger amount of heat into the outside air through the lungs.

Regulating body temperature is one example of homeostatic mechanisms at work in the body. There are many others, all with the similar objective of keeping the internal environment of the body stable. Ex-

actly how these mechanisms work is the focus of the study of physiology. It is a topic we will be returning to throughout the body of this book.

Negative Feedback

Most systems of the body maintain homeostasis by the operation of **negative feedback** mechanisms. The term *negative* refers to a mechanism that reverses a response back to a normal state. An example of negative feedback occurs when the salt concentration in the bloodstream rises above normal, as often happens after one eats a meal with a high salt content (Fig. 1–9). In response to the rise in salt, the kidneys increase their rate of removing salt from the blood and excreting it with the urine. As a result, the urine becomes concentrated with the excess salt, and the salt concentration in the blood returns to a normal level. The cycle is completed when the kidneys decrease their rate of salt removal from the blood, returning their level of activity to normal. Our previous example of temperature regulation is a negative feedback mechanism also.

Positive Feedback

Positive feedback mechanisms are quite rare in the healthy body. The term *positive* means that when a change from the normal state occurs, the mechanism

Figure 1–9
Negative feedback. Salt levels are restored by the kidney, which responds to an increase in salt content of the blood by increasing salt excretion.

What is the goal of a negative feedback mechanism?

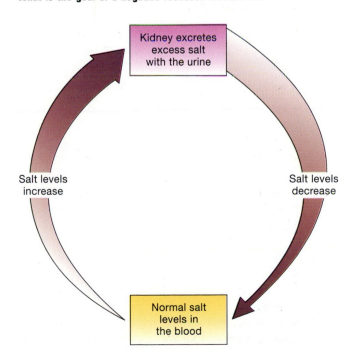

Kidney excretes
excess salt
with the urine

Salt levels
increase

Salt levels
decrease

Normal salt
levels in
the blood

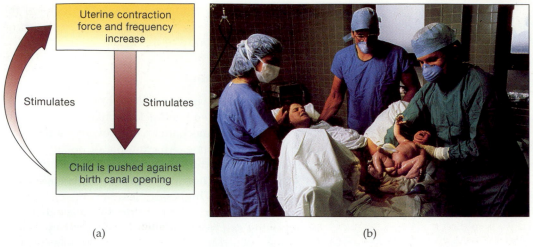

(a)   (b)

Figure 1–10
Positive feedback. (a) An increase in uterine-contraction force and frequency is
further stimulated by the pressure of the fetus against the birth-canal opening.
(b) A photograph of childbirth in progress.

How may positive feedback systems be nonbeneficial?

promotes the change yet further. Unlike negative feedback, positive feedback does not restore a body function to a normal level. Therefore, in some cases, positive feedback can lead to a "vicious circle" of continuing deterioration of homeostasis. However, there are some positive feedback mechanisms that operate under normal conditions. One example is the contraction of the uterus during birth (Fig. 1–10). In this example, increased contractions of the uterus push the baby against the opening to the birth canal (called the *cervix*). Pressure against the birth canal sends signals to the brain to stimulate further contractions. This positive feedback continues until the child passes completely through the canal.

CONCEPTS CHECK

1. What is the definition of homeostasis?

2. What is an example of a homeostatic mechanism?

3. How does a negative feedback mechanism operate?

HEALTH AND DISEASE

The disruption of homeostasis leads to an impairment of normal function. This condition is called disease.

Optimum health is a state in which body structure and function provide the individual with a complete sense of physical and emotional well-being. It is the result of all body parts' functioning in a cooperative manner to maintain homeostasis. Any reduction of this ideal state is regarded as **disease.** All diseases disrupt the sensitive balance that is achieved with homeostasis. From this broad definition, we can see that disease can vary from a minor change so small that it does not yet cause detectable pain to a severe, life-threatening illness that destroys all sense of well-being. The task of the physician is to identify the disease in its early stages so that effective action can be taken to restore health. Health is restored when the mechanisms that operate homeostasis have returned to their normal functioning states.

In most cases, a disease causes certain structural changes in the body, called **lesions,** that can be identified by a trained professional. A lesion that is recently acquired and expected to last for a short time is called **acute,** and one that is of longer duration is termed **chronic.** Lesions are often accompanied by certain sensations, such as weakness, dizziness, or pain. These sensations are referred to as **symptoms.** The process of identifying a disease from its symptoms and lesions is called a **diagnosis,** and is the first step in the treatment of the condition. A physician who specializes in identifying diseases is called a *pathologist.*

Classification of Disease

Diseases may be classified into groups that share similar developmental histories or cause similar lesions. These categories provide us with a basis for understanding the nature of human disease. From this foundation, we shall, in subsequent chapters, discuss the nature of individual diseases that are relevant to today's world. These discussions will for the most part appear in boxed, colored sections apart from the basic text.

HEALTH CLINIC

Infectious Agents: A Source of Many Diseases

Infectious agents are a variety of microscopic organisms that may invade the body for the benefit of their own survival. Soon after invasion, they often cause a disruption of homeostasis that leads to disease. The infectious agents include viruses, bacteria, fungi, and protozoans (see the figure).

Viruses (Latin *virus* = slimy, poisonous liquid) are not true cells, but are particles that are much smaller than cells and are unable to perform any of the characteristic activities of living cells without assistance. Structurally, they consist of an inner core that contains only one type of nucleic acid (RNA or DNA) surrounded by a protein coat. To reproduce, a virus must invade a living cell. The cell that is invaded is called the *host*. During invasion, a virus penetrates the cell's outer membrane and inserts its nucleic acid into the DNA of the cell. The viral nucleic acid then directs the host's metabolic machinery to synthesize new viruses. Once a certain number of virus particles have been synthesized, they burst out of the host to

infect other cells of the body. Damage to the host occurs during the release of viral particles, which causes cell death. This cycle of virus reproduction is common among all known viruses, although the time between cell penetration and release of new particles varies. Some of the more well-known examples of viruses include the herpes simplex viruses, which cause *oral* and *genital herpes*; poliovirus, which causes *polio*; the hepatitis viruses, which cause serious infections of the liver; influenza viruses, which cause the *flu*; measles virus; and the virus that causes *AIDS*, human immunodeficiency virus (HIV).

Bacteria are single-celled organisms that lack a nucleus. Unlike viruses, each bacterium is a true cell, for the functions of organization, metabolism, movement, excitability, and growth and reproduction all occur. Many types of bacteria are enclosed within a rigid cell wall in addition to the cell membrane, and all are larger in size than viruses but smaller than human cells.

(continued)

Infectious agents. (a) A color-enhanced electron micrograph of a single virus *(red)* that has been chemically stimulated to release its DNA *(green strands)*. Magnification 7500x. (b) A photomicrograph of a colony of the spherically shaped bacteria that cause scarlet fever, known as *Streptococcus scarlatinae.* Magnification 500x. (c) A color photomicrograph of the fungus *Candida albicans,* which causes infections of various parts of the body including the skin, mouth, vagina, bronchi, and lungs. Magnification 250x. (d) An electron micrograph of the protozoan *Giardia,* which contaminates drinking water and can cause dysentery. Magnification 1200x.

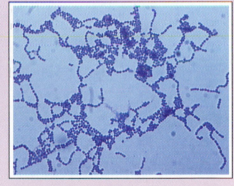

(a)

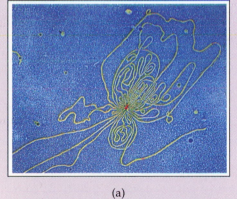

(b)

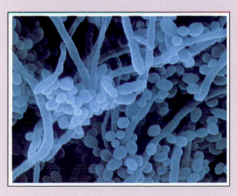

(c)

(d)

Health Clinic (continued)

Approximately 90% of the known bacteria are not harmful to humans in their normal concentrations, and many types are actually beneficial. For example, the bacterium *E. coli* is a common inhabitant of our intestinal tract and plays a beneficial role in providing us with vitamin K. However, if *E. coli* gains entry into other body areas, it can cause serious infections. Thus, even beneficial bacteria may cause serious diseases if they are not controlled effectively. There are basically two ways that a bacterium may cause disease: by directly invading a healthy tissue and killing its cells; or by releasing a damaging substance called a *toxin*. Some bacteria that invade tissues release toxins as well. A toxin can be an extremely powerful poison. For example, it has been estimated that only about one ounce of purified toxin from the bacterium that causes *diphtheria* is enough to kill everyone in New York City. The toxin produced by the bacterium that causes the disease *botulism (Clostridium botulinum)* is, incredibly, 400 times as lethal as diphtheria toxin!

Some of the many diseases caused by bacterial infection include *tetanus, botulism, diphtheria, cholera, leprosy, toxic shock syndrome, bacterial pneumonia,* and *syphilis.* Each of these diseases can cause death if not treated. The most common form of treatment for bacterial diseases is the use of antibiotics. Many antibiotics are obtained from colonies of fungi that produce these chemicals naturally in order to compete with bacteria. More recently, they are also obtained by genetic manipulation of fungi to produce greater varieties than exist naturally. Antibiotics are effective as a method of treatment because they inhibit some necessary physiological process of the bacteria while doing little harm to the patient. Although antibiotics are effective against bacteria, they have no effect on viruses.

The **fungi** are a group of simple organisms that include yeasts, molds, and mushrooms. They may be single-celled, as yeasts are, or multicellular. All fungi obtain energy by absorbing organic material from a host, which may be living or dead. Although most fungi are not harmful to humans, some varieties may become so if their populations are allowed to grow unchecked. Circumstances that permit harmful fungal growth within the body include an impairment of the immune response, and a disruption in the balance between bacterial and fungal populations that normally keep each other in check. The diseases that result from fungal infection include *candidiasis, ringworm, athlete's foot,* and *histoplasmosis.* Their nature of infection is through direct invasion of tissue. Once within tissue, the fungi destroy individual cells in order to obtain nourishment from them. Treatment of fungal infections with various antifungal antibiotics is effective.

A **protozoan** is a single-celled organism that is capable of moving about to obtain food, which it ingests in order to obtain energy. Most protozoans are similar in size to human cells. Like bacteria and fungi, most protozoans are not harmful to humans. However, the few that are can often cause serious diseases. For example, *malaria* is caused by a protozoan that infects blood, called a *plasmodium.* It afflicts over 100 million people throughout the world and causes about 1 million deaths each year. Other examples include *amebiasis,* which is a serious infection of the intestinal tract, and *toxoplasmosis,* which is an infection caused by a small protozoan common in household cats that can be passed to humans. If a pregnant woman becomes afflicted with toxoplasmosis, the infection may spread to the newborn child she carries and cause severe injuries that often lead to congenital defects.

Congenital Diseases

Congenital diseases arise at some time before birth. They can be inherited from parents or be caused by a disease-causing agent that crosses the placental barrier. In most cases, the genetic code that determines what we are is altered. Examples include Down's syndrome (formerly called mongolism), which is the most common congenital disorder, with an incidence of 1 in 600 births; Tay-Sachs disease, caused by the genetic inability to synthesize a certain enzyme; and congenital heart disease that often results when the German measles virus is carried by the mother during pregnancy.

Immunological Diseases

Diseases in the immunological category involve a reaction by the body to an invasion by foreign substances. The defensive reactions involve a series of responses by special cells of the blood called *white blood cells. Infection,* or invasion of the body by bacteria, viruses, fungi, and protozoa, is the most important cause of disease in humans. In many cases the body responds by a process called *inflammation,* which is an important method of directing white blood cells to the site of infection. Unfortunately, inflammation causes uncomfortable symptoms such as sore throat, pain from swollen tissues, and coughing. The common cold, allergies, pneumonia, and AIDS are examples of immunological diseases.

Metabolic Diseases

Recall that metabolism is the process by which body cells produce and use energy. If this important function is disturbed, the energy required by the body will not be available and health will quickly fail. All diseases potentially disrupt the energy balance in the body, upsetting homeostasis. However, certain diseases affect metabolism directly. These are metabolic diseases, and include diabetes and other disturbances to glands of the endocrine system. They also include physical injury, or *trauma*, which results in fluid loss that upsets the metabolic balance within cells.

Neoplastic Diseases

Important functions of healthy cells include normal growth and reproduction. If these functions become abnormal, the cells develop into lesions called *tumors* that threaten the normal activities of tissues and or-

gans. The various types of cancer arise in this manner, and serve as examples of neoplastic disease.

CONCEPTS CHECK

1. What is the relationship between health, disease, and homeostasis?

2. What are the four types of human disease?

CHAPTER SUMMARY

DIVISIONS OF STUDY

The study of the human body is divided into two areas of discipline: anatomy and physiology.

Anatomy Anatomy is the study of structure. It asks three basic questions: What does it look like? Where is it located? and, How does it relate to other body parts? Anatomy is divided into several areas of discipline:

Gross Anatomy The study of structures visible without the aid of a microscope.

Microanatomy The study of structures that require the use of a microscope.

Systematic Anatomy The study of structures organized into systems, such as the digestive system or the nervous system.

Regional Anatomy The study of structures organized into body regions, such as the head, the neck, or the trunk.

Physiology Physiology is the study of body function. It asks the basic question, How does it work? Its theme is how the body maintains a relatively stable internal environment to optimize body activities.

BASIC TERMINOLOGY

The language that is used to name body parts and their locations is based on Latin and Greek word parts. It is universal in its application.

Directional Terms The terms that are used to describe the relative locations of structures are listed in Table 1–1.

Body Planes Planes are imaginary flat surfaces. They are used to section the body in different directions for the purpose of viewing internal structures. The major types of planes include the following:

Sagittal Extends along the long axis of the body, dividing it into right and left portions.

Frontal Extends along the long axis of the body, dividing it into anterior and posterior portions.

Horizontal Extends perpendicular to the long axis of the body, dividing it into superior and inferior portions.

STRUCTURAL LEVELS OF ORGANIZATION

The body is organized by a series of building-block components:

Chemical Level Chemicals are substances that cannot be simplified further under natural conditions. They are composed of atoms, which are bonded together to form ions and molecules. The important large molecules in the body are proteins, carbohydrates, fats, and nucleic acids.

Cellular Level The cell is the basic structural and functional unit of life.

Tissue Level A tissue is a group of similar cells that, when combined, perform a common function.

Organ Level Organs are structures that have distinct structures and functions. They are composed of two or more different tissue types.

System Level A system is a combination of at least one organ and its associated structures. When the components of a system are combined, they perform a general function.

The Complete Organism The organism is a complete individual, composed of all the structural building blocks at the lower levels.

THE BODY PLAN

The basic structural plan of the body consists of numerous regions. The large internal areas contain cavities, which house organs.

Body Regions The major body regions are the head, neck, trunk, upper appendages, and lower appendages. These are further divided into the regions indicated in Table 1–3.

Body Cavities The major body cavities include the following:

The dorsal cavity contains a cranial cavity that houses the brain, and a vertebral canal that houses the spinal cord.

The ventral cavity contains the thoracic cavity superior to the diaphragm, and the abdominopelvic cavity inferior to the diaphragm.

The thoracic cavity contains two pleural cavities that surround the lungs, the pericardial cavity around the heart, and the mediastinum, which contains the heart, its major vessels, the esophagus, the trachea, and other nearby structures in the center of the thorax.

The abdominopelvic cavity contains the abdominal cavity in its larger superior portion, and the pelvic cavity in the inferior portion. The abdominal cavity contains the liver, stomach, pancreas, spleen, and small intestine; the pelvic cavity contains the internal reproductive organs and the lower end of the large intestine.

Abdominopelvic Regions   The abdominopelvic cavity is divided by a series of invisible lines into nine regions for the purpose of providing an aid for identifying internal sources of surface pain. Alternatively, it may be divided into four regions, or quadrants, by another series of lines.

THE CHARACTERISTICS OF LIFE

All cells that are alive are characterized by several functions. These functions are necessary to sustain the living state.

Organization   Living cells contain smaller units that have structural organization. These units obey physical laws to organize themselves into the basic unit of life, the cell. Organization is required to support functions.

Metabolism   The ways in which the cell obtains and uses energy constitute metabolism. Metabolism consists of anabolism, which is the process of obtaining energy to build more complex structures; and catabolism, by which energy locked inside chemical bonds is freed and made available to do work.

Movement   The movement of molecules must occur in living cells so that essential substances required to support life, particularly oxygen and nutrients, can enter the cell. Also, waste products from catabolism must exit the cell to prevent their poisoning effects.

Excitability   The ability of a living cell to perceive a change in the environment and respond to it is excitability. This is important if the cell is to survive during periods of change.

Growth and Reproduction   All cells grow, or increase in size because of anabolic activities, at some stage in their life cycle. Reproduction is the development of two new cells from an original parent cell.

HOMEOSTASIS: THE BALANCE OF LIFE

In order to survive, the body as a whole must be able to maintain a relatively stable internal environment despite fluctuations in the external environment. In many ways, the body operates homeostatic mechanisms similar to the operation of a thermostat in a house.

Negative Feedback   Mechanisms that restore the body to normal levels of function.

Positive Feedback   Mechanisms that promote a condition farther from normal levels of function.

HEALTH AND DISEASE

Health is an ideal state in which the body is properly maintained by the processes of homeostasis. Disease is defined as any alteration of homeostasis. Health and disease are therefore relative terms; we are constantly fluctuating between the states of optimal health and serious illness. The types of diseases are varied; they are categorized into four groups:

Congenital Diseases   Diseases that arise before birth. They may be inherited or acquired.

Immunological Diseases   Diseases that involve the invasion of a substance foreign to the body.

Metabolic Diseases   Diseases that upset the metabolic balance of cells directly. These include trauma.

Neoplastic Diseases   Diseases that involve the proliferation of cells that divide without the proper controls.

KEY TERMS

abdominopelvic (p. 11)	catabolism (p. 13)	homeostasis (p. 6)	physiology (p. 4)
acute (p. 18)	chronic (p. 18)	mediastinum (p. 12)	pleural (p. 11)
anabolism (p. 13)	diaphragm (p. 11)	metabolism (p. 13)	thoracic (p. 11)
anatomy (p. 4)	excitability (p. 16)	pericardial (p. 11)	

QUESTIONS FOR REVIEW

OBJECTIVE QUESTIONS

1. Which of the following divisions of study focus on how the body maintains itself despite changes in the external environment?
   a. gross anatomy   c. physiology
   b. systematic anatomy   d. microanatomy

2. The trunk is inferior to which body part?
   a. head   c. knee
   b. back   d. left foot

3. The brain within the cranial cavity lies _____ relative to the bones of the skull.
   a. deep   c. superficial
   b. distal   d. anterior

4. The right elbow is distal to the:
   - a. left elbow
   - b. right shoulder
   - c. digits of the right hand
   - d. left kneecap

5. When the body is hanging upside down, the thorax is always: _____ to the diaphragm.
   - a. anterior
   - b. inferior
   - c. superior
   - d. medial

6. The body plane that divides the body into right and left portions is called the:
   - a. frontal plane
   - b. horizontal plane
   - c. sagittal plane
   - d. jet plane

7. The basic structural and functional unit of life is the:
   - a. molecule
   - b. protein
   - c. tissue
   - d. cell

8. A building block of the body that is composed of a group of similar cells that combine to perform a certain function is called a(n):
   - a. tissue
   - b. molecule
   - c. system
   - d. organ

9. The body cavity that lies superior to the diaphragm and contains the heart is the:
   - a. ventral cavity
   - b. mediastinum
   - c. pericardial cavity
   - d. all of the above

10. The abdominopelvic cavity:
    - a. lies superior to the diaphragm
    - b. lies within the trunk region
    - c. contains the lungs
    - d. does not contain organs

11. The characteristic function of living cells that involves the ways in which the body breaks apart molecules to release energy is called:
    - a. homeostasis
    - b. metabolism
    - c. movement
    - d. excitability

12. The ability of a cell to perceive a change in the environment and respond to it is called:
    - a. anabolism
    - b. growth
    - c. excitability
    - d. organization

13. The ability of the human body to change its functions in order to compensate for a change in the environment, for the purpose of maintaining internal stability, is called:
    - a. disease
    - b. metabolism
    - c. homeostasis
    - d. excitability

14. Any state in which the internal balance of the body becomes disturbed is called:
    - a. health
    - b. disease
    - c. homeostasis
    - d. symptom

15. A disease that is characterized by the presence of foreign substances as the primary cause is called:
    - a. congenital
    - b. metabolic
    - c. immunological
    - d. neoplastic

ESSAY QUESTIONS

1. List the six levels of structural organization and discuss how they must interrelate to support the survival of the complete organism.

2. Describe the plan of the body by identifying the locations of the body regions, cavities, and major organs.

3. Define homeostasis, and describe why it is a key factor in maintaining health. Provide an example, not described in the text, of a homeostatic mechanism at work.

ANSWERS TO ART LEGEND QUESTIONS

Figure 1–1
Can you identify a body part that is proximal to the hand? Body parts proximal to the hand include the wrist, arm, elbow, and shoulder.

Figure 1–2
How are the hands oriented in the anatomical position? The palms of the hands are faced anteriorly, with fingers spread.

Figure 1–3
A sagittal plane divides the body into what parts? A sagittal plane divides the body into right and left parts.

Figure 1–4
What is the organized combination of two or more different types of tissue into a single structure called? Two or more tissues may be organized to form an organ.

Figure 1–5
What is the location of the axilla relative to the thigh (in directional terms)? The axilla is superior to the thigh.

Figure 1–6
What major organs occupy the thoracic cavity? The heart, the two lungs, and the thymus occupy the thoracic cavity.

Figure 1–7
Can you identify the two organs that occupy much of the epigastric region? The liver and stomach are the primary occupants of the epigastric region.

Figure 1–8
What are the responses to heat that your body uses to maintain homeostasis? Vasodilation of blood vessels in the skin, perspiration, and increased respiration rate are the primary responses to heat.

Figure 1–9
What is the goal of a negative feedback system? A negative feedback system seeks to reverse a response back to a normal state.

Figure 1–10
How may positive feedback systems be nonbeneficial? Positive feedback systems may promote a response away from a normal state.

THE CHEMICAL BASIS OF THE BODY

LEARNING OBJECTIVES

After studying this chapter, you should be able to:

1. Define element, atom, molecule, and compound and distinguish between them.

2. Describe how chemical bonds are formed.

3. Distinguish between ionic and covalent bonds.

4. Describe a chemical reaction that results in the formation of new products.

5. Distinguish between organic and inorganic compounds.

6. Identify the important features of water.

7. Describe salts, acids and bases, and define pH.

8. Distinguish among carbohydrates, lipids, proteins, and nucleic acids on the basis of their chemical structures and the roles they play in the body.

The most basic building blocks of all substances in the universe, whether living or not, are nonliving units called **atoms** and **molecules.** The properties that make each of them unique are the basis for the variation we see in the world around and within us: whether a rock crystal, a sycamore tree, or a human brain. In this chapter we will learn about the nature of atoms and molecules as a basis for understanding the structure and functions of the human body.

THE COMPOSITION OF MATTER

Matter is the stuff of life. Its most basic unit is the atom, which combines chemically with other atoms to form molecules.

Matter is the basic material of the universe. It is anything that occupies space and has mass (mass is the amount of matter an object contains; it has weight when it is subjected to the force of gravity, as on the surface of the earth). It may occur in any of three forms: gas, liquid, or solid. A familiar example is water: it exists in the gaseous state as steam, in the liquid state as fluid water, and in the solid state as ice (Fig. 2–1). Regardless of the form it takes, all matter is composed of an amazingly small variety of basic units that may combine in different ways to form an almost infinite variety of substances.

Atoms and Chemical Elements

Atoms are the smallest units of matter that have their own distinct set of properties. They are classified according to these properties into the various types of **chemical elements.** Each chemical element is composed of atoms that share the same characteristics but differ from the atoms of any other element. For example, all atoms of the element sodium have the same characteristics, which differ from those of oxygen, gold, uranium, and any other element. Each chemical element may therefore be distinguished from other elements on the basis of its atoms.

Figure 2–1
The three states of matter. This geyser and its surrounding area in Yellowstone National Park demonstrate the three states of H_2O: steam (a gas), water (a liquid), and snow (a solid).

Table 2–1	ELEMENTS OF THE HUMAN BODY		
Element	Symbol	Atomic Number	% Body Mass
Oxygen	O	8	65
Carbon	C	6	18
Hydrogen	H	1	10
Nitrogen	N	7	3
Calcium	Ca	20	1.5
Phosphorus	P	15	1.0
Potassium	K	19	0.4
Sulfur	S	16	0.3
Sodium	Na	11	0.2
Magnesium	Mg	12	0.1
Chlorine	Cl	17	0.1
Total			**99.6**
Iron	Fe	26	Trace
Iodine	I	53	Trace
Manganese	Mn	25	Trace
Copper	Cu	29	Trace
Zinc	Zn	30	Trace
Cobalt	Co	27	Trace
Fluorine	F	9	Trace

Over the years, scientists have been able to identify 109 types of chemical elements, all of which are represented in a chart known as the *periodic table of the elements* (shown in Fig. 2–2). Of these, only about one-third are found in the human body. Table 2–1 shows the elements of the human body in their relative concentrations. Notice that the first four elements constitute over 96% of the total body weight (oxygen: 65%; carbon: 18%; hydrogen: 10%; nitrogen: 3%). When calcium and phosphorus are included, almost 99% of body weight is accounted for. The remaining 20 elements are found in very low concentrations and account for the remaining amount.

When elements are referred to in a discussion or an equation, they are often expressed in an abbreviated form that is called a **chemical symbol.** The symbol is usually taken from the first one, two, or three letters of the English or Latin name of the element. Examples of chemical symbols include O for oxygen, C for carbon, Ca for calcium, and Fe for iron (from *ferrum,* the Latin for iron).

Atomic Structure

Each atom consists of two basic regions: a **nucleus** and an **electron shell.** The nucleus is located in the center of the atom and contains two basic types of particles: **protons** and **neutrons.** Protons and neutrons have

Group
IA

																	VIIA	VIII
1 H	IIA											IIIA	IVA	VA	VIA	1 H	2 He
3 Li	4 Be											5 B	6 C	7 N	8 O	9 F	10 Ne
11 Na	12 Mg	IIIB	IVB	VB	VIB	VIIB	⎡—VIIB—⎤			IB	IIB	13 Al	14 Si	15 P	16 S	17 Cl	18 Ar
19 K	20 Ca	21 Sc	22 Ti	23 V	24 Cr	25 Mn	26 Fe	27 Co	28 Ni	29 Cu	30 Zn	31 Ga	32 Ge	33 As	34 Se	35 Br	36 Kr
37 Rb	38 Sr	39 Y	40 Zr	41 Nb	42 Mo	43 Tc	44 Ru	45 Rh	46 Pd	47 Ag	48 Cd	49 In	50 Sn	51 Sb	52 Te	53 I	54 Xe
55 Cs	56 Ba	57 *La	72 Hf	73 Ta	74 W	75 Re	76 Os	77 Ir	78 Pt	79 Au	80 Hg	81 Tl	82 Pb	83 Bi	84 Po	85 At	86 Rn
87 Fr	88 Ra	89 **A	104 Unq	105 Unp	106 Unh	107 Uns	108 Uno	109 Une									

*Lanthanide Series

58 Ce	59 Pr	60 Nd	61 Pm	62 Sm	63 Eu	64 Gd	65 Tb	66 Dy	67 Ho	68 Er	69 Tm	70 Yb	71 Lu

**Actinide Series

90 Th	91 Pa	92 U	93 Np	94 Pu	95 Am	96 Cm	97 Bk	98 Cf	99 Es	100 Fm	101 Md	102 No	103 Lr

Figure 2–2
The periodic table of the elements. The elements found in the human body are highlighted.

about the same mass, and make up almost the entire mass of the atom. The electron shell is the region surrounding the nucleus and is composed of one or more swiftly moving particles known as **electrons.** The electrons provide an extremely small contribution to the mass of an atom—each is only 1/1800 the mass of a proton (Fig. 2–3)!

The electron shell represents an approximation of the location of the electron or electrons that orbit about the nucleus. You may have seen a diagram showing the *planetary model* of the atom as in Figure 2–4a before. This is a highly simplified model of atomic structure that depicts electrons moving about the nucleus in fixed, circular orbits like the planets orbiting

Figure 2–3
Structure of an atom. This atom of oxygen contains eight electrons in the electron shell that orbit about the nucleus. The nucleus contains eight protons and eight neutrons.

What particles make up the nucleus of an atom?

ⓟ Proton
ⓝ Neutron
● Electron

Oxygen atom

Figure 2–4
The two ways of representing an atom. (a) The planetary model of the carbon atom. (b) The orbital model of the same atom. Although the orbital model is more accurate, it is more complex because of its three-dimensional nature.

Is an atom three-dimensional in nature?

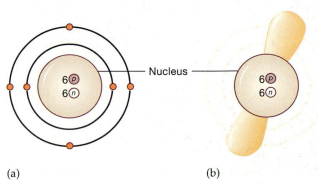

Nucleus

(a)

(b)

the sun. However, this is not accurate, because it is not possible to determine the exact location of an electron at any given time, since electrons jump around the space surrounding the nucleus in unknown trajectories at nearly the speed of light. A more accurate model, called the *orbital model,* is sometimes used to describe electron movement. In this model the region surrounding the nucleus that a given electron or electron pair may probably be found within is indicated. This region is called an **electron orbital** (Fig. 2–4b). You may be better able to appreciate the spatial relationship between an electron orbital and a nucleus by imagining an atom with a single proton and electron enlarged to the size of a football field. An atom of this size would have a nucleus represented by a solid lead ball no larger than a golf ball in the exact center of the field. The electron would be represented by a tiny mosquito buzzing randomly about, keeping mostly within the limits of the field. Although the orbital model is a more accurate picture of the atom, we will use the planetary model for most of the remaining descriptions of atomic structure for the sake of simplicity.

Electric Charge of the Atom

Scientists have known for many years that the number of protons equals the number of electrons in an atom. They have also discovered that protons and electrons exhibit a property that is known as **electric charge.** Protons have a positive (+) charge, and electrons have a negative (−) charge. Neutrons have no electric charge. Since the number of protons equals the number of electrons in an atom, the positive charges equal the negative charges. The atom is thus *electrically neutral.* For example, a carbon atom has six protons and six electrons, and therefore no electric charge. This is a general principle that applies to all atoms.

In some cases, an atom may gain or lose an electron. If this occurs, the atom becomes electrically charged and is henceforth called an **ion.** Thus, all atoms are electrically neutral and all ions are electrically charged. We will discuss later how ions are formed and the important roles they play.

Atomic Number and Weight

As we have now seen, the atoms that make up one type of element are different from the atoms that make up another type of element. For example, the atoms that form the element carbon differ from the atoms that make oxygen. What makes atoms different? Simply, the *numbers* of protons, neutrons, and electrons that are present. Each element contains a particular number of protons and electrons in each of its atoms. The number of protons in each atom is a characteristic feature of the element, and is called its **atomic number** (Fig.

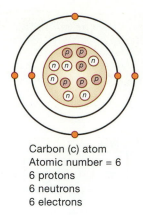

Carbon (c) atom
Atomic number = 6
6 protons
6 neutrons
6 electrons

Figure 2–5
The carbon atom. The electrical charges balance with equal numbers of electrons (−) and protons (+). This atom is thus electrically neutral.

What is the electrical charge on a neutron?

2–5). Another characteristic feature of an element is its **atomic weight.** It is the mass of all atomic particles within an atom, and is usually represented as the sum of the protons and neutrons in the nucleus of an atom within the element (because the electrons contribute a very small mass to the total atomic weight).

Molecules and Compounds

Atoms do not exist only as isolated units, but often occur in combinations of two or more atoms to form more complex units of matter called **molecules.** Molecules have their own set of properties, which are quite different from the properties of the atoms that make them up.

A molecule may be formed by the combination of two or more atoms from the same element. For example, two atoms from the element hydrogen (H) may combine to form a molecule of hydrogen gas (H_2). Similarly, two atoms from the element oxygen (O) may combine to form a molecule of oxygen gas (O_2).

A molecule may also be formed by the combination of two or more different kinds of atoms. These molecules are called **compounds.** When two atoms of the element hydrogen combine with one atom of the element oxygen, a compound called water (H_2O) is formed. The combination of a single atom of hydrogen and four atoms of carbon forms the compound methane (CH_4).

CONCEPTS CHECK

1. What is the relationship between matter, elements, and atoms?

2. What are the four most common elements that occur in the body?

3. What is the arrangement of protons, neutrons, and electrons within an atom?

4. How are atoms, molecules, and compounds related?

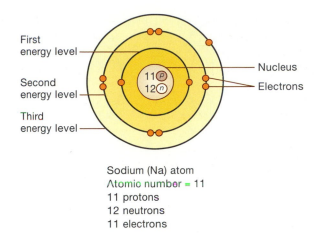

Sodium (Na) atom
Atomic number = 11
11 protons
12 neutrons
11 electrons

Figure 2–6
The sodium atom. Three energy levels contain electrons: the first level contains two, the second contains eight, and the third contains one electron.

How many energy levels would an atom have if its shell contained only two electrons?

CHEMICAL BONDS

Chemical bonds are formed between atoms when the electrons in their outer energy levels are gained, lost, or shared.

When two or more atoms combine to form a molecule, they are held together by an invisible force of attraction that is called a **chemical bond.** A chemical bond is not a particle of matter that creates a physical bridge linking two atoms together, but is an energy relationship between the electrons and protons of the atoms that are joining together.

Electrons and Chemical Bonding

We have seen from our earlier description of the orbital model of atoms that electrons exist within a region of space that surrounds the nucleus. This space is called an electron orbital. To take this discussion a step further, the electron orbitals themselves are confined to a space that is determined by the amount of energy the electrons have. This region of space is called an **electron shell.** The electrons with the lowest amounts of energy exist in the shell that is closest to the nucleus, while those with higher energy exist in the shells that are farther from the nucleus. Because each electron shell represents a different **energy level,** these terms are used interchangeably.

The number of electron shells an atom contains depends on the number of electrons that travel about its nucleus, since each shell has the maximum number of electrons it can hold. For example, the first shell of all atoms, which has the lowest energy level and is located closest to the nucleus, holds a maximum of only two electrons. An atom that contains three electrons must therefore have two shells: the first shell would hold its maximum of two electrons, and the second shell would hold the remaining electron. In general, electrons fill the lower energy shells to their maximum capacity first. Any additional electrons travel in shells that correspond to higher energy levels (Fig. 2–6).

The electrons in the shells with the highest energy level determine many of the important properties of the atom. One property is the atom's *stability,* or relative tendency to change into a different form. If an atom contains a number of electrons that fills the outer electron shell to its capacity, the atom will not tend to change into a different form. The atom is in a stable form that is called **inert.** An example of an inert atom is helium, which has two electrons that fill its only electron shell (Fig. 2–7). Conversely, an atom that contains a number of electrons that does not fill the outer shell has a strong tendency to change into a different form. This type of unstable atom is called **reactive.** The hydrogen atom is reactive because it contains only one electron, which is one less than is necessary to fill the first energy level. Reactive atoms seek to fill their outer energy levels in order to become more stable.

Figure 2–7
The hydrogen (*left*) and helium (*right*) atoms. Hydrogen is a reactive atom because its single electron does not fill the first electron shell, which has a capacity of two electrons. It therefore has a tendency to change into a different form by reacting with other atoms. Helium, on the other hand, is inert because its two electrons fill the first electron shell. This provides it with a high degree of stability.

Is the helium atom electrically neutral?

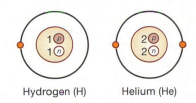

Hydrogen (H)   Helium (He)

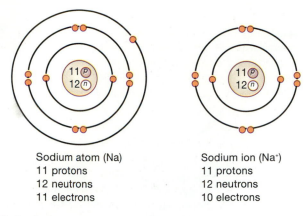

Sodium atom (Na)
11 protons
12 neutrons
11 electrons

Sodium ion (Na⁺)
11 protons
12 neutrons
10 electrons

Figure 2–8
The sodium atom (*left*) and the sodium ion (*right*). The sodium ion has lost an electron, with the result that it contains one more proton than the number of electrons. It thus carries a positive charge.

Why doesn't the sodium atom gain seven electrons instead of losing one?

Types of Chemical Bonds

As we have seen, reactive atoms have a tendency to change because their outer energy levels are not filled with the maximum number of electrons. One common type of change that occurs between two or more reactive atoms is the formation of a chemical bond. A chemical bond is formed when a reactive atom fills its outer energy level with electrons from another reactive atom, bringing both atoms close together. There are two major ways in which this may happen: by gaining or losing electrons in the outer energy levels of the bonding atoms, or by sharing the electrons in the outer energy levels.

Ionic Bonds

Recall that atoms are electrically neutral because the numbers of protons and electrons are equal, balancing the electrical charge on the atom. However, this balance is changed in reactive atoms when they gain or lose electrons. If an atom gains an electron, its electrons outnumber its protons, providing it with an electrical charge that is overall negative. If an atom loses an electron, it acquires a positive charge. The resulting particle with a negative or positive charge is called an **ion**. The relationship between atoms and ions is shown in Figure 2–8.

An **ionic bond** occurs when a positive ion and a negative ion are held together by the forces that are exerted by particles of opposite charge. These forces of attraction are called *electrostatic*. An example of an ionic bond is seen in ordinary table salt, NaCl (sodium chloride), and is illustrated in Figure 2–9. A sodium atom contains 11 protons and 11 electrons, providing it with a single electron orbiting in the third electron shell (there are two electrons in the first shell, eight in the second shell, and one in the third shell). It has a tendency to lose this electron to another atom, since the third electron shell requires eight electrons to be filled, and it is far more efficient to lose a single electron than to gain seven. A chlorine atom contains seventeen protons and seventeen electrons, providing it with seven electrons in its outer electron shell (two in the first shell, eight in the second shell, and seven in the third). Chlorine has a tendency to gain an electron, since it takes less energy to add one electron to fill its outer shell than to lose seven electrons. Consequently, the single electron from the sodium atom is taken through electrostatic force by the chlorine atom. When this oc-

Figure 2–9
Formation of an ionic bond. Sodium loses its outer electron to chlorine, resulting in the formation of two electrically charged ions (Na⁺ and Cl⁻) that bond to one another to form the compound NaCl. The force of attraction that pulls the electron from sodium to chlorine and holds the ions together is an electrostatic force, and the bond is called an *ionic bond*.

Which substance in this figure is a molecule?

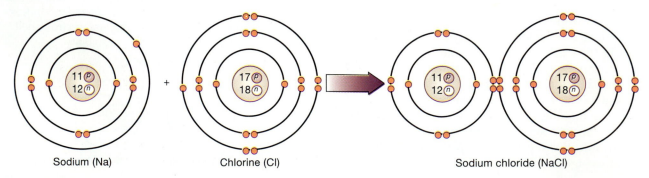

Sodium (Na)   Chlorine (Cl)   Sodium chloride (NaCl)

curs, the two atoms are joined together to form a compound of NaCl.

In our example of an ionic bond, the NaCl that is formed is a solid crystal that we recognize as table salt. When NaCl is dropped into water, the ionic bond is broken and the molecule breaks apart, or *ionizes*. The result of NaCl ionization is the presence of an ion of sodium, which has a positive charge and is written Na^+, and an ion of chlorine, which has a negative charge and is written Cl^-. In general, positively charged ions are called **cations** (KAT-ī-onz), and negatively charged ions are termed **anions** (AN-ī-onz). The process of ionization is important to body function because many of the chemical reactions that occur in the body require the presence of cations and anions in specific concentrations.

Covalent Bonds

A second way that atoms can fill their outer orbitals is by sharing electrons with other atoms. The result is a molecule that is held together by **covalent** (kō-VĀ-lent) **bonds**. For example, a hydrogen atom has one electron in its first orbital. In this form it is very unstable and seeks to fill its only orbital with any available electron. If another hydrogen atom is present, the two will combine to form the gas hydrogen (H_2). They form a bond by sharing their single electrons, thereby filling each atom's orbital (Fig. 2–10a). Because each atom contributes a single electron to the bond, it is called a *single covalent bond*. A single covalent bond is expressed as a single line between atoms (H—H).

Another example of a covalent bond occurs between oxygen atoms, each of which contains eight electrons (two in the first orbital, six in the second). This is shown in Figure 2–10b. In this case, the outer orbital requires two electrons to be filled. When O_2 gas is formed, two of the electrons in the outer orbital of each atom are shared. Because two electrons from each atom are involved, it is called a *double covalent bond*. A double covalent bond is expressed as a double line between atoms (O=O).

Atoms that are in abundance in the body form covalent bonds quite easily. Hydrogen, oxygen, and carbon atoms bond with each other and with other atoms by covalent bonds to form the structural basis of body organization.

Hydrogen Bonds

Hydrogen bonds are weak bonds that are formed between hydrogen atoms and a molecule with a weak negative charge. They are similar to ionic bonds, since the attractive force is electrostatic, but are much weaker. Hydrogen bonds become important when they form in large numbers to provide strength to large

Figure 2–10

Formation of covalent bonds. (a) Two hydrogen atoms become more stable when they share electrons, forming a single molecule of hydrogen (H_2). (b) In a similar manner, two oxygen atoms achieve greater stability when they share two pairs of electrons. This results in the formation of a molecule of oxygen (O_2).

How does a covalent bond differ from an ionic bond?

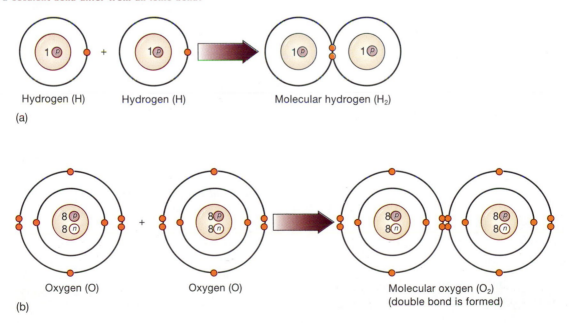

HEALTH CLINIC

Clinical Uses of Atomic Particles

The atoms of a particular element generally contain the same numbers of protons and neutrons in their nuclei. However, some elements contain atoms that have a different number of neutrons from most other atoms of the same element. These atoms are called **isotopes.** For example, a hydrogen nucleus normally contains only one proton and no neutrons. One of the isotopes of hydrogen, however, contains one proton and one neutron, and is therefore heavier than normal hydrogen atoms. Water containing this hydrogen isotope, known as *deuterium,* is called *heavy water* because of this additional mass.

Some isotopes are stable, and have similar chemical properties to their more common atoms. However, some isotopes have unstable nuclei. These unstable isotopes lose neutrons or protons at measurable rates, and are called **radioisotopes.** The emission of subatomic particles releases energy in the form of ionizing radiation. The loss of subatomic particles is called *radioactive decay.*

The radiation released by radioactive decay can penetrate and destroy tissues. While it is well documented that the effects of radiation can lead to disease, including cancer, this destructive effect can be used to our benefit. For example, rapidly dividing cells are more sensitive to the destructive effects of radiation than are cells that divide more slowly. Therefore, radiation is often used to destroy rapidly dividing cancer cells. If the radiation treatment is effective, the cancer cells are killed with only minor destruction of healthy tissue.

Radiation is also used in diagnosis of diseases. For example, radioisotopes may be used to detect problems associated with organ functions because they can be traced as they pass through the body. Compounds labeled with radioisotopes of cobalt are often used to monitor the intestinal absorption of certain vitamins.

The use of X-rays is another application of radiation that is common in the diagnosis of certain diseases. X-rays are a form of radiation resulting from the loss of energy that occurs when electrons move from a higher energy level to a lower energy level. X-rays pass readily through soft tissues, but cannot pass through dense tissues as easily. As a result, the X-rays penetrating soft tissues expose a photographic film, producing dark areas on the film. The dense tissues blocking X-ray penetration show white on the film because of its underexposure (see the figure). X-rays are routinely used to examine dense structures such as bone and teeth, especially for suspected bone fractures and dental examinations. Mammograms are low-energy X-rays of the breast. They can be used to detect breast tumors because the tumors are slightly more dense than normal tissue. Because ionizing radiation in the form of radioisotopes, X-rays, and mammograms can cause cell damage in prolonged or high doses, your exposure frequency should be minimized.

More sophisticated techniques that utilize X-rays or other forms of electromagnetic energy are currently available to aid in the diagnosis of disease. These techniques

---

molecules. They also play an important role in determining the three-dimensional shapes of large molecules, such as DNA.

CHEMICAL REACTIONS

Atoms and molecules react with one another when chemical bonds are formed or broken. New products result.

A chemical reaction is the process of forming or breaking chemical bonds between atoms. Its end result is the formation of new chemical combinations. There are two basic types of chemical reactions: synthesis and decomposition.

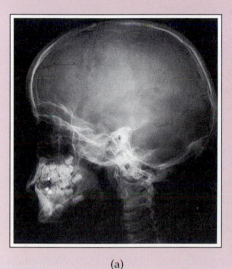

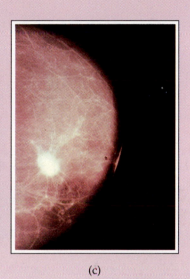

(a)   (b)   (c)

Applications of atomic particles. (a) X-ray of the head and neck regions. (b) This example is a color-enhanced x-ray of a normal right hand. (c) Mammogram. The solid particle within the breast is a primary tumor.

are collectively called *diagnostic imaging* and include computed axial tomography (CAT), positron-emission tomography (PET), and magnetic-resonance imaging (MRI), which were discussed in the previous chapter.

Synthesis Reactions

The combination of atoms, ions, or molecules to form larger molecules is called a **synthesis reaction.** It requires the formation of new bonds between the combining units, which are called **reactants.** The new molecule that is formed has different properties from the reactants; it is called the **end product.** Synthesis reactions are expressed in the formula:

$$A + B \rightarrow AB$$

Synthesis reactions are important processes that are involved in the growth of body structures and the repair of worn or damaged parts. When these reactions occur in the cell they are referred to as *anabolism,* which will be described in more detail in Chapter 17.

Decomposition Reactions

The breakdown of molecules into simpler molecules, atoms, or ions is called a **decomposition reaction.** It requires the chemical bonds that bind the reactants to be broken. Decomposition reactions are symbolized in the formula:

$$AB \rightarrow A + B$$

Decomposition reactions are also important processes that occur in the body. They are involved in the chemical breakdown (*digestion*) of food for energy to fuel body activities. When these reactions occur in the cell they are called *catabolism,* which will also be discussed further in Chapter 17.

CHEMICAL COMPOUNDS OF THE CELL

Inorganic compounds do not contain chains of carbon atoms. Water is the most important inorganic compound in the body. Organic compounds contain carbon, and form the primary structural and functional units of the body.

As we have learned from Chapter 1, the cell is composed of basic units of matter that are not living: atoms and molecules. These chemicals are primarily in the form of compounds. The chemical compounds that compose cells are divided into two categories: inorganic compounds and organic compounds.

Inorganic Compounds

Inorganic compounds generally do not contain chains of carbon atoms and are usually held together by ionic bonds. Important inorganic compounds in the body include water, salts, and most acids and bases.

Water

The most important inorganic compound in the body is water. It is by far the most abundant, making up about 60% to 70% of total body mass. Its importance to body structure and function is due to the following properties:

1. *Water is a universal solvent.* A **solvent** is a medium, such as liquid or gas, that dissolves—or breaks into smaller particles—a material known as a **solute.** Water has the capability of dissolving more types of solutes than any other known solvent. This feature makes it the desired medium for the chemical reactions of the body.

2. *Water is an important transport medium.* The small size of the water molecules makes it possible for liquid water to transport tiny solutes, such as Na^+ and Cl^-, as well as larger compounds and cells, throughout the body. For example, the liquid portion of blood is composed of more than 90% water, and is responsible for transporting vital materials such as sugar, vitamins, and oxygen between the bloodstream and body cells.

3. *Water has a high heat capacity.* Water absorbs and releases heat quite slowly and does not change in temperature readily. This feature helps the body regulate internal temperature by enabling heat to be transmitted from muscle cells to the surface of the body without raising the temperature of body fluids beyond a critical level.

4. *Water is an effective lubricant.* Water is released by cells to reduce friction between moving body parts. This occurs in the cavities surrounding the heart, lungs, and abdominal organs, where the organs move about constantly. Water also lubricates the joints between opposing bones, and when released by salivary glands it lubricates food for its easy passage through the digestive tract.

Salts

A salt is a compound that dissolves, or *ionizes*, when placed in a solvent such as water. The products of this type of decomposition reaction are ions (Fig. 2–11a). Because the ions from a salt are capable of conducting an electric current, they are referred to as **electrolytes.** The common electrolytes that play important roles in the body include Na^+ (sodium), Cl^- (chlorine), K^+ (potassium), Ca^{2+} (calcium), Mg^{2+} (magnesium), PO_4^{3-} (phosphate), CO_3^{2-} (carbonate), and HCO_3^- (bicarbonate).

Acids and Bases

An **acid** is a molecule that releases one or more hydrogen ions (H^+) when it ionizes in water. If the acid ionizes completely to release all of the hydrogen ions it contains, it is called a *strong* acid. Weak acids ionize partially, releasing only some of their hydrogen ions into water. An example of a reaction involving an acid is the decomposition reaction for hydrochloric acid (HCl) when it is in water (Fig. 2–11b). HCl is a strong acid that releases essentially all of its hydrogen ions:

$$HCl \rightarrow H^+ + Cl^-$$

A **base** is a molecule that reduces the concentration of hydrogen ions in a solution. When it is in water, it ionizes to release negatively charged ions that tend to combine with the positively charged hydrogen ions that may be present. Like strong acids, strong bases ionize completely; weak bases ionize partially. An example of a reaction involving a strong base is the decomposition reaction for sodium hydroxide, which occurs when it is in water (Fig. 2–11c):

$$NaOH \rightarrow Na^+ + OH^-$$

One product of this reaction, the hydroxyl ion (OH^-), has a strong tendency to form an ionic bond with a hy-

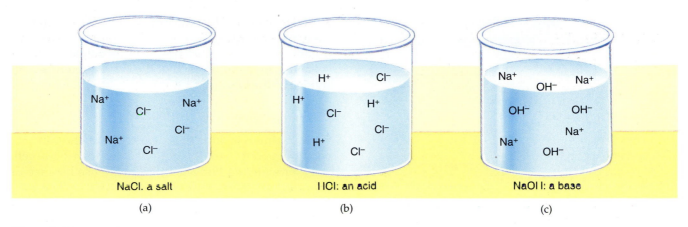

Figure 2–11
Salts, acids, and bases. (a) A salt is a substance that ionizes in a solvent. The ions
that form when NaCl is dropped into water are Na^+ and Cl^-. (b) An acid also
ionizes in water, resulting in the release of H^+ as one of its products. (c) A base is
a substance that removes H^+ from solution, as NaOH does when its OH^-
combines with H^+ to form H_2O.

How does an ion differ from an atom?

drogen ion. Once this bond is formed, the hydrogen
ion is removed from the solution to form water:

$$H^+ + OH^- \rightarrow H_2O$$

Acids and bases play a vital role in body activities.
Our bodies require these molecules in order to survive.
However, the relative strengths of acids and bases vary
depending on the degree of ionization that occurs. If
the degree of ionization is great, the body is subjected
to potentially serious damage. Fortunately, our bodies
have homeostatic mechanisms that keep acid and base
concentrations stable. These mechanisms are called
buffer systems. A buffer system reacts with strong
acids and bases and converts them to weak acids and
bases that have less of an effect.

Measuring Acids and Bases: The pH Scale
The concentrations of acids and bases are measured by
determining the relative number of hydrogen ions in
a volume of water. On the **pH scale,** values for hydrogen-ion concentration range from 0 to 14 and are
expressed in chemical units called *moles per liter* (Fig.
2–12). The midpoint on the scale is 7, and is referred
to as the neutral point. Notice from Table 2–2 that dis-
tilled water is neutral. A solution that contains many
H^+ has a value less than 7 on the scale, and is called
acidic. Conversely, a solution that contains few H^+ has
a value greater than 7, and is referred to as *alkaline*
(*basic*). Each whole number on the scale represents a
tenfold difference in H^+ concentration from the adjacent whole number. For example, white wine has a pH
value of about 3, and tomato juice has a pH of 4. This

Figure 2–12
The pH scale. pH is determined by the concentration of
H^+ in a solution. As the H^+ concentration declines, it is
replaced by OH^- and the solution becomes more basic. A
pH of 7 is neutral because the solution contains equal
numbers of H^+ and OH^-.

Is a solution with a pH of 6.0 acidic or basic?

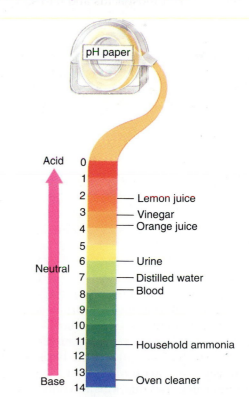

Table 2–2 pH OF COMMON FLUIDS

Substance	pH
Hydrochloric acid in stomach	1.0–3.0
Lemon juice	2.2–2.4
Vinegar	2.4–3.4
Grapefruit juice	3.0–3.3
Orange juice	3.0–4.0
Urine	4.8–8.4
Milk	6.3–7.5
Saliva	6.5–7.5
Drinking water	6.5–8.0
Distilled water	7.0
Blood plasma	7.3–7.5
Semen	7.4–7.5
Egg white	7.6–8.0
Milk of magnesia	10.0–11.0
Saturated lime	12.0

means that white wine contains approximately ten times more H^+ than tomato juice.

CONCEPTS CHECK

1. What type of chemical bond is usually involved in inorganic compounds?

2. What are four properties of water that are vital to the body?

3. What is an electrolyte?

4. What ion is released from an acid that determines its relative strength?

Organic Compounds

Organic compounds are the group of molecules that serve as the building blocks of most living structures and control most functions. They each contain one or more carbon atoms. Carbon is the essential component because of its unique properties of size and electron distribution that enable it to form covalent bonds with other carbon atoms, and with many other atoms such as oxygen, hydrogen, nitrogen, and phosphorus. This enables organic compounds to form long chains, branched configurations, and ring structures using the carbon atoms as a skeleton. The variation that we see in carbon-containing molecules is the basis for the incredible diversity of life on our planet.

The four categories of organic compounds that provide a significant contribution to human body structure and function include carbohydrates, lipids, proteins, and nucleic acids. We will also consider the organic compound adenosine triphosphate (ATP), since it is involved in each of the energy-requiring activities of the body.

Carbohydrates

Carbohydrates include the sugars and starches, and account for about 2% of the mass of an average body cell. They provide the body with an easily used source of energy with which to fuel body activities. They may also be used as a food reserve, to be called upon at a later time for energy needs. To a lesser extent, they are also used in combination with other molecules for body structure.

Carbohydrate molecules contain carbon, hydrogen, and oxygen, with hydrogen and oxygen frequently in a ratio of 2 to 1. This ratio is also found in the water molecule (H_2O), and is demonstrated by the formulas of the sugar molecules glucose ($C_6H_{12}O_6$) and sucrose ($C_{12}H_{22}O_{11}$). The arrangement of the molecules is in chains or rings of carbon atoms with hydrogen and oxygen atoms attached (Fig. 2–13). The number of carbon atoms varies with the type of carbohydrate. There are three types: monosaccharides, disaccharides, and polysaccharides.

Monosaccharides (mon'-ō-SAK-a-rīdz) are also called *simple sugars*, since they contain the smallest number of carbon atoms (from three to seven). They form the building blocks for the larger carbohydrate molecules. Examples of simple sugars include glucose, fructose, and galactose. Glucose is the preferred source of energy of the cell; its structure is shown in Figure 2–13.

Disaccharides (dī-SAK-a-rīdz) are formed when two monosaccharides are joined together. During the process of disaccharide formation, a water molecule is

Figure 2–13
The structure of glucose. (a) Glucose may form a straight chain of carbon atoms, or (b) its carbon atoms may form a more stable ring structure.
What is a simple sugar also known as?

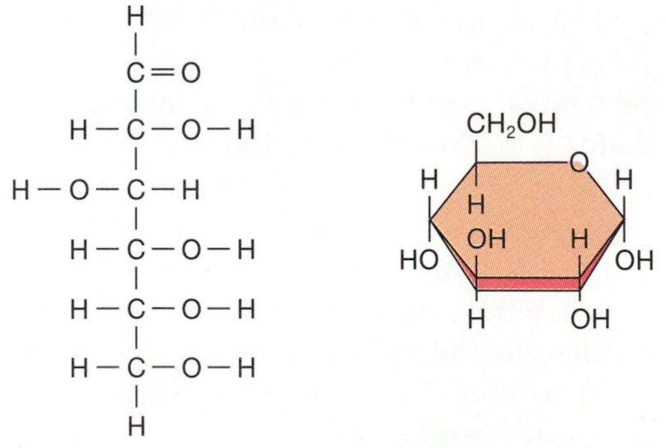

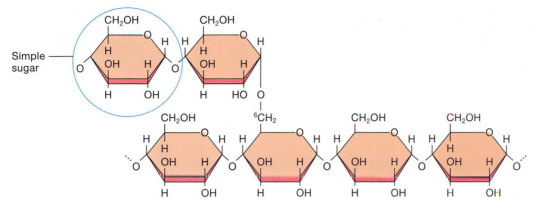

Figure 2–14
Polysaccharide structure. A polysaccharide consists of long chains of simple sugars
that are chemically bonded together.

What three atoms may be found in all carbohydrates?

removed. This reaction, often called a *dehydration (or condensation) reaction,* is a very common type of synthesis reaction performed by the body. Important disaccharides include sucrose, or table sugar (glucose + fructose); lactose, found in milk (glucose + galactose); and maltose, which is also called *malt sugar* (glucose + glucose).

Polysaccharides (pol'-ē-SAK-a-rīdz) are long chains of simple sugars that are chemically bonded together (Fig. 2–14). These large carbohydrates lack the sweetness that characterizes the simple and double sugars, and are used by the body to store energy. A type of polysaccharide that is an important source of food is starch, which is synthesized by plants for their energy storage. It is composed of large numbers of glucose subunits. Glycogen (GLĪ-kō-jen) is the polysaccharide that is synthesized by animals. Like starch, it is composed of many glucose subunits as well. Glyco-

gen is stored primarily in skeletal muscle cells and liver cells, where it remains until body needs require its decomposition into glucose molecules.

Lipids

Lipids are organic molecules that do not dissolve in water, but do dissolve in organic solvents such as alcohol, chloroform, and ether. Like carbohydrates, they are composed of carbon, hydrogen, and oxygen, but they do not have the 2 to 1 ratio of hydrogen and oxygen (they contain much less oxygen). They include a variety of substances that are vital to body structure and function, among them fats, phospholipids, and steroids.

Fats are also known as **triglycerides** (trī-GLIS-er-īdz). They are composed of two types of building blocks, fatty acids and glycerol (GLI-ser-ol), that are in a ratio of 3 to 1 (Fig. 2–15). Fats are large molecules

Figure 2–15
The formation of triglycerides. A triglyceride, or fat, is formed by the combination
of glycerol and three fatty acids.

A chain of which atom forms the structural backbone of fats?

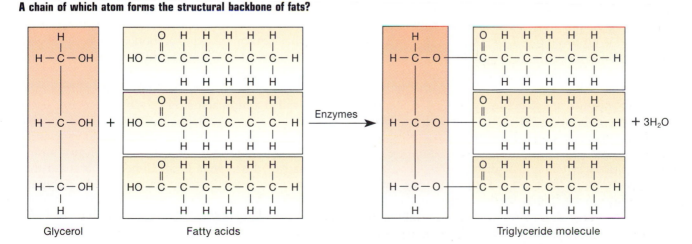

Glycerol   Fatty acids   Triglyceride molecule

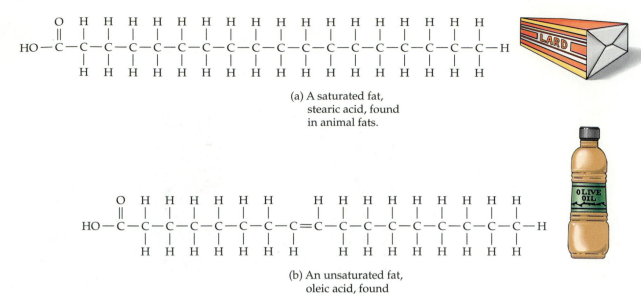

(a) A saturated fat, stearic acid, found in animal fats.

(b) An unsaturated fat, oleic acid, found in olive oil.

Figure 2–16
Saturated and unsaturated fats. (a) A saturated fat is a common animal product and is solid at room temperature. It contains only single bonds between carbon atoms. (b) An unsaturated fat is a common plant product and is liquid at room temperature. It contains double bonds between carbon atoms.

What three atoms may be found in all fats?

that must be broken down by the body before their building blocks can be absorbed and used. They are the body's most concentrated source of energy for fueling its many activities. They are also the major form of energy storage in the body, and are found mainly in fat deposits below the skin.

Fats may exist in a solid form or a liquid form at room temperature, depending on the length of the fatty acid chain and the presence of single or double bonds between carbon atoms. A fat that has a long fatty acid chain and only single bonds between carbon atoms is solid at room temperature. It is called a *saturated fat* and is typically found in animal products such as butter and meat fats. A fat that has a short fatty acid chain and contains one or more double bonds between carbon atoms is an *unsaturated fat.* Unsaturated fats are found in plants and are mostly in liquid form at room temperature. They are common in cooking oils such as olive, peanut, safflower, and corn oil (Fig. 2–16).

Phospholipids are structurally similar to fats but contain only two fatty acid chains and a phosphorus-containing group that replaces the third fatty acid chain. They are an important structural component of cell membranes, which will be discussed in Chapter 3.

Steroids are in the form of ring structures, rather than the fatty acid chains we have seen in fats and phospholipids (Fig. 2–17). Steroids are important to the body for many reasons. Cholesterol, the most well-known steroid, is required for cell membrane structure, vitamin D synthesis, and the production of steroid hormones. Cholesterol is synthesized in the liver and may be obtained from foods such as eggs, meat, and cheese. Steroid hormones are chemicals that alter cell activity in order to maintain homeostasis or influence sexual activities. They are effective because they can move through cell membranes quite easily. Hormones will be discussed in more detail in Chapter 11 when we study the endocrine system.

Proteins

Protein molecules account for 10% to 30% of the mass of a cell. They are the basic structural material of the body, and play important roles in body function as well. They are composed of carbon, oxygen, hydrogen, and nitrogen atoms, and many contain sulfur and phosphorus atoms.

All proteins are composed of building-block molecules called **amino acids** (Fig. 2–18). There are 20 different types of amino acids, each of which is capable of binding with other amino acids to form long chains. The covalent bond that is formed between two amino acids is called a *peptide bond.* Generally, a chain that contains more than 10 amino acids is called a **polypeptide,** and one that contains more than 50 is called a **protein.** The largest proteins contain more than 50,000 amino acids!

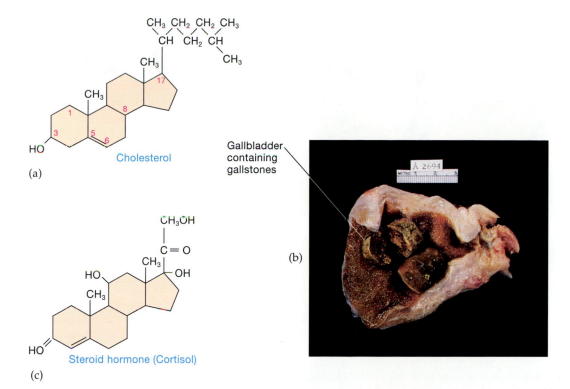

(a)

Cholesterol

(b)

Gallbladder containing gallstones

Å 2694

(c)

Steroid hormone (Cortisol)

Figure 2–17

Steroids contain interlocking rings of carbon atoms.
(a) Cholesterol is the most abundant steroid in the body.
(b) Gallstones, shown here within the gallbladder, form
from cholesterol. (c) Cortisol is an example of a steroid
hormone.

How are steroids useful to the body?

Figure 2–18

The composition of proteins. (a) Four types of amino acids
and their structural formulas. About 20 different types
occur commonly in proteins. (b) A simple protein, or
dipeptide, is formed when two amino acids combine by
the formation of a peptide bond.

**What atoms may be found in proteins that are not commonly found
in carbohydrates and fats?**

Amino acid	Structural formulas
Alanine	
Valine	
Cysteine	
Glycine	

(a)

Peptide bond

Glycine + Alanine → Glycylalanine (a dipeptide) + H_2O

Enzyme

(b)

39

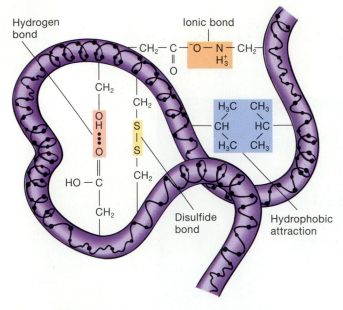

Figure 2–19
The three-dimensional structure of a large protein is stabilized by several types of chemical bonds.

What happens when the three-dimensional structure of large proteins is disrupted (during denaturation)?

Each type of protein molecule is composed of a specific sequence of amino acids. The sequence and number of amino acids determine the three-dimensional structure of the protein (Fig. 2–19), which determines the functional role of the protein in the body. Many types of proteins have important structural roles. An example of a structural protein is collagen, which reinforces bone and anchors body organs. Structural proteins are also important components of cell membranes.

Other types of proteins play important functional roles. One of the most vital types of functional proteins is the **enzyme.** Enzymes are proteins that serve as *catalysts* for body activities: they speed up chemical reactions without undergoing permanent change themselves. They are involved in nearly every chemical reaction that takes place in the body. An example of an enzyme-mediated reaction is shown in Figure 2–20. Other functional proteins include antibodies, which help provide immunity; and protein hormones, which help regulate growth and development.

Structural proteins are extremely stable molecules, but functional proteins are not; they are easily altered by exposure to excessive amounts of energy in the form of heat, radiation, electricity, and chemicals. Their alteration is a result of broken peptide and hydrogen bonds, which change the three-dimensional structure

of the protein. Once this structure is altered, the ability of the protein to carry out its intended function is affected. The process of protein alteration by external factors is called **denaturation.** An example of denaturation occurs when hemoglobin, the functional protein in red blood cells that transports oxygen and carbon dioxide, is exposed to a change in blood pH or an increase in body temperature (greater than 105°F). The region of the hemoglobin molecule that binds to oxygen, called the *active site,* is structurally changed so that it can no longer bind and carry oxygen.

Figure 2–20
Enzymes speed up chemical reactions in the body. An enzyme begins its action by binding to reacting molecules. It often "fits" to them the way a lock fits a key. Because of the influence of the enzyme, the molecules undergo rapid change. A new molecule, the product, is thereby formed. After the reaction, the unaltered enzyme can be used again.

Is an enzyme a carbohydrate, a fat, or a protein?

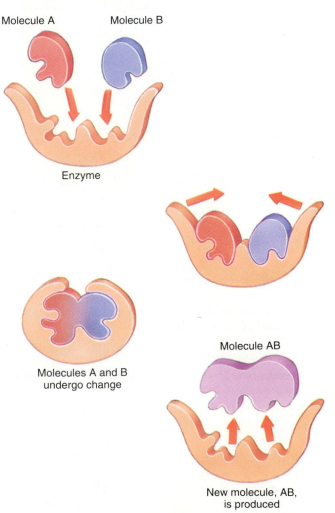

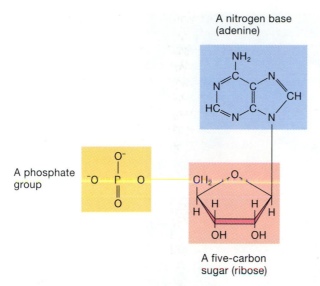

Figure 2–21

A nucleotide, the building block of nucleic acids. Each nucleotide contains a five-carbon sugar, a phosphate group, and a nitrogen base.

How does a nucleotide differ from an amino acid?

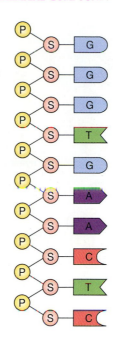

Figure 2–22

A section of the nucleic acid molecule DNA. A chain is formed when nucleotides bond to one another. Notice the sequence of bases that forms; this provides a cell with coded information. If the sequence shown carried information determining a particular trait, we would call this section of DNA a *gene*. P=phosphate, S=sugar, G=guanine, T=thymine, A=adenine, C=cytosine.

Nucleic Acids

Nucleic (noo-KLĀ-ik) **acids** are extremely large organic molecules that hold the information determining the structure and function of the cell. They typically contain carbon, hydrogen, oxygen, nitrogen, and phosphorus, which are combined to form building blocks called **nucleotides.** The atoms within each nucleotide are arranged into three basic groups: a *five-carbon sugar*, a *phosphate group*, and a *nitrogenous base* (Fig. 2–21). A nucleic acid consists of numerous nucleotides united together to form a chain that may be linear or helical (Fig. 2–22). There are two basic types of nucleic acids, both of which are found within cells: **deoxyribonucleic acid (DNA)** and **ribonucleic acid (RNA).**

DNA is a nucleic acid that contains nucleotides with the five-carbon sugar deoxyribose; the four nitrogen bases adenine, thymine, cytosine, and guanine; and a phosphate group. Its nucleotides are arranged into two strands that are twisted about each other in the form of a *double helix* (Fig. 2–23). The helical structure is stabilized by hydrogen bonds between nitrogen base pairs: adenine always pairs with thymine, and cytosine always pairs with guanine. DNA is mainly located within the nucleus of a cell, and contains the hereditary material of life called **genes.** Genes are segments of DNA molecules that determine hereditary traits and control the many activities of the cell.

RNA, the second type of nucleic acid within the cell, differs structurally from DNA in that it contains the five-carbon sugar ribose, instead of deoxyribose, and one of its nitrogen bases is uracil instead of thymine. Also, its nucleotides are arranged in a single strand, instead of the double strand found in DNA. There are several types of RNA molecules, each of which plays a different role in the cell. These roles will be explored in Chapter 3.

Adenosine Triphosphate (ATP)

Adenosine triphosphate (ATP) is a nucleotide that is found in all living organisms. It performs the essential function of capturing energy within its bonds and storing it. This energy is in a form that can be easily retrieved by the body and used to power chemical reactions that would not have proceeded easily on their own.

Structurally, ATP is similar to an RNA nucleotide molecule with two additional phosphate groups. It consists of a ribose group, an adenine base group, and—unlike the RNA nucleotide—three phosphate groups. The three phosphate groups are important, for

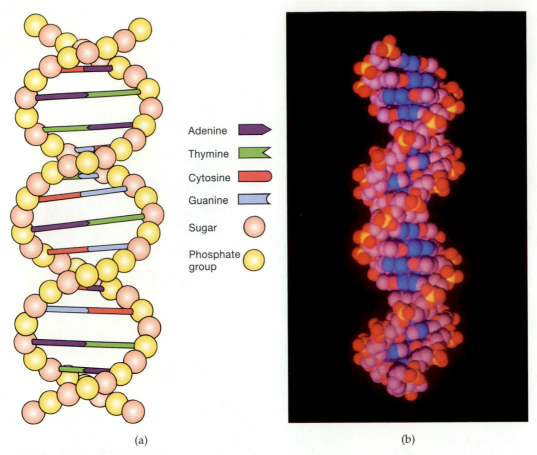

Adenine ▬▶

Thymine ◀▬

Cytosine ▬▶

Guanine ◀▬

Sugar ⬤

Phosphate group ⬤

(a)　　　　　　(b)

Figure 2–23
The double-spiral structure of DNA. (a) The base pairs form the rungs of the "ladder," and the phosphate and sugar groups form the spiraling chain. (b) A three-dimensional view of DNA.

Why is DNA an important molecule to all life on earth?

they are attached by chemical bonds that are called *high-energy phosphate bonds*. These bonds are called high-energy because, at physiologic (normal body) pH, the phosphate groups are highly charged, and so they require a large amount of energy to hold them close together. When these bonds are broken to release a phosphate group (PO_4^{2-}), energy is released that can be used immediately by the cell. The reaction is represented in the following formula:

$$ATP \rightleftharpoons ADP + PO_4^{2-} + E$$

In this formula E represents energy. Note that this reaction may go in either direction, as indicated by the arrows. This means that if energy is provided in the presence of adenosine diphosphate (ADP) and a free phosphate group (PO_4^{2-}), ATP will be synthesized. This is the mechanism for ATP generation that occurs in cells. The energy needed to drive this reaction in human cells is provided by the breakdown of glucose molecules in a process known as *cellular respiration*. Cellular respiration is one of the mechanisms of metabolism that will be discussed further in Chapter 17.

The four groups of organic compounds with their basic element and building-block components are summarized in Table 2–3.

CONCEPTS CHECK

1. What atom is a necessary component of organic molecules?

2. The atoms in organic molecules are primarily held together by what type of bond?

3. What are the major differences in chemical composition between carbohydrates, lipids, proteins, and nucleic acids?

4. What is the role of an enzyme?

5. What property of ATP makes it so important to the body?

Table 2–3   THE ORGANIC COMPOUNDS FOUND IN CELLS

Compound	Elements Present	Building Blocks
Carbohydrates	C, H, O	Simple sugars
Fats	C, H, O	Glycerol and fatty acids
Proteins	C, H, O, N, S	Amino acids
Nucleic acids	C, H, O, N, P	Nucleotides

CHAPTER SUMMARY

THE COMPOSITION OF MATTER

Matter is the "stuff of life." The fundamental particles of matter are atoms of chemical elements.

Atoms and Chemical Elements   Defined as a unit of matter with its own distinguishing properties, each element is composed of one or more similar atoms.

Atomic Structure   An atom is composed of two regions: a nucleus and an electron shell. The nucleus contains one or more protons and neutrons. Electron shells exist around the nucleus and contain one or more particles called electrons. Electrons travel within approximate areas of space called orbitals.

Electric Charge of the Atom   Protons carry a positive charge, electrons carry a negative charge, and neutrons carry no charge. The number of protons usually equals the number of electrons in an atom, making the atom electrically neutral.

Atomic Number and Weight   The number of protons and electrons provides an atom with its distinguishing features. The number of protons determines its atomic number, and the sum of the masses of protons, neutrons, and electrons determines its atomic weight.

Molecules and Compounds   When atoms are combined, they form particles with their own properties called *molecules*. When atoms that are not similar combine, they form a type of molecule called a *compound*.

CHEMICAL BONDS

The combination of atoms to form molecules is made possible by the formation of a chemical bond.

Electrons and Chemical Bonding   Electrons are confined within shells that encircle the nucleus of an atom. Each shell exists at a particular energy level. The electrons in the outermost energy level are involved in chemical bonding in atoms that are reactive.

Types of Chemical Bonds   Bonds may be formed when atoms gain or lose electrons, or when electrons are shared between atoms.

Ionic Bonds   Bonds that are formed when atoms gain or lose electrons. The force of attraction is called *electrostatic*.

Covalent Bonds   Bonds that are formed when atoms share electrons.

Hydrogen Bonds   Bonds that are formed by hydrogen atoms with molecules that have a negative charge.

CHEMICAL REACTIONS

A reaction results in a change in the nature of the particles that are involved, due to the formation or breaking of chemical bonds.

Synthesis Reaction   A reaction that involves the formation of a bond, resulting in the creation of a new, larger molecule.

Decomposition Reaction   A reaction that causes a bond to break, resulting in smaller particles.

CHEMICAL COMPOUNDS OF THE CELL

Cells are composed of inorganic and organic compounds.

Inorganic Compounds   Compounds that do not contain chains of carbon atoms. Many are held together by ionic bonds.

Water   The most abundant molecule in the body, water plays vital roles as a universal solvent, a transport medium, a good temperature-retaining medium, and a lubricant.

Salts   When placed in water, salts ionize to form charged particles called *ions*. Because ions conduct electricity, they are often called *electrolytes*. Electrolytes play important roles in the body.

Acids and Bases   An acid is a molecule that releases hydrogen ions into a solution, and a base captures or removes hydrogen ions from a solution.

Measuring Acids and Bases: The pH Scale   This is a method of comparing relative strengths, or degrees of ionization, of acids and bases.

Organic Compounds   Organic compounds contain carbon atoms and form the basis for living structures and functions.

Carbohydrates   Sugars and starches, with the general formula CH_2O. They provide much of the energy used by cells.

Lipids   Fats, phospholipids, and steroids. They provide an efficient storage molecule for energy, and structural components of membranes and other structures. They dissolve only in organic solvents. Fats are composed of fatty acids and glycerol.

Proteins   Composed of building-block units called *amino acids*, proteins form a major component of body structure. They also play important functional roles in the body, including catalysis of chemical reactions by enzymes.

Nucleic Acids   Composed of building-block units called nucleotides, nucleic acids include DNA and RNA. They play an important role in controlling the activities within cells, and in determining heredity.

Adenosine Triphosphate (ATP)   ATP is similar to a nucleotide in structure. It contains a unique type of bond that traps energy. When this bond is broken in a chemical reaction, it releases energy that is available to power cell activities. ATP is a recyclable molecule.

KEY TERMS

anion (p. 31)	decomposition (p. 33)	enzyme (p. 40)	solute (p. 34)
atom (p. 26)	denaturation (p. 40)	inorganic (p. 34)	solvent (p. 34)
cation (p. 31)	electrolyte (p. 34)	molecule (p. 28)	synthesis (p. 33)
compound (p. 28)	element (p. 26)	organic (p. 36)	triglyceride (p. 37)

QUESTIONS FOR REVIEW

OBJECTIVE QUESTIONS

1. A chemical element is composed of similar:
   a. molecules
   b. types of bonds
   c. atoms
   d. all of the above

2. Electrons exist within an approximate region of the atom that is called:
   a. nucleus
   b. neutron
   c. electron orbital
   d. energy level

3. Water is an example of a(n):
   a. atom
   b. compound
   c. element
   d. ion

4. The outer energy level of a reactive atom is always filled with the maximum number of electrons it can hold.
   a. True
   b. False

5. The particles that are involved in the formation of a chemical bond between two atoms are:
   a. outer energy-level electrons
   b. protons only
   c. neutrons and protons
   d. inner energy-level electrons

6. A type of chemical bond that is formed when electrons are shared, as occurs in organic compounds, is called:
   a. ionic
   b. covalent
   c. hydrogen
   d. none of the above

7. A chemical reaction that results in the formation of a new product by the bonding of reactants is a(n):
   a. synthesis reaction
   b. double covalent bond
   c. decomposition reaction
   d. inorganic reaction

8. Water is an inorganic compound important to the body for many reasons. Which of the following is not a feature of water?
   a. a solvent for many materials
   b. ionizes readily in solution
   c. has a high heat capacity
   d. has lubricating properties

9. A compound that ionizes completely in water by releasing hydrogen ions is called a(n):
   a. strong base
   b. weak acid
   c. strong acid
   d. electrolyte

10. The preferred energy source of the body is a six-carbon molecule called:
    a. sodium chloride
    b. glucose
    c. enzyme
    d. triglyceride

11. An example of a lipid is:
    a. RNA
    b. disaccharide
    c. cholesterol
    d. enzyme

12. Proteins are composed of building-block molecules called:
    a. glucose
    b. amino acids
    c. enzymes
    d. nucleotides

13. Enzymes are catalysts that speed up many of the chemical reactions of the body. Their building-block subunits are chemically bonded together by a bond called:
    a. peptide bond
    b. covalent bond
    c. ionic bond
    d. both (a) and (b)

14. A nucleic acid that is in the arrangement of two long strands wound into a double-helix configuration is characteristic of:
    a. ATP
    b. RNA
    c. DNA
    d. all nucleotides

15. ATP is an important molecule to the body because of its property of:
    a. ionization
    b. synthesis into large molecules
    c. energy storage
    d. heat conductance

16. If a DNA molecule contains a thymine nucleotide base on one chain, the other chain of the double helix will contain a corresponding:
    a. adenine
    b. guanine
    c. cytosine
    d. uracil

ESSAY QUESTIONS

1. Describe the structure of the atom. In your description, include both the planetary model and the orbital model of a representative atom.
2. What are the relationships between atoms, elements, molecules, compounds, and ions? Distinguish between these chemical terms.
3. Discuss the phenomenon of chemical bonding and distinguish between ionic and covalent bonds.
4. Describe the structural and functional differences between the four classes of organic compounds important to the body (carbohydrates, lipids, proteins, and nucleic acids).

ANSWERS TO ART LEGEND QUESTIONS

Figure 2–1
No question.

Figure 2–2
No question.

Figure 2–3
What particles make up the nucleus of an atom?   Protons and neutrons make up the nucleus.

Figure 2–4
Is an atom three-dimensional in nature?   Yes, the atom has a three-dimensional structure.

Figure 2–5
What is the electrical charge on a neutron?   A neutron has no electrical charge.

Figure 2–6
How many energy levels would an atom have if its shell contained only two electrons?   It would have only one energy level.

Figure 2–7
Is the helium atom electrically neutral?   Yes, the helium atom is electrically neutral (its two electrons fill the first shell).

Figure 2–8
Why doesn't the sodium atom gain seven electrons instead of losing one?   It is more energy-efficient to lose one electron than to gain seven.

Figure 2–9
Which substance in this figure is a molecule?   NaCl is a molecule.

Figure 2–10
How does a covalent bond differ from an ionic bond?   A covalent bond shares electrons, while an ionic bond involves a transfer of electrons.

Figure 2–11
No question.

Figure 2–12
How does an ion differ from an atom?   An ion contains an electric charge, whereas an atom is electrically neutral.

Figure 2–13
Is a solution with a pH of 6.0 acidic or basic?   A solution with pH 6.0 is acidic.

Figure 2–14
What is a simple sugar also known as?   A simple sugar is also called a *monosaccharide.*

Figure 2–15
What three atoms may be found in all carbohydrates?   Carbon, hydrogen, and oxygen are common to all carbohydrates.

Figure 2–16
A chain of which atom forms the structural backbone of fats?   The carbon atom forms the backbone of fats.

Figure 2–17
What three atoms may be found in all fats?   Carbon, hydrogen, and oxygen are common to all fats.

Figure 2–18
How are steroids useful to the body?   Steroids are necessary for cell membrane structure, vitamin D synthesis, and steroid hormones.

Figure 2–19
What atoms may be found in proteins that are not commonly found in carbohydrates and fats?   Nitrogen is found in all proteins, and some proteins contain sulfur and phosphorus; these are not generally found in carbohydrates and fats.

Figure 2–20
What happens when the three-dimensional structure of large proteins is disrupted (during denaturation)?   Stabilizing bonds are broken during denaturation, resulting in a loss of protein structure and, consequently, function as well.

Figure 2–21
Is an enzyme a carbohydrate, a fat, or a protein?   An enzyme is a protein.

Figure 2–22
How does a nucleotide differ from an amino acid?   A nucleotide contains a five-carbon sugar and a phosphate group, neither of which is contained in an amino acid.

Figure 2–23
No question.

Figure 2–24
Why is DNA an important molecule to all life on earth?   DNA determines the structural composition and number of proteins that are synthesized by each cell.

CELLS: THE BASIS OF LIFE

LEARNING OBJECTIVES

After studying this chapter you should be able to:

1. Describe the compartments within and surrounding a cell.

2. Describe the structure of the plasma membrane, and indicate how it may be modified to accommodate special functions.

3. Identify the functional roles of the plasma membrane.

4. Describe the ways in which materials move across the plasma membrane.

5. Distinguish between the cellular organelles on the basis of their structures and functions.

6. Explain why the nucleus is regarded as the control center of the cell.

7. Identify the structural components of the nucleus.

8. Describe the process of protein synthesis and the importance of the genetic code.

9. Distinguish between the two types of nuclear division, mitosis and meiosis.

10. Describe the sequential process of mitosis.

11. Describe the current theories that attempt to explain why aging occurs.

The cell is the most basic structural and functional unit of life. As such, it plays a key role in the structure and functions of the human body. In terms of structure, our body form and physical appearance are determined by individual cells and the ways in which they are arranged. Functionally, every one of the 75 trillion or so cells that make up our bodies is like a separate social organism: it struggles to survive in order to carry out functions that are ultimately important to the survival of the whole. If cells' normal function is disturbed by disease, the whole body may be affected. Whether as individual units or in organized groups, the normal functioning of cells is absolutely essential for the body to maintain health.

In this chapter we shall examine the structural parts of the healthy cell and many normal cell activities. Two important activities to be discussed are protein synthesis and cell division, both of which are controlled by the genetic material that lies within the cell. This will be followed by a discussion on current theories that seek to explain why the body ages. But first, let's examine some important terms used to describe the areas outside and inside the cell.

THE ENVIRONMENT OF THE CELL

The body is organized into two major compartments filled mostly with water: one outside the cells of the body, and one within the cells.

The human body consists of about 65% water, much of which is contained within cells. This fluid-filled area is known as the **intracellular environment**.[1] The remainder of body water lies outside cells, often providing a sort of bath that surrounds and supports them. This area is called the **extracellular environment**.[2] These compartments are illustrated in Figure 3–1.

Extracellular Environment

The area outside a cell contains various substances in addition to water. They include a mixture of dissolved gases, salts, food particles, and cellular **products**. A product is any material that is manufactured, or *synthesized*, by mechanisms inside the cell and released to the extracellular environment by a process called **secretion**. Some of the more common products are proteins, hormones, and vitamins. The mixture of water, products, and other substances in the extracellular environment forms a slightly thickened, syruplike liquid known as **extracellular fluid (ECF)**.

There are two types of ECF in the body. One type is located within blood vessels and the chambers of the heart and is called **plasma**. Plasma provides a liquid medium for the transport of substances in the blood. The other type of ECF is located between the cells of the body. It is called **interstitial fluid**[3] and provides a pathway for substances that are en route to and from neighboring cells. Plasma and interstitial fluid make the passage of substances possible, by providing a liquid "freeway" through which materials can travel. In some areas of the body, however, the extracellular environment contains abundant strands of protein that form a dense matting, called **matrix**, with little interstitial fluid leaking through it. This type of arrangement provides structural support for body parts but restricts the free passage of materials.

The compartment of the extracellular environment that lies between adjacent cells is called the **intercellular environment**. Where cells are arranged close together, this space often contains molecular bridges that connect adjacent cells. The bridges may also provide a means of direct communication between the cells.

Intracellular Environment

The substance of a cell is called **protoplasm**.[4] It is composed of chemicals that include water, proteins, carbohydrates, fats, nucleic acids, and electrolytes. When these basic building blocks are organized in such a way as to carry out activities characteristic of living things—organization, metabolism, excitability, movement, growth, and reproduction—protoplasm is said to be *alive* (see p. 13 in Chapter 1 for a discussion of these activities if you don't recall them).

The portion of the protoplasm that separates extracellular space from the space inside the cell is the **plasma membrane** (cell membrane), which is shown in Figure 3–1. The plasma membrane surrounds and contains the rest of the cell like a sack, creating a space inside. This internal space is known as the **intracellular environment**. The fluid mixture of water, proteins, and other chemicals within the intracellular environment is called **intracellular fluid**, or **ICF**.

In addition to the plasma membrane, there is a second major component of protoplasm: a thickened, gel-like fluid called **cytoplasm**,[5] which usually occupies most of the intracellular environment. The cytoplasm contains many small, highly organized structures that play important functions in the cell, known as **organelles**.

A third major component of protoplasm is a large, oval-shaped structure known as the **nucleus**. Like cytoplasm, the nucleus also lies within the intracellular environment. It regulates the activities of the cell.

The intracellular environment is the site where most internal cell functions occur. Activities such as energy production, energy storage, and synthesis of new products for growth or secretion are examples of these important functions.

---

[1]Intracellular: *intra-* = within + *cellular* = of the cell.

[2]Extracellular: *extra-* = outer + *cellular*.

[3]Interstitial: Latin *inter-* = between + *-stitial* = standing (referring to cells).

[4]Protoplasm: Greek *proto-* = first + *plasma* = formable material.

[5]Cytoplasm: Greek *cyto-* = cell + *plasma*.

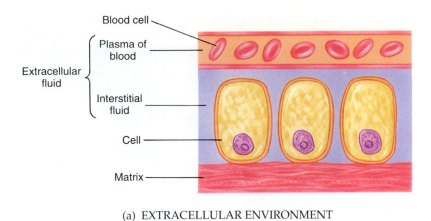

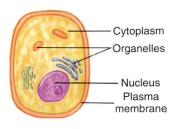

(a) EXTRACELLULAR ENVIRONMENT   (b) INTRACELLULAR ENVIRONMENT

Figure 3–1
The compartments associated with the cell. (a) The extracellular environment lies
outside the cell, and (b) the intracellular environment is located inside the cell.

What fluid is located in the extracellular environment?

CONCEPTS CHECK

1. When is protoplasm considered to be alive?

2. What is the area enveloped by the plasma membrane called?

3. What are the three main components of a cell?

CELL STRUCTURE AND FUNCTION

Human cells share the common features of plasma membrane, cytoplasm, and nucleus structure, which operate in harmony to provide cellular functions.

Nearly all human cells are too small to be seen without the use of a microscope. They vary considerably in size, ranging from the ovum (the female sex cell), which has a diameter of about 1000 micrometers (μm),* to the sperm cell (the male sex cell), which measures only about 50 μm in diameter. Cell shape also varies among different types of cells (Fig. 3–2). For the most part, this variation in size and shape is a reflection of the different roles that cells play in the body.

Although the cells of the body exhibit a great deal of variation in structure, they share a number of important features in common. For example, we have already learned that they generally contain three major components: the plasma membrane, the cytoplasm, and the nucleus. For the purpose of describing cell structure in more detail, a model of a "generalized cell" is used, which is a composite of features common to most cells. This is shown in Figure 3–3. As you study this figure and the following discussion, keep in mind that the cell presented is not a real human cell but a composite model provided to help you visualize the important features of human cells.

Plasma Membrane

The plasma membrane separates the extracellular and intracellular compartments of the body. It regulates the movement of materials between these compartments.

The plasma membrane envelops the cell completely, creating a barrier that separates the intracellular environment from the extracellular environment. Its dynamic, fluid structure enables the cell to maintain homeostasis despite changes that might occur in the environment that surrounds it. Homeostasis is maintained by the plasma membrane's ability to regulate the movement of materials into and out of the cell.

Structure

The plasma membrane is composed of approximately equal amounts (by weight) of lipid and protein molecules, with a very small amount of carbohydrates.

---

*A micrometer is a unit of measure in the metric system. It is equal to one-millionth of a meter; it would take 25,400 micrometers to equal one inch!

(Text continues on page 52)

Figure 3–2

The shape and size of a cell reflect its function in the body. (a) The egg cell, or oocyte, is the largest cell of the body, while the sperm cell produced by males is one of the smallest. Magnification 4000x. (b) Epithelial cells that are columnar, like those shown, often absorb or secrete substances. Magnification 360x. (c) Nerve cells contain branches that enable them to communicate with other cells. Magnification 500x.

Why aren't all cells the same shape and size?

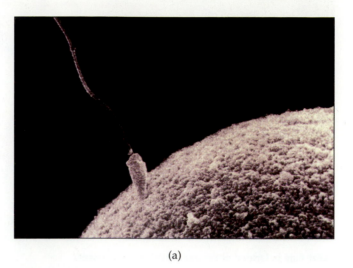

(a)

Figure 3–3

Cell structure. (a) A model of a cell illustrates the plasma membrane, the nucleus, and the major cytoplasmic organelles. (b) An electron micrograph of a human cell looks quite different, because it is a one-dimensional photograph of a sectioned cell and the organelles are not as clearly seen. Magnification 1200x.

Where is the cytosol in the cell model shown?

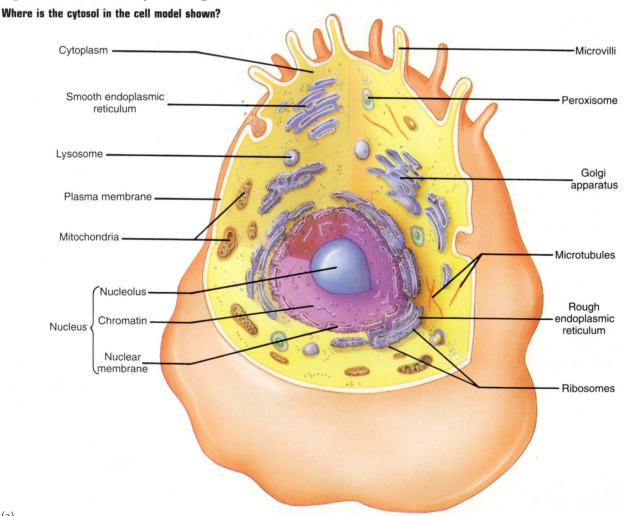

Cytoplasm

Smooth endoplasmic reticulum

Lysosome

Plasma membrane

Mitochondria

Nucleus {
  Nucleolus
  Chromatin
  Nuclear membrane
}

Microvilli

Peroxisome

Golgi apparatus

Microtubules

Rough endoplasmic reticulum

Ribosomes

(a)

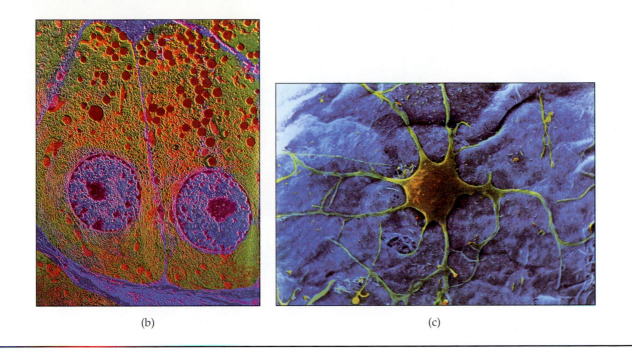

(b)   (c)

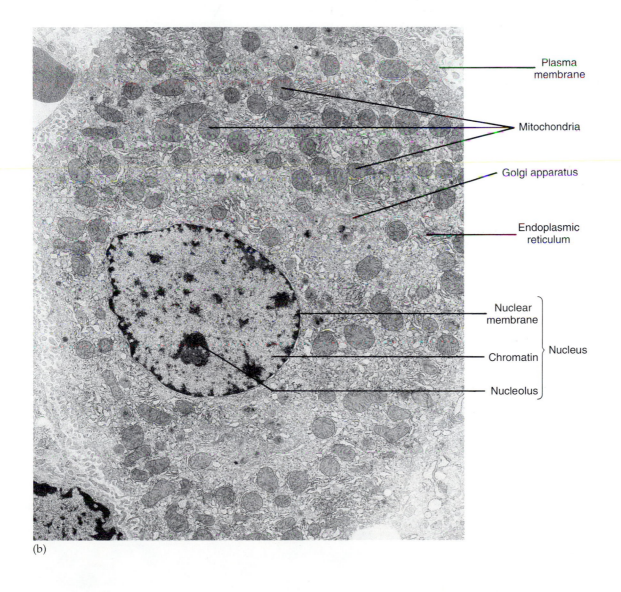

Plasma
membrane

Mitochondria

Golgi apparatus

Endoplasmic
reticulum

Nuclear
membrane

Chromatin   Nucleus

Nucleolus

(b)

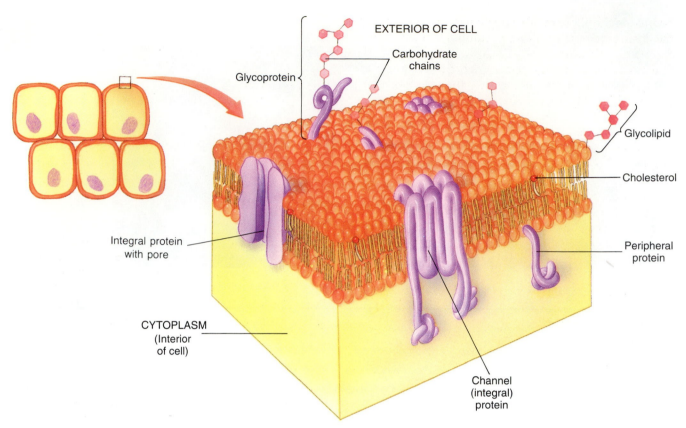

Figure 3–4
The plasma membrane. This section of a cell's plasma membrane illustrates the
lipid bilayer and both types of proteins (peripheral and integral).

What function is served by integral proteins that provide membrane channels?

These molecules are arranged to form an extremely thin barrier, only 650 to 1000 nanometers (nm), or less than one-millionth of an inch thick. This barrier is not complete, for very small molecules such as water are able to pass through freely while others may be transported across by mechanisms within the plasma membrane. The barrier is, therefore, referred to as a selectively permeable membrane barrier (a freely permeable membrane permits *all* material to pass).

The lipid component of the plasma membrane includes two types of molecules: **phospholipids** and **cholesterol**. Phospholipid molecules each contain two parts that have different characteristics: a "head" portion that is attracted to water, or is **hydrophilic**,[6] and thereby borders the watery extracellular environment; and a "tail" portion that repels water, or is **hydrophobic**,[7]

thereby bordering the interior of the membrane. As a result of their interaction with water, phospholipid molecules form a double-layered arrangement that is called the **lipid bilayer** (Fig. 3–4). In the temperature range in which the cell normally lives, the lipid bilayer is in the liquid state. This provides the membrane with a consistency that is very fluid, much like the vegetable oil that is used in cooking. Cholesterol molecules lie within the lipid bilayer and provide a stabilizing influence by increasing its mechanical strength. The lipid nature of the bilayer provides an effective barrier for the cell, since it prevents the unregulated passage of molecules that do not dissolve in oil, such as amino acids and sugars. It does, however, permit the passage of molecules that do dissolve in oil, such as oxygen, carbon dioxide, and certain hormones.

The proteins that form about one-half the composition of the plasma membrane are of two types: **peripheral** and **integral**. Peripheral proteins are attached to the outside and inside surfaces of the plasma membrane. Integral proteins, which are more numerous, have at least some portion of their structures within the lipid bilayer. Integral proteins may lie partially em-

[6]Hydrophilic: Greek *hydro-* = water + *-philic* = liking, attracted toward.

[7]Hydrophobic: *hydro-* + *-phobic* = fearing, repelled from.

bedded in the bilayer with a portion extending through the inside or outside surface, although most extend completely through (see Fig. 3–4). Integral proteins that extend through the bilayer serve as channels for the transport of materials passing through the membrane. These channels may serve as openings, or "pores," for the unregulated passage of small molecules such as water, or they may be selective channels that permit passage of only certain substances. For example, nerve and muscle cells contain "ion channels" in their membranes that regulate the movement of ions required for their normal functioning.

Peripheral and integral proteins that communicate through the outside surface of the bilayer may contain an attached carbohydrate molecule. A complex molecule that is formed from this union is known as a **glycoprotein**. Glycoproteins serve as receptors for substances that must distinguish between different types of cells, such as hormones, growth factors, and antibodies. Glycoproteins also signal nutrient molecules floating in the extracellular environment, directing them to the plasma membrane for their incorporation into the cell.

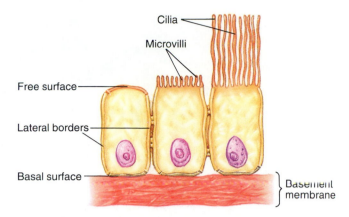

Figure 3–5
Features associated with stationary cells. Their basal surface is attached to a basement membrane, and their lateral borders have junctions that interconnect adjacent cells. The free surfaces frequently contain modifications of the plasma membrane, such as microvilli or cilia.

What benefit is provided to the cell by microvilli?

CONCEPTS CHECK

1. Why is the plasma membrane referred to as a "selectively permeable membrane" barrier?

2. Describe the arrangement of phospholipid molecules in the lipid bilayer.

3. What is the functional significance of integral proteins that extend through the membrane?

Modifications of Plasma Membranes

In many cells of the body the plasma membrane contains structures that enable the cell to perform special functions. This is the case with most cells that do not move about the body but remain in one place throughout their existence. These cells may be found lining the digestive tract, or forming the substances of the liver and kidneys, or forming the walls of the blood vessels, or in many other areas of the body. The plasma membranes of these stationary cells are divided into four regions (Fig. 3–5): a **free surface**, which is exposed to an open space, or lumen; two **lateral borders** that are adjacent to neighboring cells on each side; and a **basal surface** that is opposite the free surface. The basal sur-

face is usually attached to a matrix of protein, called the **basement membrane**. Plasma membrane modifications frequently occur on the free surface of the cell.

The free surface of the plasma membrane in some cells is extensively folded to form many tiny, slender projections called **microvilli**[8] (Figs. 3–5 and 3–6). Each microvillus is enveloped by the plasma membrane and contains a small amount of cytoplasm within it. The presence of microvilli increases the area of the plasma membrane's free surface, thereby increasing the amount of material that can pass across it. Consequently, microvilli are common in cells that absorb materials, such as the cells lining the small intestine, and in cells that release materials, such as excretory cells in the kidneys. These cells have enormous numbers of microvilli—as many as 3000 per cell.

Another plasma membrane modification along the free surface of certain cells is a collection of **cilia**[9] (Figs. 3–5 and 3–6). Cilia, like microvilli, are slender projections of the plasma membrane that extend into the lumen. Unlike microvilli, however, they contain an organized arrangement of proteins called *microtubules*, which provide structural support. Cilia are also quite a bit longer than microvilli. Although cilia increase the surface area of the plasma membrane, they do not ab-

---

[8]Microvilli: Greek *micro-* = small + Latin *villi* = shaggy hairs (plural form).

[9]Cilia: Latin *cilia* = eyelashes.

Figure 3–6
Cilia and microvilli. This electron micrograph shows the free (*top*) surface of a group of cells. Some of the cells contain long, hairlike processes called *cilia*, while others contain much shorter processes, the *microvilli*. Magnification 4110x. (*From R.G. Kessel and R.U. Kardon*, Tissues and Organs: a text-atlas of scanning electron micrographs *courtesy of Professor R.G. Kessel.*)

sorb materials. Rather, they move about in a coordinated, whiplike manner in order to move body fluids. Cilia are common among cells lining the respiratory tract, which is layered with a sticky fluid called *mucus* that is moved upward through the tract by ciliary action.

A **flagellum** is a modification of the plasma membrane that is normally found as a single, long process. Like cilia, flagella also contain microtubules that whip about. Flagella provide propulsion for cells such as sperm cells.

CONCEPTS CHECK

1. With what area of the cell are microvilli and cilia normally associated?

2. Distinguish between microvilli and cilia on the basis of their structures and functions.

Functions

We have seen that the plasma membrane is a dynamic, vital component of the cell. It provides an important physical barrier that encloses the cell contents and separates them from the extracellular environment. It contains receptors that permit substances to distinguish between cells, and directs nutrients to the plasma membrane for their incorporation into the cell. Modifications of its basic structure also occur, allowing for a diversification of specialized functions. Finally, the plasma membrane regulates the passage of materials into and out of the cell. Its selectively permeable nature permits the movement of certain substances, such as water, yet restricts the movement of others. The various ways in which materials cross the plasma membrane are important considerations in the study of body function upon which we will now focus.

The different mechanisms that transport material across the plasma membrane are distinguished by the nature of energy that is used to power their movement. They may be grouped into two categories: **passive processes** and **active processes**. Passive processes are powered by *kinetic energy*, which is a force that causes molecules to move about randomly. Passive processes do not require energy from the cell. Active processes, on the other hand, demand an input of energy from the cell. This energy is supplied by the high-energy chemical bonds of adenosine triphosphate (ATP) molecules. Energy is released to power the movement of materials across the plasma membrane when the ATP bonds are broken in chemical reactions.

Passive Processes

DIFFUSION The process of **diffusion** occurs whenever molecules move from a region of higher concentration to a region of lower concentration by their kinetic energy. Movement will continue until the molecules achieve an even distribution, or **equilibrium**. A difference in concentration from one region to another is called a **concentration gradient**, which may be compared to a hill: the region of low concentration is the bottom of the hill and the region of high concentration is the top. During diffusion, molecules move passively down the gradient hill. Examples of diffusion can be found in everyday experiences: dropping sweetener into a cup of coffee, spraying a scented aerosol into the air of a stuffy room, and opening a window to let fresh air replace stale air are just a few. An example of diffusion in the body is the movement of oxygen molecules from the tiny air sacs in the lungs to the blood vessels that surround the air sacs. The diffusion of a molecule through the plasma membrane of a cell is shown in Figure 3–7a. Notice that in this example, a

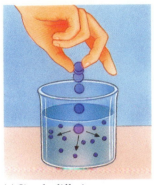

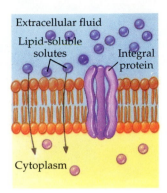

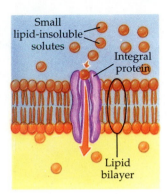

(a) Simple diffusion

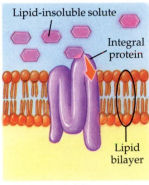

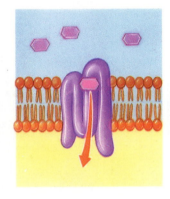

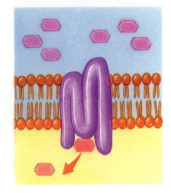

(b) Facilitated diffusion

Figure 3–7
Two mechanisms of diffusion across a plasma membrane. (a) The first box demonstrates the simple diffusion of a particle dissolving in a beaker of water; once the particle breaks apart, its subunits migrate from a region of high concentration to a region of low concentration until they are equally distributed. Diffusion may occur across a plasma membrane in one of two ways: by the passage of lipid-soluble molecules through the lipid bilayer, or by the passage of lipid-insoluble molecules through a pore formed by an integral protein. In either case, movement is achieved without assistance. (b) Facilitated diffusion is similar, but it requires the assistance of integral proteins because of the insolubility of the molecules.

Can diffusion occur against a concentration gradient?

molecule that is soluble (that is, can dissolve) within the plasma membrane is capable of diffusing directly through it. On the other hand, a molecule that is not soluble within the plasma membrane (is lipid-insoluble) may diffuse through it by passing through a pore formed by an integral (channel) protein.

FACILITATED DIFFUSION Certain large molecules that are needed by the cell for energy and growth are often more numerous outside the cell than in it. This establishes a concentration gradient that favors their movement into the cell. However, they cannot pass through the plasma membrane without assistance, because of

their large size and molecular composition. In **facilitated diffusion**, passage is made possible by the assistance of integral proteins in the plasma membrane (recall that integral proteins may extend completely through the plasma membrane). An example of this passive process occurs in the movement of the sugar molecule, glucose, into the cell (Fig. 3–7b). Glucose concentration is normally greater in the extracellular environment, but glucose is unable to move across the plasma membrane unassisted. Its movement into the cell is mediated by integral proteins that combine with it, allowing it to pass through the lipid bilayer and into the cytoplasm.

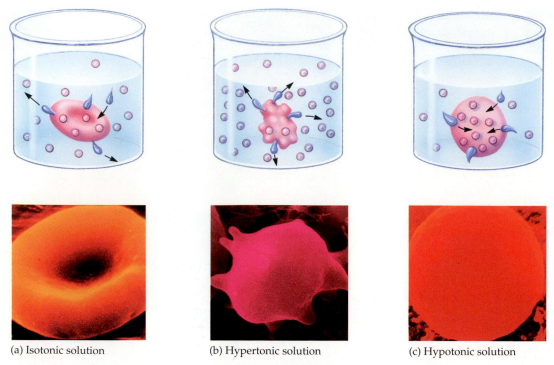

(a) Isotonic solution   (b) Hypertonic solution   (c) Hypotonic solution

Figure 3–8
Osmosis. (a) In an isotonic solution, no net movement of water occurs, and the cell remains the same. The cell is in equilibrium with the solution. (b) When the cell is placed in a hypertonic solution, the higher solute concentration draws water out of the cell, causing it to shrivel. (c) When the cell is placed in a hypotonic solution, the lower solute concentration forces water to flow into the cell, causing it to swell.

Does water move into or out of a cell placed in a hypertonic solution?

OSMOSIS   The passive process of **osmosis**[10] occurs whenever two compartments containing water, each with a different amount of dissolved particles, or solutes, are separated by a barrier that permits the movement of water molecules but excludes the movement of solutes. Osmosis is defined as the movement of water molecules (only) across a selectively permeable membrane from a region of higher water concentration to a region of lower water concentration. Water concentration is determined by the amount of solutes present: the higher the solute concentration, the lower the water concentration, and vice versa. As in diffusion, molecules move in response to a concentration gradient until a state of equilibrium is achieved.

The force exerted by the movement of water molecules is known as **osmotic pressure**. As you may suspect, the strength of this force is dependent on the magnitude of the concentration gradient: the greater the difference in water concentration between two compartments, the greater the osmotic pressure. This

force is an important factor in living systems, since the transport of substances across the plasma membrane, between cells, and between body compartments is influenced by the movement of water and the osmotic pressure that is generated.

In the cell, water molecules are able to move quickly through channels in integral proteins in the membrane. This feature makes osmosis an important factor in cell survival, as can be illustrated by the red blood cell (Fig. 3–8). If we were to place a red blood cell in a beaker filled with water containing the same concentration of solutes as the cell, water molecules would flow across the plasma membrane in both directions at equal rates. This occurs because the concentration of water molecules is the same on both sides of the plasma membrane—it is in a state of equilibrium. In this case, the extracellular fluid is said to be an **isotonic solution**[11] when compared with the intracellular fluid. If the cell is placed in a different beaker that contains a *greater* concentration of solutes than the intracellular fluid, more water will move out of the cell

---

[10]Osmosis: Greek *osmos* = act of pushing + *-osis* = a condition of.

[11]Isotonic: Greek *iso-* = equal + *tonos* = tension, force.

than in and the extracellular fluid is called a **hypertonic solution**.[12] This condition will cause the cell to dehydrate and shrink in size—a state known as *crenation*. The opposite effect, which we see in a **hypotonic solution**,[13] occurs when the cell is placed in a beaker containing a *smaller* concentration of solutes than the intracellular fluid. In this situation, more water moves into the cell than out, causing the cell to swell, or to *hypertrophy*, and eventually burst if the osmotic pressure is great enough.

Osmosis is an important factor among all cells in the body. It is one reason why the solute concentration of interstitial fluid in the extracellular environment must remain as constant as possible. This is illustrated by the fact that in some areas of the body where it is not possible to maintain a constant extracellular solute concentration, there are mechanisms that serve to protect cells. For example, urine passing through the urinary tract may at times carry large amounts of solutes, and at other times carry very little (exactly why this is so will be examined in Chapter 18). The cells lining the inner wall of the tract are protected from the ravages of these extremes by a protective coat of mucus, which is secreted by special cells in the tract wall. The mucus prevents direct contact between the cells and urine.

FILTRATION As we have just seen, the force that causes the movement of molecules in diffusion, facilitated diffusion, and osmosis is generated by the presence of a difference in numbers of molecules between two regions, or a concentration gradient. In **filtration**, the force that drives molecules is generated by a difference in the numbers of collisions that occur among molecules between two regions. This is called a **pressure gradient**. Filtration is defined as the movement of molecules across a selectively permeable membrane from a region of high pressure to a region of low pressure. In the cell, filtration involves mainly small molecules that are physically pushed through the plasma membrane by the presence of a pressure gradient. It occurs in the kidneys, where blood pressure forces small molecules such as water, small ions, and the waste product urea between the plasma membranes of the thin cells that form the walls of the blood vessels and kidney tubules.

Active Processes

ACTIVE TRANSPORT As an active process, **active transport** requires the use of energy supplied by the cell in the form of ATP. It is defined as the transport of sub-

stances against a concentration gradient across a selectively permeable membrane. In the cell, it requires the assistance of integral proteins, called *carriers*, in the plasma membrane (Fig. 3–9). Active transport is an important means of maintaining the appropriate levels of certain ions inside the cell and other ions outside the cell in spite of opposing concentration gradients. Large molecules that are needed for energy or growth functions of the cell, such as amino acids, may also be carried into the cell, where their numbers are greater than in the extracellular environment.

CYTOSIS The transport of large volumes of materials and large particles across the plasma membrane occurs

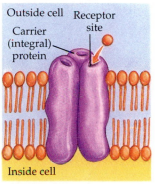
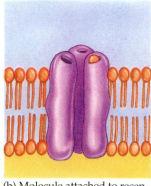

(a) Molecule attaches to receptor site

(b) Molecule attached to receptor site; channel opens

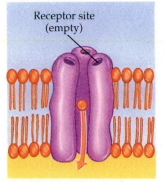
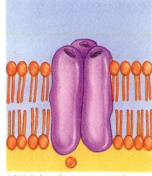

(c) Molecule transported through carrier protein; channel closes behind molecule

(d) Molecule transported to inside of cell; channel closed

Figure 3–9
Active transport. (a) Active transport begins when an integral carrier protein binds to a molecule. The chemical reaction of binding causes the protein to change shape, forming a channel through which the molecule can pass. (b) As the protein changes shape, the molecule is released from the receptor site and begins its journey through the channel. (c) The molecule is forced through the channel against a concentration gradient and enters the cytoplasm. Energy is provided by ATP. (d) Once the molecule enters the inside of the cell, the carrier protein closes the channel.

Can active transport proceed when there is no energy available?

[12]Hypertonic: *hyper-* = above normal, excessive + *tonos*.

[13]Hypotonic: *hypo-* = below, deficient + *tonos*.

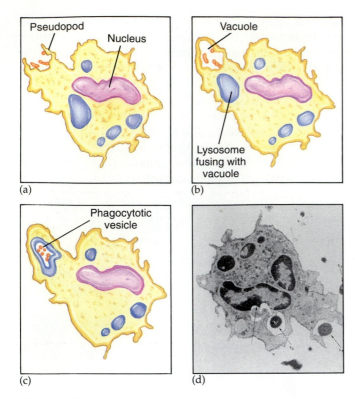

Pseudopod
Nucleus

Vacuole

(a)
(b)

Phagocytotic
vesicle

Lysosome
fusing with
vacuole

(c)
(d)

Figure 3–10
Phagocytosis—an example of endocytosis. White blood cells ingest foreign particles and cell debris by this process. (a) It involves the movement of a cell "arm," or pseudopod, around the particles. (b) The pseudopod eventually surrounds and engulfs the particles, enclosing them within a vacuole in the cytoplasm. (c) Once within the cytoplasm of the cell, the particles are digested by enzymes released from lysosomes. (d) Different stages in phagocytosis are captured by the electron micrograph.

How is pinocytosis similar to phagocytosis?

by an active process known as **cytosis**. There are two types of cytosis, which differ in the direction of material transport. **Endocytosis**[14] is the import of materials into the cell, and **exocytosis**[15] is the export of materials out.

Endocytosis brings large volumes of materials or large particles into the cell by extending a segment of the plasma membrane around the material and enclosing it (Fig. 3–10). In one type of endocytosis called **phagocytosis**[16] (fag′-ō-sī-TŌ-sis), a portion of the cell's

---

[14]Endocytosis: Greek *endo-* = within, inner + *cytosis* = condition of the cell.

[15]Exocytosis: *exo-* = outer + *cytosis.*

[16]Phagocytosis: *phago-* = eating + *cytosis.*

---

plasma membrane is pushed outward toward the targeted particle by streaming cytoplasm. This forms an armlike projection called a *pseudopod,*[17] which surrounds the particle and engulfs it. Once inside the cell, the plasma membrane around the particle forms a membrane sac called a *phagocytotic vesicle,* where the particle is digested by enzymes. Phagocytosis transports only solid particles into the cell; hence the designation, which means "cell-eating." It is a specialized process that few cells are capable of. It is most notable in white blood cells of the body's immune response as the method they use to remove dead cells and foreign material such as bacteria. A second type of endocytosis, called **pinocytosis** (pin′-ō-sī-TŌ-sis) or "cell-drinking," involves the bulk movement of fluid into the cell. Unlike phagocytosis, this process involves only a slight folding of the plasma membrane around a droplet of fluid. Pinocytosis occurs in many cells of the body, including cells that line blood vessels and kidney tubule walls.

Exocytosis is the process the cell uses to export bulk amounts of material. It may be used to release large amounts of waste products, or to release molecules that have been manufactured in the cytoplasm and are to play a role elsewhere in the body. In either case, the bulk materials are initially packaged within membrane-enclosed organelles in the cytoplasm and then transported to the plasma membrane (Fig. 3–11). Here, the organelle membrane fuses with the plasma

---

[17]Pseudopod: Greek *pseudo* = false + *pod-* = foot.

Figure 3–11
Exocytosis. During this process, waste materials or newly manufactured products are packaged into bundles within membrane-bound vesicles. The vesicles migrate to the plasma membrane and fuse with it. Once fusion is complete, an opening is created that enables the materials to escape the cytoplasm and enter the extracellular space.

Does exocytosis require the presence of energy?

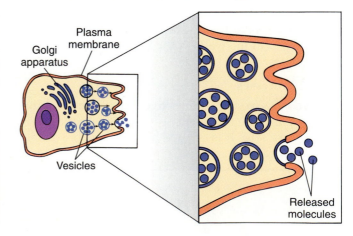

Golgi
apparatus

Plasma
membrane

Vesicles

Released
molecules

membrane, releasing the materials into the extracellular environment. The removal of waste products through exocytosis is also known as **excretion**, and the exocytosis of manufactured molecules is known as **secretion**. Secretion is the method by which important substances such as digestive enzymes, mucus, hormones, antibodies, and other products are released into the extracellular environment.

The various ways in which material may move across the plasma membrane are summarized in Table 3–1.

CONCEPTS CHECK

1. What is the difference between diffusion and facilitated diffusion?

2. When a cell is placed in a hypertonic solution, in which direction will water flow across the plasma membrane?

3. Why is active transport an active process?

4. Describe the process of phagocytosis.

Cytoplasm and Its Organelles

Organelles are membrane-enclosed structures in the cytoplasm that serve specific functions, providing the cell with a division of labor.

Cytoplasm is the protoplasmic material that is located between the plasma membrane and the nucleus. It is composed of a thickened, gel-like fluid that is known as the **cytosol**,[18] which consists of water, many types of proteins, and small amounts of fats, carbohydrates, and electrolytes. Suspended within the cytosol are numerous compartments called **organelles**,[19] which are bounded by a selectively permeable membrane. There are five major types of organelles in the cells of the body: **endoplasmic reticulum, Golgi apparatus, mitochondria, lysosomes**, and **peroxisomes**. As we will

---

[18]Cytosol: *cyto-* = cell + *-sol* = solution.

[19]Organelle: *organ-* = body + *-elle* = tiny.

Table 3–1 MOVEMENT OF MATERIALS ACROSS THE PLASMA MEMBRANE

Process	Energy Source	Method of Movement	Examples
Diffusion	Kinetic energy of molecular motion	Along a concentration gradient	Movement of oxygen, carbon dioxide, and fats through the lipid bilayer
Facilitated diffusion	Kinetic energy of molecular motion	Along a concentration gradient with the aid of a membrane protein carrier	Movement of glucose into a nutrient-hungry cell
Osmosis	Kinetic energy of molecular motion	Water movement along a concentration gradient	Movement of water molecules through pores in a membrane
Filtration	Kinetic energy of a pressure gradient	Along a pressure gradient	Movement of fluids across membranes forming capillary walls
Active transport	Cellular energy (ATP)	Against a concentration gradient with the work of a membrane protein	Movement of glucose and ions against a gradient
Cytosis			
Endocytosis	Cellular energy (ATP)	Bulk transport of substances into the cell	Phagocytosis of dead cells and bacteria by white blood cells; pinocytosis of fluid by various cells
Exocytosis	Cellular energy (ATP)	Bulk transport of substances out of a cell	Secretion of hormones by thyroid gland; secretion of digestive enzymes by pancreas cells

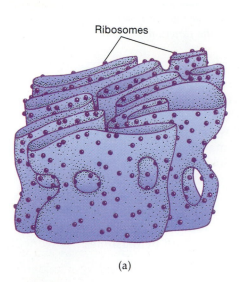

Ribosomes

(a)

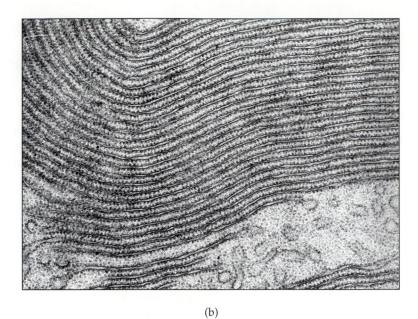

(b)

Figure 3–12
Endoplasmic reticulum. (a) In this illustration of rough ER, note the presence of spherical ribosomes that are attached to the membranous folds. (b) Ribosomes may be observed in an electron micrograph of rough ER as tiny dots lining the membranes.

What is the function of the rough ER?

see, each of these organelles contains a particular set of enzymes that perform a specific function for the cell. This provides the cell with an efficient means of separating functions, or a division of labor.

Endoplasmic Reticulum

The **endoplasmic reticulum (ER)**[20] is a series of branching and rejoining hollow tubules located throughout the cytoplasm (Fig. 3–3). Its walls consist of a single continuous sheet of lipid bilayer membrane that encloses a hollow space in the center, forming a closed sac. The long, continuous sac may be folded upon itself to form a series of flattened sections, or it may form a meshwork of tubules. In either form, the ER provides a physical barrier that separates molecules belonging in the cytosol from those that do not. The ER serves as a transportation network for newly synthesized molecules. It also contains enzymes that play an important role in the synthesis of macromolecules such as proteins and lipids, which are used to build other cellular organelles.

When the ER is viewed through the powerful electron microscope, its two types can be distinguished:

the **rough ER (RER)**, which contains small, spherical granules known as **ribosomes**[21] attached to the cytoplasmic side of the membrane (Fig. 3–12); and the **smooth ER (SER)**, which does not contain ribosomes. The rough ER is always continuous with the membrane surrounding the nucleus of the cell. Its ribosomes are often present in large numbers that form chains called *polyribosomes*. Ribosomes may also be found dissociated from the rough ER throughout the cytoplasm. In cells that synthesize large amounts of protein molecules such as antibody-secreting white blood cells, the rough ER is abundant. By contrast, cells that are active in lipid synthesis, such as liver cells, contain large amounts of smooth ER. Thus, smooth ER appears to play a role in lipid synthesis while rough ER is active in protein synthesis.

Golgi Apparatus

The **Golgi** (GŌL-jē) **apparatus** was first described in 1898 by Camillo Golgi, an Italian anatomist. It is normally found near the nucleus and is composed of numerous flattened, disc-shaped sacks called *cisternae*[22]

---

[20]Endoplasmic reticulum: *endo-* = within, inner + *-plasmic* = having form; *reticulum* = little network.

[21]Ribosome: *ribos* = five-carbon sugar + *-some* = body.

[22]Cisternae: Latin *cisterna* = hollow vessel.

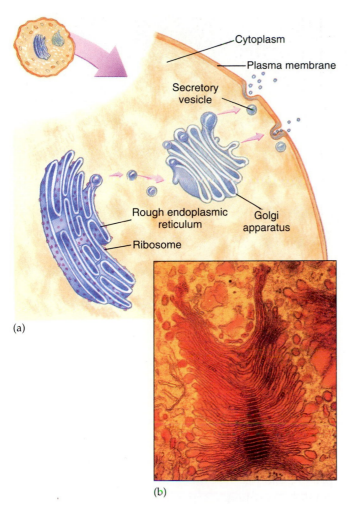

(a)

(b)

Figure 3–13

Golgi apparatus. (a) The Golgi apparatus is usually located between the rough ER and the plasma membrane of a cell. Vesicles that are in the process of exocytosis are shown migrating toward the plasma membrane and fusing with it. (b) This electron micrograph of the Golgi apparatus demonstrates the disc-shaped cisternae and the secretory granules.

What role does the Golgi apparatus play in exocytosis?

that are bound by a single lipid bilayer membrane (Figs. 3–3 and 3–13). The Golgi apparatus contains enzymes and other substances that alter newly synthesized proteins and carbohydrates originating in the ER. In many cases, proteins and carbohydrates are combined to form glycoproteins, which are used in the construction of new plasma membranes. Other alterations prepare molecules for their secretion out of the cell. As the molecules approach their moment of secretion, they are concentrated into portions of the Golgi apparatus that pinch off from the cisternae to form pack-

ages called *secretory vesicles*. During the process of exocytosis, the secretory vesicles migrate toward the plasma membrane and eventually fuse with it, releasing the packaged molecules to the exterior. As you may suspect, cells that are highly active in secreting enzymes, hormones, antibodies, and other protein-containing substances contain large numbers of Golgi apparatus.

Mitochondria

Often called the "energy powerhouse" of the cell, **mitochondria**[23] (mī'-tō-KON-drē-a) contain large amounts of enzymes that break down nutrient molecules in order to supply the cell with energy. This catabolic process is known as *cellular respiration*; it will be discussed in Chapter 17.

Mitochondria are spherical or sausage-shaped structures that may change from one of these shapes to the other (Fig. 3–14). They are bound by two layers of lipid bilayer membrane: the outer membrane is selectively permeable and forms an envelope around the structure; the inner membrane is folded into numerous convolutions called *cristae*, which extend into the interior of the mitochondrion. The cristae increase the surface area of the membrane, providing attachment sites for metabolic enzymes. Mitochondria are the only cellular organelles with two sets of bilayer membranes (the nucleus also contains two sets of bilayer membranes, but is not considered an organelle).

Within the space enveloped by the inner membrane of the mitochondrion is a space called the *matrix*, which contains a highly concentrated mixture of enzymes and DNA. The presence of DNA provides a mitochondrion with the ability to synthesize some of its own proteins and, under the direction of the nucleus of the cell, duplicate itself. In cells that utilize large amounts of energy, such as muscle cells, mitochondria are generally found in abundance largely because of their ability to self-duplicate during times of high energy need.

Lysosomes

In most cells, **lysosomes**[24] (LĪ-sō-sōmz) appear as spherical structures that are randomly distributed throughout the cytoplasm (Fig. 3–3). They are bound by a single lipid bilayer membrane and contain large numbers of hydrolytic enzymes. These enzymes are

[23]Mitochondria: Greek *mitos* = thread + *chondria* = grain (plural form).

[24]Lysosome: Greek *lys-* = to break + *-some* = body.

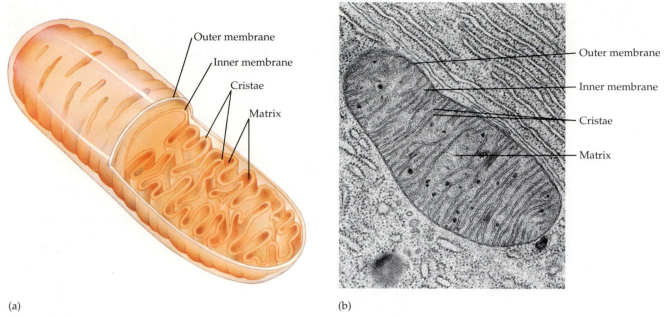

(a)   (b)

Figure 3–14
Mitochondrion. (a) The mitochondrion contains enzymes that perform cellular respiration for the production of energy along its inner membrane. Although shown here as sausage-shaped, it takes on many shapes within a cell. (b) This electron micrograph reveals the inner membrane structure of a mitochondrion. Can you identify the prominent organelle immediately surrounding it?

What essential cellular activity is performed within mitochondria?

used to break apart intracellular particles and nutrient molecules, or, in the case of a phagocytic white blood cell, ingested microorganisms and dead body cells. Because of these activities, lysosomes are often called "the digestive system of the cell."

Lysosomes also aid in the recycling process that follows cell death. When the cell dies, its lysosomes burst open by a yet-undiscovered mechanism. Hydrolytic enzymes are released immediately into the cytoplasm, where they proceed to digest the components of the cell, rendering them into a simpler form that can be reused by other cells. This function has provided lysosomes with the titles "self-destruct bags" and "suicide packets."

Another function of lysosomes is the release of enzymes into the extracellular environment by the process of exocytosis. This occurs in certain cells that break down materials located outside the cell, such as bone-destroying osteoclasts and white blood cells at the site of an infection.

Peroxisomes

Like lysosomes, **peroxisomes** are spherical structures that are distributed randomly in the cytoplasm and are bound by a single lipid bilayer membrane (Fig. 3–3).

Like the mitochondria, they contain enzymes that break down molecules when oxygen is present. Peroxisome enzymes are important to the cell because they detoxify various molecules. For example, the ethyl alcohol that is present in alcoholic beverages is toxic to cells when present in large amounts. However, almost half of the alcohol that is consumed in a glass of wine is converted by peroxisomal enzymes to a chemical called *acetaldehyde*, which can be metabolized and thereby removed as a toxic substance. Peroxisomal enzymes also play a role in the digestion of fatty acids, aiding the mitochondria in their metabolic activities.

Cytoskeleton

The **cytoskeleton** is located within the cytoplasm but is not an organelle, since it is not enveloped by a selectively permeable membrane. It is composed of proteins that are organized to provide the structural framework of the cell (Fig. 3–15). Cytoskeletal elements include **microtubules**, which are long, hollow tubes that help maintain the shape of a cell. Small branches of microtubules, known as *microtrabecular strands*, form a vast network throughout the cytoplasm. Microtubules and their branches also provide structural support for cilia and flagella, and form structures called

HEALTH CLINIC

Tay-Sachs Disease and Lysosomes

Lysosomes are important for the normal functioning of cells, since their digestive activities help to regulate concentrations of molecules within the cytoplasm. Impairment of their function results in the often fatal accumulation of substances. **Tay-Sachs disease** is an example of lysosomal impairment. It is an inherited disease with a current expression in the United States of about 1 in 1000 births. It occurs predominantly in families of Eastern European Jewish origin, and is characterized by progressive mental and physical retardation and early death. It is caused by

a genetically controlled deficiency of one of the enzymes normally released by lysosomes in nerve cells of the brain. This enzyme, called *hexosaminidase A*, normally digests lipid molecules in the cytoplasm of cells. When it is deficient, the lipids accumulate in the cells, causing them to alter in function and eventually die. Neurological functions quickly become impaired, and the afflicted child usually dies between the ages of 2 and 4 years. Currently, there is no cure for this tragic disease.

---

centrioles that may play a role in cell division (to be discussed later in this chapter).

Microfilaments are another type of cytoskeletal element. They are thin and threadlike, and are usually arranged in bundles. Microfilaments have been shown to play an important role in cellular movement, and

Figure 3–15
Cytoskeleton. This model of the cytoskeleton lattice is about 200,000 times its actual size. It is composed mainly of microtubules that traverse the cytoplasm throughout.

What is the major role of the cytoskeleton?

Endoplasmic reticulum
Plasma membrane
Ribosomes
Mitochondrion
Microtubule
Microtrabecular strands

are particularly apparent in muscle cells, where they are called *myofilaments*.

CONCEPTS CHECK

1. What is the nature of the fluid portion of cytoplasm, and what is it called?

2. Compare and contrast the structures and main functions of the five organelles.

3. What is the function of the cytoskeleton?

Nucleus

The nucleus is the "control center" of the cell because it contains DNA, which, by the use of genes, controls the synthesis of proteins.

The **nucleus**[25] is the largest, most conspicuous structure in the cell. It is spherical in shape and is usually located near the center of the cell (Figs. 3–3 and 3–16). It is often referred to as the "control center" of the cell, since it contains the genetic material that determines cell structure and regulates cell activity. It consists of

---

[25]Nucleus: Latin *nucleus* = kernel.

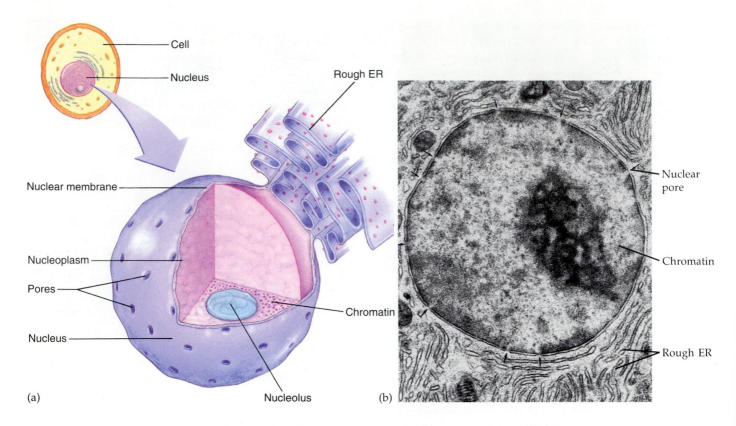

Figure 3–16
Nucleus. (a) The nucleus is enveloped by a double membrane, the nuclear membrane, which contains many large pores. In some areas it is continuous with the ER. (b) This electron micrograph shows a sectioned nucleus, revealing its chromatin within.

What purpose is served by the large pores in the nuclear membrane?

a **nuclear membrane** that envelops an internal region filled with nuclear sap known as **nucleoplasm**.

Nuclear Membrane

The envelope surrounding the nucleus is the nuclear membrane. It provides a selectively permeable barrier that separates the nucleoplasm from the cytoplasm. Structurally, it consists of two layers of lipid bilayer membrane that enclose a minute space in between. Penetrating both layers are small pores, which permit the movement of large molecules between the nucleoplasm and the cytoplasm during protein synthesis.

Nucleoplasm

The nucleoplasm (NOO-klē-ō-plazm) is a gel-like fluid similar in consistency to the cytosol found in cytoplasm. It is located within the limiting nuclear membrane. Immersed within it are structures that change form during certain stages of the cell's life cycle. They include one or more **nucleoli**, and **chromatin**, which are described next.

Nucleoli

Nucleoli[26] are small, spherical structures that are prominent in the nucleoplasm during the stage of the cell cycle known as *interphase* ("between cell divisions"). They are not visible during cell division. Structurally, they are composed of protein and RNA, which are not confined by a limiting membrane. It is thought that nucleoli may be the site of ribosome synthesis, since ribosomes have been shown to move from the nucleoli across the nuclear membrane and into the cytoplasm.

Chromatin

In the interphase nucleus, chromatin appears as tiny granules that are diffused throughout the nucleoplasm. During cell division, the chromatin organizes into larger structures called chromosomes,[27] which can

---

[26]Nucleoli: *nucle-* = kernel + *-oli* = small (plural form).

[27]Chromosome: Greek *chromo-* = color + *-some* = body.

be viewed easily under the compound microscope. A chromosome is believed to be composed of a single DNA molecule. The genetic units, or **genes**, are segments of the DNA molecule. The primary function of the nucleus—the regulation of protein synthesis—is controlled by DNA and its genes.

The parts of a cell that have been discussed are summarized in Table 3–2, where their structures, locations, and primary functions are briefly described.

CONCEPTS CHECK

1. Why is the nucleus referred to as the "control center" of the cell?

2. What are the components of the nucleus?

3. What is the structural relationship between chromatin, chromosomes, and DNA?

Table 3–2   THE PARTS OF THE CELL: A SUMMARY OF STRUCTURE AND FUNCTION

Cell Part	Structure	Location	Primary Functions
Plasma membrane	About 45% lipid, 50% protein, with a small amount (5%) carbohydrate; lipids are arranged in a bilayer with proteins attached or extended through	Envelops entire cell	Provides a barrier for the cell from the extracellular environment; regulates movement of materials across
Cytoplasm	A mixture of water, salts, and organic molecules that combine to form a thick fluid and organized structures called *organelles*	The region between plasma membrane and nucleus	Region of cellular metabolism and other activities that maintain the cell
Cytosol	Thick, gel-like fluid composed mostly of water and proteins	Surrounds organelles	Transport of substances between organelles and other cell structures
Organelles	Formed bodies	Suspended within cytosol	Varied activities that maintain the cell
Endoplasmic reticulum	Branching tubules that may be flattened or tubular; may attach small, round ribosomes	Twists and turns throughout cytoplasm; often extends between nucleus and plasma membrane	Provides a pathway for substances through cytoplasm; the site of protein synthesis (RER) and lipid synthesis (SER)
Golgi apparatus	Series of flattened, disc-shaped sacs	Usually near nucleus with secretory vesicles near plasma membrane	Prepares and packages cellular products for exocytosis
Mitochondria	Spherical or sausage shaped - with inner folded membrane	Randomly through-out cytoplasm	Site where nutrients are broken down and ATP is produced during the process of cellular respiration
Lysosomes	Small, spherical structures	Randomly through-out cytoplasm	Contain enzymes that break down materials brought into cell, or release enzymes out of cell to digest material
Peroxisomes	Small, spherical structures	Randomly through-out cytoplasm	Contain enzymes that detoxify certain harmful materials

(Continued)

Table 3-2 *(continued)*			
Cell Part	**Structure**	**Location**	**Primary Functions**
Cytoskeleton	Tiny tubules or bundles of fibers that are not membrane-bound; composed of protein	Forms a skeleton between organelles and other structures; may extend outward to support cilia and flagella	Provides structural support and movement
Nucleus	Large structure composed of a double layer of lipid membrane that envelops inner region containing nucleic acids	Often located in center of cell	Contains DNA, which regulates protein synthesis and cell division; also contains RNA
Nuclear membrane	Double layer of lipid bilayer membrane with large pores	Surrounds nucleoplasm	Regulates movement of materials between cytoplasm and nucleoplasm
Nucleoplasm	A gel-like fluid that consists of water, nucleic acids, and proteins	Contained within the nuclear membrane	Site of nucleic acids, DNA, and RNA
Nucleoli	Small, spherical structures containing RNA and proteins; visible only during interphase	Randomly within the nucleoplasm	Site of RNA synthesis
Chromatin	Granular, threadlike material composed of DNA and proteins	Diffuse throughout the nucleoplasm	DNA contains the genes, the genetic material

Protein Synthesis

The genetic code in the DNA of cells determines body structure and function due to the important structural and functional role of proteins.

You may be aware from previous courses in biology that DNA provides the basis for heredity from parents to offspring. Thus, it largely determines our physical appearance, our body structure, our ability to resist disease, the way in which we age, much of our behavior—in short, who and what we are as individuals. Exactly how can a tiny molecule play such a profound role in our lives? The answer is basically quite simple, although the mechanism is not: *DNA regulates the synthesis of proteins in the cell.* That is, DNA determines the exact structure of each new protein, and the quantity that is to be synthesized. This has a vital im-

pact on the body because of the crucial role played by proteins in cell structure and function. For example, some proteins contribute to the structure of organelles and the plasma membrane. Many other proteins are enzymes and hormones, which determine the rates and directions of the chemical reactions in the cells and elsewhere in the body. Thus, proteins provide a significant contribution to the structure and the basis for function in the cell. Since the cell is the basic unit of life, we can conclude that "what we are" is largely determined by our proteins, which in turn are determined by DNA.

Protein synthesis involves two processes, **transcription** and **translation**. These processes will be discussed briefly in this section to provide you with a basic understanding of how protein synthesis is achieved by the cell. But first, we have to "crack" the genetic code before we can see how DNA regulates protein synthesis.

The Genetic Code

How does DNA regulate protein synthesis? Basically, it provides a message that directs the sequence of amino acids forming the polypeptide chain (recall from Chapter 2 that a polypeptide is a protein consisting of more than ten amino acids). The message is derived from the structure of the DNA molecule. In Chapter 2 we learned that DNA resembles a long, twisted ladder comprising nucleotide subunits that are repeated along the length of the molecule (Fig. 2–24). Each nucleotide is composed of a sugar group, phosphate groups, and one of four nitrogen bases (adenine, thymine, cytosine, or guanine). The sugar and phosphate groups from each nucleotide form the uprights on one side of the DNA ladder. The rungs are formed by nitrogen bases that pair between opposing nucleotides.

The message of DNA is contained in the specific order of the nitrogen base pairs that occur along the length of the molecule. The base pairs are arranged in groups of three. Each group of three provides a single message that codes for a particular amino acid. This message is called the *triplet code*. The different combinations of base pairs that form the triplet codes for the 20 existing amino acids have been identified, and some are shown in Table 3–3. A section of DNA that contains all the triplet base pairs to code for one complete polypeptide chain is known as a **gene**. The message that a gene provides in order to produce a particular polypeptide chain is the **genetic code**.

Table 3–3	A SELECTION OF MATCHING DNA BASE TRIPLETS, mRNA CODONS, AND AMINO ACIDS	
DNA Base Triplet	**mRNA Codon**	**Amino Acid**
TAC	AUG	"Start" message
AAA	UUU	Phenylalanine
ACA	UGU	Cysteine
GGG	CCC	Proline
GCT	CGA	Arginine
TTT	AAA	Lysine
TGC	ACG	Tyrosine
CTC	GAG	Aspartic acid
AGG	UCC	Serine
ATT	UAA	"Stop" message

Transcription

We have learned that DNA located in the nucleus of the cell regulates protein synthesis, but if you recall, proteins are actually manufactured at the ribosomes in the cytoplasm. How can this be? Genetic research has informed us that a type of RNA, called **messenger RNA (mRNA)**, serves as an intermediary between DNA in the nucleus and ribosomes of the cytoplasm. Two other types of RNA, called **transfer RNA (tRNA)** and **ribosomal RNA (rRNA)**, are also involved in protein synthesis, as we shall discover shortly.

Transcription is the process by which all types of RNA are formed. During transcription, the triplet code sequence in a section of DNA is copied, or *transcribed*, onto an RNA molecule within the nucleus (Fig. 3–17). This is made possible by an unraveling of a section of DNA, which exposes nucleotide bases. RNA nucleotides floating free in the nucleoplasm combine specifically to exposed complementary DNA bases, and with the assistance of enzymes, the nucleotides then combine to form an RNA molecule.

A newly transcribed RNA molecule containing a sequence of triplet bases that corresponds to the triplet code of the unraveled section of DNA is called messenger RNA (mRNA). Each triplet is called a **codon**. Because each codon corresponds precisely to a section of DNA, the sequence of nucleotide bases forming the genetic code is preserved in the mRNA molecule. Once it is formed, the mRNA molecule exits the nucleus through pores in the nuclear membrane and passes into the cytoplasm, where it associates with ribosomes anchored to the rough ER or existing unattached in the cytosol.

Transfer RNA (tRNA) also arises from transcription, but its structure differs from that of mRNA. The sequence of nucleotide bases in tRNA contains small differences from the DNA molecule, resulting in its formation into a folded structure (Fig. 3–17). Once formed, tRNA exits the nucleus through pores in the nuclear membrane and passes into the cytoplasm. Here, it performs the role of transporting free amino acids to ribosomes.

Like tRNA, ribosomal RNA (rRNA) is transcribed from DNA with small differences in the nucleotide base sequence. These differences result in its formation into a structure that is more spherical in shape (Fig. 3–17). Once formed, rRNA is believed to combine with proteins in the nucleolus to form a ribosome. Ribosomes exit the nucleolus and nucleus through large pores. In the cytoplasm the ribosomes may attach to the membranous walls of endoplasmic reticulum, or they may exist unattached throughout the cytoplasm. Ribosomes provide an attachment site for mRNA during protein synthesis.

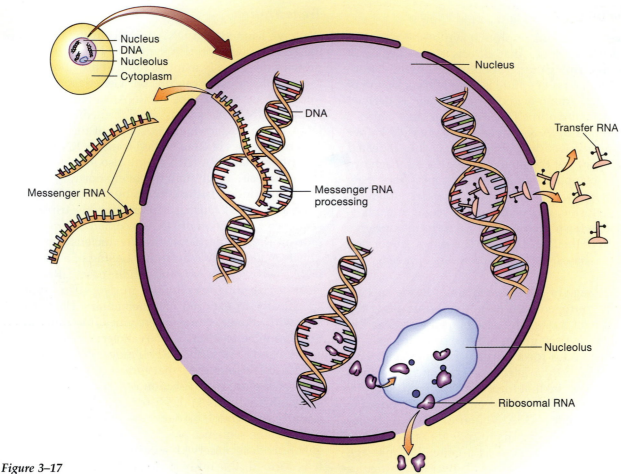

Figure 3–17

The process of transcription. During this process the triplet code sequence in a section of DNA is copied, or transcribed, onto a messenger RNA molecule within the nucleus. Transfer RNA and ribosomal RNA are also transcribed at this time. These molecules are transported out of the nucleus and into the cytoplasm, where they are made ready for the next process of protein synthesis, translation.

How is a genetic code for a particular protein passed from DNA to mRNA?

Translation

The interpretation of the codon message in mRNA into a specific sequence of amino acids occurs during the process of **translation**, which takes place entirely within the cytoplasm. Translation is described schematically in Figure 3–18. In the first step of translation, the mRNA attaches to a ribosome. This is soon followed by the movement of tRNA molecules, which begin to transfer amino acids that are existing free within the cytoplasm to the ribosome site. Recall that there are 20 different types of amino acids; each type combines with a specific type of tRNA. When the tRNA–amino acid complex arrives at the ribosome, its triplet base, called the **anticodon**, combines with a complementary codon on the mRNA. The ribosome then moves along the mRNA, and the next tRNA–amino acid complex moves into position. The two amino acids that have remained attached to the tRNA molecules are joined by a chemical bond with the assistance of enzymes, and the first tRNA detaches from the mRNA. This process is repeated to form a growing peptide chain, until a **stop codon** is encountered on the mRNA, stopping the process. The completed polypeptide is then liberated from the ribosomal complex.

In summary, as the ribosome moves along the mRNA strand, it *translates* the nucleotide base sequence into an amino acid sequence. The polypeptide continues to grow as amino acids brought by tRNA molecules link together by chemical bonds until the stop message is encountered. The process of protein synthesis may be summarized by the following formula:

$$DNA \rightarrow RNA \rightarrow protein$$

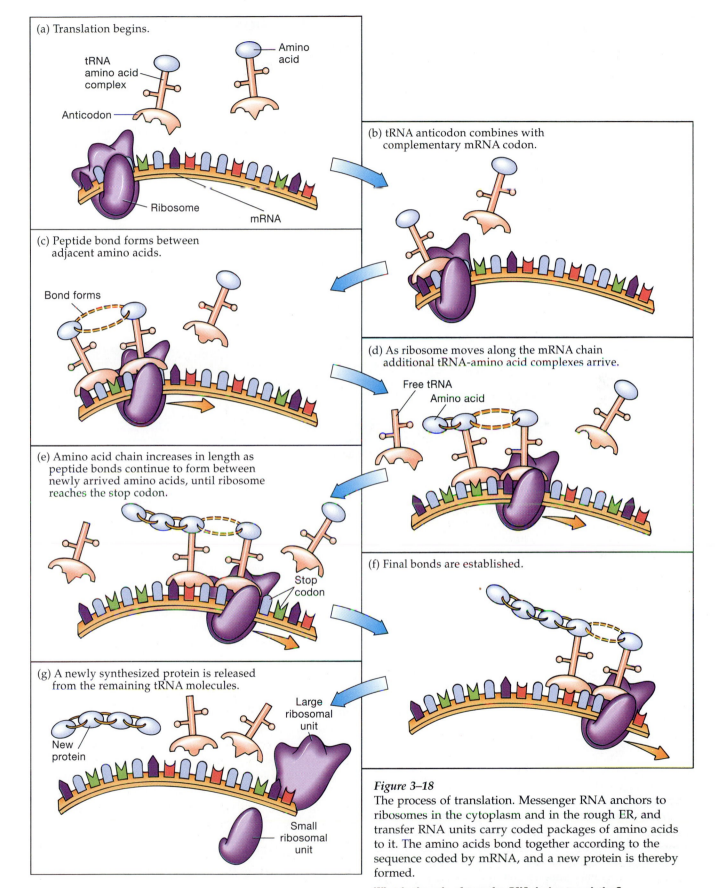

(a) Translation begins.

tRNA amino acid complex

Amino acid

Anticodon

Ribosome

mRNA

(b) tRNA anticodon combines with complementary mRNA codon.

(c) Peptide bond forms between adjacent amino acids.

Bond forms

(d) As ribosome moves along the mRNA chain additional tRNA-amino acid complexes arrive.

Free tRNA

Amino acid

(e) Amino acid chain increases in length as peptide bonds continue to form between newly arrived amino acids, until ribosome reaches the stop codon.

Stop codon

(f) Final bonds are established.

(g) A newly synthesized protein is released from the remaining tRNA molecules.

Large ribosomal unit

New protein

Small ribosomal unit

Figure 3–18
The process of translation. Messenger RNA anchors to ribosomes in the cytoplasm and in the rough ER, and transfer RNA units carry coded packages of amino acids to it. The amino acids bond together according to the sequence coded by mRNA, and a new protein is thereby formed.

What is the role of transfer RNA during translation?

This is the central dogma of a growing field of biology known as molecular genetics. Within this formula lies the explanation for the structures and functions of all life on earth.

CELL DIVISION

Cell replacement and body growth are accomplished by mitosis and cytokinesis. Sex-cell production is accomplished by meiosis.

Cells reproduce by the process of cell division. It is the method by which old or diseased cells are replaced, growth of the body occurs, and sex cells for reproduction of the organism are produced. When a cell divides in a normal fashion, two major events occur: the division of the nucleus, called *nuclear division*; and the division of the cytoplasm, called *cytoplasmic division*. The result of both events is the splitting of the parent cell to produce two independent daughter cells. There

HEALTH CLINIC

Inherited Diseases and a New Hope

All cells manufacture proteins by the processes of transcription and translation. If the genetic code is altered in a section of DNA, the synthesis of proteins in that cell will be affected. In diseases that are inherited, the DNA in the cells of the body does not provide the proper message for protein synthesis. The unfortunate result is disease caused by the deficiency of an enzyme or a structural protein that normally plays a significant role in the body. An important example of an inherited disease is *phenylketonuria*, which is caused by a deficiency of an enzyme (phenylalanine hydroxylase) that is required for normal metabolism of the amino acid phenylalanine. This disease results in mental retardation if not corrected. Other examples are *Tay-Sachs disease* (see the previous Health Clinic box); *achondroplasia*, which inhibits normal bone growth at the ends of long bones, resulting in severe stunting of growth (dwarfism) and very short limbs; *sickle cell anemia*, in which a structural protein in red blood cells is synthesized abnormally, resulting in sickle-shaped cells that are not capable of transporting enough oxygen; *hemophilia*, in which a protein required for normal blood coagulation is deficient, resulting in uncontrolled bleeding following minor injuries; and *cystic fibrosis*, a lethal disease that causes an incorrect channel protein to be produced in body cells, leading to a severe accumulation of mucus in the lungs, pancreas, and intestines.

Fortunately, molecular geneticists have begun an unprecedented attempt to identify the entire genetic code of humans—the so-called *human genome*. This is an enormous project, since it involves unraveling the 2 meters of DNA that is packed into a human cell and contains over 100,000 individual genes—some with as many as 10,000 base pairs each! Ultrafast computers are being used, and some scientists estimate that the human genome may be known by 1998. Once this information is available and the nature of gene action is more completely understood, individual human genes that are the sources for many diseases may be manipulated to correct the deficiencies.

Although this may sound a bit like a science fiction fantasy, great strides have already been made toward this end. For example, the genes responsible for causing *cystic fibrosis*, *neurofibromatosis*, and *fragile X-linked mental retardation* have been completely mapped and identified, and progress is rapid on mapping the genes that cause *schizophrenia*, a form of *breast cancer*, and a form of *diabetes*! It is the hope of the scientific community that in the years to come, an emerging technology now coming of age as a result of work in the manipulation of bacterial DNA, known as *genetic engineering*, will be capable of manipulating, and thereby correcting, these errant genes and many others.

are two important ways nuclear division may proceed: through mitosis or meiosis.

Mitosis

Mitosis[28] is the division of the cell nucleus into two genetically identical daughter nuclei. Prior to this process, the cell has undergone a period of cytoplasmic growth, resulting in a doubling of the cell's original size. The DNA has also changed during this period, having duplicated itself to provide the cell with a double set of chromosomes (a total of 92, in humans). The process of DNA duplication is called **replication**. This period of cytoplasmic growth and DNA replication is known as **interphase**,[29] for it occurs between mitoses.

The process of mitosis divides the replicated DNA into two identical sets that are packaged inside two newly formed nuclei. When this is accompanied by cytoplasmic division, or **cytokinesis**[30] (sī'-tō-kī-NĒ-sis), the result is the production of two daughter cells that are each about half the size of the parent cell. Each daughter cell contains an identical copy of the parent cell's genetic material. Mitosis occurs as a continuous sequence of events, most of which can be observed with a compound microscope. It is divided into a series of phases in order to make it easier to understand. The phases are shown in Figure 3–19 and are described in the following paragraphs.

[28]Mitosis: Greek *mitos* = thread + *-osis* = a condition of.

[29]Interphase: *inter-* = between + *phase* = stage or period of time.

[30]Cytokinesis: Greek *cyto-* = cell + *kinesis* = movement.

Figure 3–19
Cell division. This series of illustrations depicts the processes of nuclear division (mitosis) and cytoplasmic division (cytokinesis) that occur between interphases. During interphase, the chromosomes are not visible, and a nuclear membrane envelops the nuclear chromatin. As cell division begins and the cell enters prophase, the nuclear membrane gradually disappears and the chromosomes become visible as sets of chromatids. During metaphase, the chromosomes move to the central region of the cell. Anaphase is marked by the separation of chromosomes, which is followed by their movement toward opposite ends of the cell (to the spindle poles) during telophase. A nuclear membrane soon envelops the nuclear material at each pole. Cytokinesis completes division, and is marked by the development of a cleavage furrow dividing the cytoplasm in half, resulting in the formation of two independent cells.

At which stage of mitosis do chromosomes migrate to the center of the cell?

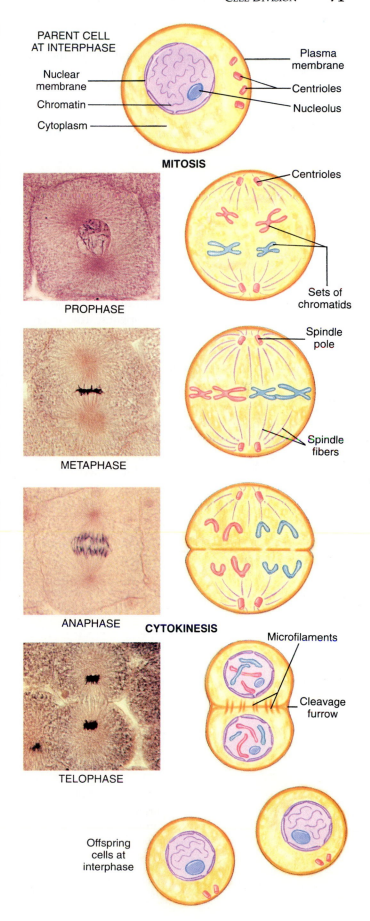

Late Interphase

In the **late interphase** stage of the cell cycle, the nucleus begins preparing for mitosis. Initially, the chromatin thickens to form chromosomes that gradually become visible as elongated structures irregularly arranged within the nucleus. Centrioles, which are composed of microtubules and are located in the cytoplasm, begin to migrate to opposite poles of the cell.

Prophase

Prophase[31] is the first stage of mitosis. During this period the chromosomes continue to thicken until they become clearly visible as two longitudinal halves, called **chromatids**, that are connected to each other at a constricted region called the **centromere**. Also, centrioles complete their migration to opposite poles, and radiating microtubules develop around each centriole. As the nuclear membrane disappears, the microtubules begin to form a network of bundles that extend from one side of the cell to the centromere of each chromatid. These fibers collectively form the **spindle**.

Metaphase

During **metaphase**,[32] the spindle continues to expand, pushing chromosomes to the center of the cell. By the end of this stage the nuclear membrane has completely disappeared, the spindle is well developed, and the chromosomes are lined up along the center of the cell.

Anaphase

In **anaphase**,[33] the third stage of mitosis, the chromatids are separated by the shrinking of spindle fibers and are pulled toward opposite poles of the cell. During late anaphase, cytokinesis begins with the development of a cleavage furrow (a line of division) through the cytoplasm.

Telophase

Telophase[34] is the final stage of mitosis. Early in this stage the two groups of chromatids reach opposite poles of the cell. Upon their arrival, the chromatids, now referred to as *chromosomes*, clump together, and a nuclear membrane forms around each set. The spindle soon disappears and the nucleoli within each newly formed nucleus reappear. The cleavage furrow that first appeared during anaphase deepens, and eventually divides the cytoplasm completely to form two independent daughter cells. The chromosomes in the daughter cells soon uncoil to resume the granular chromatin state, and the new cells enter the interphase stage.

Mitosis and cytokinesis are the methods by which most body cells are replaced to make up for cell death caused by normal wear and tear, aging, and disease. These processes are very important to body health, since many cells have a limited life span and must be replaced on a continual basis by cells that have identical features to the original ones. In fact, over 1 trillion cells die and must be replaced each minute in a healthy adult! However, certain specialized cells are not capable of dividing, and thus cannot replace themselves (for example, nerve cells, heart muscle cells, and skeletal muscle cells). When these cells die, their functions either are taken over by neighboring cells or are lost.

Mitosis and cytokinesis also produce body growth. From the first division of the fertilized egg into two identical cells through adulthood, each division that results in growth of the individual is generated by these processes.

Meiosis

Meiosis[35] is the division of a parent cell into genetically different daughter cells, each of which contain only one-half the number of chromosomes of the parent. The resulting daughter cells in meiosis are the sex cells, or **gametes**, which are produced in only one site in the body—in the sex organs, or **gonads**. They are produced for the purpose of reproduction of the individual. Because gametes contain only one-half the normal amount of genetic material necessary to maintain a cell (23 chromosomes), they are short-lived. When a female gamete (ovum) successfully fuses with a male gamete (sperm) during fertilization, a cell with a complete set of chromosomes (46) results and the life of a new individual begins. This complete cell is called a **zygote**.

---

[31]Prophase: *pro-* = before + *phase* = stage or period of time.

[32]Metaphase: *meta-* = after + *phase*.

[33]Anaphase: *ana-* = up, toward, or apart + *phase*.

[34]Telophase: *telo-* = end, final + *phase*.

[35]Meiosis: Greek *meion* = less + *-osis* = condition of.

HEALTH CLINIC

Cancer: The Misguided Cell

Despite its importance to the health of the cell, the DNA within the nucleus is not protected by special structural means. A host of substances are capable of penetrating the plasma and nuclear membranes, causing damage that may impair DNA function. For example, viruses and certain bacteria can penetrate the plasma membrane to invade the environment of the nucleus, often attacking the DNA itself. Environmental sources of radiation such as ultraviolet light and X-rays penetrate easily to cause damage. Certain chemicals such as benzene and formaldehyde have also been shown to damage DNA.

The nature of the damage resulting from penetration of these substances is fragmentation of DNA, which impairs the genetic code. If the DNA damage is not repairable by enzymes, the damage becomes irreversible and results in disease. When a cell's DNA is damaged beyond its ability to repair, the protein synthesis machinery either will shut down, so that cell death will soon follow, or will accelerate beyond any normal rate. If the latter is the case, uncontrollable cell growth and division will occur and the normal function of the cell will come to a halt. Unfortunately, the cells with damaged DNA usually pass this "mutated" DNA to their daughter cells. A *mutation* is a permanent chemical change in DNA. Genes that have become mutated sufficiently to upset their normal activity are called *oncogenes*. Oncogenes are genes within the genome (the gene population in normal body cells) that are believed to cause cancer. For example, many oncogenes have been shown by investigators to code for abnormal proteins. Many have also been shown to release growth factors that can stimulate unregulated cell growth and division. If cells containing oncogenes are not stopped by the body's immune response, they will form a diseased conglomerate of mutant cells called a **tumor**.

There are two major types of tumors that may arise, which differ in their rates of growth and how they spread. **Benign tumors** grow slowly and remain localized. They cause damage by physically pushing healthy tissues aside, and are usually not life-threatening. **Malignant tumors**, on the other hand, are life-threatening because they grow very rapidly and can spread throughout the body to vital organs. The process by which malignant tumors spread is called *metastasis*; it involves the transport of mutated cells through lymphatic channels and the bloodstream to distant body sites. Malignant tumors collectively are more commonly known as **cancer**, which is currently the leading cause of death in the United States. How may we increase our chances of avoiding this dreaded disease? We will learn more about tumors and cancer in the next chapter, including how to identify early signs and what substances to avoid.

Meiosis consists of two consecutive nuclear divisions, called *meiosis I* and *meiosis II*. Each division is divided into phases similar to those described for mitosis. Also similar to mitosis, meiosis I is preceded by a doubling of chromosomes (replication). But in prophase of meiosis I, an event not seen in mitosis occurs. The chromosomes migrate to their replicated twins (called their *homologous*[36] chromosomes) and align with them. This matched pairing of homologous chromosomes is called **synapsis**. During synapsis, the free ends of the chromosomes contact one another at one or more points, allowing for the exchange of genetic material from one chromosome to the other. This interaction is called *crossing over*; it provides the opportunity for random mixing of genes between a pair of homologous chromosomes. Soon afterward, during metaphase of meiosis I, the attached pairs of homologous chromosomes line up in a random manner at the center of the cell. This random orientation causes an additional scrambling of genetic information. The random mixing of genetic information during this phase, combined with the phenomenon of crossing over, are

---

[36]Homologous: *homo-* = similar, same + *logous* = pertaining to

the primary causes of the genetic variation that we see in all populations of organisms.

Once meiosis I is complete, two daughter cells that differ from each other genetically have arisen from a single parent cell. Meiosis II begins immediately, allowing no time for DNA to replicate again. Its phases mirror mitosis in every way, and result in the production of two daughter cells from each of the two cells beginning meiosis II. Each daughter cell, or gamete, contains only 23 chromosomes, and is genetically different from all other gametes.

To summarize the process of meiosis, we can conclude that it accomplishes two important tasks: (1) it provides for the genetic variation in populations by mixing genetic information during meiosis I; and (2) it produces cells with one-half the number of chromosomes to serve as gametes for reproduction.

CONCEPTS CHECK

1. What process must DNA undergo before mitosis can begin?

2. Chromatids become visible during which stage of mitosis?

3. Cells that die and must be replaced are produced by what two processes?

4. What provides for the genetic variation we see within populations?

CELLS AND AGING

Aging in our bodies is a result of destructive changes that occur in cells with time.

As time proceeds and we grow older, certain discernible changes occur in the body. They are most noticeable as graying and loss of the hair, wrinkling of the skin, an accumulation and redistribution of body fat, loss of agility and coordination, decreased muscle tone and mass, and even some slowing of mental processes. These are signs of the **aging process**, which actually begins soon after sexual maturity is reached. Why must this happen to us? Aging is the result of a reduction of function in cells that leads to their eventual death. The answer, therefore, may be found at the level of the cell.

The signs that we associate with aging are caused by certain irreversible changes within cells, a condition known as **cellular senescence**. A number of theories have been offered to account for these changes. One theory suggests aging may be "turned on" by the genetic code. Cells that continually divide in the body have been experimentally shown to "run out of gas" after a predetermined period of time; that is, they gradually lose the ability to divide. In these experiments, cells removed from the body and grown in culture divided a certain number of times and then stopped. Furthermore, cells removed from older individuals divided fewer times than cells taken from younger people. When cells were taken from different animals and cultured, the average number of divisions varied with the species. These results suggest that the life span of cells may be programmed within the genetic code of the DNA molecule. After a predetermined time or number of divisions, an "aging gene" in the cell may direct the cell to shut down its functions.

Another possible cause of reduced cell function with increasing age is the so-called "free-radical" theory. A free radical is a molecule of oxygen that contains free electrons, which are capable of damaging proteins inside the cell. As an individual increases in age, the number of free radicals in the cytoplasm increases. The result is an increasing impairment of cell function and possible damage to DNA, followed eventually by cell death. Free radicals have been shown to be produced by environmental factors such as ionizing radiation and chemicals (many of which are in the foods we eat, the water we drink, and the air we breathe). On the other hand, substances called *antioxidants* inhibit the production of free radicals; these include vitamin C, vitamin E, and the mineral selenium. A growing body of literature supports the notion that free radicals influence the aging process.

Currently, there are other theories that seek to explain the process of aging. For example, the activity of white blood cells in destroying what are thought to be healthy body cells (called the *autoimmune response*) is under investigation because it produces symptoms similar to those caused by aging. Also, it is possible that some as yet unknown factor that triggers aging in body cells is released by the brain. Regardless of which theory or combination of theories proves to be correct, we know that the aging process occurs first at the level of the cell because of its limited life span and limited ability for division. These limitations on the cell result in limitations on the functioning of tissues, organs, systems, and ultimately the whole organism.

CHAPTER SUMMARY

THE ENVIRONMENT OF THE CELL

There are two areas associated with the cell: the extracellular and intracellular environments.

Extracellular Environment The area outside the cell, composed of fluid that contains products manufactured by the cell, dissolved gases, salts, food particles, and water. The fluid is called extracellular fluid (ECF), which consists of plasma in the blood vessels and heart, and of interstitial fluid between all body cells. Interstitial fluid makes transport of materials between cells and other body regions possible. The area between cells forms the intercellular environment.

Intracellular Environment The area within a cell, confined by the cell's plasma membrane and containing intracellular fluid (ICF), the nucleus, and the cytoplasm with its organelles. The plasma membrane, nucleus, and cytoplasm are units of protoplasm that make up the three parts of the cell.

CELL STRUCTURE AND FUNCTION

Most cells are too small to be seen with the unaided eye.

Plasma Membrane The protoplasmic unit that envelops the cell, which helps to maintain the cell's homeostasis by regulating movement of material into and out of the cell.

Structure The plasma membrane is composed of fat and protein, with a small amount of carbohydrate. The fats include phospholipids and cholesterol, and compose the fluid lipid bilayer. The proteins may be peripheral (attached to the bilayer surface) or integral (associated with the internal part of the bilayer). Integral proteins often form channels that materials may use to pass through the bilayer. Some proteins form a complex with a carbohydrate, called a *glycoprotein*.

Modifications of Plasma Membranes Many stationary cells in the body have plasma membranes that exhibit a structural modification. Usually occurring on the free surface of the cell (exposed to an open space), they include microvilli, cilia, and flagella. Microvilli are tiny cytoplasmic projections that increase the surface area of the membrane for added absorption or excretion. Cilia are larger, hairlike projections that are supported by microtubules. Their whiplike movement propels body fluids. Flagella are much longer projections that are normally found as a single process, providing the cell with motility.

Functions The primary function of the plasma membrane is regulation of movement of material into and out of the cell. This may be performed passively, when no cellular energy is necessary, or actively, when there is an input of energy.

Passive Processes

Diffusion The movement of molecules from a region of high concentration to a region of low concentration.

Facilitated Diffusion The movement of molecules from a region of high concentration to a region of low concentration across a selectively permeable membrane that requires the aid of integral proteins.

Osmosis The movement of water molecules from a region of high water concentration (low solute concentration) to a region of low water concentration (high solute concentration) across a selectively permeable membrane.

Filtration The movement of molecules from a region of high pressure to a region of low pressure across a selectively permeable membrane.

Active Processes

Active Transport The transport of substances from a region of low concentration to a region of higher concentration across a selectively permeable membrane.

Cytosis The bulk transport of large volumes of substances across a plasma membrane. It includes endocytosis, which carries material into the cell: phagocytosis, or "cell eating"; and pinocytosis, or "cell drinking." Cytosis also includes exocytosis, which transports material out of the cell. Secretion is the release of manufactured products, and excretion is the release of waste material.

Cytoplasm The protoplasmic material between the plasma membrane and the nucleus. It contains a thick fluid called *cytosol* and numerous organelles bounded by a selectively permeable membrane. The organelles provide a division of labor for the cell.

Endoplasmic Reticulum A highly branched structure comprising a bilayer membrane that forms a flattened sac or tube. It provides a transportation network for molecules. If it contains attached ribosomes, it is called *rough ER* and serves as the site for protein synthesis. If no ribosomes are present, it is called *smooth ER* and may synthesize lipids and complex carbohydrates.

Golgi Apparatus Composed of a series of flattened, disc-shaped tubes bounded by a bilayer membrane. It contains enzymes to complete the synthesis of materials, and exports them to the extracellular environment by exocytosis.

Mitochondria Spherical or sausage-shaped structures that are composed of two layers of lipid bilayer membrane. The outer layer envelopes the mitochondrion, and the inner layer forms shelves upon which are attached enzymes. Mitochondria are the "energy powerhouses" of the cell, since their enzymes perform catabolism in the process of cellular respiration.

Lysosomes Spherical structures that contain hydrolytic enzymes. They are the "digestive system" of the cell, since they digest substances brought into the cell. They also break down cell structures following cellular death, and may release enzymes to break down other materials outside the cell.

Peroxisomes Spherical structures resembling lysosomes, but containing oxidative enzymes that break down toxic materials introduced into the cell.

Cytoskeleton Not an organelle, because it is not bounded by a bilayer membrane. The cytoskeleton provides the structural framework of the cell. It includes microfilaments and microtubules.

Nucleus The largest structure of the cell; the "control center" of the cell, because its main function is the regulation of protein synthesis.

Nuclear Membrane A double layer of lipid bilayer membrane that encloses the nucleus and separates it from the cytoplasm. It contains large pores.

Nucleoplasm A gel-like fluid confined within the nuclear membrane. It contains nucleoli and chromatin during the period of the cell cycle between cell divisions.

Nucleoli Small, spherical structures composed of protein and RNA. They may be the site for RNA synthesis, and disappear when cell division begins.

Chromatin Tiny granules in the nucleoplasm that form into larger structures called *chromosomes* at the onset of cell division. They are composed of DNA, which contains the genetic units known as *genes*.

Protein Synthesis The main function of the DNA in the nucleus of the cell; the process that largely determines structure and function of the body. It is divided into two processes: transcription and translation.

The Genetic Code The code that determines body structure and function is within the structure of the DNA molecule. It is contained in the order of nitrogen base pairs that occur along its length, which are arranged into groups of three. Each particular triplet base set keys to one of 20 amino acids. The section of DNA that contains enough codons to code for all amino acids composing a single protein is called a *gene*.

Transcription The process in which DNA passes its message to messenger RNA, which exits the nucleus and attaches to a ribosome in the cytoplasm.

Translation The process in which the base sequence in mRNA is translated into an amino acid sequence to form a protein. This is accomplished by matching triplet base groups in the mRNA, called *codons*, to triplet base groups in the tRNA, called *anticodons*, that are also carrying a particular amino acid. The amino acids form a protein when they combine chemically as they are transported by the tRNAs. The entire process of protein synthesis may be summarized in the formula:

$$DNA \rightarrow RNA \rightarrow protein$$

CELL DIVISION

Growth of the body, replacement of cells, and production of sex cells require the process of cell division. This includes two events: the division of the nucleus, or nuclear division; and the division of the cytoplasm, or cytoplasmic division.

There are two ways in which nuclear division occurs: by mitosis and by meiosis.

Mitosis The division of the nucleus into two genetically identical daughter nuclei. When it is accompanied by cytoplasmic division, or cytokinesis, two daughter cells result from an original parent cell, each with the same genetic information and about one-half the volume of the parent. Prior to mitosis during interphase, DNA is duplicated, or replicated, and cell growth has doubled the cell volume. Mitosis may be divided into the following phases:

Late Interphase During this period the chromosomes and centrioles become visible, and DNA replication is complete.

Prophase Each chromosome thickens to form a pair of chromatids that are attached at the centromere. The centrioles migrate to opposite poles of the cell as the nuclear membrane disappears, and the spindle is formed.

Metaphase The spindle fibers expand, pushing the chromosomes to the center of the cell.

Anaphase Chromatids are separated at the centromeres and pulled toward opposite poles of the cell. Cytokinesis begins with the development of a cleavage furrow.

Telophase Sister chromatids reach opposite poles of the cell, and a nuclear membrane forms around each group. The spindle disappears, and the cleavage furrow widens to eventually split the cell in half. Two daughter cells, each with a complete set of genetic information, are thus formed.

Mitosis is the method by which dead cells are replaced and body growth occurs.

Meiosis The division of a parent cell into genetically different and haploid (containing one-half the chromosome number) daughter cells. It is the method by which sex cells, or gametes, are produced for the purpose of sexual reproduction. Meiosis consists of two distinct division sequences, called meiosis I and meiosis II. Meiosis I provides for genetic variation by the crossing over of genetic information during synapsis, and by the random mixing that occurs when homologous chromosomes line up. Meiosis II proceeds in a manner similar to mitosis.

CELLS AND AGING

The aging process is made apparent by changes that occur in the body with increasing time. These changes are the result of reduced function in cells that leads to their eventual death. This condition of the cell is known as *cellular senescence* and is not yet well understood. One theory suggests that aging is caused by alterations in the genetic code, which may cause changes in the cell as age progresses. Another theory suggests that free radicals that damage cells accumulate with time to cause aging. Other theories have also been presented, but much further research is needed before the riddle behind aging is solved.

KEY TERMS

concentration gradient (p.54)
crenation (p.57)
diffusion (p.54)
equilibrium (p.54)
excretion (p.59)
filtration (p.57)

hydrophilic (p.52)
hydrophobic (p.52)
hypertonic (p.57)
hypotonic (p.57)
hypertrophy (p.57)
interstitial fluid (p.48)

isotonic (p.56)
plasma (p.48)
pressure (p.57)
product (p.48)

protoplasm (p.48)
secretion (p.48)
senescence (p.74)

QUESTIONS FOR REVIEW

OBJECTIVE QUESTIONS

1. Which of the following environments contain products that have been secreted by the cell?
   a. intracellular environment
   b. cytoplasm
   c. extracellular environment
   d. nucleoplasm
2. The endoplasmic reticulum (ER) is a component of the:
   a. intercellular environment
   b. matrix
   c. extracellular environment
   d. cytoplasm
3. The fluid located in the extracellular environment that transports materials between cells is called:
   a. plasma
   b. interstitial fluid
   c. cytoplasm
   d. blood
4. Phospholipids in the plasma membrane form a bilayer that is very:
   a. thick
   b. solid and hard
   c. fluid
   d. impermeable
5. The primary function of proteins in the lipid bilayer is to provide structural support for the plasma membrane.
   a. True
   b. False
6. Certain materials may pass through the plasma membrane through pores or ion channels provided by:
   a. glycoproteins
   b. cholesterol
   c. phospholipids
   d. integral proteins
7. A modification of plasma membrane structure that increases its surface area for increased absorption or excretion is:
   a. microvilli
   b. flagella
   c. cilia
   d. secretory vesicles
8. The passage of material across the plasma membrane along a concentration gradient, with the aid of integral proteins, is a process called:
   a. diffusion
   b. active transport
   c. facilitated diffusion
   d. endocytosis
9. When a cell is placed in an environment that is hypertonic to the cell, water will move in a net direction:
   a. out of the cell
   b. both ways equally
   c. into the cell
   d. none of the above
10. The process by which a white blood cell engulfs a large substance, such as a dead cell or a bacterium, is called:
    a. exocytosis
    b. phagocytosis
    c. filtration
    d. active transport
11. Which of the following types of cells would you suspect contain the largest number of mitochondria?
    a. skin cells
    b. muscle cells
    c. fat cells
    d. bone cells
12. Which of the following organelles is (or are) contained within a limiting lipid bilayer membrane?
    a. endoplasmic reticulum
    b. mitochondria
    c. lysosomes
    d. all of the above
13. The organelle that is the site of protein synthesis is:
    a. rough ER
    b. mitochondrion
    c. peroxisome
    d. nucleus
14. In the genetic code, a nitrogen base sequence in DNA codes for a particular sequence of:
    a. amino acids
    b. genes
    c. sugar molecules
    d. phosphate groups
15. Which of the following molecules transfer the genetic code from DNA to the cytoplasm during protein synthesis?
    a. mRNA
    b. rRNA
    c. tRNA
    d. a polypeptide
16. The nucleus is often called the "control center" of the cell because it:
    a. controls cytosis
    b. controls protein synthesis
    c. produces ATP
    d. controls catabolism
17. Nuclear division that occurs for the replacement of damaged and diseased body cells and for body growth is called:
    a. mitosis
    b. meiosis
    c. cytokinesis
    d. telophase

18. The division of a cell nucleus into two genetically different and unequal daughter nuclei describes the process of:
    a. mitosis
    b. meiosis
    c. cytokinesis
    d. anaphase

19. The aging process in the body can be traced directly to:
    a. loss of function in cells
    b. poor diet habits
    c. lack of skin care
    d. poor blood circulation

ESSAY QUESTIONS

1. Describe the mechanisms that permit the movement of each of the following substances into the cell: (a) water; (b) glucose; (c) oxygen; (d) protein.
2. Identify the primary functions of each cellular organelle.
3. Describe the passage of a protein molecule from its origin on a portion of the rough endoplasmic reticulum to its secretion out of the cell.
4. Explain why the nucleus is regarded as the "control center" of the cell.
5. Diagram the processes of translation and transcription.
6. Describe the changes that occur in a cell during the phases of mitosis.

ANSWERS TO ART LEGEND QUESTIONS

Figure 3–1
What fluid is located in the extracellular environment? Interstitial fluid is located in the extracellular environment.

Figure 3–2
Why aren't all cells the same shape and size? Different shapes and sizes are necessary in order to meet the demands of body functions.

Figure 3–3
Where is the cytosol in the cell model shown? The cytosol is located in the cytoplasm between the limiting membranes of organelles.

Figure 3–4
What function is served by integral proteins that provide membrane channels? Integral proteins permit the movement of substances through the plasma membrane.

Figure 3–5
What benefit is provided to the cell by microvilli? Microvilli increase the absorptive surface area of the plasma membrane, thus increasing the amount of absorption that can occur.

Figure 3–7
Can diffusion occur against a concentration gradient? No, diffusion can occur only along a concentration gradient.

Figure 3–8
Does water move into or out of a cell placed in a hypertonic solution? Water moves out of a cell placed in a hypertonic solution.

Figure 3–9
Can active transport proceed when there is no energy available? No, active transport requires energy in the form of ATP.

Figure 3–10
How is pinocytosis similar to phagocytosis? Both pinocytosis and phagocytosis involve the mass transport of materials into the cell, which requires energy.

Figure 3–11
Does exocytosis require the presence of energy? Yes, exocytosis requires energy in the form of ATP.

Figure 3–12
What is the function of the rough ER? Rough ER contains ribosomes, which provide an attachment site for mRNA during protein synthesis.

Figure 3–13
What role does the Golgi apparatus play in exocytosis? The Golgi apparatus packages substances in secretory vesicles, which transport the substances to the plasma membrane. The vesicles fuse with the plasma membrane during exocytosis.

Figure 3–14
What essential cellular activity is performed within mitochondria? The generation of available energy through the process of cellular respiration is performed by enzymes within the mitochondria.

Figure 3–15
What is the major role of the cytoskeleton? The cytoskeleton provides structural support for organelles and assists in cellular movement.

Figure 3–16
What purpose is served by the large pores in the nuclear membrane? The large pores permit the movement of mRNA by the process of transcription.

Figure 3–17
How is a genetic code for a particular protein passed from DNA to mRNA? The genetic code is copied onto mRNA by the process of transcription.

Figure 3–18
What is the role of tRNA during translation? Transfer RNA transports amino acids from the cytoplasm to the site of protein synthesis (attachment site between mRNA and a ribosome).

Figure 3–19
At which stage of mitosis do chromosomes migrate to the center of the cell? During metaphase the chromosomes migrate to the center of the cell.

TISSUES

CHAPTER OUTLINE

LEARNING OBJECTIVES

After studying this chapter you should be able to:

1. Define the term *tissue* and indicate the four primary types of tissue found in the body.

2. Describe the structural characteristics of epithelial tissue and indicate its common functions.

3. Distinguish between the various types of epithelial tissue and provide an example of where each is found in the body.

4. Describe the structural characteristics of connective tissue that distinguish it from epithelial tissue.

5. Distinguish between the types of connective tissue on the basis of their structures and functions.

6. Describe the characteristics of muscle tissue and indicate the three types.

7. Describe the organization of nerve tissue and indicate its primary function.

8. Define the term *membrane* and distinguish between the types of membranes found in the body.

9. Identify the roles membranes play in the body.

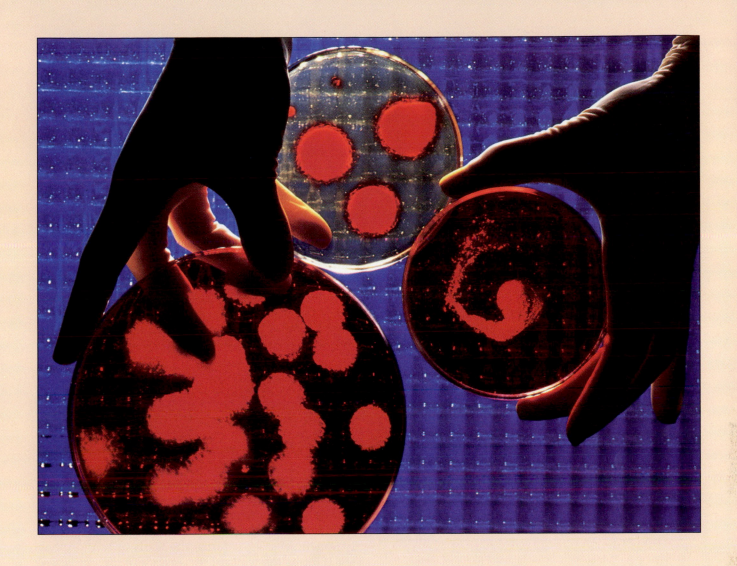

Cells are organized packets of chemicals that provide the basis for structure and function. However, they cannot meet the body's needs as individual isolated units. Functions such as protection, support of body parts, secretion or absorption of large amounts of material, and body movement require an effort by large numbers of cells within combined, organized units. A **tissue** provides this organization; it is a group of similar cells sharing a common origin that are united to perform a particular function. A tissue also includes the extracellular fluid and cellular products that exist between the cells, which are commonly referred to as *intercellular material*.

There are four primary types of tissues in the human body: **epithelial**, **connective**, **muscle**, and **nerve**. These tissues are combined in various ways to form the organs of the body, and may also be combined in a somewhat simpler manner to form **membranes**. In this chapter we will examine the four types of tissues and the natures of the membranes they may form.

81

EPITHELIAL TISSUE

Epithelial tissue is composed of cells packed close together. They perform mainly protective, absorptive, and secretory functions.

Epithelial tissue consists of cells that are closely joined together with little or no intercellular material between adjacent cells. The cells are so tightly packed that blood vessels cannot penetrate between them; epithelium is therefore an **avascular**[1] tissue. It covers body surfaces, lines the insides of body cavities and organs, and forms glands. It also comprises the functional cells within certain organs, such as the liver and kidneys. In different areas of the body it performs different functions: secretion, absorption, and protection are its major roles. Epithelial tissue may be classified into two basic categories: covering and lining epithelium, and glandular epithelium.

Covering and Lining Epithelium

Covering and lining epithelial tissue generally covers external body surfaces and lines the inner walls of cavities and organs. Its cells are stationary, anchored by their basal surfaces to a thin layer of intercellular material that was identified in the previous chapter as the *basement membrane*, which connects the epithelium

---

[1]Avascular: *a-* = without + *vascular* = pertaining to blood vessels.

---

to an underlying region of connective tissue. The opposite surface of the epithelium, called the *free surface*, is open to the lumen of a body cavity or organ, or to the external environment.

Classification of covering and lining epithelium is based on variations in cell shape and the layered arrangement of cells. Three general shapes can be recognized, although variations of these shapes exist. They are flat, or **squamous**; cube-shaped (measuring equally in height and width), or **cuboidal**; and cylindrical (greater in height than width), or **columnar**. The arrangement of cells may be in a single layer, called **simple**, or in multiple layers, called **stratified**. The most common types of covering and lining epithelium include simple squamous, simple cuboidal, simple columnar, stratified squamous, pseudostratified columnar, and transitional; these are described in the following paragraphs and summarized in Table 4–1.

Simple Squamous

Simple squamous epithelium consists of flattened cells that are arranged in a single layer: that is, all cells contact the basement membrane and have a free surface (Fig. 4–1). Its cells have a centrally located nucleus, and their close arrangement resembles that of a tiled floor when viewed from above (called a *top view*). This epithelial sheet is found in the body where the layer of cells must be thin to permit efficient diffusion of materials. It lines the inside walls of blood vessels and lymphatic vessels, it forms the walls of tiny capillaries, it forms the walls of the air sacs in the lungs, and it forms the linings of the body cavities.

Table 4–1 COVERING AND LINING EPITHELIAL TISSUE

Type of Epithelium	Structural Characteristics	Example in the Body
Simple squamous	Single layer of flattened cells	Lines inside wall of blood vessels; forms walls of capillaries and lung air sacs
Simple cuboidal	Single layer of cube-shaped cells	Forms walls of ducts in skin glands and kidney tubules
Simple columnar	Single layer of cylindrical cells	Lines inside walls of stomach, intestines
Stratified squamous	Multiple layers of cells, with cells along free edge flattened in shape	Superficial layer of the skin, mouth, and throat
Pseudostratified columnar	Single layer of irregularly shaped cells that appear multiple, often with cilia	Lines inside walls of larynx, trachea, and bronchi
Transitional	Multiple layers of spherical or irregularly shaped cells	Lines inside walls of urinary bladder and ureter

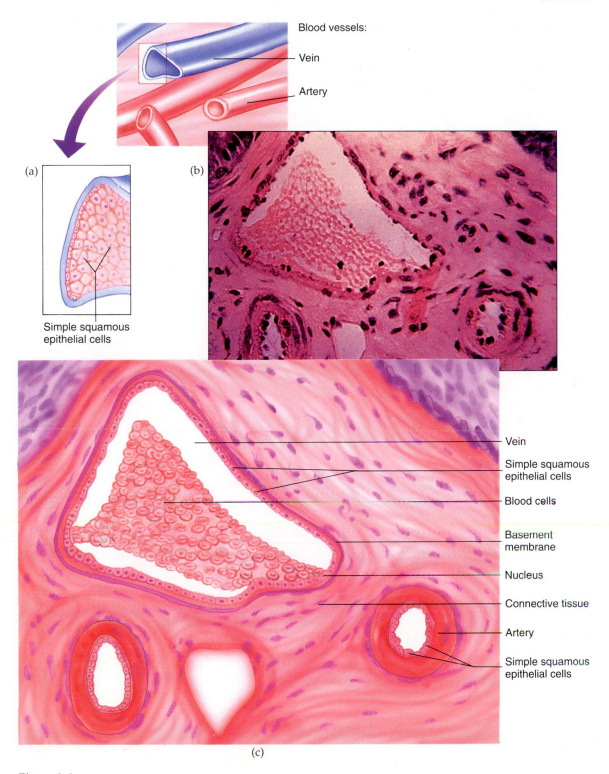

Blood vessels:

Vein

Artery

(a)

(b)

Simple squamous
epithelial cells

Vein

Simple squamous
epithelial cells

Blood cells

Basement
membrane

Nucleus

Connective tissue

Artery

Simple squamous
epithelial cells

(c)

Figure 4–1
Simple squamous epithelium. (a) A single layer of flattened cells may be found
lining the inner walls of blood vessels. Note how close together the cells are in this
top view. (b) A photomicrograph of a section through several blood vessels and
surrounding tissue. Simple squamous epithelium lines the inner wall of these
vessels. This section is illustrated in (c). Magnification 250x.

How many cell layers do you see in simple squamous epithelium?

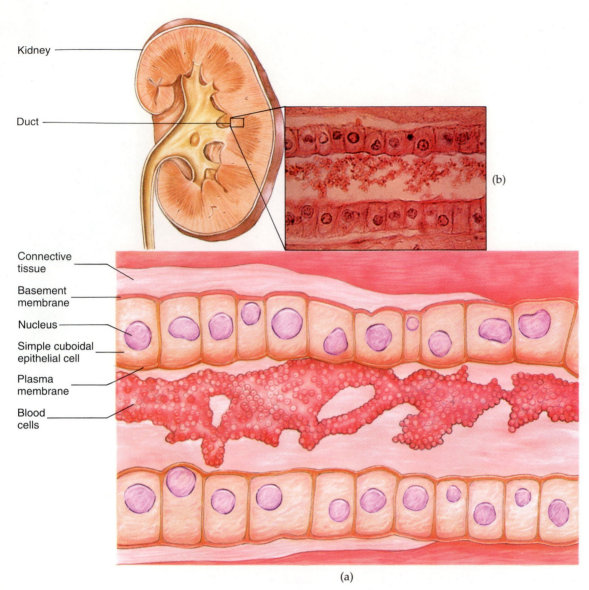

Kidney

Duct

(b)

Connective tissue

Basement membrane

Nucleus

Simple cuboidal epithelial cell

Plasma membrane

Blood cells

(a)

Figure 4–2
Simple cuboidal epithelium. (a) The tissue may be found lining tubules and ducts in the kidney, and its cells are cube-shaped and are arranged in a single layer. The blood cells in the duct space are an abnormal finding; their presence usually indicates kidney disease. (b) A photomicrograph of a sectioned duct from a kidney that is lined with simple cuboidal epithelium. Magnification 250x.

Where in the body would you expect to find simple cuboidal epithelium?

Simple Cuboidal

Composed of a single layer of cube-shaped cells, **simple cuboidal epithelium** resembles closely fitted polygons (Fig. 4–2). Its cells have a centrally located nucleus and often contain cilia and/or microvilli along their free border. This tissue commonly forms the walls of small tubes, or *ducts,* that carry secretions from one region of the body to another. It is found in the kidneys, in the liver, and in many glands.

Simple Columnar

Simple columnar epithelium consists of a single layer of elongated, cylindrical cells whose nuclei typically lie near the basement membrane (Fig. 4–3). Its cells fre-

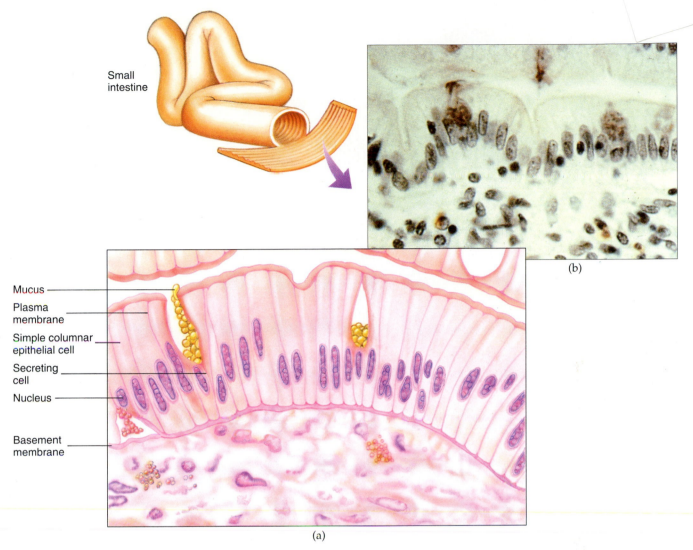

Small intestine

Mucus

Plasma membrane

Simple columnar epithelial cell

Secreting cell

Nucleus

Basement membrane

(a)

(b)

Figure 4–3
Simple columnar epithelium. (a) The cells are cylindrical and in a single layer. Some cells are specialized for secretion. (b) A photomicrograph of a small section of the small intestine, which is lined with simple columnar epithelium. Magnification 250x.

What is the shape of the individual cells in simple columnar epithelium?

quently secrete a product, so their cytoplasm often contains an abundance of rough ER and Golgi apparatus. It is found lining the inside wall of the uterus and of digestive organs such as the stomach and small intestine. In the small intestine, its free border contains microvilli, which aid in absorption of food particles.

Stratified Squamous

Stratified squamous epithelium is a multiple-layered arrangement of cells that commonly covers areas of the body that must resist constant wear and tear (Fig. 4–4). In general, stratified epithelium is categorized on the basis of the shape of the surface layer of cells only, so in this tissue the surface layer of cells is squamous. The deepest layers contain cuboidal or columnar cells that become flattened as they approach the surface of the structure they cover. This protective tissue forms the outer layer of the skin and dips in at all the openings of the body to protect them from abrasion.

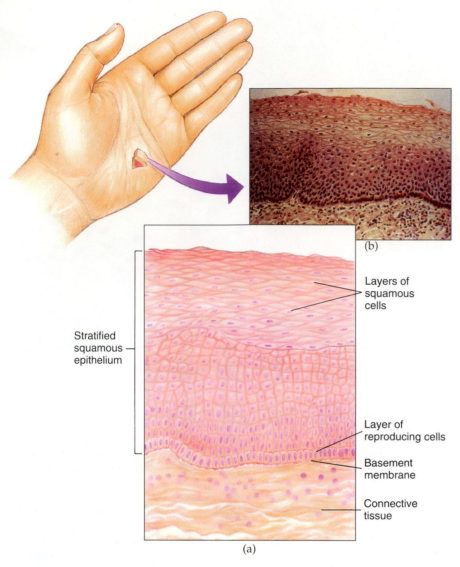

Layers of
squamous
cells

Stratified
squamous
epithelium

Layer of
reproducing cells

Basement
membrane

Connective
tissue

(a)

(b)

Figure 4–4
Stratified squamous epithelium. (a) Although the cells near the basement
membrane appear cylindrical or cube-shaped, the ones near the tissue free surface
are flattened. (b) A photomicrograph of a section of skin, which is covered with a
thick layer of stratified squamous epithelium. Magnification 50x.

Where in the body might you find stratified squamous epithelium?

Pseudostratified Columnar

Pseudostratified[2] (soo'-dō-STRAT-i-fīd) **columnar epithelium** is a curious arrangement of cells that appears multiple-layered but in reality is not (Fig. 4–5). This illusion is created by the lack of uniformity among cell shapes and locations of nuclei. We know that this tissue contains a single layer of cells because each cell makes contact with the basement membrane, although not all cells reach the free surface. Often, the cells contain long cilia that create currents for the movement of mucus. Where this occurs, the tissue is referred to as *pseudostratified ciliated columnar (PSCC)* epithelium.

[2]Pseudostratified: *pseudo-* = false + *stratified* = layered.

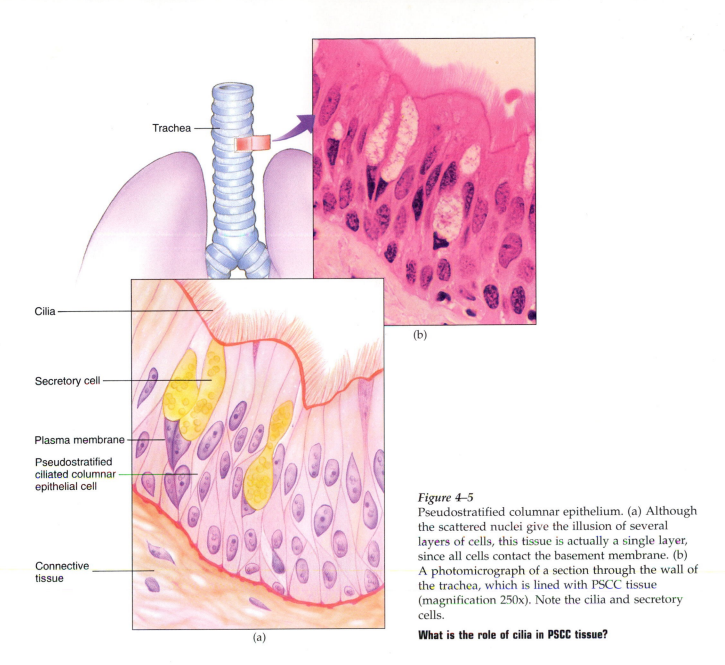

Trachea

Cilia

Secretory cell

Plasma membrane

Pseudostratified ciliated columnar epithelial cell

Connective tissue

(a)

(b)

Figure 4–5
Pseudostratified columnar epithelium. (a) Although the scattered nuclei give the illusion of several layers of cells, this tissue is actually a single layer, since all cells contact the basement membrane. (b) A photomicrograph of a section through the wall of the trachea, which is lined with PSCC tissue (magnification 250x). Note the cilia and secretory cells.

What is the role of cilia in PSCC tissue?

This tissue lines parts of the respiratory tract, such as the trachea and bronchi.

Transitional

Transitional epithelium consists of a multiple-layered arrangement of cells of cubelike or irregular shape (Fig. 4–6). This tissue is much like crepe paper: the individual cells have the capacity to expand in size and recoil, giving the tissue the remarkable ability to stretch (a feature known as *elasticity*) and to return to its original shape (a feature known as *extensibility*). Transitional epithelium lines the insides of organs that require elasticity and extensibility for normal function, such as the urinary bladder and the ureters (the tubes that travel from the kidneys to the bladder).

CONCEPTS CHECK

1. What are the four primary types of tissue in the human body?

2. How does the arrangement of cells in epithelial tissue restrict the presence of blood vessels?

3. How is covering and lining epithelium categorized?

4. Where may each of the listed types of covering and lining epithelium be found in the body?

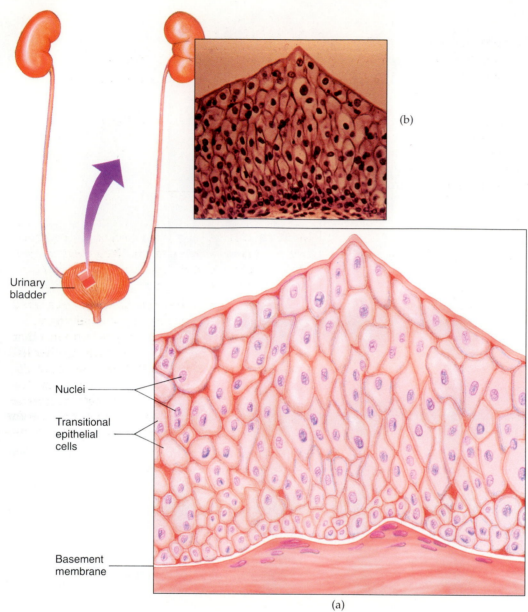

Urinary
bladder

Nuclei

Transitional
epithelial
cells

Basement
membrane

(b)

(a)

Figure 4–6
Transitional epithelium. (a) The cells form a multiple-layered arrangement. When
the tissue is stretched, the cells flatten in shape. (b) A photomicrograph of a
section through the wall of the urinary bladder. Magnification 100x.

How does transitional epithelium differ from stratified squamous epithelium?

Glandular Epithelium

Glandular epithelium consists of closely packed cells
that are highly specialized to manufacture and secrete
products. A particular type of glandular epithelium is
simply referred to as a **gland**. There are many varieties
of glands in the body, ranging from simple, single-
celled glands that secrete mucus in the digestive tract
to complex glands like the pituitary gland, which controls many important body activities. Regardless of

their variation, all glands have one thing in common:
their primary function is the secretion of one or more
products.

There are many ways to classify the various types
of glands in the body. However, all but one are beyond
the scope of this book. The classification scheme we
will learn is based on the way in which glandular products are distributed to appropriate areas of the body.
In this scheme all glands fall into one of two categories:
exocrine and endocrine.

Exocrine Glands

Exocrine glands empty their products into ducts. The ducts transport the product onto the body surface or into a cavity, where the product will provide some benefit. Sweat glands and oil glands in the skin, salivary glands, and single-celled mucous glands are examples of exocrine glands (Fig. 4–7). Most exocrine glands are supplied by blood vessels and nerves that are closely associated with them, and they are supported by connective tissue. Consequently, they are composed of more than one type of tissue and are thus regarded as organs as well as glands.

Endocrine Glands

Endocrine glands secrete their products into the extracellular space, where the products diffuse into the bloodstream. The bloodstream transports the products for their distribution throughout the body. The pituitary gland at the base of the brain, the thyroid gland in the neck, and the adrenal glands on top of each kidney are examples of endocrine glands. Because these glands also contain other types of tissue, such as connective and nervous, we may refer to them correctly as organs as well. In fact, they collectively form one of the systems of the body—the endocrine system—and will be studied in more detail in Chapter 11.

CONNECTIVE TISSUE

Connective tissue consists of a vast amount of intercellular material secreted by interspersed cells. It supports and protects body parts and manufactures blood cells.

Connective tissue is composed of widely scattered cells that lie within a large amount of nonliving intercellular material. It functions mainly to support other body structures, and is the "glue" that keeps other tissues and organs in place. The cells of connective tissue are of two types: one produces and maintains the intercellular material, and the other protects the tissue from infection. The intercellular material that is produced is composed of a mixture of sugar-protein mol-

Figure 4–7

Examples of exocrine glands. (a) The parotid salivary gland in the cheek region is the largest exocrine gland in the body. (b) A cross section through the skin shows a sweat gland and an oil gland associated with a hair root; both glands are exocrine glands. (c) A section through the lining of the trachea reveals secretory cells in the PSCC tissue. These cells are single-celled glands, called *goblet cells.*

What common feature do all exocrine glands share?

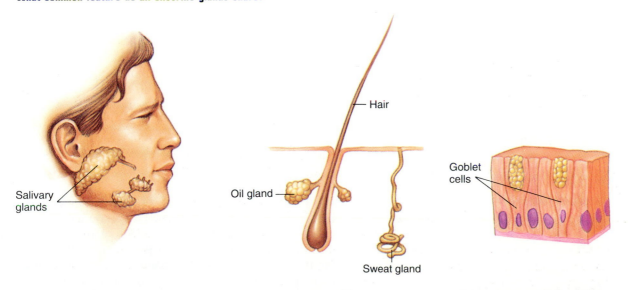

(a)   (b)   (c)

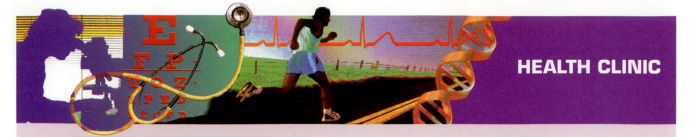

When the Body Lacks Sufficient Collagen

Collagen is used extensively by the body for tissue repair and structural support. Its importance to health is illustrated by the effects on the body resulting from *collagen disease*. This is a group of diseases that have in common their effect upon connective tissues, which is primarily to reduce the quantity and quality of collagen. One example is a disease called *scurvy*, which is generally caused by poor nutrition that results in a low collagen production rate. Its symptoms include the inability of wounds to heal, the rupture of small blood vessels, and the inability of bones and teeth to grow or maintain themselves. Scurvy can be treated by administering vitamin C, which is a necessary building block for collagen formation.

Other collagen diseases that are more serious and difficult to treat include *systemic lupus erythematosus (SLE)* and *rheumatic fever*. SLE is a disease of unknown origin that causes severe swelling, or inflammation, of blood vessels, membranes, and nerves throughout the body. If it affects vital organs, such as the heart or kidneys, it can cause death. Rheumatic fever also causes severe swelling, and usually occurs in young school-age children. It is caused by bacteria and may affect the brain, heart, and joints, where it can cause permanent damage.

---

ecules and interstitial fluid, which is known as *ground substance*, and several types of *protein fibers*.

Collagenous fibers are the most abundant type of protein fiber in the intercellular material. They are thick, wavelike strands composed of a protein called **collagen**,[3] which is flexible but has great tensile strength (that is, it resists stretching). These features are important in many areas of the body—for example, in tendons, which must connect muscle to bone, and in organs that must remain stationary in a body cavity. Collagen is, therefore, widely distributed throughout the body. In fact, it is the most abundant protein—in an average individual it makes up nearly 10% of total body weight! Collagen is also used extensively by the body for tissue repair. For example, it is the main component of scar tissue that forms over a wound during the healing process (if you have a scar, notice how little it stretches but how tightly it binds the skin together; these are characteristic features of collagen).

Elastic fibers are a second type of protein fiber in the intercellular material, and are composed of the protein **elastin**. Elastin is not as strong as collagen and forms much thinner fibers, but it has the important properties of elasticity and extensibility (remember—

these properties provide the ability to stretch and return to the original shape). You can test for the presence of elastin in your skin by pinching yourself and watching as the skin returns to its original shape. Scars in your skin lack this elasticity, because scar tissue does not contain elastic fibers.

Reticular fibers are the third type of protein fiber common to connective tissue. They are composed of the protein **reticulin**, which resists physical stress despite its thin, branching shape. In most connective tissues, reticular fibers are not abundant.

An important characteristic of connective tissue (except in cartilages) is the presence of blood vessels that pass through the intercellular material; that is, it is a **vascular** tissue. This ample blood supply provides the tissue with important capabilities for growth and repair. In fact, connective tissue is the first tissue to be formed when any organ of the body is damaged. The early formation of scar tissue—which is connective tissue rich in collagen without the elastic fibers—over superficial wounds in the skin is an example of this repair function in progress.

There are several types of connective tissues in the body. They may be distinguished on the basis of the density of proteins in their intercellular material, and on the nature of the cells that produce the intercellular material. They include connective tissue proper, cartilage, bone, and blood-forming tissue, and are summarized in Table 4–2.

---

[3]Collagen: Greek *colla* = glue + *-gen* = producer.

TABLE 4-2   DISTINGUISHING FEATURES AND FUNCTIONS OF CONNECTIVE TISSUE

Type of Connective Tissue	Structural Features	Primary Functions
Connective tissue proper	The fibroblast cell produces an intercellular material consisting of fluid ground substance, interstitial fluid, and fibers.	Anchors and supports body parts, confers immunity, stores fat, connects tissues
Loose connective tissue (areolar tissue)	Fibers are not abundant. All three types of fibers are present. In addition to fibroblasts there are phagocytic cells (macrophages).	Anchors body parts; confers immunity
Adipose tissue	Fibroblasts are specialized to store fat; intercellular material is minimal.	Stores fat
Dense connective tissue	Fibers are abundant.	Supports and connects body parts
Dense regular	Fibers are arranged parallel to one another.	Connects muscle to bone (tendons) and bone to bone (ligaments)
Dense irregular	Fibers are not in parallel array, but form an irregular network.	Supports organs and attaches them to each other and to other structures
Cartilage	The chondrocyte cell produces the matrix, composed of a thickened ground substance and fibers.	Forms a firm or elastic connection between structures
Hyaline cartilage	Matrix is dominated by ground substance and reinforced with collagenous fibers.	Forms an elastic joint surface between opposing bones; forms a template for growing bone
Elastic cartilage	Matrix is dominated by elastic fibers and supported by firm ground substance.	Forms an elastic frame for the ears, nose, and epiglottis
Fibrocartilage	Matrix is dominated by collagenous fibers with a thick ground substance.	Forms a shock-absorbing pad between opposing bones in the spinal column
Bone	The osteocyte cell maintains a hard matrix composed of mineral salts and collagenous fibers.	Provides structural "mainframe" support for the body, since it forms the skeleton
Compact bone	Matrix is deposited in layers called *lamellae*, which encircle a central canal. Surrounding the central canal and embedded in lamellae are osteocytes. This constitutes the osteon, or haversian system.	Forms the dense part of bone and provides structural integrity
Spongy bone	Matrix is deposited in thin, bony plates called *spicules*, between which are spaces filled with red bone marrow.	Forms the interior framework of bone; houses blood-forming tissue
Blood-forming tissue and blood	Specialized stem cells produce the intercellular material, composed of newly formed blood cells and protein. Blood consists of formed elements and plasma.	Manufactures blood cells; blood is a medium of transport for a host of substances

Connective Tissue Proper

Connective tissue proper is a group of connective tissues found throughout the body. In each type, the cell that produces the intercellular material is called a **fibroblast.**[4] This tissue may be categorized into three general groups, which differ in the amounts and types of fibers present in the matrix: loose connective tissue, adipose tissue, and dense connective tissue.

Loose Connective Tissue

Loose connective tissue is the most widespread of all connective tissues. Generally, its intercellular material is composed of all three types of fibers, which are in a loose, disorganized network surrounded by a fluid ground substance (Fig. 4–8a). This provides a semifluid material that has the consistency of a soft gel.

Interspersed within the abundant intercellular material of loose connective tissue are relatively small numbers of cells. The most numerous cells are fibroblasts, which produce the intercellular material. Loose connective tissue also contains phagocytic white blood cells, known as *macrophages*. The macrophages are particularly numerous during an infection, pointing to the important role they play in the immune response.

Loose connective tissue provides a structural anchor to body parts by way of its numerous fibers. It is found between the skin and muscle layers of the body, on the surfaces of most organs, and filling in spaces between organs and other body parts. Loose connective tissue is also known as **areolar tissue.**[5]

Adipose Tissue

Adipose tissue is composed mainly of specialized fibroblasts, called *adipose cells* or *adipocytes*,[6] which are large, spherical cells containing a deposit of fat. The intercellular material surrounding the adipose cells is minimal, with mainly reticular fibers that interweave about the cells (Fig. 4–8b). Adipose tissue stores energy as fat, forms an insulating padding between organs, and acts as a shock-absorbing cushion.

The fat stored within adipose cells is in the form of molecules called *triglycerides*. Their relative abundance within adipose tissue is determined by the energy requirements of the body. Factors such as exercise, nutrition, and certain hormones influence the body's energy requirements and volume of fat reserves.

Dense Connective Tissue

Dense connective tissue contains protein fibers that are packed tightly together with little space in between for other substances. It thus contains fewer cells and less ground substance than loose connective tissue, but more protein fibers. Collagenous fibers are in abundance, but elastic and reticular fibers are present as well. There are two types of dense connective tissue, which differ in the relative abundance of fibers: dense regular connective tissue, and dense irregular connective tissue.

Dense Irregular Connective Tissue

The fibers of dense *irregular* connective tissue are not in a parallel arrangement, but rather branch extensively to form a dense matting of protein (Fig. 4–8c). The fibroblasts are distributed randomly throughout the tissue, and there is again a minimal amount of ground substance. It is located in the deep layer of the skin, called the *dermis*, and forms an external wrapping around bones and cartilage.

Figure 4–8 ▶

Areolar, adipose, and dense irregular connective tissues. (a) Loose connective tissue. This type of connective tissue contains numerous fibroblasts, which secrete the collagenous, elastic, and reticulin fibers. They also produce the ground substance. A photomicrograph of areolar tissue that binds organs together is also shown; magnification 250x. (b) Adipose tissue. This connective tissue contains fibroblasts that are modified for energy storage. They are packed close together, allowing little space between cells. The cytoplasm and nucleus of each cell are squeezed against the plasma membrane to permit max-imum space for the fat droplet. A photomicrograph of adipose tissue below the skin layer is also shown; magni-fication 250x. (c) Dense irregular connective tissue. This tissue is densely packed with collagenous fibers that extend in various directions. Also shown is a photomicro-graph of the lower layer of skin (the dermis), which is composed mainly of this type of connective tissue; magnification 100x.

What cells produce the proteins in loose connective tissue?

---

[4]Fibroblast: Latin *fibro-* = fiber + Greek *-blast* = in the process of forming.

[5]Areolar: Latin *areol-* = little area + *-ar* = referring to.

[6]Adipocytes: Latin *adipo-* = presence of fat + Greek *-cyte* = cell.

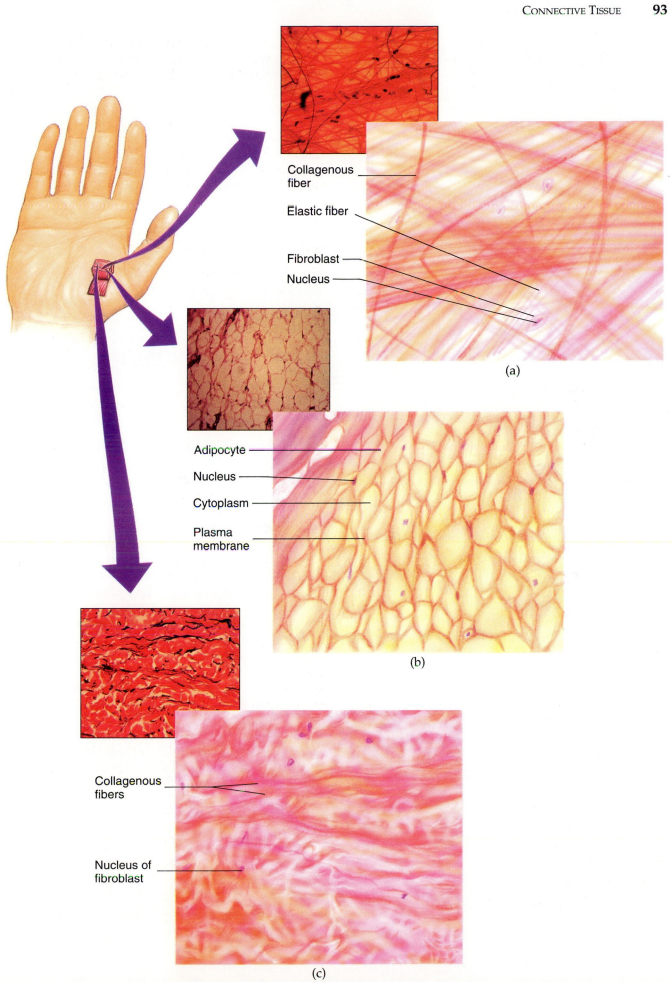

Collagenous fiber

Elastic fiber

Fibroblast
Nucleus

(a)

Adipocyte

Nucleus

Cytoplasm

Plasma membrane

(b)

Collagenous fibers

Nucleus of fibroblast

(c)

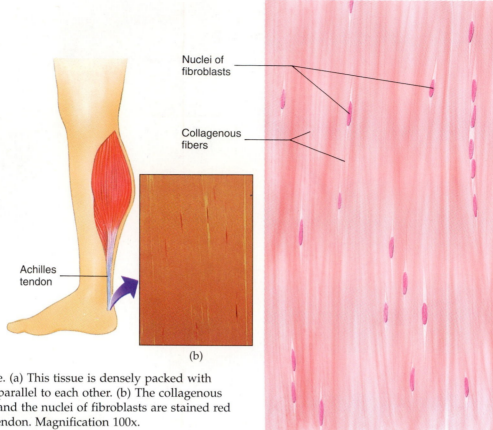

Nuclei of fibroblasts

Collagenous fibers

(b)

(a)

Figure 4–9
Dense regular connective tissue. (a) This tissue is densely packed with collagenous fibers that extend parallel to each other. (b) The collagenous fibers are stained orange-pink and the nuclei of fibroblasts are stained red in this photomicrograph of a tendon. Magnification 100x.

What protein provides strength to tendons and ligaments?

Dense Regular Connective Tissue

Dense *regular* connective tissue contains densely packed fibers that extend parallel to each other (Fig. 4–9). Its fibroblasts are distributed in a linear fashion beside groups of fibers, and there is very little ground substance. Dense regular connective tissue is extremely resistant to the forces of physical stress, and is the primary component of tendons and ligaments.

Cartilage

Cartilage forms a harder, more solid structure than that of connective tissue proper. It contains a very dense and firm intercellular material that is composed of many protein fibers within a thickened, gel-like ground substance. Its dense arrangement of fibers and thickened ground substance is known as **matrix**. The matrix is maintained by cartilage cells, called **chondrocytes**,[7] that lie embedded within small spaces or chambers called **lacunae**.[8] The chondrocytes obtain nourishment by the diffusion of materials across the matrix from a vascular layer of dense connective tissue surrounding the cartilage, called the **perichondrium**.[9] There are three types of cartilage that are present in an adult, which differ in the types of fibers that dominate in the matrix. They are hyaline, elastic, and fibrocartilage.

CONCEPTS CHECK

1. What type of cell produces the intercellular material in connective tissue proper?

2. What distinguishes loose connective tissue from dense connective tissue?

3. What are the functions of areolar tissue and adipose tissue?

4. How may dense regular and dense irregular connective tissues be distinguished?

---

[7]Chondrocyte: Greek *chondro-* = cartilage, gristle, or granule + -*cyte* = cell.

[8]Lacunae: Latin *lacuna* = small lake.

[9]Perichondrium: Greek *peri-* = around + *chrondr-* = cartilage + -*ium* = material.

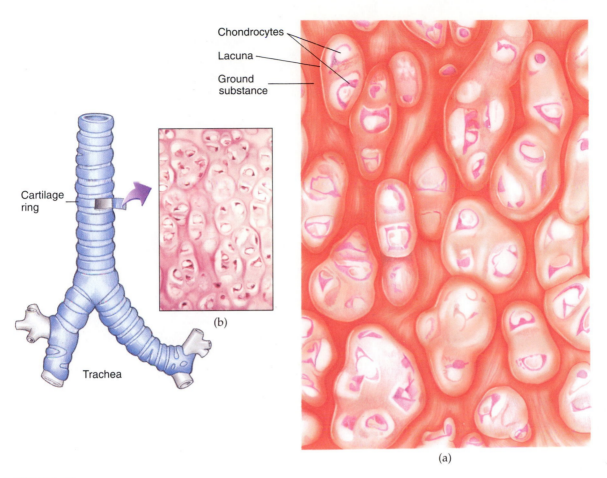

Chondrocytes
Lacuna
Ground
substance

Cartilage
ring

Trachea

(b)

(a)

Figure 4–10
Hyaline cartilage. (a) Chondrocytes embedded within lacunae are surrounded by
the abundant ground substance. Collagenous fibers are present, but they are not
visible. (b) A photomicrograph of a section through hyaline cartilage from the
trachea. Magnification 250x.

Where in the body is hyaline cartilage found?

Hyaline Cartilage

Hyaline cartilage[10] is a bluish-white, almost opaque
cartilage that is the most abundant of the three types
(Fig. 4–10). Its matrix is dominated by the presence
of protein-sugar molecules called *chondroitin sulfate*,
which are the major components of the ground substance. The fibers that form a part of the matrix are
composed of collagen, but they cannot be seen under
the microscope with usual staining methods, because
of the masking effect of chondroitin sulfate. The chondrocytes of hyaline cartilage, which are embedded
within lacunae, are sparsely distributed throughout
the extensive matrix.

Hyaline cartilage is located in the upper portion of
the respiratory tract; at the ends of bones, where it
forms movable joints; and at the ends of the ribs. It
also forms most of the skeleton of the fetus, and precedes bone formation in most bones of a growing child.

Elastic Cartilage

A yellowish cartilage characterized by the presence of
elastic fibers that dominate the matrix, **elastic cartilage**
is a firm but flexible tissue. When it is viewed under
the microscope, thin, wavy elastic fibers may be visible as they course through the matrix surrounding the
chondrocytes inside their lacunae (Fig. 4–11). The
ground substance of elastic cartilage also contains
chondroitin sulfate, but this does not dominate the matrix as it does in hyaline cartilage.

Elastic cartilage forms the supportive framework
for the ears, the end of the nose, and the small lid
over the opening to the larynx in the throat, called
the *epiglottis*.

[10]Hyaline: Greek *hyal-* = glass + *-ine* = like, resembling.

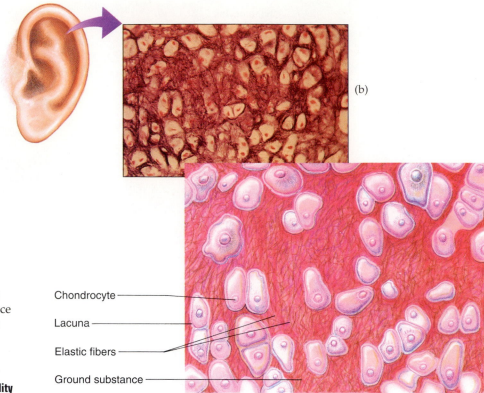

Figure 4–11
Elastic cartilage. (a) Elastic fibers are visible in the ground substance of this cartilage. (b) Elastic fibers are the black, wavy lines in the matrix of this photomicrograph from the ear. Magnification 250x.

What protein provides the stretchability that characterizes elastic cartilage?

Chondrocyte

Lacuna

Elastic fibers

Ground substance

(a)

Fibrocartilage

Fibrocartilage consists of a solid but flexible matrix characterized by the predominance of thick collagenous fibers. When viewed under a microscope, the collagenous fibers appear as dark, wavy lines that weave about the chondrocytes inside their lacunae (Fig. 4–12).

Fibrocartilage is often found in close association with hyaline cartilage in joints, such as the knee, where it acts as a shock absorber. Another example is the intervertebral discs, which are padded joints, composed mostly of fibrocartilage, that lie between opposing vertebrae in the spinal column.

CONCEPTS CHECK

1. How does cartilage differ from loose connective tissue?

2. How do chondrocytes obtain nourishment?

3. What is the prominent component of the matrix in hyaline cartilage, elastic cartilage, and fibrocartilage?

Bone

Bone is a connective tissue whose intercellular material is filled with mineral salts and collagenous fibers, providing it with a hard and durable structure that exceeds that of all other tissues. Because of its dense nature, the intercellular material of bone is called **matrix**. The matrix is maintained by bone cells called **osteocytes**.[11] Like the chondrocytes of cartilage, the osteocytes lie embedded within chambers called **lacunae**. Osteocytes obtain nourishment by diffusion from blood vessels that penetrate through the hard matrix. The blood vessels originate in a membrane that surrounds the bone, known as the **periosteum**.[12] There are two types of bone tissue, compact bone and spongy bone, which we will briefly define in the following paragraphs and describe in more detail in Chapter 7.

Compact Bone

The matrix of **compact bone** is densely packed with deposits of mineral salts and collagen (Fig. 4–13). Its

[11]Osteocyte: *osteo-* = bone + *-cyte* = cell.

[12]Periosteum: *peri-* = around + *osteum* = pertaining to bone.

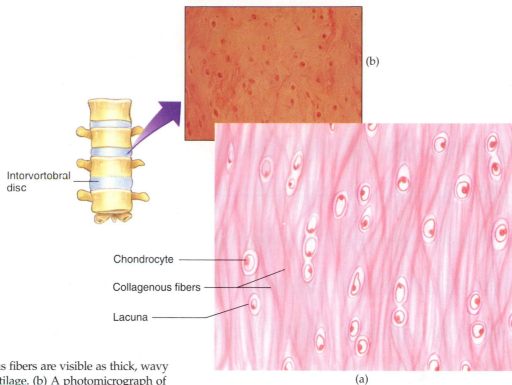

Chondrocyte

Collagenous fibers

Lacuna

(b)

(a)

Figure 4–12
Fibrocartilage. (a) Collagenous fibers are visible as thick, wavy lines in the matrix of this cartilage. (b) A photomicrograph of fibrocartilage taken from an intervertebral disc. Magnification 250x.

What cells produce the dense ground substance of cartilage?

Intorvortobral disc

matrix is so dense, in fact, that nourishment must be provided to osteocytes by way of blood vessels that extend through the bone in canals. These canals, called **osteonic canals** (or **haversian canals**), are surrounded by the hard matrix, which is laid down by bone cells into a series of thin, orderly layers called **lamellae**. Lamellae usually form concentric patterns that encircle osteonic canals. Nutrients reach the osteocytes by diffusing across the barrier formed by lamellae by way of tiny channels, known as **canaliculi**. In the outer layers of compact bone, however, there are no osteonic canals, and lamellae form a solid matrix.

In a bone, each individual osteonic canal with its concentric lamellae and osteocytes inside lacunae constitutes a cylindrical unit known as an **osteon**, or **haversian system**. Osteons are cemented together by other types of lamellae to form the substance of compact bone, which covers the surfaces of all bones in the body.

Spongy Bone

The matrix of **spongy bone** is not as densely packed as that of compact bone. Rather, it contains spaces that are filled with blood-forming tissue, called **red marrow**. The hard region of matrix is composed of mineral salts and collagen that form small, thin plates called **spicules** (or **trabeculae**). The spicules fuse together during their formation to form an irregular network, leaving spaces in between that fill with marrow. The osteocytes inside lacunae are located within the spicules. In general, spongy bone tissue forms the interiors of bones.

Blood-Forming Tissue and Blood

Blood-forming tissue manufactures the cellular components of blood. This tissue contains basically three types of components: stem cells that produce blood cells; newly formed blood cells that are in the process of maturing; and a small amount of protein. Because it lacks collagen and mineral salts, blood-forming tissue is the softest type of connective tissue in the body.

There are two types of blood-forming tissue. One type, which occurs inside cavities within spongy bone, was already identified as **red marrow**. This tissue, often referred to as **hematopoietic tissue**,[13] initiates the production of all blood cells (red blood cells, white

---

[13]Hematopoietic: Greek *hemato-* = blood + *-poietic* = forming.

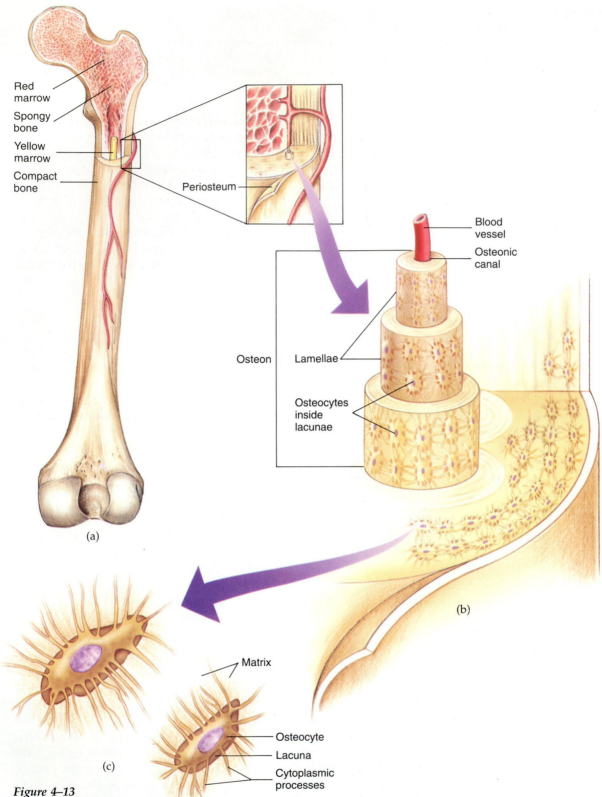

Red
marrow

Spongy
bone

Yellow
marrow

Compact
bone

Periosteum

(a)

Osteon

Blood
vessel

Osteonic
canal

Lamellae

Osteocytes
inside
lacunae

(b)

Matrix

Osteocyte

Lacuna

Cytoplasmic
processes

(c)

Figure 4–13

Bone. (a) Bone tissue may be compact or spongy. Both types are shown in this partial
section of a human bone. Red marrow, a type of blood-forming tissue, is located within
spongy bone spaces. Compact bone forms the dense edges of a bone, while spongy bone
lines the internal cavity and fills its ends. (b) Compact bone is a composite of subunits
called *osteons*. (c) Within each osteon are numerous osteocytes. Each osteocyte is embedded
within a lacuna and is surrounded by hard matrix.

What materials provide bone with its matrix?

blood cells, and platelets). A second type of blood-forming tissue is called **lymphoid tissue**; it is found throughout the body in lymph nodes, in the tonsils of the throat, in the spleen, and in the thymus of children. It is the maturation site of two types of white blood cells, called *lymphocytes* and *monocytes*.

Blood is a type of connective tissue also (Fig. 4–14). Its living cells (red blood cells, white blood cells, and platelets) are called **formed elements** and are surrounded by a fluid matrix known as **plasma**. The fibers of the matrix are dissolved protein molecules that become visible during clotting. Blood acts as the transport vehicle for the cardiovascular system, carrying respiratory gases, nutrients, wastes, and many other substances throughout the body.

CONCEPTS CHECK

1. What two products of osteocytes are the primary components of bone matrix?

2. How do compact bone and spongy bone differ in structure?

3. Where is blood-forming tissue located?

MUSCLE TISSUE

Muscle tissue consists of closely arranged cells with little intercellular material. Proteins within the cells enable contraction to occur, which produces movement.

Muscle tissue consists of specialized cells that contain molecular filaments of protein within their cytoplasm. These cells are specialized to cause the cell to shorten, or contract. The proteins are mainly myosin and actin, and are arranged in parallel bundles: during contraction, the actin filaments slide over the myosin filaments. The result of muscle cell contraction is the production of movement in the body. There are three types of muscle tissue, each of which is found in a different area of the body. They are skeletal muscle, smooth muscle, and cardiac muscle.

Skeletal Muscle

Skeletal muscle is the type of muscle tissue that is attached to bones. It is the primary tissue of the muscular system. The attachment of skeletal muscle to bones is by way of bands of dense regular connective tissue known as **tendons**. Skeletal muscle tissue is located

Figure 4–14
Blood. (a) Although it is very watery, blood is a type of connective tissue. Its matrix is fluid and is called *plasma*. (b) A photomicrograph of blood. Magnification 800x.

Where do blood cells originate?

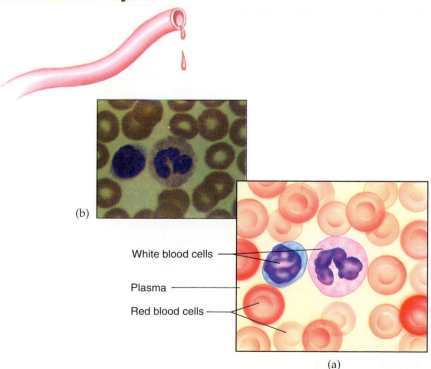

(b)

White blood cells

Plasma

Red blood cells

(a)

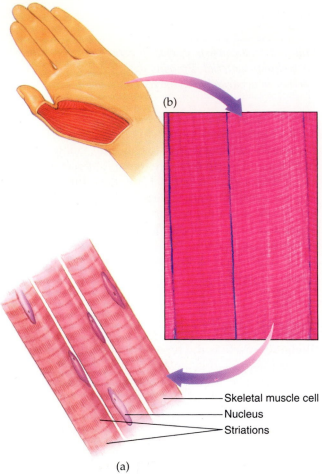

Skeletal muscle cell
Nucleus
Striations

(a)

Figure 4–15
Skeletal muscle. (a) Each skeletal muscle cell is called a *fiber*, and it contains many nuclei. The cells in the illustration are not complete, since skeletal muscle cells are as long as the muscle itself. The striations are dark lines formed by the internal arrangement of protein fibers. (b) A photomicrograph of skeletal muscle. Magnification 250x.

Where in the body are skeletal muscles located?

deep under the skin layer, and its contraction is under conscious (voluntary) control. Its function is the production of body movement. When viewed under the microscope, its cells are observed to exhibit alternating light and dark bands that extend at right angles to the length of the cell. These bands are called *striations* (Fig. 4–15). The striations reveal the way in which the protein filaments are arranged.

Smooth Muscle

Smooth (visceral) muscle forms part of the walls of blood vessels and visceral organs (such as the stomach, intestines, urinary bladder, and uterus). Its con-

traction helps propel substances that move through these tubes and chambers. Unlike skeletal muscle, smooth muscle tissue is not under our conscious control (it is involuntary) and does not contain striations in the cytoplasm of its cells (Fig. 4–16a). Striations are not present because the protein filaments are more diffuse and are not organized into tightly packed bands or groups.

Cardiac Muscle

Cardiac muscle is found in the walls of the heart, where it is called the **myocardium**.[14] Its coordinated contractions push blood out of the heart and through the body. Cardiac muscle tissue is not under conscious control (it is involuntary), but its cells contain striations that are visible with a microscope (Fig. 4–16b). Between adjacent cardiac cells the plasma membrane is thickened, forming a special junction called an **intercalated disk**.

NERVE TISSUE

Muscle tissue consists of neurons, which conduct rapid signals, and neuroglia, which provide support.

Nerve tissue is characterized by its well-developed properties of *conductivity* and *excitability*, which provide it with the ability to send and to carry electrochemical signals throughout the body. It consists of two types of cells: **neurons** and the **neuroglia** (Fig. 4–17). Neurons are highly specialized cells that conduct electrochemical signals very rapidly. Neuroglia are more numerous than neurons in organs such as the brain and spinal cord, and generally serve to maintain and support the neurons.

CONCEPTS CHECK

1. What special property characterizes muscle tissue, and what is its function?

2. Where may skeletal muscle, smooth muscle, and cardiac muscle be found in the body?

3. What specialized properties are characteristic of nerve tissue?

[14]Myocardium: *myo-* = muscle + *-cardium* = pertaining to the heart.

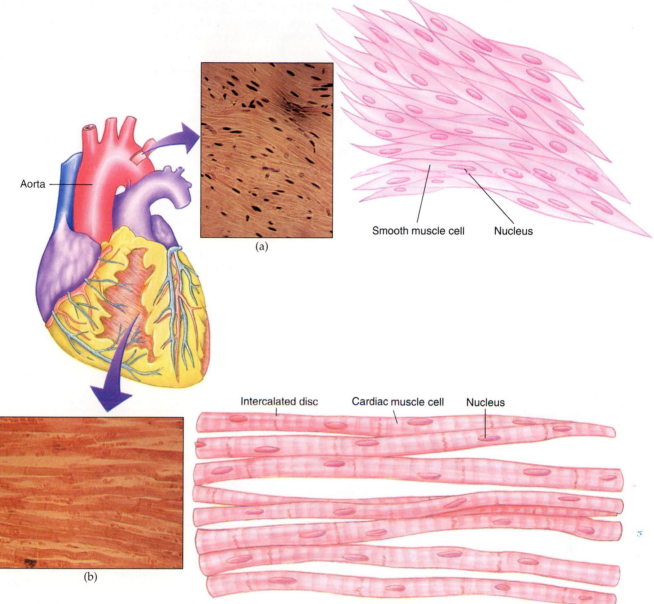

Figure 4–16

Smooth muscle and cardiac muscle. (a) Smooth muscle tissue contains cells that are spindle-shaped. The photomicrograph is of a section of smooth muscle in the wall of the aorta; magnification 250x. (b) Cardiac muscle tissue contains cells that have squared edges. Where adjacent cells meet there is a thickening of the plasma membrane, called the *intercalated disc.* A photomicrograph of cardiac tissue reveals its features; magnification 250x.

What features of cardiac muscle can you find that are not present in smooth muscle tissue?

MEMBRANES

Membranes are a simple combination of tissues that include connective tissue and, in most cases, epithelial tissue to form a thin sheeting.

A **membrane** is the simplest combination of tissues in the body to form a functional unit. It consists of connective tissue that is usually associated with epithelial tissue, and contains other structures passing through it such as blood vessels, lymphatic vessels, and nerves. Membranes structurally divide areas of the body or organs, line the internal surfaces of hollow organs and body cavities, and anchor organs to other structures. In general, they provide support and protection for body structures that transport substances throughout the body.

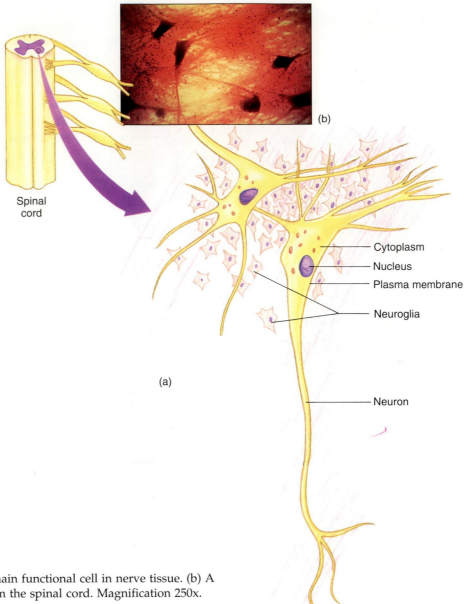

Figure 4–17
Nerve tissue. (a) The neuron is the main functional cell in nerve tissue. (b) A photomicrograph of nerve tissue from the spinal cord. Magnification 250x.

What special feature distinguishes nerve tissue from all other tissues?

Most membranes in the body contain a layer of epithelial cells that are associated with connective tissue. These membranes are termed **epithelial membranes** and include three types found in different areas of the body: mucous membranes, which line passageways leading to the outside of the body; serous membranes, which line body cavities and cover certain organs; and the cutaneous membrane, which constitutes the skin. A fourth membrane, the synovial membrane, does not contain a layer of epithelium; it lines the cavities of the joints. These membranes are diagrammed in Figure 4–18.

Cutaneous Membrane

The **cutaneous membrane** is also known as the *skin*. It constitutes the primary organ of the integumentary system, which will be discussed in Chapter 6. It is shown in Figure 4–18a.

Serous Membranes

Serous membranes line the internal surfaces of the thoracic and abdominopelvic cavities. They also cover most organs that lie within these cavities, and may bind them to each other and to the body walls. Exam-

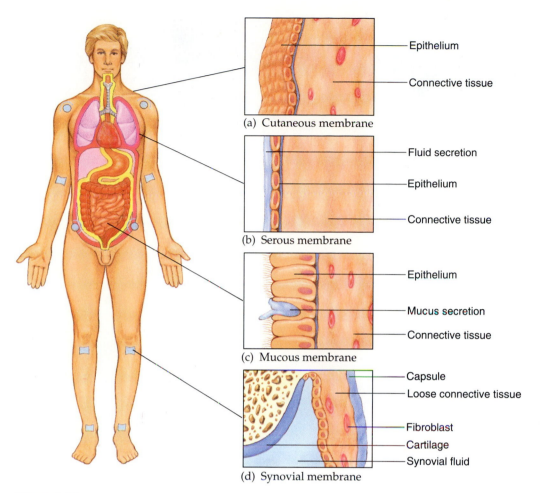

Figure 4–18
Membranes. (a) The cutaneous membrane is the skin. (b) Serous membranes contain epithelium that secretes a watery fluid. They cover the heart and lungs, and line the abdominopelvic cavity. (c) Mucous membranes contain epithelium that secrete mucus. They line the openings to the body, the digestive tract, and the respiratory tract. (d) Synovial membranes line joint cavities and do not contain an epithelial layer.

With the exception of synovial membranes, of what two tissues are membranes composed?

ples of serous membranes include the **pericardium**, which surrounds the heart; the two **pleurae**, which surround each lung; and the **peritoneum**, which lines the abdominal cavity and covers abdominal organs. The epithelium that lines the outer surfaces of serous membranes secretes a clear, watery fluid that provides lubrication (Fig. 4–18b).

Mucous Membranes

Mucous membranes line the internal walls of the digestive tract from the mouth to the anus, and the respiratory tract from the nasal cavity to the air sacs of

the lungs. Mucous membranes also line the ureters, urinary bladder, and urethra of the urinary tract, and all organs of the reproductive tract. The epithelium that covers their outer surface secretes **mucus**, which is a sticky, thick liquid composed of water and carbohydrates (Fig. 4–18c). The purposes of mucus production are to trap foreign particles for their removal out of the body, to maintain a moist internal environment, and to protect cells that would otherwise be exposed to the potentially harmful liquids that pass through (for example, stomach acids and urine salts).

(*Text continues on page 107*)

HEALTH CLINIC

Tissues, Tumors, and Cancer

A **tumor**, or **neoplasm**, is an overgrowth of cells to form a tissue that has no useful purpose to the body. It becomes a threat to health when it physically replaces healthy tissue and competes with surrounding tissues (see the photomicrographs). A tumor originates in a single cell that, as a result of damage to the DNA, has lost its ability to regulate growth and division. An alteration of chromosomes that affects the DNA is called a **mutation**. Most mutations are caused by exposure to environmental agents, called **carcinogens**. A carcinogen may be a source of ionizing radiation (either ultraviolet light from the sun or X-rays) or a chemical. One of the more well-known groups of carcinogens is found in cigarette tobacco, which is the

documented cause of 90% of all *lung cancers* and may influence the development of other cancers as well.

A mutation may also be caused by a virus. Experiments performed on animals and on human cells grown in culture have shown that certain viruses are able to enter a cell and change the genetic code by incorporating their own genetic material into that of the cell. The result is a mutant cell that manufactures proteins that are different from those of normal cells. If this mutant cell evades detection by white blood cells of the immune system and is stimulated by the viral genes to divide more rapidly than normal, a tumor will develop. An example of a virus-induced tumor occurs in the disease *adult T-cell leukemia-*

A tumor arises from normal tissue. As the tumor grows, it competes with healthy tissue by overcrowding it and reducing its ability to perform normal functions. (a) A healthy bronchial passageway to the lungs is lined with ciliated cells. The cilia sweep dust and other foreign particles away from the lungs. Magnification 4000x. (b) Tumor cells (*green*) invade the bronchial wall and crowd out normal cells lining the bronchus. Magnification 3000x.

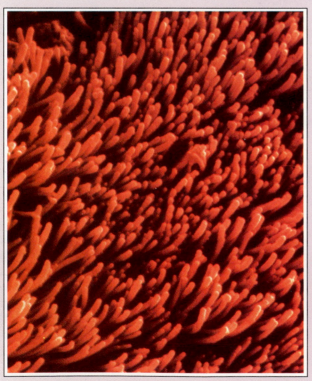

(a)

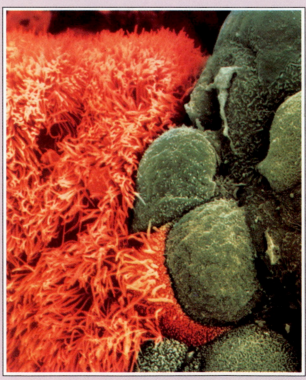

(b)

lymphoma (ATLL), which has been shown to be caused by a virus called *human T-cell leukemia-lymphoma virus type 1 (HTLV-1)*.

There are two major types of tumors, which differ in their rates of growth and the natures of their origins. **Benign tumors** grow slowly and remain localized. **Malig-**

nant tumors grow rapidly and infiltrate into surrounding tissues, often spreading into lymphatic and blood vessels that can transport them to distant sites in the body.

A benign tumor is a slowly growing tissue that causes damage to healthy tissues by pushing them aside. It often arises from an area of the body that has received re-
(continued)

Incidences of diagnosed cancers and deaths caused by cancers in men and women in the United States. These figures represent 1993 estimates and exclude the most common of all cancers, nonmelanoma skin cancers.

Leading Sites of Cancer Incidence and Death – 1993 Estimates

Cancer Incidence by Site and Sex*		Cancer Deaths by Site and Sex*	
Male	**Female**	**Male**	**Female**
Prostate 165,000	Breast 182,000	Lung 93,000	Lung 56,000
Lung 100,000	Colon & Rectum 75,000	Prostate 35,000	Breast 46,000
Colon & Rectum 77,000	Lung 70,000	Colon & Rectum 28,800	Colon & Rectum 28,200
Bladder 39,000	Uterus 44,500	Pancreas 12,000	Ovary 13,300
Lymphoma 28,500	Lymphoma 22,400	Lymphoma 11,500	Pancreas 13,000
Oral 20,300	Ovary 22,000	Leukemia 10,100	Lymphoma 10,500
Melanoma of the Skin 17,000	Melanoma of the Skin 15,000	Stomach 8,200	Uterus 10,100
Kidney 16,800	Pancreas 14,200	Esophagus 7,600	Leukemia 8,500
Leukemia 16,700	Bladder 13,300	Liver 6,800	Liver 5,800
Stomach 14,800	Leukemia 12,600	Brain 6,600	Brain 5,500
Pancreas 13,500	Kidney 10,400	Kidney 6,500	Stomach 5,400
Larynx 10,000	Oral 9,500	Bladder 6,500	Multiple Myeloma 4,600
All Sites 600,000	All Sites 570,000	All Sites 277,000	All Sites 249,000

*Excluding basal and squamous cell skin cancer and carcinoma in situ.   Source: American Cancer Society, *Cancer Facts & Figures – 1993*

Health Clinic (*continued*)

peated injury, and can originate in any tissue of the body. Fortunately, a benign tumor can usually be completely removed through surgery. Unless they are allowed to grow unchecked, benign tumors do not pose a serious threat to life.

Malignant tumors are a life-threatening disease because they grow very rapidly and can spread throughout the body to vital organs. The process by which a malignancy spreads to other areas of the body is called *metastasis*. In this process, groups of cells from the tumor infiltrate into lymphatic vessels and are carried toward lymph nodes, where they may settle to form secondary tumors. From the lymph nodes the malignant cells can be carried throughout the body to distant sites. The tumor may also infiltrate the bloodstream, where its cells may be carried elsewhere. If the malignant tumor is not destroyed quickly by the body's immune response or by medical intervention, it may become established in vital organs and interfere with their functions, eventually killing the afflicted person.

Malignant tumors collectively are more commonly known as **cancer**. Cancer is currently the leading cause of death in the United States (it replaced heart disease as the leading killer in 1990); some form of cancer will eventually develop in about 1 in every 4 persons. Cancer currently accounts for nearly 18% of all deaths in the developed nations of the world. Despite over 20 years of intensive research, the incidences of many types of cancer (such as *malignant melanoma* and *breast cancer*) are on the increase.

Every known cancer falls into one of three main categories: carcinoma, sarcoma, and leukemia. **Carcinomas** arise from epithelial tissue and are among the greatest threats to health. For example, *lung carcinoma* is the most common cancer in men, and *mammary carcinoma* (breast cancer) is the most common cancer in women; both have a high incidence of death. **Sarcoma** is the type of cancer that arises from most types of connective tissues. It is a rapidly spreading disease because of the ample supply of blood and lymphatic vessels in the affected tissues. A common type of this cancer is *osteosarcoma*, or cancer of bone. **Leukemia** arises from blood-forming tissues, such as red marrow or lymph nodes, and results in the production and distribution of blood cells that cannot carry out normal functions. In advanced stages of leukemia, the production of abnormal cells outstages the production of healthy cells, causing symptoms of anemia (lack of oxygen delivered to body cells) and a lowered resistance to infection. In recent years, advances in medicine have dramatically improved diagnosis and treatment of most leukemias. If detected early, most forms of this deadly cancer now have a high probability of a successful cure. The major types of cancer are listed in the accompanying table according to their anticipated 1993 incidences and death rates, by primary location (site) in the body and by sex.

How may we increase our chances of avoiding these dreaded diseases? The American Cancer Society has published a list of early warning signs that are possible indicators of a developing cancer. The society has concluded, after studying millions of cancer patients, that, if identified and treated at an early stage, most forms of cancers (about 89%) may be forced into a state of remission (in which their growth stops and tumor size diminishes). These early warning signs are provided in the accompanying table. Additionally, your risk of developing cancer of any type can be minimized by heeding the following advice, which is recommended by the U.S. Surgeon General and the American Cancer Society:

1. Do not use tobacco. All forms of tobacco use (including smoking and chewing) have been shown to cause cancer.
2. Avoid prolonged exposure to the sun without protection by a sunscreen or sun block.
3. Avoid unnecessary exposure to X-rays.
4. Avoid high-fat, smoked, and cured foods. Increase the fiber content of your diet.
5. Examine your body on a regular basis: your skin for black moles or sensitized areas; if male, your testes for possible lumps; if female, your breasts for possible changes. Any unusual change in your body, such as recurrent pain, blood in the stools, or persistent swelling, should be reported immediately to a qualified physician.

AMERICAN CANCER SOCIETY EARLY WARNING SIGNS

1. Change in bowel or bladder habits
2. A sore that does not heal
3. Unusual bleeding or discharge
4. Thickening or lump in the breast or elsewhere
5. Indigestion, or difficulty when swallowing
6. Obvious change in a wart or mole, or the sudden appearance of a black mole
7. Nagging cough or continued hoarseness

HEALTH CLINIC

Membranes and Immunological Diseases

Membranes play an important role in the development of diseases caused by foreign agents, including bacteria, viruses, fungi, and protozoans. They are often the first structures to be affected by an infectious agent and may serve as pathways for the spread of disease, since they are widespread, interconnected, and often vascular. The usual response to the invasion of an infectious agent is *inflammation*, in which the walls of capillaries supplying the affected region open to permit the rapid movement of white blood cells and interstitial fluid out of the bloodstream and into the infected area. This results in the accumulation of fluid in the extracellular space to produce swollen membranes, a condition commonly known as *edema*. Edema in the mucous membranes produces ex-cessive discharge of mucus and swollen passages, as we find in the common cold that affects the head region, and in *colitis*, which affects the large intestine. Edema in serous membranes often causes the inhibition of fluid secretion and enlargement of the membranes, resulting in friction and pain when opposing membranes rub against each other. This occurs in the diseases *pericarditis*, which affects the pericardium; *pleurisy*, which affects the lungs; and *peritonitis*, which affects the peritoneum. Edema in the cutaneous membrane is evidenced by swelling in the infected area of the skin and the region that surrounds it. Edema in a synovial membrane produces an enlargement of the joint cavity, which is a common ailment resulting from physical injury as well as from an infection.

Synovial Membranes

Synovial membranes line the inside walls of cavities that surround certain joints, such as the joints in the knee, elbow, and shoulder. Synovial membranes do not contain epithelial tissue and are therefore not regarded as epithelial membranes. They are composed of loose connective tissue that contains a small amount of fat. The cells within this tissue secrete a clear, watery fluid called *synovial fluid* that lubricates the opposing bones of the joint as they move. It also nourishes the cartilage at the ends of the bones with oxygen and nutrients (Fig. 4–18d).

CONCEPTS CHECK

1. What is a membrane, and where may membranes generally be found in the body?

2. What roles are played by the mucus secreted by mucous membranes?

3. What are the three primary serous membranes, and where are they located?

4. How do synovial membranes differ from mucous and serous membranes?

CHAPTER SUMMARY

EPITHELIAL TISSUE

Cells that are closely packed with little intercellular material between them. Because of the cell density, no blood vessels can penetrate the tissue and it is thus avascular (without blood vessels).

Covering and Lining Epithelium Covers body surfaces and lines the walls of cavities and organs. It is categorized on the basis of cell shape and arrangement.

Simple Squamous Flattened cells that form a single layer. It commonly lines blood vessels and lymphatic vessels, forms the walls of capillaries, and forms the walls of air sacs in the lungs.

Simple Cuboidal Cube-shaped cells in a single layer. It forms the walls of ducts and kidney tubules.

Simple Columnar Cylindrical cells that form a single layer. It lines the walls of digestive-tract organs and of the uterus.

Stratified Squamous A multiple-layered arrangement of cells with the surface layer composed of flattened cells. It is usually protective in function, and it forms the outer layer of the skin.

Pseudostratified Columnar A single layer of irregularly shaped cells that visually appears to be in a multiple-layered arrangement. It lines the respiratory tract, where it contains cilia.

Transitional   A multiple-layered arrangement of cubelike or irregularly shaped cells that can expand and contract. It lines the urinary bladder and ureters.

Glandular Epithelium   Closely packed cells that are specialized to manufacture and secrete products. It forms glands, and may be categorized on the basis of how their products are distributed.

Exocrine Glands   Glands that empty their products into ducts, which channel the secretions to the body surface or into a cavity. They include salivary glands, sweat glands, and oil glands.

Endocrine Glands   Glands that secrete their products into the extracellular space, where they diffuse into the bloodstream. They include the pituitary gland, the thyroid gland, and the adrenal glands. The endocrine glands are also organs, since they consist of more than one type of tissue; and together they compose the endocrine system.

CONNECTIVE TISSUE

Composed of widely scattered cells that lie within a large amount of intercellular material. The intercellular material is mainly produced by the cells and usually consists of two things: a ground substance of sugar-protein molecules and interstitial fluid, and protein fibers. The fibers are of three types: collagenous fibers, elastic fibers, and reticular fibers. Connective tissue is highly vascular.

Connective Tissue Proper   In this type of connective tissue, the intercellular material is produced by cells known as *fibroblasts* and is dominated by protein fibers.

Loose Connective Tissue   Also called areolar tissue. The intercellular material consists of fibers not in abundance but arranged in a loose network, producing a soft, gel-like material. It contains all three fiber types, as well as macrophages that fight infection. It provides a structural anchor for body parts.

Adipose Tissue   Contains specialized fibroblasts known as *adipocytes* or *adipose cells*, which store large amounts of fat.

Dense Connective Tissue   The intercellular material contains fibers that are packed tightly together, and thus this tissue has fewer cells and less ground substance but more fibers than loose connective tissue.

Dense Regular Connective Tissue   The fibers extend parallel to one another. This tissue resists physical stress and is found in tendons and ligaments.

Dense Irregular Connective Tissue   The fibers branch to form a dense matting. It is found in the deep layer of the skin.

Cartilage   Contains a firm, dense, intercellular material owing to the thickened ground substance or fibers. The intercellular material is called *matrix*; it is produced and maintained by cells called *chondrocytes*. These cells lie in chambers surrounded by matrix known as *lacunae*. The chondrocytes receive nourishment from the perichondrium.

Hyaline Cartilage   The matrix is dominated by chondroitin sulfate in the ground substance and contains collagen. It is located in the upper respiratory tract tubes and at the ends of bones and forms the skeleton of the fetus.

Elastic Cartilage   The matrix is dominated by elastic fibers, providing the cartilage with elastic properties. It is found in the ears, at the end of the nose, and in the epiglottis.

Fibrocartilage   The matrix is dominated by collagenous fibers. It is found as a component of certain joints, such as the intervertebral discs.

Bone   The intercellular material, or matrix, consists of mineral salts and collagen, which is maintained by cells known as *osteocytes*. The osteocytes lie surrounded by hard matrix within lacunae.

Compact Bone   The matrix consists of closely packed deposits of mineral salts, which are laid down in mostly concentric layers known as *lamellae*. In the center lies a tube, the osteonic canal, which serves as a passageway for blood vessels. Combined, this concentric arrangement constitutes the osteon, or haversian system.

Spongy Bone   The matrix consists of small plates of mineral salts and collagen, called *spicules*, that form a network with spaces in between. The spaces are filled with blood-forming tissue, called *red bone marrow*.

Blood-Forming Tissue and Blood   A tissue that manufactures blood cells. Its intercellular material contains stem cells, newly formed blood cells, and protein. It consists of two types: red bone marrow, or hematopoietic tissue, found in spongy bone; and lymphoid tissue, found in lymph nodes, tonsils, the spleen, and the thymus in young children. Red bone marrow initiates production of all blood cells. Lymphoid tissue produces two types of white blood cells, known as *monocytes* and *lymphocytes*. As a whole, the blood consists of formed elements suspended in a fluid matrix known as *plasma*. Blood provides for the transport of many materials throughout the body.

MUSCLE TISSUE

Consists of cells specialized to shorten, or contract. Their coordinated contraction produces movement of body parts. The two main proteins in the cytoplasm of the cells involved in contraction are myosin and actin.

Skeletal Muscle   Muscle that is attached to bones by way of bands of dense connective tissue known as *tendons*. It is under conscious control of the brain and contains striations. The function of skeletal muscle is to provide body movement by moving bones.

Smooth Muscle   Muscle that forms part of the walls of blood vessels and visceral organs. It is not under conscious control and does not contain striations. Smooth muscle propels material as it passes through body tubes, and alters the sizes of organs.

Cardiac Muscle   Muscle that forms the walls of the heart. It is not under conscious control but does contain striations. Cardiac muscle contraction helps propel blood through the body's system of vessels.

NERVE TISSUE

A tissue characterized by the well-developed properties of conductivity and excitability. It consists of two types of cells, neurons and neuroglia. Neurons conduct electrochemical signals very rapidly, and neuroglia support and maintain the neurons.

MEMBRANES

The simplest combination of tissues in the body that forms a functional unit. Membranes consist of connective tissue usually associated with epithelial tissue, and are highly vascular. Those that contain epithelium are termed *epithelial membranes*. Membranes divide areas of the body or organs, line hollow organs and cavities, and anchor organs.

Cutaneous Membrane Also known as the skin; it will be studied in Chapter 6.

Serous Membranes Line the internal surfaces of the thoracic and abdominopelvic cavities and cover the surfaces of many of their organs. They include the pericardium, around the heart; the two pleurae, each of which surrounds one lung and also lines the same side of the thoracic cavity wall; and the peritoneum, which lines the abdominal cavity and covers most of its organs. The epithelium secretes a watery fluid that provides lubrication.

Mucous Membranes Line the internal walls of the digestive tract, the respiratory tract, the reproductive tract, and the ureters, urinary bladder, and urethra of the urinary tract. The epithelium may secrete mucus, which traps foreign particles, maintains a moist environment, and forms a protective layer over cells.

Synovial Membranes Line the walls of cavities that surround certain joints. They lack an epithelial layer. The cells secrete a fluid, synovial fluid, which lubricates the joints and nourishes joint cartilage.

KEY TERMS

avascular (p. 82)
cartilage (p. 94)
chondrocyte (p. 94)
conductivity (p. 100)

contraction (p. 99)
elasticity (p. 87)
epithelial (p. 82)
excitability (p. 100)

extensibility (p. 87)
fibroblast (p. 92)
membrane (p. 101)
neuroglia (p. 100)

neuron (p. 100)
osteocyte (p. 96)
osteon (p. 97)
vascular (p. 90)

QUESTIONS FOR REVIEW

OBJECTIVE QUESTIONS

1. The type of tissue that usually consists of cells suspended in abundant intercellular material with numerous blood vessels is:
   a. epithelial tissue
   b. connective tissue
   c. muscle tissue
   d. nerve tissue
2. An example of an avascular tissue is:
   a. dense connective tissue
   b. bone
   c. simple cuboidal epithelium
   d. adipose tissue
3. An epithelial tissue that contains a single layer of elongated, cylindrical cells is:
   a. simple squamous
   b. simple columnar
   c. pseudostratified columnar
   d. simple cuboidal
4. Tissue that covers body regions exposed to a large degree of wear and tear often contains more than one cell layer. An example of this type of tissue is:

   a. skeletal muscle
   b. stratified squamous epithelium
   c. pseudostratified columnar epithelium
   d. simple cuboidal epithelium
5. A group of similar cells that secrete a product into the extracellular space is called:
   a. neuroglia
   b. exocrine gland
   c. endocrine gland
   d. simple squamous epithelium
6. Collagen is a protein that is particularly abundant in:
   a. covering and lining epithelium
   b. dense regular connective tissue
   c. smooth muscle
   d. adipose tissue
7. Areolar loose connective tissue serves to:
   a. support body weight
   b. bind structures together
   c. produce body movement
   d. form joints between bones

8. Osteocytes embedded within chambers that are surrounded by a matrix composed of mineral salts and collagen may be found in:
   a. hyaline cartilage
   b. bone
   c. tendons and ligaments
   d. areolar tissue

9. The fibers that form a thick protein matting in dense irregular connective tissue include:
   a. collagenous fibers   c. elastic fibers
   b. reticular fibers   d. myosin fibers

10. Chondrocytes secrete a matrix that contains a thickened ground substance in which type of tissue?
    a. compact bone   c. cartilage
    b. dense connective tissue   d. skeletal muscle

11. A bluish-white, abundant form of cartilage that forms the ends of bones and precedes bone formation in most bones of growing children is called:
    a. hyaline cartilage   c. fibrocartilage
    b. elastic cartilage   d. none of the above

12. In which type of connective tissue is the matrix deposited in thin, orderly layers called *lamellae* that encircle a central tube known as the *osteonic canal*?
    a. spongy bone   c. blood-forming tissue
    b. compact bone   d. hyaline cartilage

13. A tissue that contains cells specialized for contraction and is connected to bones for the production of their movement is called:
    a. skeletal muscle   c. nerve tissue
    b. cardiac muscle   d. spongy bone

14. The simplest combination of tissues in the body that divides body regions or organs, lines internal surfaces, and anchors organs is a(n):
    a. organ   c. membrane
    b. system   d. body cavity

15. A membrane that secretes a watery fluid that prevents friction from developing during the movement of body organs is a:
    a. cutaneous membrane   c. serous membrane
    b. mucous membrane   d. all of the above

ESSAY QUESTIONS

1. Compare and contrast the main types of epithelial tissue on the basis of their general structural and functional characteristics.

2. Discuss why epithelial tissue is avascular and connective tissue is highly vascular.

3. Describe the composition of the matrix found in areolar tissue, dense connective tissue, cartilage, and bone. How does its structural nature relate to its functional role?

4. What body function is muscle tissue specialized to perform? Why can't epithelial or connective tissue perform this function?

5. How do mucous membranes differ in body location and in function from serous membranes?

A N S W E R S   T O   A R T   L E G E N D   Q U E S T I O N S

Figure 4–1
How many cell layers do you see in simple squamous epithelium? One cell layer is present in simple squamous epithelium.

Figure 4–2
Where in the body would you expect to find simple cuboidal epithelium? Simple cuboidal epithelium may be found lining tubules and ducts, such as those within the kidneys.

Figure 4–3
What is the shape of the individual cells in simple columnar epithelium? Cells in simple columnar epithelium are cylindrical in shape.

Figure 4–4
Where in the body might you find stratified squamous epithelium? Stratified squamous epithelium is found in the superficial layer of skin, and it lines body openings to the exterior.

Figure 4–5
What is the role of cilia in PSCC tissue? The cilia in PSCC tissue move about to create a current of mucus.

Figure 4–6
How does transitional epithelium differ from stratified squamous epithelium? Transitional epithelium consists of cells that are each roughly spherical in shape, whereas the superficial layers of cells in stratified squamous epithelium are flattened.

Figure 4–7
What common feature do all exocrine glands share? All exocrine glands secrete their products into ducts, which empty into body cavities or onto the body surface.

Figure 4–8
What cells produce the proteins in loose connective tissue? Fibroblasts secrete proteins in loose connective tissue.

Figure 4–9
What protein provides strength to tendons and ligaments? Collagen provides strength in tendons and ligaments.

Figure 4–10
Where in the body is hyaline cartilage found? Hyaline cartilage is located in most joints between opposing bones, in the nose,

and between the ends of most ribs and the sternum. It also forms most of the skeleton in a fetus.

Figure 4–11
What protein provides the stretchability that characterizes elastic cartilage? Elastin provides the stretchability feature of elastic cartilage.

Figure 4–12
What cells produce the dense ground substance of cartilage? Chondrocytes produce and maintain cartilage matrix, including the ground substance.

Figure 4–13
What materials provide bone with its matrix? Mineral salts and collagen make up bone matrix.

Figure 4–14
Where do blood cells originate? Blood cells originate in red marrow and lymphoid tissue.

Figure 4–15
Where in the body are skeletal muscles located? Skeletal muscles are located between the skin and internal body cavities, and attach to bones.

Figure 4–16
What features of cardiac muscle can you find that are not present in smooth muscle tissue? Skeletal muscle contains visible striations and intercalated discs which are not present in smooth muscle.

Figure 4–17
What special features distinguish nerve tissue from all other tissues? Nerve tissue has highly developed properties of excitability and conductivity.

Figure 4–18
With the exception of synovial membranes, of what two tissues are membranes composed? Epithelial and connective tissues make up membranes (other than synovial membranes).

CHAPTER 5

ORGANS AND SYSTEMS: OVERVIEW OF THE HUMAN BODY

CHAPTER OUTLINE

LEARNING OBJECTIVES

After studying this chapter, you should be able to:

1. Distinguish between the structural levels of organization.

2. Explain the structural relationships that exist between tissues, organs, and systems.

3. Identify the basic functions of each of the 11 systems of the body.

4. Identify the primary organs that form a part of each system.

5. Discuss the structural organization of each system.

Important local functions are performed by tissues and membranes, but these cannot carry out the generalized activities that affect larger areas of the body or the body as a whole. Functions such as propelling blood at a regulated, discontinuous rate through blood vessels, receiving information from throughout the body and coordinating the proper response, or controlling fluid volume in the body by removing water or salts from the bloodstream each require a combination of tissues that are integrated into a single structural unit. This combination involves the integration of at least two of the four primary tissue types (epithelial, connective, muscle, nerve) into a distinctly formed, organized unit known as an **organ**. For example, the organ that propels blood is the heart; the organ that receives information and coordinates the response is the brain; the organs that control fluid volume are the kidneys. There are many other organs in the body, each of which performs an activity affecting a large region of the body or the body as a whole.

Organs are, in turn, combined into larger units of structural organization known as **systems**. In a system, each organ and its associated structures share a common, general function that directly affects the body's ability to survive or reproduce. For example, the respiratory system contains the trachea, the bronchi, and the lungs. Each of these organs performs a particular function: the trachea transports air between the throat and the bronchi; the bronchi transport air between the trachea and the smaller tubes within the lungs; the lungs are the site of gas diffusion between the air and the bloodstream. Their common function when integrated as the respiratory system is to bring oxygen from the external environment into the bloodstream, and to carry the waste product, carbon dioxide, in the opposite direction. If either the trachea, bronchi, or lungs were removed without replacement, the function of the respiratory system would cease and survival of the individual would not be possible.

There are 11 systems in the human body, to which you were briefly introduced in Chapter 1. In this chapter each system and the organs it contains are examined in more detail. This study represents an overview of the human body, as it concentrates on the general aspects of structure and function in an effort to present the "wide picture" of body organization. It is provided as preparation for studying subsequent chapters that examine the body, system by system, in further detail.

INTEGUMENTARY SYSTEM

The integumentary system consists of the skin and numerous accessory organs. It helps maintain homeostasis.

The integumentary system covers the surface of the body, providing it with a protective, highly sensitive barrier against the external environment. Its primary organ is the **skin**, or **integument**, which is also known as the *cutaneous membrane* (Fig. 5–1).

The skin consists of two layers: the superficial layer, or **epidermis**, which is composed of epithelium that forms a waxy protective coat of dead cells and protein over the surface of the body; and the deep layer, the **dermis**, which consists of connective tissue that supports and maintains the epidermis. Deep to the skin is a layer of fat and areolar tissue, called the **hypodermis**, which anchors the skin to the deeper-lying structures of the body and provides an insulating blanket and food reserve.

Embedded within the dermis are numerous accessory organs, including **blood vessels**, **receptors**, and **glands**. Blood vessels provide the dermis with an ample supply of blood, making it possible for the skin to grow and regenerate itself. Receptors consist of nerve endings in association with other tissues, and are capable of detecting changes in the environment. If the changes are great enough, receptors may initiate an awareness of fine touch, pressure, heat and cold, and pain. The glands in the dermis are of two types: sweat glands, which secrete sweat onto the surface of the skin; and oil glands, which secrete a lubricating oil at the base of a hair.

The integumentary system performs a number of vital functions. The protective layer formed by the epidermis provides a physical barrier that guards underlying organs against the external environment, prevents fluid loss, and helps maintain internal body temperature. The dermis and accessory organs provide a means of communication with the external environment, remove metabolic waste products by way of sweat glands, and also help maintain internal body temperature. These functions point to the overall role of the integumentary system: the maintenance of homeostasis.

CONCEPTS CHECK

1. What is the relationship between organs and systems?

2. What are the organs of the integumentary system?

3. In what ways does the integumentary system help maintain homeostasis?

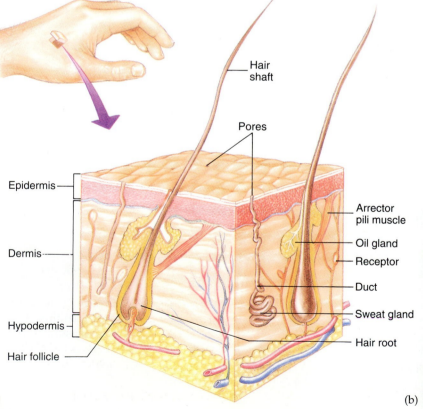

(a)

(b)

Figure 5–1
Integumentary system. (a) This cutaway view of a section of skin shows its major components. Although the skin as a whole is a single organ, it contains many smaller organs within: blood vessels, receptors, and glands. (b) A section of human skin; magnification 100x.

What tissue makes up the epidermis of the skin?

Labels for figure (a): Hair shaft, Pores, Epidermis, Dermis, Hypodermis, Hair follicle, Arrector pili muscle, Oil gland, Receptor, Duct, Sweat gland, Hair root

SKELETAL SYSTEM

The skeletal system consists of bones, joints, and associated connective tissues. Its supportive nature makes body movement possible.

The skeletal system consists of **bones**, joints between opposing bones, and the connective tissues associated with them. Each individual bone is regarded as an organ, since the bones form independently of one another during the developing years. The bones, joints, and associated tissues form the supportive frame of the body, or **skeleton**.

The skeleton contains 206 bones (Fig. 5–2), each of which falls into one of two categories: the **axial skeleton**, which contains the bones that lie along or near the vertical axis of the body; and the **appendicular skeleton**, with the bones that are lateral to the vertical axis. The axial skeleton consists of the bones of the skull, the vertebral column, and the thoracic cage. The appendicular skeleton contains the bones of the shoulder (or pectoral) girdle, the upper limbs, the hip (or pelvic) girdle, and the lower limbs.

The joints, or **articulations**, are junctions between bones. They consist of one of several types of connecting material that extends between opposing bones, and may contain a cavity as well. Depending on the nature of the connecting material and the presence of

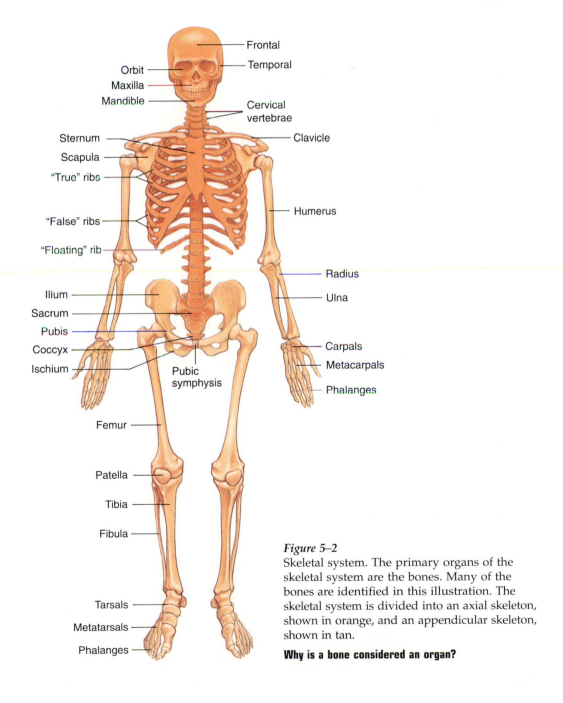

Figure 5–2
Skeletal system. The primary organs of the skeletal system are the bones. Many of the bones are identified in this illustration. The skeletal system is divided into an axial skeleton, shown in orange, and an appendicular skeleton, shown in tan.

Why is a bone considered an organ?

a cavity, they permit varying amounts of movement. Supporting and strengthening the joints are ligaments, which extend from one bone to the next across the joint; and tendons and muscles, which surround certain joints.

The functions of the skeletal system include support of soft body tissues and organs; protection of organs that are surrounded by bone, such as the brain and spinal cord; storage of mineral salts that are required for normal body functions (particularly calcium and phosphorus); and providing the site for blood cell formation, which occurs within the red marrow of spongy bone tissue. The skeletal system also provides attachment sites for muscles by way of tendons, making body movement possible.

MUSCULAR SYSTEM

The organs of the muscular system are the muscles that attach to bones. Contraction of the muscles provides body movement.

The movement of body parts to produce body movement is provided by the muscular system. It is composed of the type of muscle tissue that is attached to bones—skeletal muscle—and its associated connective tissue. Skeletal muscle tissue and connective tissue are organized to form the more than 500 **muscles** of the body, each of which is a distinctly separate organ (Fig. 5–3).

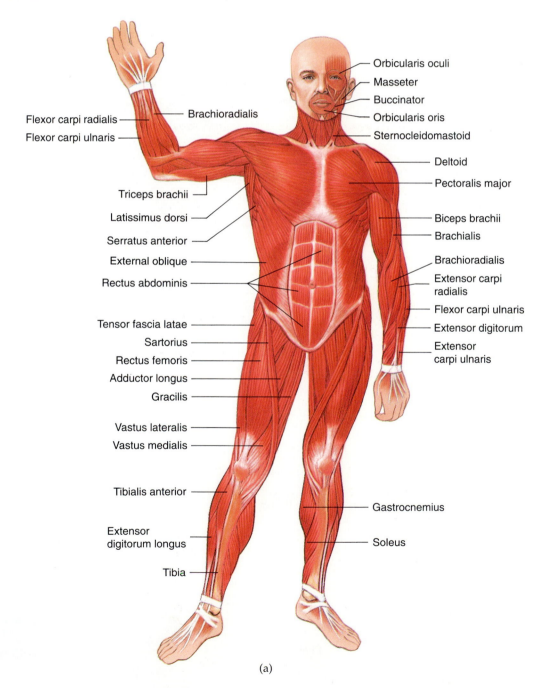

(a)

A single muscle consists of cells packaged into discrete units, known as *bundles*, which are highly specialized to contract. The coordinated, simultaneous contraction of cells within a bundle causes it to shorten. When all bundles within a muscle shorten at the same time, the muscle as a whole contracts. Muscles are normally attached at their opposite ends to two different bones: movement results when a contracting muscle pulls one of its attached bones toward the other, more stationary bone.

The connective tissue that is associated with a muscle envelops the muscle's individual cells, the bundles, and the muscle itself. It also extends beyond the main body of the muscle to form a tendon, which forms a firm attachment with a bone. Within the connective tissues are blood vessels and nerves, providing the muscle with a source of nourishment and energy, removing the waste products of metabolism, and providing nerve stimulation that begins the process of contraction within the cells.

The primary function of the muscular system is the production of skeletal movement. Other roles include the maintenance of body form, or posture, and support of the skeleton. It also functions in the generation of heat, and is thereby involved in the maintenance of body temperature—a homeostatic function.

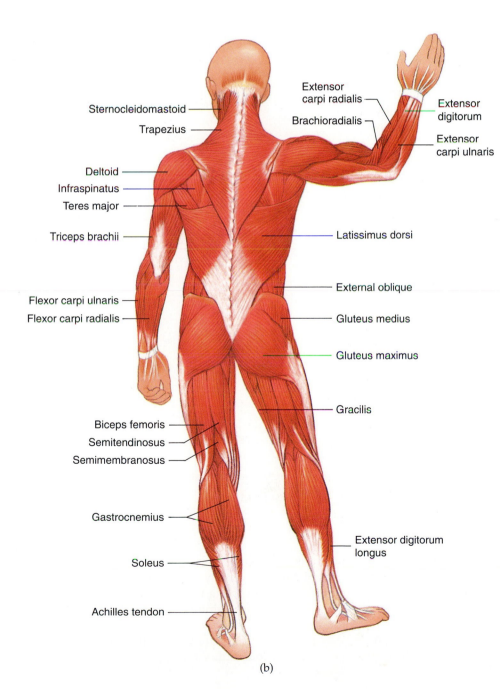

Sternocleidomastoid

Trapezius

Deltoid

Infraspinatus

Teres major

Triceps brachii

Flexor carpi ulnaris

Flexor carpi radialis

Biceps femoris

Semitendinosus

Semimembranosus

Gastrocnemius

Soleus

Achilles tendon

Extensor carpi radialis

Brachioradialis

Extensor digitorum

Extensor carpi ulnaris

Latissimus dorsi

External oblique

Gluteus medius

Gluteus maximus

Gracilis

Extensor digitorum longus

(b)

Figure 5–3
Muscular system. The muscles that move the body are the organs that make up the muscular system. Many of the major muscles are identified here. (a) Anterior view. (b) Posterior view.

What organs are muscles attached to?

NERVOUS SYSTEM

The nervous system consists of the brain, spinal cord, nerves, and special sense organs, and is classified on the basis of structure and function. It maintains homeostasis by way of rapid impulses.

The nervous system provides a means of communication between the body and the external environment. Its organs include the **brain**, the **spinal cord**, the **nerves** that extend between the brain and spinal cord and other areas of the body, and **special sense organs** that monitor changes in the external environment, such as the eye and ear (Fig. 5–4). Each organ is composed of nerve tissue that is supported by connective tissue coverings; it receives nourishment and waste removal by way of a large supply of blood vessels.

The organs of the nervous system are structurally classified into two categories: the **central nervous system (CNS)**, comprising the brain and spinal cord; and the **peripheral nervous system (PNS)**, which consists of the nerves that extend between the CNS and the structure in which they terminate (a visceral organ, a gland, or a muscle) or originate.

The nervous system may also be classified on the basis of functional differences, such as the direction information is traveling. This classification scheme uses the CNS as a point of reference, although its components are also included. It consists of an **afferent**, or **sensory**, portion, which carries information toward the CNS; and an **efferent**, or **motor**, portion, which carries information away from the CNS.

The efferent portion is further divided into the **somatic nervous system** and the **autonomic nervous system**. The somatic component stimulates skeletal muscle contraction and is under conscious control: if you choose to move an arm or a leg, you can do so at will because the somatic nervous system follows your mental command. All other structures in the body receiving commands from the CNS, such as glands, visceral organs, blood vessels, and the heart, have nerve connections to the autonomic nervous system, which is

not under conscious control. You cannot consciously control the diameter of your blood vessels or keep your stomach from growling when you are hungry, because their smooth muscle linings are innervated by the autonomic nervous system.

As a coordinated unit, the components of the nervous system maintain the body in a state of stability,

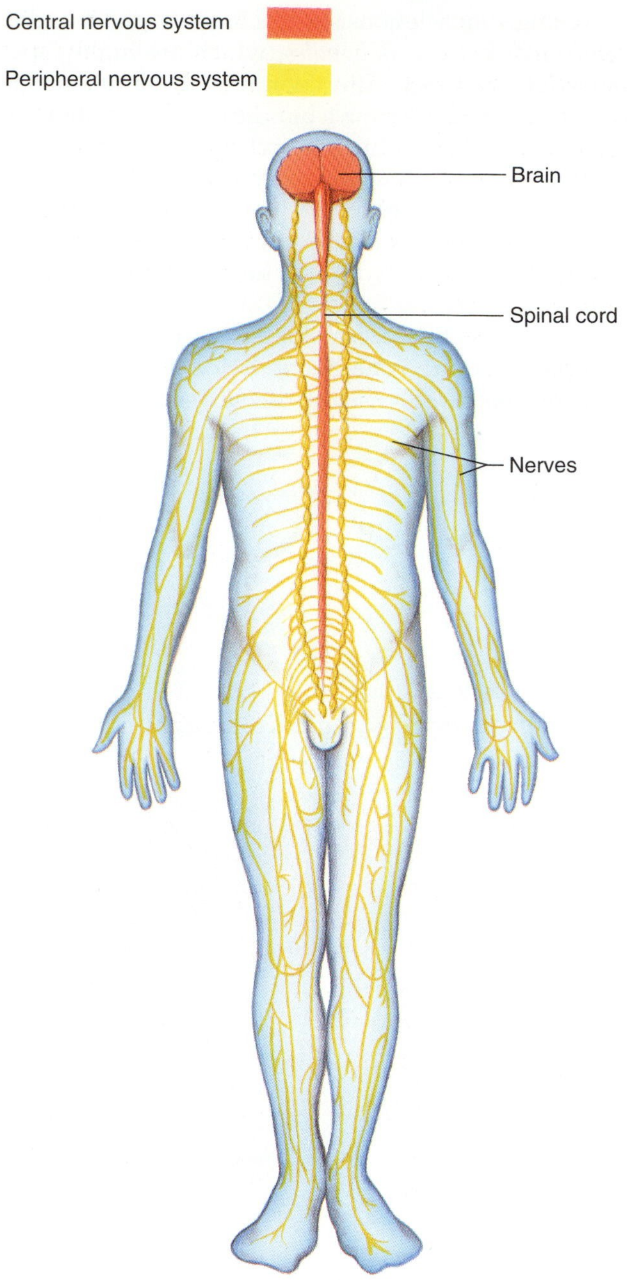

Central nervous system

Peripheral nervous system

Brain

Spinal cord

Nerves

Figure 5–4
Nervous system. The organs of the nervous system include the brain, spinal cord, special sense organs, and nerves.

Of what tissue are the organs of the nervous system primarily composed?

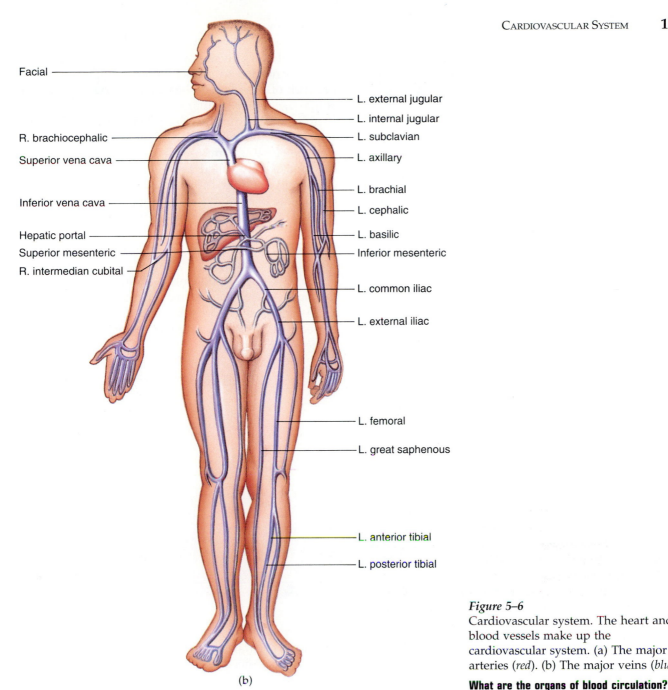

Facial

L. external jugular
L. internal jugular
L. subclavian
R. brachiocephalic
L. axillary
Superior vena cava

L. brachial
Inferior vena cava
L. cephalic

Hepatic portal
L. basilic
Superior mesenteric
Inferior mesenteric
R. intermedian cubital

L. common iliac

L. external iliac

L. femoral

L. great saphenous

L. anterior tibial

L. posterior tibial

(b)

Figure 5–6
Cardiovascular system. The heart and blood vessels make up the cardiovascular system. (a) The major arteries (*red*). (b) The major veins (*blue*).

What are the organs of blood circulation?

CARDIOVASCULAR SYSTEM

The organs of the cardiovascular system include the heart and blood vessels. This system provides transportation for the various substances carried by the blood.

Every living cell requires basic materials to survive. In the case of human cells, the basic materials are oxygen, used to decompose molecules for the release of energy stored in their chemical bonds; and nutrients, used to provide the building blocks required for the synthesis of new molecules. Also, waste materials from the cell's metabolic activities may be toxic if allowed to linger, so they must be removed continuously to ensure the cell's survival. How are the basic materials brought to every living cell and waste materials removed? A system of tubes carries a liquid containing these substances, propelled mainly by a muscular organ, to every corner of the body. This is the cardiovascular system.

The organs of blood circulation are the **blood vessels** and the **heart** (Fig. 5–6). Blood vessels that transport blood away from the heart are called **arteries**, and those that transport blood toward the heart are called **veins**. As arteries extend from the heart they become smaller in diameter and branch extensively. Eventually, their branches dwindle to a microscopic size and become extremely thin-walled. These tiny tubes are called **capillaries**, and are the site of the exchange of materials between the bloodstream and the interstitial fluid. Capillaries form an extensive network in every organ, for they must be close to every living cell to enable materials to diffuse in and out adequately. As blood continues to flow, it passes through capillaries into vessels that become larger and fewer as they progress toward the heart. These larger vessels, the veins, deliver blood back to the heart.

The heart provides most of the propulsion needed to move blood through the vessels. It is a hollow, muscular organ composed mainly of cardiac muscle tissue with supporting connective tissue. Internally it contains four chambers: two upper (superior), thin-walled **atria** that collect blood as it enters the heart; and two lower (inferior), thick-walled **ventricles** that provide the force needed to propel blood through the body's network of vessels. The heart and blood vessels constitute the cardiovascular system; when blood and the lymphatic structures are included, they are often referred to collectively as the *circulatory system*.

The blood is composed of cells and cell-like particles, collectively called **formed elements**, suspended in a liquid medium known as **plasma**. The formed elements are of three types: red blood cells, or **erythrocytes**, which have special properties allowing them to transport oxygen and carbon dioxide gases; white blood cells, or **leukocytes**, which protect the body from infection and disease; and platelets, or **thrombocytes**, which help form blood clots for minimizing fluid loss during an injury. The plasma is a clear, yellowish fluid that is composed mostly of water. Suspended or dissolved within it are proteins, nutrients, gases, enzymes, and hormones.

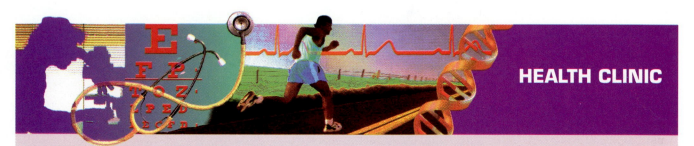

HEALTH CLINIC

Systemic Diseases

A systemic disease is a condition that affects the whole body, instead of a localized region or body part. The name *systemic* is used because such a condition is capable of afflicting numerous body systems, such as the digestive, cardiovascular, and endocrine systems, all at once. It often arises from a single area of the body, but its spread to other sites results in widespread symptoms. Because systemic diseases are so widespread, they are often life-threatening. An example of a systemic disease is *systemic lupus erythematosus (SLE)*, which results in widespread swelling of membranes, tissues, and entire organs. It is a type of collagen disease (see the box on page 90) whose cause is still not known. Another systemic disease is *tuberculosis (TB)*, which is currently on the increase in the United States. Usually associated with the lungs, this bacteria-caused disease may establish itself in any one of a number of different organs throughout the body. The bacterial colonies form hard plaques in living tissue, destroying its normal function. TB is spread by airborne particles of sputum or other body fluids and may be difficult to treat if not arrested early, because of the bacteria's resistance to many types of antibiotics. Yet another example of a systemic disease is *cystic fibrosis (CF)*, which affects all exocrine glands throughout the body. Cystic fibrosis is a hereditary disease whose symptoms occur early in life. The inherited gene alters the normal amount of sodium that passes through plasma membranes, resulting in the excessive secretion of thick mucus. The mucus blocks ducts of the pancreas, preventing the secretion of digestive enzymes and thereby interfering with nutrition. It also blocks the trachea and bronchi of the respiratory system, often leading to lung collapse. Most deaths from CF occur as a result of respiratory failure. Many other diseases may be of a systemic nature; among them are other types of hereditary diseases, many forms of cancer, immune-suppressive diseases such as AIDS, diseases caused by dietary imbalances, and most forms of allergies.

LYMPHATIC SYSTEM

The lymphatic system contains lymphatic vessels, lymph nodes, the spleen, the thymus, and the tonsils. It provides immunological protection from foreign particles.

The lymphatic system is, in some ways, similar to the cardiovascular system, for it, too, consists of a system of tubes that carry fluid through the body. However, it has no source of propulsion, and it transports fluid in one direction only: toward the heart. The fluid, called **lymph**, is a yellowish, watery fluid that contains no red blood cells or platelets. Lymph is transported from lymphatic capillary networks within most body organs, where it originates, to the heart through a system of vessels that are separate from those of the cardiovascular system. Lymph is cleansed of foreign particles and toxins as it is carried back to the heart.

The primary organs of the lymphatic system include **lymphatic vessels** that transport the lymph; numerous **lymph nodes** that are each about the size of a small pea and are located along the lymphatic vessel route; the **spleen**, located lateral to the stomach; the **thymus gland**, which lies near the heart; and the **tonsils**, in the throat (Fig. 5–7). The lymph nodes, spleen, thymus, and tonsils are composed of a specialized type of tissue, called *lymphoid tissue*, which manufactures two particular types of white blood cells: monocytes and lymphocytes. These cells are very important components of the body's defense arsenal.

The primary function of the lymphatic system is to protect the body from the harmful effects of microorganisms, toxins, and dead or diseased body cells. This is performed in basically two ways: by stationary cells that compose the structure of lymphatic organs, which filter the harmful materials out of lymph as it passes toward the heart; and by cells that travel through the bloodstream, lymph, and interstitial fluid, actively pursuing the harmful materials and destroying them. The stationary cells are referred to as the *mononuclear phagocytic system*, and the active cells play the primary roles in the *immune response*. In both cases, the cells are various types of white blood cells.

CONCEPTS CHECK

1. What is the primary function of the cardiovascular system?

2. How do arteries, veins, capillaries, and lymphatic vessels differ functionally?

3. What is the primary function of the lymphatic system?

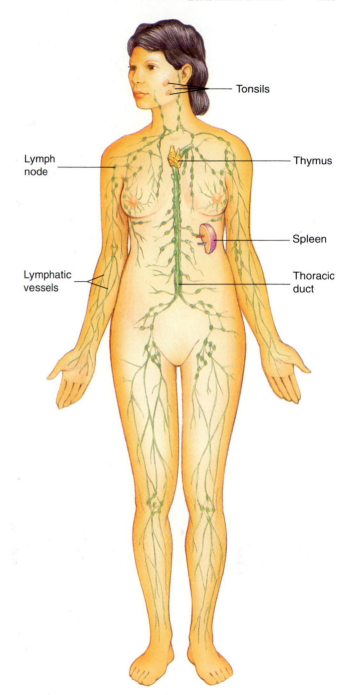

Figure 5–7
Lymphatic system. The vessels of the lymphatic system may be found in most body areas. They drain lymph into the bloodstream. Other lymphatic organs include the pea-sized lymph nodes, the tonsils, the thymus, and the spleen.

What fluid is carried by lymphatic vessels?

RESPIRATORY SYSTEM

The respiratory system consists of the lungs and the tubes that transport air to and from them. It provides for the exchange of gases between the exterior and the bloodstream.

All cells in the human body require oxygen to survive. The oxygen must be supplied on a continuous basis, for if its levels drop too much, the cell's metabolic machinery will shut down and the cell will perish for want of energy to power its life-sustaining reactions. Oxygen is readily available from the external environment, for it is a component of the air we breathe. However, in order for it to become available to every cell, it must be transported into the body and find its way into the bloodstream.

Conversely, every cell must rid itself of the gas, carbon dioxide, which is produced during metabolism as a waste material. As carbon dioxide levels rise during metabolic activity, it leaves the cell and enters the bloodstream by simple diffusion. Once in the bloodstream, carbon dioxide may dissolve or combine with other molecules to form carbonic acid and bicarbonate ions. If these materials are allowed to accumulate in the bloodstream, they will reach toxic levels and threaten homeostasis. The body must, therefore, have a way of removing carbon dioxide from the bloodstream and of eliminating it from the body completely.

How may oxygen travel from the external environment into the bloodstream, and carbon dioxide from the bloodstream to the external environment? By way of the respiratory system, which consists of a series of tubes that channel air and a highly specialized surface area that permits rapid diffusion of gases with the bloodstream.

The respiratory system is functionally divided into two components: a **conducting zone** and a **respiratory zone** (Fig. 5–8). The conducting zone consists of the tubes that transport air between the external environment and the lungs, and includes organs such as the **nose**; the throat, or **pharynx**; the voicebox, or **larynx**; the windpipe, or **trachea**; and the two **bronchi**, which each pass into one of the lungs. The respiratory zone consists of the **lungs**, where diffusion occurs between the air supply and the bloodstream. Gas exchange in the lungs occurs between the thin walls of microscopic sacs called **alveoli** and the thin walls of capillaries that surround them.

Figure 5–8
Respiratory system. The organs of the respiratory system transport air to and from the tiny air sacs within the lungs, where gas exchange with the bloodstream occurs. They include the nose, pharynx, larynx, trachea, bronchi, and lungs.

What respiratory organs carry air molecules to and from the lungs?

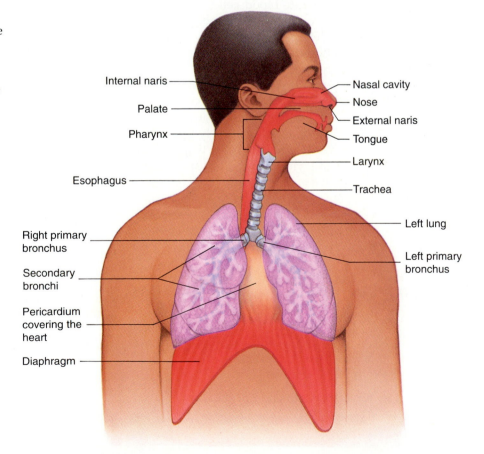

Internal naris — Nasal cavity — Nose — Palate — External naris — Pharynx — Tongue — Larynx — Esophagus — Trachea — Left lung — Right primary bronchus — Left primary bronchus — Secondary bronchi — Pericardium covering the heart — Diaphragm

DIGESTIVE SYSTEM

The organs of the digestive system include the mouth, salivary glands, esophagus, stomach, small intestine, pancreas, liver, gallbladder, and large intestine. This system prepares food particles for their entry into the bloodstream.

In addition to oxygen, cells require nutrients in order to survive. These compounds provide cells with energy-containing molecules and the raw materials necessary for growth, repair, reproduction, and the production of new substances. Nutrients are obtained from the food we eat, which is broken down by a se-ries of stages into a form small enough to pass through capillary walls for its transport by the bloodstream. The breakdown, or **digestion**, of food particles into smaller forms for the nourishment of body cells is the primary function of the digestive system.

The digestive system is a long, continuous tube, composed of numerous organs, that extends through the body from the mouth to the anus, receiving ducts from specialized organs along the way (Fig. 5–9). The digestive process begins in the **mouth**, which breaks apart the food initially and receives secretions from the **salivary glands** that start the process of carbohydrate digestion. Swallowed food then passes briefly through the **pharynx** before entering the **esophagus**, a long,

Figure 5–9
Digestive system. The combined functions of digestive organs provide nutrients for transportation to body cells.

Within which digestive organ does nutrient absorption take place?

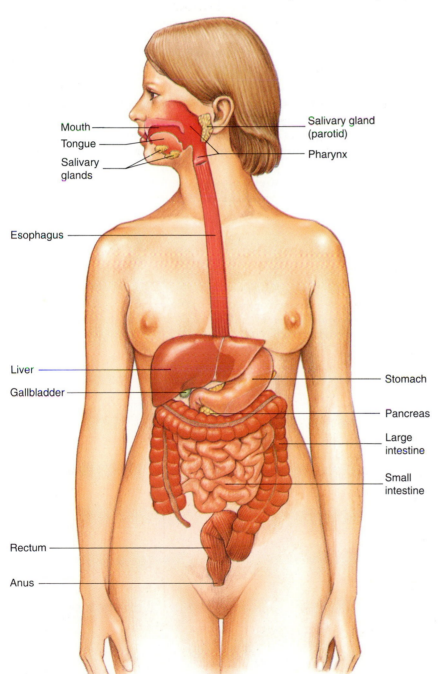

Mouth
Tongue
Salivary glands
Salivary gland (parotid)
Pharynx
Esophagus
Liver
Gallbladder
Stomach
Pancreas
Large intestine
Small intestine
Rectum
Anus

muscular tube that extends through the neck and thoracic cavity and penetrates the diaphragm just before it unites with the **stomach**. The stomach is a J-shaped enlargement of the digestive tube where swallowed material is converted to a liquid mixture and where protein digestion begins by the action of enzymes, but where very little absorption into the blood occurs.

From the stomach, the liquefied food passes into the long, twisting **small intestine**, where digestion proceeds and is eventually completed. The **liver** and **gallbladder** provide the small intestine with a yellowish-green liquid, called *bile*, which aids in the digestion of fats. The liver also removes nutrients from the bloodstream and converts them to chemical energy, energy-storage molecules, or other useful substances and returns them to the bloodstream for their distribution throughout the body. The **pancreas** provides the small intestine with a liquid filled with enzymes that aid in the digestion of protein, carbohydrates, and fats. The small intestine is lined with a specialized epithelium that serves as the primary site of nutrient absorption into the bloodstream.

All material that is not absorbed by the small intestine passes into the **large intestine**, where water is removed and returned to the bloodstream. The solidified waste material is then eliminated from the body by passing through the terminal opening, the anus.

CONCEPTS CHECK

1. Within which respiratory organ is the site of gas exchange between the external air and the bloodstream?

2. What are the organs of the digestive system, and what are their roles in digestion and absorption?

URINARY SYSTEM

The urinary system contains the kidneys, ureters, urinary bladder, and urethra, and maintains homeostasis by regulating fluid content and volumes in the body.

The organs of the urinary system include the two **kidneys**, against the posterior wall of the trunk; the two **ureters** (YOO-re-terz), which extend from the kidneys downward; the **urinary bladder**, at the base of the abdominopelvic cavity, which receives the two ureters; and the **urethra** (yoo-RĒ-thra), which extends from the bladder to open to the exterior (Fig. 5–10).

Figure 5–10
Urinary system. The kidneys and ureters are located behind the digestive organs against the back wall of the abdominopelvic cavity. The urinary bladder is at the base of the pelvic cavity, and the urethra penetrates the body wall to open to the exterior. (a) The urinary system in the female. (b) The urinary system in the male.

Which urinary organs transport urine from the kidneys to the bladder?

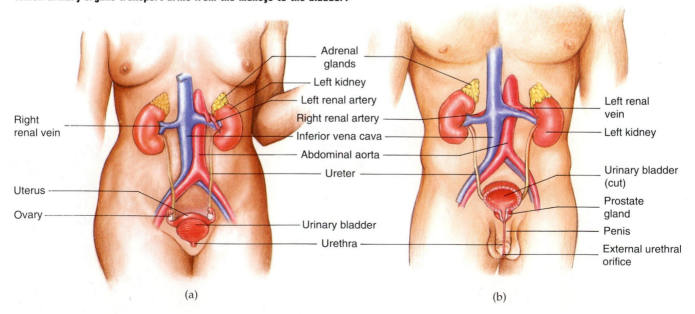

Adrenal glands
Left kidney
Left renal artery
Right renal artery
Inferior vena cava
Abdominal aorta
Ureter

Right renal vein

Uterus
Ovary

Urinary bladder
Urethra

Left renal vein
Left kidney

Urinary bladder (cut)
Prostate gland
Penis
External urethral orifice

(a)

(b)

The primary functions of the urinary system are important in maintaining homeostasis. They include regulating the water, salt, and acid-base balance in body fluids, and removing from the bloodstream nitrogen-containing waste materials that form during cellular metabolism. These functions are performed by the kidneys, which are specialized to receive blood and modify it in order to stabilize the internal balance of the body despite changes that occur in body-fluid content, blood pressure, and rate of metabolism. The resulting waste that is produced by kidney function is in the form of liquid urine. The remaining urinary organs serve to transport and temporarily store urine, and to eliminate it from the body.

REPRODUCTIVE SYSTEM

The reproductive system contains organs that differ between males and females. It provides a means of continuing the species.

The reproductive system provides a means of ensuring continued survival of the species. It is the only system that is not primarily concerned with homeostasis of the individual organism. The reproductive system is also the only system in which the organs differ significantly between males and females (Fig. 5–11).

In males, the organs that produce sex cells are the paired **testes**. Other organs transport and maintain the sex cells, or *sperm*, and include the two **epididymides** (ep'-i-DID-i-mī-dēz'; **epididymis** is the singular form) located with the testes inside the external skin-covered sac known as the **scrotum**; the paired **ductus deferentes**, or **vasa deferentia** (singular **ductus** or **vas deferens**), which extend from each epididymis upward through the abdominal wall and into the abdomino-pelvic cavity, where they wind around the urinary bladder to unite with the urethra; the **urethra**, extending from the urinary bladder to open to the exterior; and the structure that projects from the body and contains much of the urethra through its center, the **penis**.

Several glands provide secretions into the reproductive fluid, or semen, which prolongs survival of the sperm cells and improves reproductive success. These are the two **seminal vesicles** behind the bladder, the **prostate gland** surrounding the urethra near its union with the bladder, and the two **bulbourethral** (bul'-bō-yoo-RĒ-thral) **glands** near the origin of the penis.

The female organs that produce sex cells are the paired **ovaries**, supported by connective tissue structures against the posterior wall of the pelvic cavity. Normally, one of the two ovaries releases a single sex cell, or *ovum*, at approximately 28-day intervals. The ovum is first swept into one of the two **uterine (fallopian) tubes**, where fertilization may occur. Eventu-

Figure 5–11
Reproductive system. (a) The male reproductive system produces sperm and conveys them to the outside by way of a series of tubes and ducts. (b) The female system produces the egg and provides a site for support of the embryo during pregnancy.

Do all reproductive organs normally differ between males and females?

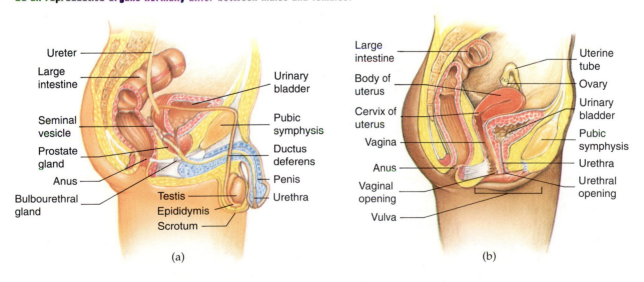

(a) (b)

Table 5–1   THE SYSTEMS OF THE BODY

System	Organs	Primary Functions
Integumentary	Skin Accessory organs	Protects from radiation, microorganisms, fluid loss; helps maintain body temperature
Skeletal	Bones	Supports, and protects vital organs, stores mineral salts, serves as attachment site for muscles, forms blood cells
Muscular	Muscles	Moves bones, maintains posture, supports skeleton, generates heat
Nervous	Brain Spinal cord Nerves Special sense organs	Receives, interprets, and responds to sensations in order to maintain homeostasis; monitors and controls body activities
Endocrine	Pituitary gland Thyroid gland Parathyroid glands Adrenal glands Pancreas Gonads Thymus Pineal gland	Maintains long-term homeostasis
Cardiovascular	Heart Blood vessels	Transports substances throughout the body
Lymphatic	Lymphatic vessels Lymph nodes Spleen Thymus Tonsils	Protects from foreign particles, toxins, and dead or diseased body cells
Respiratory	Nose Pharynx Larynx Trachea Bronchi Lungs	Provides for gas exchange between the exterior and the bloodstream

ally the ovum—or, if fertilized, the *zygote*—is swept by currents into the next organ, the **uterus**. The uterus is the site of embryo implantation and development, and of *menstruation*. Inferior to the uterus is the **vagina**, which opens to the exterior. The external genital organs are collectively referred to as the **vulva**.

Table 5–1 summarizes the 11 systems of the body by listing the organs and the major functions that characterize each system.

CONCEPTS CHECK

1. What functions are performed by the kidneys?

2. What are the reproductive organs in males?

3. What are the reproductive organs in females?

System	Organs	Primary Functions
Digestive	Mouth Salivary glands Pharynx Esophagus Stomach Pancreas Liver Gallbladder Small intestine Large intestine	Prepares food particles for their entry into the bloodstream
Urinary	Kidneys Ureters Urinary bladder Urethra	Maintains homeostasis by regulating water, salt, and acid-base balance in the bloodstream; removes nitrogen-containing waste materials
Reproductive		Provides for procreation to sustain the species
Male	Testes Epididymides Scrotum Ductus deferentes Urethra Penis Accessory glands	
Female	Ovaries Uterine tubes Uterus Vagina Vulva	

CHAPTER SUMMARY

Organs are composed of tissues of more than one type that combine to form a distinct structure with a particular function. The function meets a requirement of the body that is of a broader scope than the needs of a local region. A system is composed of organs and their associated structures, which perform a function necessary for the survival of the whole organism, or for reproduction.

INTEGUMENTARY SYSTEM

Primary organ is the skin, which is composed of two distinct layers: the superficial epidermis, composed of epithelium; and the deep dermis, composed of connective tissue. Embedded within the dermis are accessory organs that include glands, receptors, and blood vessels. The main functions of the integumentary system include protection from the external environment, prevention of fluid loss, maintenance of internal body temperature, excretion of metabolic wastes, and communication with the external environment—all of which point to the overall function of maintaining homeostasis.

SKELETAL SYSTEM

Composed of bones, which are regarded as true organs, joints, and their associated connective tissue. When combined, they make up the framework of the body, the skeleton. The skeleton contains 206 different bones, which are divided into two categories: the axial skeleton, which lies along or near the body midline; and the appendicular skeleton, which lies lateral. The joints permit variable degrees of movement between bones, depending on the type of connective tissue that binds them and whether or not a cavity is present. The skeletal system provides support for soft body parts and protects certain organs, stores minerals (mainly calcium and phosphorus), serves as a site for blood cell formation, and provides a firm attachment for muscles during body movement.

MUSCULAR SYSTEM

The organs are the muscles, usually attached to bones, which contain connective tissue coverings, blood vessels, and nerves as well as skeletal muscle tissue. The muscles are specially structured to shorten, or contract, and thus produce movement when the muscle is firmly attached to bones by way of tendons. In addition to providing movement of the skeleton, the muscular system also maintains posture and supports the skeleton, and generates heat that can be used to help regulate body temperature.

NERVOUS SYSTEM

Composed of organs containing nerve tissue, connective tissue, and blood vessels. It includes the brain, spinal cord, nerves, and special sensory organs. It is structurally divided into the central nervous system (CNS), which includes the brain and spinal cord; and the peripheral nervous system (PNS), which includes the nerves. Functionally, the nervous system is divided into two systems on the basis of the direction the impulse travels: an afferent, or sensory, portion carries impulses to the CNS; an efferent, or motor, portion conducts impulses away from the CNS. The efferent portion is further divided into the somatic division, which is under conscious control and stimulates skeletal muscle; and the autonomic division, which is not under conscious control and innervates glands, smooth muscle, and cardiac muscle. The primary function of the nervous system is the regulation of homeostasis, which it performs by receiving information regarding a change in the environment, transporting the information rapidly by way of electrochemical impulses, interpreting the information, and initiating responses by making changes in the body.

ENDOCRINE SYSTEM

Composed of glands which do not contain ducts and which secrete hormones. The endocrine glands include the pituitary gland, the thyroid gland, the parathyroid glands, the adrenal glands, the endocrine cells within the pancreas, and the gonads (testes in the male, ovaries in the female), which also produce the sex cells. The thymus, pineal gland, stomach, small intestine, and kidneys also play a role in endocrine activities. Each of these glands is an organ, for all are composed of connective tissue, blood vessels, and nerves as well as glandular epithelium. The products they secrete are collectively known as *hormones*; they modify a target cell by changing its metabolic or protein-synthesis machinery. This modification serves to help the body maintain homeostasis in concert with other body systems.

CARDIOVASCULAR SYSTEM

The heart and blood vessels are the organs of the cardiovascular system, which provides a means of transporting materials required for cell maintenance and normal body activities. The heart provides most of the propulsion, and contains four chambers: two upper atria, and two lower ventricles. The vessels carry the blood away from the heart by way of arteries, and back to the heart by way of veins. In between are tiny, thin-walled capillaries, which are the sites of exchange between the bloodstream and interstitial fluid. The blood is composed of formed elements and plasma. The formed elements include red blood cells, or erythrocytes; white blood cells, or leukocytes; and platelets, or thrombocytes. When the heart, blood vessels, blood, and lymphatic structures are considered as a unit, it is referred to as the *circulatory system*.

LYMPHATIC SYSTEM

Composed of organs that include the lymphatic vessels, lymph nodes, spleen, thymus, and tonsils. The lymphatic vessels transport a fluid, known as *lymph*, from capillary networks within organs toward the heart. Lymph eventually rejoins the blood circulation. As lymph is transported, it is filtered of foreign particles and toxins by white blood cells. There are two functional components of the lymphatic system, both of which serve the general function of protecting the body from infection and disease: the mononuclear phagocytic system, consisting of stationary cells within lymphatic organs that filter lymph as it passes through; and the cells of the immune response, which consists of mobile white blood cells that actively pursue foreign particles through the bloodstream, interstitial fluid, and lymph.

RESPIRATORY SYSTEM

Oxygen is a necessary ingredient for metabolism in cells, and carbon dioxide is a metabolic waste material that can upset homeostasis if allowed to accumulate in body fluids. The respiratory system transports these gases between the external environment and the bloodstream. It consists of two portions: a conducting portion, which transports gases between the exterior and the lungs; and a respiratory portion, which permits gas exchange. The site where gas diffusion occurs within the lungs is between the thin-walled sacs known as *alveoli* and the thin walls of capillaries that surround the alveoli.

DIGESTIVE SYSTEM

In addition to oxygen, cells also require nutrients. Nutrients are needed as sources of energy, growth, reproduction, and the production of new materials. Nutrients are obtained from the food we eat, which is broken down, or digested, by the digestive system into particles small enough to enter the bloodstream. The digestive system consists of a series of organs that form a long tube extending from the mouth to the anus and accessory organs that communicate with the tube. The organs are the mouth; the salivary glands, which empty saliva into the mouth to begin carbohydrate digestion; the pharynx; the esophagus; the stomach, where protein digestion begins and food is converted to a liquid; the small intestine, where digestion continues and is completed and where absorption into the bloodstream occurs; the liver and gallbladder, which provide bile for the breakdown of fats in the small intestine; the pancreas, which sends a variety of enzymes important in digestion to the small intestine; and the large intestine, which releases the solid waste material to the exterior.

URINARY SYSTEM

The organs are the kidneys, the ureters, the urinary bladder, and the urethra. The kidneys help maintain homeostasis by regulating the salt-water balance and pH in body fluids, and by removing nitrogen-containing waste materials that result from metabolism. Liquid urine results from these activities, and is transported, temporarily stored, and eliminated by the remaining urinary organs.

REPRODUCTIVE SYSTEM

Primary function is to continue the species, and it is the only system not concerned with homeostasis. Its organs differ between males and females. In males, the organs include the testes, which produce the sex cells, or sperm; the epididymis located with each testis inside the external sac, called the scrotum; the ductus deferens, which extends from each epididymis into the pelvic cavity to unite with the urethra; and the urethra, which passes through the penis to open to the exterior. Accessory organs provide secretions to support the sperm, and include the seminal vesicles, prostate gland, and bulbourethral glands.

In females, the organs include the ovaries, which normally release a single sex cell, or ovum, about once every 28 days; the uterine tubes; the uterus, which is the site of embryo implantation and development and the site of menstruation; and the vagina, which opens to the exterior. The external genitalia are referred to as the vulva.

KEY TERMS

alveoli (p. 124)
capillary (p. 122)
digestion (p. 125)

erythrocyte (p. 122)
gland (p. 114)
hormone (p. 120)

leukocyte (p. 122)
lymph (p. 123)
menstruation (p. 128)

receptor (p. 114)
system (p. 113)
thrombocyte (p. 122)

QUESTIONS FOR REVIEW

OBJECTIVE QUESTIONS

1. A level of structural organization that performs a role affecting a region greater than its immediate location, and is composed of more than one type of tissue, is a(n):
   a. cell
   b. organ
   c. tissue
   d. membrane

2. The integumentary system forms a barrier over the body that helps protect against:
   a. loss of fluids
   b. external temperature changes
   c. physical injury
   d. all of the above

3. The blood vessels, glands, and receptors of the integumentary system may be found:
   a. in the dermis or hypodermis
   b. above (superficial to) the dermis
   c. in the epidermis
   d. on the surface of the skin

4. The system of the body that provides support of soft body parts, mineral storage, and a site for blood cell formation is the:
   a. integumentary system
   b. muscular system
   c. skeletal system
   d. digestive system

5. The organs of the muscular system are the:
   a. bones and joints
   b. muscles
   c. muscle bundles
   d. tendons and ligaments

6. The skeletal system is not involved in the production of movement that is performed by muscle contraction.
   a. True
   b. False

7. The brain and spinal cord are components of the:
   a. peripheral nervous system
   b. central nervous system
   c. endocrine system
   d. autonomic nervous system

8. The nervous system is capable of detecting a change in the environment and responding rapidly to that change by way of:
   a. the release of hormones
   b. the afferent nervous system
   c. an electrochemical impulse
   d. none of the above

9. The nervous system is the only system of the body involved in maintaining homeostasis.
   a. True
   b. False

10. The pituitary gland, thyroid gland, and adrenal glands are organs of the:
    a. digestive system
    b. endocrine system
    c. nervous system
    d. none of these; they are not organs

11. Oxygen and nutrients are transported to cells from other regions of the body by way of the:
    a. respiratory system
    b. lymphatic system
    c. cardiovascular system
    d. digestive system

12. Ventricles of the heart push blood initially through _____, which transport it away from the heart.
    a. atria   c. capillaries
    b. arteries   d. veins

13. Formed elements and plasma are components of the:
    a. extracellular environment   c. blood
    b. interstitial fluid   d. lymph

14. Tiny particles and fluid can diffuse out of the bloodstream into the interstitial fluid across the thin walls of the:
    a. capillaries   c. heart
    b. veins   d. arteries

15. The function of the lymphatic system is:
    a. to supply cells with oxygen
    b. to protect from infection and disease
    c. to supply cells with lymph
    d. to channel fluid through the heart

16. The site of gas exchange between the external air supply and the bloodstream is within the:
    a. lungs   c. heart
    b. trachea   d. throat

17. The process of digestion begins in the:
    a. trachea   c. mouth
    b. small intestine   d. salivary glands

18. The organ that receives secretions from the liver, gallbladder, and pancreas and is the site of nutrient absorption into the bloodstream is the:
    a. stomach   c. small intestine
    b. large intestine   d. esophagus

19. The organ that helps maintain homeostasis by regulating the water-salt balance and pH of body fluids and removes nitrogen-containing waste materials is the:
    a. kidney   c. brain
    b. ureter   d. small intestine

20. The system that is mainly concerned with continuing the species and that does not play a role in homeostasis is the:
    a. excretory system
    b. digestive system
    c. reproductive system
    d. cardiovascular system

21. The ovaries, uterine tubes, and uterus are organs that form a part of the:
    a. male reproductive system   c. excretory system
    b. female reproductive system   d. digestive system

ESSAY QUESTIONS

1. Describe what a failure of the cardiovascular system would do to the ability of the nervous system, endocrine system, and digestive system to function correctly.

2. Explain why the respiratory system is needed for a person to survive. What happens when it fails?

ANSWERS TO ART LEGEND QUESTIONS

Figure 5–1

What tissue makes up the epidermis of the skin? Epithelium makes up the epidermis of the skin.

Figure 5–2

Why is a bone considered an organ? A bone is an organ because it is an organized arrangement of two or more types of tissues that perform a general function.

Figure 5–3

What organs are muscles attached to? Muscles are attached to bones.

Figure 5–4

Of what tissue are the organs of the nervous system primarily composed? Nervous system organs are mainly composed of nerve tissue.

Figure 5–5

What products are secreted by all endocrine glands? Hormones are secreted by all endocrine glands.

Figure 5–6

What are the organs of blood circulation? The organs that circulate blood include the arteries, the veins, and the heart.

Figure 5–7

What fluid is carried by lymphatic vessels? Lymph is carried within lymphatic vessels.

Figure 5–8

What respiratory organs carry air molecules to and from the lungs? The nose, pharynx, larynx, trachea, and bronchi carry air to and from the lungs.

Figure 5–9

Within which digestive organ does nutrient absorption take place? Nutrient absorption takes place within the small intestine.

Figure 5–10

Which urinary organs transport urine from the kidneys to the bladder? The ureters carry urine from the kidneys to the bladder.

Figure 5–11

Do all reproductive organs normally differ between males and females? Yes, all reproductive organs differ between males and females.

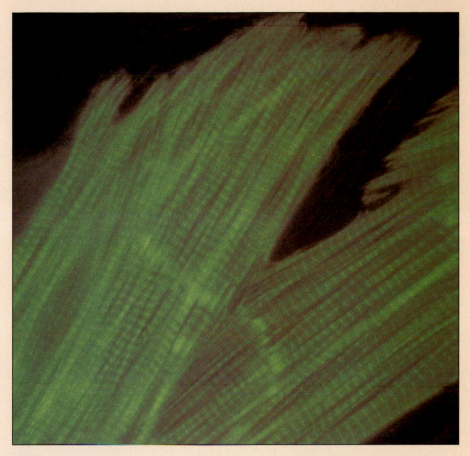

SYSTEMS THAT COVER, SUPPORT, OR MOVE THE BODY

In previous chapters, we saw that the body is organized by levels of building-block components, which include chemicals, cells, tissues, organs, and systems. From this point onward, the most complex level, the systems, will be studied in finer detail. This begins our systematic approach to the study of the human body. The study will include a close inspection of system structure to show how a system's chemicals, cells, tissues, and organs are interwoven to form functional parts; and of system physiology to show the importance of each major component in maintaining our health. Each of the 11 systems is examined separately in the chapters that follow to allow a clear picture to be drawn of its structure and function.

Unit 2 begins our study of body systems by presenting three systems that cover, support, or move the body. The system that covers the body, the integumentary system, opens the unit in Chapter 6. The support system—the skeletal system—is presented in Chapter 7; and the muscular system, which moves the body, follows in Chapter 8.

THE INTEGUMENTARY SYSTEM

LEARNING OBJECTIVES

After studying this chapter, you should be able to:

1. Identify the basic functions of the integumentary system.

2. Distinguish between the two layers of skin on the basis of structure and function.

3. Identify the layers of the epidermis, and describe the changes that occur in the cells as they are pushed toward the surface.

4. Describe the structure and function of the hypodermis.

5. Discuss the role of the integumentary system in regulating body temperature.

The integumentary system covers the body like gift wrap. It has an outer surface area of about 2 square meters (20 square feet) in the average adult, and accounts for about 15% of total body weight. Its major organ is the skin, but it also includes smaller accessory organs that lie within or extend through the skin, such as hair, glands, receptors, and blood vessels. The major functions of this vital system include the following:

1. **Protection**—The physical barrier formed by the skin protects against the loss of body fluids, body damage due to physical injury or ultraviolet light, and invasion by microorganisms.

2. **Regulation of body temperature**—The skin provides an insulating barrier against the external environment and contains sweat glands whose secretions provide evaporative cooling. The skin also contains a blood supply that can be regulated to help cool or warm the body when needed.

3. **Communication**—Sensory receptors located within the skin react to stimuli, such as heat, cold, touch, and pain, and relay this information to the spinal cord and brain.

4. **Excretion of wastes**—Sweat glands in the skin release small amounts of metabolic waste materials through tiny pores onto the skin surface.

5. **Vitamin D production**—Vitamin D is produced by skin cells when they become exposed to ultraviolet light. This vitamin makes possible the absorption of calcium through the digestive tract. Calcium is needed for bone growth and repair, muscle contraction, and all nervous system functions.

THE SKIN

The skin is the primary organ of the integumentary system. It is divided into the epidermis and the dermis.

The skin is one of the larger and more dynamic organs of the body. It is the primary organ of the integumentary system, and plays an important role in each of the functions listed in the opening of this chapter. It consists of two distinct layers (Fig. 6–1): a thin, superficial layer of epithelium called the **epidermis**[1], and a thicker, deep layer of connective tissue known as the **dermis**. Because the skin contains one layer of epithelium and one of connective tissue, it is regarded as a membrane in addition to its status as an organ. As you may recall from Chapter 4, skin is called the *cutaneous membrane*.

Underlying the deeper layer of skin, the dermis, is a region that provides a connection between the skin and the muscle layer of the body. This is the **hypodermis**[2] (also called the *superficial fascia* or *subcutaneous layer*).

Epidermis

The epidermis is a region of stratified squamous epithelium that is in a continual state of replacement. Skin color is a function of melanocytes, which lie within the epidermis.

The epidermis is a dynamic sheet of cells that forms a waterproof, protective wrap over the body's surface. It is in a state of constant change, for it must renew itself continually to replace the millions of cells that are

[1]Epidermis: Greek *epi-* = upon, on top of + *dermis* = skin.

[2]Hypodermis: Greek *hypo-* = beneath + *dermis*.

Figure 6-1
Section of the human skin. The epidermis is lifted from the underlying dermis in one corner of this section to show how these two layers interdigitate.

What accessory organs can you identify in this illustration?

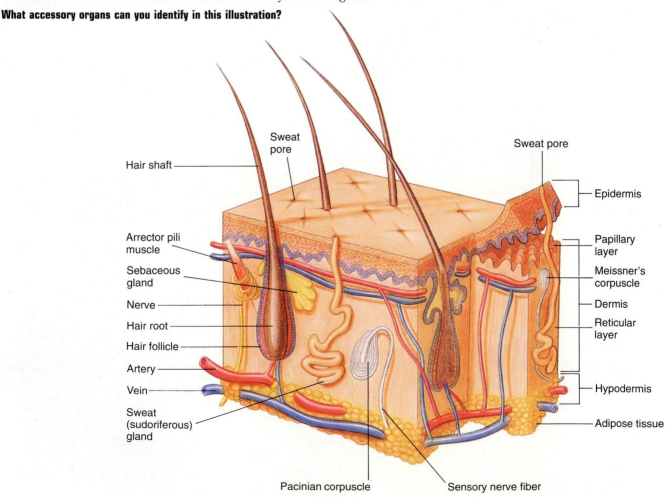

Hair shaft · Sweat pore · Sweat pore · Epidermis · Arrector pili muscle · Papillary layer · Sebaceous gland · Meissner's corpuscle · Nerve · Dermis · Hair root · Reticular layer · Hair follicle · Artery · Vein · Hypodermis · Sweat (sudoriferous) gland · Adipose tissue · Pacinian corpuscle · Sensory nerve fiber

worn away and lost by the activities of everyday life. It is composed of **stratified squamous epithelium**, which typically consists of a basal layer of living, actively dividing columnar cells that push older cells toward the surface as new cells are produced. The older cells gradually flatten and die as they near the skin surface, and their cytoplasmic contents are replaced by a tough, waterproof protein called **keratin** to form an almost impenetrable protective barrier. The layers of the epidermis can be observed on a stained microscope slide of the skin (Fig. 6–2), and are commonly categorized into the following divisions (from deepest layer to most superficial layer):

Stratum basale[3]—the deepest layer of the epidermis. It consists of a single layer of columnar cells capable of continued cell division (by mitosis). The newly produced cells are pushed toward the skin surface as more cells are manufactured. As they approach the surface, they become distanced from the nourishing blood supply located in the dermis and begin to die (recall that there are *no* blood vessels in epithelium!).

---

[3]Stratum basale: Latin *stratum* = layer + *basale* = bottom or base.

Stratum spinosum[4]—a multiple-layered arrangement of cuboidal cells. The cells contain molecular bridges that connect them to adjacent cells. This gives them a "spiny" or "prickly" appearance, from which their name is derived. The nuclei of the cells are often darkened (a condition called *pyknosis*), which is an early sign of cell death. Their fate is sealed because the nutrients and oxygen in interstitial fluid have become exhausted before the fluid is able to reach them by diffusion.

Stratum granulosum[5]—three to five rows of partially flattened cells whose cytoplasm contains small granules. The granules are proteins that are in the process of transforming into the waterproofing protein keratin.

Stratum lucidum[6]—present only in the thick skin of the palms of the hands and soles of the feet. It

---

[4]Stratum spinosum: Latin *stratum* + *spinosum* = spiny.

[5]Stratum granulosum: *stratum* + *granulosum* = containing small grains.

[6]Stratum lucidum: *stratum* + *lucidum* = clear or transparent.

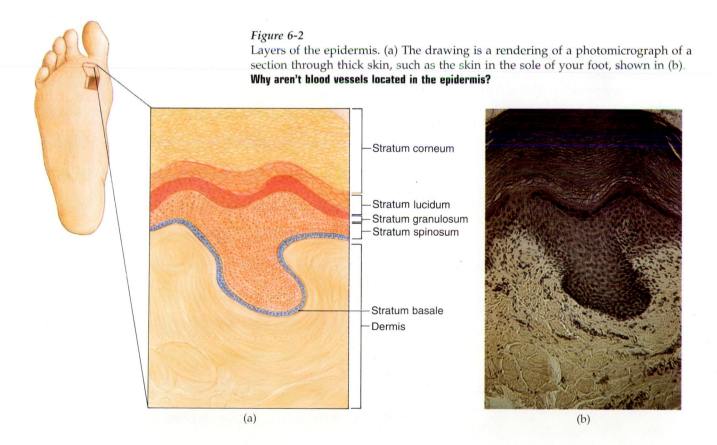

Figure 6-2
Layers of the epidermis. (a) The drawing is a rendering of a photomicrograph of a section through thick skin, such as the skin in the sole of your foot, shown in (b). **Why aren't blood vessels located in the epidermis?**

Stratum corneum

Stratum lucidum
Stratum granulosum
Stratum spinosum

Stratum basale
Dermis

(a) (b)

consists of three to four rows of flattened, dead cells that are mostly transparent. The process of keratin formation continues within these cell "ghosts."

Stratum corneum[7]—the most superficial layer. It consists of 20 to 50 rows of flattened, dead cells. The dead cells are constantly being lost ("sloughed off") by normal wear and tear, and are replaced by deeper layers. The entire process of cell production in the stratum basale, migration to the surface, and sloughing off takes about three weeks to one month. Each cell "ghost" in the stratum corneum contains keratin, which is important in protecting the skin against water loss. Although the level of protection by keratin is important, the barrier it forms is not completely impermeable. For example, after soaking in water, the keratin layer softens, increasing its permeability to water. As a result, water tends to move outward from your skin cells (by osmosis), causing your skin to "wrinkle" temporarily, especially after a long soak such as a hot bath.

Skin Color

The amount of color in the skin is determined by special cells that lie between the dividing cells of the stratum basale. They are called **melanocytes**[8] and secrete a dark-colored protein, or pigment, called **melanin** (Fig. 6–3). The greater the amount of melanin in your skin, the darker your skin color. In other words, the color of your skin is not determined by the number of melanocytes that are present, which is about the same among individuals of similar size, but by the amount of melanin they produce. Melanin production is regulated by the genetic code within the DNA of melanocytes, so skin color for the most part is an inherited trait. To a lesser degree, it is also regulated by a hormone secreted by an endocrine gland at the base of the brain, the pituitary gland.

Despite its careful regulation, melanin production can be modified by exposing the skin to ultraviolet light. Prolonged exposure to this type of light stimulates melanocyte activity, leading to increased melanin secretion in the epidermis and to darker skin. This process is called *tanning*. The function of melanocyte activity is to protect body cells from the harmful effects of ultraviolet light, which tends to break apart the DNA molecule, causing mutations. Protection is made possible by melanin's ability to absorb ultraviolet light,

Figure 6-3
A melanocyte and melanin. Melanocytes are normally found between the cells of the stratum basale, and between the basale and the dermis.

What benefit is provided by melanin in the skin?

thus reducing the likelihood that its harmful rays will damage cells lying deeper in the body. Unfortunately, excessive exposure of the skin to ultraviolet light may overwhelm the protective function of melanin and result in sunburn or, much worse, cause mutations in melanocytes, leading to the most lethal form of skin cancer, *malignant melanoma*.

Skin color may also receive contributions from another pigment, carotene. It is present in the stratum corneum and in the dermis, and its effects are most apparent among people of Asian origin. It gives a yellowish hue to the skin.

The pinkish skin of Caucasian people contains small amounts of melanin and carotene, and is primarily influenced by the presence of blood within blood vessels of the dermis. An increase in blood flow to the skin causes the color to become more pink or even red, and a decrease causes the skin to become a pale shade of pink.

CONCEPTS CHECK

1. What are the five major functions of the integumentary system?

2. Why do cells die as they are pushed farther from the stratum basale?

3. What factors determine the color of your skin?

---

[7]Stratum corneum: *stratum + corneum* = horny.

[8]Melanocytes: Greek *melano-* = dark-colored + *-cyte* = cell.

HEALTH CLINIC

Skin Cancer

The incidence of malignant tumors of the skin is currently on the rise in the United States. In most cases they arise from overexposure to ultraviolet light, underlining the importance of keeping your skin protected when out of doors. Malignant tumors account for about 35% of all newly reported skin cancer cases, and can disrupt body homeostasis to the extent of causing death. There are three frequently encountered types of skin cancers that everyone should be informed about:

Squamous cell carcinoma—a noninvasive cancer that usually arises from flattened cells of the epidermis. It occurs on parts of the bodies of fair-skinned individuals that are repeatedly exposed to sunlight. The lesions develop and grow quickly ($\frac{1}{2}$ cm per week on the average). They appear as small, red, conical, hard nodules that break open soon after forming. Metastasis usually follows soon after the squamous cell carcinoma lesions appear. This cancer is treated by excision (surgical removal) or X-ray irradiation.

Basal cell carcinoma—a cancer that occurs mostly on parts exposed to sunlight. The tumor arises from cells of the basal layer of the epidermis, and grows slowly (1 to 2 cm per year). The lesions often appear as red, waxy nodules in the skin. Although this type of cancer rarely metastasizes, the lesions can grow deeply to invade underlying vital organs if they are neglected. Treatment is by excision.

Malignant melanoma—the most life-threatening cancer that arises from the skin. The original tumor originates in melanocytes located in the basal layer of the epidermis. It may appear on the skin as a diffuse, discolored, tender region, or as a small, discolored nodule. A melanoma often starts as a small, mole-like growth that increases in size, changes color, becomes ulcerated, and bleeds easily from a slight injury. It metastasizes quickly, spreading first to nearby lymph nodes before reaching vital organs. In the United States each year, over 30,000 new cases are reported and about 4000 patients die of this disease. Since 1973, the incidence rate of melanoma has increased steadily about 4% each year. Treatment is effective only if this rapidly developing cancer is detected and treated early. Malignant melanoma is usually treated by widespread excision and X-ray irradiation.

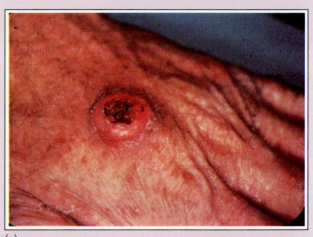
(a)

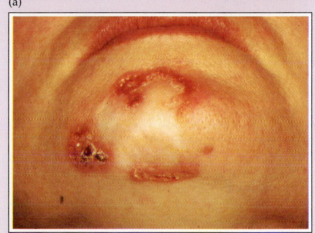
(b)

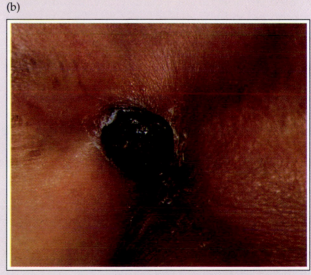
(c)

Skin cancer. (a) Squamous cell carcinoma. (b) Basal cell carcinoma. (c) Melanoma in an advanced stage (sternal region of a female).

Dermis

The dermis is composed of connective tissue, which contains a large supply of blood vessels and houses the accessory organs of the integumentary system.

The dermis is a region of connective tissue that underlies the epidermis. Its cells are not crowded close together like the epithelium of the epidermis, but are scattered far apart. In a healthy individual, the intercellular material of the dermis contains a large amount of collagen that forms a dense mat, giving it the consistency of a firm, wet sponge.

The blood vessels that extend through the dermis supply the stratum basale of the epidermis, as well as its own cells, with nourishment and waste removal. Blood supply to your skin is very important—some have estimated that nearly one-quarter of the total blood volume in the body may be present in the dermis of the skin. Not only does the blood nourish cells of the skin, but it also helps regulate body temperature with changes in blood vessel volume that are controlled by the brain. This important function will be discussed further at the end of this chapter.

The dermis is structurally divided into two areas: a superficial area adjacent to the epidermis, called the **papillary region**,[9] and a deep, thicker area known as the **reticular region**.[10] The papillary region is composed of loose (areolar) connective tissue. It is named for its fingerlike projections, or *papillae*, that extend toward the epidermis (see Fig. 6–1). The papillae provide the dermis with a "bumpy" surface that interdigitates with the epidermis, strengthening the connection between the two layers of skin. In the palms, fingers, soles, and toes, the influence of the papillae projecting into the epidermis forms contours in the skin surface. These are called **friction ridges,** because they help the hand or foot to grasp by increasing friction. Friction ridges occur in patterns that are genetically determined and are therefore unique to the individual, making it possible to use fingerprints or footprints as a means of identification.

The reticular region lies deep to the papillary region and is usually much thicker (see Fig. 6–1). It is composed of dense irregular connective tissue, and receives its name from the dense concentration of collagenous, elastic, and reticular fibers that weave throughout it. These proteins fibers give the dermis its properties of strength, extensibility, and elasticity.

---

[9]Papillary: Latin *papilla* = nipple.

[10]Reticular: Latin *reticulum* = little network.

Skin wrinkles that accompany aging and excessive ultraviolet light exposure are a result of a change in the production or quality of these proteins and a corresponding loss of their properties. Also located within or extending through the reticular region are the numerous accessory organs of the integumentary system, including the roots of hair, sebaceous glands, sweat glands, receptors, nails, and blood vessels.

CONCEPTS CHECK

1. How is the dermis different from the epidermis?

2. What roles do blood vessels play in the dermis?

3. How can you distinguish between the two areas of the dermis?

ACCESSORY ORGANS

Accessory organs are located mostly within the dermis, although they do not originate there. They play a variety of roles, including protection, communication, and excretion.

Although the accessory organs lie embedded within the reticular region of the dermis, they have originated elsewhere during embryonic development. At this early stage in life, groups of cells from the epidermis migrate into the dermis and begin the process that eventually leads to the formation of hair and nails, or sweat glands and sebaceous glands. These accessory organs may therefore be referred to as *epidermal derivatives*. Other accessory organs such as receptors and blood vessels arrive in the dermis from other areas of the body through normal growth processes.

Hair

Hair protects the skin from injury that may be caused by sunlight and foreign particles. It is supported by epithelial cells that form a downward extension of the epidermis into the dermis, called the **hair follicle** (Fig. 6–4). Each hair contains two portions: the part surrounded by the hair follicle is the **root**, and the part that extends away from the body surface is the **shaft**. The hair follicle and root are composed mostly of living cells that obtain nutrition from a cluster of blood vessels at their base, or **bulb**. The growth of hair is similar to replacement of epidermis: its only growing re-

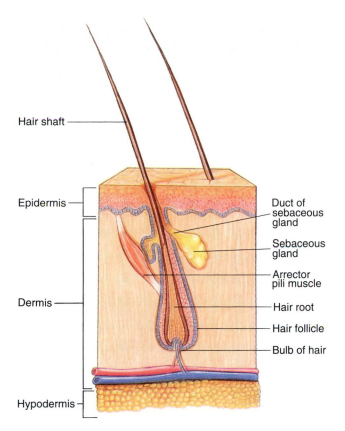

Figure 6-4
The structure of hair. Hair grows upward from the bulb, where it receives its blood supply.

How are the living cells of a hair nourished?

gion is a deep layer of cells that push newly produced cells toward the surface. In hair, these developing cells are located in the bulb and so are adjacent to the blood supply. As cells mature they are pushed toward the surface and eventually die, and their cytoplasm is replaced by the protein keratin. The shaft is thus made up entirely of dead cells composed mainly of keratin.

Hair grows at a normal rate of about 1 mm every 3 days. It will continue to grow as long as the dividing cells in the follicle are healthy. Shaving or cutting the hair has no effect on growth rate, since the portion being cut is the shaft, which consists only of nonliving cells full of protein. Also, the loss of hair during combing or brushing does not mean you are "going bald"—this is a normal process in the daily cycle of hair growth and replacement (about 100 hairs are lost and replaced each day in a normal adult scalp). However, if the dividing follicle cells become diseased as a result of infection, trauma, emotional stress, drugs, ionizing radiation, an inherited condition, or the influence of certain hormones, hair that is normally lost will not be replaced. If this affects scalp hairs, baldness (*alopecia*)

will result. For example, the male hormones, called *androgens*, have been shown to inhibit follicle activity, leading to a break in the cycle of hair growth, loss, and replacement. As a result, hair that is lost is not replaced. This leads to "male pattern baldness," which is genetically determined.

Hair color (yellow, brown, or black) is provided by melanin, which is present in various concentrations to produce different shades. As one increases in age, the amount of melanin produced is reduced, causing the hair to turn to gray or even white. An increased accumulation of air within the hair shaft with age also promotes the loss of color.

Associated with hair follicles are two important structures. A small, narrow band of smooth muscle that extends at an angle from the follicle to the papillary region of the dermis is called the **arrector pili**[11] (a-REK-tor PĪ-lē). The contraction of this muscle pulls the hair to a more vertical position when you are cold or frightened. It produces a small amount of heat from the muscle, increases the insulating effect of hair (which, as you might guess, is more important among fur-bearing mammals), and presses against the sebaceous gland, forcing oil out onto the skin surface. The contraction of many such muscles produces the "goose bumps" you see on your skin when cold. The sebaceous gland is a second structure associated with each hair follicle, and is described in the next few paragraphs.

Sebaceous Glands

Sebaceous glands,[12] or oil glands, are nearly always associated with hair. Each gland consists of a cluster of glandular epithelial cells that are usually connected to a nearby hair follicle by way of a short duct (see Fig. 6-4). Sebaceous glands are distributed throughout the skin except in the palms and soles, and they secrete an oily substance called **sebum**.

Sebum consists of water, fats, cholesterol, protein, and salt. It is secreted by the sebaceous gland into a short duct that empties into a small space between the hair follicle and hair root, or sometimes directly onto the skin surface (such as in the eyelids). It serves to keep hair and skin soft and pliable and provides a water-resistant layer to the skin surface. Its production and secretion are accelerated by sex hormones.

Occasionally, the ducts that carry sebum may become plugged by large amounts of sebum or dead

---

[11]Arrector pili: Latin *arrector* = something that raises, or causes to stand + *pilus* = hair.

[12]Sebaceous: Latin *sebum* = grease + *-aceous* = pertaining to.

cells, especially during puberty, when sebaceous gland activity is accelerated by sex hormones. A plugged duct or swollen gland is called a *blackhead*; it may lead to the formation of a pimple or boil if bacteria become established. Frequent cleansing of the skin helps to minimize growth of bacteria.

Sweat Glands

Sweat glands, which are also known as **sudoriferous glands**,[13] are widely distributed throughout the skin in great numbers. They secrete a watery substance simply called *sweat* or *perspiration*, which consists of water, salts, and small amounts of the metabolic waste material urea. The secretion of sweat helps maintain body temperature by cooling the body off as it evaporates and, to a lesser degree, aids the kidneys in the elimination of metabolic wastes. Each gland originates as a single tube that is tightly coiled into a ball in the dermis or hypodermis (see Fig. 6–1). As sweat is secreted, it passes from the coil into a winding duct that opens onto the skin surface by way of a **pore**.

There are two types of sweat glands, **eccrine** and **apocrine**. Eccrine glands function throughout life and are widely distributed throughout the skin. They secrete a watery sweat in response to elevated body temperatures. Apocrine glands begin functioning during puberty in response to the production of sex hormones. They secrete a thickened sweat that contains proteins, which promote the growth of microorganisms normally found on the skin. Since the microorganisms produce odoriferous materials, sweat from apocrine glands may have a strong odor. Apocrine glands are active primarily during periods of emotional stress, and are more numerous in the armpits and groin.

Nails

Nails are formed from the compressed outer layer of the epidermis. They are thus composed of keratin, and serve to protect the ends of fingers and toes from injury, as well as to help us pick up small objects and grip the floor as we walk. The visible surface of the nail is the **nail body** (Fig. 6–5). The **cuticle**, or **eponychium** (ep´-ō-NIK-ē-um), is a flap of stratum corneum that overlies the proximal edge of the nail. Deep to the cuticle is the **nail root**, which contains an active stratum basale known as the **nail matrix**. The nail matrix pushes new cells outward as they are produced. The new cells soon die and become compressed to form new nail material. A small part of the nail matrix can be seen through the nail body as a light-colored crescent, called the **lunula** (LOO-nyoo-la). Because the formation of keratin requires calcium, a deficiency of this important mineral in your diet is often expressed through the development of thin, brittle nails.

---

[13]Sudoriferous: Latin *sudor* = sweat + *ferre* = to bear.

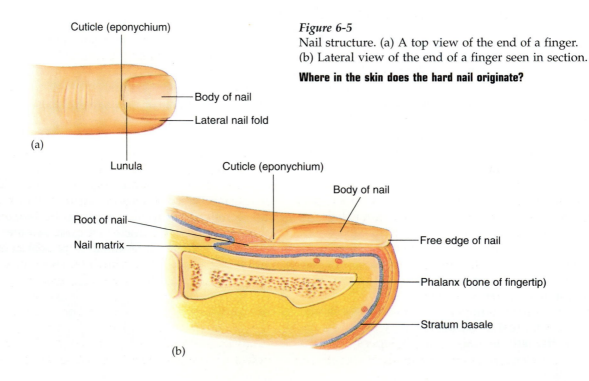

(a)

Figure 6-5
Nail structure. (a) A top view of the end of a finger. (b) Lateral view of the end of a finger seen in section.

Where in the skin does the hard nail originate?

(b)

Receptors

In the skin, **receptors** usually consist of the distal ends of nerve cells that are wrapped in a capsule of connective tissue. These nerve cells carry impulses toward the brain and are found throughout the body, in addition to the millions present in the skin. Receptors are similar to antennae, for they receive information about the world. This information is sent over nerves to the brain at incredible speeds (about 40 meters per second!), where it is interpreted as *sensations* such as cold or heat, pressure, fine touch, and pain. Sensations tell the body about changing conditions so that the brain can make the necessary changes in order to maintain homeostasis. The skin thus functions as an important sense organ. The various types of receptors found in the skin include Pacinian corpuscles, which respond to pressure changes, Meissner's corpuscles, which respond to slight changes in pressure to detect fine touch, and a variety of nerve endings that respond to excessive temperature and pressure changes for the detection of pain (see Fig. 6–1).

CONCEPTS CHECK

1. Why do you lose a hair every time a follicle dies?

2. What product does a sebaceous gland secrete, and what is its function?

3. How may the two types of sweat glands be distinguished?

4. What general function is performed by skin receptors?

HYPODERMIS

The hypodermis lies deep to the skin. It is composed of loose connective tissue and adipose tissue, and provides an insulating barrier for deeper structures.

The hypodermis is not usually considered a component of the skin organ, so it is described separately here. It is a region deep to the dermis that connects the skin to underlying body structures. The connection is by way of strands of collagen that extend from the dermis through the hypodermis to the muscle layer.

The hypodermis is composed of **adipose tissue** and **loose connective tissue**, and varies greatly in thickness over different areas of the body. It insulates deep tissues from extremes of hot or cold, and provides a shock-absorbing cushion as well as a reserve for energy storage.

HOMEOSTASIS: TEMPERATURE REGULATION

The integumentary system plays a key role in the homeostatic mechanism that regulates body temperature.

Our internal body temperature is kept within a narrow range, averaging about $37 \pm 1°C$ ($98.6 \pm 1.5°F$) during good health. It is very important for this range to be maintained for body functions to continue, since the enzymes that perform work such as moving molecules across plasma membranes, transporting substances, and catalyzing chemical reactions are functional only within this narrow range. If the temperature drops too low, enzyme activity slows and may stop altogether; if the temperature rises too high, enzymes break apart, or **denature.** Either event results in a breakdown of body functions. Thus, body temperature must be maintained within this narrow range for health to continue. Regulation of temperature is an important part of the overall homeostasis of the body.

However, keeping the body's temperature stable is not a simple task; it requires a balancing act between the amount of heat produced internally by cellular activities (heat is a by-product of metabolism) and the amount of heat gained from or lost to the external environment. This is where the integumentary system enters the scene. As we shall see, our skin plays a major role in regulating body temperature to within the narrow limits required.

Suppose you are walking through the desert during a sunny afternoon when the air temperature is $40°C$ (about $105°F$). A sequence of homeostatic events takes place in an effort to keep your body temperature below that of the air (Fig. 6–6 on page 146). Heat-sensitive receptors in the skin sense the external temperature and relay this information quickly to the brain. The brain immediately stimulates sweat glands to increase secretion, and beads of sweat form on your body. The brain also signals blood vessels in the dermis to dilate, or relax and expand, increasing blood flow to the skin. Excessive heat carried by the blood escapes through the skin to the air. As the sweat evaporates, your skin surface is cooled, and its lowered tem-

HEALTH CLINIC

Skin Repair

The skin provides considerable protection from mechanical injury, chemical hazards, and bacterial invasion because the epidermis is relatively thick and covered with keratin. Secretions from sebaceous glands and sweat glands also benefit this protective barrier. In the event of an injury that damages the skin's protective barrier, the body triggers a response called **inflammation,** which sends fluids carrying phagocytic white blood cells to the injury site. Once the invading microorganisms have been brought under control, the skin proceeds to heal itself. The remarkable ability of the skin to heal even after considerable damage has occurred is largely due to the presence of stem cells in the dermis and cells in the stratum basale of the epidermis, all of which can generate new tissue.

When an injury extends through the epidermis into the dermis, bleeding occurs and the inflammatory response begins (see the illustration). Clotting mechanisms in the blood are soon activated, and a clot, or **scab,** forms within several hours. The scab temporarily restores the integrity of the epidermis and restricts the entry of microorganisms. Soon after the scab is formed, cells of the stratum basale begin to divide by mitosis and migrate to the edges of the scab. About one week after the injury, the edges of the wound are pulled together by contraction. Although

Steps in wound healing. (a) Initial bleeding due to damaged blood vessels in the dermis. Inflammation (swelling, redness, and pain) of the tissue surrounding the wound soon follows, which brings phagocytic white blood cells to the injury site. (b) Establishment of the scab that has resulted from blood clot formation. Epithelial cells migrate around the scab and proliferate. (c) Continued proliferation of epithelial cells, and production of new intercellular material by fibroblasts in the dermis. (d) Replacement of the scab by scar tissue.

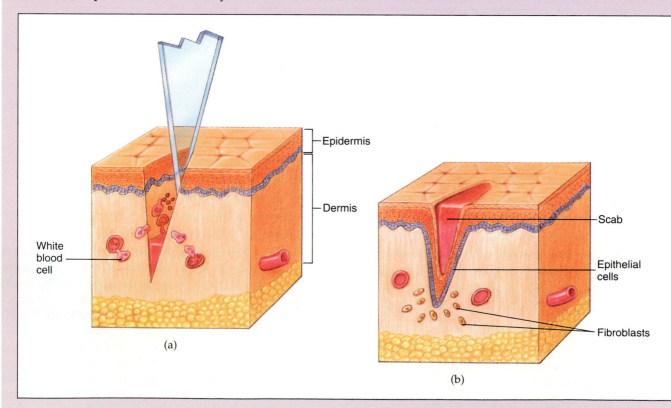

the mechanism of contraction is uncertain, it is an important part of the healing process when damage has been extensive. In a major injury, if epithelial cell migration and tissue contraction cannot cover the wound, suturing the edges of the injured skin together, or even replacement of lost skin with skin grafts, may be required to restore the skin.

As epithelial cells continue to migrate around the scab, the dermis is repaired by the activity of stem cells. These active cells produce collagenous fibers and ground substance. Blood vessels soon grow into the dermis, restor-ing circulation. If the injury is very minor, the epithelial cells eventually restore the epidermis once the dermis has been regenerated.

In injuries that are not minor, the repair mechanisms are unable to restore the skin to its original condition. The repaired region contains an abnormally large number of collagenous fibers, and relatively few blood vessels. Damaged sweat and sebaceous glands, hair follicles, muscle cells, and nerves are seldom repaired. They are usually replaced by the fibrous tissue. The result is the formation of an inflexible, fibrous **scar tissue**.

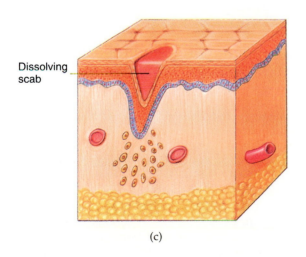

Dissolving scab

(c)

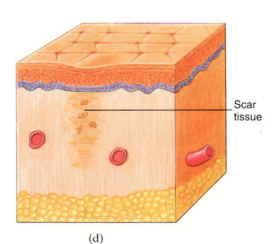

Scar tissue

(d)

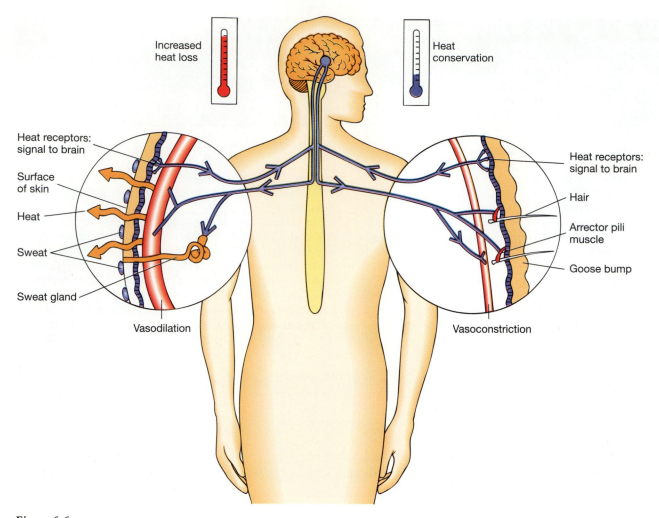

Figure 6-6
The role of the skin in temperature regulation. To increase heat loss during hot days or heavy exercise, blood vessels in the dermis undergo dilation (vasodilation), allowing more blood near the surface, where heat can escape from the body. Sweat glands are activated to secrete, cooling the body surface. The skin conserves heat by constriction of the blood vessels (vasoconstriction) in the dermis, which reduces blood flow and heat loss. These responses are controlled by the brain, which monitors temperature by heat receptors in the skin.

Why is the regulation of body temperature important to the body's homeostasis?

perature helps to further cool the blood. The cooled blood travels to other body parts, and your internal temperature lowers in response.

When the body is exposed to extremely cold temperatures, other responses occur in the skin to stabilize internal temperature. For example, the brain stimulates blood vessels in the skin to constrict, reducing the volume of heat-carrying blood flow and thereby reducing the amount of heat lost through the skin. The

brain may also stimulate skeletal muscles throughout the body to contract slightly (causing shivering) and the arrectores pilorum in the skin to contract (causing "goose bumps"). These contractions increase muscle metabolism and generate greater quantities of heat as a by-product, and your internal temperature rises in response.

Other systems of the body are also involved in temperature regulation. The cardiovascular, respiratory, and nervous systems play important roles. For example, during an increase in body temperature the heart beats faster, pushing more blood from the deeper areas of the body to the skin. Breathing rate also increases, which carries out more heat as we expire. And, as we have seen for the skin in the examples in this section, during any change in internal temperature the brain is the vital organ that interprets the change and orchestrates the response.

CONCEPTS CHECK

1. What are the functions of the hypodermis?

2. Why must internal body temperature be maintained within a limited range?

3. What are the roles performed by receptors, sweat glands, and blood vessels in the integumentary system during overheating?

CLINICAL TERMS AND DISEASES OF THE INTEGUMENTARY SYSTEM

Any disease of the skin affects its ability to perform one or more of its functions. A serious disease can upset homeostasis and thus have profound effects on the body's overall health. In fact, a loss of about one-third of the skin (as occurs in serious burns) will usually lead to death if medical treatment is not readily available. Most diseases of the skin are caused by microorganisms that are normally present on the skin in great numbers but do not become pathogenic (disease-causing) unless the epidermis is penetrated and the immune response impaired. For example, population densities of bacteria on the skin may be as high as 3 million per square centimeter in a healthy individual! Some of the more common disorders of the skin are as follows:

Acne Vulgaris A common ailment whose symptoms include edema (fluid leakage into body spaces), which forms skin eruptions; or pimples, which usually occur on the face, back, and shoulders. The development of pimples is the result of excessive production of sebum and its breakdown by bacteria into waste products that cause irritation. Since sebum production is accelerated by sex hormones, acne is most common during puberty, when sex hormone production is high. Treatment includes frequent cleansing, antibiotics, a chemical called *benzyl benzoate*, and exposure to ultraviolet light. If acne lesions cause scarring, a procedure known as *dermabrasion* may be used to improve the skin's appearance. In this procedure the skin is frozen and anesthetized (numbed), then carefully smoothed with fine sandpaper or motor-driven abrasive brushes.

Boils and Carbuncles A boil, or furuncle, forms in the skin when a hair follicle develops a deep-seated bacterial infection, called an **abscess**. The causative organism is *Staphylococcus aureus*, so a boil is a type of "staph infection." A carbuncle is several boils developing from adjacent follicles and joining to form a single mass with numerous drainage points. If a boil or carbuncle is not treated, it may spread to other tissues and can have serious consequences. Treatment is usually by the administration of antibiotics.

Burns Regarded as a type of metabolic disease, since their effect is to traumatize the metabolic balance that cells must maintain in order to survive. Burns may be caused by any form of heat-producing energy, including fire, chemicals, radiation, and electricity. The specific effect of heat on body tissue is the denaturing of proteins. Depending on the severity of the burn, protein destruction can lead to cell death and loss of body fluids and electrolytes. The loss of fluids and electrolytes can be life-threatening, for it can lead to a critical condition of the body known as *shock*. Shock generally occurs whenever significant amounts of body fluids are lost, and kidney and heart failure usually follow if the fluid and electrolyte balance is not restored quickly. Burns of over 50% of the body surface are often fatal, especially in infants and people over the age of 60, who are more sensitive to their harmful effects.

Burns are classified according to the depth of tissue damage. The least severe is a *first-degree burn*, in which tissue damage is limited to the superficial layers of the epidermis. A typical first-degree burn causes redness; a mild sunburn is a common example. A *second-degree burn* usually causes redness and blistering, as the epidermis and dermis are both damaged. However, the damage is not severe enough to prevent the skin from healing itself quickly. The most severe burn is the *third-degree burn*, which completely destroys the epidermis, dermis, and accessory organs within a region of skin. Usually, the edges

(continued)

of a third-degree burn cannot regenerate sufficient new tissue, so the surgical implantation of skin taken from other areas of the body or from skin culturing is necessary to restore the damaged tissues (a procedure known as *grafting*).

Dermatitis   A general term to describe an inflammation of the dermis. Inflammation produces symptoms of redness (erythema) and swelling (edema) in the affected area of skin, and is sometimes followed by excessive scaling of the epidermis (called *desquamation*), the formation of vesicles (fluid-filled bumps), and secondary infection by bacteria. Dermatitis may be caused by contact with an irritating substance (contact dermatitis), or may have a more complex origin relating to a congenital condition (atopic dermatitis or eczema, and sebhorrheic dermatitis). In all cases the symptoms can be lessened by treating the affected skin with ointment containing corticosteroid (in the form of hydrocortisone), which is a hormone normally produced by the adrenal glands that reduces the symptoms of inflammation.

Herpes Simplex   An immunological disease that is caused by a virus. It can be spread to others by direct contact with the lesion; in other words, it is *contagious*. The primary symptom is a recurrent outbreak of vesicles on the skin and mucous membranes of the mouth or genitals, causing burning and stinging sensations. The virus that attacks the mouth region is called herpes simplex type 1, or oral herpes, and the lesions are better known as cold sores. The virus attacking the genital region is herpes simplex type 2, or genital herpes, and is a type of venereal disease; that is, it is spread by sexual intercourse or genital contact. The outbreaks caused by either virus often follow periods of body stress resulting from fever, sunburn, indigestion, fatigue, menstruation, or nervous tension. Scientists have shown that the herpes virus appears to incubate within the nuclei of nerve cells. When the ability of the immune system to respond is lowered, the virus travels along nerves to mucous membranes, where it develops quickly, causing lesions and pain. Treatment is currently limited to the administration of *acyclovir* onto lesions and the use of symptomatic pain relievers, although some evidence suggests the amino acid isoleucine may be effective in reducing outbreaks for some individuals.

Herpes Zoster   More commonly known as *shingles*. It is an acute immunological disease caused by a virus that is very similar to the one that causes herpes simplex. Recently, scientists have found that it is a form of the contagious childhood disease chickenpox (varicella), although it is most common among individuals over the age of 50 years. Its primary symptom is the formation of burning vesicles that erupt through the skin over nerve tracts, usually along the trunk of the body. Treatment is limited to symptomatic pain relievers, and serious sufferers are often hospitalized for continuous treatment of lesions. Symptoms usually diminish in about two weeks.

Pediculosis   An immunological disease that is caused by infestation of parasites on the scalp, trunk, or groin area. The parasites are small, bloodsucking animals called *lice*, which are related to insects, spiders, and mites. Pediculosis occurs among people who live in crowded conditions, but it can infect anyone, since contact with infested toilet seats can spread pubic lice (also known as *crabs*) and contact with infested hairbrushes or combs can spread head lice. Its primary symptom is itching of the affected area, and in most cases, the lice or their egg sacs (known as *nits*) can be viewed as white specks. It is commonly treated successfully with lotions or dips containing an insecticide that kills the lice.

Psoriasis   A congenital disease of the skin that afflicts about 3% to 4% of the world population. It is characterized by the presence of small, round skin elevations that are covered with flaky skin. This symptom is the result of an accelerated rate of mitosis in epidermal cells that is influenced by hormonelike substances in the blood called *prostaglandins*. An individual who suffers from this common inherited disease often experiences periodic "flares" that are associated with hormones, poor nutrition, and emotional stress. The flares may last weeks, months, or even years, with lifelong recurrences. Treatment is through administration of corticosteroid ointments, proper diet, and stress management.

Ringworm   An infection of the skin by a fungus. The disease is also known as *tinea*, and its symptoms include ringed patches of redness or vesicles with a clearing center. It causes minor itching, and if it occurs in the scalp the lesions cause balding. When tinea affects the groin area it is commonly called *jock itch*, and when it affects the feet it is called *athlete's foot*. Treatment of this acute disease is through a variety of drugs that have proven toxic to the fungus.

Tumors A disease resulting in the formation of a group of cells that play no functional role in the body. In most cases involving the skin they arise from overexposure to ultraviolet light from the sun, which induces mutations in normal cells, transforming them into nonfunctional cells. Tumors that involve the skin may be *benign*, or slow-growing and noninvasive; or *malignant*—fast-growing, invasive, and able to spread to other sites (metastasize). Benign tumors include freckles, which are small areas of excess melanin in the basal layer of the epidermis, and moles (also called *nevi*), which are elevated regions of the epidermis. The cells that form moles, called *nevus cells*, may or may not be pigmented with melanin.

Moles must be watched carefully, for if they begin to increase in size, darken, or bleed, they may be in the process of transforming into a malignant form and should be checked immediately by a physician. Malignant tumors of the skin include squamous cell carcinoma, basal cell carcinoma, and malignant melanoma. These forms of cancer were discussed previously in this chapter.

Warts An elevation of the skin caused by a virus that has penetrated the epidermis. There are at least seven different types of warts, all of which are contagious by direct physical contact. They normally persist for 2 to 18 months.

CHAPTER SUMMARY

The major organ of the integumentary system is the skin. The system also includes smaller accessory organs such as the hair, glands, receptors, and blood vessels. The skin's major functions are protection, regulation of body temperature, communication through sensory perception, excretion of metabolic wastes, and production of vitamin D.

THE SKIN

The skin plays a major role in the functions of the integumentary system. It comprises two layers, the epidermis and the dermis.

Epidermis The superficial layer of the skin, composed of stratified squamous epithelium that is in a state of continuous cell replacement. The deepest layer, the stratum basale, contains cells that divide, pushing the older cells toward the skin surface. The next layers in sequence include the stratum spinosum, stratum granulosum, stratum lucidum, and stratum corneum. As the cells approach the uppermost layer, the corneum, they die and their cytoplasm is replaced by the waterproofing protein, keratin.

Skin Color Melanocytes are cells located between the dividing cells of the stratum basale that secrete a dark protein pigment called *melanin*. The more melanin produced, the darker the skin. The rate of melanin production is an inherited trait, but it can be increased by exposure to ultraviolet light to cause tanning. The function of melanin is protection of the skin from the harmful effects of ultraviolet light.

Dermis The deep region of the skin beneath the epidermis, composed of dense irregular connective tissue, which contains a large blood supply. The dermis is divided into a superficial papillary region and a deep reticular region. The accessory organs lie embedded within the dermis.

ACCESSORY ORGANS

Hair, nails, and glands originate in the epidermis. Receptors originate in nerve tissue.

Hair Hair is produced by epithelial cells that form the hair follicle. The root is the portion of the hair that is submerged in the dermis and surrounded by the follicle, and the shaft extends away from the body surface. Hair growth is similar to growth of the epidermis, in that new cells are continually produced near the base and are pushed upward, where they eventually die and become filled with keratin (in the shaft). Associated with hair follicles are small muscles, the arrector pili, and sebaceous glands.

Sebaceous Glands Sebaceous glands secrete an oily substance called *sebum* into the space between the hair follicle and the root or sometimes directly onto the skin surface. Sebum keeps hair and skin soft and pliable, and provides a waterproof layer.

Sweat Glands Also known as sudoriferous glands. These secrete sweat to help maintain body temperature and to help eliminate metabolic wastes. Each gland is a coiled ball that releases sweat through a long duct and out the pore. There are two types of sweat glands, eccrine and apocrine. Eccrine glands release a watery sweat to lower body temperature, and apocrine glands release a more thickened sweat once puberty is reached.

Nails   Nails are formed from the outer layer of the epidermis and are, therefore, composed of keratin. The cuticle lies over the proximal edge of the nail, and the nail root lies deep to the cuticle and produces new cells.

Receptors   The distal ends of nerve cells that carry impulses toward the brain. Receptors are found in all regions of the skin, in the dermis. They sense changes in the external environment and relay this information to the brain for processing.

HYPODERMIS

A region of tissue between the dermis and the muscle layer of the body. It is composed of loose connective tissue and adipose tissue. The hypodermis insulates, provides a shock-absorbing cushion, and serves as a reserve for energy storage.

HOMEOSTASIS: TEMPERATURE REGULATION

An important function of the integumentary system is the regulation of body temperature within the narrow limits defined by enzyme activity. In this way, the integumentary system helps maintain homeostasis. Mechanisms for regulation include sensing temperature changes by receptors in the skin, sweating and dilation of blood vessels to cool the blood and reduce body temperature, and constriction of blood vessels and muscle contraction to warm the blood and increase body temperature.

KEY TERMS

apocrine (p. 142)	epidermis (p. 136)	pigment (p. 138)	sebaceous (p. 141)
denature (p. 143)	hypodermis (p. 136)	pyknosis (p. 137)	sebum (p. 141)
dermis (p. 136)	melanoma (p. 138)	receptor (p. 143)	sudoriferous (p. 142)
eccrine (p. 142)	papillae (p. 140)		

QUESTIONS FOR REVIEW

OBJECTIVE QUESTIONS

1. Which of the following is an accessory organ of the skin?
   a. stratum corneum
   b. collagen
   c. hair
   d. arrector pili
2. The portion of the skin that must renew itself to replace lost cells is the:
   a. hypodermis
   b. epidermis
   c. dermis
   d. sebaceous gland
3. The layer of the epidermis that consists only of dead cells that have been replaced by keratin is the:
   a. stratum basale
   b. stratum spinosum
   c. stratum granulosum
   d. stratum corneum
4. Melanocytes protect the skin from ultraviolet light damage by producing:
   a. epithelial cells
   b. melanin
   c. keratin
   d. mucus
5. A cut through the skin of your finger does not produce bleeding. This implies that the cut did not penetrate the:
   a. dermis
   b. stratum basale
   c. epidermis
   d. stratum corneum
6. The dermis is composed of cells that lie interspersed within a dense intercellular material of protein fibers. This type of tissue is called:
   a. dense irregular connective tissue
   b. smooth muscle tissue
   c. stratified squamous epithelium
   d. cartilage
7. To permanently prevent a hair from growing, one must:
   a. cut the hair at the shaft
   b. destroy the hair follicle
   c. cut the hair at the skin surface
   d. remove the keratin
8. Skin and hair are kept from becoming brittle and dry by the secretion of:
   a. sebum
   b. perspiration
   c. sweat
   d. follicle cells
9. The integumentary system aids the regulation of body temperature by:
   a. sweating when overheated
   b. constriction of blood vessels in the dermis when cold
   c. sensing changes in air temperature
   d. all of the above
10. A tumor that arises from the melanocytes of the skin and spreads rapidly is:
    a. a freckle
    b. malignant melanoma
    c. basal cell carcinoma
    d. a mole

ESSAY QUESTIONS

1. Describe the normal process of epidermal replacement, indicating the changes that take place in the cells as they are pushed toward the surface. Also indicate why continual replacement is necessary to maintain health.
2. Explain why hair, sebaceous glands, sweat glands, and receptors are classified as accessory *organs* and not just accessory *structures* to the skin.
3. Discuss the importance of the large blood supply that is carried by the dermis. Include in your discussion the cells it nourishes and the ways in which it helps maintain homeostasis.

ANSWERS TO ART LEGEND QUESTIONS

Figure 6–1
What accessory organs can you identify in this illustration? Hair, sebaceous glands, sweat glands, and receptors are included in this illustration.

Figure 6–2
Why aren't blood vessels located in the epidermis? Blood vessels are unable to squeeze into the tight arrangement of cells within the epidermis.

Figure 6–3
What benefit is provided by melanin in the skin? Melanin absorbs ultraviolet light, thus reducing damage the rays might otherwise cause to deeper cells.

Figure 6–4
How are the living cells of a hair nourished? The living cells of a hair receive nourishment from the blood supply located at the base of the bulb.

Figure 6–5
Where in the skin does the hard nail originate? The hard nail originates from the nail root, which lies deep to the cuticle.

Figure 6–6
Why is the regulation of body temperature important to the body's homeostasis? Body temperature must be maintained in order to provide an optimum environment for all living cells. Any extreme change in body temperature can lead to the damage of important enzymes and structural proteins, which would, in turn, damage cells. If large numbers of cells are affected, the body's overall homeostasis becomes threatened.

THE SKELETAL SYSTEM

LEARNING OBJECTIVES

After studying this chapter, you should be able to :

1. Identify and describe the five functions of the skeletal system.

2. Distinguish between long bones, short bones, flat bones, and irregular bones, and provide examples of each.

3. Identify the parts of a typical long bone.

4. Describe the inorganic and organic components of bone tissue, and distinguish between the three types of bone cells on the basis of their functions.

5. Describe the microscopic structure of compact bone, and compare it with that of spongy bone.

6. Explain the process by which intramembranous bones are formed.

7. Describe how endochondral bones are formed from cartilage.

8. Describe the process of bone growth in length and width.

9. Define the role of bone remodeling in maintaining homeostasis.

10. Identify the 206 bones of the skeleton and their major features.

11. Distinguish between the three types of joints in the body and provide examples of each.

12. Describe the structure of a synovial joint and indicate the types of synovial joints.

13. Define the movements that are possible at synovial joints.

The bones and joints that form the skeletal system are composed mainly of two types of supportive connective tissue: bone tissue and cartilage. Dense connective tissue, blood-forming tissue, and nerve tissue also combine to provide a strong but flexible, stable yet dynamic, basis for body structural support. Although it is common to think of the skeleton as providing the body with support only, its functions go far beyond that. The more important ones include the following:

Support—The function for which the skeleton is well known. Its strong, rigid nature enables it to serve as a structural frame that supports other body structures.

Protection—Some bones physically surround internal body organs, such as the cranial bones around the brain and the rib cage around the heart and lungs. The hard quality of bone provides a partial shield against damage to these and other organs.

Aid in movement—Bones provide a place of attachment for skeletal muscles, enabling coordinated movement to occur. The rigid nature of bones enables them to serve as levers for attached muscles to pull against during contraction.

Blood cell formation—Blood cells are manufactured by a blood-forming connective tissue called *red marrow* that resides within bone tissue. This manufacturing process is known as *hematopoiesis*.

Storage—Bone tissue is the storehouse and main reserve for two important minerals, calcium and phosphate, which are needed for muscle contraction, nerve cell function, and movement of materials across plasma membranes.

BONE STRUCTURE

Bones come in a variety of shapes, but share a number of common features. They are composed mainly of bone tissue, which is a mixture of organic and inorganic materials that are highly organized to form either compact bone or spongy bone.

One of the common misconceptions about the human body held by many people is their perception of bones. Bones are often regarded as rigid, inflexible, lifeless structures that break like dry wooden sticks if bent too far. On the contrary, bones are complex organs that are tough but flexible in a healthy individual and are in a state of constant change. In this section we will examine the structural nature of this most prominent organ of the skeletal system.

Types of Bone

When bones are categorized on the basis of their general shape, we find they can be grouped into four basic types. These types include long bones, short bones, flat bones, and irregular bones (Fig. 7–1). **Long bones** are greater in length than in width. This structure is designed to absorb stress from body weight. Long bones are found in the arms, forearms, hands, thighs, legs, and feet. **Short bones** are about equal in length and width, forming a shape that is roughly similar to a cube. Examples of short bones are in the wrists and ankles. **Flat bones** are thin and flat, as in the cranium of the head, the ribs, and the sternum. **Irregular bones** have complex shapes that do not fit into any of the previous three categories. They include the vertebrae and bones of the face.

Figure 7–1
Types of bones. (a) Long bones, such as the humerus of the arm, are greater in length than in width. (b) Short bones, such as the carpals of the wrist, are about the same in length and width. (c) Flat bones of the skull and elsewhere have broad, flat surfaces. (d) Irregular bones, such as a vertebra, have complex shapes.

How does a long bone differ from a flat bone?

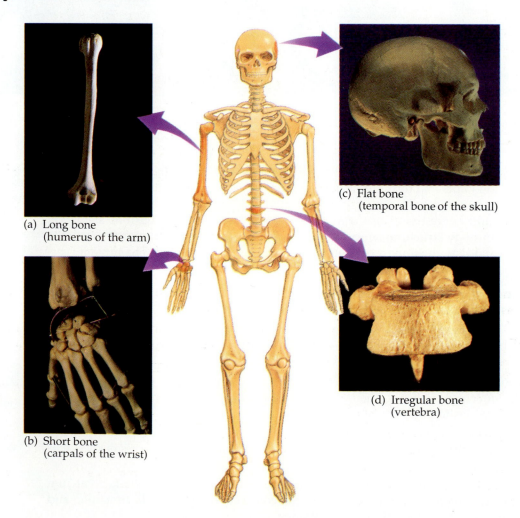

(a) Long bone
(humerus of the arm)

(b) Short bone
(carpals of the wrist)

(c) Flat bone
(temporal bone of the skull)

(d) Irregular bone
(vertebra)

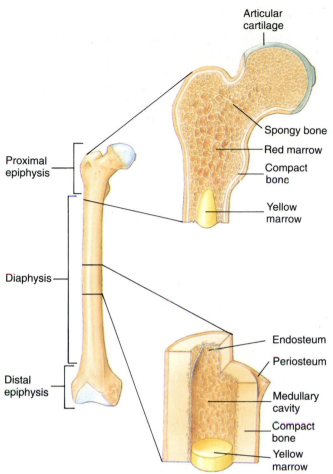

Articular cartilage

Spongy bone

Red marrow

Compact bone

Yellow marrow

Proximal epiphysis

Diaphysis

Distal epiphysis

Endosteum

Periosteum

Medullary cavity

Compact bone

Yellow marrow

Figure 7–2
Parts of a long bone. This is a drawing of a femur, or thigh bone, with its proximal end (the proximal epiphysis) and a part of its middle region (the diaphysis) sectioned along the frontal plane.

What functions are served by the periosteum?

Parts of a Long Bone

When describing the areas of a particular bone that can be viewed without the aid of a microscope, it is convenient to use a long bone as an example. This is shown in Figure 7–2. The long central shaft is known as the **diaphysis**[1] (dī-A-fi-sis). It lies between the extreme ends of the bone, which are called **epiphyses**[2] (e-PIF-i-sēz). Each epiphysis forms a joint, or *articulation*, with another bone. Along its outer surface where the joint occurs is a thin layer of hyaline cartilage, called the **articular cartilage**.

A sheet of dense connective tissue envelops the bone, except in areas where the articular cartilage is present. This important tissue is called the **periosteum**[3] (par'-ē-OS-tē-um). It is firmly attached to the bone's outer surface by special protein fibers, and contains a

large supply of blood vessels. It also contains a type of bone cell, the *osteoblast*, that is active in bone growth and repair. The functions of the periosteum include bone nourishment, attachment to ligaments and tendons, and bone growth and repair.

Sectioning through the long bone allows you to view its internal features, which are also shown in Figure 7–2. Upon first glance, you may notice there is a distinct difference between the epiphyses and diaphysis internally. This is due to the arrangement of two types of bone tissue within these two areas: tightly packed **compact bone** forms the walls of the diaphysis, and a lattice network of **spongy bone** occupies the epiphyses. A thin layer of compact bone covers the surfaces of the epiphyses. The many small spaces within the spongy-bone network are filled with blood-forming connective tissue known as **red marrow**. In the diaphysis, the compact bone borders a central chamber that is continuous with many of the spongy-bone channels. This large chamber is called the **medullary cavity**. It is filled with a second type of marrow, **yellow marrow**, which is rich in fatty tissue for energy storage. Lining the interior surface of the medullary cavity and extending into spongy bone spaces is a thin membrane known as the **endosteum**[4] (en-DŌ-stē-um).

Bone Composition

Bone is composed of bone tissue, cartilage, dense connective tissue, blood-forming tissue, blood vessels, and nerves. By far its primary component is bone tissue, which is a hard, strong, flexible tissue that is in a state of constant change. As you may recall from Chapter 4, bone tissue consists of a vast, hardened matrix with embedded cells interspersed throughout.

Bone tissue contains inorganic and organic materials. The inorganic components are mineral salts made from calcium phosphate and calcium carbonate to form mineral crystals, known as *hydroxyapatite*. These crystals compose two-thirds the total weight of bone and provide it with its hardness and limited flexibility. They form a matrix so dense that blood vessels cannot penetrate; during bone development, canals are provided for the passage of vessels and are maintained

[1]Diaphysis: Greek *dia-* = through + *phys-* = growth + *-is* = presence of (a growth that is present throughout).

[2]Epiphysis: *epi-* = on top of, upon + *physis* (a growth that is present upon).

[3]Periosteum: *peri-* = around + *osteum* = presence of bone.

[4]Endosteum: *endo-* = inside of, or within + *osteum*.

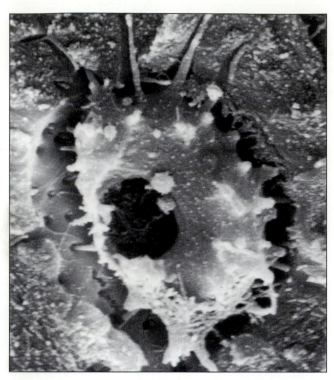

Figure 7–3
The osteocyte. This electron micrograph of bone tissue shows the osteocyte in the center, lying within its lacuna. Surrounding the lacuna is hardened bone matrix. *(From R.G. Kessel and R.H. Kardon, Tissues and Organs: a text–atlas of scanning electron microscopy. Micrographs courtesy of Professor R.G. Kessel.)*

throughout life. Organic materials include mainly collagen, which is produced by bone cells and serves to reinforce and further strengthen the matrix, and the cells themselves.

There are three types of bone cells in living bone. Recent investigations have found that each type actually represents a particular stage in the life cycle of a single, generalized bone cell. **Osteoblasts**[5] represent a youthful stage in the life of a bone cell, and are usually found on the surfaces of bone tissue (in the periosteum). They actively produce matrix (mineral salts and collagen) and remain unrestrained by it. **Osteocytes**[6] are mature bone cells that have become trapped within chambers (*lacunae*) surrounded by matrix. Osteocytes may produce additional matrix but at a reduced rate. An example of an osteocyte is shown in Figure 7–3. **Osteoclasts**[7] are bone cells that wander throughout bone tissue, secreting a substance that dissolves the mineral salt crystals of the matrix. It is thought that osteoclasts may arise from osteoblasts under the influence of hormones.

Microscopic Structure of Bone

The bone that is formed by the organization of its inorganic and organic components is not a solid, rock-like crystal structure, but rather a porous material with pores containing living cells and blood vessels. As we saw in the discussion of a long bone, regions of a single bone often vary in density of the matrix and may fall under the classification of spongy bone or compact bone, depending on the degree of matrix porosity.

Compact bone is the type of bone tissue that contains a dense matrix. It is thicker in the diaphysis than in the epiphyses of long bones, and in flat bones it is deposited on the internal and external surfaces. When compact bone is viewed under a microscope, its osteocytes can be seen within spaces called **lacunae** (Fig. 7–4). The lacunae lie between thin sheets of matrix called **lamellae**,[8] which are layered in concentric circles around a series of canals that extend parallel to the long axis of the bone. These canals are called **osteonic canals** or **haversian canals**. Passing through the canals are blood vessels, which provide nutrients to the osteocytes. The nutrients diffuse from the osteonic canals through tiny channels that extend across the solid barrier formed by the lamellae. These channels, called **canaliculi**,[9] permit the osteocytes to communicate to each other as well as to the osteonic canals. Each osteonic canal, with the lamellae, osteocytes, and canaliculi surrounding it, is called an **osteon** or **haversian system**.

The blood vessels of the osteons are interconnected with each other and with a rich network of blood vessels in the periosteum and endosteum. The connections are by way of a network of blood vessels that extend perpendicular to the bone's long axis. These blood vessels pass through canals in the bone matrix called **Volkmann's canals**.

Spongy bone contains many more pores and spaces within its substance than does compact bone. Spongy bone consists of numerous thin plates of bone, known as **trabeculae**[10] (tra-BEK-yoo-lē), that interconnect with one another, leaving spaces in between. The spaces are filled with **red marrow**, which is the site of blood-cell formation (hematopoiesis). Like the matrix of compact bone, the trabeculae contain osteocytes that lie within lacunae. However, there is no need for a se-

[5]Osteoblast: *osteo-* = bone + *-blast* = sprout or formation.

[6]Osteocyte: *osteo-* + *-cyte* = cell.

[7]Osteoclast: *osteo-* + *-clast* = breaking or destruction.

[8]Lamellae: Latin *lamella* = little plate.

[9]Canaliculi: *canal-* = channel or stream + *-culi* = tiny.

[10]Trabeculae: plural form of Latin *trabecula* = little beam.

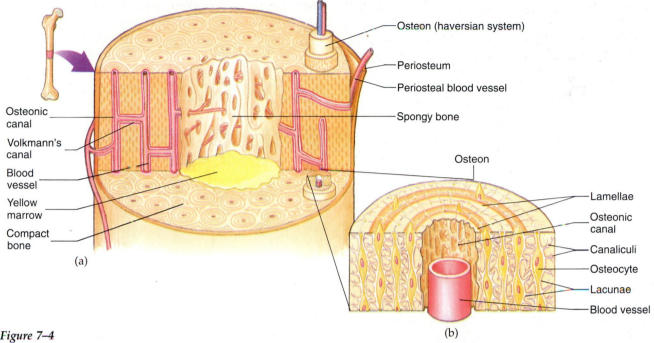

Osteon (haversian system)

Periosteum

Periosteal blood vessel

Spongy bone

Osteonic canal

Volkmann's canal

Blood vessel

Yellow marrow

Compact bone

(a)

Osteon

Lamellae

Osteonic canal

Canaliculi

Osteocyte

Lacunae

Blood vessel

(b)

Figure 7–4
Microscopic structure of bone. This diagram illustrates the organization of compact bone. (a) This slice through a long bone's diaphysis shows the relationships between the periosteum, the compact-bone layer, and the spongy-bone layer. Note how the compact-bone layer is made up of numerous osteons (haversian systems). (b) Individual osteocytes, lacunae, and lamellae can be seen in a more highly magnified view of a single osteon. (c) A photomicrograph of an osteon seen at cross section (magnification 100x).

What are the components of an osteon?

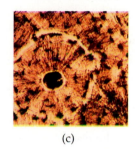

(c)

ries of channels cutting through dense matrix as we have seen in compact bone, because of the thin structure of the trabeculae. In spongy bone, the osteocytes receive nourishment from the rich blood supply in the red marrow by way of short canaliculi that extend between the trabecular surface and the bone cells.

CONCEPTS CHECK

1. What are the five functions of the skeletal system?

2. What are the parts of a typical long bone that can be identified without a microscope?

3. How do osteocytes receive nourishment in compact bone?

4. How does spongy bone differ structurally from compact bone?

BONE DEVELOPMENT AND GROWTH

Bones begin developing at an early stage in life in two ways: from embryonic membranes, or from cartilage. After birth, bones grow in two directions, lengthwise and widthwise, until body growth stops. Throughout adulthood, bone remodeling continues in order to provide a recycling of bone materials.

The process of bone development begins during the first two months of prenatal life. During this early time, special precursor cells migrate to areas of the embryonic body from which bones will arise, such as the head, the thorax, and the limb buds. Once established, the precursor cells begin the development of bone in one of two ways. In one method, cells develop between thin sheets of connective tissue, or embryonic membrane. Bones that arise in this manner are known as intramembranous bones.[11] In the second method, a template of cartilage is produced, which is later transformed into bone. Bones originating from cartilage are called **endochondral bones**.[12] In both methods of bone development, the osteoblasts secrete the new bone matrix composed of collagen and mineral salts. The secretion of new bone matrix is called **ossification.**

Intramembranous Bones

The development of intramembranous bones begins soon after the formation of embryonic membranes, about the fifth week of life. At this time, precursor cells transform into cells called osteoblasts that are capable of secreting collagen and mineral salts. The osteoblasts cluster in groups along the embryonic membranes and become active, secreting new matrix around themselves. Soon, thin plates of bone are produced, which interconnect to form spongy bone tissue. Eventually, the osteoblasts cement themselves within the newly hardened matrix and become trapped within lacunae. When this occurs, they reduce their rate of secretion and thereby become osteocytes. The membrane surrounding the bone, the periosteum, is formed by cells of the embryonic membrane that lie outside the developing bone. Once established, osteoblasts within the periosteum produce a layer of compact bone over the surface of the newly formed spongy bone, completing the development process. Bones that develop in this manner include the flat bones of the skull, the mandibles (lower jaw), and the clavicles (collarbones).

Endochondral Bones

Bones that form from cartilage include all but the flat bones of the skull, mandible, and clavicles. Their development begins around the sixth week of life when clusters of precursor cells transform into cartilage-producing cells, called **chondroblasts**. These early chondroblasts produce hyaline cartilage in areas where bone is to form. The mass of cartilage provides a model, or template, for the development of bone.

The cartilage model expands rapidly at first until it resembles the shape of the bone that will replace it (Fig. 7–5). Blood vessels soon penetrate it, causing some chondroblasts to enlarge and die while others transform into osteoblasts. The increased flow of blood

[11]Intramembranous: *intra-* = within + *membranous* = pertaining to a membrane.

[12]Endochondral: Greek *endo-* = within + *chondral* = pertaining to cartilage.

also imports osteoblasts from other areas. The death of chondroblasts leads to a destruction of cartilage, forming caverns inside the model. In a long bone, for example, this occurs first in the center of the diaphysis, where it is called the **primary ossification center**. Once caverns are established, osteoblasts migrate into these spaces and produce spongy bone tissue. At the same time, other osteoblasts in the newly formed periosteum deposit a thin layer of compact bone along the surface of the model.

Soon after the primary ossification center is established, more blood vessels penetrate the model at the epiphyses. In our long bone example these are known as the **secondary ossification centers**. The conversion of cartilage to bone begins again, as cartilage is destroyed and osteoblasts produce spongy bone tissue in replacement.

As the bone nears its final stages of early development, the space formed by cartilage destruction in the center of the bone expands to form the medullary cavity, and the walls of the diaphysis thicken with compact bone. Eventually, spongy bone fills the epiphyses and a thin layer of compact bone encircles them. Between the epiphyses and the diaphysis, a narrow band of cartilage called the **epiphyseal** (ep'-i-FIZ-ē-al) **plate** remains. Also, a thin layer of cartilage, the articular cartilage, remains at the end of each epiphysis, where a joint will form.

Bone Growth

Once an endochondral bone has formed, it can increase in size in two directions: length and width. Lengthwise expansion, or **interstitial growth**, is accomplished at the epiphyseal plate. This region of cartilage contains chondroblasts that produce new cells and new intercellular material, which are pushed toward the epiphysis. At the opposite end of the plate, near the diaphysis, the older cartilage is converted to bone. As a result, the diaphysis of the growing bone increases its length while the epiphyseal plate regenerates itself.

A long bone of a growing child will increase in length as long as the chondroblasts in the epiphyseal plate are active. They will continue to be active until the growth-controlling factors of the body, the genetic code and hormones released by the pituitary and other glands, stop their activity. Once chondroblast activity stops, the epiphysis and the diaphysis permanently fuse, and lengthwise growth becomes no longer possible. Along the area of fusion between epiphysis and diaphysis, a visible line of ossification called the **epiphyseal line** is formed in the bone.

A growing long bone normally undergoes an increase in width, called **appositional growth**, while it

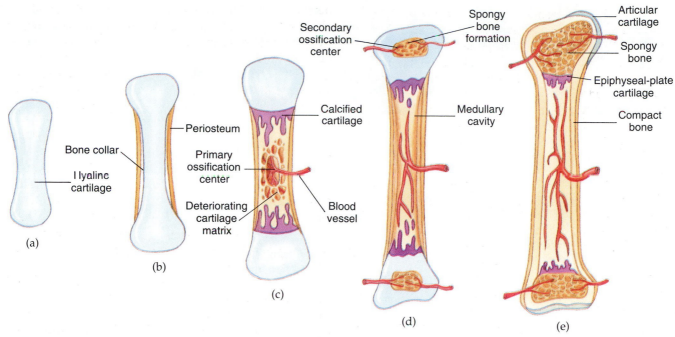

Figure 7–5
Endochondral bone formation. (a) An endochondral bone begins as a model of
hyaline cartilage in an embryo. (b) It soon forms a collar of early bone around its
midsection. (c) The primary ossification center forms as cartilage is replaced by
spongy bone. Once formed, it proceeds to expand as more blood vessels penetrate
the area. (d) A central cavity is formed, and secondary ossification centers begin to
emerge. (e) Ossification continues until spongy bone fills the epiphyses and
compact bone surrounds the entire structure. Hyaline cartilage remains in the
epiphyseal plates and the articular cartilages (the epiphyseal plates provide for
lengthwise bone growth until young adulthood).
Why does the entrance of a blood vessel into the early cartilage model begin ossification?

increases in length. This is accomplished by osteoblasts
in the periosteum, which deposit new compact bone
along the outer surfaces of the diaphysis. At the same
time, the medullary cavity in the center expands by the
destruction of bone along the inner surfaces. This is
performed by osteoclasts, which secrete enzymes that
dissolve the bone matrix. Appositional growth is usually an unequal process as more bone is deposited than
removed, providing a continually thicker and stronger
bone.

Bone Remodeling

Once an individual reaches a certain age, the genetic
code commands the pituitary gland to stop the growth
process, and growth in height comes to a halt as the
epiphyseal plate disappears. However, this does not

mean the bones become inactive. On the contrary, the
process of osteoblast deposition of new bone and reabsorption of bone by osteoclasts continues throughout life. In fact, every gram of bone tissue in your body
will have been recycled at least twice by the time you
reach the age of 60 years! This process is known as
bone remodeling.

Bone remodeling varies in activity from bone to
bone, and even from bone area to bone area. For example, some bones or bone areas, such as the distal
part of the femur, are replaced every five to six months,
whereas other bones are replaced much less frequently.
Remodeling does not take place on every bone at the
same time; at any one time, most bones are not involved in remodeling. In general, the bone areas that
receive the greatest stress or injuries undergo more frequent remodeling. In fact, remodeling is an important
part of the bone repair process.

HEALTH CLINIC

Bone Fractures and Repair

When a bone is required to accept more stress than it is able, the result is often a break, or **fracture**. There are several types of fractures, which are classified on the basis of the extent and nature of the breakage (see the illustration). A *complete fracture* occurs when the bone is broken completely through. If the break does not extend through the bone, it is termed an *incomplete* or *greenstick fracture*. A break that is not marked by a tear in the skin is a *closed* or *simple fracture*, while one in which the skin is pierced is called a *compound fracture*. In some cases, a bone may fragment during the breakage; when it does, the result is called a *comminuted fracture*.

A fracture may heal if the broken ends of the bone make contact with one another. If this does not occur during the injury, medical intervention is necessary to bring the broken ends together. The healing process begins with bleeding and inflammation (increased blood flow, swel-

ling, and infiltration of white blood cells) of the fracture site. Bleeding soon produces a blood clot. Within about 48 hours of the injury, the blood clot softens by the action of white blood cells, producing a mass of protein fibers called a **procallus**. Over the next several days, fibroblasts arrive at the injury site and secrete a dense connective tissue network, which replaces the procallus. Chondroblasts and osteoblasts also begin to arrive at the site, having originated from the nearby periosteum and endosteum of the injured bone. By the end of the first week, clusters of newly formed cartilage and bone appear throughout the injury site. Gradually, inflammation subsides and additional bone is laid down by the osteoblasts. After several weeks, the injury site is occupied by a bone mass called the **osseous callus**. The osseous callus acts as an internal splint, cementing the open ends of the fracture together. Once the osseous callus has undergone remodeling to restore bone structure, the injured bone is healed.

X-ray photographs of common bone fractures.

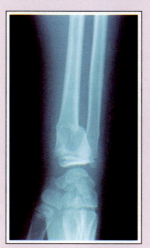

A greenstick fracture

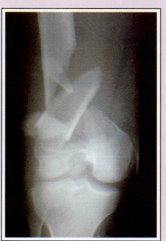

A comminuted fracture

A compound fracture

The cyclic generation of new bone and its reabsorption occurs to help maintain homeostasis of the body. It provides a method for removing the vital minerals calcium and phosphate from the blood supply—if they are in excess—by storing them in new bone tissue. It also removes these minerals from storage and returns them to the blood during times of low supply. Finally, bone remodeling provides an important method of healing bone tissue that has been damaged by disease or fracture.

CONCEPTS CHECK

1. Where does the primary ossification center occur in endochondral bone development?

2. What is the role of the epiphyseal plate during bone growth?

3. Why is bone remodeling an important activity in adult bones?

ORGANIZATION OF THE SKELETON

The skeleton provides a frame for the body, and includes all bones and joints. It is organized into two divisions, axial and appendicular. Bones contain important features that are associated with their particular functions.

The 206 bones of the human body are organized into a single, interconnected frame known as the **skeleton**.

It is divided into two portions: an **axial skeleton**, which contains the bones that lie within the midvertical axis of the body; and the **appendicular skeleton**, which contains the bones that lie outside the midvertical axis and are associated with the appendages. Before we take a more detailed look at the skeleton, let's first examine its general arrangement, described in the following outline and shown in Figure 7–6.

Figure 7–6
The skeletal system. The axial skeleton is shown in dark orange, and the appendicular skeleton is in tan.

Why is the vertebral column considered part of the axial skeleton?

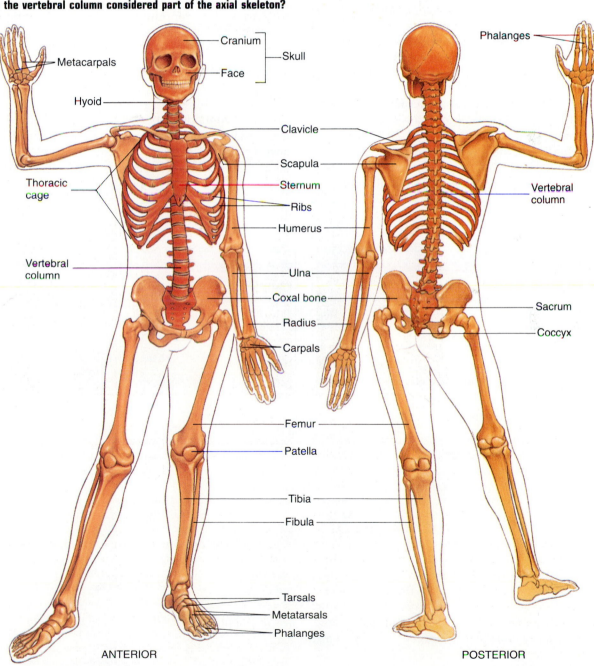

ANTERIOR

POSTERIOR

I. **Axial Skeleton**
  1. **Skull:** the set of bones of the head region. It consists of the bones of the **cranium** and the **facial bones**.
  2. **Hyoid bone:** a small bone in the anterior part of the neck.
  3. **Vertebral column:** the backbone. It consists of the **vertebrae** and **sacrum**.
  4. **Thoracic cage:** the bones in the chest region. It includes the **sternum** and the 12 pairs of **ribs** that are attached to the vertebral column.

II. **Appendicular Skeleton**
  1. **Pectoral girdle:** a frame connecting the upper limbs to the axial skeleton. It contains a **scapula** and a **clavicle** on either side of the body.
  2. **Upper limbs:** the upper appendages, each of which consists of a **humerus**, a **radius**, an **ulna**, **carpals**, **metacarpals**, and **phalanges**.
  3. **Pelvic girdle:** the lower frame connecting the lower limbs to the axial skeleton. It contains two **coxal** (hip) **bones**. Together with the lower part of the vertebral column (the sacrum), they form the **pelvis**.
  4. **Lower limbs:** the lower appendages, each of which include a **femur**, a **patella**, a **tibia**, a **fibula**, **tarsals**, **metatarsals**, and **phalanges**.

Surface Features of Bones

Bone surfaces contain structural features that distinguish one bone from another but do not vary much from person to person. For example, at the base of everyone's skull is a large opening, the foramen magnum, which is a feature only the occipital bone has. The features of bones suggest their relationships to skeletal system functions. They include: projections and depressions that may serve as attachment points for ligaments or tendons, to strengthen the bone, or for articulation with other bones; openings to permit the penetration of blood vessels or nerves; and spaces for housing body structures. A list of surface features is provided in Table 7–1, which you should use as an aid to learning the important features in boldface type that follow.

Bones of the Axial Skeleton

Skull

The skull contains 22 bones, which, for the most part, are closely adjoined by rigid, narrow joints known as **sutures**. In the skull of an adult, sutures appear as jagged lines between adjacent bones (Figs. 7–7 and 7–9 through 7–11).

Table 7–1   SURFACE FEATURE TERMINOLOGY IN BONES

Term	Description	Example
Condyle	Literally, "knuckle"; a large, rounded prominence	Occipital condyles at the base of the skull
Facet	A smooth articular surface	On thoracic vertebrae for rib attachment
Fissure	A narrow opening or cleft	Orbital fissure of the sphenoid bone
Foramen	An opening or hole through bone	The foramen magnum at the base of the skull
Fossa	A depression or groove	Glenoid cavity of the scapula
Process	Any projection from the surface of a bone	Styloid process of the temporal bone
Spine	A narrow or pointed projection	Spine of the scapula
Trochanter	A large, blunt process	Greater trochanter of the femur
Tubercle	A small, rounded process	Greater tubercles on the proximal end of the humerus
Tuberosity	A rounded, elevated area of a bone that is usually roughened	Deltoid tuberosity of the humerus

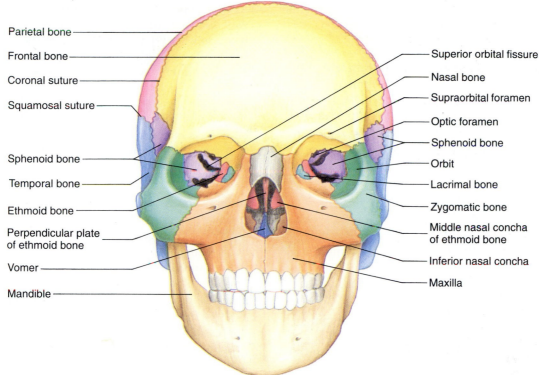

Figure 7–7
The skull, anterior view.
Is the frontal bone a flat bone or a long bone?

Within the substance of the skull are chambers that are lined with mucous membranes and are filled with air, called **sinuses** (Fig. 7–8). They connect with the nasal cavity to drain fluids, reduce the weight of the skull, and resonate sound from the voice. Inflammation of the mucous membranes lining the sinuses, which may be caused by an allergy or an infection, is called *sinusitis*. If the amount of swelling is sufficient

to block the drainage channels into the nasal cavity, a buildup of fluid pressure can lead to a sinus headache. Five bones of the skull contain sinuses: the frontal bone, the ethmoid, the sphenoid, and the two maxillary bones.

The bones of the skull include 8 bones of the cranium and 13 smaller bones of the face known as facial bones. Another bone of the skull is the single lower

Figure 7–8
The sinuses of the skull. (a) Anterior view, superimposed on the face. (b) Lateral view of the head sectioned along the midsagittal plane.
What cavity in the head do the skull sinuses drain into?

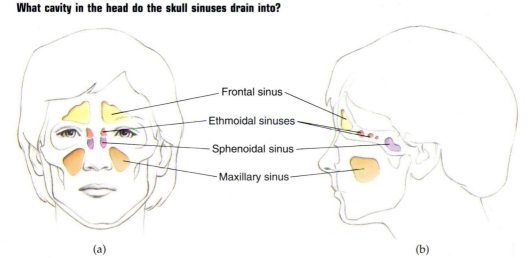

Frontal sinus
Ethmoidal sinuses
Sphenoidal sinus
Maxillary sinus

(a)   (b)

jawbone, the mandible, which hangs from attachments to the cranium.

Cranium

The cranium encloses and protects the brain within the space it creates, and provides an attachment site for muscles of the scalp, lower jaw, neck, and back. It consists of the following eight bones (refer to Figs. 7–7, 7–9, 7–10, and 7–11):

Frontal bone—the large bone forming the anterior part of the skull above the eyes, or forehead (Fig. 7–7). Notice the large eye sockets, whose roofs are formed by the frontal bone. These sockets are called **orbits**. Also note the small hole above each orbit. This is the **supraorbital foramen**, through which pass blood vessels and nerves. Within the frontal bone are the **frontal sinuses**, one above each orbit near the midline (shown in Figs. 7–8 and 7–11).

Parietal bones—two bones, right and left, forming much of the lateral aspects of the cranium (Fig. 7–9). They meet at the top of the skull at the **sagittal suture**, and unite with the frontal bone at the **coronal suture**.

Occipital bone—a thick bone forming the posterior wall and floor of the cranium (Figs. 7–9 and 7–10). It meets with the parietal bones at the **lambdoidal** (lam-DOY-dal) **suture**. Its most prominent feature is the **foramen magnum**, a large opening through its inferior surface. The foramen magnum permits passage of the spinal cord as it extends between the cranial cavity and the vertebral canal. Bordering both sides of this opening are rounded processes called **occipital condyles**, which articulate (form joints) with the first vertebra (the atlas) for head movement.

Temporal bones—two bones on either side of the cranium below the parietal bones (Figs. 7–9 and 7–10). Each temporal bone unites with a parietal bone along a **squamosal suture**. Along the inferior margin of each temporal bone are several important features. The opening is the **external auditory meatus**, which leads toward the inner parts of the ear. Anterior to this opening is a depression called the **mandibular fossa**, which provides an articular (joint) surface for the mandible. A bridgelike extension of bone that projects anteriorly is the **zygomatic process**, which

Figure 7–9
The skull, lateral view.
How many bones of the cranium are visible in this lateral view?

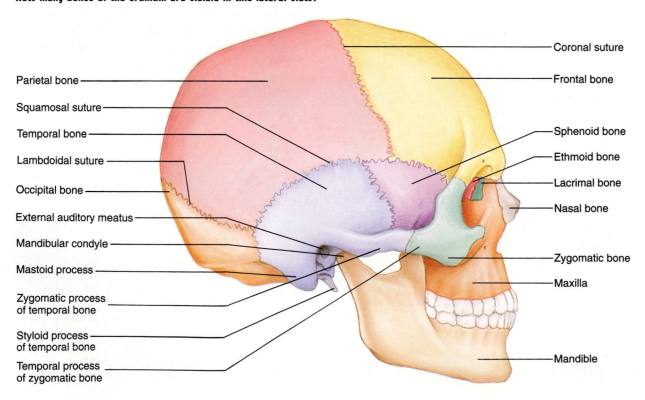

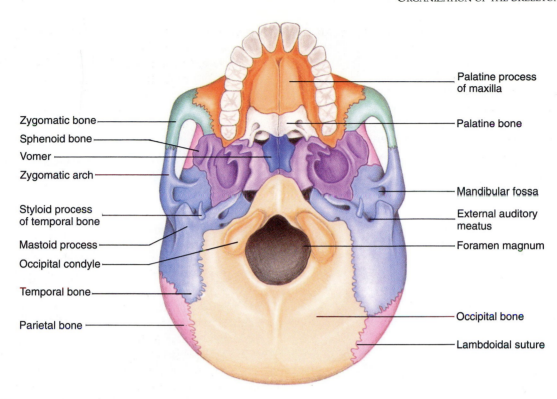

Zygomatic bone

Sphenoid bone

Vomer

Zygomatic arch

Styloid process
of temporal bone

Mastoid process

Occipital condyle

Temporal bone

Parietal bone

Palatine process
of maxilla

Palatine bone

Mandibular fossa

External auditory
meatus

Foramen magnum

Occipital bone

Lambdoidal suture

Figure 7–10
The skull, inferior view with the lower jaw removed.
What structure extends through the foramen magnum in life?

joins the zygomatic bone to form the cheekbone, or **zygomatic arch**. Projecting downward from its origin below the external auditory meatus is a narrow, pointed projection called the **styloid process**.[13] It serves as an anchorage for muscles of the tongue and pharynx. Posterior to the styloid process is a rounded projection called the **mastoid process**, which provides an attachment for several muscles of the neck.

Sphenoid bone—a single butterfly-shaped bone, the sphenoid (SFĒ-noyd) is wedged between other bones in the anterior margin of the cranium. Parts of it form the lower lateral walls (Fig. 7–9) and floor (Fig. 7–10) of the cranium, and posterior walls of the orbits (Fig. 7–7). When this bone is viewed by looking through the orbits, a round hole that penetrates the medial wall may be seen. This is the **optic foramen**, through which passes the optic nerve on its way to the brain. The two widened cracklike openings lateral to the optic foramen are the **superior** and **inferior orbital**

fissures, which transmit blood vessels and nerves. Within the cranial cavity a part of the sphenoid bone projects upward to form a saddle-shaped process called the **sella turcica**[14] (SEL-a TER-si-ka), which houses the pituitary gland (see Fig. 7–11). Inside the bone itself are two small spaces known as the **sphenoidal sinuses** (see Fig. 7–8).

Ethmoid bone—a small bone anterior to the sphenoid bone (portions of this mostly internal bone can be seen in Figs. 7–7, 7–9, and 7–11). Portions of the ethmoid bone form sections of the cranial floor, orbital walls, and nasal-cavity walls. The cranial floor segment is called the **cribriform plate**, which divides the cranial cavity from the nasal cavity (Fig. 7–11). Projecting upward into the cranial cavity from the cribriform plate is a thin process that resembles a rooster's comb, called the **crista galli**,[15] which provides a point of attachment for membranes that enclose the brain. Projecting

[13]Styloid: *styloid* = stakelike or spearlike.

[14]Sella turcica: *sella* = saddle + *turcica* = Turkish (Turkish saddle).

[15]Crista galli: Latin *crista* = crest, or comb + *gallus* = rooster.

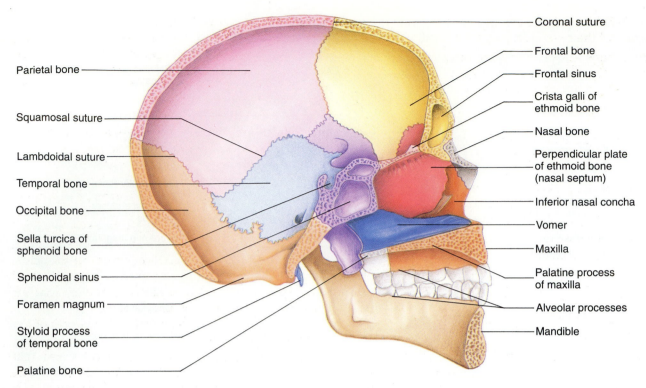

Parietal bone

Squamosal suture

Lambdoidal suture

Temporal bone

Occipital bone

Sella turcica of sphenoid bone

Sphenoidal sinus

Foramen magnum

Styloid process of temporal bone

Palatine bone

Coronal suture

Frontal bone

Frontal sinus

Crista galli of ethmoid bone

Nasal bone

Perpendicular plate of ethmoid bone (nasal septum)

Inferior nasal concha

Vomer

Maxilla

Palatine process of maxilla

Alveolar processes

Mandible

Figure 7–11
The skull, lateral view of a midsagittal section.
What bones form the nasal septum?

downward from the plate is another thin process, the **perpendicular plate**, which forms most of the nasal septum dividing the nasal cavity into right and left halves. The main body of the ethmoid forms much of the lateral walls of the nasal cavity, and contains thin, scroll-like projections extending into the cavity, called the **superior** and **middle nasal conchae** (Fig. 7–7). Within the ethmoid body are numerous spaces called the **ethmoidal sinuses** (Fig. 7–8).

Facial Bones

The bones of the facial skeleton include 13 immovable bones and a movable lower jaw. They support the face and provide attachments for muscles that control facial expressions and move the jaw (refer to Figs. 7–7, 7–9, 7–10, and 7–11).

Maxillary bones—two bones on each side of the face that form the upper jaw (Figs. 7–7 and 7–9). Portions of them help form the floor of the orbits, the roof of the mouth, and the walls and floor of the nasal cavity. The largest of the sinuses, the **maxillary sinuses**, are located within these bones (Fig. 7–8). They drain directly into the nasal cavity. Also, the upper set of teeth articulate with the

maxillary bones at projections called **alveolar processes**. The anterior roof of the mouth is actually a horizontal projection of each maxillary bone called the **palatine process**, which normally fuses along the midline before birth to form most of the *hard palate* (Fig. 7–10). If the fusion is not complete, an opening remains between the mouth and the nasal cavity. This condition is called a *cleft palate* and is often associated with a *cleft lip*, which is a split in the upper lip. A cleft palate and cleft lip can be corrected with surgery.

Palatine bones—two L-shaped bones that are located posterior to the maxillary bones (Figs. 7–10 and 7–11). Their horizontal portions form the posterior roof of the mouth and floor of the nasal cavity, and their vertical portions help form the lateral walls of the nasal cavity.

Zygomatic bones—two bones on each side of the face that form part of the cheekbones and orbits (Figs. 7–7 and 7–9). Each zygomatic bone contains a **temporal process** that extends to the temporal bone on the side of the head. It unites with the zygomatic process, forming the prominence of the cheekbone known as the **zygomatic arch**.

Nasal bones—two small, rectangular bones that meet at the midline to form the bridge of the nose

(Fig. 7–7) between the orbits. The larger, remaining portion of the nose is formed by cartilage plates.

Lacrimal bones—two small, thin bones that are similar in size and shape to small fingernails. They are posterior and lateral to the nasal bones, and form part of the orbit's medial walls (Figs. 7–7 and 7–9).

Vomer—a single bone located along the midline within the nasal cavity (Figs. 7–7 and 7–11). Along its superior margin it unites with the perpendicular plate of the ethmoid, forming the *nasal septum*. The nasal septum divides the nasal cavity into right and left chambers.

Inferior nasal conchae—two thin, scroll-like bones attached to the lateral walls of the nasal cavity (Fig. 7–7). They are located below the superior and middle conchae of the ethmoid, and together they form the three shelves of the nasal cavity through which air is channeled. The three conchae are also called *turbinates*.

Mandible—the single lower jaw, which articulates with the temporal bones (Figs. 7–7 and 7–9). It is the only movable bone of the skull. The mandible has a **mandibular condyle**, which articulates with the mandibular fossa of the temporal bone. The **alveolar process** of the mandible, like that of the maxillary bones, is an arch containing sockets for teeth.

Hyoid Bone

The single hyoid bone is a unique part of the skeleton, since it does not articulate with any other bone. It is located in the neck region below the mandible, where it is suspended from the styloid process of the temporal bone by ligaments and muscles. The horseshoe-shaped hyoid bone supports the tongue and provides attachment for some of its muscles.

Vertebral Column

The **vertebral column** acts as a strong, flexible rod that supports the trunk while permitting anterior, posterior, rotational, and lateral movements. It extends from the base of the skull to the pelvis (Fig. 7–12) and is composed of a series of irregular bones known as **vertebrae**. Between each two vertebrae is a compressible mass of fibrocartilage, called an **intervertebral disc**. This is the structure that is damaged during certain back injuries, when it is referred to as a "slipped (herniated) disc." The vertebrae are held in place by connections to the intervertebral discs, intervertebral ligaments, and the deep muscles of the back. As a complete unit, the vertebral column provides protection to the spinal cord, which extends through a central canal that forms from openings through the vertebrae. This central canal extends through the entire length of the vertebral column and is known as the **vertebral canal**.

The vertebral column typically contains 33 vertebrae, which are divided into 6 regions in the following way: 7 **cervical vertebrae** in the neck region; 12 **thoracic vertebrae** in the thorax; 5 **lumbar vertebrae** in the lower trunk; 5 **sacral vertebrae** fused into a single bone called the **sacrum**; and an average of 4 (ranging from 3 to 5 in different individuals) **coccygeal vertebrae** fused into one bone called the **coccyx**. When the sacrum and coccyx are considered as individual bones, the 33 vertebrae of the column make up 26 actual bones. As we will see in a moment, the vertebrae of one region have certain structural features that set them apart from those of other regions.

Vertebral regions correspond to the curves of the column, which you can see in a lateral view (Fig. 7–12a). These curves include a **cervical curve** that bends to the front (anteriorly), a **thoracic curve** that bends backward (posteriorly), a **lumbar curve** that bends back to the front, and a **sacral curve** that bends to the back. The cervical and lumbar curves are called *concave curves*, and the thoracic and sacral curves are called *convex curves*. These curves lend an added strength to the column, help maintain balance during upright posture, and help absorb shocks during running, jumping, and walking. A newborn baby has only a single, convex curvature. The cervical curve develops as the child is able to hold up its head, and the lumbar curve develops as the child learns to stand and walk (Fig. 7–12b).

A Typical Vertebra

The cervical, thoracic, and lumbar vertebrae are basically similar in form (Fig. 7–13). For example, they each have a single drum-shaped portion called the **body**. The body is the weight-bearing part of the vertebra and attaches to the intervertebral discs. Extending from the body is the **vertebral arch**, which completes the bony circle around the large opening, called the **vertebral foramen**, through which passes the spinal cord. The vertebral arch contains: two narrow bridges that attach to the body, called the **pedicles**; two plates that extend from the pedicles backward, called the **laminae**; a single posterior projection known as the **spinous process**; two lateral projections extending from the laminae, called **transverse processes**; and four smaller upward- and downward-facing projections called **superior** and **inferior articulating processes**. The spinous processes and transverse processes pro-

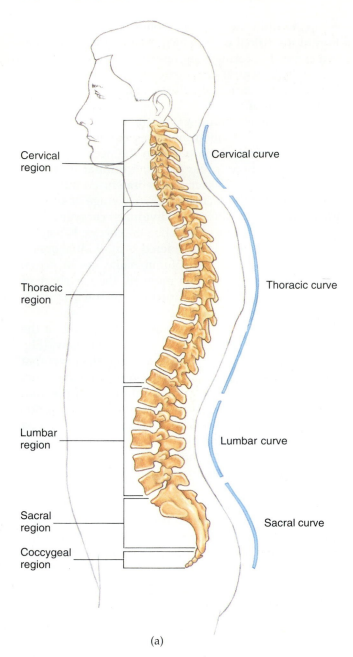

Cervical region

Cervical curve

Thoracic region

Thoracic curve

Lumbar region

Lumbar curve

Sacral region

Sacral curve

Coccygeal region

(a)

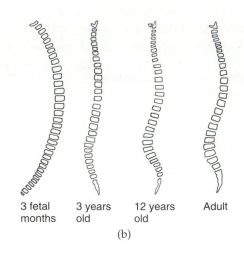

3 fetal months

3 years old

12 years old

Adult

(b)

Figure 7–12
The vertebral column. (a) Location of the vertebral column in the body. (b) Development of the vertebral curves.
The vertebrae of which region articulate with the ribs?

vide attachments for muscles and ligaments, and the articulating processes unite with vertebrae above and below.

Cervical Vertebrae

The seven vertebrae of the neck region provide support to the head. They differ from vertebrae of other regions in two major ways: they are smaller and lighter, and they contain a small hole through each transverse process, called the **transverse foramen**. This opening permits passage of arteries extending to the brain.

The first two cervical vertebrae have unique features that are specialized to permit movement of the head (Fig. 7–14). The first vertebra, called the **atlas**, articulates with the occipital condyles of the skull. It has no body, and contains especially large superior articulating processes with smooth, cartilage-covered surfaces called **facets**. Its articulation with the head permits up-and-down motion, such as when nodding "yes." The second vertebra is the **axis**. It contains a body with a prominent tooth-like **odontoid process** (or **dens**), which projects upward through the ring of the

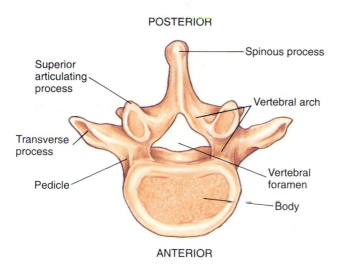

Figure 7–13
The features of a typical vertebra, top (superior) view.
What bone articulates with a superior articulating process?

Figure 7–14
The first two cervical vertebrae, as viewed from the top (superior view): (a) the atlas; (b) the axis. (c) Photograph of the atlas and axis in their articulated positions (posterior view; the sketch is an anterior view).
Which of these two vertebrae articulates with the skull?

atlas. When you turn your head from side to side, as in shaking the head "no," the atlas pivots around the odontoid process.

Thoracic Vertebrae

The twelve thoracic vertebrae are larger than cervical vertebrae and contain long, pointed spinous processes that project downward (Fig. 7–12). They are the only members of the vertebral column that articulate with ribs. Consequently, they are unique among all vertebrae in that they have smooth facets on the sides of their bodies and on the transverse processes where the ribs attach.

Lumbar Vertebrae

The five lumbar vertebrae are larger and thicker than cervical and thoracic vertebrae, particularly at the bodies (Fig. 7–12). This reflects the added stress placed upon them by the increasing amounts of body weight they must support.

Sacrum

The sacrum is a large, triangular bone that forms the posterior part of the pelvis (Fig. 7–15). As we have already learned, it is the result of the fusion of five vertebrae. The spinous processes from these fused bones are still present as the **median sacral crest**, and the two superior processes that articulate with the fifth lumbar

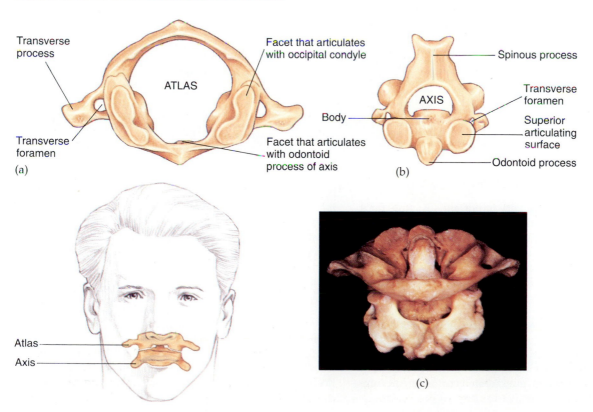

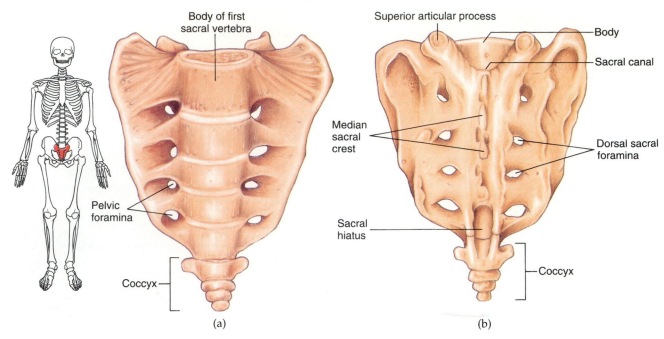

(a)

(b)

Figure 7–15
The sacrum: (a) anterior view; (b) posterior view.
What structures extend through the sacral foramina?

vertebra are the **superior articular processes**. The vertebral canal that continues from the lumbar region is the **sacral canal**, which opens at the opposite end of the sacrum through the **sacral hiatus**. The four pairs of openings alongside the median sacral crest are the **dorsal sacral foramina**, and those on the anterior side are called **pelvic foramina**. Through these foramina pass nerves and blood vessels.

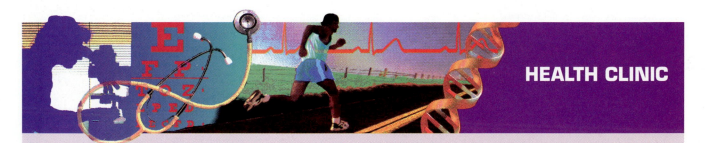

HEALTH CLINIC

Disorders of the Vertebral Column

The vertebral column is a very important structure, for it is responsible for protecting the spinal cord and for supporting much of the weight and movement of the body. Consequently, if something is amiss with the spine it produces wide-ranging effects. For example, certain congenital diseases bring disabling disfigurements or even death when the vertebral column is involved. *Spina bifida* is one such condition, in which the child is born with deformed or missing laminae. In extreme cases, the spinal cord is exposed, producing a water-filled sac in the back that, if broken during birth, usually causes paralysis or death. Other diseases include abnormal curvatures of the spine. For example, *lordosis* is an exaggerated curving in the lumbar region (swayback), and *kyphosis* is an exaggerated curving out of the thoracic region. These conditions are often aggravated by poor posture, obesity, or pregnancy and are a source of low back pain. *Scoliosis* is the lateral deviation of the spine from the midline. It is extremely common (found in 2% of all people over the age of 14 in the United States) and only rarely causes noticeable symptoms. More serious cases of scoliosis can be progressive, however, leading to severe pain, disfigurement, and possible damage to the heart and lungs. These cases may be treated with electrical stimulation of the muscles at selected areas of the curve, which tends to correct the scoliosis when successful.

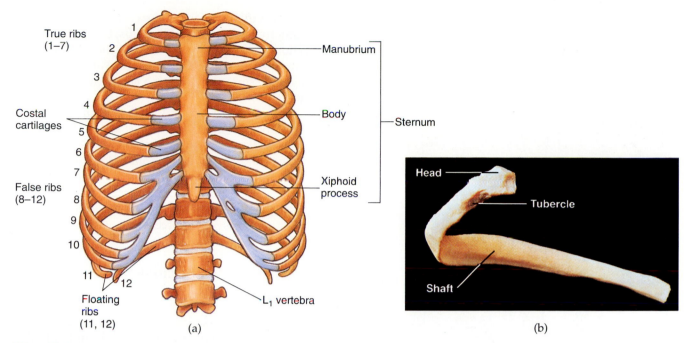

True ribs (1–7)

1
2
3
4
5
6
7

Costal cartilages

False ribs (8–12)

8
9
10
11
12

Floating ribs (11, 12)

Manubrium

Body

Sternum

Xiphoid process

L₁ vertebra

(a)

Head

Tubercle

Shaft

(b)

Figure 7–16
(a) The thoracic cage, anterior view. (b) Photograph of a rib, posterior view. *(From* Color Atlas of Human Anatomy, 2nd Edition *by R.M.H. McMinn & R.T. Hutchings, © 1988 Year Book Medical Publishers, Inc.)*

How do true ribs differ from false ribs?

Coccyx

The coccyx is a series of three to five small, fused or partially fused bones. The coccyx is attached to the inferior end of the sacrum by ligaments (Fig. 7–15).

Thoracic Cage

The thoracic vertebrae, sternum, and ribs form the thoracic cage (Figs. 7–16 and 7–17). The cage is a conical, basket-shaped structure that forms a partial enclosure around the organs of the chest and supports the shoulder girdle and upper limbs.

Sternum

The **sternum**, or breastbone, is a flat, narrow bone located along the vertical midline of the chest (Figs. 7–16 and 7–17). It consists of three parts: a superior **manubrium**; a larger, middle **body**; and a small, inferior **xiphoid process**. The sternum articulates with the two clavicles at its superior end, and with the ribs by way of cartilages along its lateral borders.

Ribs

There are usually 12 pairs of **ribs** in every individual. Each pair attaches to a thoracic vertebra and curves around to the front of the thorax, stopping short of the sternum. Attachment to the sternum is by way of a

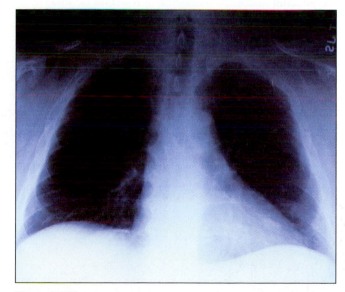

Figure 7–17
X-ray photograph of the thoracic cage.

band of hyaline cartilage, called the **costal cartilage**. Look closely at Figures 7–16 and 7–17, and notice that the costal cartilages do not form a direct link between every rib and the sternum. The first seven pairs of ribs, known as **true ribs**, are the ones that connect to the

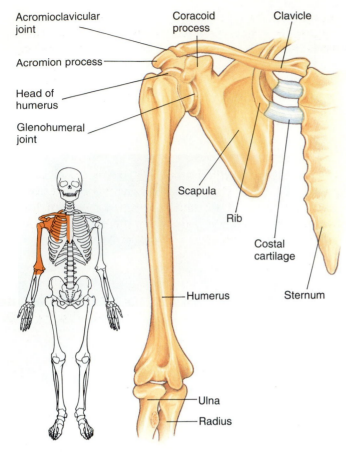

Figure 7–18
The pectoral girdle. This drawing of the right pectoral girdle shows an anterior view and includes the sternum and the humerus that connect to it.
What two bones articulate with the clavicle?

sternum directly by means of the costal cartilages. The remaining five pairs, called **false ribs**, have an indirect connection or no connection at all. The last two (or sometimes three) pairs of false ribs are also known as **floating ribs**, since they lack the costal cartilage completely.

A typical rib has several distinguishing features that set it apart from other bones (Fig. 7–16b). It contains a long, flattened **shaft** that curves around in a semicircle. The **head** of a rib is an enlargement at its posterior, or proximal, end that articulates with the body of one or two thoracic vertebrae. Near the head is a small projection called a **tubercle**, which articulates with the transverse process of a vertebra.

Bones of the Appendicular Skeleton

Pectoral Girdles

The **pectoral girdles** (shoulder girdles) provide a connection between the axial skeleton and the upper limbs (Fig. 7–18). Each girdle contains two bones: a clavicle, or collarbone; and a scapula, or shoulder blade. Together these bones support the arms and attach to muscles that move the arms.

Clavicles

The **clavicles** are slender, rodlike bones that resemble the letter S in shape. Each one extends horizontally between the sternum and the scapula at the base of the neck (Fig. 7–18). At its union with the scapula is one of two joints that form the shoulder, called the

Figure 7–19
The scapula. (a) Anterior, (b) lateral, and (c) posterior views of the left scapula.
What bone articulates with the scapula at the glenoid cavity?

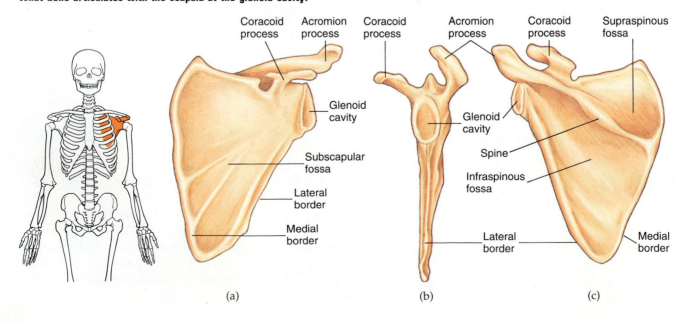

(a)   (b)   (c)

acromioclavicular joint. The clavicles help to hold the shoulder in place during movement of the arm.

Scapulae

The two **scapulae** are located lateral to each clavicle on either side of the upper back. Each one is triangular in shape with thin, broad surfaces and two prominent processes (Fig. 7–19). Its surfaces, edges, and processes provide important attachment sites for muscles that move the shoulder and upper arm. The anterior surface, which is called the **subscapular fossa**, is slightly concave and is quite smooth. The posterior surface contains a prominent ridge called the **spine** that divides it into unequal portions: the superior portion is the **supraspinous fossa**; and the larger, inferior portion is the **infraspinous fossa**. The lateral end of the spine forms one of the major processes, called the **acromion process**, which unites with the clavicle at the acromioclavicular joint. The other large process, the **coracoid process**, arises anterior to the acromion and

curves laterally. It provides attachments for muscles of the arm and chest. Between the two processes, along the lateral edge of the scapula, is an oval depression called the **glenoid cavity**. It receives the head of the humerus (upper arm bone), forming the second joint of the shoulder, which is called the **glenohumeral joint**.

Upper Limbs

The upper limbs contain a total of 60 bones, which support the arms, wrists, hands, and fingers of each side. Each limb includes the humerus of the upper arm, the radius and ulna of the lower arm, the metacarpals in the wrist, the carpals in the hand, and the phalanges of the fingers.

Humerus

The **humerus** is the prominent long bone of the arm that extends from the shoulder to the elbow (Fig. 7–20).

Figure 7–20
The humerus. (a) Anterior and (b) posterior views of the right humerus.
What joint do the capitulum and trochlea help to form?

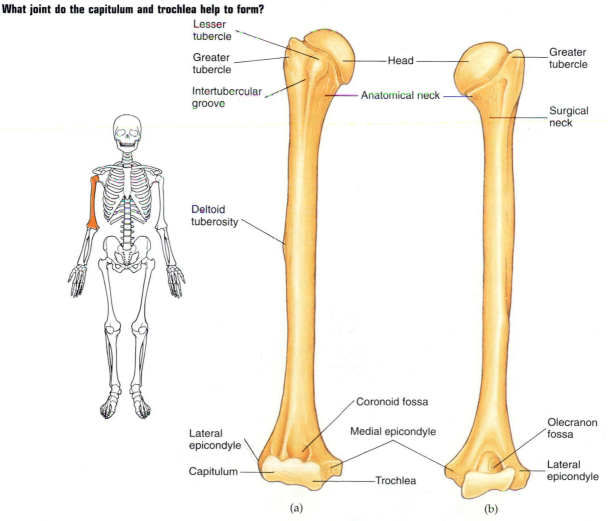

It contains many important features, which are presented next, beginning with the proximal end at the shoulder and proceeding toward the distal end at the elbow.

At the proximal end of the humerus is a smooth, round surface known as the **head**. It articulates with the glenoid cavity of the scapula, forming the glenohumeral joint of the shoulder. Just distal to the head is a narrow circular groove known as the **anatomical neck**. This groove separates the head from two roughened projections that lie below: the **greater tubercle** on the lateral side, and the **lesser tubercle** on the anterior side. Both tubercles provide attachments for the muscles that move the arm at the shoulder. Between these two extends a narrow channel called the **intertubercular groove**. A region named the **surgical neck** is located distal to the tubercles; it is so called because fractures commonly occur there. Distal to the surgical

neck, near the middle of the shaft on the lateral side, is a roughened elevation known as the **deltoid tuberosity**. It provides an attachment for the deltoid muscle of the shoulder.

The distal end of the humerus contains four prominent processes and two depressions, all of which play roles in the formation of the elbow joint. Two smooth, somewhat rounded processes lie in the center of the distal end: the **capitulum**[16] (ka-PIT-yoo-lum) on the lateral side, which articulates with the radius; and the **trochlea**[17] (TRŌK-lē-a) on the medial side, which articulates with the ulna. The two processes that form the lateral and medial borders of the distal end are the **epicondyles**. They provide attachments for muscles

[16]Capitulum: Latin *capit-* = head + *-ulum* = little.

[17]Trochlea: *trochlea* = presence of a pulley.

Figure 7–21
The ulna and radius. These two bones of the right forearm are shown in their articulating position, anterior view. The head of the radius articulates with the radial notch of the ulna, and the head of the ulna articulates with the ulnar notch of the radius.

What process fits into the trochlear notch of the ulna in life?

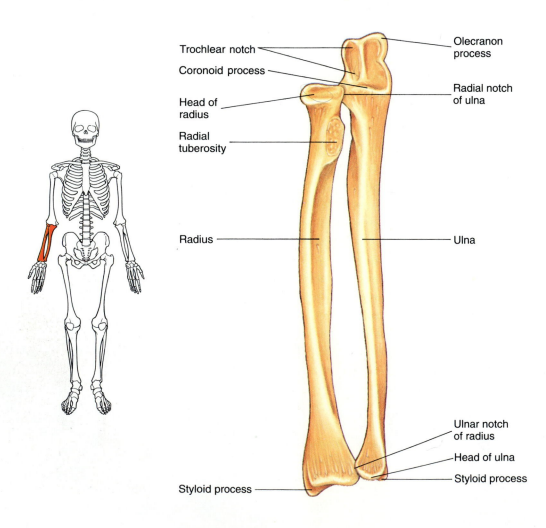

and ligaments of the elbow. Between the epicondyles on the anterior side is the **coronoid fossa**, a depression that receives a process of the ulna. On the posterior side is the second depression, the **olecranon** (ō-LE-kra-non) **fossa**, which receives a process of the ulna when the arm is straightened.

Radius

The **radius** is the lateral bone of the forearm (Fig. 7–21). You can identify it on your own forearm as the bone that is always in line with the thumb. It contains a disc-shaped **head** on its proximal end that articulates with the capitulum of the humerus and a notch of the ulna (called the radial notch). Just distal to the head is a roughened surface called the **radial tuberosity**, which provides an attachment for the large biceps brachii muscle of the arm. At the distal end of the radius, a pointed projection called the **styloid process**[18] provides attachments for ligaments of the wrist.

Ulna

The **ulna** is located medial to the radius in the forearm (Fig. 7–21). At its proximal end is a large projection known as the **olecranon process**. This process forms

[18]Styloid: *styloid* = stakelike or spearlike.

the bony tip of your elbow and contacts the olecranon fossa of the humerus when you extend (straighten) your arm at the elbow. The **coronoid process** is a smaller projection that lies distal to the olecranon process. Between these two processes is a semicircular depression called the **trochlear (semilunar) notch**, which receives the trochlea of the humerus. The distal end of the ulna contains a small **head** that is separated from the wrist by a fibrocartilage disc. A pointed **styloid process** extends downward to provide attachment to ligaments of the wrist.

Hand

The bones of the hand consist of the carpals, which form the wrist; the metacarpals, which support the body of the hand; and the phalanges, which support the fingers (Fig. 7–22).

The **carpal bones** include eight small bones that are closely bound to each other by ligaments. The bones are arranged in two transverse rows, four to a row, and are indicated by name in Figure 7–22. As a complete unit, the carpal bones articulate with the radius and the ulnar fibrocartilage disc, and with the metacarpals.

The five **metacarpal bones** form the skeleton of the palm of the hand. Each bone consists of a flattened

Figure 7–22
The bones of the hand. (a) The carpals, metacarpals, and phalanges of the left hand are drawn in this posterior view. (b) Photograph of the hand skeleton.

Which bones form the body of the hand?

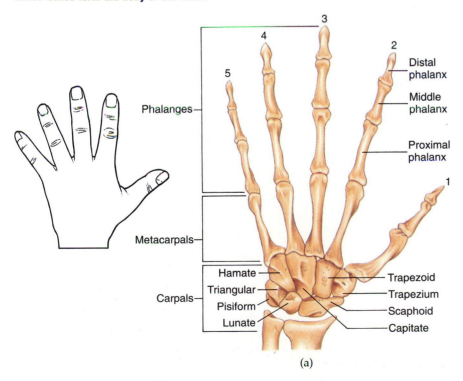
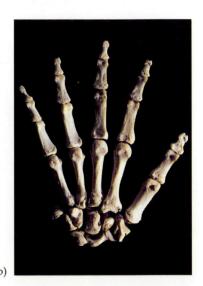

(a)   (b)

base, an elongated shaft, and a distal, rounded head. The heads of the bones form the knuckles of a clenched fist. The bones are numbered 1 through 5, beginning with the metacarpal of the thumb. The metacarpals articulate with the carpal bones on their proximal ends and with the phalanges on their distal ends.

The **phalanges** support the fingers, or digits, of the hand. They number 14 in each hand: two in the thumb (proximal and distal) and three in each remaining finger (proximal, middle, and distal).

Pelvic Girdle

The **pelvic girdle** provides a strong, durable frame for supporting the lower limbs, which must carry the weight of the body. It consists of the two large **coxal** (pelvic or hip) **bones** (Figs. 7–23 and 7–25), which unite with one another anteriorly and with the sacrum posteriorly to form a ringed structure called the **pelvis**. The large superior opening in the center of the pelvis is called the **pelvic inlet**. The pelvises of males and females have important differences, which relate to the function of childbirth in females. These are shown in Figure 7–24 and summarized in Table 7–2. The pelvis provides support for the upper parts of the body, attachment to lower limbs, and protection for organs of the lower trunk and forms the skeletal part of the birth canal in females.

Coxal Bones

In a newborn infant, each coxal bone consists of three separate bones: a superior bone, the ilium; an inferior and anterior bone, the pubis; and an inferior and pos-

Table 7–2	PELVIC GIRDLE DIFFERENCES BETWEEN MALES AND FEMALES	
Feature	**Male**	**Female**
Orientation	Tilted backward	Tilted forward
Pelvic inlet	Narrow, heart-shaped	Wide, oval-shaped
Pubic arch	Less than 90° angle	Broad, greater than 90° angle
Sacrum	Narrow and long	Wide and short with greater curvature
Coccyx	Immovable	Movable
Bone thickness	Thick and heavy	Thin and light

terior bone, the ischium. As the child grows, the three bones begin to fuse together until they merge completely in the adult, where they become subdivisions of a single bone (Fig. 7–25). The area of fusion of the three early bones is marked by a deep, cup-shaped depression known as the **acetabulum**[19] (a-se-TAB-yoo-lum). This cavity is located on the lateral side of each coxal bone and is the socket for the head of the femur.

The **ilium** (IL-ē-um) is the largest of the coxal bone's three subdivisions. It contains a large, flattened portion that flares outward to form the bony ridge of the hip. The superior margin of this ridge is called the

---

[19]Acetabulum: *aceta-* = acid, such as vinegar; a vinegar cup.

Figure 7–23
The pelvic girdle, anterior view.
In the female pelvis, which opening forms the walls of the birth canal?

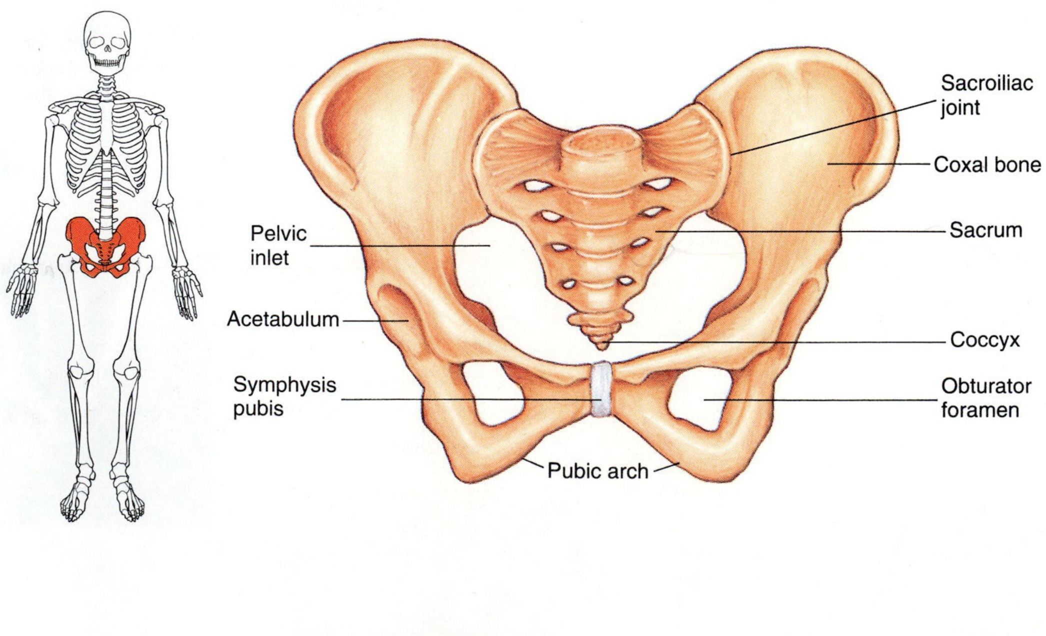

Pelvic inlet

Acetabulum

Symphysis pubis

Pubic arch

Sacroiliac joint

Coxal bone

Sacrum

Coccyx

Obturator foramen

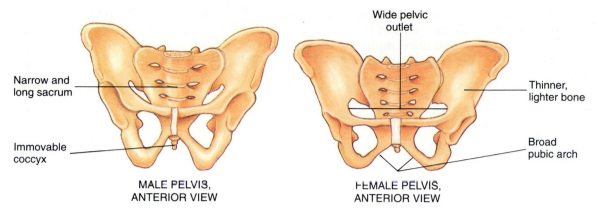

Narrow and long sacrum

Immovable coccyx

Wide pelvic outlet

Thinner, lighter bone

Broad pubic arch

MALE PELVIS, ANTERIOR VIEW

FEMALE PELVIS, ANTERIOR VIEW

Figure 7–24
Comparison of the male and female pelvises, anterior views.
What differences can you identify between female and male pelvises?

iliac crest. Along its posterior margin is a roughened surface, the **auricular surface**, where the ilium unites with the sacrum to form the **sacroiliac joint**. The ilium also contains numerous processes, called spines, that provide attachment sites for muscles of the hip and thigh. The most prominent spine is the **anterior superior iliac spine**, which is located at the anterior end of each ilium.

The **ischium** (IS-kē-um) forms the inferior and posterior portion of the coxal bone. Its shape resembles the letter L. It contains three major features: a projection near its superior border with the ilium, called the **ischial spine**; a roughened, elevated surface at the angle of the L that supports your body weight while you are sitting, known as the **ischial tuberosity**; and a flattened region that fuses with the pubis, called the **ramus**.

Figure 7–25
The right coxal bone, lateral view.
What are the three fused bones that form the adult coxal bone?

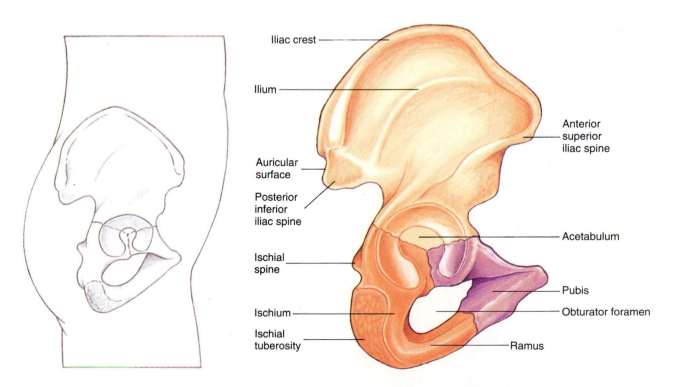

Iliac crest

Ilium

Auricular surface

Posterior inferior iliac spine

Ischial spine

Ischium

Ischial tuberosity

Anterior superior iliac spine

Acetabulum

Pubis

Obturator foramen

Ramus

The **pubis** (PYOO-bis) is the inferior and anterior part of the coxal bone. It unites with the pubis of the opposite coxal bone to form a joint called the **symphysis pubis**. The inferior angle formed by the two coxal bones in the pubic region is known as the **pubic arch**. The large opening that is formed by the curvatures of the pubis and ischium is the **obturator foramen**, which is the largest foramen in the skeleton. The term *obturator* means something that closes off or stops up and refers to the fact that the foramen is actually filled with a fibrous membrane, with portions of muscles, and with blood vessels and nerves passing through it.

Lower Limbs

The bones of the lower limb support the thigh, leg, ankle, and foot. The bones include the large femur in the thigh, the patella at the knee, the tibia and fibula in the foreleg, the tarsals and metatarsals of the foot, and the phalanges of the toes (refer to Fig. 7–6). Like the upper limbs, the lower limbs also contain a total of 60 bones.

Femur

The femur is the longest and heaviest bone in the body (Fig. 7–26). It extends from its union with the coxal bone at the hip joint, to the knee. At its proximal end it contains a large, ball-shaped **head** that projects medially to fit into the acetabulum of the coxal bone. From the head a constricted area known as the **neck** extends to the main shaft. Two large processes border the base of the neck: the upper, lateral **greater trochanter**[20] and the lower, medial **lesser trochanter**. These provide attachment sites for muscles of the hip and thigh. Extending down the shaft of the femur on its posterior side is a narrow ridge, called the **linea aspera**, which also provides for muscle attachment.

[20]Trochanter: Greek *trochanter* = one who runs.

Figure 7–26
The right femur: (a) anterior view; (b) posterior view.
The head of the femur articulates with what bone?

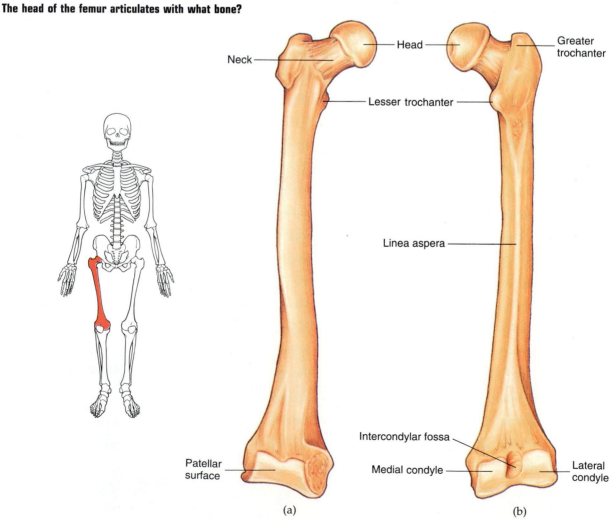

(a)   (b)

The proximal end of the femur contains two prominent, rounded processes that articulate with the tibia of the foreleg. These are the **lateral** and **medial condyles**. Between them is a depression that is apparent on the posterior side, called the **intercondylar fossa**.[21] On the anterior side is a smooth, slight depression known as the **patellar surface**, which articulates with a separate bone that forms the kneecap, or **patella**. The patella is located within a large tendon that wraps over the anterior side of the knee (known as the *quadriceps tendon*).

Tibia

The **tibia** is the larger of the two bones that support the leg (Fig. 7–27). It is located on the medial side, and extends between the knee and the ankle. Its proximal end is expanded into a **lateral condyle** and a **medial condyle**. Their smooth, concave surfaces articulate with the condyles of the femur. Distal to the condyles

on the anterior side is a roughened surface known as the **tibial tuberosity**, which provides an attachment site for a large ligament called the *patellar ligament*.

The distal end of the tibia contains a pointed process, the **medial malleolus**[22] (ma-LĒ-o-lus), which forms the bony ridge you can feel on the medial side of your ankle. The lateral side is smooth for its articulation with the fibula. The inferior surface of the tibia's distal end is slightly concave for its articulation with a large bone of the foot, forming the ankle joint.

Fibula

The **fibula** is a thin, twisted bone located lateral to the tibia in the leg (Fig. 7–27). The **head** of the fibula is at its proximal end, and articulates with the lateral condyle of the tibia. The distal end has a projection called the **lateral malleolus**, which you can feel as a large bump on the lateral side of your ankle. It articulates with a large bone of the foot, the talus.

---

[21]Intercondylar fossa: *inter-* = between + *condylar* = having a rounded surface + *fossa* = depression.

[22]Malleolus: Latin *malleus* = hammer + *-olus* = little.

Figure 7–27
The two bones of the left leg, the tibia and fibula (anterior view).
Is the position of the fibula lateral or medial to the tibia?

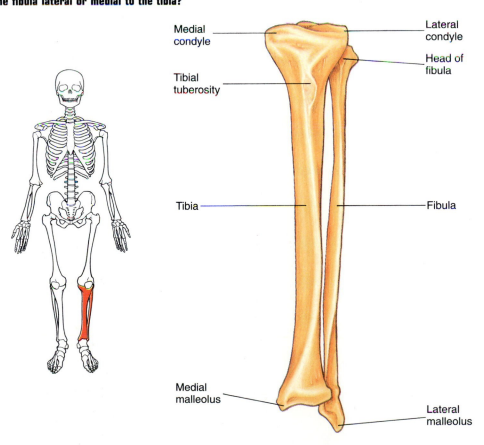

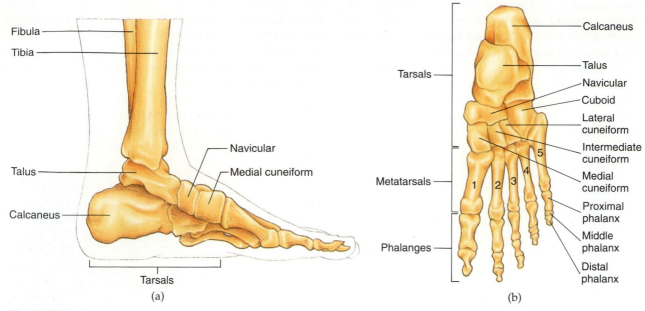

Figure 7–28
The bones of the left foot: (a) medial view; (b) superior view.
What prominent bone of the foot forms the heel?

Foot

The foot contains 26 bones that constitute the ankle; the body of the foot, or instep; and the toes (Fig. 7–28). The ankle consists of seven **tarsal bones**, the most prominent of which are the **talus** and the **calcaneus**. The talus is the only bone that articulates with the tibia and fibula, so it must bear the weight of the body briefly before shifting it to other bones of the ankle. Below the talus is the calcaneus, or heel bone, which is the largest and strongest of the bones of the foot. It helps support the weight of the body and serves as an attachment site for muscles of the leg and foot. The remaining five bones of the ankle are anterior to the calcaneus; they are identified by name in Figure 7–28.

The instep consists of five **metatarsal bones**, numbered 1 to 5 from the medial to the lateral side. Like the metacarpals of the hand, each metatarsal comprises a proximal base, a shaft, and a distal head. The heads form the ball of the foot. Between the calcaneus and the ball of the foot are the **arches**, which elevate the midsection of the foot slightly above ground. The arches are maintained by ligaments that extend between the tarsals and metatarsals; they help provide lift during walking.

The toes are supported by the **phalanges**, which are similar to those of the fingers. There are two phalanges in the great (big) toe, and three in each of the four other toes.

CONCEPTS CHECK

1. What are the bones that form the facial skeleton?

2. What are the major features of a typical vertebra?

3. What bones form the upper limb?

4. What are the major features of the coxal bones?

5. What bones form the lower limb?

JOINTS

Joints are the junctions between bones. There are three types, which differ in the nature of the material that connects the bones together and in the degree of movement that is permitted.

Joints, or **articulations**, are junctions between opposing bones. Depending largely on the nature of the material that connects two bones together, they permit variable degrees of movement. Basically, a joint may be *immovable* (permit no movement between bones), be *slightly movable* (permit a little movement), or be *freely* *movable* (permit almost unrestricted movement). Joints may be categorized into three groups on the basis of their binding materials. These are fibrous joints, cartilaginous joints, and synovial joints.

Fibrous Joints

A **fibrous joint** consists of fibrous (dense) connective tissue between articulating bones. The bones are typically held quite closely together, and little or no movement is allowed. An example of a fibrous joint is a **suture**, which you have observed between the flat bones of the cranium (Fig. 7–29). At birth, sutures are still in the process of formation, and the bones of the cranium are joined by fibrous membranes. These early joints are called **fontanels**, which you may recognize as "soft spots" in the head of a recently born infant (Fig. 7–30). Fontanels permit some movement between bones to allow compression of the head as it passes through the birth canal. The fontanels soon close as the cranial bones grow together; they are replaced by immovable sutures.

In addition to sutures, two other types of fibrous joints may be found in the skeleton. **Syndesmoses** (sin'-dez-MŌ-sēz) are fibrous joints that separate the bones by some distance; the bones are supported by ligaments. An example of a syndesmosis is found between the distal ends of the tibia and fibula in the leg. **Gomphoses** (gom-FŌ-sēz) are fibrous joints consisting of pegs fitted closely into sockets and kept in place by ligaments. The joint between a tooth and its socket in the jaw is an example of a gomphosis.

Cartilaginous Joints

A joint that binds opposing bones together with cartilage is a **cartilaginous joint** (Fig. 7–31). In some cases, hyaline cartilage forms a relatively stiff, firm connection between bones such as in the costal cartilages associated with the ribs. Fibrocartilage is the binding material in other cartilaginous joints, such as the symphysis pubis of the pelvic girdle and the intervertebral discs between vertebrae of the vertebral column. Intervertebral discs receive further support from ligaments that extend between vertebrae, reducing the chances of injury to the vertebral column. Cartilaginous joints are slightly movable, owing to the limited degree of flexibility permitted by cartilage.

Figure 7–29
Fibrous joints. The flat bones of the cranium are separated by tight-fitting joints composed of fibrous (dense) connective tissue, called *sutures*. (a) Lateral view of the skull. (b) Photograph of sutures in an adult skull.

Does a suture permit much movement between articulating bones?

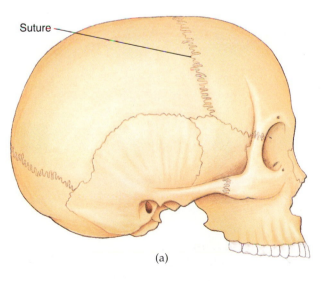

(a)

(b)

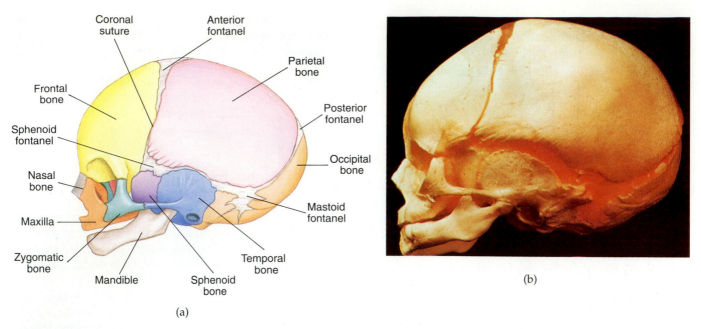

(a)

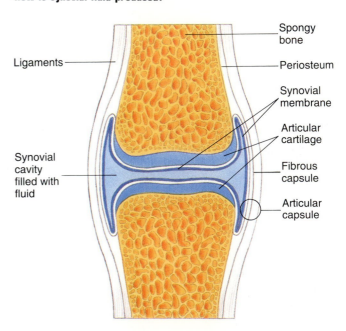

(b)

Figure 7–30

The skull of a fetus or newborn. (a) Lateral view. The fontanels connect the flat bones of the cranium. (b) Photograph of a fetal skull. *(From* Color Atlas of Human Anatomy, 2nd Edition *by R.M.H. McMinn & R.T. Hutchings, © 1988 Year Book Medical Publishers, Inc.)*

Of what type of tissue are the fontanels composed?

Synovial Joints

Most joints of the body are **synovial joints**, occurring between the numerous bones of the limbs. They include the joints of the shoulder, knee, elbow, and wrist; joints between phalanges; joints of the jaw; and many others.

Synovial joints permit a greater range of movement than fibrous and cartilaginous joints; they are regarded as freely movable. This is made possible by the presence of a fluid-filled space between opposing bones, instead of thick binding material (Fig. 7–32). The space is called a **synovial cavity**; it is enclosed within a tubular capsule known as the **articular capsule**. The

Figure 7–31

Cartilaginous joints. The bodies of adjacent vertebrae are separated by slightly movable joints composed of cartilage, called *intervertebral discs*.

Of what type of cartilage are the intervertebral discs composed?

Figure 7–32

A typical synovial joint.

How is synovial fluid produced?

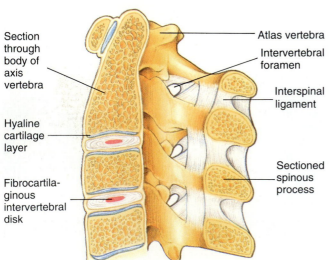

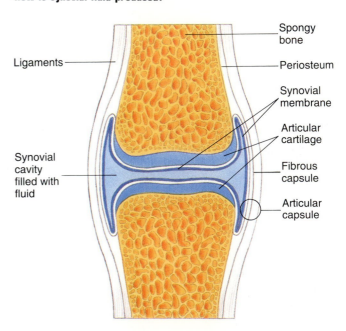

Back Injury and Intervertebral Discs

Back injuries are commonly caused by bending over to lift heavy objects. When this occurs, too much stress has been applied to the vertebral column. The result is often a condition known as a *herniated disc*, or slipped disc. The herniated disc is an intervertebral disc that has been damaged. The nature of damage usually involves a tearing of the outer ring of fibrocartilage, known as the *annulus fibrosus*, and a subsequent extrusion of the gelatinous inner core of the disc (called the *nucleus pulposus*). The extruded nucleus pulposus may press on the spinal cord or on spinal nerves exiting from the cord, causing numbness, excruciating pain, or even destruction of nerve tissue. The X-ray shown in the illustration demonstrates a protrusion of the nucleus pulposus. Herniated discs are often treated with bed rest, traction, and pain-killing drug therapy. If this is not successful, a more radical treatment may follow, including surgery that removes part of the vertebra (called a *laminectomy*) or fusion of vertebral elements.

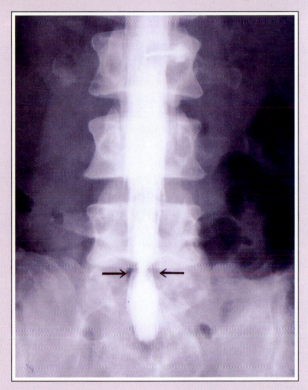

Herniated disc. (a) X-ray of the lower vertebral column in a patient with a herniated disc. The arrows point to the herniation pressing against the spinal cord. (b) Extrusion of the nucleus pulposus—which is shown pressing against the spinal cord—is the cause of the herniated disc.

(a)

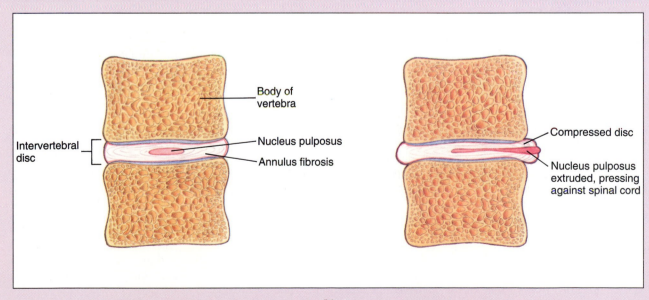

Body of vertebra

Intervertebral disc

Nucleus pulposus

Annulus fibrosis

Compressed disc

Nucleus pulposus extruded, pressing against spinal cord

(b)

outer layer of the articular capsule is continuous with the periostea of the articulating bones, and is subsequently very tough. The outer layer is called the **fibrous capsule**; its strength is further reinforced by ligaments extending from bone to bone that surround the capsule. The inner layer is a thin layer of loose connective tissue known as the **synovial membrane**, which secretes a thick, clear fluid (**synovial fluid**) into the synovial cavity. Synovial fluid provides a shock-absorbing liquid cushion between opposing bones, and a source of lubrication to reduce friction during movement. The surfaces of the articulating bones that are exposed to the synovial cavity are lined with smooth hyaline cartilage (the **articular cartilage**).

Some synovial joints contain accessory structures that improve movement. The knee, for example, has discs of fibrocartilage that subdivide the synovial cavity into two or more separate cavities (Fig. 7–33). These discs are called **menisci** (me-NIS-kī) and serve to increase the stability of the joint. Associated with certain joints are **bursae** and **tendon sheaths**, which lie outside the joint and act as "ball bearings" by reducing friction. Bursae are flattened sacs lined with synovial membrane and containing synovial fluid. They lie between bone and tendons, muscles, or ligaments of ma-

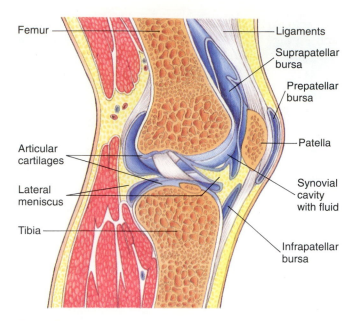

Figure 7–33
The knee joint, a specialized type of synovial joint. It is shown sectioned through the midsagittal plane in lateral view.

What benefits are provided by the presence of bursae and menisci in a synovial joint?

SPORTS CLINIC

Sports Injuries

Injuries to the body during athletic activities have increased dramatically over recent years, since people are becoming more interested in physical fitness and are participating more in sports. The most common injuries involve the joints.

Torn cartilage is an injury usually associated with the knee. It is a shearing or tearing of one of the two menisci within the knee's synovial cavity, usually on the medial side. A sudden pop or crack signals the damage, and its symptoms include pain at the joint, cracking sounds during walking, and locking (the inability to fully extend the knee). Treatment of the torn meniscus is

surgical removal of the loose piece of cartilage using an instrument known as an *arthroscope*. A more severe form of injury to the knee is a complete tearing of three of the ligaments that lie within the knee joint (medial collateral, lateral collateral, and anterior cruciate ligaments), which may accompany the torn cartilage. This injury is often called the "terrible triad" and usually proves devastating to future activities.

Joint dislocations usually involve the shoulder. The shoulder is supported by a "cuff" of four muscles that surround the shoulder joint, called the *rotator cuff*. When the arm is raised and suddenly forced back-

ward, the rotator cuff tears and the head of the humerus dislocates from its joint. Relocation of the humerus and immobilization of the shoulder for some time are required for healing.

Sprains are caused by exaggerated bending or twisting movements that stretch or tear the ligaments and tendons supporting the synovial joint. Damage to blood vessels and nerves may occur, causing the development of bruises and local pain. Swelling of the damaged joint usually follows, causing further pain and often slowing healing.

jor joints, such as the shoulder, to reduce friction. Tendon sheaths are basically elongated bursae that are wrapped about tendons in the hand and foot.

Types of Synovial Joints

Synovial joints are classified according to the physical placement of one bone in relation to another. There are six types, which are listed as follows and shown in Figure 7–34.

Gliding—Also called a *plane joint*. It contains articulating surfaces that are nearly flat. They are found between some wrist and ankle bones, between the clavicle and sternum, and between the articular processes of adjacent vertebrae. They permit sliding movements.

Hinge—The convex surface of one bone fits into the concave surface of another. Examples include the elbow, the knee, and the joints between the phalanges. Hinge joints permit movement back and forth, like the motion of a single-hinged door.

Pivot—The cylindrical surface of one bone rotates within a ring formed by another bone. The joint between the axis and atlas of the vertebral column is an example of a pivot joint. Movement is limited to rotation about a central axis.

Condyloid—An oval-shaped condyle of one bone fits into a cavity of another bone. This occurs between the proximal carpal bones of the wrist and the distal ends of the radius and ulna. Condyloid joints allow back-and-forth movement and circular movement.

Saddle—The convex surface of one bone fits into the concave surface of another. Other contours of the two bones also complement the fit. An example of a saddle joint is between a carpal (the trapezium) and the metacarpal of the thumb. These joints

Figure 7–34
The different types of synovial joints allow varying degrees of movement. The arrows indicate the range of movement that is normally permitted.
What is an example of a hinge joint?

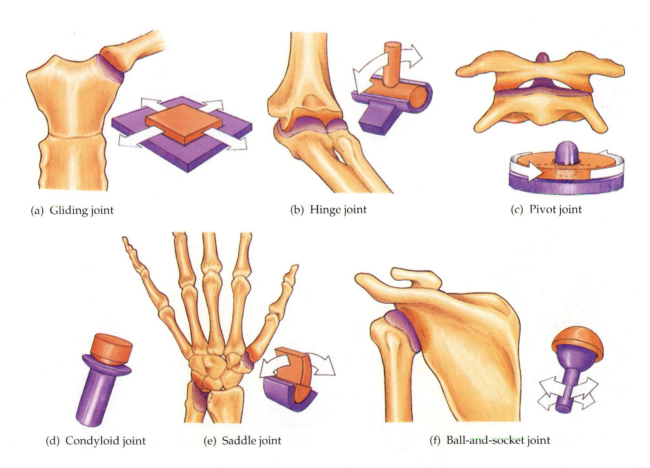

(a) Gliding joint   (b) Hinge joint   (c) Pivot joint

(d) Condyloid joint   (e) Saddle joint   (f) Ball-and-socket joint

allow back-and-forth, side-to-side, and some pivotal movements.

Ball and Socket—The ball-shaped process of one bone fits into the cup-shaped socket of another bone. The hip and shoulder joints are examples of this type of joint, which permits more movement than any other type of joint in the body.

Types of Movements at Synovial Joints

The movement of a bone at a joint is provided by the contraction of skeletal muscles. Typically, a muscle is attached by tendons to two bones. When a muscle contracts, or shortens, one bone is pulled toward the other bone, which is being held stationary by other muscles. The point of attachment of the tendon to the movable bone is called the **insertion**, and the attachment site of the tendon to the stationary bone is called the **origin**. The nature of the movement that occurs depends on the physical fit of the two articulating bones, the tightness of ligaments, and the muscles that are attached by tendons. The movement is called an **action**. There are numerous types of actions; the most important are listed as follows and shown in Figure 7–35.

Flexion—a decrease in the angle between two bones, such as bending the arm at the elbow or the leg at the knee. A special type of flexion occurs with the foot at the ankle: known as **dorsiflexion**, it occurs when the top of the foot is moved toward the ankle.

Extension—an increase in the angle between two bones, as in straightening the arm at the elbow or the leg at the knee. **Hyperextension** increases the angle beyond any normal anatomical position. A special type of extension occurs with the foot at the ankle: known as **plantar flexion**, it occurs when the foot is positioned as in standing on the toes.

Abduction—movement away from the vertical midline of the body, as in raising the thigh or arm horizontally from the side.

Adduction—movement toward the vertical midline, as when returning the arm or thigh to the side.

Circumduction—movement of a limb so that its distal end forms a circular pattern. An example is the circular movement of a finger while the hand remains still.

Rotation—movement of a part around a central axis. Twisting the head from side to side (nodding "no") is an example of rotation.

Pronation—turning the hand to point the palm downward, or posterior.

Supination—turning the hand to point the palm upward, or anterior.

Eversion—turning the foot to point the sole outward.

Inversion—turning the foot to point the sole inward.

Protraction—moving a body part straight outward from the midline. An example is sticking the chin outward.

Retraction—the opposite of protraction. It involves returning the body part straight back to the midline.

CONCEPTS CHECK

1. How do fibrous, cartilaginous, and synovial joints differ in structure and function?

2. What are the major components of a synovial joint?

3. What specific movements may be permitted by synovial joints?

HOMEOSTASIS

The skeletal system supports homeostasis of the body by providing skeletal support and protection, blood-cell production, and mineral recycling.

The skeletal system performs several important roles in the maintenance of homeostasis. In fact, each of its major functions has a direct impact on the body's internal stability. For example, skeletal support and protection of the body's soft tissues is vital for maintaining health. The importance of these functions becomes clear when they are impeded, as occurs when one suffers a fractured limb, cranium, or rib cage. The consequences of these injuries can include immobilization or physical damage to a vital organ such as the brain or a lung. In either case, homeostasis can be seriously disrupted.

In a healthy body, the skeletal system supports homeostasis on a continual basis by producing blood cells. As we have learned, blood cells are manufactured within red marrow. Because the functions of blood are vital to homeostasis, the normal production of these

Abduction

Adduction

Eversion

Inversion

Supination

Pronation

Figure 7–35
Types of movements.
(Figure continues on
next page)

Flexion

Extension

Hyperextension

Figure 7–35
Types of movements.
(continued)

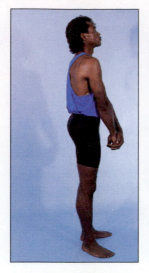

Protraction   Retraction

Circumduction

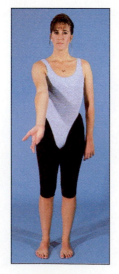

Rotation   Flexion   Extension

cells is necessary for survival. When red marrow becomes damaged by disease, blood homeostasis may be severely affected. For example, in the disease *leukemia*, the ability of red marrow to manufacture normal quantities of healthy white blood cells becomes impaired because of the development of dysfunctional marrow tissue. Large numbers of nonfunctional white blood cells result, which also reduces the number of red blood cells that can be produced because of overcrowding. As a result, the ability of the blood to defend against infection and transport oxygen to vital organs becomes impaired. If not treated, the disruption of homeostasis by leukemia usually results in death. Current treatments for leukemia include bone marrow transplants, radiation therapy, and drug therapy.

The role of the skeletal system in storing mineral salts also supports the body's homeostatic balance. The minerals that are stored mainly include calcium and phosphate, but potassium and several micronutrients are stored in bone as well. When the body is in need of them, these minerals may be released by the action of osteoclasts during bone remodeling for use elsewhere in the body. When these minerals are in abundant supply, osteoblasts may bind them into new bone matrix. In this way, the loss of minerals valuable to the body is minimized. This is particularly true for calcium, which is essential for the normal functioning of skeletal muscles, the heart, and the brain. The levels of calcium ions in the blood are continually monitored by the hypothalamus of the brain, and fluctuations in its concentrations are kept minimal by the regulated release of hormones (parathyroid hormone by the parathyroid glands and calcitonin by the thyroid gland) that control osteoclast activity and bone remodeling.

CLINICAL TERMS OF THE SKELETAL SYSTEM

Achondroplasia An inherited disease involving the conversion of cartilage to bone during bone development. This results in dwarfism with usually a normal-size trunk and head but very short extremities. There is no known treatment.

Gout Also called *gouty arthritis*; a type of acute arthritis that is caused by the accumulation of uric acid crystals in the synovial fluid of joints, usually in the lower limbs. The precipitation of the crystals, which are a waste product from protein metabolism, causes pain and inflammation. Treatment usually includes anti-inflammatory drugs.

Osteitis Deformans (Paget's Disease) A slowly progressive bone disease that is characterized by unregulated bone destruction followed by bone deposition, causing gross deformity of bones and bone pain. It occurs among 10% of people in their eighties or older, but the cause is yet unknown.

Osteoarthritis A form of arthritis in which one or many joints undergo degenerative changes, including body sclerosis (death of osteoblasts), loss of articular cartilage, uncontrolled growth of bone and cartilage within the joint, and inflammation of the synovial membrane of the joint. It is the most common form of arthritis, and its sufferers experience pain, stiffness, enlargement, and deformity of the afflicted joint. Its cause is unknown, but may include genetic, metabolic, mechanical (physical injury), and endocrine factors. Treatment includes rest of the afflicted joint, heat, and anti-inflammatory drugs; and in severe cases, joint surgery becomes necessary.

Osteomalacia and Rickets Metabolic diseases caused by a deficiency of vitamin D, which is necessary for calcium absorption from the intestinal tract into the bloodstream. Rickets is the result of this deficiency in children, where it causes weakened bones that are manifested by bowed limbs and a soft skull. Osteomalacia occurs in adults, and causes thinning and softening of bones leading to skeletal deformities. Both diseases can be treated with adequate calcium and vitamin D intake.

Osteomyelitis An infection of bone most commonly caused by the bacterium *Staphylococcus aureus*. Infection may be acquired through an open fracture or other trauma, or may spread from other tissues through the bloodstream. Treatment includes antibiotic drugs and minor surgery to drain the infected area.

Osteoporosis A decrease in total bone mass that usually affects postmenopausal women and el-

(continued)

derly men and women. It results in an increase in fractures, poor posture, and difficulty in walking. Osteoporosis represents an imbalance between bone formation and bone resorption. It can be counterbalanced by mild exercise, vitamin D and calcium intake, and, in some cases, estrogen therapy.

Rheumatoid Arthritis   A disease involving the degeneration of cartilage and dense connective tissue in a synovial joint and its replacement by bone. Ultimately, the joint becomes completely fused with bone. This form of arthritis is regarded as an autoimmune disease, as a person's own immune system destroys the synovial tissues. It is not yet understood why this occurs, and there is no cure. Treatment is limited to pain-killing and anti-inflammatory drugs.

Tumors of Bone   Tumors of bone tissue are relatively rare, but when they do occur they can be extremely painful, since they expand within an enclosed space. Most types undergo malignant change, so they are often life-threatening. The most common is malignant myeloma, which originates in the bone marrow. Other types include osteochondroma and osteogenic sarcoma, both of which are most frequent among adolescents aged 10 to 20 years.

CHAPTER SUMMARY

The skeletal system consists of bones and joints, which are composed mostly of bone tissue and cartilage. The system performs five major functions: support, protection, aid to movement, blood-cell formation, and storage of minerals.

BONE STRUCTURE

Bones are dynamic, somewhat flexible organs.

Types of Bone   All bones fall into four categories, which are based on their general shapes.
Long Bones   Greater in length than in width.
Short Bones   About equal in length and width.
Flat Bones   Thin and flat.
Irregular Bones   Usually a combination of shapes.

Parts of a Long Bone   A typical long bone contains the following parts that are visible without the aid of a microscope:
Diaphysis   The long, central shaft.
Epiphysis   The extreme ends.
Articular Cartilage   A thin layer of hyaline cartilage on the outer surfaces of the epiphyses where they form a joint.
Periosteum   A sheet of fibrous connective tissue that covers the outer surface of bone, except on the articular surface.
Compact Bone   A type of bone tissue that forms the walls of the diaphysis.
Spongy Bone   The bone tissue that makes up the inner part of the epiphysis.
Red Marrow   A connective tissue within spongy bone that manufactures blood cells.
Medullary Cavity   The central, internal space within the diaphysis of the bone.
Yellow Marrow   A fatty tissue within the medullary cavity, for energy storage.
Endosteum   A thin membrane lining the inner wall of the medullary cavity.

Bone Composition   Bone is composed mostly of bone tissue. Bone tissue contains inorganic and organic components.
Inorganic   Includes mineral salts of calcium phosphate and calcium carbonate to form crystals of hydroxyapatite.
Organic   Includes collagen in the matrix, and three types of cells: osteoblasts, which produce matrix; osteocytes embedded within lacunae; and osteoclasts, which dissolve bone matrix.

Microscopic Structure of Bone   Bone tissue is either compact or spongy.
Compact Bone   Organized into concentric units called *osteons* or *haversian systems*. Each osteon contains a single, central canal called the *osteonic canal* (or *haversian canal*); concentric layers of hardened matrix, called *lamellae*; canals that extend longitudinally between the osteonic canal and the blood supply, called *Volkmann's canals*; and tiny canals that extend between the osteonic canal and cells embedded within lacunae, called *canaliculi*.
Spongy Bone   More porous than compact bone, it consists of numerous bony plates with spaces in between filled with red marrow. The cells are embedded within lacunae in the bony plates.

BONE DEVELOPMENT AND GROWTH

The process of bone development begins before birth and proceeds in two different ways: from embryonic membranes and from cartilage.

Intramembranous Bones   Bones that develop from embryonic membranes. Spongy bone develops first, forming thin, bony plates in the center of the bone. Soon compact bone forms along the outer surfaces. Intramembranous bones are the flat bones of the skull, the mandibles, and the clavicles.

Endochondral Bones Bones that develop from cartilage. Each begins as a cartilage model; then a blood vessel penetrates into the center of the diaphysis (the primary ossification center). The flow of blood causes cartilage cells to swell and die, leaving caverns. The blood flow also brings in bone-producing cells. Bone matrix is produced by the newly arrived osteoblasts, and more blood vessels penetrate the epiphyses (the secondary ossification centers). Soon the cartilage is replaced by bone, except at the epiphyseal plate and the articular cartilages.

Bone Growth The bones of a growing child increase in length (interstitial growth) at the epiphyseal plate, where cartilage cells continually regenerate and die to be replaced by bone. Bones expand in width (appositional growth) through production of compact bone by osteoblasts in the periosteum, while the medullary cavity expands through destruction of the inner wall by osteoclasts.

BONE REMODELING

The continuous recycling of bone occurs as osteoblasts produce new matrix and osteoclasts destroy old. This provides for mineral availability to other organs, and a method for bone repair following injury.

ORGANIZATION OF THE SKELETON

The 206 bones of the skeleton are divided into axial and appendicular divisions.

Surface Features of Bones Bones contain features on their surfaces that correspond to their functions, including projections, depressions, and openings. Many share a common terminology.

Bones of the Axial Skeleton The axial skeleton includes the bones of the skull, the hyoid bone in the neck, the vertebral column, and the thoracic cage.
Skull Divided into the cranium, housing the brain, and the facial skeleton.
Cranium Includes the frontal bones, the parietal bones, the occipital bone, the temporal bones, the sphenoid bone, and the ethmoid bone.
Facial Bones Include the maxillary bones, the palatine bones, the zygomatic bones, the nasal bones, the lacrimal bones, the vomer, the inferior nasal conchae, and the mandible.
Hyoid A small, horseshoe-shaped bone in the neck region. It does not contact any other bone.
Vertebral Column Includes the 33 vertebrae, which form 26 bones. The spine is divided into regions: cervical, thoracic, lumbar, the sacrum, and the coccyx.
Thoracic Cage Includes the 12 pairs of ribs and the sternum. Most ribs are attached to the sternum by way of cartilages called the *costal cartilages*.

Bones of the Appendicular Skeleton The appendicular skeleton consists of the pectoral girdle, the upper limbs, the pelvic girdle, and the lower limbs.
Pectoral Girdles Include the clavicles and the scapulae.

Upper Limbs Include the humerus, radius, ulna, carpals, metacarpals, and phalanges on each side.
Pelvic Girdle Include the coxal bones and the sacrum. The coxal bones each contain an ilium, an ischium, and a pubis.
Lower Limbs Include the femur, tibia, fibula, tarsals, metatarsals, and phalanges on each side.

JOINTS

Joints are junctions between bones. They are classified into three categories, based on the nature of the material binding the bones together.

Fibrous Joints Joints that are bound by fibrous connective tissue. They permit little or no movement, and include sutures.

Cartilaginous Joints Joints that are bound by cartilage. They permit little movement, and include costal cartilages and intervertebral discs.

Synovial Joints Joints that contain a joint cavity within an articular capsule, which binds the bones together. The cavity is called a *synovial cavity* and contains synovial fluid secreted by an inner lining of synovial membrane. Menisci, bursae, and tendon sheaths are sometimes associated with synovial joints. They permit a wide range of movement.
Types of Synovial Joints Classification is based on the physical fit between two bones, which also determines the types of movements permitted.
Gliding Articulating surfaces are nearly flat, allowing a sliding movement (as between some wrist and ankle bones).
Hinge A convex surface of one bone fits into a concave surface of another, allowing back-and-forth movements (as in the elbow).
Pivot The cylindrical surface of one bone fits into a ring formed by another, allowing a twisting rotation (as in axis and atlas).
Condyloid The oval-shaped condyle of one bone fits into a cavity of another, permitting back-and-forth and circular movements (as in the wrist).
Saddle A convex surface fits into a concave surface, allowing back-and-forth, circular, and pivotal movements (as in the metacarpal of the thumb with the carpal).
Ball and Socket A ball-shaped process fits into a cup-shaped socket, allowing the widest range of movement (as in the hip joint).
Types of Movements at Synovial Joints Movements are caused by skeletal muscles pulling on bones while stabilizing other bones. The point of attachment of a muscle tendon to a more stationary bone is the origin, and the attachment site of a tendon to a more movable bone is the insertion. The actual movement is the action, of which there are numerous types.
Flexion Decrease in the angle between two bones. Dorsiflexion occurs at the ankle only.
Extension Increase in the angle between two bones. Plantar flexion occurs at the ankle only.

Abduction   Movement away from the vertical midline.

Adduction   Movement toward the vertical midline.

Circumduction   Movement of the distal end to form a circular pattern.

Rotation   Movement of a part around a central axis.

Pronation   Turning the hand to point the palm downward or posterior.

Supination   Turning the hand to point the palm upward or anterior.

Eversion   Turning the foot to point the sole outward.

Inversion   Turning the foot to point the sole inward.

Protraction   Moving a body part straight outward from the midline.

Retraction   Moving a body part straight toward the midline.

HOMEOSTASIS

The skeletal system helps maintain homeostasis by the support and protection of vital soft organs, the production of blood cells in the red bone marrow, and the provision of mineral storage.

KEY TERMS

action (p. 186)
articulation (p. 180)
canaliculi (p. 156)
chondroblast (p. 158)
condyle (p. 162)
diaphysis (p. 155)

endosteum (p. 155)
epiphysis (p. 155)
foramen (p. 162)
fossa (p. 162)
hematopoiesis (p. 153)

hydroxyapatite (p. 155)
insertion (p. 186)
lamellae (p. 156)
notch (p. 162)
origin (p. 186)

osteoblast (p. 156)
osteoclast (p. 156)
osteocyte (p. 156)
osteon (p. 156)
periosteum (p. 155)

QUESTIONS FOR REVIEW

OBJECTIVE QUESTIONS

1. Which of the following is *not* a major function of the skeletal system?
   a. support
   b. regulation of body temperature
   c. blood-cell formation
   d. storage of mineral salts

2. A type of bone that is thin and flat, such as the bones that form the cranium, is regarded as a(n):
   a. long bone
   b. irregular bone
   c. flat bone
   d. heavy bone

3. The extreme ends of a long bone that contain spongy bone are called:
   a. diaphysis
   b. epiphyses
   c. articular cartilage
   d. periosteum

4. The sheet of fibrous connective tissue that envelops bone and provides attachment for ligaments and tendons is the:
   a. periosteum
   b. synovial membrane
   c. endosteum
   d. diaphysis

5. An organic component of bone tissue that actively produces collagen and hydroxyapatite is the:
   a. osteoblast
   b. osteoclast
   c. calcium carbonate
   d. chondroblast

6. Nutrients are supplied to osteocytes within lacunae by way of tiny channels that extend from the osteonic canal to lacunae through:
   a. red marrow
   b. canaliculi
   c. transverse communicating channels
   d. the medullary cavity

7. Bony plates with spaces in between filled with red marrow characterize which type of tissue?
   a. compact bone
   b. long bone
   c. hyaline cartilage
   d. spongy bone

8. The flat bones of the skull, the mandibles, and the clavicles arise during the developmental stage from:
   a. cartilage
   b. other bones
   c. embryonic membranes
   d. all of the above

9. During endochondral bone development, the primary ossification center arises as a result of:
   a. penetration by a blood vessel
   b. cartilage production
   c. expansion of the cartilage model
   d. osteoblast destruction

10. Growth in length of a long bone is achieved by the conversion of cartilage to bone at the:
    a. primary ossification center
    b. periosteum
    c. epiphyseal plate
    d. secondary ossification center

11. The interconversion of minerals between bone and the bloodstream in the skeleton of an adult is known as:
    a. homeostasis   c. bone growth
    b. bone remodeling   d. none of the above

12. The axial skeleton consists of the skull, hyoid bone, vertebral column, and:
    a. pectoral girdle   c. thoracic cage
    b. pelvic girdle   d. clavicles

13. The "soft spots" in the skull of an infant are regions where the sutures have not yet closed, and are called:
    a. fontanels   c. sinuses
    b. foramen   d. plane joints

14. The coronal suture is the joint between which two bones?
    a. frontal and temporal c. parietal and frontal
    b. temporal and sphenoid d. maxillary and frontal

15. The bone that contains the prominent foramen magnum and articulates with the atlas is the:
    a. axis   c. occipital bone
    b. parietal bone   d. mandible

16. The zygomatic process, styloid process, and mastoid process may all be found on which bone?
    a. temporal bone   c. zygomatic bone
    b. occipital bone   d. sphenoid bone

17. The nasal septum is formed by the perpendicular plate of the ethmoid, the cartilages, and the:
    a. nasal bones   c. vomer
    b. inferior nasal conchae d. maxillary bones

18. The large bone that forms the thigh and articulates with the pelvic girdle at the hip joint is the:
    a. humerus   c. tibia
    b. femur   d. sternum

19. The large, superior expanded portion of the coxal bone is the:
    a. ilium   c. sacrum
    b. ischium   d. pubis

20. Which type of joint is regarded as freely movable?
    a. fibrous   c. cartilaginous
    b. synovial   d. suture

21. Which joint is correctly matched with the type of joint indicated?
    a. sagittal suture—cartilaginous
    b. symphysis pubis—fibrous
    c. intervertebral disc—synovial
    d. knee—synovial

22. The shoulder joint is an example of which type of synovial joint?
    a. condyloid   c. pivot
    b. ball-and-socket   d. hinge

23. Which movement occurs at the shoulder when you raise your arms to the side of your body and form a T?
    a. flexion   c. abduction
    b. pronation   d. retraction

ESSAY QUESTIONS

1. Describe the process of endochondral bone formation, and compare it with intramembranous bone formation.

2. Compare and contrast the three types of joints, indicating the range of movements that each type permits and why.

3. Consider how the body might appear without a skeleton. Describe this appearance, and discuss why survival on land would not be possible in this circumstance.

A N S W E R S   T O   A R T   L E G E N D   Q U E S T I O N S

Figure 7–1

How does a long bone differ from a flat bone?   A long bone is greater in length than in width, while a flat bone's distinguishing feature is its flat shape.

Figure 7–2

What functions are served by the periosteum?   The periosteum serves as a point of attachment for ligaments and tendons; it provides for bone nourishment by way of its many blood vessels; and its osteoblasts are active in bone growth and repair.

Figure 7–3

What materials are secreted by the osteocyte?   The osteocyte secretes bone matrix, which includes mainly mineral salts (hydroxyapatite crystals) and collagen.

Figure 7–4

What are the components of an osteon?   An osteon consists of a circular arrangement of plates of bone matrix, called *lamellae*. Between the plates are lacunae, which contain osteocytes. Tiny channels, called *canaliculi*, penetrate through the lamellae and interconnect the lacunae. At the center of the lamellae is a central canal called an osteonic canal, which contains blood vessels and nerves.

Figure 7–5

Why does the entrance of a blood vessel into the early cartilage model begin ossification?   The entrance of a blood vessel brings in a fresh supply of blood, causing the chondroblasts to expand and either burst or transform into osteoblasts. The blood also transports osteoblasts into the site. Once the osteoblasts have arrived, they begin secreting new matrix.

Figure 7–6
Why is the vertebral column considered part of the axial skeleton?
The vertebral column is located along the central (midline) axis of the body.

Figure 7–7
Is the frontal bone a flat bone or a long bone?   The frontal bone is a flat bone.

Figure 7–8
What cavity in the head do the skull sinuses drain into?   The sinuses within the skull drain into the nasal cavity.

Figure 7–9
How many bones of the cranium are visible in this lateral view?
All six bones of the cranium are visible (frontal, parietal, occipital, temporal, sphenoid, and ethmoid bones).

Figure 7–10
What structure extends through the foramen magnum in life?
The spinal cord extends through the foramen magnum.

Figure 7–11
What bones form the nasal septum?   The vomer and the perpendicular plate of the ethmoid form the nasal septum.

Figure 7–12
The vertebrae of which region articulate with the ribs?   The thoracic vertebrae articulate with the ribs.

Figure 7–13
What bone articulates with a superior articulating process?   The vertebra that is adjacent superiorly articulates with a superior articulating process.

Figure 7–14
Which of these two vertebrae articulates with the skull?   The atlas articulates with the skull.

Figure 7–15
What structures extend through the sacral foramina?   Nerves and blood vessels extend through the dorsal sacral foramina and the pelvic foramina.

Figure 7–16
How do true ribs differ from false ribs?   True ribs are attached to the sternum directly (without branching) by way of costal cartilages, whereas false ribs either are attached by way of branching costal cartilages or are not attached at all (as in floating ribs).

Figure 7–18
What two bones articulate with the clavicle?   The sternum (at the medial end) and the humerus (at the lateral end) articulate with the clavicle.

Figure 7–19
What bone articulates with the scapula at the glenoid cavity?
The humerus articulates with the scapula at the glenoid cavity.

Figure 7–20
What joint do the capitulum and trochlea help to form?   The capitulum and trochlea of the humerus help form the elbow joint.

Figure 7–21
What process fits into the trochlear notch of the ulna in life?
The trochlea of the humerus fits into the trochlear notch, which forms much of the elbow.

Figure 7–22
Which bones form the body of the hand?   The metacarpal bones form the body of the hand.

Figure 7–23
In the female pelvis, which opening forms the walls of the birth canal?   The pelvic inlet forms the walls of the birth canal.

Figure 7–24
What differences can you identify between female and male pelvises?   Refer to Table 7-2 for a complete list of differences.

Figure 7–25
What are the three fused bones that form the adult coxal bone?
The ilium, ischium, and pubis form the adult coxal bone.

Figure 7–26
The head of the femur articulates with what bone?   The head of the femur articulates with the acetabulum of the coxal bone (to form the hip joint).

Figure 7–27
Is the position of the fibula lateral or medial to the tibia?   The fibula is lateral to the tibia.

Figure 7–28
What prominent bone of the foot forms the heel?   The calcaneus forms the heel.

Figure 7–29
Does a suture permit much movement between articulating bones?
No, a suture permits very little or no movement between articulating bones.

Figure 7–30
Of what type of tissue are the fontanels composed?   Fontanels are composed of fibrous (dense) connective tissue.

Figure 7–31
Of what type of cartilage are the intervertebral discs composed? The intervertebral discs are composed of fibrous cartilage.

Figure 7–32
How is synovial fluid produced?   Synovial fluid is secreted by cells within the synovial membrane.

Figure 7–33
What benefits are provided by the presence of bursae and menisci within a synovial joint?   Bursae reduce friction within a joint, and menisci help to stabilize a joint.

Figure 7-34
What is an example of a hinge joint?   The elbow, the knee, and the joints between the phalanges are all examples of hinge joints.

THE MUSCULAR SYSTEM

LEARNING OBJECTIVES

After studying this chapter, you should be able to:

1. Indicate the primary functions of muscles.

2. Describe the connective tissues associated with muscle.

3. Identify and describe the microscopic components of skeletal muscle tissue.

4. Identify the components of the neuro-muscular junction.

5. Explain the sliding-filament mechanism of muscle contraction.

6. Describe, in their proper order of occurrence, the events leading to muscle contraction.

7. Indicate the roles of ATP in muscle contraction, and how this energy is supplied.

8. Describe the phenomenon of oxygen debt and how it may lead to fatigue and cramping.

9. Define threshold stimulus, and relate it to the concept of the all-or-none response.

10. Using a myogram, compare and contrast the different types of muscle contractions.

11. Define *origin* and *insertion*, and describe how muscle contraction produces movement by way of group actions.

12. Identify the major muscles on the basis of their locations, origins, insertions, and actions.

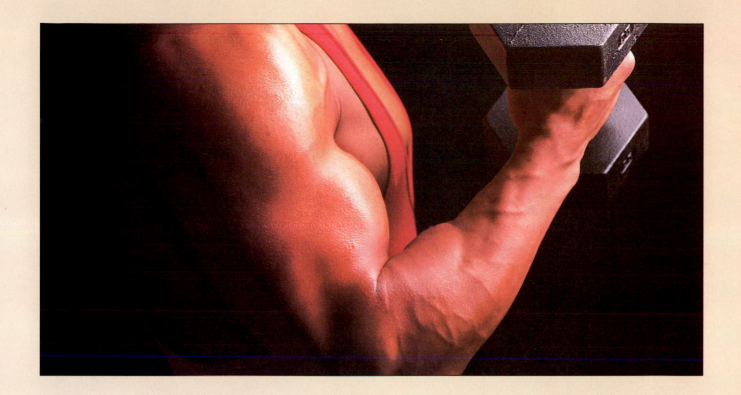

Muscle tissue consists of specialized muscle cells that are each endowed with four important properties. Although other cells of the body exhibit some of these properties, the combination of all four distinguishes the cells of muscle tissue from the cells of other tissues. **Contractility** is the ability of a cell to shorten in length. When the cell is connected to other structures, it produces a force that can result in movement. **Excitability** is the ability to receive and respond to a stimulus. A stimulus is an environmental change, such as temperature, pH, or, as we will see in the case of muscle, the presence of a chemical called a *neurotransmitter*. **Extensibility** is the ability to increase in length (that is, to stretch or extend). Muscle cells shorten in length when they contract and stretch when they relax. Finally, **elasticity** is the ability of a cell to return to its resting form after it has contracted or stretched.

As you will recall from Chapter 4, there are three types of muscle tissue in the body: skeletal muscle, smooth muscle, and cardiac muscle. The muscular system is made up only of skeletal muscle tissue (along with its associated connective tissues, nerve tissue, and blood supply). Smooth muscle tissue, on the other hand, belongs to any system of which it forms a part; for example, the smooth muscle tissue forming the walls of the stomach is part of the digestive system. Cardiac muscle tissue is found only in the heart wall. It is, therefore, part of the cardiovascular system. The three types of muscle tissue and an organ containing each are illustrated in Figure 8–1.

Skeletal muscle tissue forms the bulk of the organs in the muscular system. These organs are the **muscles** (or **skeletal muscles**) and account for about 40% to 50% of your total body weight. All 500 or so muscles of the body share three major functions:

Movement—The production of movement of body parts is the major function of muscles. Each motion relies on the integrated functioning of bones, joints, nerves, and nearby muscles to achieve a coordinated movement.

Support—The rigid connections between muscles and bones help to hold the body in an upright posture and strengthen the skeletal frame.

Heat production—Muscle produces heat as a by-product of movement. The heat serves to help maintain normal body temperature. Heat production is one important way the muscular system helps maintain homeostasis.

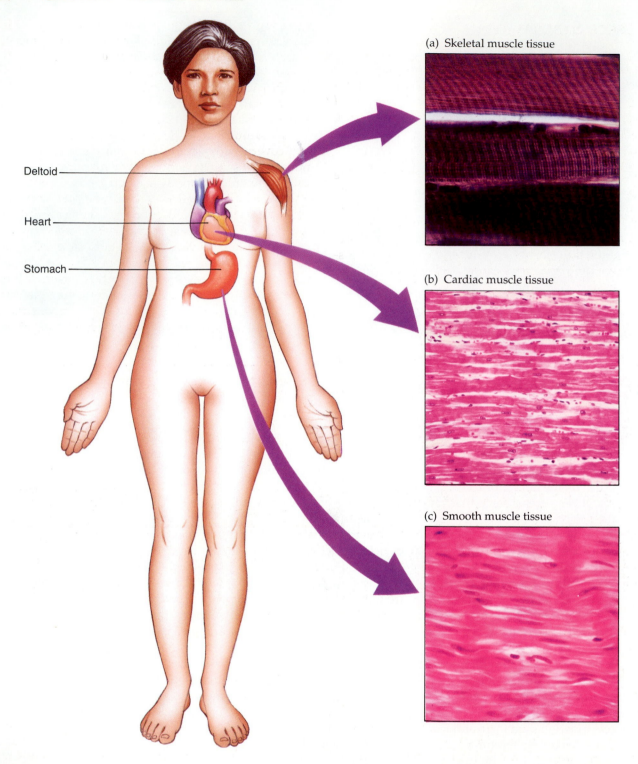

(a) Skeletal muscle tissue

(b) Cardiac muscle tissue

(c) Smooth muscle tissue

Deltoid

Heart

Stomach

Figure 8–1
The three types of muscle tissue. (a) Skeletal muscle is attached to bones and is the major component of the muscles forming the muscular system. (b) Cardiac muscle forms the walls of the heart. (c) Smooth muscle is found in the walls of visceral organs and blood vessels.
Which type of muscle tissue is the main component of the organs of the muscular system?

MUSCLE STRUCTURE

The connective tissue of muscle is primarily fascia, with which it is closely integrated. The muscle itself consists mainly of muscle cells, whose protein subunits are arranged in a special way to permit contraction. Associated with each muscle is a nerve supply.

A muscle is an organ that usually reaches from one bone to another. It is composed mainly of skeletal muscle tissue, but connective tissue and nerve tissue also form important parts. In this section we will look at muscle structure by examining its associated connective tissues, the structural organization of skeletal muscle tissue, and the connections between muscle and nerve tissue.

Connective Tissues of Muscle

The most abundant connective tissue associated with muscle is **fascia** (FASH-ē-a). In general, the term *fascia* refers to a sheet or broad band of dense connective tissue that may occupy the narrow space between skin and muscle or may surround muscle and other organs. In Chapter 6 we learned that **superficial fascia** (the hypodermis) is beneath the skin. A second type of fascia, known as **deep fascia**, is the type that is associated with muscles. The deep fascia of muscle surrounds a muscle, and thereby serves to support it and hold it together into a single unit. Like other forms of fascia, the deep fascia of muscle also provides an important route for passage of blood vessels and nerves.

In addition to deep fascia are other connective tissue coverings associated with muscle (Fig. 8–2). The outermost covering envelops the entire muscle and is called the **epimysium**;[1] it is deep fascia. Deep to it is a thinner layer of dense connective tissue known as the **perimysium**.[2] The perimysium penetrates into the muscle dividing it into compartments, called bundles or *fascicles* (FA-si-kulz), of skeletal muscle cells. Each individual muscle cell is enveloped by the deepest, thinnest portion of dense connective tissue, the **endomysium**.[3] Each of these three coverings transmits blood vessels and nerves to muscle components, as well as providing an interconnecting network of protein fibers that structurally support muscle.

[1]Epimysium: *epi-* = upon, on top of + *-mysium* = presence of muscle.

[2]Perimysium: *peri-* = around + *mysium*.

[3]Endomysium: *endo-* = within + *mysium*.

SPORTS CLINIC

Tendon Injuries

The protein fibers that form a tendon are braided with one another, forming a strong, durable attachment to bone or another muscle. Where a tendon attaches to a bone, there is usually a rough tubercle, line, or ridge on the bone to strengthen the connection. The integrity of this connection is very important, for it must withstand great amounts of tension as the muscle contracts. During normal activity, such as walking, lifting small objects, and swinging your arms, a tendon usually experiences only about 25% of the maximal stress it can withstand. However, very rapid, unexpected stretches of a tendon can cause it to rupture. For example, stepping into a hole in the ground unexpectedly while strolling or jogging may exceed the tension limits of the Achilles tendon attaching to the heel, resulting in its rupture. Another example is loading an excessive amount of weight on a weight machine and struggling to lift it before allowing your muscles to warm up gradually.

The price to pay for a ruptured tendon can be great. For example, if it is torn completely, surgery is necessary to reconnect the torn ends, and the joint and muscle group it supports must be immobilized (usually by casting) for an extended period of time. In some cases, total recovery may not be achievable, since the formation of scar tissue often impedes the range of movement and strength of contraction. For tears that are not complete, the healing process occurs more rapidly and successfully. *Tendinitis* is usually not associated with the rupture of an afflicted tendon, but refers instead to a condition of local pain and inflammation of a tendon due to its overuse. It occurs in athletes who repeatedly use the muscle to which the tendon is attached.

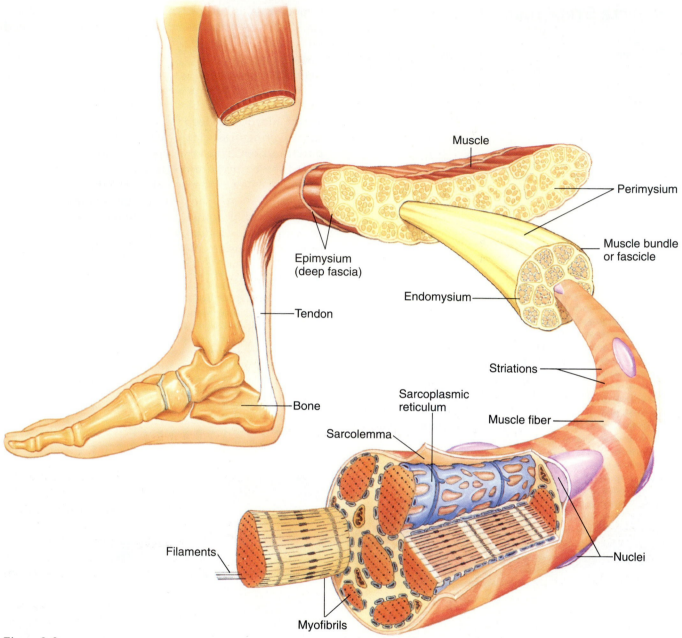

Figure 8–2

Structure of a muscle. A muscle is supported by several layers of connective tissue, which are made visible in this cutaway view. The muscle itself is composed of smaller units, called *muscle bundles* (or *fasciculi*). Each muscle bundle is composed of numerous muscle fibers, each of which contains a number of cylindrical subunits known as *myofibrils*. Myofibrils are made up of protein subunits called *filaments*.

Which layer of connective tissue surrounds individual muscle fibers?

As a muscle nears its point of attachment to a bone or another muscle, the three layers of connective tissue converge to form a single band or broad sheet that extends away from the fleshy part of the muscle. A thick band of dense connective tissue that forms most connections between muscle and bone is called a **tendon**. Fibers from the tendon converge into the periosteum of the bone to strengthen this important connection. A broad sheet of dense connective tissue extending away from the fleshy part of some muscles is known as an **aponeurosis**. An aponeurosis may attach a muscle to a bone or, in some cases, a muscle to another muscle.

Other connective tissues associated with muscle include loose connective tissue (areolar tissue) and adipose tissue. These tissues are usually located external to the epimysium and serve to support its attachment to other bones or muscles, to provide for energy storage, and to cushion the muscle from physical injury.

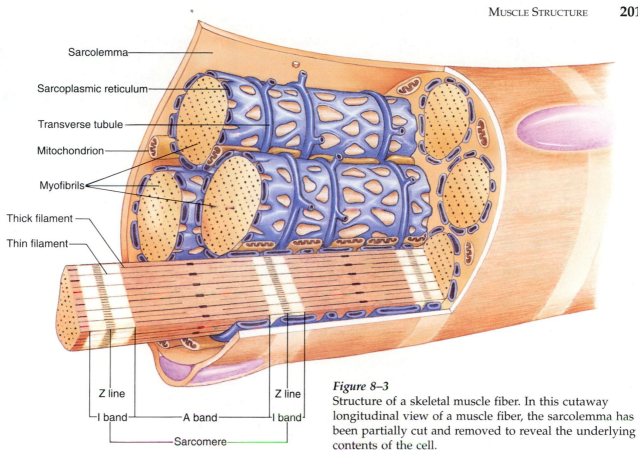

Sarcolemma

Sarcoplasmic reticulum

Transverse tubule

Mitochondrion

Myofibrils

Thick filament

Thin filament

Z line

I band

A band

Z line

I band

Sarcomere

Figure 8–3
Structure of a skeletal muscle fiber. In this cutaway longitudinal view of a muscle fiber, the sarcolemma has been partially cut and removed to reveal the underlying contents of the cell.

What structures occupy most of the space within the sarcoplasm?

Microscopic Structure of Muscle

A single cell of skeletal muscle tissue is called a **muscle fiber**, owing to its extremely long, cylindrical shape (some fibers of the thigh are over 0.5 meter in length, yet less than 0.1 millimeter in diameter!). They are typically the length of the muscle of which they form a part, and contain many nuclei (a feature called *multinucleate*). The plasma membrane of each fiber is called a **sarcolemma**,[4] and the cytoplasm is called **sarcoplasm**[5] (Figs. 8–2 and 8–3). These alternative terms for "plasma membrane" and "cytoplasm" are used because of the unique nature of skeletal muscle cell structure, which you are about to discover.

Skeletal muscle fibers are specialized to shorten, or contract, and return to their original position when they relax. Their internal structure reflects these specializations. Within the sarcoplasm are hundreds of mitochondria, which furnish (in the form of ATP) about 95% of the energy necessary for contraction. Also within the sarcoplasm is a membranous sac similar to the endoplasmic reticulum of other cells, called the **sarcoplasmic reticulum**.[6] It functions in the storage of calcium, which is needed for muscle contraction. Between adjacent sacs is a tube, called the **transverse (T) tubule**, which unites with the sarcolemma. The transverse tubules help activate the calcium-transport mechanism

in the early stages of contraction. Most of the internal space deep to the sarcoplasmic reticula is occupied by cylindrical cords of protein, called **myofibrils**,[7] which lie parallel to each other as they extend throughout the length of the fiber. Myofibrils contain two types of protein filaments: **thick filaments**, composed of the protein myosin; and **thin filaments**, composed of the proteins actin, troponin, and tropomyosin.

The arrangement of thick and thin filaments is in a distinct, consistent manner. Scientists have been able to map their arrangement, which is shown in Figures 8–3 and 8–4. Basically, the arrangement consists of alternating bands of thick and thin filaments along the length of the myofibril. In regions where the thick and thin filaments overlap, it is known as the **A band**. The term *A band* is used because this region of overlap appears dark (anisotropic) when viewed in a microscope. In regions where only thin filaments occur, it is called **I band**. The term *I band* is used because the presence of only thin filaments provides a microscopic view that

[4]Sarcolemma: Greek *sarco-* = flesh + *lemma* = husk, outer shell.

[5]Sarcoplasm: *sarco-* + *plasm* = matter.

[6]Sarcoplasmic reticulum: *sarcoplasmic* = of fleshy matter + *reticulum* = tiny network.

[7]Myofibrils: *myo-* = muscle + *fibril* = tiny fiber.

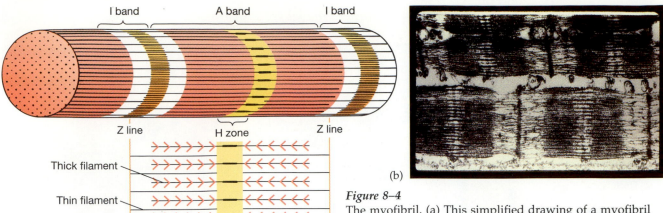

(a)

(b)

Figure 8–4
The myofibril. (a) This simplified drawing of a myofibril shows the highly organized pattern of thick and thin filaments to form the A band, I band, and H zone of the sarcomere. (b) An electron micrograph of skeletal muscle. Can you identify the Z lines, A bands, and I bands in this photograph?

What protein fibers make up the light bands in a skeletal muscle fiber?

is light (isotropic). The alternating bands of dark and light regions form the striations that are visible through the microscope. The electron micrograph in Figure 8–4b shows this pattern clearly.

Thin strands of protein extending perpendicular to the length of the myofibril intersect the thin filaments at regular intervals. These protein strands form **Z lines**, which line up with Z lines in other myofibrils within the same fiber. The segment of a myofibril between two Z lines is called a **sarcomere**.[8] Each sarcomere contains half of two I bands and a central A band. Within the A band is a region that is less dense in its center. This is the **H zone**, where there is no overlap of thin filaments. Observe the diagram of a segment of myofibril in Figure 8–4, and take special note that we can describe a myofibril as being composed of successive sarcomeres arranged in a linear fashion.

Nerve Supply

The outsides of most plasma membranes contain more positively charged ions than negatively charged ions, providing them with a net positive charge. The inside of a plasma membrane has a negative charge, due to its greater number of negatively charged ions. This difference in charges across a plasma membrane causes a small voltage difference or potential, which is called **resting membrane potential**. We will discuss this topic in more detail in Chapter 9.

When a nerve cell or muscle cell is stimulated, its resting membrane potential is changed for a very brief period of time. The stimulus causes ion channels in the membrane to open, allowing positively charged ions (sodium ions) to rush into the cell. This causes the inside of the plasma membrane where ion channels are opened to become positively charged. Soon the ion channels close, and sodium ions are transported back to the outside of the cell. Once this is complete, the resting membrane potential is restored. This brief reversal of charges across the plasma membrane is called an **action potential**.

The nerve cells that carry action potentials to skeletal muscle fibers are called **motor neurons**. The motor neuron originates from the spinal cord or brain and extends through the body until it terminates at junctions with many muscle fibers. The functional unit consisting of a single motor neuron and the numerous muscle fibers it stimulates is called a **motor unit** (Fig. 8–5).

The site where the motor neuron and each muscle fiber unite is highly specialized for transferring the action potential from nerve tissue to muscle tissue (Fig. 8–6). At this site the muscle fiber looks a bit different: its plasma membrane (the sarcolemma) is recessed to form a pocket called a **synaptic cleft**, and its sarcoplasm contains many mitochondria. The sarcolemma lining the synaptic cleft is highly folded, and is called the **motor end plate**. A terminal branch of the motor neuron fits into the synaptic cleft, leaving a small gap between the two membranes. This special site, which includes the terminal end of a motor neuron, a motor end plate, and the narrow space in between, is referred to as the **neuromuscular junction**.

Within the cytoplasm of the motor neuron at its terminal end are many tiny sacs, called **synaptic vesicles**. These vesicles contain **acetylcholine** (as'-ē-til-

---

[8]Sarcomere: *sarco-* = flesh + *-mere* = segment.

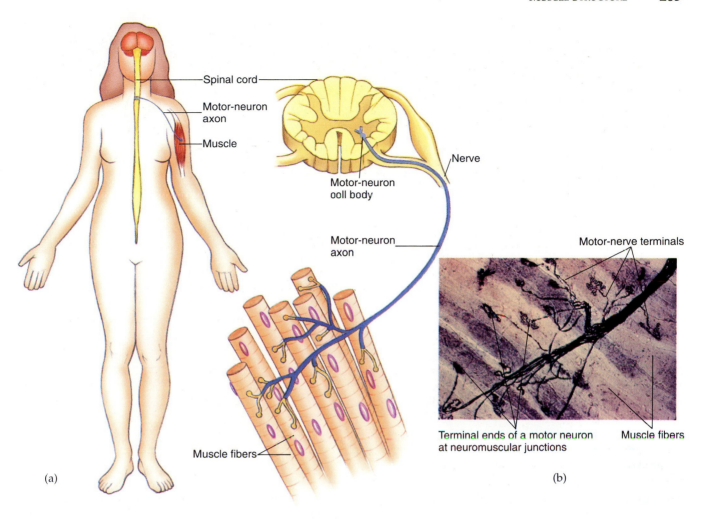

Spinal cord

Motor-neuron axon

Muscle

Nerve

Motor-neuron ooll body

Motor-neuron axon

Motor-nerve terminals

Muscle fibers

Terminal ends of a motor neuron at neuromuscular junctions

Muscle fibers

(a)

(b)

Figure 8–5
The motor unit. (a) A motor unit consists of a single motor neuron and its connections to numerous skeletal muscle fibers. (b) A photomicrograph of a motor unit.

How many motor neurons are in a single motor unit?

KŌ-lēn), or **ACh**, which is a type of **neurotransmitter**. A neurotransmitter is a chemical that carries a signal from one nerve terminal to another neuron or a muscle cell.

When an action potential arrives at the terminal end of the motor neuron, the synaptic vesicles are stimulated to release acetylcholine. This chemical diffuses across the synaptic cleft to the motor end plate of the muscle fiber. Once it reaches the motor end plate, acetylcholine binds to receptor molecules in the muscle cell membrane, causing the ion channels in the membrane to open. As a result of the open channels, sodium ions flow into the cell, initiating an action potential in the muscle cell, which causes the cell to contract. Exactly how the muscle cell contracts is the topic of our next section.

CONCEPTS CHECK

1. What layer of connective tissue transmits tiny blood vessels to individual muscle cells?

2. What is a cylindrical cord of protein that extends the length of the muscle fiber called?

3. What is the segment of protein between two adjacent Z lines called?

4. What structure does acetylcholine first contact once it leaves the motor neuron and diffuses across the synaptic cleft?

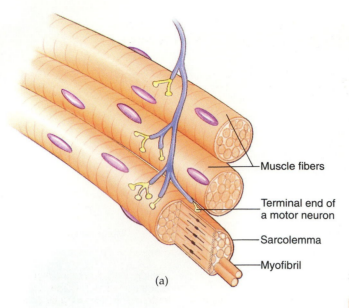

- Muscle fibers
- Terminal end of a motor neuron
- Sarcolemma
- Myofibril

(a)

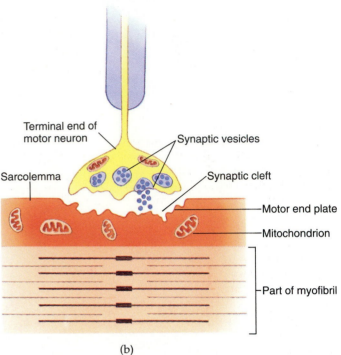

- Terminal end of motor neuron
- Synaptic vesicles
- Sarcolemma
- Synaptic cleft
- Motor end plate
- Mitochondrion
- Part of myofibril

(b)

Figure 8–6
The neuromuscular junction. (a) The terminal end of a motor neuron contacts the sarcolemma of a skeletal muscle fiber. (b) In this cutaway close-up, we can see that the contact between neuron and muscle is separated by a narrow space, called the *synaptic cleft*. Notice the neurotransmitters (represented by blue dots) diffusing across the synaptic cleft.

What neurotransmitter is packaged within the motor neuron synaptic vesicles?

PHYSIOLOGY OF MUSCLE CONTRACTION

Muscle contraction is accomplished when the sarcomeres of muscle fibers shorten in length. This movement requires a stimulus, calcium ions, and energy in the form of ATP.

In a motor unit, a single motor neuron stimulates an average of 150 muscle fibers (the actual range is from 25 to as many as 3000), which contract simultaneously to provide a smooth contraction. How does an individual muscle fiber contract, once it receives the stimulus? The contraction process is achieved by the sliding action of the thin filaments inward toward the H zones, causing each sarcomere along the length of the myofibril to shorten. When all myofibrils shorten simultaneously, the muscle fiber shortens. This concept is called the **sliding filament mechanism** and is described in more detail in the following paragraphs.

The Fiber at Rest

Before a stimulus is received, the muscle fiber is in a state described as "at rest." In this state most available calcium ions are being stored within the sarcoplasmic reticulum. ATP molecules are chemically bound to thick filaments (myosin proteins), and the thin fila-

ments are intact with all three proteins (actin, troponin, and tropomyosin). The fiber is ready to contract should a stimulus occur.

Role of the Stimulus

The release of large amounts of acetylcholine into the synaptic cleft signals the beginning of the events leading to muscle contraction. Acetylcholine release provides the stimulus that is required for muscle contraction to begin. Once acetylcholine contacts the motor end plate of the muscle fiber, an action potential is generated that travels along the sarcolemma, down the T-tubule membranes, and through the membranes of the sarcoplasmic reticulum. The sarcoplasmic reticulum membrane is altered by the action potential, resulting in an increase in its permeability to calcium ions. The ions are thereby released into the sarcoplasm and diffuse to the myofibrils.

Muscle Contraction

Once calcium ions reach the myofibrils, they bind to troponin molecules in the thin filaments (Fig. 8–7). This causes actin and troponin molecules to undergo a change in shape, resulting in the exposure of binding

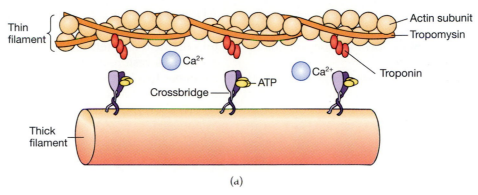

Thin filament

Actin subunit

Tropomysin

Ca²⁺

Ca²⁺

Troponin

Crossbridge

ATP

Thick filament

(a)

Figure 8–7
A model of the sliding filament mechanism of muscle contraction. (a) At rest. The thin filament is intact, and ATP is bound to the myosin thick filament. (b) Calcium ions bind to the thin filament, displacing troponin and exposing the binding site on actin. (c) As a result, the myosin cross bridge attaches to actin. (d) The splitting of ATP provides the energy to move the myosin cross bridges, which slide the thin filament past the thick filament. (e) As another ATP molecule binds to the myosin thick filament, the cross bridge detaches from actin and returns to its original angle.

What is the role of calcium ions in muscle contraction?

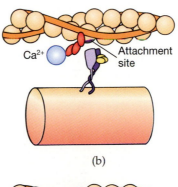

Ca²⁺

Attachment site

(b)

Calcium ions bind to the thin filament, changing shape of troponin, and exposing binding site on actin subunit.

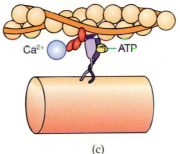

Ca²⁺

ATP

(c)

Crossbridge attaches to the actin subunit on the thin filament.

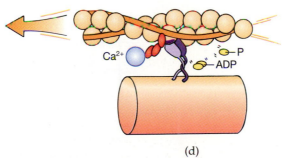

Ca²⁺

P

ADP

(d)

ADP and P are released producing energy that is used to shift the crossbridge and slide the thin filament.

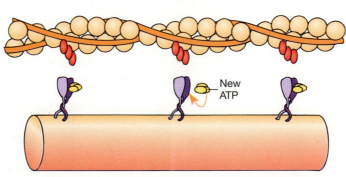

New ATP

(e)

New ATP attaches, breaking attachment of crossbridge. Crossbridge bends back to original angle.

sites on the thin filaments. Once these new sites become available, portions of the thick filaments, called **cross bridges**, bind to them, forming connections between thin and thick filaments.

About the same time the cross bridge connections occur, calcium ions activate the breakdown of ATP that is bound to the thick filaments. In this reaction, myosin plays the role of an enzyme by catalyzing the breakdown of ATP into $ADP + PO_4^{2-}$ and energy. Part of the energy released is used to move cross bridges, and part is released as heat. The heat released during muscle contraction increases body temperature, which is why your body becomes warm during exercise. In fact, about 80% or more of your body heat is produced by muscle contraction.

Movement of the cross bridges causes the thin filaments to be drawn toward the center of the sarcomere. This mechanism begins as each cross bridge tilts inward. Once the tilting action is complete, another ATP molecule binds to myosin, causing the thick filament's cross bridge to break from its attachment to the thin filament and return to its original angle. Once its position is restored, the cross bridge re-forms its attachment to a binding site on the thin filament, but in a position closer to the sarcomere's center. The cycle of cross bridge formation, movement, and release occurs again and again, resulting in a continual movement of thin filaments toward the center of the sarcomere. Consequently, the Z lines of the sarcomere are drawn together and the sarcomere shortens (Fig. 8–8). The simultaneous shortening of sarcomeres in a myofibril causes it to shorten as well, leading to the contraction of the muscle fiber.

When a person dies, ATP becomes unavailable because of the shutdown of the cell's metabolic activities. The cross bridges that have formed are, therefore, not released, and a condition of muscular rigidity called **rigor mortis** occurs.

Figure 8–8
The sliding filaments of the sarcomere cause muscle contraction. (a) The sarcomere at rest. (b) Once calcium ions are made available, the myosin cross bridges attach to actin. ATP is utilized to shift the cross bridges; the shifting moves the thin filaments inward. When all thin filaments move across the thick filaments simultaneously, the Z lines are drawn inward and the sarcomere shortens to produce a state of contraction. (c) Complete contraction is achieved once the thin filaments have slid inward as far as they may go.

What is the role of ATP during muscle contraction?

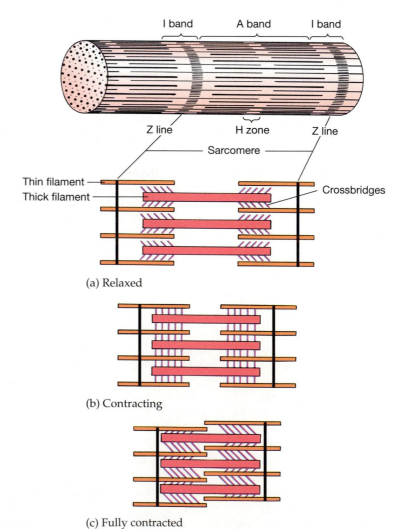

I band   A band   I band

Z line   H zone   Z line

Sarcomere

Thin filament
Thick filament

Crossbridges

(a) Relaxed

(b) Contracting

(c) Fully contracted

Return to Rest

When the action potential passing down the motor neuron stops, acetylcholine release comes to a halt. However, the stimulus does not end until all remaining molecules of acetylcholine on the motor end plate are inactivated. Their inactivation is accomplished by an enzyme in the sarcolemma called *acetylcholinesterase*, or *AChE* (also known as *cholinesterase*). Immediately following acetylcholine inactivation, calcium ions are returned to the sarcoplasmic reticulum by active transport, which requires additional energy in the form of ATP. With the absence of calcium ions, the original shape of the thin filaments is restored. The sudden lack of binding sites on the thin filaments causes their attachments to cross bridges to break, and the thin filaments slide back to their original position in the sarcomere. Within a fraction of a second, the muscle fiber relaxes. The muscle fiber is readied for the next stimulus as ATP is quickly regenerated by mitochondria.

Energy for Contraction

Energy is required for contraction in many ways. The three most direct uses of energy, as we have just seen, include: (1) the mechanical movement of cross bridges, (2) the breakage of cross bridge attachments from thin filaments, and (3) the return of calcium to the sarcoplasmic reticulum. Energy is made available for use when the high-energy bonds in ATP are broken. In this reaction, the breakdown of ATP yields ADP + PO_4^{2-} + energy.

ATP is manufactured by mitochondria within muscle fibers during cellular respiration, in which sugar molecules (mainly glucose) are degraded for the release of energy that can be used to drive other reactions. This energy is temporarily stored within ATP, according to the formula

$$ADP + PO_4^{2-} + energy \longrightarrow ATP$$

In a muscle fiber at rest, ATP is quickly produced by mitochondria in this manner and stored within the cell for the next contraction cycle.

Once contraction begins, the stored ATP is used up in a matter of seconds. During strenuous activity or prolonged contractions, its rapid use quickly exceeds its rate of production. Thus, other sources of energy are required. One source is a high-energy molecule called **creatine phosphate**, which can be stored within muscle fibers for longer periods than ATP and is usually 4 to 6 times more abundant in muscle than ATP. The breakdown of creatine phosphate begins shortly after contraction begins, and the energy that is released is used to regenerate ATP that can be made available within a fraction of a second.

If we were to rely on stored ATP and creatine phosphate for all our muscles' energy needs, our muscles would not be able to sustain a contraction for longer than 15 seconds. Therefore, more energy sources must be available. One additional source is **glycogen**, which is a storage form of glucose present in muscle and in the liver. The breakdown of glycogen releases large amounts of glucose, which yields enough ATP after it is broken down to sustain contractions for several minutes. When muscle activity is called upon for a longer period of time, such as in prolonged and strenuous exercise, fat molecules are utilized. Fat contains a greater concentration of potential energy than any other molecule in the body.

Oxygen Debt

Oxygen is required for muscle contraction because it is used during cellular respiration to synthesize ATP (recall that cellular respiration is the breakdown of glucose molecules by enzymes to produce ATP and occurs in the mitochondria of cells). During strenuous exercise, the production of ATP reaches a maximum rate in muscle fibers, and large amounts of oxygen are required to maintain this rate. However, after several minutes of exercise the respiratory and cardiovascular systems cannot bring in enough oxygen to meet the demands, and oxygen levels become depleted as a result. This temporary lack of oxygen availability is called **oxygen debt**. It leads to an accumulation in muscle fibers of a substance called *lactic acid*, which is a by-product of cellular respiration. The presence of lactic acid in muscle produces the soreness you may experience after strenuous exercise. Lactic acid continues to be produced in muscle cells until the oxygen required for its further metabolism becomes available. The oxygen debt is eventually repaid when additional oxygen is brought into the body, which is made possible by the rapid and deep breathing that accompanies heavy exercise. Once additional oxygen becomes available, lactic acid is metabolized within the muscle cells, although muscle soreness may persist for hours because of the slow conversion rate of this metabolic reaction.

Among highly trained athletes, the rate at which lactic acid is produced in skeletal muscle cells during heavy activity is about one-half that of untrained individuals. This reduced rate is a benefit of training, but it may be genetically influenced as well. Thus, trained athletes are capable of achieving greater levels of activity before lactic acid production and the development of oxygen debt begins. This is why trained athletes do not become short of breath as easily as untrained individuals.

In muscles that are exercised strenuously for a prolonged period of time, the oxygen debt may lead to

muscle fatigue. Fatigue is the inability of a muscle to contract in a normal manner. It is caused by changes in the muscle that occur as a result of the accumulation of lactic acid, such as a decrease in pH, which may render it unresponsive to stimulation. In some cases, cramps may follow fatigue. A **cramp** occurs when a muscle contracts spasmodically without relaxing. This painful condition is usually caused by an insufficient amount of ATP to properly return calcium ions to the sarcoplasmic reticulum, preventing the muscle fiber from relaxing.

CONCEPTS CHECK

1. Why is a stimulus required for a muscle to contract?

2. What two functions are performed by calcium ions in muscle contraction?

3. In what three ways are ATP molecules used in muscle contraction?

4. Why does strenuous exercise lead to oxygen debt and fatigue?

SMOOTH MUSCLE AND CARDIAC MUSCLE

Smooth muscle tissue and cardiac muscle tissue differ from skeletal muscle tissue not only in structure, but also in the way they contract.

We have seen that skeletal muscle tissue is the primary component of the body's muscles, which move our appendages. The strength that can be achieved through muscle contraction is due to the highly organized structure of each muscle fiber, and the combined contractions of numerous fibers. In smooth and cardiac muscle tissues, the shape of individual cells and the internal arrangement of proteins differ from those of skeletal muscle tissue, providing for contractions that differ as well (Table 8–1).

A cardiac muscle cell has a single nucleus and is roughly rectangular in shape, with branches that contact adjacent cells. At the contact sites are thickenings of the plasma membranes, called **intercalated discs**. These connections facilitate stimulus conduction between cardiac cells, allowing the cardiac cells to function as a unit. Internally, cardiac muscle cells contain

Table 8–1   COMPARISON OF MUSCLE TISSUES

Factor	Skeletal	Smooth	Cardiac
Location	Attached to skeleton	In walls of hollow organs	In walls of heart
Nature of control	Voluntary	Involuntary	Involuntary
Shape of fibers	Long, cylindrical	Spindle-shaped	Cylindrical, branching
Striations (presence of sarcomeres)	Present	Absent	Present
Speed of contraction	Most rapid	Slowest	Intermediate
Strength of contraction	Greatest	Least	Intermediate
Ability to remain contracted	Least	Greatest	Intermediate

The three types of muscle tissue.

(a) Skeletal muscle fibers

(b) Smooth muscle fibers

(c) Cardiac muscle fibers

Muscle Hypertrophy and Disuse Atrophy

Heavy exercise and good nutrition can lead to increased muscle mass. When muscles enlarge in size, they are said to *hypertrophy*. At the present time, most researchers agree that actual enlargement of a muscle is mainly due to an increase in the size of individual muscle fibers. In each fiber, the mitochondria increase in number and new filaments of actin and myosin are added to the sarcomeres. Eventually, the addition of new protein may lead to the de-velopment of additional myofibrils. Hypertrophy of a muscle provides for an increased force of contraction, as the strength of muscle contraction is directly related to the diameter of the muscle fibers.

If muscles are not kept active but become immobilized for an extended period of time, such as during enforced bed rest or loss of nervous stimulation, muscles begin to lose their protein mass. As a result, muscle fibers decrease in size. This condition is called *disuse atrophy*, and can result in a decrease of muscle strength at the surprising rate of 5% each day. As atrophy progresses in long-term disuse, most of the muscle tissue is replaced by fibrous connective tissue, making muscle rehabilitation nearly impossible. In short-term disuse (less than about two years of immobilization), exercise is usually successful in returning muscle fibers to their original size.

thick and thin filaments that are arranged into sarcomeres. The presence of alternating light and dark bands in the sarcomeres produces striations, as we have seen in skeletal muscle cells. As a result of the regular organization of proteins, cardiac muscle contraction is forceful, although slightly less than skeletal muscle contraction. Also, cardiac muscle tissue does not develop an oxygen debt and does not fatigue. Its contraction is autorhythmic, which means it does not require an external stimulus to begin the contraction cycle.

Smooth muscle cells are relatively small and are spindle-shaped, with a single nucleus in each cell. Internally, they contain a protein distribution different from that found in skeletal muscle cells. There are no troponin and fewer actin fibers in the thin filaments, and the filaments are not organized into sarcomeres. Consequently, smooth muscle cells are not striated. Also, no T tubules or sarcoplasmic reticula are present. Smooth muscle cells contract more slowly than skeletal muscle and cardiac muscle cells, and with less force. However, they have the greatest ability to sustain a contraction relative to skeletal and cardiac cells. Like cardiac cells, they do not develop an oxygen debt. Under most circumstances, smooth muscle cells require an external stimulus to contract, which can be in the form of nerve stimulation or hormone stimulation.

MUSCULAR RESPONSES

A muscle fiber responds to a stimulus of sufficient strength by contracting. The nature of contraction of the muscle as a whole may vary according to the number of motor units responding, the frequency of stimuli received, and how tension is applied.

We have learned in this chapter that a stimulus from a motor neuron may cause a muscle fiber to contract. However, our nervous systems are capable of varying the strength and frequency of the stimulus. How does this affect the muscle's response? We will explore this question and others in this section.

All-or-None Response

The weakest stimulus that can initiate a contraction is called a **threshold stimulus**. Any stimulus that is weaker will not cause even a partial contraction; this is called a **subthreshold stimulus**. When a muscle fiber receives a stimulus of threshold strength or greater, it will contract to its complete capacity. In other words, a muscle fiber does not partially contract; it either does it all the way, or not at all. This feature of muscle fibers is called the **all-or-none response**.

The all-or-none response may seem contradictory at first thought, as we are capable of adjusting the strength of a handshake, the power of a kick, the length of a jump: in other words, the strength of muscle contractions. How is this possible if muscle fibers can contract only to their fullest extent? Recall our discussion on the motor unit: a single motor neuron stimulates an average of 150 muscle fibers (see Fig. 8–6). All fibers in a motor unit are stimulated to contract simultaneously, so each motor unit follows the all-or-none response. A whole muscle comprises many motor units, each receiving its own motor neuron that has a distinct threshold stimulus. If the muscle were to receive a minor series of stimuli from the brain, for example, only the motor units with low thresholds would respond and the muscle would contract slightly. Additional levels of stimulus strength would stimulate motor units with higher thresholds, causing the muscle to contract with greater force. As the intensity of the stimulation increases, more motor units are activated until the muscle is contracting with maximal force. This adding of motor units as stimulus strength increases is known as **recruitment**. Recruitment of motor units in a muscle provides us with the ability to vary the force of our muscle contractions.

Types of Muscle Contractions

In addition to stimulus strength, the frequency of stimuli received by the muscle may also vary. A change in stimulus frequency has an immediate effect on the nature of muscle contraction, producing the different types that are described in the following paragraphs.

Twitch

A **twitch contraction** is a rapid response to a single stimulus that is slightly over threshold. It lasts only about one-tenth of a second before the muscle returns to rest. It may be demonstrated in the muscle of a laboratory animal, in which the muscle is removed and stimulated electrically and its contraction measured and plotted as shown in Figure 8–9. This type of plot is known as a **myogram**. Notice from the myogram the delay of contraction after the stimulus is applied. This is called the **latent period** and is thought to represent the time required for calcium ions to be released, myosin to be activated, and cross bridge attachment to occur. The latent period is followed by the **period of contraction**, which is indicated on the myogram as the upward tracing. During this phase the muscle is pulling at its attachments, causing itself to shorten. The downward tracing that follows is the **period of relaxation**, during which the muscle returns to its original length.

The twitch contraction is the basic unit of muscle contraction. It is not the usual method of body movement, because of the small number of fibers that contract. As you are about to see, all other contractions are basically a combination of twitches.

Treppe

When a muscle is stimulated in such a way as to produce single twitches that rapidly follow each other, the first few twitch contractions progressively increase in force. This phenomenon is known as **treppe**, or the *staircase effect* (Fig. 8–10a). Some investigators have suggested that treppe occurs to enable a muscle to warm up prior to reaching a full contraction.

Figure 8–9
A myogram recording of a single muscle twitch. A single stimulus causes a slight muscle contraction, which is measured as the force of contraction per unit of time.

What mechanism is represented by the latent period?

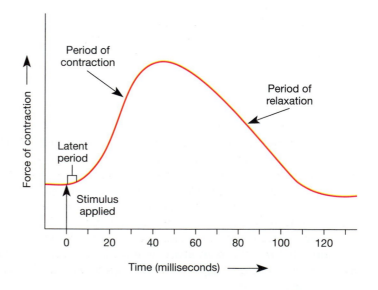

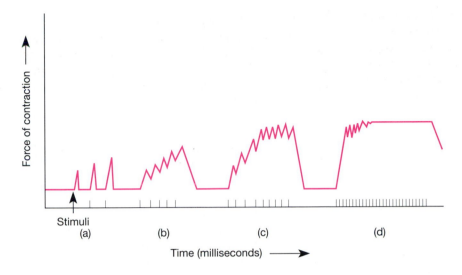

Figure 8–10
Types of muscle contractions. When the frequency of stimuli increases, the nature of muscle contraction changes. (a) Treppe. If a muscle is allowed to relax between stimuli, the contractions will be simple twitches that increase slightly in strength. (b) Wave summation. If a muscle is not allowed to relax between stimuli, contractions increase considerably in strength. (c) Incomplete tetanus. If the frequency of stimuli increases further, summation will reach a maximum value. (d) Complete tetanus. Should the frequency of stimuli increase yet further, no relaxation will occur between stimuli, and the twitches will fuse completely at a maximum value.

Which of these types of contractions is used by skeletal muscle in the performance of an activity such as walking?

Wave Summation

A stimulus that is great enough to cause a muscle to contract lasts for a definite period of time. When a muscle receives a second stimulus before the first contraction cycle is complete, the second contraction will be stronger than the first. This phenomenon is called **wave summation** and is shown in the myogram in Figure 8–10b. The force of muscle contraction provided by wave summation may be 4 times as great as that achieved by a series of twitches.

Tetanus

If a series of stimuli bombard a muscle before each contraction cycle can reach completion, say, at a rate of about 20 to 30 per second, the wave summation will reach a maximum value and be sustained until the stimuli stop. This feature of muscle is called **incomplete tetanus** and is shown in Figure 8–10c. During incomplete tetanus, partial relaxation occurs between stimuli. If the stimuli increase their frequency yet further, say, to 35 to 50 per second, the sustained contraction will be maintained without any relaxation whatsoever. This state of muscle contraction is called **complete tetanus** (Fig. 8–10d). Complete tetanus represents a fusion of twitches and provides a forceful, sustained contraction. You may recognize the term *tetanus* from the disease caused by bacteria that gain entrance to the body through a puncture wound, a disease also known as *lockjaw*. *Tetanus* is used to describe this disease because one of its symptoms is severe cramping with continuous tetanic contractions in the muscles of the body, including those of the jaw.

Muscle contraction by tetanus provides the usual means of body movement. Depending on the strength of the contraction needed, it may be provided by incomplete tetanus, complete tetanus, or a combination of both. For example, walking requires a high frequency of stimuli from the brain and spinal cord to the muscles of the legs, which respond by a series of sustained contractions. Running and jumping require a greater frequency of stimuli, resulting in a temporary state of complete tetanus to produce a forceful series of contractions.

Tetanus also provides the type of contraction necessary to maintain **muscle tone**. Muscle tone is a series of sustained contractions by a small number of fibers that causes the muscle to tighten slightly but does not cause body movement. Its importance in maintaining posture is made evident when muscle tone is suddenly lost, such as when a person loses consciousness and the body collapses.

Isotonic and Isometric

Isotonic and isometric contractions are actually two types of tetanic contractions. They are terms that are frequently used to describe two different effects of **tension** on muscles. Tension is the force exerted by a muscle contraction; it requires the use of energy.

An **isotonic contraction**[9] produces movement as the muscle pulls an attached structure, usually a bone, toward a more stationary structure. The tension that is applied is held constant until the muscle relaxes. This is the usual means of producing body movement and is recognized as an active form of exercise that provides greater muscle enlargement and endurance.

Isometric contraction[10] produces muscle tension, but the muscle does not shorten. No body movement results, although energy is still used to provide the tension. An example of an isometric contraction occurs when you push against an immovable object, such as a wall. Your muscles contract and tense, but no body movement results. Isometric exercises are often used by athletes to strengthen joints, and may be used to burn energy (calories) while minimizing the enlargement of muscle mass.

CONCEPTS CHECK

1. How does a muscle fiber respond when it is subjected to stimuli that gradually increase in strength?

2. How may recruitment increase the strength of muscle contraction?

3. What is the response of a muscle when it receives stimuli that increase in frequency?

4. What is the difference between isotonic and isometric contractions?

---

[9]Isotonic: Greek *iso-* = same, constant + *tonic* = pertaining to strength or tension.

[10]Isometric: *iso-* + *metric* = length, measurement.

PRODUCTION OF MOVEMENT

Movement occurs when a muscle contracts, pulling a movable bone toward a more stationary bone. In most movements, many muscles are involved and each plays one of several possible roles.

Muscle contraction produces a wide range of body movements. The nature of each movement is determined by many factors, such as how the muscle forms its attachments, the structure of the joint, and the interactions of nearby muscles.

Origin and Insertion

Muscles produce movement by pulling on their attachments, which are usually tendons that are attached to bones. Most muscles extend from one bone to another, crossing over the joint between the two opposing bones. In these muscles, one end is attached by tendons to a bone that is held somewhat immovable or stationary, while the opposite end is attached by tendons to a bone that is movable. The point of attachment to the more stationary bone is called the muscle's **origin**, and the attachment to the more movable bone is the muscle's **insertion**. Contraction of the muscle causes movement when the insertion is pulled toward the origin (Fig. 8–11). In the muscles of the limbs, the origins are proximal and the insertions are distal.

Group Actions

For a particular movement to occur in a smooth, coordinated fashion, more than one muscle must be involved. For example, when you wish to raise your arm above your head, a simple command to contract a single muscle in the shoulder would be insufficient. Rather, your brain signals to a group of muscles to contract at appropriate intervals, and your arm rises. The coordinated response of a group of muscles in order to bring about a body movement is called a **group action**.

The muscles within a group action perform specific roles in order to accomplish the desired movement. They include **prime movers**, which cause the desired action; **antagonists**, which must relax during the action; **synergists**, which steady the movement; and **fixators**, which stabilize the origin of the prime mover.

For an example of these roles, let's use flexion of the arm at the elbow joint as the intended movement. In order to flex your arm at the elbow, the biceps brachii muscle on the anterior side of the humerus must contract to pull the forearm toward the humerus. Because this muscle causes the movement (flexion), it is called the prime mover. The muscle opposite to the

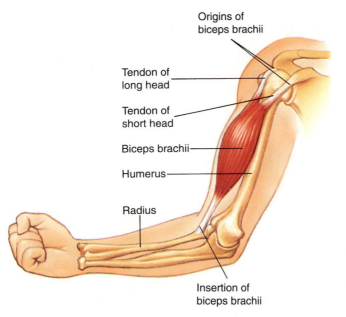

Origins of
biceps brachii

Tendon of
long head

Tendon of
short head

Biceps brachii

Humerus

Radius

Insertion of
biceps brachii

Figure 8–11
Origin and insertion of a muscle. The muscle shown is the biceps brachii, located in the arm.
Which is the movable bone in this example?

biceps, the triceps brachii on the posterior side of the humerus, must relax and yield to the movement. The triceps is thus the antagonist in this action. Other muscles of the arm adjacent to the biceps and numerous muscles of the forearm contract during flexion to keep the arm and elbow joint steady. These muscles are the synergists for this movement. Finally, muscles of the chest, back, and shoulder that have origins near the shoulder joint contract to keep it stable, permitting a greater force to be applied by the biceps. These are the fixators. If our example were a different movement, say, adduction of the arm at the shoulder, these muscles would each play a different role in accomplishing the movement.

SPORTS CLINIC

The Effects of Exercise on Muscle

Exercise physiologists have identified three categories of training regimens, each of which provides a different response to the body. In most cases, athletic training involves elements of all three. These categories are motor-skill development (learning and coordination), endurance, and strength.

Motor-skill development occurs whenever you learn a new skill that requires a coordinated action, such as typing or bowling. It occurs under a regime of repetition that demands relatively little energy expenditure. The learning component occurs primarily in the brain (learning is discussed in Chapter 10).

Endurance represents submaximal, sustained muscular efforts, such as are required for marathon running or aerobic dancing. Exercising for endurance results in an increased ability of all involved motor units to metabolize high-energy-sustaining molecules, such as fat and glycogen. There is also a limited increase in muscle hypertrophy. Perhaps the most significant benefit of endurance exercise, however, is its effect on other systems. The demand on muscles to increase their metabolic capacity places an increased load on the cardiovascular and respiratory systems, which leads to an improved blood supply to muscles, an increased stroke volume of the heart (greater strength of cardiac contraction), and increased lung capacity.

Muscle strength can be increased by repetitious, major physical efforts that involve many motor units. Strength represents brief, maximal muscular efforts, such as are required for weight lifting. Exercising for strength induces the synthesis of additional myofibrils, resulting in hypertrophy of the active muscle cells. It also provides an enhanced ability of muscle cells to utilize energy-containing molecules such as glycogen, glucose, and creatine phosphate. The increased stress that accompanies exercising for strength also induces growth of tendons and bones.

MAJOR MUSCLES OF THE BODY

The major muscles provide for movement of all movable bones of the body. They are distributed throughout, and each one's name corresponds to its appearance, location, action, or relationship to other structures.

Throughout the remainder of this chapter, we will focus on the names of the major muscles of the body. They are organized for study on the basis of the body parts they move, beginning with the head region and moving downward. For each muscle identified, the origin, insertion, and primary action are provided in the accompanying table. Also, a "map" showing the locations of the body's muscles on the anterior and posterior sides is provided in Figures 8–12 and 8–13 for your reference.

As you prepare to begin your study of the muscles, glance through the tables briefly and notice the names of muscles. They are in scientific nomenclature, although some have been simplified a bit. In other words, each one is named in a way that is descriptive of its appearance, location, action, or relationship to other structures, using Latin or Greek roots to form the words. For example, the term *biceps brachii* literally means "two heads (points of origin) in the arm," which adequately describes this muscle of the upper arm that

has two heads, or points of origin. Another example is the large muscle of the chest region, the *pectoralis major*, which translates to "large chest." The translations of muscle terms are provided as footnotes to aid you in your study of the muscles.

Muscles of the Head and Neck

The major muscles associated with the head and neck include muscles that provide facial expressions, muscles involved in chewing, and muscles in the neck that move the head.

Muscles of Facial Expression

Several muscles of the head act on facial structures to provide us with the ability to change our facial expressions (Fig. 8–14). The **frontalis** of the forehead raises the eyebrows, while the **occipitalis** located on the lower back of the head pulls the scalp backward. Surrounding each eye is the circular **orbicularis oculi**[11] (or-bi'-kyoo-LAR-is O-kyoo-lī) which closes the eyelids and squinches the eyes. Another circular muscle, the **orbicularis oris**,[12] surrounds the mouth. Together with the **buccinator**[13] (BYOO-si-nā-ter), these "kissing" muscles pucker the mouth. Because the buccinator also flattens the cheeks, as when whistling or blowing a trumpet, it is also referred to as the trumpeter's muscle. Smiling is accomplished mainly by the **zygomaticus**[14] (zī-gō-MA-ti-kus) muscles by raising the corners of the mouth. The muscles of facial expression are summarized in Table 8–2.

[11]Orbicularis oculi: Latin *orbis* = orb or sphere + *oculi* = of the eye.

[12]Orbicularis oris: *orbis* + *oris* = of the mouth.

[13]Buccinator: *buccinator* = trumpeter.

[14]Zygomaticus: *zygos* = cheek.

Table 8–2   MUSCLES OF FACIAL EXPRESSION

Muscle	Origin	Insertion	Action
Frontalis	Occipital bone	Skin around the eye	Raises the eyebrows
Occipitalis	Occipital bone	Epicranial aponeurosis	Pulls scalp posteriorly
Orbicularis oculi	Maxillary and frontal bones around the orbit	The eyelid	Closes eye
Orbicularis oris	Muscles surrounding the mouth	Skin at the corner of the mouth	Closes and protrudes lips; shapes lips during speech
Buccinator	Maxilla and mandible	Orbicularis oris	Compresses the cheeks as in whistling, blowing, and sucking
Zygomaticus	Zygomatic bone	Skin and muscle at the corner of the mouth	Raises the corner of the mouth, as in smiling

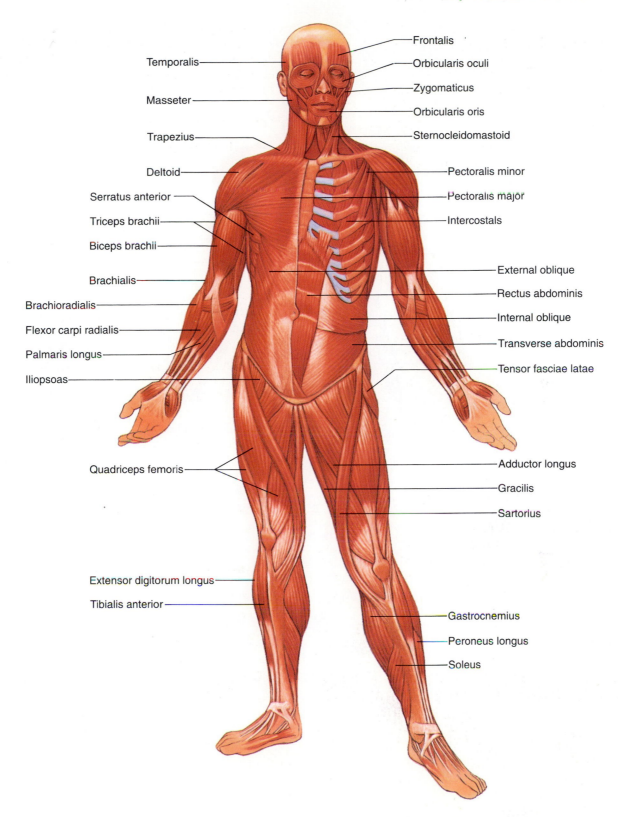

Temporalis

Masseter

Trapezius

Deltoid

Serratus anterior

Triceps brachii

Biceps brachii

Brachialis

Brachioradialis

Flexor carpi radialis

Palmaris longus

Iliopsoas

Quadriceps femoris

Extensor digitorum longus

Tibialis anterior

Frontalis

Orbicularis oculi

Zygomaticus

Orbicularis oris

Sternocleidomastoid

Pectoralis minor

Pectoralis major

Intercostals

External oblique

Rectus abdominis

Internal oblique

Transverse abdominis

Tensor fasciae latae

Adductor longus

Gracilis

Sartorius

Gastrocnemius

Peroneus longus

Soleus

Figure 8–12
Major muscles of the body, anterior view.

**Is the deltoid muscle an organ, a tissue, or a
membrane?**

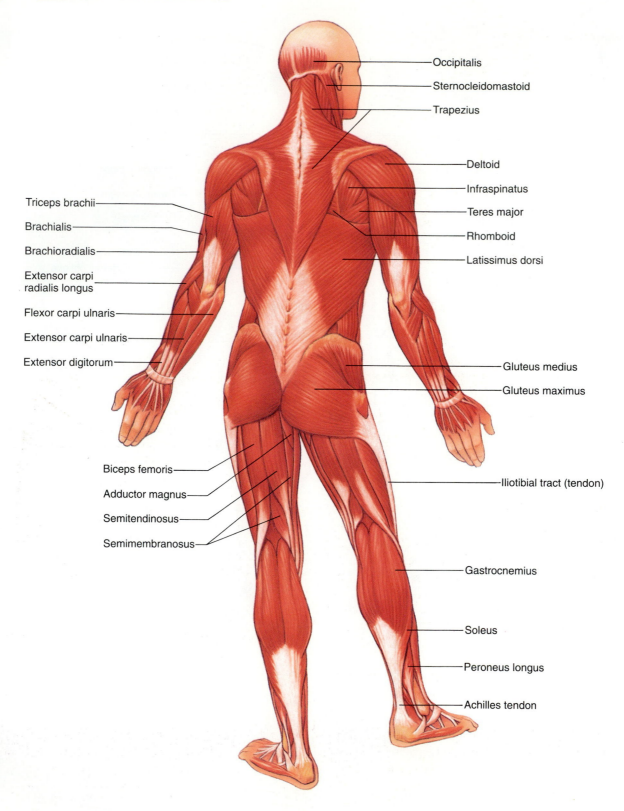

Figure 8–13
Major muscles of the body, posterior view.
What muscles of the back can you identify in this illustration?

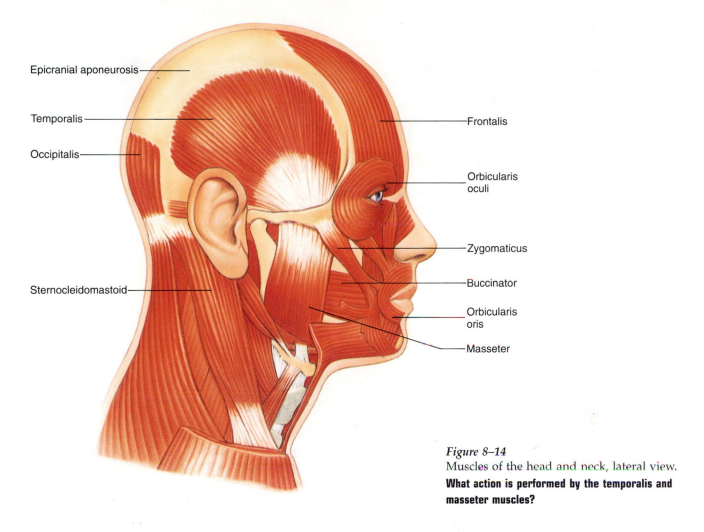

Epicranial aponeurosis

Temporalis

Occipitalis

Sternocleidomastoid

Frontalis

Orbicularis oculi

Zygomaticus

Buccinator

Orbicularis oris

Masseter

Figure 8–14
Muscles of the head and neck, lateral view.

What action is performed by the temporalis and masseter muscles?

Muscles of Mastication

The major muscles involved in chewing, or mastication, include two of the most powerful muscles of the body (Fig. 8–14). The **masseter**[15] muscle extends from the zygomatic process to the mandible. It closes the mouth, and can be felt on the side of the jaw during mastication. The **temporalis** muscle occupies most of the temporal bone, and also closes the mouth.

Neck Muscle

The primary muscle of the neck that moves the head is the **sternocleidomastoid**[16] (ster-nō-klī-dō-MAS-

toyd) muscle. This pair of narrow muscles is easily seen on the anterior and lateral sides of the neck (Figs. 8–12 and 8–14). From the anterior view, the origin of the two sternocleidomastoids forms a V at the base of the neck. The muscles that move the jaw and head are summarized in Table 8–3.

[15]Masseter: *masseter* = one who chews.

[16]Sternocleidomastoid: *sterno-* = breastbone + *cleido-* = clavicle + *mastoid* = breast-resembling process (on the temporal bone).

Table 8–3	MUSCLES THAT MOVE THE LOWER JAW AND HEAD		
Muscle	**Origin**	**Insertion**	**Action**
Masseter	Zygomatic process of temporal bone and zygomatic arch	Mandible	Closes the mouth by elevating the mandible
Temporalis	Temporal bone	Mandible	Closes the mouth by elevating the mandible
Sternocleidomastoid	Manubrium of sternum, and clavicle	Mastoid process of temporal bone	Contraction of both sides flexes the head; contraction of one side rotates the head

SPORTS CLINIC

When You Pull or Tear a Muscle

A pulled or torn muscle sometimes results from strenuous exercise, especially if a "warm-up" period was not included prior to the exertion. It results in local pain in the affected muscle and swelling in the region surrounding the muscle, often to the point of severely restricting movement. These symptoms are caused by internal bleeding, or *hemorrhaging*, of the damaged muscle into the muscle bundle. Although this sounds quite serious, most pulled muscles are not, and healing occurs quickly with adequate treatment. Immediate treatment for pulled or torn muscles is a formula known by many as "RICE": rest, ice, compression, and elevation. This procedure is intended to reduce hemorrhaging and inflammation as much as possible. Ice massage and gentle stretching for several days are suggested in order to help restore the muscle to its normal state.

Upper-Limb Muscles

The muscles of the upper limb include those that attach and support the pectoral girdle (the scapula and clavicle) and upper limb, and those that move the arm, forearm, and hand.

Muscles of the Pectoral Girdle

The connection of the pectoral girdle to the thorax is supported by a group of muscles. These muscles attach the scapula to the thorax and move the scapula. They are shown in Figures 8–13, 8–15, and 8–16. The largest muscle in this group is the **trapezius**[17] (tra-PĒ-zē-us). It is a diamond-shaped, flat muscle of the upper back that extends from the base of the occipital bone and the spines of the first 18 vertebrae to the acromion process and spine of each scapula. Other muscles of the pectoral girdle include the **levator scapulae**[18] (SKA-pyoo-lē), **rhomboids** (ROM-boydz), **serratus** (ser-Ā-tis) **anterior**,[19] and **pectoralis minor**.[20] For the most part, the muscles of the pectoral girdle act as fixators to hold the scapula firmly in place when the muscles of the arm contract. They also move the scapula in different positions, which increases the range of arm movement. The pectoral girdle muscles are described further in Table 8–4.

[17]Trapezius: Greek *trapezius* = table-like.

[18]Levator scapulae: Latin *levator* = one who raises, or elevates + *scapulae* = of the shoulder blade.

[19]Serratus: Latin *serratus* = sawlike.

[20]Pectoralis minor: Latin *pectoralis* = of the breast + *minor* = smaller.

Table 8–4 MUSCLES THAT MOVE THE PECTORAL GIRDLE

Muscle	Origin	Insertion	Action
Trapezius	Occipital bone and spines of the cervical and thoracic vertebrae	Acromion and spine of the scapula	Elevates and rotates the scapula; adducts the scapula; depresses the shoulder; extends the head
Levator scapulae	First four cervical vertebrae	Scapula	Elevate and adduct the scapula; flex the head to either side
Rhomboids	Seventh cervical and first five thoracic vertebrae	Scapula	Adduct the scapula to "square the shoulders"; rotate scapula as in paddling a canoe
Serratus anterior	The first eight ribs	Scapula	Abducts the scapula (pulls it forward), and rotates it
Pectoralis minor	Ribs 3 through 5	Scapula, at its coracoid process	Draws the scapula forward and downward; elevates the ribs

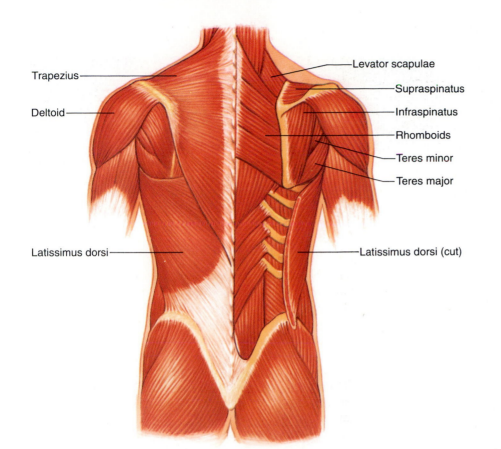

Trapezius

Deltoid

Latissimus dorsi

Levator scapulae

Supraspinatus

Infraspinatus

Rhomboids

Teres minor

Teres major

Latissimus dorsi (cut)

Figure 8–15
Muscles of the trunk, posterior view. Superficial muscles on the right side (the trapezius and latissimus dorsi) are removed to reveal underlying muscles.

What is the main role provided by the muscles of the pectoral girdle?

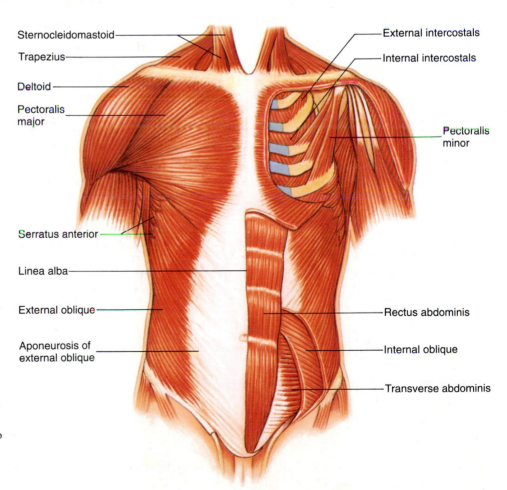

Sternocleidomastoid

Trapezius

Deltoid

Pectoralis major

Serratus anterior

Linea alba

External oblique

Aponeurosis of external oblique

External intercostals

Internal intercostals

Pectoralis minor

Rectus abdominis

Internal oblique

Transverse abdominis

Figure 8–16
Muscles of the trunk, anterior view. Superficial muscles on the left side (pectoralis major and external oblique) are removed to reveal underlying muscles.

Where does the pectoralis major insert, and what are its major actions?

219

Muscles That Move the Arm

The attachment of the arm to the pectoral girdle is supported and strengthened by the **pectoralis major, latissimus dorsi** (la-TI-si-mis DOR-sī), and **deltoid**[21] (DEL-toyd) muscles. The pectoralis major forms the upper chest (Figs. 8–12 and 8–16). Its insertion at the greater tubercle of the humerus enables it to flex, adduct, and medially rotate the arm. The latissimus dorsi is a large, flat muscle of the lower back, which inserts at the intertubercular groove of the humerus (Figs. 8–13 and 8–15). Often called the "lat" by athletes, this powerful muscle extends, adducts, and medially rotates the arm. It is also referred to as the swimmer's muscle, since a swimmer uses its three motions during the power stroke of the crawl. The deltoid muscle wraps over the shoulder and is the major abductor of the upper limb (Figs. 8–15 through 8–18). The deltoid is a common site for the administration of injections.

A group of four muscles serves to further strengthen the shoulder joint during arm movements (Figs. 8–17 and 8–18). Called the *rotator cuff* muscles, they are the **subscapularis**[22] (sub-SKAP-yoo-lar'-is), **supraspinatus**[23] (sū'-pra-spi-NĀ-tus), **infraspinatus** (in'-fra-spi-NĀ-tus), and **teres** (TAYR-ēz) **minor**[24] muscles. Each of these muscles originates on the scapula, and their tendons blend with the fibrous capsule of the shoulder joint to form a cap over the proximal end of the humerus. A rotator cuff injury involves damage to one or more of these muscles or their tendons. The **teres major** is a small muscle that also crosses over the shoulder joint, but it does not contribute to the rotator cuff structure. The muscles that move the arm are described further in Table 8–5.

Muscles That Move the Forearm

The muscles moving the forearm can be divided into anterior and posterior compartments. The largest posterior muscle is the **triceps brachii**[25] (Fig. 8–17). It is the primary extensor of the forearm. The smaller **brachioradialis** (brā'-kē-ō-rā-dē-AL-is) helps the anterior muscles flex the forearm (Figs. 8–19 and 8–20). The anterior muscles are the primary flexors of the forearm, and include the large **biceps brachii** and the deeper **brachialis** (Fig. 8–18).

In addition to the actions of flexion and extension, the forearm is also capable of supination (turning the forearm so that the palm is up) and pronation (turn-

---

[21]Deltoid: *deltoid* = triangular.

[22]Subscapularis: Latin *sub-* = below, underneath + *scapularis* = of the shoulder blade.

[23]Supraspinatus: *supra-* = above + *spinatus* = pertaining to the spine (of the scapula).

[24]Teres: Latin *teres* = rounded.

[25]Triceps brachii: *tri-* = three + *-ceps* = head + *brachium* = arm.

Table 8–5   MUSCLES THAT MOVE THE ARM

Muscle	Origin	Insertion	Action
Pectoralis major	Clavicle, sternum, costal cartilages of first six ribs	Greater tubercle of humerus	Flexes the arm; adducts and medially rotates the arm
Latissimus dorsi	Spines of lower six thoracic vertebrae, lumbar vertebrae, lower ribs, and iliac crest	Intertubercular groove of humerus	Extends the arm; adducts and medially rotates the arm; pulls the shoulder downward and back
Deltoid	Acromion and spine of scapula, and clavicle	Deltoid tuberosity of humerus	Abducts the arm; aids in extending and flexing the humerus
Subscapularis	Scapula, anterior surface	Lesser tubercle of humerus	Rotates the arm medially
Supraspinatus	Scapula, posterior surface	Greater tubercle of humerus	Abducts the arm
Infraspinatus	Scapula, posterior surface below spine	Greater tubercle of humerus	Rotates the arm laterally
Teres major	Scapula	Lesser tubercle of humerus	Extends, adducts, and medially rotates the arm
Teres minor	Scapula	Greater tubercle of humerus	Rotates the arm laterally with the infraspinatus

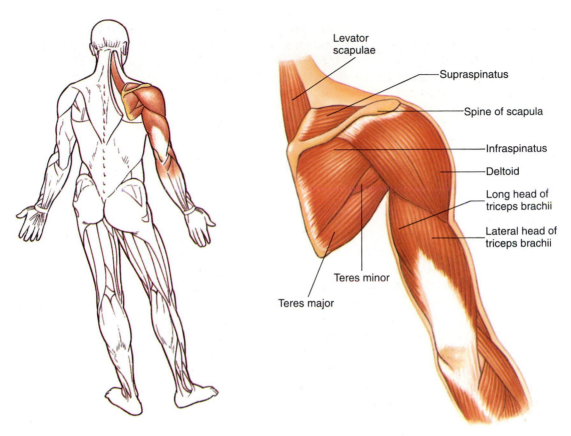

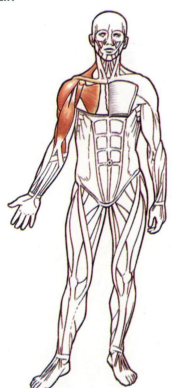

Figure 8–17 (above)
Muscles of the right shoulder and arm, posterior view. The trapezius has been removed in this drawing.

What four muscles (three of which are shown in this figure) make up the rotator cuff?

Figure 8–18 (below)
Muscles of the right shoulder and upper arm, anterior view. The pectoralis major is reflected and the rib cage is removed in this drawing.

What action is provided by the biceps brachii and brachialis muscles?

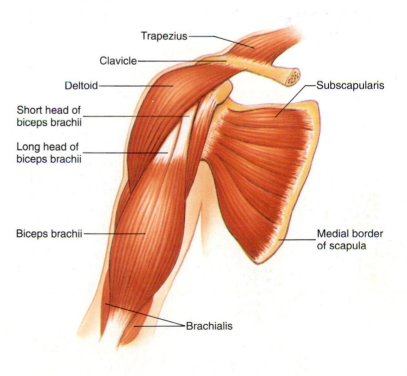

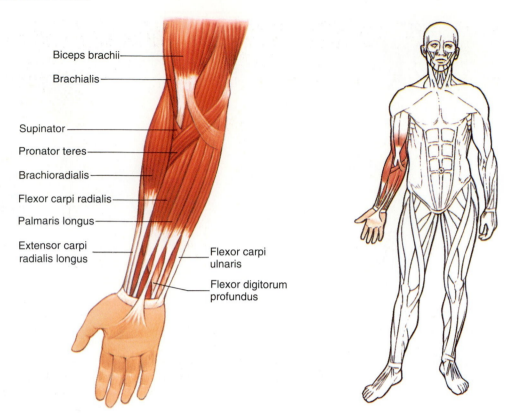

Figure 8–19
Muscles of the right forearm, anterior view.

What muscles extend the hand?

Figure 8–20
Muscles of the right forearm, posterior view.

Where does the triceps brachii insert?

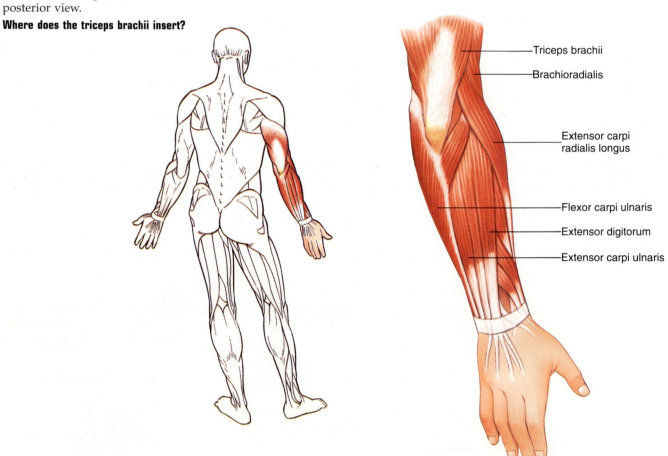

Table 8-6 MUSCLES THAT MOVE THE FOREARM

Muscle	Origin	Insertion	Action
Biceps brachii	Two heads of origin on the scapula	Radial tuberosity of the radius	Flexes the forearm at the elbow; supinates the hand
Brachialis	Shaft of humerus	Coronoid process of ulna	Flexes the forearm
Brachioradialis	Distal end of humerus	Base of styloid process of radius	Flexes the forearm
Triceps brachii	Three heads of origin on the scapula and humerus	Olecranon process of ulna	Extends the forearm
Supinator	Distal end of the humerus and proximal end of the ulna	Proximal end of radius	Supinates the forearm
Pronator teres	Distal end of the humerus and coronoid process of the ulna	Shaft of radius	Pronates the forearm

ing the palm down). Supination is accomplished by the biceps brachii and a small muscle that crosses the elbow, the **supinator** (SOO-pi-nā-ter). Pronation is performed by the contraction of another small muscle that crosses the elbow, the **pronator teres** (PRŌ-nā-ter TAYR-ēz). The muscles that move the forearm are further described in Table 8-6.

Muscles That Move the Hand and Fingers

The muscles that move the hand and fingers also include anterior and posterior compartments. They occupy the forearm, and are shown in Figures 8–19 and 8–20. There are actually about 20 of these muscles, but we will examine only a few of the superficial ones. For the most part, the muscles on the anterior side provide flexion of the wrist and digits (bones of the finger), while those on the posterior side cause extension. The

tendons of these muscles can be seen through the surface of the forearm when the muscles contract.

The **flexor carpi radialis** (FLEKS-er KAR-pē rā'-dē-AL-is) is located on the anterior side of the forearm. It flexes and abducts the wrist. The **flexor carpi ulnaris** (ul-NAYR-is) is located medial to the flexor carpi radialis, and performs the same action. The **palmaris longus** (pal-MAYR-is LONG-us) is located between the two, and assists them in flexing the wrist. The **flexor digitorum profundus** (di-ji-TOR-um prō-FUN-dus) lies along the anterior surface of the ulna and flexes the digits of each finger.

The **extensor carpi radialis longus** is located on the posterior side of the forearm. Together with the **extensor carpi ulnaris**, it extends the wrist. The **extensor digitorum**, as its name suggests, extends the digits of each finger. The muscles moving the hand and fingers are further described in Table 8–7.

Table 8-7 MUSCLES THAT MOVE THE HAND AND FINGERS

Muscle	Origin	Insertion	Action
Flexor carpi radialis	Distal end of humerus	Second and third metacarpals	Flexes and abducts the wrist
Flexor carpi ulnaris	Distal end of humerus and olecranon process of ulna	Carpal and metacarpal bones	Flexes and adducts the wrist
Palmaris longus	Distal end of humerus	Fascia of the palm	Flexes the wrist
Flexor digitorum profundus	Anterior surface of ulna	Distal phalanges in fingers 2 through 5	Flexes the distal phalanges of each finger
Extensor carpi radialis longus	Distal end of humerus	Second metacarpal	Extends and abducts the wrist
Extensor carpi ulnaris	Distal end of humerus	Fifth metacarpal	Extends and adducts the wrist
Extensor digitorum	Distal end of humerus	Middle and distal phalanges in fingers 2 through 5	Extends the digits of each finger

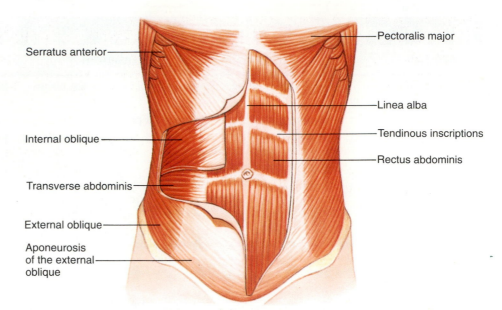

Figure 8-21
Muscles of the abdominal wall.

What is the common action of the external oblique, internal oblique, and transverse abdominis muscles?

Trunk Muscles

Muscles of the trunk include the muscle groups that move the vertebral column, muscles of the thorax, and muscles of the abdominal wall.

Muscles That Move the Vertebral Column

The muscles that move the vertebral column enable us to bend, twist, and stretch the trunk. They are very strong, for they provide much of the support to keep the posture erect when a person is standing or sitting upright. The muscles moving the vertebral column are organized into a series of muscle groups, the most prominent of which is the **erector spinae** (see Table 8–8). Located on each side of the back, deep to the trapezius and latissimus dorsi muscles, they are important in keeping the back straight and the body erect. Other muscle groups lie deeper, and are largely responsible for flexion, extension, abduction, and rotation of the vertebral column.

Muscles of the Thorax

The muscles of the thorax are mainly involved in the process of breathing. The **external intercostals**, for example, elevate the ribs during inspiration (Fig. 8–16). This helps to expand the thorax. The **internal intercostals** depress the ribs during forced expiration (exhaling forcefully). A third thoracic muscle, the **diaphragm** (DĪ-a-fram), is located internally between the thoracic and abdominal cavities. When this dome-shaped muscle contracts, it pushes down into the abdominal cavity to expand the thoracic cavity volume, which results in inspiration. The muscles of the thorax are further described in Table 8–8.

Muscles of the Abdominal Wall

The muscles forming the anterior wall of the abdomen flex and rotate the vertebral column and compress the abdominal cavity (Fig. 8–21). By forming the wall of the abdomen, they also form a shield that protects

Muscle	Origin	Insertion	Action
Erector spinae	Vertebrae, pelvis	Superior vertebrae, ribs	Extends vertebral column
External intercostals	Ribs	Rib below the rib of origin	Elevate ribs for inspiration
Internal intercostals	Ribs	Rib above rib of origin	Depress ribs for forced expiration
Diaphragm	Interior of body wall	Central tendon of diaphragm	Expand the thorax by compressing the abdomen for inspiration

Table 8–8 MUSCLES OF THE VERTEBRAL COLUMN AND THORAX

Table 8–9 MUSCLES OF THE ABDOMINAL WALL

Muscle	Origin	Insertion	Action
Rectus abdominis	Pubis and symphysis pubis	Xiphoid process and costal cartilages of fifth to seventh ribs	Flexes the vertebral column, which thereby compresses the abdomen
External oblique	Lower eight ribs	Iliac crest and linea alba	When both sides contract, aids rectus abdominis; when one side contracts, aids back muscles in trunk rotation and flexes the vertebral column laterally
Internal oblique	A large aponeurosis of the lower back, iliac crest, and costal cartilage of lower ribs	Linea alba and pubis	Same as external oblique
Transverse abdominis	Aponeurosis of lower back, iliac crest, and costal cartilage of lower ribs	Linea alba and pubis	Same as external oblique

the abdominal visceral organs. In a person who is relatively well muscled and has little body fat, a vertical line can be seen extending from the sternum to the navel. This line is formed by a ridge of connective tissue, called the **linea alba** (LIN-ē-a AL-ba). On each side of the linea alba is a long, flat muscle called the **rectus abdominis**[26] (REK-tus ab-DOM-i-nis). Crossing the rectus abdominis at three or more locations are **tendinous inscriptions**, which give the abdominal wall of a well-muscled person the appearance of segmentation.

Lateral to the rectus abdominis are three layers of muscle. From superficial to deep, they are the **external oblique, internal oblique**, and **transverse abdominis**. The fibers of these three muscles are oriented in opposite directions of one another. When all three contract completely, they compress the abdomen. When the same side (right or left) of all three contracts, the muscles of the back are aided in lateral flexion of the vertebral column. The four muscles of the abdominal wall are further described in Table 8–9.

Lower-Limb Muscles

The muscles of the lower limb include those that move the thigh, those that move the leg, and those that move the foot and toes.

Muscles That Move the Thigh

Many of the muscles that move the thigh at the hip joint are located in the pelvic region (Figs. 8–22 through 8–24). They typically originate from the coxal bones

and insert on the femur. A group of muscles on the anterior side, collectively called the **iliopsoas**[27] (il-ē-ō-SŌ-as), flexes the thigh. The muscle on the lateral side of the pelvis is the **tensor fasciae latae** (TEN-ser FA-shē-ē LA-tē), which abducts, flexes, and medially rotates the thigh. The tensor fasciae latae receives its name from the fact that it tenses a thick band of fascia on the lateral side of the thigh, called the *iliotibial-tract tendon*. By so doing, it can abduct the thigh. On the posterior side of the pelvis are the gluteal muscles, the largest of which is the **gluteus** (GLOO-tē-us) **maximus**.[28] It makes up the bulk of the buttocks and extends the thigh when the thigh is flexed at a 45° angle. The **gluteus medius** is located just superior and lateral to the maximus; it is a common site for injections. It abducts and medially rotates the thigh.

In addition to the pelvic muscles identified previously, some muscles located in the thigh also attach to the coxal bones and can thereby cause movement of the thigh. Many of these thigh muscles have a dual role: they also move the leg at the knee. The thigh muscles that move the thigh as their primary action include the **adductor longus, adductor magnus**, and **gracilis**.[29] Each of these three muscles is located on the medial

[26]Rectus abdominis: *rectus* = vertical, straight + *abdominis* = of the abdomen.

[27]Iliopsoas: *ilio-* = pertaining to the ilium + *psoas* = loin.

[28]Gluteus maximus: *gluteus* = of the buttocks + *maximus* = greatest.

[29]Gracilis: *gracilis* = slender.

side of the thigh (Fig. 8–22). Their contraction draws the thigh toward the midline of the body, or adducts it. The remaining thigh muscles perform a dual role, and are included in the following discussion on muscles that move the leg. The muscles that move the thigh are described further in Table 8–10.

Muscles That Move the Leg

The muscles that move the leg are located in the thigh. Many of these muscles have dual roles, as they also move the thigh. They are divided into two groups: the anterior thigh muscles, which extend the leg and flex the thigh; and the posterior thigh muscles, which flex the leg and extend the thigh. The medial thigh muscles, which adduct the thigh, were described previously.

The anterior thigh muscles include the **sartorius**[30] and a group of four muscles known as the **quadriceps femoris**[31] (Figs. 8–22 and 8–24). The sartorius is a long, narrow muscle that extends from the ilium to the proximal end of the tibia. It is referred to as the "tailor's muscle," since it permits sitting in a cross-legged po-

[30]Sartorius: Latin *sartor* = tailor.

[31]Quadriceps femoris: Latin *quadri-* = four + *-ceps* = head + *femoris* = of the femur.

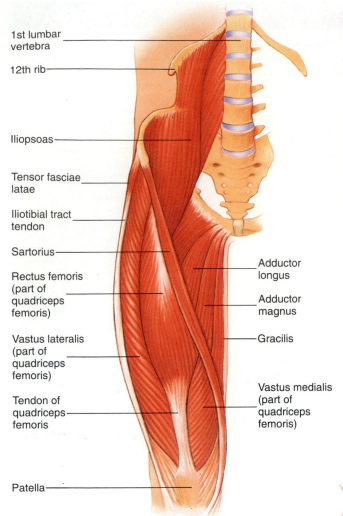

Figure 8–22
Muscles of the right thigh and pelvis, anterior view.

The adductor leg muscles insert upon what long bone?

Labels for Figure 8–22:
- 1st lumbar vertebra
- 12th rib
- Iliopsoas
- Tensor fasciae latae
- Iliotibial tract tendon
- Sartorius
- Rectus femoris (part of quadriceps femoris)
- Vastus lateralis (part of quadriceps femoris)
- Tendon of quadriceps femoris
- Patella
- Adductor longus
- Adductor magnus
- Gracilis
- Vastus medialis (part of quadriceps femoris)

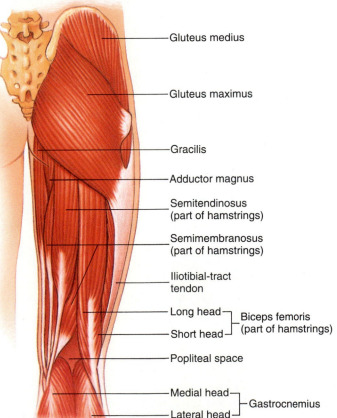

Figure 8–23
Muscles of the right thigh and pelvis, posterior view.

The gluteus muscles originate from what bone?

Labels for Figure 8–23:
- Gluteus medius
- Gluteus maximus
- Gracilis
- Adductor magnus
- Semitendinosus (part of hamstrings)
- Semimembranosus (part of hamstrings)
- Iliotibial-tract tendon
- Long head
- Short head
- Biceps femoris (part of hamstrings)
- Popliteal space
- Medial head
- Lateral head
- Gastrocnemius

Table 8–10 MUSCLES THAT MOVE THE THIGH

Muscle	Origin	Insertion	Action
Iliopsoas	Iliac fossa and lumbar vertebrae	Lesser trochanter of femur	Flexes and medially rotates the thigh
Tensor fasciae latae	Iliac crest of ilium	Tibia by way of fascia of the thigh	Abducts, flexes, and medially rotates the thigh
Gluteus maximus	Ilium, sacrum, and coccyx	Posterior surface of femur and fascia of the thigh	Extends the thigh
Gluteus medius	Ilium	Greater trochanter of femur	Abducts and medially rotates the thigh
Adductor longus	Pubis and symphysis pubis	Posterior surface of femur	Adducts, flexes, and laterally rotates the thigh
Adductor magnus	Ischial tuberosity	Posterior surface of femur	Adducts the thigh; anterior part flexes the thigh, posterior part extends the thigh
Gracilis	Pubis	Medial surface of tibia	Adducts the thigh; flexes and medially rotates the leg

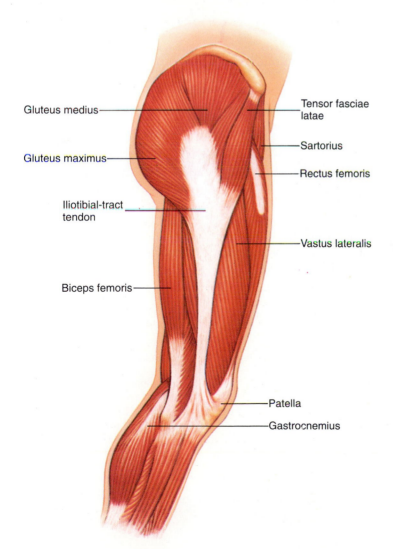

Gluteus medius

Gluteus maximus

Iliotibial-tract tendon

Biceps femoris

Tensor fasciae latae

Sartorius

Rectus femoris

Vastus lateralis

Patella

Gastrocnemius

Figure 8–24
Muscles of the right thigh and pelvis, lateral view.

What are the four muscles of the quadriceps femoris group (two of which are shown in this figure)?

sition by flexing the thigh and leg and rotating the leg laterally, which is a position tailors used to occupy for long periods at a time. The quadriceps femoris are the primary extensors of the leg. Each of the four muscles in this group arises from a different head, or point of origin (hence the four heads forming the term *quadriceps*), and the muscles insert by way of a common tendon. The tendon inserts into the patella (kneecap) before continuing as the patellar tendon into the tibial tuberosity.

The posterior thigh muscles are often referred to as **hamstring muscles** (Fig. 8–23). In general, they flex the leg and extend the thigh. Their tendons of insertion can be easily felt on the posterior side of the knee, particularly when the knee is slightly bent. The hamstrings acquired their name from the butcher shop, where the tendons of insertion in pigs were used to suspend hams during curing. A relatively common, yet serious, injury occurs in athletics when one or more hamstring muscles or their tendons become stretched beyond their limits, causing a tear. This is known as a "pulled hamstring." The muscles that move the leg are described in more detail in Table 8–11.

Muscles That Move the Foot and Toes

The muscles moving the foot and toes are located in the leg, with their tendons of insertion extending into the foot. They can be divided into three groups: anterior, posterior, and lateral. As with our study of the

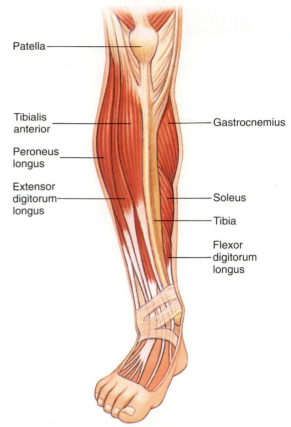

Figure 8-25
Muscles of the right leg, anterior view.

What is the common action of the anterior muscles of the leg?

Table 8–11	MUSCLES THAT MOVE THE LEG		
Muscle	**Origin**	**Insertion**	**Action**
Quadriceps femoris			
Rectus femoris	Ilium and margin of acetabulum	Patella and tibial tuberosity by way of the common tendon	Extends the leg at the knee and flexes the thigh at the hip
Vastus lateralis	Greater trochanter and posterior surface of the femur	Same as rectus femoris	Extends the leg
Vastus medialis	Medial surface of the femur	Same as rectus femoris	Extends the leg
Vastus intermedius	Anterior and lateral surface of the femur	Same as rectus femoris	Extends the leg
Hamstring group			
Biceps femoris	Two heads of origin: 1. Ischium 2. Linea aspera of femur	Proximal ends of the fibula and tibia by way of a common tendon	Flexes and rotates the leg laterally; extends the thigh
Semitendinosus	Ischium	Medial surface of tibia	Extends the thigh and flexes the leg
Semimembranosus	Ischium	Proximal end of tibia	Extends the thigh and flexes the leg

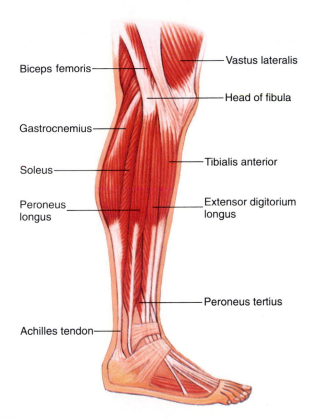

Figure 8–26
Muscles of the right leg, lateral view.
What two muscles share their insertion at the calcaneus bone by way of the Achilles tendon?

of this muscle everts the foot and assists the posterior leg muscles in plantar flexion. The smaller **peroneus tertius** also everts the foot, but it assists the anterior muscles in dorsiflexion. The muscles of the leg are described further in Table 8–12.

forearm muscles, we will examine only the major superficial muscles.

The anterior muscles of the leg are extensors. They provide dorsiflexion of the foot and extension of the toes. The two primary anterior leg muscles are the **tibialis anterior** and **extensor digitorum longus** (Fig. 8–25).

The superficial muscles on the posterior side of the leg form the bulge of the calf. They include the **gastrocnemius**[32] (gas'-trok-NĒ-mē-us) and the **soleus**[33] (SŌ-lē-us) (Figs. 8–25 through 8–27). At their distal ends, they insert by way of a common tendon on the calcaneus (heel). This common tendon is known as the **Achilles tendon**. Both muscles are flexors, providing plantar flexion of the foot.

The lateral muscles of the leg include the peroneus (payr-ō-NĒ-us) muscles, the most prominent of which is the **peroneus longus**[34] (Fig. 8–26). The contraction

Figure 8–27
Muscles of the right leg, medial view.
What is the action of the tibialis anterior?

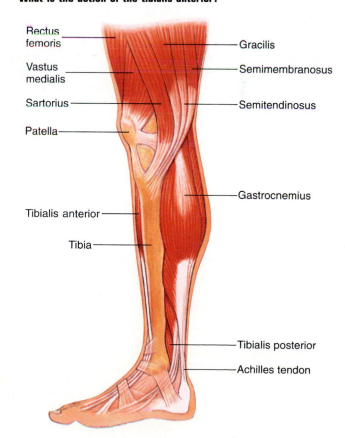

---

[32]Gastrocnemius: Greek *gaster* = stomach + *kneme* = lower leg.

[33]Soleus: Latin *soleus* = pertaining to the sole of the foot.

[34]Peroneus longus: *peroneus* = pinlike + *longus* = long.

Table 8-12 MUSCLES THAT MOVE THE FOOT AND TOES

Muscle	Origin	Insertion	Action
Tibialis anterior	Proximal two-thirds of tibia	Tarsal bone (cuneiform) and first metatarsal	Dorsiflexion; inverts foot
Extensor digitorum longus	Proximal end of tibia, anterior surface of fibula	Second and third phalanges of toes 2 to 5	Dorsiflexion; everts foot
Gastrocnemius	By two heads at distal end of femur	Calcaneus by way of Achilles tendon	Plantar flexion; flexion of the leg at the knee
Soleus	Proximal ends of tibia and fibula	Calcaneus by way of Achilles tendon	Plantar flexion
Peroneus longus	Proximal ends of tibia and fibula	Tarsal and metatarsal bones	Everts foot; plantar flexion
Peroneus tertius	Distal surface of fibula	Fifth metatarsal	Dorsiflexion; everts foot

SPORTS CLINIC

The Truth About Muscle-Building Steroids

Anabolic steroids that are taken to improve muscle mass are synthetically produced versions of the male sex hormone testosterone. Normally available in pill form, these chemicals are controlled substances that are legally available through prescription sources only. Often called "roids" in locker room jargon, their illegal use has been accelerating in recent years among individuals concerned about their lack of weight or muscle mass. While the sustained use of steroids can increase weight gain and muscle mass, they may also accelerate a temporary ability to recuperate from high-intensity training sessions. Their popularity among athletes is well documented, and is the reason why athletes competing in the Olympic Games are required to undergo blood testing prior to competition.

In recent years, evidence that steroids are highly addictive, potentially dangerous drugs has been accumulating. For example, steroids have been shown to affect naturally occurring opioids and neurotransmitters in the brain, which are also involved in heroin, cocaine, and alcohol abuse. In some individuals, a euphoric experience is reached by "stacking" the dose taken (taking 10 times a normal dose or more). Steroid abuse causes behavioral changes, often by increasing the assertive nature of the individual and turning his or her focus to steroid-related activities. Some reports of increased aggression have also been documented, with steroid addicts flying into a "roid rage" and causing harm to themselves and others. Also, in many cases, the sex drive becomes markedly reduced. Evidence is now growing that indicates permanent kidney damage and increased cancer incidence may be the price to pay for steroid abuse.

Steroid dependency is becoming increasingly common, requiring painful withdrawal symptoms that are similar to those for heroin and cocaine. According to one recent report, about 14 percent of steroid users get hooked on the muscle-expanding hormones. The criterion that can be used to identify symptoms of steroid dependency is the presence of at least three of the following: (1) more steroids often taken than intended; (2) a desire to control or reduce steroid use, which is not realized; (3) a large amount of time spent on steroid-related activities; (4) replacement of daily activities by steroid use; (5) continued, prolonged use despite recognition of problems caused by steroids; (6) tolerance of increasing doses; (7) frequent feelings of anxiety or intoxication when not using steroids; and (8) the use of steroids to relieve anxiety symptoms.

HOMEOSTASIS

The muscular system helps maintain homeostasis in two different ways: by providing body movement, and by producing heat for temperature regulation. The movement of body parts is a necessary activity for survival. Not only is it the only means we have of manipulating our environment, bringing food to our mouths, and chewing and swallowing the food, but skeletal muscle movement also keeps blood circulating through the body by preventing it from pooling in large veins. The contraction of muscles surrounding blood vessels pushes against the blood vessel walls, pressing blood toward the heart. When skeletal muscle activity is inhibited, as occurs in injuries that cause paralysis, life expectancy becomes limited because of insufficient blood circulation to vital organs. This is why it is important for bedridden patients to be massaged and turned periodically.

Temperature regulation is a homeostatic function that is provided by several organ systems, including the integumentary system, the cardiovascular system, the nervous system, and the muscular system. The role of the muscular system in maintaining internal body temperature is important, for the contraction of skeletal muscle is the primary means of heat generation by the body. Approximately 80% of the energy produced during metabolism for muscle contraction is in the form of heat. In certain conditions, this heat must be eliminated to prevent excessive internal temperatures. For example, during times of abundant heat, such as during a warm, active day, the body must remove the excess heat produced by muscle contraction. As we have learned in previous chapters, this is done by increasing blood flow to the skin's surface, increasing the rate of perspiration, and increasing the breathing rate. In other conditions, such as when the weather is cold and body activity is minimal, additional body heat must be generated in order to keep the internal temperature constant. This is done by the contraction of small muscles in the skin (the arrectores pilorum muscles) and by shivering. Ideally, the contraction of these muscles provides the heat necessary to keep the internal temperature from dropping, and thereby maintains the homeostatic balance.

CLINICAL TERMS OF THE MUSCULAR SYSTEM

Charley horse An injury to muscle resulting in its tearing, leading to intense, prolonged pain. Bleeding into adjacent tissues usually follows, forming subdermal discoloration (bruising). This is a common injury in contact sports, and frequently occurs at the quadriceps of the thigh.

Electromyography A recording of the electrical events involved in muscle contraction. This data is useful in diagnosing neuromuscular problems, and is obtained by attaching surface electrodes or inserting a needle electrode into the muscle and recording electrical changes.

Muscular dystrophy A group of genetically transmitted diseases characterized by progressive atrophy of muscles without a corresponding loss of nerve tissue. All forms of this disease exhibit a loss of strength with increasing disability and deformity. At the present time there is no cure.

Myalgia A condition of muscle pain that accompanies many infectious diseases, such as influenza (the "flu"), measles, rheumatic fever, toxoplasmosis, and others. It may also be caused by metabolic disturbances and certain drugs.

Myasthenia gravis A chronic disease resulting in weakness and early fatigue of muscles. It is the result of a defect in the conduction of nerve impulses at the neuromuscular junction due to a deficiency of acetylcholine.

Myoma A common benign tumor of the muscle forming the middle wall of the uterus.

Myositis Inflammation of muscle tissue. Myositis may be caused by infection, injury, or infestation by parasites.

Myotonia A condition in which a muscle or muscle group does not readily relax after contraction.

Shin splints A common term for the irritation of the anterior tibialis muscle of the leg. This occurs following excessive exercise, and leads to inflammation of the muscle that is complicated by the tight fascia surrounding it.

Torticollis A condition in which the neck is twisted following injury to the sternocleidomastoid muscle on one side; also called *wryneck*. The head tilting becomes fixed to one side. Exercise that stretches the affected muscle corrects the condition.

Muscle tissue exhibits the properties of contractility, excitability, extensibility, and elasticity. The muscular system is composed mainly of skeletal muscle tissue. Its organs, the muscles, perform the functions of generating body movement, supporting the body, and producing heat.

MUSCLE STRUCTURE

Muscle is composed of skeletal muscle tissue, connective tissues, and nerve tissue.

Connective Tissues of Muscle   Fascia is dense connective tissue that includes two forms: superficial fascia beneath the skin, and deep fascia around organs, including muscles. When associated with muscle, connective tissue appears in three layers: epimysium, which surrounds the whole muscle and is deep fascia; perimysium, which divides the muscle into bundles, or fascicles; and endomysium, which envelops individual muscle cells. Other connective tissues include areolar tissue and fat.

Microscopic Structure of Muscle   Skeletal muscle is highly specialized to contract. Each cell is called a *fiber*, and contains the following components:
   Sarcolemma   The plasma membrane.
   Sarcoplasm   The cytoplasm.
   Sarcoplasmic Reticulum   Similar to the ER of other cells; a saclike organelle that stores calcium ions.
   Transverse Tubules   Tubes that intersect adjacent sarcoplasmic reticula. They play a role in calcium transport.
   Myofibril   A cylindrical cord of protein in the sarcoplasm.
   Filaments   Protein molecules that make up the myofibril. Two types are recognized:
      Thick Filaments   Composed of the protein myosin.
      Thin Filaments   Composed of the proteins actin, troponin, and tropomyosin.
   Sarcomere   A segment of the myofibril, whose thick and thin filaments are divided into regions:
      A Band   Where thick and thin filaments overlap (dark).
      I Band   Where only thin filaments occur (light).
      H Zone   Where only thick filaments occur (light), in the center.

Nerve Supply   A motor neuron provides a stimulus to the muscle fiber. One motor neuron with the muscle fibers it innervates is called a *motor unit*. The region between the motor neuron and the sarcolemma of the muscle fiber is called the *neuromuscular junction*. It consists of the following:
   Synaptic Cleft   A recessed pocket of the sarcolemma that receives the motor neuron.
   Motor End Plate   The region of the sarcolemma at the synaptic cleft.
   Synaptic Vesicles   Small sacs within the terminal end of the motor neuron that contain acetylcholine, a neurotransmitter. Acetylcholine is released as an action potential reaches the neuron terminal end. It diffuses across the synaptic cleft until it contacts the motor end plate.

PHYSIOLOGY OF MUSCLE CONTRACTION

A muscle contracts when the thin filaments slide across the thick filaments toward the center of each sarcomere. This is the sliding filament mechanism of contraction, and includes the following sequence:

The Fiber at Rest   Calcium ions are stored within the sarcoplasmic reticulum, ATP molecules are bound to the thick filaments, and thin filaments are intact.

Role of the Stimulus   Acetylcholine is released into the synaptic cleft and binds to receptor molecules in the motor end plate of the muscle fiber. An action potential is stimulated in the sarcolemma, which passes along the membrane until reaching the sarcoplasmic reticulum. Once there, the action potential causes the sarcoplasmic reticulum to release its stored calcium ions.

Muscle Contraction   Calcium ions diffuse into the sarcoplasm, where they bind to thin filaments. This frees binding sites on actin molecules. The heads of myosin thick filaments attach to the exposed binding sites, forming cross bridge connections. The cross bridges shift their angle, causing thin filaments to slide over thick filaments. This shifting requires energy in the form of ATP, which is bound to the myosin molecules. Once shifting is complete, another ATP molecule provides the energy to break the bond between the thick-filament cross bridge and the thin filament.

Return to Rest   ATP provides the energy for enzymes to return calcium ions to the sarcoplasmic reticulum. The lack of calcium ions causes the thin filaments to return to their original shape, closing off the binding site to cross bridges. The thin filaments then slide back to their original positions in the sarcomere.

Energy for Contraction   Energy is provided in the form of ATP. Because it is used up very quickly, other, longer-term sources of energy are available: glucose, creatine phosphate, glycogen, and fats release additional ATP following their metabolic conversion.

Oxygen Debt   Oxygen is required for contraction because it is needed to synthesize ATP. After strenuous exercise, more energy is used than can be provided in the presence of oxygen, and products (lactic acid) accumulate in muscle. Rapid breathing eventually restores the oxygen that is needed. Oxygen debt may lead to muscle fatigue, in which the muscle becomes unable to contract because of an accumulation of lactic acid and subsequent unavailability of ATP.

SMOOTH MUSCLE AND CARDIAC MUSCLE

Smooth and cardiac muscle tissues differ from skeletal muscle tissue in structure and in the manner of contraction.

MUSCULAR RESPONSES

Different factors alter the muscular response. They include variations in stimulus strength, stimulus frequency, and tension.

All-or-None Response The weakest stimulus that can initiate contraction is called a *threshold* stimulus. Once this is reached, the muscle contracts all the way. This is called the *all-or-none response.*

Types of Muscle Contractions Contractions vary according to the stimulus frequency that is received.
Twitch A rapid response to a single stimulus that is slightly over threshold. When measured on a myogram, a twitch consists of a latent period, a period of contraction, and a period of relaxation. A twitch is the basic unit of muscle contraction.
Treppe When a muscle receives successive stimuli, the strength of the contraction increases slightly with each stimulus. Treppe is also called the *staircase effect.*
Wave Summation When a muscle receives a second stimulus before the first is complete, the contraction will be stronger.
Tetanus When successive stimuli are received before contractions are completed, the muscle will reach a maximum contraction strength and be sustained. If some relaxation occurs between contractions, the response is called *incomplete tetanus.* If no relaxation occurs, it is called *complete tetanus.* Incomplete and complete tetanus are the types of contractions that are used to produce most body movements.
Isotonic and Isometric Two types of tetanic contractions that differ in the way in which tension is applied. Tension is the force exerted by a muscle contraction. Isotonic contraction produces body movement with tension, and isometric contraction does not produce body movement with the application of tension.

PRODUCTION OF MOVEMENT

Origin and Insertion The site of attachment of a muscle tendon to a stationary bone is called the muscle's *origin*, and the site of attachment to a movable bone is the muscle's *insertion.*

Group Actions For a smooth, coordinated movement to occur, more than one muscle is involved. The interaction of numerous muscles to bring about movement is called a *group action.* Four roles of muscles are recognized:
Prime Mover The muscle that causes the desired action.
Antagonist The muscle that relaxes and yields to the prime mover.
Synergist The muscle(s) that steady the movement by stabilizing the joint and moving bones.
Fixator The muscle(s) that stabilize the bone and joint at the muscle's origin.

MAJOR MUSCLES OF THE BODY

Muscles of the Head and Neck
Muscles of facial expression include the frontalis and occipitalis, the orbicularis oculi and orbicularis oris,

the buccinator, and the zygomaticus. Muscles of mastication (chewing) include the masseter and temporalis. The primary muscle moving the head is the sternocleidomastoid.

Upper-Limb Muscles
Muscles of the pectoral girdle include the trapezius, pectoralis minor, levator scapulae, rhomboids, and serratus anterior. Muscles moving the arm include the pectoralis major, latissimus dorsi, deltoid, subscapularis, supraspinatus, infraspinatus, teres major, and teres minor. Muscles moving the forearm include the triceps brachii and brachioradialis on the posterior side of the arm, and the biceps brachii and brachialis on the anterior side of the arm. The supinator and pronator teres also move the forearm and cross the elbow. Muscles that move the hand and fingers are located in the forearm and include the flexor carpi radialis, flexor carpi ulnaris, palmaris longus, flexor digitorum profundus, extensor carpi radialis longus, and extensor digitorum.

Trunk Muscles
Muscles that move the vertebral column include the erector spinae group and the deep muscles of the back. The primary muscles of the thorax include the external intercostals, internal intercostals, and diaphragm. They are involved in the breathing process. The muscles of the anterior abdominal wall are the rectus abdominis, external oblique, internal oblique, and transverse abdominis.

Lower-Limb Muscles
The primary muscles that move the thigh include the iliopsoas on the anterior side of the pelvis, the tensor fasciae latae on the lateral side of the pelvis and hip, the gluteus maximus and gluteus medius on the posterior side of the pelvis and hip, and the adductor longus, adductor magnus, and gracilis on the medial side of the thigh. The muscles that move the leg also move the thigh. They include the anterior group of thigh muscles, known as the *quadriceps femoris,* and the posterior group of thigh muscles, known as the *hamstrings.* The quadriceps femoris includes the rectus femoris, vastus lateralis, vastus intermedius, and vastus medialis. The hamstrings include the biceps femoris, semitendinosus, and semimembranosus. The major superficial muscles of the leg that move the foot and toes include the tibialis anterior and extensor digitorum longus on the anterior side of the leg, the gastrocnemius and soleus on the posterior side, and the peroneus longus and peroneus tertius on the lateral side.

HOMEOSTASIS

The muscular system helps maintain homeostasis in two ways:

Production of Movement Contraction of muscles not only provides for movement of body parts, but also assists the flow of blood through blood vessels.

Production of Heat Contraction of muscles produces heat, which can be used to help maintain the body temperature at a constant rate.

antagonist (p. 212)   fiber (p. 200)   myofibril (p. 200)   sarcomere (p. 202)
endomysium (p. 199)   fixator (p. 212)   neurotransmitter (p. 203)   sarcoplasm (p. 200)
epimysium (p. 199)   glycogen (p. 207)   origin (p. 212)   sarcoplasmic reticulum (p. 200)
fascia (p. 199)   insertion (p. 212)   perimysium (p. 199)   synergist (p. 212)
fascicle (p. 199)   isometric (p. 212)   recruitment (p. 210)   tetanus (p. 211)
fatigue (p. 208)   isotonic (p. 212)   sarcolemma (p. 200)   threshold stimulus (p. 209)

QUESTIONS FOR REVIEW

OBJECTIVE QUESTIONS

1. The layer of connective tissue that divides a muscle into bundles, or fascicles, is the:
   a. perimysium
   b. epimysium
   c. endomysium
   d. ectomysium

2. The band of connective tissue that extends from a muscle to attach a bone is called a(n):
   a. ligament
   b. tendon
   c. aponeurosis
   d. perimysium

3. The membranous sac within the sarcoplasm of a muscle fiber that stores calcium ions is called:
   a. endoplasmic reticulum
   b. sarcoplasmic reticulum
   c. sarcolemma
   d. mitochondrion

4. A cylindrical cord of protein that extends the length of the muscle fiber is the:
   a. tendon
   b. thick filament
   c. myofibril
   d. sarcomere

5. The H zone is a region of thick filaments in the center of a(n):
   a. A band
   b. sarcomere
   c. motor unit
   d. muscle fiber

6. A motor unit consists of a single motor neuron and the many _____ it stimulates.
   a. muscles
   b. muscle fibers
   c. fascicles
   d. myofibrils

7. In a neuromuscular junction, the surface of the sarcolemma is highly folded. This region is called the:
   a. motor end plate
   b. synaptic vesicles
   c. synaptic cleft
   d. motor unit

8. According to the sliding-filament mechanism, thin filaments:
   a. break in half
   b. completely disappear
   c. slide toward the H zone
   d. move from one fiber to the next

9. In a muscle fiber at rest, calcium ions are:
   a. stored within synaptic vesicles
   b. stored in sarcoplasmic reticulum
   c. in high concentrations in the sarcoplasm
   d. none of the above

10. ATP provides the energy for:
    a. moving cross bridges
    b. disconnecting cross bridge attachments
    c. returning calcium ions to the SR
    d. all of the above

11. The accumulation of lactic acid in muscle following strenuous exercise is due to an insufficient amount of:
    a. ATP
    b. oxygen
    c. glycogen
    d. creatine phosphate

12. The weakest stimulus that can initiate a muscle contraction is called a(n):
    a. threshold stimulus
    b. twitch
    c. subthreshold stimulus
    d. incomplete tetanus

13. A muscle either contracts all the way or does not contract at all. This phenomenon is called the:
    a. threshold stimulus
    b. tetanic contraction
    c. all-or-none response
    d. isometric exercise

14. A muscle increases its strength of contraction as additional _____ are recruited.
    a. sarcomeres
    b. motor units
    c. myofibrils
    d. stores of ATP

15. A muscle contraction that lasts only a fraction of a second in response to a single stimulus is called a(n):
    a. isotonic contraction
    b. wave summation
    c. twitch contraction
    d. complete tetanus

16. Body movement and muscle tone are accomplished by:
    a. twitch contractions
    b. isometric contractions
    c. tetanic contractions
    d. all of the above

17. The site of attachment of a muscle tendon to a movable bone during an action is called the muscle's:
    a. insertion
    b. synergist
    c. origin
    d. fixation

18. In a group action, the muscle that reduces its tension and yields to the movement is called the:
    a. antagonist
    b. synergist
    c. prime mover
    d. fixator

19. Which muscle allows you to bend your arm at the elbow and raise a glass from the tabletop to your lips for drinking?
    a. latissimus dorsi
    b. extensor carpi ulnaris
    c. biceps brachii
    d. triceps brachii

20. Which muscle allows you to extend your lower leg and kick a ball lying on the ground?
    a. biceps femoris
    b. sartorius
    c. gluteus maximus
    d. quadriceps femoris

21. Which muscle allows you to raise your heel and stand up on tiptoe?

    a. gastrocnemius   c. tibialis anterior

    b. semitendinosus   d. gracilis

ESSAY QUESTIONS

1. Describe the sequence of events involved in muscle contraction, from the "at rest" state to the "return to rest" state. Include the roles of calcium and ATP.

2. Explain how oxygen debt develops during exercise, and how it may lead to fatigue and cramping. How may an athlete prolong the use of muscles and thereby delay the oxygen debt?

3. Describe how a twitch contraction can lead to complete tetanus. Put the concept of tetanus into everyday experiences; for example, how does tetanus play a role in jumping, or kicking a ball?

4. Describe how recruitment of motor units can lead to increased force of muscle contraction.

ANSWERS TO ART LEGEND QUESTIONS

Figure 8–1

Which type of muscle tissue is the main component of the organs of the muscular system? Skeletal muscle tissue is the main component of muscles.

Figure 8–2

Which layer of connective tissue surrounds individual muscle fibers? The endomysium surrounds muscle fibers.

Figure 8–3

What structures occupy most of the space within the sarcoplasm? Myofibrils occupy much of the sarcoplasm.

Figure 8–4

What protein fibers make up the light bands in a skeletal muscle fiber? Actin, troponin, and tropomyosin, which constitute the thin filaments, make up the light bands.

Figure 8–5

How many motor neurons are in a single motor unit? One motor neuron is in a single motor unit.

Figure 8–6

What neurotransmitter is packaged within the motor neuron synaptic vesicles? Acetylcholine is the neurotransmitter in motor neuron synaptic vesicles.

Figure 8–7

What is the role of calcium ions in muscle contraction? Calcium ions bind to the thin filament, thereby freeing binding sites for the attachment of thick-filament cross bridges.

Figure 8–8

What is the role of ATP during muscle contraction? ATP provides the energy to shift cross bridges once they become attached to thin filaments. Its energy is also required to break the cross bridge attachment and to power enzymes that remove calcium from the sarcoplasm.

Figure 8–9

What mechanism is represented by the latent period? The latent period is thought to represent the time it takes for calcium ions to enter the sarcoplasm, myosin to be activated, and cross bridge attachment to occur.

Figure 8–10

Which of these types of contractions is used by skeletal muscle in the performance of an activity such as walking? Primarily, complete tetanus is used by skeletal muscle to perform an activity such as walking.

Figure 8–11

Which is the movable bone in this example? The radius is the movable bone (the bone of insertion) during contraction of the biceps brachii.

Figure 8–12

Is the deltoid muscle an organ, a tissue, or a membrane? The deltoid muscle, as are all muscles, is an organ.

Figure 8–13

What muscles of the back can you identify in this illustration? The trapezius, infraspinatus, teres major, rhomboid, and latissimus dorsi are all muscles of the back that are visible.

Figure 8–14

What action is performed by the temporalis and masseter muscles? Both the temporalis and masseter muscles close the mouth by elevating the jaw (to chew).

Figure 8–15

What is the main role played by the muscles of the pectoral girdle? The pectoral girdle muscles move the scapula.

Figure 8–16

Where does the pectoralis major insert, and what are its major actions? The pectoralis major inserts on the greater tubercle of the humerus, and it flexes, adducts, and medially rotates the arm.

Figure 8–17

What four muscles (three of which are shown in this figure) make up the rotator cuff? The subscapularis, supraspinatus, infraspinatus, and teres minor muscles make up the rotator cuff.

Figure 8–18

What action is provided by the biceps brachii and brachialis muscles? The biceps brachii and brachialis muscles flex the arm at the elbow.

Figure 8–19
What muscles extend the hand?   The extensor carpi radialis longus and the extensor carpi ulnaris extend the hand at the wrist.

Figure 8–20
Where does the triceps brachii insert?   The triceps brachii inserts at the olecranon process of the ulna.

Figure 8–21
What is the common action of the external oblique, internal oblique, and transverse abdominis muscles?   The external oblique, internal oblique, and transverse abdominis muscles flex the vertebral column, thereby compressing the abdomen.

Figure 8–22
The adductor leg muscles insert upon what long bone?   The adductor muscles of the leg insert on the femur.

Figure 8–23
The gluteus muscles originate from what bone?   The gluteus muscles originate from the coxal bones of the pelvis.

Figure 8–24
What are the four muscles of the quadriceps femoris group (two of which are shown in this figure)?   The quadriceps femoris group includes the rectus femoris, vastus lateralis, vastus medialis, and vastus intermedius muscles of the thigh.

Figure 8–25
What is the common action of the anterior muscles of the leg?   The anterior leg muscles extend the foot (dorsiflexion) and toes.

Figure 8–26
What two muscles share their insertion at the calcaneus bone by way of the Achilles tendon?   The gastrocnemius and soleus muscles both insert at the calcaneus bone by way of the Achilles tendon.

Figure 8–27
What is the action of the tibialis anterior?   The tibialis anterior extends (dorsiflexion) and inverts the foot.

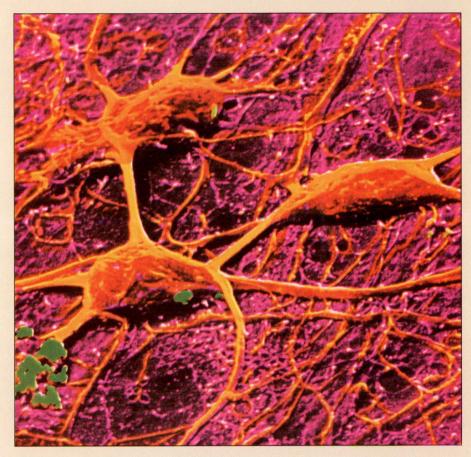

SYSTEMS THAT CONTROL BY COMMUNICATION

The task of keeping the body healthy and alive is a job performed by every one of its many trillions of cells. Like the players in a great symphony orchestra, each cell must perform its role for the good of the whole. The conductor of this great orchestra is the nervous system, which provides control and coordination for the cells' many activities. Its performance is achieved by way of rapid nerve impulses, which can speed information throughout all areas of the body. A second system assists the nervous system in control, but it uses a different method of communication. The endocrine system controls cell activities by the release of chemicals that must circulate through the bloodstream, and is therefore much slower in bringing about a response. Regardless of which system, nervous or endocrine, is playing the major role in control, the goal is the same: maintaining the harmony of homeostasis—the stable internal environment of the body—to support survival by the achievement of health.

The two systems that control body activities by way of communication, the nervous and endocrine systems, are grouped together in this unit. Three chapters are included: Organization of the Nervous System (Chapter 9), The Special Senses and Functional Aspects of the Nervous System (Chapter 10), and The Endocrine System (Chapter 11).

ORGANIZATION OF THE NERVOUS SYSTEM

LEARNING OBJECTIVES

After studying this chapter, you should be able to:

1. Identify the divisions of the nervous system.

2. Distinguish both structurally and functionally between the neuroglia and neurons.

3. Describe the structure of a neuron, and distinguish between myelinated and unmyelinated fibers.

4. Identify the different types of neurons.

5. Describe the events involved in maintaining a resting potential, and in initiating an action potential.

6. Describe the structure of a synapse, and explain how an impulse is transmitted from cell to cell.

7. Identify the protective coverings of the brain and spinal cord.

8. Describe the structure of the spinal cord, and distinguish between gray and white matter.

9. Describe the conduction pathways of the spinal cord and the reflex arc.

10. Identify the ventricles of the brain, and explain how cerebrospinal fluid is produced and circulated.

11. Distinguish between the parts of the brain on the basis of structural and functional differences.

12. Identify the organs of the PNS.

13. Distinguish between cranial nerves and spinal nerves, and describe how spinal nerves branch.

14. Distinguish between the somatic and autonomic nervous systems on the basis of their effectors and pathways.

15. Distinguish between the sympathetic and parasympathetic autonomic divisions.

The nervous system is the control center and communication network of the body. It is called the control center because it monitors the state of the body, processes the information received, and initiates the response necessary to maintain internal stability, or homeostasis. Its role in monitoring the body is called its **sensory** function, for it can perceive, or "sense," changes as they occur. Its ability to process the sensory information it has received is called its **integrative** function, for processing involves combining ("integrating") pieces of information before arriving at a desired response. Its ability to initiate a response is called its **motor** function, since the response causes a change to occur that is often associated with body movement or secretion of a product.

Each of the three functions of the nervous system is performed by means of a vast communication network. This network is made up of the organs of the nervous system, which include the brain, spinal cord, nerves, and sensory organs. These organs communicate with each other and with other organs, such as muscles, glands, and visceral organs, by way of nerve impulses. Nerve impulses travel very rapidly, making split-second adjustments to changes in the body that are necessary to keep the internal environment stable.

A balanced and controlled internal environment is vital to life itself. The function of maintaining this balance, or homeostasis, is primarily performed by the nervous system, since only it can sense changes in the body, integrate the information, and initiate a motor response to a change very rapidly. Thus, by controlling the homeostatic balance of our bodies, the nervous system plays a key role in our basic survival.

DIVISIONS OF THE NERVOUS SYSTEM

The nervous system consists of the central nervous system, which includes the brain and spinal cord, and the peripheral nervous system, containing the nerves.

The nervous system is divided into two major categories, the central nervous system and the peripheral nervous system, each with its organs or subdivisions (Fig. 9–1).

The **central nervous system (CNS)** includes the **brain** and **spinal cord**. These organs are located along the vertical midline of the body, and are therefore "central" in location. All sensory nerve impulses arrive in the CNS, and all motor nerve impulses originate from the CNS.

The **peripheral nervous system (PNS)** comprises the nerves that lie outside, or peripheral to, the CNS. These nerves, which include two types, convey all sensory and motor impulses that travel between the CNS and the rest of the body. The **cranial nerves** originate from the brain and extend to the head and neck, and the **spinal nerves** arise from the spinal cord to supply the body below the head. Cranial and spinal nerves give rise to smaller nerves that branch extensively throughout the body.

The nerves that arise from cranial and spinal nerves will serve one of two subdivisions of the PNS: the **somatic system** or the **autonomic system**. The somatic system contains sensory nerves that serve sensory receptors in the skin and sensory organs of the head, and motor nerves that stimulate skeletal muscles; that is, the parts of your body that make you aware of the world around you or that you can consciously control. For this reason, the somatic system is often referred to as the *voluntary* nervous system. The nerves of the autonomic system consist mainly of motor nerves that carry impulses to visceral organs, blood vessels, and glands; in other words, structures that are under your body's unconscious or "automatic" control. The autonomic nervous system is often referred to as the *involuntary* nervous system because of its role in unconscious activities.

The autonomic system is further subdivided into two categories: the **sympathetic division**, which plays an active role during conditions of stress (such as injury, emotional excitement, and strenuous physical activity); and the **parasympathetic division**, which is active during conditions of normal organ functioning.

CONCEPTS CHECK

1. What are the organs of the central nervous system?

2. What nerves arise from the spinal cord?

3. What body structures receive their nerve supply from the autonomic system?

NERVE TISSUE

The primary structural and functional unit of nerve tissue is the neuron. Each neuron is specially adapted to form and conduct impulses and transmit them from cell to cell.

As we learned in Chapter 3, nerve tissue is one of the four primary types of tissue in the body. It differs from any other tissue in that it is highly specialized to respond to changes in the environment surrounding it,

Figure 9–1
Organization of the nervous system. The arrows represent cranial nerves (connecting to the brain) and spinal nerves (connecting to the spinal cord). Structures colored yellow are within the CNS, those colored red are part of the PNS, and those in orange are body organs that are not part of the nervous system.

What are the organs of the central nervous system?

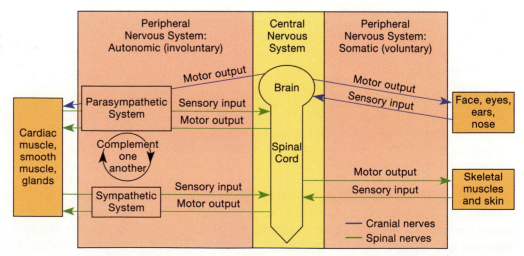

240

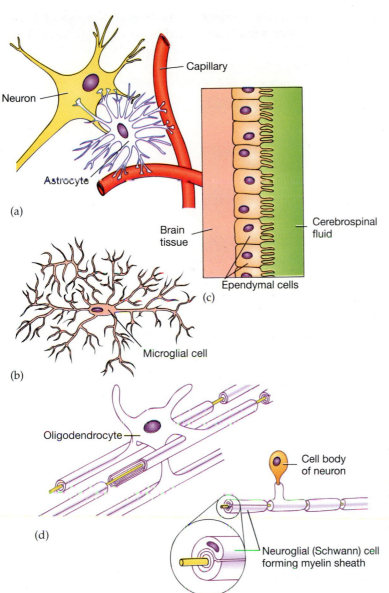

Figure 9–2
Types of neuroglia. (a) Astrocytes provide a framework for the fragile brain, connecting neurons and blood vessels together. (b) Microglial cells wander throughout the CNS, phagocytizing foreign debris and dead cells. (c) Ependymal cells line cavities within the CNS. (d) Oligodendrocytes wrap around axons in the CNS, forming an insulating barrier known as the *myelin sheath.* (e) Schwann cells wrap around long axons in the PNS, providing an insulating myelin sheath.
Which neuroglial cells are found in the CNS?

and to conduct impulses. It consists of two categories of cell types: **neurons**, which are the responsive cells that conduct impulses at great speeds; and **neuroglia** (noo-RŌG-lē-a), which serve to support and maintain neuronal structure and function.

Neuroglia

Neuroglia, or simply glial cells, form about 90% of the substance of the brain and spinal cord, and a small part of peripheral nerves. In the brain and spinal cord they provide a structural framework for the more fragile neurons, produce a supportive fatty sheath around neuron axons, remove unwanted materials by phagocytosis, and aid in the circulation of cerebrospinal fluid

(Fig. 9–2). Recent experimental evidence suggests that certain types of neuroglia in the brain may also aid neurons in impulse conduction and in the formation of junctions between neurons. In peripheral nerves, the neuroglia produce a supportive fatty sheath around many neuron axons. The various types of neuroglia, with their locations and functions, are listed in Table 9–1.

Neurons

Despite the abundance of neuroglia in the central nervous system, neurons are considered the primary structural and functional unit of nerve tissue. They carry this title because they perform the functions re-

Table 9–1   SUPPORTING CELLS OF NERVE TISSUE: THE NEUROGLIA

Type of Cell	Location	Function
Astrocytes	Brain and spinal cord	Anchor neurons to blood capillaries; control the flow of ions around neurons; may aid neurons in impulse conduction
Ependymal cells	Line cavities in brain and spinal cord	Help form and circulate cerebrospinal fluid
Microglia	Brain and spinal cord	Phagocytize invading microorganisms and dead nerve tissue
Oligodendrocytes	Brain and spinal cord	Provide insulating coverings around CNS axons, forming a myelin sheath
Schwann cells	Large nerves of the peripheral nervous system	Provide insulating coverings around large PNS axons

quired by the nervous system as a whole: they sense changes in the environment, integrate the information, and carry out a motor response, all by means of the rapid conduction of impulses. They are found in a variety of shapes and sizes, which relate to the particular roles they play (Fig. 9–3).

Structure

Although neurons vary in shape and size, each exhibits general features that are common to all others. These features include the division of the cell into three portions: a cell body, numerous dendrites, and a single axon (Fig. 9–4).

Figure 9–3
Shapes of neurons. Many shapes and sizes of neurons exist, some of which are represented in this drawing.
What three structural features do all neurons have in common?

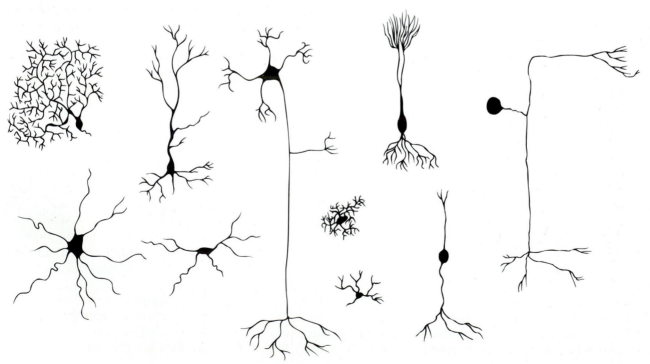

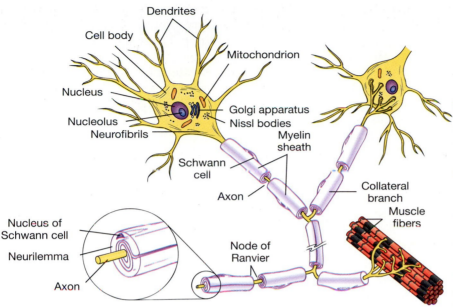

Figure 9–4
Neuron structure. A typical neuron, like the motor neuron shown here, contains dendrites, a cell body, and an axon. Branches from the axon are called *collaterals*. The cell body contains the nucleus and much of the cytoplasm. The close-up is a section of the myelin sheath. The Schwann cell wraps around the axon to form multiple layers of insulation. Its nucleus and most of its cytoplasm remain on the outer layer, forming the neurilemma.

In what direction do impulses travel as they pass through the axon?

The **cell body** is the portion of the neuron that is more typical of other body cells. It consists of a mass of cytoplasm enveloped by a plasma membrane, with a nucleus containing a prominent nucleolus. Other organelles are also present, including mitochondria, lysosomes, and Golgi apparatus. In addition, there are components not found in other cell types, including small, membranous sacs called **Nissl bodies**, which are similar to rough endoplasmic reticula. Ribosomes are attached to the surfaces of Nissl bodies, indicating that protein synthesis takes place on these structures. A network of fine threads similar to microtubules, called **neurofibrils**, is also characteristic. In addition, the cytoplasm of a mature neuron lacks mitotic spindle fibers, rendering the neuron incapable of cell division.

Dendrites[1] are thin, branching extensions originating from the cell body. Each is enveloped by the plasma membrane, and they usually contain many of the cytoplasmic contents found in the cell body. In most neurons, the dendrites are short, highly branched, and quite numerous. They provide many surfaces for receiving impulses from adjacent neurons. Once impulses are received, dendrites conduct them in one direction only: *toward* the cell body.

The **axon** is a highly specialized part of the neuron that conducts impulses *away from* the cell body. It is often called a **nerve fiber**. Like the dendrites, the axon is surrounded by a plasma membrane, and its cytoplasm contains many of the contents of the cell body. There is only one axon to every neuron, although axons may give off side branches (called **collaterals**). Also, many small branches often occur at the axon's

terminal end, each with a specialized ending that communicates with another cell.

The axons of many neurons in the peripheral nervous system extend for great distances between the spinal cord and muscles and skin, some for as long as 1 meter. These axons are partially enclosed within sheaths composed of Schwann cells, a type of neuroglia. The Schwann cells are wound tightly around the axon, forming multiple layers of plasma membrane with very little cytoplasm in between. Because the plasma membrane of the Schwann cells contains an unusually high proportion of lipid molecules, the sheath that is formed is rich in fat and provides a high insulation and nutrition value for the axon. This white, fatty insulating barrier is called the **myelin** (MĪ-e-lin) **sheath**. The cytoplasm and nuclei of the Schwann cells remain on the outer layer of the myelin sheath, forming what is called the **neurilemma**.[2] Look carefully at Figure 9–4, and notice that the myelin sheath is not continuous throughout the length of the axon. Schwann cells do not contact one another; there are gaps between them where the myelin sheath is absent. These gaps are called **nodes of Ranvier**[3] (ran-vē-Ā) and aid in impulse conduction in a very interesting way, which will be discussed later in this chapter.

Axons that are somewhat shorter in length may also be associated with insulating cells. For example,

[1]Dendrites: Greek *dendr-* = tree
[2]Neurilemma: Greek *neur-* = nerve + *lemma* = sheath.
[3]Ranvier: after the French physician Louis Ranvier.

smaller axons of peripheral nerves may be enclosed by Schwann cells but lack the thick myelin sheath. These axons are called **unmyelinated fibers**, while long axons with the myelin sheath are termed **myelinated fibers**. Also, some axons of the brain are provided with a myelin sheath that is similar to the one produced by Schwann cells, but is produced by oligodendrocytes instead. These axons are therefore myelinated fibers as well. In general, groups of myelinated fibers appear white because of the presence of the myelin sheaths and are therefore called **white matter**. Conversely, groups of unmyelinated fibers and neuron cell bodies appear as **gray matter**.

Types of Neurons

As was stated earlier, neurons come in many shapes and sizes. They also perform different specific functions. On the basis of their differences in structure and function, the various types of neurons may be categorized in the following way.

On the basis of *structural* differences, there are three major types of neurons: multipolar, bipolar, and unipolar (Fig. 9–5). **Multipolar neurons** have many dendrites arising from the cell body, and a single axon. The neurons that carry impulses from the CNS to skeletal muscles are multipolar neurons. **Bipolar neurons** have a single dendrite and a single axon arising from the cell body, each extending in opposite directions. They are found in special sensory areas, such as in the eyes, ears, and nose. **Unipolar neurons** contain a single nerve fiber extending from the cell body. A short distance from the cell body the fiber splits into two branches: one branch extends to the spinal cord and serves as the axon, while the other branch extends to the peripheral part of the body and serves as the dendrite. The neurons that carry impulses from skin receptors to the spinal cord are unipolar neurons.

On the basis of *functional* differences, again there are three major types of neurons: sensory, association, and motor. **Sensory neurons** (afferent neurons) carry nerve impulses from a peripheral part of the body to the CNS. Sensory neurons are unipolar in structure. **Association neurons** (interneurons) are located within the CNS. They form links between neurons, and provide a means of relaying impulses from one region of the brain or spinal cord to another. Association neurons are usually multipolar in structure. **Motor neu-**

Figure 9–5
Structural types of neurons:
(a) multipolar, (b) bipolar, and
(c) unipolar. The red arrow indicates the direction of impulse conduction.

What are the three functional distinctions of neurons?

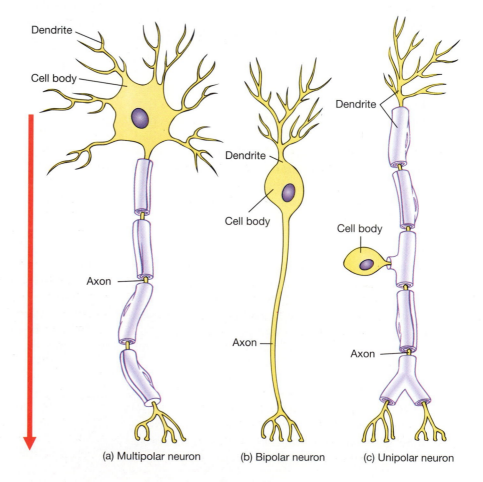

(a) Multipolar neuron   (b) Bipolar neuron   (c) Unipolar neuron

Neuron Regeneration

At about 3 years of age, the cytoplasm of every neuron in your body loses its mitotic apparatus and the cell is no longer able to divide. Unlike the cells of your skin, once a neuron cell body is damaged or destroyed past this age it will not be replaced by a newly produced cell. A neuron that is destroyed is, under normal circumstances, permanently lost.

However, a damaged neuron can be repaired by cellular mechanisms if the damage is not too extensive. For example, damage to myelinated axons in the PNS can often be repaired if the cell body remains intact. Regeneration of cytoplasmic materials is accomplished mainly by Nissl bodies, which migrate to the damaged area. This new growth is supported by Schwann cells, which form a nourishing and protecting tube around the damaged axon. Thus, peripheral nerves that have been partially severed or crushed can regenerate under some circumstances, restoring at least partial function.

For more severe cases of nerve injury that exceed the cellular mechanisms of recovery, recent research into the field of spinal cord injuries is beginning to blossom with discoveries of potential therapies. For example, in one series of studies on laboratory rats, neuroscientists found that transplants of Schwann cells onto the severed ends of axons can coax the damaged nerves to grow toward one another. In another very promising approach, scientists are exploring the use of genetically engineered cells that can produce a protein known as *nerve growth factor (NGF)*, which has been shown to stimulate complete axon growth. In another approach, investigators have found a different protein, normally present in the myelin sheath, that has growth-inhibitory properties. When this protein is neutralized with genetically engineered antibodies, severed axons have been capable of growing as quickly as 1 millimeter per day! In yet another approach, administration of a steroidal drug called *methylprednisolone* within hours of a spinal cord injury has been shown to minimize paralysis. Presumably, this drug works by reducing the breakdown of plasma membranes in cells that have been crushed or severed, thus also reducing the quantity of liberated free radicals that kill or damage nearby healthy cells. Within the next several years, it is hoped that these findings and others will significantly reduce the incidences of accidental paralysis that result from nerve damage.

---

rons (efferent neurons) carry nerve impulses from the CNS to parts of the body capable of responding, such as muscles or glands. The general term for parts that respond to stimuli is **effectors**. Motor neurons are multipolar in structure.

Function

The primary function of a neuron is the generation of a nerve impulse. In many ways, the nerve impulse is similar to the flow of electricity through an insulated wire. In this section we shall see how the nerve impulse flows through the cell, and how it is passed from cell to cell.

Resting Membrane Potential

A potential difference occurs when there is a separation of charge between two points. We can observe an example of a potential difference in a battery, where there is a flow of electrons between the positive pole and the negative pole. The electron flow between the poles creates a voltage, which has a value that can be measured.

A potential difference that may be measured in voltage also occurs across the plasma membranes of all living cells in the body. In a cell, the plasma membrane separates a difference in charge between the extracellular environment and the intracellular environment. This difference in charge is produced by an unequal distribution of ions across the membrane, which is maintained by integral proteins in the plasma membrane that form guarded openings into the cell known as **ion channels**. The ion channels regulate the movement of ions across the plasma membrane. Ion channels are not peculiar to neurons; they may also be found in other cell types, such as muscle cells, where they perform similar functions.

The ions that have the greatest influence in creating a difference in charge across the membranes of most cells are sodium (Na^+) and potassium (K^+). When a neuron is at rest (that is, when it is not con-

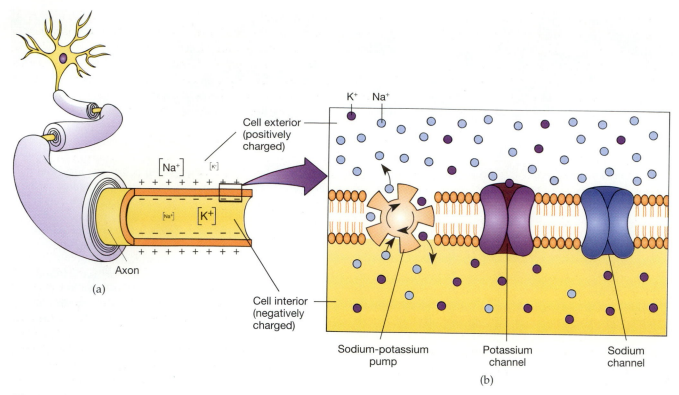

Figure 9–6
The resting membrane potential. (a) The relative sizes of ions in this illustration (shown as [Na⁺] and [K⁺]) indicate their concentrations relative to one another. A net positive charge is present on the outer surface of the plasma membrane, and a net negative charge is present on the inner surface. (b) The unequal distribution of ions (the blue dots are Na⁺, and the violet dots are K⁺) is maintained primarily by the sodium-potassium pump in the plasma membrane, which transports 3 Na⁺ out of the cell for every 2 K⁺ transported back into the cell. It is also assisted by the leakage of some potassium ions outward. At rest, the sodium- and potassium-ion channels are closed, as they are shown here. During an action potential, these ion channels open to allow for a rapid flow of ions across the membrane.
How is the high sodium ion concentration maintained on the outside of the membrane?

ducting impulses), there are about 10 times more sodium ions outside the cell than inside, while the distribution of potassium ions is roughly the reverse (Fig. 9–6). How is this uneven distribution of ions made possible? Primarily by an ion channel mechanism known as the **sodium-potassium pump**, which uses up large amounts of energy to continually transport sodium ions out of the cell and potassium ions inward. Other factors are involved as well, as we are about to discover.

As the sodium-potassium pump forces sodium ions out of the cell and potassium ions in, a concentration gradient is established across the plasma membrane. This gradient tends to cause the diffusion of sodium ions back into the cell and the diffusion of potassium ions out. If conditions for uninterrupted diffusion were ideal, we would eventually find an equal distribution of ions across the membrane, and no net charge. However, two factors are involved that disrupt equal diffusion rates. First, the sodium-potassium pump transports an uneven ratio of ions. For every three sodium ions pumped out of the cell, two potassium ions are pumped inward. This uneven ratio establishes an electrical gradient across the membrane, in which more positive ions are maintained on the outside of the cell. Second, the plasma membrane is more permeable to potassium ions than to sodium ions. Consequently, potassium ions tend to leak back out of the cell (by diffusion) at a rate greater than sodium ions tend to leak in.

As a result of the combined effects of sodium being pumped out and potassium leaking out, the outside of the plasma membrane tends to accumulate positive ions, giving it a positive charge. The inside of the membrane has a negative charge, due to the low quantities of positive ions, the presence of chloride ions (Cl⁻), and the presence of negatively charged proteins inside the cell that are unable to cross over. The potential difference resulting from this uneven distribution of ions in a resting neuron is called the **resting membrane potential**. A membrane that exhibits the resting membrane potential is said to be in a *polarized* state.

Like other forms of potential difference, the resting membrane potential can be measured in terms of voltage. For example, the potential on the inside of a polarized membrane is −70 millivolts (mV). This is very small; a single flashlight battery ("D" cell) carries a potential that is over 1500 times greater (1.5 volts). Also, the actual number of ions that flow across the membrane is extremely small; so small, in fact, that the intracellular and extracellular concentrations of sodium and potassium ions are almost equal. However, sufficient numbers of potassium ions leak through the membrane to produce the difference in the distribution of ions that establishes the resting potential.

Action Potential

A characteristic of neurons is their ability to respond to a change in the environment around them, a feature called *excitability*. They are able to respond when the change, or stimulus, becomes great enough to alter the resting membrane potential of a particular region of the membrane. The result is a rapid change in that particular region in membrane permeability to sodium ions, which diffuse inward through newly opened sodium-ion channels. This sudden change in membrane permeability to sodium ions, followed by the flow of sodium ions into the cell, causes the electrical charge on the inside of the membrane to change to a more positive value. When this occurs, the membrane in the region near the point of stimulus has become *depolarized*. For example, we learned previously that the potential on the inside of a polarized membrane (that is, the resting membrane potential) is −70 millivolts. The sudden flow of sodium ions into the cell causes this charge to increase to +30 millivolts. This increase in charge during depolarization is shown in the graph in Figure 9–7. Thus, during depolarization the charge on the inside of the membrane has changed from a negative value to a positive value. The charge on the outside of the membrane also changes in response, but from positive to negative.

Soon after a region of the membrane has depolarized, the resting membrane potential becomes restored, or *repolarized*. Repolarization involves the rapid diffusion of potassium ions outward across newly opened potassium ion channels. At the same time as potassium ions begin to move out of the cell, the flow of sodium ions into the cell slows due to the closing of the sodium ion channels. The result is the net movement of positive ions out of the cell, which provides the outer membrane with a positive charge once again (and the inner membrane with a negative charge).

The depolarization of a region of plasma membrane, followed quickly by its repolarization, is called an **action potential** (Fig. 9–8), which is otherwise known as a *nerve impulse*. The chemical events associated with the action potential, including the flow of sodium ions inward followed by the flow of potassium ions outward, occurs extremely rapidly: about 1/1000 of a second. Also, action potentials may occur in rapid

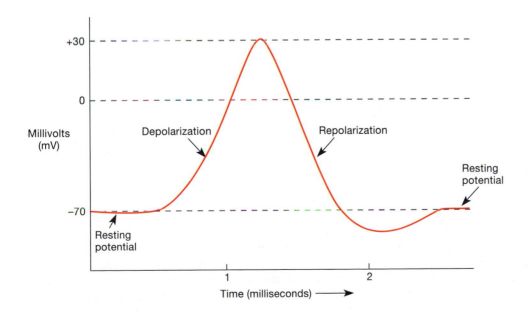

Figure 9–7
The action potential. The graph plots the change in electrical charge that occurs across a plasma membrane during an action potential. The depolarization phase increases the charge from its resting potential (−70 mV) to +30 mV. It is quickly followed by repolarization, which eventually returns the charge to rest. The action potential takes a little more than 1 millisecond to complete.

The rapid movement of what ions into the cell leads to depolarization?

Figure 9–8
Depolarization and
repolarization. (a) The resting
potential maintains positive
charges outside and negative
charges along the inside of the
plasma membrane.
(b) Depolarization occurs as
the membrane permeability to
sodium ions changes,
permitting their rapid
diffusion inward. As a result,
charges on the membrane are
reversed. (c) The action
potential is propagated as
depolarization spreads along
the membrane. (d) Soon after
depolarization, membrane
channels open to permit the
rapid diffusion of potassium
outward, restoring the positive
charge to the outside of the
membrane. The membrane
has thus repolarized.

**What two changes in the plasma
membrane lead to repolarization?**

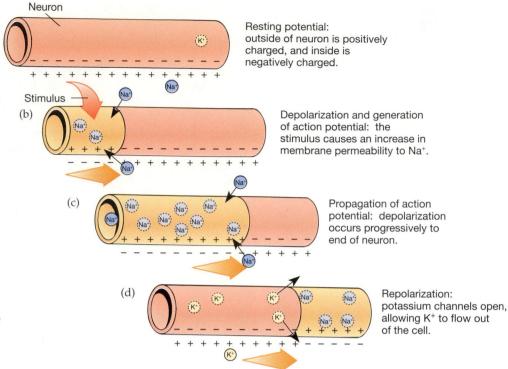

Resting potential:
outside of neuron is positively
charged, and inside is
negatively charged.

Depolarization and generation
of action potential: the
stimulus causes an increase in
membrane permeability to Na⁺.

Propagation of action
potential: depolarization
occurs progressively to
end of neuron.

Repolarization:
potassium channels open,
allowing K⁺ to flow out
of the cell.

succession, since the number of ions that flow through the gated channels during a given action potential is extremely small. This allows for the rapid generation of many action potentials before the sodium-potassium pump must restore the ion concentrations to their original levels. Although many types of cells exhibit resting membrane potentials, only neurons and muscle cells are capable of producing action potentials.

Conduction of the Nerve Impulse

Once an action potential is stimulated, it begins a journey along the length of the neuron. The starting point of this journey is the highly localized region of the membrane where the stimulus was applied. When the inside of the cell becomes more positive with respect to the outside during depolarization, an electrical gradient is created between the point of stimulus and adjacent regions of the membrane. As a result, an ionic current inside the cell begins to increase membrane permeability to sodium ions nearby. This current causes a reversal of ion distribution, or depolarization, to the adjacent region, which is shortly followed by repolarization (Fig. 9–8). The reversal of ions to the adjacent region of the membrane repeats itself over and over again, and the nerve impulse is conducted from the point of stimulation to the opposite end of the neuron. Thus, nerve impulse conduction is actually a wave of ionic reversals, leading to brief negative charges on

the outside and positive charges on the inside, that moves along the surface of the neuron membrane.

The speed at which a nerve impulse is conducted is extremely rapid, but does vary according to the presence or absence of the myelin sheath. In the preceding discussion of impulse conduction, we assumed that the impulse travels unimpeded. However, this is the case only in unmyelinated fibers, since the myelin sheath in myelinated fibers blocks the continuous flow of ions. How, then, is the impulse conducted in myelinated fibers? By a process known as **saltatory conduction**, in which the impulse actually jumps across the myelin sheath from one node of Ranvier to another node of Ranvier, much like the way in which a flat skipping stone thrown into water jumps from one point of contact with the water to the next (Fig. 9–9). Since the impulse jumps great distances (2 to 3 mm) as it travels from node to node, the speed of conduction is much greater than that of impulses that must travel step by step along the fiber. Myelinated fibers are therefore the fastest conductors in the body, carrying impulses at speeds up to 130 meters per second (over 300 miles per hour!), whereas unmyelinated fibers do not exceed a speed of 10 meters per second. Rapid saltatory conduction is a feature that is important to homeostasis, since it provides a means for split-second responses to emergency situations that often arise in the body.

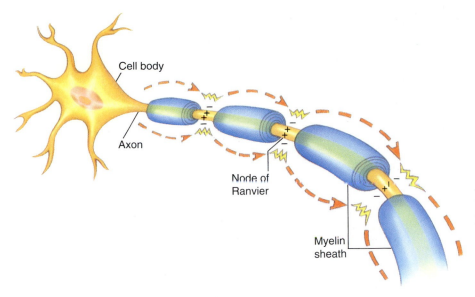

Cell body

Axon

Node of
Ranvier

Myelin
sheath

Figure 9–9
Saltatory conduction. An action potential is propagated along a myelinated fiber by jumping from one node of Ranvier to another.
What is the benefit of saltatory conduction?

All-or-None Response

Like the contraction of muscle fibers, the conduction of a nerve impulse occurs in an **all-or-none** fashion. In other words, if a stimulus is strong enough to cause an action potential, the impulse will be conducted along the entire length of the neuron at a maximum strength. The minimum strength of a stimulus required to initiate an action potential is referred to as the **threshold stimulus**. Increasing the strength of the stimulus beyond its threshold level will have no effect on the impulse whatsoever. It's a bit like blowing up a bridge: once enough dynamite has been ignited to cause it to fall, adding more will not make it fall more. Also, a stimulus weaker than threshold, called **subthreshold**, cannot initiate an impulse. However, a series of subthreshold stimuli that are quickly applied to the neuron may have a cumulative effect that can lead to an action potential. This phenomenon is called **summation**.

Transmission of Impulses from Cell to Cell

The junction between adjacent neurons is called a **synapse**[4] (Fig. 9–10). The impulse must cross this junction in order to pass from neuron to neuron. The neuron that sends the impulse to the synapse is called the **presynaptic neuron**, and the one that receives the impulse across the synapse is the **postsynaptic neuron**. The route of transmission is always in one direction, from the presynaptic neuron to the postsynaptic neuron.

Typically, the axon of the presynaptic neuron ends at a round bulb, called the **synaptic end bulb**, which contains numerous **synaptic vesicles**. The synaptic

vesicles contain chemicals, called **neurotransmitters**, that relay the impulse across the synapse. Investigators have identified about 50 different neurotransmitters, some examples of which include acetylcholine (ACh), norepinephrine, dopamine, serotonin, glutamate, and gamma aminobutyric acid (GABA). The postsynaptic neuron usually has a concave surface that forms a gap, called the **synaptic cleft**, across which neurotransmitters must diffuse in order to contact the postsynaptic membrane.

A nerve impulse is transmitted across a synapse in the following way. When a nerve impulse arrives at the synaptic end bulb of a presynaptic neuron, calcium ion channels in the plasma membrane open briefly, and calcium ions flow into the end bulb from the surrounding interstitial fluid. Their presence causes the fusion of synaptic vesicles with the presynaptic plasma membrane, and neurotransmitters are released into the synapse by exocytosis. These chemicals diffuse across the gap through the interstitial fluid until they contact the plasma membrane of the postsynaptic neuron. Once contact is made, the result is either excitation or inhibition of the postsynaptic neuron, depending on the nature of the neurotransmitter and its receptor. Whichever its effect may be, it is extremely brief because the neurotransmitter is immediately inactivated by special enzymes or transported away by other enzymes. This limits the effect of neurotransmitters to a fraction of a second, and readies the synapse for the next transmission.

Excitatory versus Inhibitory Transmissions

One effect of neurotransmitters upon a postsynaptic neuron is to increase its membrane permeability to sodium ions, causing an action potential. This is called an **excitatory transmission**. Generally, the release of a

---

[4]Synapse: Greek *synapsis* = a coming together, a union.

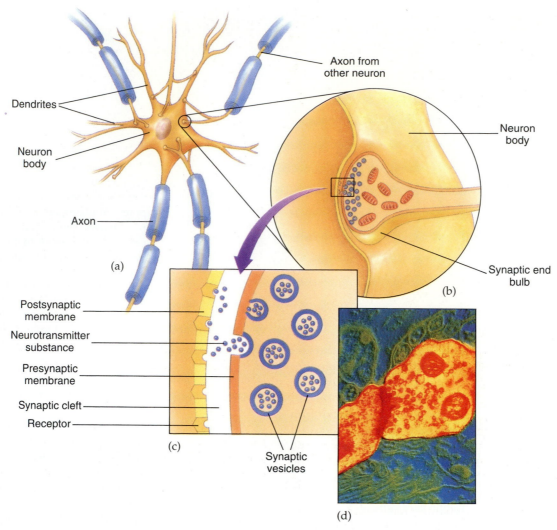

Figure 9–10

Impulse transmission. (a) An impulse moves from one neuron to the next across a gap junction known as the *synapse*. (b) Closer view of a synapse. (c) In this close-up of a synapse, the neurotransmitter substances are released from the presynaptic membrane and diffuse across the synaptic cleft to the postsynaptic membrane, which responds as a receptor. (d) False-color electron micrograph of a synapse (the darkened band) between adjacent neurons.

What is the chemical that carries an impulse across a synapse by diffusing from the presynaptic membrane to the postsynaptic membrane?

certain volume of neurotransmitter molecules from numerous presynaptic end bulbs is required to cause the action potential. As each molecule of neurotransmitter is received, the postsynaptic membrane becomes a bit more excited as more sodium ions are allowed to flow into the cell. At this point, the postsynaptic membrane becomes prepared for additional stimuli that will trigger an action potential. This state of partial polarization is known as **facilitation**. Once the postsynaptic membrane has received an additional volume of excitatory neurotransmitters, the action potential occurs and the nerve impulse begins its journey through the

cell. Neurotransmitters that are excitatory include acetylcholine and norepinephrine.

Neurotransmitters may also increase the postsynaptic membrane permeability to potassium ions. This is called an **inhibitory transmission**, since it lowers the chance that a nerve impulse will be transferred across a synapse successfully. It begins when a neurotransmitter contacts the postsynaptic membrane and opens the potassium ion channels. As a result, potassium flows out of the cell. Chloride ion channels are also opened, allowing chloride ions (Cl^-) to move into the cell. Sodium ion permeability is not affected. Conse-

Chemicals and Their Influences on Nervous Function

Certain chemicals are essential for the proper functioning of the nervous system. These are the neurotransmitters, of which about 50 different types are currently known. Neurotransmitters are synthesized by neurons within their cytoplasm from raw materials delivered by the bloodstream or other body fluids.

Chemicals that influence nervous activity may also be introduced into the body. These foreign chemicals may be used to alter normal function, or used to correct a deficiency that causes abnormal function. For the most part, their mechanism of action upon nerve tissue is either by stimulation of neural activity or by inhibition. These chemicals or drugs fall into one of several major categories: stimulants, depressants, antidepressants, psychedelics, analgesics, and antianxiety drugs.

Stimulants generally increase the synaptic transmission of impulses, especially in the brain. This tends to produce a sense of increased energy, elevate mood, and decrease appetite, but it also increases irritability and anxiety. Stimulants include amphetamines ("uppers," "whites"), methedrine ("speed"), caffeine, cocaine, and nicotine (in tobacco). Long-term use of stimulants often leads to addiction.

Depressants cause an inhibition of impulses at the synapse, or they may alternatively block receptors of neurotransmitters such as norepinephrine and acetylcholine. In general, high doses can result in extreme depression leading to death. Examples of depressants include anesthetics (chloroform, ether, and toluene), antianxiety drugs (sleeping pills, tranquilizers), ethyl alcohol (in whiskey, beer, and wine), barbiturates (phenobarbital, Seconal, and Nembutal), and opiates (heroin, morphine, and codeine). Many of these depressants are addictive.

Antidepressants increase the levels of norepinephrine in the brain, reversing the effects of psychological depression. They include a family of drugs called *dibenzepines*, which are used to balance the depressive turns in patients suffering from manic-depressive disorder (a form of schizophrenia).

Psychedelic drugs affect the role of the neurotransmitter serotonin in the brain, although the exact nature of their effect remains unknown. Small doses of psychedelic drugs alter perception and mood, while larger doses produce distortions of perception called *hallucinations*. Psychedelic drugs include cannabis (marijuana), LSD (lysergic acid diethylamide), and mescaline.

Analgesics are a class of drugs that relieve pain, presumably by interfering with the transmission of pain impulses to the cerebral cortex. They include aspirin, acetominophen (Tylenol), and ibuprofen (Nuprin, Advil). Certain depressants also serve as analgesics, because of their inhibition of pain. They are the opiates (heroin, morphine, and codeine), which are also highly addictive.

---

quently, positive ions accumulate on the outside of the cell, establishing a positive charge, while negative ions provide a negative charge along the inside of the plasma membrane. When this occurs, the membrane is said to be *hyperpolarized*. Neurotransmitters that are inhibitory include GABA, endorphins, and enkephalins, which selectively inhibit impulses generated by pain stimuli. These compounds serve as the body's natural painkillers.

Processing at the Synapse

Recall that the postsynaptic membrane is located at the dendrites and cell body of a single neuron. This membrane may receive thousands of presynaptic end bulbs originating from thousands of presynaptic neurons. Some neurotransmitters arriving across the synapse from these numerous end bulbs may have an excitatory effect, while others may have an inhibitory effect. The overall effect on the postsynaptic membrane is determined by the sum of their effects. For example, if most of the neurotransmitters are excitatory, their effect will be accumulated and result in an action potential. However, if there is a greater number of inhibitory neurotransmitters at a given time, the action potential will be blocked.

By the complex processing of excitatory and inhibitory neurotransmitters at the synapse, the nervous system is provided a means of regulating nerve impulses. The synapse therefore plays an important role in body activities that require this important regulation. For example, homeostasis requires impulses to be channeled to specific areas of the body, such as skeletal muscles to produce heat during cold weather; much of this channeling is made possible by the processing of impulses at synapses. Also, recent investigations into brain function (a growing field called *neurophysiology*) suggest that the complex processing occurring at brain synapses provides us with our ability to store and retrieve information, create visual images in the mind, and develop or modify behavior. These activities require a physical change in the organization of the brain at the level of synapses, which is a feature of the brain known as *plasticity*. The functional activity of the brain will be discussed in more detail later in this chapter and in Chapter 10.

CONCEPTS CHECK

1. What is a myelinated nerve fiber, and why does it conduct impulses faster than unmyelinated fibers?

2. What are the structural and functional types of neurons?

3. How is an action potential generated in a resting neuron?

4. What is the difference between an excitatory transmission and an inhibitory transmission?

THE CENTRAL NERVOUS SYSTEM

The central nervous system is the center of integrative functions for the nervous system. It consists of the brain and spinal cord.

In past years the central nervous system (the brain and spinal cord) was compared to a telephone switchboard, with its incoming relays, complicated switching networks, and outgoing wire connections. More recently, it has often been described as a vastly complex computer, although no computer yet exists that can match the integrative abilities of the brain. Whichever comparison is made, the central nervous system remains as one of the great wonders of nature, for it is within it that your thoughts, memory, and personality reside. In this section the structure of the spinal cord and brain is presented. We will also look at some of their functional relationships, but the more complex integrative functions will be examined in Chapter 10.

The Spinal Cord

The spinal cord is protected by bone, fluid, and membranes. It is composed of gray and white matter, and serves as a conduction pathway between the brain and nerves and as a reflex center.

The spinal cord is a long, slender bundle of nerve tissue that extends about 42 cm (17 inches) from the base of the brain to the first or second lumbar vertebra (Fig. 9–11). It emerges through the foramen magnum at the base of the skull and travels throughout its length within the center of the vertebral canal. The spinal cord

Figure 9–11
The spinal cord. To expose the spinal cord and show its important features, the vertebral arches of the vertebral column have been removed and the dura mater cut.
What two components of the nervous system does the spinal cord connect?

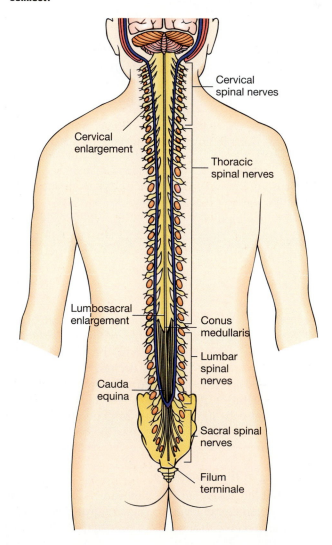

Cervical spinal nerves

Cervical enlargement

Thoracic spinal nerves

Lumbosacral enlargement

Conus medullaris

Cauda equina

Lumbar spinal nerves

Sacral spinal nerves

Filum terminale

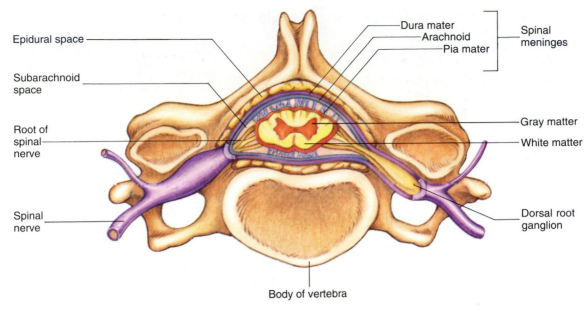

Epidural space

Subarachnoid space

Root of spinal nerve

Spinal nerve

Dura mater
Arachnoid
Pia mater
Spinal meninges

Gray matter
White matter

Dorsal root ganglion

Body of vertebra

Figure 9–12
The spinal cord in cross section. The spinal cord is shown in place within the vertebral foramen of a vertebra.

What is the outermost meningeal layer?

provides a two-way connection between the brain and the peripheral nerves.

Protective Coverings

The spinal cord is a soft, fragile structure that must be protected to avoid physical injury that can be life-threatening. Its protection is provided by the vertebral column that surrounds it, by fluid that provides a liquid cushion, and by several layers of membrane known as **meninges**[5] (me-NIN-jēz).

The three meningeal membranes are the outermost dura mater, the middle arachnoid, and the inner pia mater (Figs. 9–12 and 9–13). The **dura mater**[6] (DYOO-ra MĀ-ter) forms a thick, tough sheath around the spinal cord. Along its outer margins it is not attached to the vertebral column, but instead borders a rather large space called the **epidural space**. This area is actually filled with fat and areolar connective tissue, providing an insulative and protective barrier between the dura mater and the vertebral column. Internal to the dura mater is the middle membrane, the **arachnoid**[7] (a-RAK-noyd), whose collagen fibers form a matting that looks much like a thick cobweb. The innermost membrane, the **pia mater**[8] (PĒ-a MĀ-ter), is attached

to the outer surface of the spinal cord and is quite thin and delicate. Between the arachnoid and pia mater is a space, called the **subarachnoid space**, which is filled with fluid that originates from the brain. This clear, colorless fluid is **cerebrospinal fluid (CSF)**. It circulates around and through the spinal cord and brain, providing a shock-absorbing liquid cushion and transporting vital materials.

All three meningeal membranes extend as a unit well beyond the inferior end of the spinal cord, so their covering of the cord is complete. At the opposite end of the cord, the three membranes continue uninterrupted to the brain, which they also surround and protect. The relationships between cerebrospinal fluid, the three meninges, and the brain will be further discussed later.

Structure

The spinal cord is in the shape of a long cylinder that is slightly flattened on its anterior and posterior surfaces (Figs. 9–11 and 9–12). It consists of a continuous series of 31 segments, arranged in a vertical, linear fashion. Each segment of the cord gives rise to a pair of prominent nerves, called **spinal nerves**, by way of **spinal roots**. The spinal nerves relay information between the spinal cord and the peripheral areas of the body, and are components of the peripheral nervous system.

Along the length of the cord are two areas that are slightly thickened because of the abundance of spinal

[5]Meninges: Greek *meninx* = membrane.

[6]Dura mater: Latin *dura* = tough + *mater* = mother.

[7]Arachnoid: Greek *arachne* = spider + *-oid* = similar to.

[8]Pia mater: *pia* = gentle, delicate + *mater* = mother.

HEALTH CLINIC

Lumbar Puncture

Since the spinal cord ends at the second lumbar vertebra level and the meningeal layers continue beyond that point to about the second sacral vertebra, the subarachnoid space below the second lumbar vertebra is not associated with the spinal cord. Therefore, there is no danger of damaging the spinal cord in this region. This is where cerebrospinal fluid is removed for clinical testing. This procedure is called a *lumbar puncture* or *spinal tap*, and is

performed when there are indications of infection or trauma to nerve tissue in the cord or brain. It involves the penetration of the skin, lower back muscles, dura mater, and arachnoid by a syringe needle, and the withdrawal of fluid from the subarachnoid space into the syringe. This procedure may be accompanied by a temporary, severe headache resulting from the sudden drop in cerebrospinal fluid pressure.

nerves serving the appendages: the **cervical enlargement** in the neck region, and the **lumbosacral enlargement** in the lower back (Fig. 9–11). The cervical enlargement contains nerves that supply the upper appendages, and those of the lumbosacral enlargement supply the lower appendages.

The spinal cord tapers to a point at the level of the second lumbar vertebra, where it ends as the **conus medullaris**[9] (KŌ-nus med′-yoo-LAR-is) (Fig. 9–11). From this point extend numerous spinal nerves, which continue downward through the vertebral canal for some distance before exiting outward. This collection of spinal nerves is called the **cauda equina**[10] because of its resemblance to a horse's tail. An extension of the pia mater also continues beyond the spinal cord, and beyond the limits of the other meninges all the way to the back of the coccyx. This membrane is called the **filum terminale**.

When the spinal cord is sectioned across its horizontal plane (a cross section) and viewed from above, its flattened shape from back to front can be seen (Figs. 9–12 and 9–13). The two grooves that mark these surfaces partially divide the cord into right and left portions. They are the deep **anterior median fissure** and the shallower **posterior median sulcus**. The cross-sectional view of the spinal cord also reveals its internal features, which include gray matter and white matter.

Gray Matter

Recall that gray matter is nerve tissue that consists of unmyelinated fibers and neuron cell bodies. Neuroglia

---

[9]Conus medullaris: Latin *conus* = cone + *medullaris* = pertaining to the core, or spinal cord.

[10]Cauda equina: Latin *cauda* = tail + *equinus* = of a horse.

Figure 9–13
Structure of the spinal cord. (a) This anterior, slightly tilted view shows the connections of the cord with three pairs of spinal nerves. (b) Cross section of the spinal cord and its meninges.

How does the posterior gray horn of the spinal cord differ functionally from the anterior gray horn?

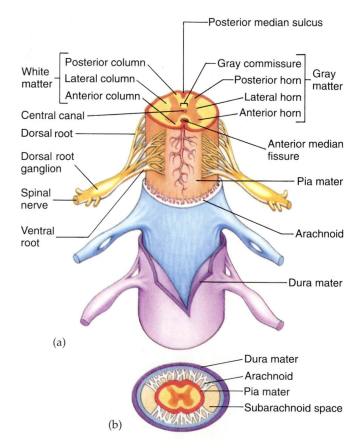

also forms a part of gray matter. In the spinal cord, gray matter is located in the center of the cord in a pattern that resembles the shape of the letter H (Figs. 9–12 and 9–13). The two upper "arms" of the H are called **posterior horns**, and the two lower "legs" are the **anterior horns**. Between them on either side is a small projection of gray matter called the **lateral horn**, which may be found only on the thoracic and first two lumbar segments of the cord. A horizontal bar of gray matter connects the right and left sides of the H together; this is the **gray commissure**. In its center is the **central canal**, which extends throughout the length of the cord and carries cerebrospinal fluid.

There is a functional distinction between the three types of gray horns of the cord. The anterior gray horns contain cell bodies of motor neurons, whose fibers exit the cord with spinal nerves and extend to skeletal muscles. The posterior gray horns contain the terminal endings of sensory neurons. They originate in sensory regions of the skin, muscles, and visceral organs, and their cell bodies lie just outside the spinal cord in clusters called spinal **ganglia**[11] (clusters of neuron cell bodies located in the brain are known as **nuclei**). Lateral gray horns contain cell bodies of certain motor neurons of the autonomic nervous system. Thus, anterior gray horns carry a motor function, posterior gray horns carry a sensory function, and lateral gray horns are autonomic in function.

White Matter

White matter in the spinal cord, as in all nerve tissue that is white, is composed mostly of myelinated fibers. In the cross section of the cord (Figs. 9–12 and 9–13), notice that it surrounds the central gray matter. It is divided into three regions, known as the **anterior, posterior**, and **lateral columns** (funiculi). Each column consists of longitudinal bundles of myelinated fibers that represent major nerve pathways extending up and down the spinal cord. These pathways are called **nerve tracts**.

Functions

The spinal cord serves two major functions: as a conduction pathway for impulses traveling between the brain and peripheral nerves, and as a reflex center.

Conduction Pathways

The conduction pathway provided by the spinal cord has mainly two-way traffic: up toward the brain, and down away from the brain. This traffic is kept separated, like a two-lane highway, by two distinct nerve tracts. The **ascending tracts** carry sensory information to the brain, and the **descending tracts** carry motor information away from the brain. Both the ascending and descending tracts are further divided into more localized groups of myelinated fibers, which are named according to their common points of origin and destination. By way of these tracts, the spinal cord ties together the many pieces of information received from sensory neurons, the integrative powers of the brain, and the responding capabilities of the motor neurons.

Reflex Center

The spinal cord also serves as a reflex center for the nervous system. A **reflex** is an extremely rapid way of responding to an emergency situation. It is the simplest pathway an impulse may take, since it involves but a small number of neurons. Basically, it provides a rapid response because the impulse does not travel all the way to the higher levels of the brain; it is routed within the spinal cord or lower brain level directly to a motor neuron within a fraction of a second. This means you do not have to think about reacting; your body does it automatically for you.

The **reflex arc** is the conduction pathway for a reflex action. It begins with a **receptor** at the end of a sensory neuron. A receptor is a specialized dendrite of a sensory neuron that has the ability to generate an action potential in response to a particular change in its environment. Once the receptor is stimulated by the sudden change—for example, a pain receptor stimulated by heat at the end of your finger (see Fig. 9–14)—the nerve impulse generated travels quickly along the **sensory neuron** to the CNS, which is, in this example, the spinal cord. Once inside the CNS, the impulse is transmitted to one or more **association neurons**, which make up the **reflex center** that quickly processes the information by routing it to the appropriate **motor neurons**. The association neurons also have connections with other parts of the nervous system, explaining why you can feel pain from the burn and think about how best to treat it (after the emergency response is made). The motor neurons then conduct the impulses to the **effectors**, in this case the skeletal muscles of your arm, to bring about the response.

In our example of a reflex arc, we used what is called the **withdrawal reflex**. This type of reflex is protective because its rapid response time minimizes the extent of an injury. There are other types of reflexes, including the knee jerk, or **patellar reflex**, which involves only two neurons (the sensory neuron transmits directly to a motor neuron within the spinal cord) and is used frequently by health professionals to diagnose nervous disorders. Withdrawal and patellar reflexes are called **somatic reflexes**, because the effectors are skeletal muscles. A second major group is categorized as **visceral reflexes**, which affect smooth and cardiac muscles to cause automatic responses such as heart

---

[11]Ganglia: Greek *ganglion* = knot.

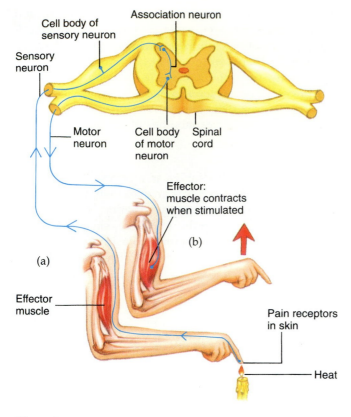

Cell body of
sensory neuron

Association neuron

Sensory
neuron

Motor
neuron

Cell body
of motor
neuron

Spinal
cord

Effector:
muscle contracts
when stimulated

(b)

(a)

Effector
muscle

Pain receptors
in skin

Heat

Figure 9–14
The reflex arc. (a) A reflex arc originates from sensory receptors and travels along a sensory neuron to the posterior gray horn of the spinal cord. Within the cord, the impulse is conveyed to an association neuron, which routes the impulse to the anterior gray horn and a motor neuron. (b) The motor neuron carries the impulse to a skeletal muscle effector to produce the response.

Why does a reflex arc bring about a more rapid response than a coordinated, skilled movement?

rate, breathing, vomiting, sneezing, and coughing. Regardless of the particular type, all reflexes are important in maintaining homeostasis because of their ability to provide a rapid response to an environmental change.

CONCEPTS CHECK

1. What are the meningeal layers protecting the spinal cord and brain?

2. How is gray matter functionally organized in the spinal cord?

3. What are the conduction pathways in the spinal cord?

4. What are the components of a reflex arc?

The Brain

The brain is the integrative center of the nervous system. It consists of several regions whose components perform specific functions, providing a division of labor in the brain.

The brain is far more massive and complex than the spinal cord. It is one of the largest organs of the body, and contains about 100 billion neurons in addition to the abundant neuroglia. It is divided into three major regions: the **forebrain**, the **midbrain**, and the **hindbrain**, shown in Figure 9–15. These divisions are based on the embryological development of the brain, which parallels that of other mammals. The forebrain is by far the largest division of the brain, containing the massive, highly complex **cerebrum**. It also contains the smaller **diencephalon**, located inferior to the cerebrum. The midbrain is quite small and is located below the diencephalon. The hindbrain occupies the inferior part of the brain and contains the **pons, medulla oblongata,** and **cerebellum.** As you can see in Figure 9–15, the combination of the midbrain, pons, and medulla forms yet another division of the brain, called the **brain stem.**

Protective Coverings

As in the spinal cord, the nerve tissue that makes up the brain is very soft and fragile, requiring continuous protection from several sources. It receives its protection from the flat bones of the cranium, cerebrospinal fluid, and the meninges. Also like the cord, the meninges of the brain consist of an outer **dura mater**, a middle **arachnoid**, and an inner **pia mater**, with a **subarachnoid space** present between the arachnoid and pia mater layers (Fig. 9–16). Recall that the meninges of the brain are continuous with those of the spinal cord.

Cerebrospinal Fluid and Ventricles of the Brain

Cerebrospinal fluid (CSF) is a clear, colorless fluid that circulates within and around the spinal cord and brain. It functions as a liquid cushion for these delicate organs, preventing them from crashing against the walls of the cranium or vertebral column during sudden body movements and supporting their weight. CSF also helps to nourish the brain and remove metabolic waste materials, in addition to the nourishment that is provided by the extensive capillary network.

Much of the nervous system's CSF is contained within cavities located inside the brain. These cavities are called **ventricles**[12] and are continuous with the cen-

---

[12]Ventricle: Latin *ventri-* = belly, chamber + *-cle* = little.

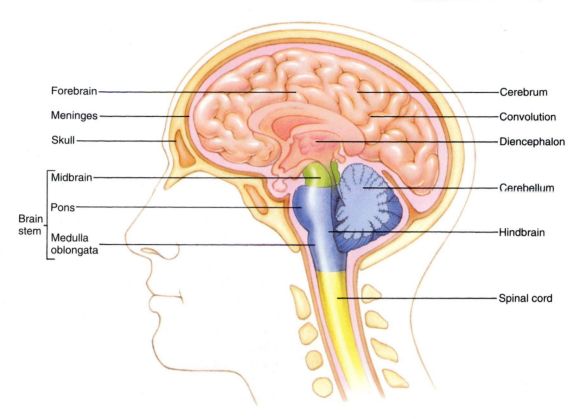

Forebrain —
Meninges —
Skull —
Midbrain —
Pons —
Brain stem {
Medulla oblongata —

— Cerebrum
— Convolution
— Diencephalon
— Cerebellum
— Hindbrain
— Spinal cord

Figure 9–15
The brain. In this lateral view of a sagittally sectioned brain, its primary divisions are indicated in color (forebrain in pink, midbrain in green, and hindbrain in blue).

The cerebrum is a part of what region of the brain?

tral canal of the spinal cord and the subarachnoid space associated with both brain and cord (Figs. 9–16 and 9–17). There are four ventricles: two **lateral ventricles**, located in each cerebral hemisphere; a single **third ventricle** in the center of the diencephalon; and a single, narrow **fourth ventricle** between the cerebellum and the medulla oblongata. Each lateral ventricle is connected to the third ventricle by a channel called the **foramen of Monro,** and the third ventricle communicates with the fourth by a channel called the **cerebral aqueduct**.

CSF is formed within the ventricles by the filtration of blood plasma through dense networks of capillaries. These networks resemble masses of jelly within the ventricle; each network is called a **choroid plexus** (kō-ROYD PLEK-sus). The capillary walls of the choroid plexus and its supporting neuroglia form part of the **blood-brain barrier,** which selectively permits certain substances to enter, such as water, glucose, and oxygen, while blocking others. The resulting CSF is therefore similar in composition to blood plasma, but lacking in materials selected out, such as cells, bacte-

ria, and many proteins. The continual production of CSF through the choroid plexus of each ventricle creates a fluid pressure, causing a current of fluid to flow through the brain and spinal cord. CSF is eventually returned to the bloodstream by its reabsorption across fingerlike projections of the arachnoid, called **arachnoid villi,** which deliver it into a large vein on top of the brain known as the *superior sagittal sinus.*

The flow of CSF is in one direction, which begins in the lateral ventricles as it is produced from the choroid plexus. It passes slowly through the foramen of Monro into the third ventricle, then into the fourth ventricle through the cerebral aqueduct. From there it passes into the central canal and subarachnoid space of the spinal cord, and then into the cranial subarachnoid space and back into the bloodstream.

Cerebrum

The **cerebrum** is the largest structure of the brain, occupying most of the forebrain and total bulk of the brain. It is often called the "higher brain" because of

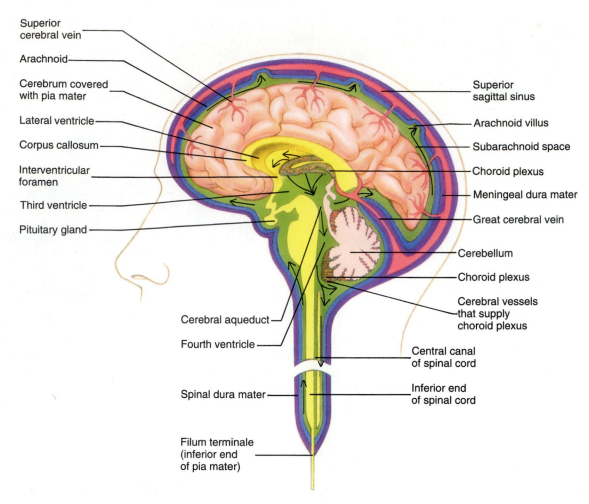

Superior
cerebral vein

Arachnoid

Cerebrum covered
with pia mater

Lateral ventricle

Corpus callosum

Interventricular
foramen

Third ventricle

Pituitary gland

Cerebral aqueduct

Fourth ventricle

Spinal dura mater

Filum terminale
(inferior end
of pia mater)

Superior
sagittal sinus

Arachnoid villus

Subarachnoid space

Choroid plexus

Meningeal dura mater

Great cerebral vein

Cerebellum

Choroid plexus

Cerebral vessels
that supply
choroid plexus

Central canal
of spinal cord

Inferior end
of spinal cord

Figure 9–16
Circulation of cerebrospinal fluid. CSF is initially produced at the choroid plexus
of the lateral ventricles. From there it passes through the third and fourth
ventricles before entering the central canal of the spinal cord and subarachnoid
space. It rejoins the blood supply as it is absorbed through the arachnoid villi.
Note the locations of the meninges and ventricles of the brain in this schematic
drawing.

Trace the pathway of cerebrospinal fluid from its origin to its reabsorption.

Figure 9–17
Ventricles of the brain. The two lateral, the third, and the
fourth ventricles are shown superimposed through the
brain in this schematic drawing.

What ventricles are connected by the cerebral aqueduct?

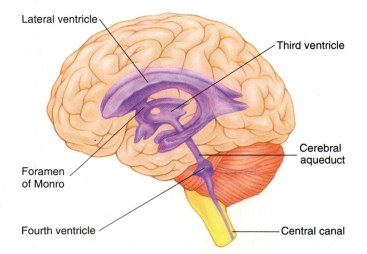

Lateral ventricle

Third ventricle

Cerebral
aqueduct

Central canal

Foramen
of Monro

Fourth ventricle

Accumulation of Fluids in the Brain

Two types of fluids are present within the cranial cavity: blood in blood vessels, and cerebrospinal fluid in ventricles and associated channels. Because the volume capacity of the cranial cavity cannot change (due to the rigid skull bones), any increase in fluid volume compresses the soft, semifluid brain tissue. For example, in a severe head injury, blood vessels that course through brain tissue may rupture, causing blood to collect within the cranial cavity. The accumulation of blood within the confined space of the cranial cavity leads to an increase in pressure, called *intracranial pressure*. As a result of increased intracranial pressure, the soft brain tissue becomes compressed. This causes neural damage and dysfunction that often leads to unconsciousness and death. Likewise, any disturbance of the CSF production and reabsorption equilibrium will re-

sult in an increased fluid volume and an increase in intracranial pressure. This disturbance can be caused by head injuries, masses such as tumors or abscesses blocking one of the flow channels, or congenital defects that restrict normal CSF circulation. Because CSF production continues, the ventricles gradually expand, pushing brain tissue against the cranial walls. The resulting pressure often leads to functional deterioration. In infants born with a congenital defect that restricts CSF circulation, the brain tissue pushes against the flexible cranium, which has not yet sutured, and the cranium expands. This condition is called *hydrocephalus*, or "water on the brain." Successful treatment usually involves the surgical installation of a shunt, which serves as a bypass to drain the excess CSF and reduce intracranial pressure.

---

its varied and complex functions in conscious thought, memory, and learning. It also receives and interprets sensations, and initiates motor responses. Structurally, it is divided into right and left portions that are called **cerebral hemispheres**.

A structural characteristic of the cerebrum is its prominent wrinkles or foldings, known as **convolutions** (Fig. 9–18). These occur as a result of rapid brain growth during embryonic development. Where the foldings project upward, they are called **gyri**, and where they extend downward to form a shallow groove, they are called **sulci**. A deep groove is known as a **fissure**. There are two major fissures: the **longitudinal fissure**, dividing the right and left cerebral hemispheres across the midline of the brain; and the **transverse fissure**, which divides the lower margin of the cerebrum from the smaller cerebellum.

The cerebrum is divided into functional regions called **lobes** (Fig. 9–18). Each lobe contains a group of functions that are distinct from those of other lobes. These functions will be discussed in Chapter 10. There are four lobes within each cerebral hemisphere: the **frontal lobe**, the **parietal lobe**, the **occipital lobe**, and the **temporal lobe**. Their locations correspond to those of the cranial bones that lie above them. The frontal and parietal lobes are separated by the **central sulcus**,

and the frontal and temporal lobes are separated by the **lateral sulcus**.

When the cerebrum is sectioned, we can observe its internal features. Notice from Figure 9–18c that each lobe of the cerebrum internally contains two shades of color: an external gray region and an internal white region. The external region is known as the **cerebral cortex** and is composed of gray matter. Although it averages only about 2 mm thick, it contains billions of cell bodies and synapses and represents the source of cerebral integrative functions. The region underlying the cerebral cortex is composed of white matter, which contains myelinated fibers extending in three major directions: from hemisphere to hemisphere across a bridge connecting the two, called the **corpus callosum**; from one region of a hemisphere to another (within the same hemisphere); and from one hemisphere to other parts of the brain.

Embedded within the white matter of each hemisphere are several masses of gray matter known as **basal ganglia**. These clusters of cell bodies, or *nuclei*, serve as relay stations for motor impulses originating in the cerebral cortex and en route to the spinal cord. They also control large, unconscious movements of skeletal muscles, such as swinging your arms while walking.

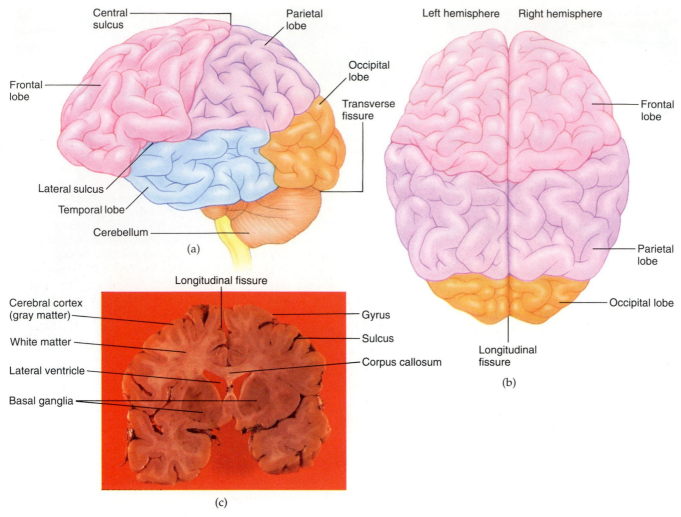

Figure 9–18

The cerebrum. (a) Lateral view. (b) Top view. (c) Photograph of a frontal section, showing some internal features of the cerebrum. Note the locations of gray and white matter.

What are the four lobes of each cerebral hemisphere?

Diencephalon

Also forming part of the forebrain along with the cerebrum, the **diencephalon**[13] is located inferior to the corpus callosum. It is composed largely of gray matter, which is organized to form two prominent structures: the **thalamus**, and the **hypothalamus**.[14] It is also associated with two endocrine glands, the **pineal** and **pituitary glands**, whose endocrine functions will be described in Chapter 11. The diencephalon may be viewed in a sagittal section of the brain, since it lies within the substance of the brain (Fig. 9–19).

Thalamus

The **thalamus** is the largest portion of the diencephalon, forming about four-fifths of its total weight. It surrounds the third ventricle just below the corpus callosum, and consists of two masses of gray matter covered by a thin layer of white matter. Each mass is on either side of the third ventricle, and the two masses are connected by a narrow bridge. Along its posterior margin is attached a small oval structure, the **pineal gland**.

The thalamus contains clusters of neuron cell bodies (nuclei), many of which serve as relay stations for

[13]Diencephalon: Greek *dia-* = through or across + *encephalos* = brain.

[14]Hypothalamus: Greek *hypo-* = below + *thalamos* = inner chamber.

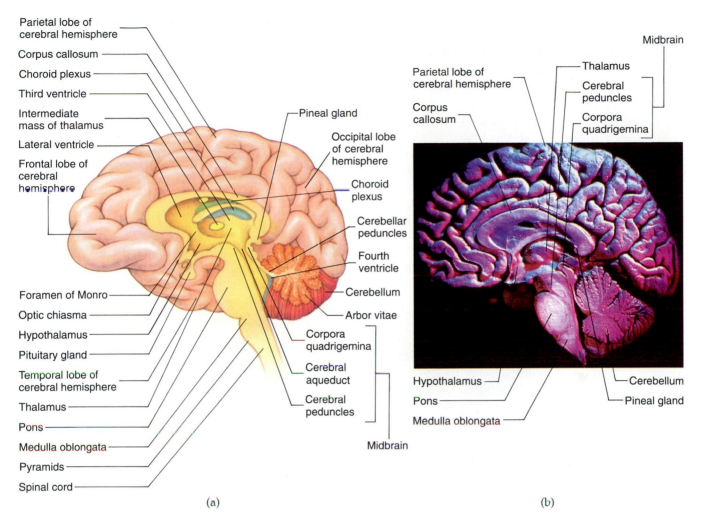

Parietal lobe of cerebral hemisphere
Corpus callosum
Choroid plexus
Third ventricle
Intermediate mass of thalamus
Lateral ventricle
Frontal lobe of cerebral hemisphere
Pineal gland
Occipital lobe of cerebral hemisphere
Choroid plexus
Cerebellar peduncles
Fourth ventricle
Foramen of Monro
Optic chiasma
Hypothalamus
Pituitary gland
Temporal lobe of cerebral hemisphere
Cerebellum
Arbor vitae
Corpora quadrigemina
Cerebral aqueduct
Cerebral peduncles
Thalamus
Pons
Medulla oblongata
Pyramids
Spinal cord
Midbrain

(a)

Parietal lobe of cerebral hemisphere
Corpus callosum
Midbrain
Thalamus
Cerebral peduncles
Corpora quadrigemina
Hypothalamus
Pons
Medulla oblongata
Cerebellum
Pineal gland

(b)

Figure 9–19
Internal structures of the brain. (a) The sagittally sectioned brain is shown in a lateral view. (b) Photograph of a sagittally sectioned brain, lateral view.
With what part of the diencephalon does the pituitary gland form a connection?

most sensory impulses on their way to the cerebral cortex for processing. Other cell bodies serve as switching stations for voluntary motor impulses leaving the cortex. Thus, the thalamus is the principal relay station for sensory impulses (except for smell) traveling to the cerebral cortex, and for involuntary motor impulses traveling outward. Additionally, an early conscious recognition of sensations takes place within the thalamus. This awareness is very primitive, for only sensations that are related to survival, such as pain, temperature, touch, and pressure, are realized here.

Hypothalamus

The **hypothalamus** is a small part of the diencephalon that lies below the larger thalamus. It is partially housed within the sella turcica of the sphenoid bone, so it is well protected. Attached to its inferior margin

by a short stalk is the **pituitary gland**. Although it is small in size, the hypothalamus is a giant in terms of function. It contains clusters of neuron cell bodies (nuclei) that control many involuntary body activities, most of which have a direct effect on homeostasis.

The hypothalamus is the control center for the autonomic nervous system. As such, it regulates all visceral activities (such as digestion, respiration, and heart rate). It also serves as an intermediary between the nervous and endocrine systems in that it stimulates or inhibits the master endocrine gland of the body, the pituitary gland. It serves as a body thermostat in that it controls internal body temperature. It also regulates food and water intake, and maintains waking and sleeping patterns. It is associated with emotion: rage, aggression, sex drives, and physiological symptoms of emotional stress. Its role in the development of emotion is tied to parts of the thalamus, cerebral cortex, basal ganglia, and other nuclei that together constitute the **limbic system**, which will be discussed further in the following chapter.

Midbrain

The **midbrain** is by far the smallest of the three major embryological divisions of the brain. It is located between the diencephalon and pons (Fig. 9–19), and consists of anterior and posterior portions that are structurally and functionally distinct. The anterior portion consists of bundles of myelinated fibers, called **cerebral peduncles**. These bundles provide a main connection for motor pathways extending between the cerebrum and cerebellum. The posterior portion contains structures known as **corpora quadrigemina**, which house nuclei that serve as reflex centers for rapid eye, head, and trunk movements.

Pons

The **pons**[15] is located directly inferior to the midbrain, where it appears as a rounded bulge (Fig. 9–19). It is the superior portion of the hindbrain, and consists of white matter and scattered masses of nuclei. The white matter provides a bridge connecting the spinal cord to the brain and parts of the brain with each other by way of nerve fibers that travel in two directions: longitudinally and transversely. Some nuclei operate with centers in the medulla oblongata in regulating the breathing rhythm, while others relay sensory impulses from peripheral nerves to the cerebral cortex.

Medulla Oblongata

The **medulla oblongata**[16] is the most inferior of the brain structures. It lies below the pons, and extends downward until it unites with the spinal cord (Fig. 9–19). When viewed from the side, the midbrain, pons, and medulla oblongata resemble a stalk supporting the cauliflowerlike cerebrum. Because of this resemblance, the combination of these three structures is often called the **brain stem**.

Like the spinal cord, the medulla is composed of white matter on the outside and gray matter on the inside. Its white matter consists of all the ascending (sensory) and descending (motor) nerve fibers that extend between the brain and spinal cord. On its posterior side the descending tracts cross to opposite sides, forming an expanded region known as the **pyramids**. As a result, fibers originating from the left side of the cerebral cortex activate muscles on the right side, and vice versa. This is also true of most sensory impulses; however, crossing of ascending tracts occurs at a pair of prominent nuclei within the medulla.

The gray matter of the medulla oblongata consists of reflex centers that play a role in visceral reflexes, consciousness, and arousal. Some of its more important visceral reflex centers include the **cardiac center**, which regulates heart rate; the **vasomotor center**, which regulates blood pressure by varying the diameter of blood vessels; and the **respiratory center**, which controls the depth and rhythm of breathing.

Cerebellum

The **cerebellum**[17] is the posterior portion of the hindbrain (Figs. 9–18 and 9–19). It is quite large, although not nearly as large as the cerebrum, which lies above it. It is divided into two **hemispheres** that are connected along the midline by the central **vermis**.[18]

Like the cerebrum, the cerebellum consists mostly of white matter, with a thin shell of gray matter, called the **cortex**, on its outer surface. The white matter forms a distinctive pattern, which to many resembles a branching tree. This pattern is called the **arbor vitae** (AR-ber VI-tē)—literally, the "tree of life"—which you can see in a sagittal section. Visible along the outer surface of the cortex are narrow ridges, or convolutions, that are much less prominent than those of the cerebrum. The upfolds are called **folia**, and the shallow downfolds are **sulci**. The **transverse fissure** completely separates the cerebellum from the cerebrum.

The cerebellum contains three paired bundles of myelinated fibers, which connect it to other parts of the brain. These are the **cerebellar peduncles**. These fibers carry sensory information from sensory organs in muscles, joints, and the inner ear to enable the brain to determine the status of voluntary motor activities, equilibrium, and balance. They also convey motor impulses that coordinate and provide precision to skeletal muscle contraction, which are initiated by the cerebrum. In this way, the cerebellum acts as an "automatic pilot" for motor responses: it regulates the performance of skeletal muscles to keep them within a narrow margin of coordinated, smooth response. Its regulation of muscle response also includes posture maintenance.

---

[15]Pons: Latin *pons* = bridge.

[16]Medulla oblongata: Latin *medulla* = central part, or spinal cord + *oblongata* = long.

[17]Cerebellum: *cerebrum* = main part of the brain + *-ellum* = little.

[18]Vermis: Latin *vermis* = worm.

CONCEPTS CHECK

1. How is cerebrospinal fluid produced and circulated through the CNS?

2. What is the functional importance of the cerebral cortex?

3. Which part of the brain controls the autonomic system and many endocrine functions?

4. What parts of the brain compose the brain stem?

THE PERIPHERAL NERVOUS SYSTEM

The peripheral nervous system provides a communication pathway for impulses traveling between the CNS and other parts of the body. It consists of nerves, ganglia, and sensory receptors.

The **peripheral nervous system (PNS)** consists of the **nerves** that extend between the skin, muscles, visceral organs, and glands and the central nervous system. It also includes other structures lying outside the CNS, such as **ganglia** and **sensory receptors**. The PNS provides a vital communication link between the CNS and other areas of the body.

As we have seen earlier in this chapter, the PNS is subdivided into two components: the **somatic system**,[19] which transports information that is involved with conscious activities; and the **autonomic system**,[20] which transports information related to activities that do not require conscious awareness. Both systems convey both sensory and motor impulses, although the autonomic system is more frequently concerned with motor activities. Their structural components are, for the most part, separate, although some structures and pathways are shared at least partially.

In our following discussion of the PNS, we will first examine its organs, the nerves, ganglia, and sensory receptors. This is followed by a look at the emergence of nerves from the brain and spinal cord—the so-called cranial nerves and spinal nerves. How these nerves branch to peripheral areas of the body will also be examined. Finally, we will study some major structural and functional features of the somatic and autonomic divisions of the PNS.

Organs of the PNS

Nerves, ganglia, and many types of sensory receptors are organs of the peripheral nervous system.

Nerves and Ganglia

A **nerve** is a true organ, for it is composed of more than one type of tissue and performs a general function—the transport of nerve impulses from one part of the body to another. You should avoid confusing the term *nerve* with *nerve fiber*, which is the axon of a single neuron. Each nerve consists of parallel bundles of nerve fibers (not all of which are myelinated) enclosed by wrappings of connective tissue (Fig. 9–20).

There are three sheaths of connective tissue associated with each nerve. The tough, fibrous outer covering around the whole nerve is called the **epineurium**.[21] Within the body of the nerve, groups of nerve fibers are wrapped together with connective tissue to form bundles, or *fascicles*. This middle wrapping is called the **perineurium**. Yet deeper within is the loose, delicate **endoneurium** that surrounds each individual nerve fiber. The endoneurium also encloses the myelin sheath, when it is present.

There are three different types of nerves, which are distinguished by the nature of the nerve fibers they contain. A nerve may contain only fibers from sensory neurons, only fibers from motor neurons, or fibers from both. Nerves with only sensory fibers are called **sensory**, or **afferent, nerves** and carry impulses toward the CNS. Nerves with only motor fibers are called **motor**, or **efferent, nerves** and carry impulses away from the CNS. Most nerves, however, contain fibers from both sensory and motor neurons and are simply called **mixed nerves.**

Ganglia are clusters of neuron cell bodies located outside the CNS (as opposed to *nuclei*, which are clusters of cell bodies within the CNS). Like nerves, ganglia are enclosed by wrappings of connective tissue.

Sensory Receptors

Sensory receptors are structures that are specialized to respond to stimuli, or changes in the environment. Many sensory receptors are the endings of dendrites from sensory neurons. These dendrites may end free within a body tissue, such as in epithelia and connective tissue, where they are very abundant. They often have tiny, knoblike swellings at their ends and respond to pain, pressure, or temperature. Some sensory re-

---

[19]Somatic: Greek *somat-* = body + *-ic* = pertaining to.

[20]Autonomic: Greek *auto-* = self + *nomicos* = regulating.

[21]Epineurium: Greek *epi-* = upon, on top of + *neurium* = presence of a nerve.

ceptors contain dendrites that are enclosed within a connective tissue capsule. These types are common in the dermis of the skin, where they respond to pressure, and in muscles and joints, where they respond to movement.

Sensory receptors may also be complex structures with numerous tissues associated with them. These are called **special sensory organs**, and include the eye, the ear, taste buds in the mouth, and olfactory organs in the nasal epithelium. The receptor cells that respond to stimuli are actually located deep within these organs. The receptor cells are connected to sensory neurons, which convey the impulse to the brain for processing. Sensory receptors and the four special sensory organs will be discussed in more detail in Chapter 10.

CONCEPTS CHECK

1. What three layers of connective tissue are associated with a nerve?

2. Why are nerves and ganglia regarded as organs?

3. What is the function of a sensory receptor?

Cranial Nerves and Spinal Nerves

The prominent nerves that merge with the CNS are the cranial nerves associated with the brain and the spinal nerves associated with the spinal cord.

Cranial Nerves

The **cranial nerves** are nerves that attach directly to the brain (Fig. 9–21). They extend through the cranium by way of various openings in the skull to supply mainly head and neck structures. There are 12 pairs of cranial nerves, each with a name and a Roman numeral. The Roman numeral corresponds to the relative point of attachment to the brain. For example, the olfactory nerve (I) attaches to the brain at a location anteriormost to other cranial nerves; the optic nerve (II) attaches from a point behind the olfactory nerve; and the oculomotor nerve (III) attaches behind the optic nerve.

The cranial nerves vary considerably in their functions. They may be sensory, motor, or mixed. Three pairs of cranial nerves contain only sensory fibers, and are closely associated with special sense organs. These

Figure 9–20
Nerve structure. The left side of this figure is an electron micrograph of a single nerve, shown in cross section. The right side is a drawing of a nerve, with its components revealed. *(From R.G. Kessel and R.H. Kardon,* Tissues and Organs: a text–atlas of scanning electron microscopy. *Micrographs courtesy of Professor R.G. Kessel.)*

How do the connective tissue wrappings of a nerve parallel those of a skeletal muscle?

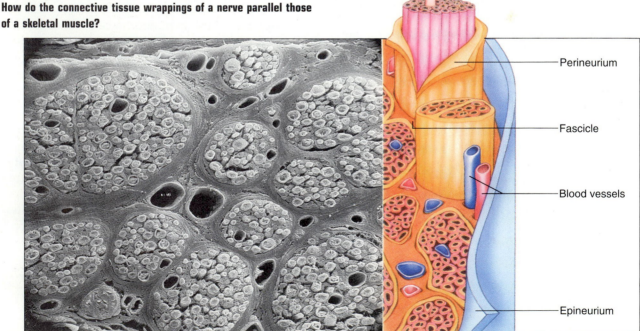

- Axon
- Myelin sheath
- Endoneurium
- Perineurium
- Fascicle
- Blood vessels
- Epineurium

are the olfactory, the optic, and the vestibulocochlear nerves, which supply the sensory receptors in the nasal chambers, the eyes, and the ears, respectively. Two pairs of nerves are entirely motor in function, and supply muscles of the head and neck. They are the spinal accessory (or accessory) and hypoglossal nerves. The remaining seven pairs of cranial nerves contain both sensory and motor fibers, and are therefore mixed nerves. The mixed nerves include the oculomotor, trochlear, trigeminal, abducens, facial, glossopharyngeal, and vagus nerves.

The 12 pairs of cranial nerves are summarized in Table 9–2, which provides the Roman numeral, name, function, and a brief description of each pair.

Table 9–2 THE CRANIAL NERVES

	Cranial Nerve	Type of Nerve	Innervation	Sensation or Action
I	Olfactory	Sensory	Sensory endings in nasal epithelium	Sense of smell
II	Optic	Sensory	Sensory cells in the retina of each eye	Sense of sight
III	Oculomotor	Mixed	Skeletal muscles of the eyeball, smooth muscles in the iris of the eye, sensory endings in eye muscles	Movement of eye, control of light entering the eye, eye muscle sensibility
IV	Trochlear	Mixed	Skeletal muscles of the eyeball, sensory endings in eye muscle	Movement of eye, eye muscle sensibility
V	Trigeminal	Mixed		
	1. Ophthalmic		Eyes, tear glands, upper eyelids, skin of the forehead	Sensation of innervated areas
	2. Maxillary		Upper mouth, skin of the face	Sensation of innervated areas
	3. Mandibular		Sensory fibers to scalp, skin of the jaw, lower mouth; motor fibers to jaw muscles and muscles in floor of the mouth	Sensation of innervated areas; movement of the jaw for chewing
VI	Abducens	Mixed	Motor and sensory fibers to muscles of the eyeball	Movement of the eye; sensation of eye movement
VII	Facial	Mixed	Sensory fibers to tongue; motor fibers to face muscles, tear glands, and salivary glands	Sense of taste; movement of muscles for facial expression; glandular secretion
VIII	Vestibulocochlear	Sensory	Inner ear	Senses of hearing and equilibrium
IX	Glossopharyngeal	Mixed	Sensory fibers to pharynx, tongue, and carotid arteries; motor fibers to pharynx and salivary glands	Sense of taste; sense of touch in throat and mouth; movement of pharynx
X	Vagus	Mixed	Motor fibers to larynx and pharynx, heart, and visceral organs of the thorax and abdomen; sensory fibers to pharynx, larynx, esophagus, and visceral organs of the thorax and abdomen	Movement associated with speech and swallowing; heart rate; peristalsis; visceral sensations
XI	Accessory	Motor	Muscles of the upper mouth, pharynx, larynx, neck, and back	Movements associated with speech and swallowing; movements of the shoulder and head
XII	Hypoglossal	Motor	Muscles of the tongue	Movement of the tongue

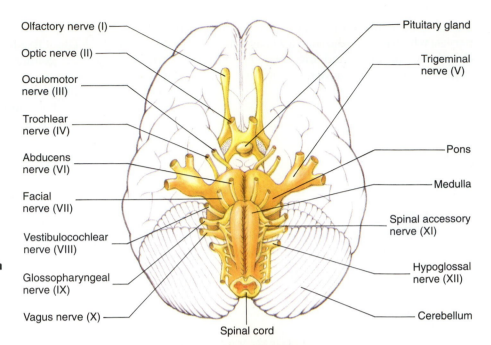

Olfactory nerve (I)
Optic nerve (II)
Oculomotor nerve (III)
Trochlear nerve (IV)
Abducens nerve (VI)
Facial nerve (VII)
Vestibulocochlear nerve (VIII)
Glossopharyngeal nerve (IX)
Vagus nerve (X)
Spinal cord

Pituitary gland
Trigeminal nerve (V)
Pons
Medulla
Spinal accessory nerve (XI)
Hypoglossal nerve (XII)
Cerebellum

Figure 9–21
Cranial nerves. The origins of the cranial nerves from the base of the brain are shown.

Which cranial nerves are sensory, which are motor, and which are mixed in function?

Spinal Nerves

Thirty-one pairs of **spinal nerves** attach directly to the spinal cord, one pair from each segment (Fig. 9–22). Each spinal nerve is mixed in function and contains thousands of nerve fibers. The spinal nerves extend from the spinal cord to branch into smaller nerves that link the neck, trunk, and limbs to the CNS. There are 8 pairs of cervical spinal nerves (numbered C_1 to C_8), 12 pairs of thoracic nerves (T_1 to T_{12}), 5 pairs of lumbar nerves (L_1 to L_5), 5 pairs of sacral nerves (S_1 to S_5), and a single pair of coccygeal nerves (C_0).

Roots of Spinal Nerves

The association between the spinal cord and spinal nerves is very close. The fibers that compose the nerves have some part of their structures within the gray matter of the cord. In the case of motor neurons, the cell bodies are located in the anterior horns. In sensory neurons, the terminal ends of the axons are located within the posterior horns. Thus, there is a separation of function within the cord and with the fibers as they merge with it. This separation continues with the fibers until they combine at the intervertebral foramina, where the spinal nerve is actually formed (Fig. 9–23).

Cervical plexus
Brachial plexus
Lumbosacral plexus

Phrenic nerve
Axillary nerve
Musculocutaneous nerve
Thoracic nerves
Radial nerve
Ulnar nerve
Median nerve
Obturator nerve
Femoral nerve
Sciatic nerve
Saphenous nerve

Figure 9–22
Spinal nerves. The spinal nerves of the right side are shown extending through the appendages, while those of the left side have been cut in this posterior view.

Are the spinal nerves sensory, motor, or mixed in function?

266

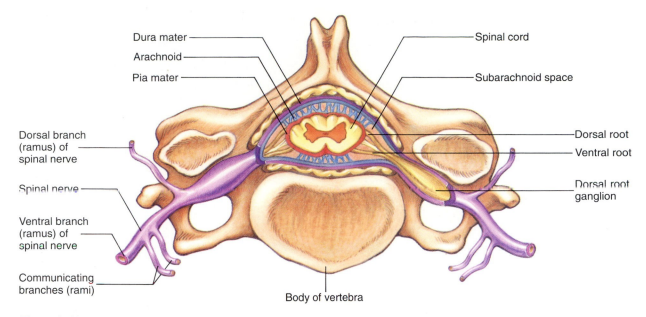

Figure 9–23
The spinal cord, spinal roots, and spinal nerves in cross section. The spinal cord is shown in its position within the vertebral canal.
What is the direction of impulses that are traveling along the dorsal root of a spinal nerve?

The fibers between the cord and the point of spinal nerve formation are referred to as a **root**. The groups of fibers that are sensory in function are called **dorsal roots**, since they unite with the posterior (dorsal) horn of the cord. The motor fibers are called **ventral roots**, since they emerge from the anterior (ventral) horn.

The dorsal root of each spinal nerve contains a large swelling just after it unites with the ventral root. This is the **dorsal root ganglion**, which contains the cell bodies of the sensory neurons whose dendrites conduct impulses from the peripheral regions of the body. The axons of these neurons extend from the ganglion to the posterior horn of the spinal cord.

Rami

The dorsal and ventral roots unite to form spinal nerves at the intervertebral foramina. Soon after its emergence from the vertebral column, each spinal nerve divides into smaller branches (Fig. 9–23). These branches are called **rami**. There are two major rami associated with each spinal nerve: the **dorsal** (posterior) **ramus**, which extends into the back to supply its muscles and skin; and the **ventral** (anterior) **ramus**, which is the larger of the two and extends forward to supply the trunk and limbs.

Two smaller rami also branch, although not from every spinal nerve. They are components of the sympathetic division of the autonomic system. They are called **communicating rami**, and are located only in the thoracic and first two lumbar segments.

Plexuses

The ventral rami from the 31 spinal nerves continue from the back region in an anterior direction. In the lower trunk region (T_2 through T_{12}), the ventral rami extend directly to supply body parts. All other ventral rami branch and recombine to form complex networks, called **plexuses**, before continuing onward. These branching networks occur in the cervical, brachial (upper chest), lumbar, and sacral regions on both sides of the spinal cord (refer to Fig. 9–22). Because of this extensive branching, individual body parts, such as the biceps brachii muscle in your arm, receive nerve impulses from more than one spinal nerve root. An advantage to this dividing and regrouping is that damage to a single root or spinal cord segment does not lead to complete motor or sensory loss in the body part that is supplied.

The **cervical plexus** is located beneath the sternocleidomastoid muscle in the anterior neck region (Fig. 9–22). It is formed by the branching of the first four cervical nerves. A pair of prominent nerves that arise from the cervical plexus are the **phrenic nerves**, which course downward from the plexus through the thorax to the diaphragm. The diaphragm receives motor impulses from these nerves, enabling breathing movements to occur. The importance of the phrenic nerves to breathing is made evident when they are severed, or when the C_3 through C_5 region is crushed or destroyed: the diaphragm becomes paralyzed and respiratory arrest occurs.

The **brachial plexus** is located between the vertebral column and the axilla (armpit) (Fig. 9–22). It is formed by the intermixing of the lower four cervical nerves and most of the first thoracic nerve. Its major branches supply the skin and muscles of the upper limbs, and are called the **musculocutaneous, ulnar, median, radial**, and **axillary nerves**.

The first four lumbar nerves and first four sacral nerves intermix on either side to form the network in the lower trunk called the **lumbosacral plexus** (Fig. 9–22). It lies deep within or on top of the iliopsoas muscle. Its branches supply the skin and muscles of the abdominal wall, pelvic wall, thighs, legs, and feet. The major branches include the **femoral nerve**, the **obturator nerve**, and the **sciatic nerve**.

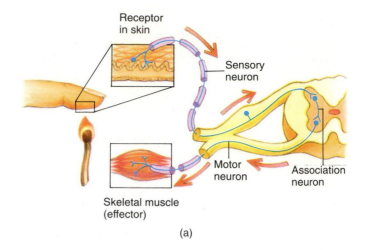

(a)

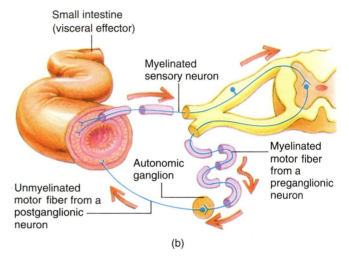

(b)

CONCEPTS CHECK

1. What is the difference between a cranial nerve and a spinal nerve?

2. Which spinal nerve root contains only motor fibers?

3. What plexus arises from the ventral rami in the upper chest region?

Functional Divisions of the PNS

The somatic and autonomic systems are functionally distinct: the somatic is related to conscious activities, and the autonomic, to unconscious activities. Although they are distinct in this regard, their pathways share portions of numerous cranial and spinal nerve routes.

Somatic System

The somatic nervous system is a part of the PNS that conveys information related to activities involving a conscious effort, such as talking, walking, and running. It is composed of nerves, ganglia, and sensory receptors that are widely distributed. Its sensory and motor pathways course throughout the body between peripheral regions and the CNS (Fig. 9–24a).

The sensory component of the somatic system contains the sensory receptors in the skin, visceral organs, and special sensory organs. It also contains the sensory nerves (and mixed nerves with sensory fibers) that convey impulses from these receptors to the CNS for processing. In the sensory pathway, there is only one sensory neuron between the sensory receptor and the CNS.

Figure 9–24
The pathways of the PNS. (a) The somatic system. In this example of a somatic pathway, a reflex arc is shown extending from a skin receptor to a muscle effector. (b) The autonomic system. In this generalized example of an autonomic pathway, the sensory impulse passes from a visceral organ to the CNS. The motor impulse is routed from the CNS through a preganglionic neuron to an autonomic ganglion, then to a postganglionic neuron. The postganglionic neuron terminates in a visceral effector (in this example, smooth muscle of the small intestine).

What types of effectors receive impulses from postganglionic neurons?

The motor component consists of motor nerves (and mixed nerves with motor neurons). The motor nerves terminate in effectors, which are always skeletal muscles. Only one type of neurotransmitter is released at synapses with skeletal muscles, acetylcholine (ACh), which can produce only one effect: excitatory, resulting in the contraction of muscle. The motor pathway consists of a single motor neuron extending between the spinal cord and the effector. The pathways of the somatic nervous system will be discussed in more detail in Chapter 10.

Autonomic System

The autonomic nervous system is the portion of the peripheral nervous system that conveys information related to activities that do not require a conscious effort to maintain, such as blood vessel diameter, stomach wall contractions, heart rate, and glandular secretions. It performs these functions automatically by reflex actions, although its name was not derived from this fact. The system was named "autonomic" years ago when scientists thought it operated independently, or "autonomously," from the brain and spinal cord. It is now understood that it does not function independently, but is regulated by centers in the brain, particularly the hypothalamus and medulla oblongata, which serve as reflex centers.

Like the somatic system, the autonomic system contains sensory and motor components. Its sensory components are sensory neurons that carry signals from receptors within the skin and visceral organs. Sensory pathways are quite simple, for the impulses pass directly to centers in the hypothalamus, brain stem, or spinal cord. These centers integrate the sensory information and route a motor impulse to smooth muscles, cardiac muscle, or glands by way of special autonomic motor pathways. The motor impulse may stimulate smooth muscle contraction, cardiac muscle contraction, or glandular secretion, or it may inhibit them, depending primarily on the nature of the neurotransmitter that is released.

The autonomic motor pathways are quite complex and include two basic routes that differ both structurally and functionally. They are called the **sympathetic division** and the **parasympathetic division**. Many visceral organs receive nerve fibers from both divisions. In these cases, the effect of an impulse from one division will counter the effect of an impulse from the other division; in other words, their effects are *antagonistic*. For example, smooth muscle in the wall of the stomach receives both sympathetic and parasympathetic fibers. Impulses from the sympathetic fibers inhibit contraction, while impulses from parasympathetic fibers stimulate contraction for digestion.

Autonomic Motor Pathways

An autonomic motor pathway contains two neurons, unlike a somatic motor pathway, which contains a single neuron between the spinal cord and a skeletal muscle (Fig. 9–24b). The first neuron begins with its cell body within the brain or spinal cord, and is called the **preganglionic neuron**. Its axon is myelinated and exits the CNS by way of a cranial or spinal nerve. Once outside the CNS, it separates from the nerve and extends to a **ganglion**, where it terminates. At the ganglion it synapses with the dendrites or cell body of the second neuron, called the **postganglionic neuron**. The axon from this neuron is unmyelinated and extends from the cell body in the ganglion to a visceral effector (smooth muscle, cardiac muscle, or a gland), where it terminates. The actual arrangement of preganglionic neurons, ganglia, and postganglionic neurons varies between the two divisions.

In the sympathetic division, the preganglionic neurons originate from the gray matter (lateral horns) of the spinal cord in only its middle segments: the 12 thoracic and first 2 or 3 lumbar segments (Fig. 9–25). The preganglionic fibers exit the cord along the ventral roots of the spinal nerves in these segments (Fig. 9–26). A short distance from their point of exit, the fibers branch away from the spinal nerves and extend through the communicating rami. They soon enter a chain of ganglia that lie in a vertical row along either side of the vertebral column, called the **sympathetic trunk ganglia**. The preganglionic fiber may terminate at the first ganglion it encounters, forming a synapse with many postganglionic neurons. The axons from these neurons return to a spinal nerve by way of the communicating rami. They travel with the spinal nerve until they reach a visceral effector.

Alternatively, the preganglionic fiber does not terminate at the first sympathetic trunk ganglion it encounters, but rather continues in a vertical direction to another ganglion in the chain. Here, it synapses with postganglionic neurons, which extend along with spinal nerves to a visceral effector.

In some cases, the preganglionic neuron does not terminate in a sympathetic trunk ganglion, but instead extends beyond it into the abdominal or the pelvic cavity. Once here, it synapses with a postganglionic neuron at a **prevertebral ganglion**. The axon from the postganglionic neuron then extends from the ganglion through a plexus before terminating at a visceral effector.

In the parasympathetic division, the preganglionic neurons originate from the brain stem and the sacral region of the spinal cord (S_2 through S_4). They are shown in Figure 9–25. Their axons emerge as part of a cranial nerve or as part of the ventral root of a spinal nerve, and extend with these nerves to end at **terminal ganglia** located within or near the visceral effectors. Postganglionic neurons originating from the terminal ganglia extend a short distance to supply the visceral effectors.

Autonomic Functions

The effect of autonomic impulses upon smooth muscle, cardiac muscle, and glands may be either to stimulate or to inhibit their activity. This dual function is primarily due to the release of two different types of neurotransmitters by the two autonomic divisions.

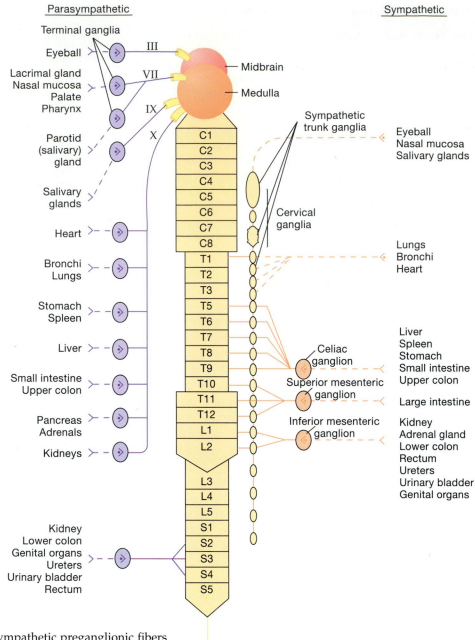

Figure 9–25
Autonomic motor pathways. The sympathetic preganglionic fibers emerge from the T_1 through L_2 segments of the spinal cord, while the parasympathetic preganglionic fibers emerge from the brain stem and the sacral region of the spinal cord. Sympathetic pathways are shown on the right side in orange, and parasympathetic pathways are on the left side in violet. The organs they innervate are listed.

How does the distribution of sympathetic preganglionic fibers differ from that of parasympathetic preganglionic fibers?

Most sympathetic postganglionic fibers release *norepinephrine* (noradrenaline), and are thus referred to as **adrenergic fibers**. Parasympathetic postganglionic fibers release *acetylcholine* and are called **cholinergic fibers**. These two neurotransmitters have opposite effects on the visceral organs. Thus, visceral organs that receive postganglionic fibers from both divisions are stimulated by one neurotransmitter and inhibited by the other. However, most visceral organs are actually controlled primarily by one division. The effects of autonomic stimulation on many visceral organs are summarized in Table 9–3.

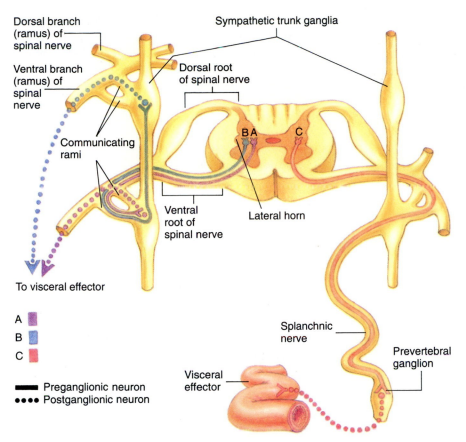

Figure 9–26
Alternate routes of the sympathetic motor pathway. (a) The preganglionic fiber may synapse in a sympathetic trunk ganglion at the same level it emerges from; (b) it may synapse in a sympathetic trunk ganglion at a different level; (c) or it may extend through the sympathetic trunk ganglion and synapse at a prevertebral ganglion. After synapsing with the preganglionic fiber at a ganglion, the postganglionic fiber extends to the effector.

What are the general effects of sympathetic stimulation?

The sympathetic division is primarily involved with processes that use energy. Its adrenergic fibers increase heart rate, increase glucose levels in the blood, channel blood flow to skeletal muscles, lungs, and heart and away from digestive organs, and inhibit digestive contractions. It prepares the body for emergency situations that require quick thinking or physical reaction, and is therefore called the "fight or flight" division.

The parasympathetic division is mainly concerned with body activities that involve a conservation of energy. Its cholinergic fibers stimulate the smooth muscles in the digestive tract and glands in the digestive wall, allowing energy-supplying foods to be digested and stored. It also inhibits heart rate. Because of these effects, it is often called the "rest-repose" division.

Table 9–3   PHYSIOLOGICAL EFFECTS OF THE AUTONOMIC NERVOUS SYSTEM

Organ	Effects of Sympathetic Stimulation	Effects of Parasympathetic Stimulation
Eye	Dilates pupil	Constricts pupil
Heart	Increases rate and strength of contraction	Decreases rate and strength of contraction
Lung	Dilates bronchioles	Constricts bronchioles
Stomach	Decreases digestive activity	Increases digestive activity
Urinary bladder	Relaxes muscles in wall	Constricts muscles in wall
Sweat glands (in skin)	Stimulates secretion	No innervation
Blood vessels in skeletal muscles	Vasodilation (increases blood flow)	No innervation
Blood vessels in skin	Vasoconstriction (decreases blood flow)	No innervation

Autonomic Regulation

The control centers of the autonomic system are located in the CNS. Within these centers, autonomic impulses are initiated and channeled through pathways that eventually lead to the visceral effectors. In visceral organs that do not receive dual innervation from both divisions, the control centers may regulate the strength of the stimulus in order to control the effect. For example, most blood vessels receive only adrenergic fibers, which stimulate the smooth muscle lining their walls. The diameter of the blood vessel can be increased (dilated) by decreasing the strength of the stimulus. In visceral organs that receive dual innervation, the control centers can reduce the stimulus strength to one division and increase it to the other to allow a dominating effect by one division. For example, reducing the stimulus strength from the cholinergic fibers to the smooth muscle in the stomach wall and increasing the stimulus strength to the adrenergic fibers will allow the adrenergic fibers to inhibit contraction, and the stomach will stop churning.

The control centers include primarily the hypothalamus and the medulla oblongata. As you may recall, the hypothalamus helps regulate body temperature, endocrine gland secretion, hunger, and thirst. It accomplishes this by controlling the strength of impulses and the route of transmission along the autonomic pathways. In a similar way, the medulla oblongata of the brain stem controls heart rate, blood vessel diameter, and breathing rate. These centers are influenced to some extent by the cerebral cortex, which becomes involved with the autonomic system when an individual becomes emotionally stressed. Autonomic pathways may be used by these higher centers to regulate emotional behavior.

CONCEPTS CHECK

1. How many neurons are required to conduct a motor impulse from the spinal cord to a skeletal muscle?

2. How do the sympathetic and parasympathetic divisions differ in their conduction pathways?

3. How do the two autonomic divisions differ in their responses?

4. What are the main controlling centers of the autonomic system?

HOMEOSTASIS

The role of the nervous system in maintaining the stability of internal systems, or homeostasis, is significant in many ways. For example, we have seen how a simple reflex arc can protect us from injury by way of the withdrawal reflex. Every other reflex that we have, including the many visceral reflexes that occur without requiring conscious effort—breathing, blinking the eyes, heart rhythm, blood vessel constriction, and others—occurs in order to keep the internal environment of the body in a stable, optimum state for body functions to proceed uninterrupted. The reflexes do so by initiating an internal response, such as skeletal muscle contraction, cardiac muscle contraction, smooth muscle contraction, or glandular secretion. If conditions should change, the reflexes can adjust their level of output in response to the change in order to maintain stability. For the most part, these reflexes are controlled by the hypothalamus and the medulla oblongata. The response time is extremely rapid, since reflexes eliminate the need to send impulses to the higher centers of the brain (the cerebral cortex) before managing the proper response. Regulation of internal activities by reflex action is the primary way in which the nervous system maintains homeostasis.

The nervous system maintains homeostasis in ways other than by reflexes, as well. The brain's higher centers, for example, can direct a response to an emergency situation on a conscious level. This enables us to utilize our memories of past experiences in order to respond in an optimum way. For example, a child learns to look both ways before crossing a street. In so doing, the child is avoiding an emergency situation by drawing on his or her learning abilities. The use of the brain's higher centers in maintaining homeostasis may also utilize problem-solving skills and other means of intelligence. The simple act of putting on a coat in cold weather is one example.

Both methods of adjusting body activities in order to maintain homeostasis rely on an important feature characteristic of the nervous system. This feature is the ability to sense changes in the environment (internal and external) and to relay this information rapidly to the CNS. This sensory function is a vital component of the mechanisms involved in homeostasis management, for the nervous system must be able to monitor changes if it is to initiate a meaningful response.

CLINICAL TERMS OF THE NERVOUS SYSTEM

Amyotrophic lateral sclerosis (ALS)   A relatively rare neurologic disorder characterized by degeneration of the motor neuron cell bodies within the anterior gray horns of the spinal cord and within the brain. It is progressive, and there is no known cause or cure. It is also known as *motor-neuron disease* and *Lou Gehrig's disease.*

Bacterial meningitis   An infection of the meninges commonly caused by the bacterium *Haemophilus influenzae.* It usually affects children under the age of 5, although it may be part of the normal flora in adults. It can cause severe headaches and fever, and can lead to encephalitis and brain damage.

Cephalalgia   The clinical term for headache. Headaches may originate from a number of causes, such as muscle spasms in the neck or head, spasms in the smooth muscles in the walls of blood vessels serving the brain, or a drop in blood glucose. Migraine headaches are the most severe form, although their cause is not yet understood.

Encephalitis   An infection of brain tissue by bacteria. It is characterized by severe headache and fever. If not treated with antibiotics quickly, this disease can cause brain damage leading to dysfunction or death.

Glioma   Tumors of the nervous system that originate from neuroglia. They are usually slow-growing and benign, although their overcrowding can lead to neuron destruction and nervous dysfunction. They are named according to the cell type they arise from, including astrocytoma, oligodendroglioma, and schwannoma.

Guillain-Barré syndrome   Also known as *idiopathic polyneuritis*, a disease characterized by a loss of myelin protecting spinal nerves and nerve roots. Its symptoms include muscular weakness beginning in the legs and spreading to the trunk. Its cause is unknown, and recovery is usually complete with respiratory assistance.

Multiple sclerosis   A progressive, fatal disease that involves the deterioration of the myelin sheath protecting axons in the brain. It is thought to be an autoimmune disease, in which the body's own white blood cells enter brain tissue and destroy functional oligodendrocytes. It may be virus-initiated, although evidence is lacking to substantiate this theory. It is characterized by a gradual loss of motor functions.

Schizophrenia   A disorder of the brain that is characterized by abnormal behavior, including social withdrawal, auditory or visual hallucinations, and an inability to distinguish reality from the dream state. It is often accompanied by a degeneration of brain tissue. Recent CAT scan studies have shown that schizophrenics typically exhibit enlarged lateral ventricles in the cerebrum, and a reduced hippocampus (a portion of the cerebrum near the ventricles). Drug therapy helps relieve symptoms, but there is no cure and the cause is not well understood.

CHAPTER SUMMARY

The nervous system is the control center and communication network of the body. It performs these functions through sensory, integrative, and motor activities.

DIVISIONS OF THE NERVOUS SYSTEM

The nervous system is divided into the central nervous system (CNS), containing the brain and spinal cord, and the peripheral nervous system (PNS), containing the nerves and ganglia. The PNS contains cranial nerves, which arise from the brain, and spinal nerves, which arise from the spinal cord, as well as their many branches. It is subdivided into the somatic system, which is under conscious control, and the autonomic system, which is under unconscious control. The autonomic system contains sympathetic and parasympathetic divisions.

NERVE TISSUE

Nerve tissue is composed of neurons, which conduct impulses, and neuroglia, which support the neurons.

Neuroglia   There are numerous types of neuroglia, all of which support and help maintain the neurons.

Neurons   These are the structural and functional units of nerve tissue, for they perform the sensory, integrative, and motor functions of the nervous system.

Structure Each neuron contains a cell body, numerous dendrites, and a single axon. An impulse travels from the cell body and dendrites to the axon, and out. The axons in large peripheral neurons may contain a fatty, supportive sheath called the *myelin sheath*; these are called *myelinated* axons. The presence of myelin makes white matter white; the lack of myelin is apparent in gray matter.

Types of Neurons Neurons are classified on the basis of structure and function. Structurally, there are three types: multipolar, bipolar, and unipolar. There are three functional types as well: sensory, association, and motor.

Function The function of neurons is the generation and transmission of nerve impulses.

Resting Membrane Potential At rest, a neuron has an electrical charge due to the uneven distribution of ions across its membrane. The outside carries a positive charge due to sodium and potassium ions, and the inside carries a negative charge due to a smaller number of potassium ions and negative ions that cannot cross the membrane.

Action Potential Neurons respond to changes in their environment; this is called *excitability*. If the change, or stimulus, is above a minimum level, then the membrane permeability changes initially for sodium ions. These ions diffuse into the cell, causing the membrane to change its electrical charge and depolarize. Immediately afterward, potassium-ion channels open to allow potassium ions to move outward, restoring the charge on the outside of the membrane to positive, or repolarizing. Depolarization followed by repolarization is an action potential, or nerve impulse.

Conduction of the Nerve Impulse The action potential spreads from one point along the length of the neuron by depolarization followed by repolarization of adjacent points. The wave of negative charges thus moves along the outside of the membrane. Propagation is more rapid in myelinated fibers because of saltatory conduction from one node of Ranvier to another.

All-or-None Response The minimal stimulus to initiate depolarization is called a *threshold stimulus*. If one stimulus is strong enough to cause an action potential, increasing its strength further will have no effect. However, subthreshold stimuli can accumulate to cause an action potential; this is called *summation*.

Transmission of Impulses from Cell to Cell Transmission occurs across a gap between neurons, called a *synapse*. The signal is carried across the synapse by neurotransmitters.

Excitatory versus Inhibitory Transmissions A transmission involving neurotransmitters that increase the postsynaptic membrane permeability to sodium ions is called an *excitatory transmission*. This requires a certain volume of neurotransmitters. The addition of neurotransmitter molecules tends to lead toward depolarization. Partial depolarization is called *facilitation*. A transmission that decreases permeability to sodium ions is called an *inhibitory transmission*.

Processing at Synapses The overall effect of neurotransmitters on the postsynaptic membrane is determined by the sum of excitatory and inhibitory effects. This is a means for providing a processing and integration of impulses.

THE CENTRAL NERVOUS SYSTEM

The central nervous system consists of the brain and spinal cord.

The Spinal Cord A long bundle of nerve tissue extending through the vertebral canal from the base of the brain to the lower back.

Protective Coverings Three components protect the cord: the vertebral column, cerebrospinal fluid (CSF), and the meninges. There are three layers of meninges: the outer dura mater, the middle arachnoid, and the inner pia mater. Between the arachnoid and pia mater is an important space filled with CSF, the subarachnoid space. The meninges of the cord are continuous with those of the brain.

Structure The cord consists of 31 segments, each containing a pair of spinal nerves. In cross section, it can be observed to contain white matter and gray matter.

Gray Matter Contains unmyelinated fibers, neuron cell bodies, and neuroglia. It occupies the center of the cord in the shape of the letter H, and consists of the posterior gray horns, the anterior gray horns, the lateral gray horns, the gray commissure connecting right and left, and a central canal filled with CSF. The posterior horns are sensory, and the anterior horns are motor.

White Matter Contains myelinated fibers, and surrounds the gray matter. It contains three regions: the anterior, posterior, and lateral columns. Longitudinal bundles of myelinated fibers extend through the white matter, forming the nerve tracts.

Functions The spinal cord functions as a conduction pathway and as a reflex center.

Conduction Pathways The spinal cord provides two-way traffic between the brain and peripheral nerves. The ascending tracts carry sensory information to the brain, and the descending tracts carry motor impulses away from the brain.

Reflex Center A spinal reflex is a rapid signal that bypasses the higher brain centers for a reduced response time. The pathway is called a *reflex arc* and involves a sensory receptor, sensory neuron, reflex center in the spinal cord or lower brain area, and motor neuron. Somatic reflexes stimulate skeletal muscle, and visceral reflexes stimulate or inhibit visceral organs.

The Brain The brain is divided into three regions: the forebrain, midbrain, and hindbrain. The forebrain contains the cerebrum and diencephalon, the midbrain is the smallest, and the hindbrain contains the pons, medulla oblongata, and cerebellum.

Protective Coverings The brain is protected by cranial bones, CSF, and the three meningeal layers (dura mater, arachnoid, and pia mater).

Cerebrospinal Fluid and Ventricles of the Brain There are four ventricles, or cavities, inside the brain—two lateral ventricles, a third ventricle, and a fourth ventricle—all of which intercommunicate by way of narrow channels. CSF is produced by filtration through capillary networks in the ventricles, called *choroid plexuses*. Its continual production rate establishes a fluid current that pushes CSF through the ventricles, around the subarachnoid space of the spinal cord, to the subarachnoid space around the brain. It is reabsorbed into the bloodstream through arachnoid villi at the top of the brain.

Cerebrum The largest structure in the brain, divided into right and left parts, called *hemispheres*. Each hemisphere is divided into functional lobes: the frontal, parietal, occipital, and temporal. Its wrinkles are called convolutions; they contain upfoldings known as *gyri* and downfoldings known as *sulci* or *fissures*. When the brain is sectioned, we can observe its outer region of gray matter, known as the *cerebral cortex*, and its inner region of white matter. Embedded inside the white matter are basal ganglia, clusters of gray matter.

Diencephalon Located below the large cerebrum. It contains two structures: the thalamus and the hypothalamus. The pineal gland and pituitary gland are connected to the diencephalon.

Thalamus The largest part of the diencephalon, it serves as the principal relay station for sensory impulses entering the brain and voluntary motor impulses traveling out.

Hypothalamus Located below the thalamus; the primary control center for the autonomic nervous system. It regulates endocrine gland activity, controls body temperature, regulates food and water intake, and is associated with emotion as part of the limbic system.

Midbrain Located between the diencephalon and the pons. It consists of cerebral peduncles on its anterior side, which provide a main connection for motor and sensory pathways; and the colliculi on the posterior side, which contains reflex centers for rapid eye, head, and trunk movements.

Pons Located below the midbrain, it bridges the cerebrum and cerebellum.

Medulla Oblongata Together with the midbrain and pons, makes up the brain stem. The medulla lies below the pons and is continuous with the spinal cord. On its posterior side the motor nerves cross, forming a bulge called the *pyramids*. Sensory nerves also cross in the medulla. The medulla contains numerous visceral reflex centers, such as the cardiac center, the vasomotor center, and the respiratory center.

Cerebellum The posterior portion of the hindbrain, divided into two hemispheres and containing narrow convolutions. Its outer region is the gray-matter cortex, and its inner is white matter with nuclei. It connects to three paired bundles of myelinated fibers, called *cere-*

bellar peduncles, which send motor impulses to skeletal muscles to coordinate contraction and provide precision of movement.

THE PERIPHERAL NERVOUS SYSTEM

The PNS consists of nerves, ganglia, and sensory receptors. These organs constitute the two divisions of the PNS, the somatic system and the autonomic system.

Organs of the PNS Nerves, ganglia, and many sensory receptors are true organs because they contain more than one type of tissue and perform a general function.

Nerves and Ganglia Each nerve consists of parallel bundles of nerve fibers wrapped by three layers of connective tissue: the epineurium, perineurium, and endoneurium. Ganglia are clusters of neuron cell bodies outside the CNS, also wrapped with connective tissue coverings.

Sensory Receptors Structures that are specialized to respond to stimuli. They may be simple endings of sensory neurons or complex special sense organs such as the eye.

Cranial Nerves and Spinal Nerves

Cranial Nerves Nerves that originate from the brain. There are 12 pairs, each with a name and Roman numeral. They may be sensory, motor, or mixed in function.

Spinal Nerves Nerves that originate from the spinal cord. There are 31 pairs, all of which are mixed in function.

Roots of Spinal Nerves Spinal nerve roots connect the spinal nerves with the spinal cord. They lie within the vertebral canal and extend from the gray horns to the intervertebral foramina. The dorsal roots contain sensory fibers, and the ventral roots contain motor fibers.

Rami Each spinal nerve divides into two main branches soon after emerging from the vertebral column: a dorsal ramus, which supplies deep back muscles, and a ventral ramus, which supplies the trunk and limbs.

Plexuses The ventral rami of spinal nerves in the cervical, brachial, lumbar, and sacral regions branch extensively and regroup to form networks, called *plexuses*.

Cervical Plexuses In the neck region, formed by C_1 through C_4.

Brachial Plexuses In the upper chest between the axilla and neck, formed by C_5 through C_8 and most of T_1.

Lumbosacral Plexuses In the lower trunk region, formed by L_1 through S_4.

Functional Divisions of the PNS

Somatic System The part of the PNS whose motor component is under conscious control. Its sensory pathways carry impulses from sensory receptors to the CNS by way of a single sensory neuron, and its motor pathways carry impulses from the CNS to skeletal muscles by way of a single motor neuron.

Autonomic System   The part of the PNS that is under unconscious, or automatic, control. Its sensory pathways lead from visceral organs to autonomic reflex centers in the hypothalamus, brain stem, or spinal cord. Its motor pathways are complex, being divided into sympathetic and parasympathetic divisions.

Autonomic Motor Pathways   All impulses travel from the CNS to a preganglionic neuron, then to a ganglion, then to a postganglionic neuron, and finally to the visceral effector. The actual pathway differs between the two autonomic divisions.

Sympathetic Division   The preganglionic fiber exits from a spinal nerve in the 12 thoracic and first two or three lumbar segments. It passes to a chain of ganglia, called the *sympathetic trunk ganglia*, to eventually terminate at one. Here it synapses with a postganglionic neuron. The neuron returns to a spinal nerve, which it transverses until it terminates at an effector. Alternatively, the preganglionic fiber may extend beyond the sympathetic trunk ganglia into the abdominal and pelvic cavities to a prevertebral ganglion. The postganglionic fiber then extends to the visceral effector from this ganglion.

Parasympathetic Division   The preganglionic fibers originate from the brain stem by way of cranial nerves and from the sacral region of the spinal cord by way of spinal nerves. They extend directly to terminal ganglia located within or near the visceral effectors. From here, postganglionic neurons extend to the effectors.

Autonomic Functions   Functions are mainly dependent on the nature of the neurotransmitters involved. There are two basic types: norepinephrine, released by adrenergic fibers of the sympathetic division; and acetylcholine, released by cholinergic fibers of the parasympathetic division. These two types produce different effects on body activities.

Sympathetic Division   Primarily involved with processes that use energy; called the "fight or flight" division.

Parasympathetic Division   Primarily involved with processes that conserve energy; called the "rest-repose" system.

Autonomic Regulation   Autonomic functions are regulated by reflex centers in the hypothalamus, brain stem, and spinal cord. Many effectors receive dual innervation (from both divisions), although only one division is usually predominant. Control is achieved by influencing the strength of the stimulus, and directing pathways.

HOMEOSTASIS

The nervous system plays a key role in maintaining internal stability. It does so mainly by reflex action, but also by use of higher brain centers. It relies on sensory information in order to initiate a meaningful response.

KEY TERMS

autonomic nervous system (p. 269)
cortex (p. 259)
effector (p. 255)
fissure (p. 259)
ganglia (p. 255)

gyri (p. 259)
meninges (p. 253)
nerve (p. 263)
neurotransmitter (p. 249)
nuclei (p. 255)
plexus (p. 267)

polarize (p. 247)
ramus (p. 267)
reflex (p. 255)
somatic nervous system (p. 268)

sulcus (p. 259)
synapse (p. 249)
transmission (p. 249)
ventricle (p. 256)

QUESTIONS FOR REVIEW

OBJECTIVE QUESTIONS

1. The three main activities performed by the nervous system to maintain homeostasis are sensory, motor, and:
   a. secretion
   b. visual
   c. integration
   d. temperature control

2. The autonomic nervous system is part of the:
   a. brain
   b. peripheral nervous system
   c. central nervous system
   d. somatic nervous system

3. The cells that support nerve tissue function and account for about 90% of the substance of the CNS are called:
   a. neuroglia
   b. Schwann cells
   c. neurons
   d. fibroblasts
4. All multipolar neurons contain a cell body, numerous dendrites, and a single:
   a. axon
   b. myelin sheath
   c. nerve fiber
   d. both a and c above
5. White matter in nerve tissue receives its coloration from:
   a. neuroglia
   b. myelin sheaths
   c. neurilemma
   d. dendrites
6. A neuron that conducts an impulse away from the CNS is called a(n):
   a. unipolar neuron
   b. motor neuron
   c. sensory neuron
   d. association neuron
7. The ion gradient across a resting neuron membrane is established because the plasma membrane is more permeable to which ion?
   a. potassium
   b. chloride
   c. sodium
   d. calcium
8. When sodium ions flow into a neuron because of a change in membrane permeability, the membrane changes its electrical charge and becomes:
   a. polarized
   b. depolarized
   c. hyperpolarized
   d. repolarized
9. A nerve impulse is conducted rapidly along a myelinated nerve fiber by a process called:
   a. saltatory conduction
   b. nervous transmission
   c. repolarization
   d. electrical conduction
10. A single subthreshold stimulus has what effect on a neuron?
    a. initiates an action potential
    b. has no effect
    c. summation
    d. depolarization
11. Neurotransmitters are released from synaptic vesicles located within:
    a. postsynaptic neurons
    b. presynaptic neurons
    c. synapses
    d. interstitial fluid
12. A transmission that causes a postsynaptic neuron to increase its membrane permeability to sodium ions is called a(n):
    a. excitatory transmission
    b. automatic transmission
    c. inhibitory transmission
    d. manual transmission
13. The meningeal membrane that is attached to the vertebral column and protects the spinal cord is the:
    a. pia mater
    b. arachnoid
    c. dura mater
    d. myelin sheath
14. The region of the spinal cord that contains cell bodies of motor neurons is the:
    a. lateral column
    b. anterior gray horn
    c. posterior gray horn
    d. central canal
15. The reflex center in a reflex arc is composed of:
    a. cell bodies of sensory neurons
    b. association neurons
    c. myelinated nerve fibers
    d. effectors

16. In the brain, the diencephalon is part of the:
    a. cerebrum
    b. forebrain
    c. thalamus
    d. brain stem
17. Cerebrospinal fluid is produced within cavities of the brain called:
    a. choroid plexuses
    b. arachnoid villi
    c. ventricles
    d. hemispheres
18. The longitudinal fissure divides the _____ into right and left hemispheres.
    a. cerebrum
    b. medulla oblongata
    c. cerebellum
    d. frontal lobe
19. The main control center for autonomic functions and endocrine gland activity is the:
    a. thalamus
    b. cerebrum
    c. hypothalamus
    d. cerebellum
20. The organs of the peripheral nervous system include ganglia, special sensory organs, and:
    a. nerve fibers
    b. effectors
    c. nerves
    d. spinal cord
21. Nerves that originate from the brain are called:
    a. spinal nerves
    b. cranial nerves
    c. plexuses
    d. mixed nerves
22. The portion of the PNS that conveys impulses related to conscious activities is the:
    a. sympathetic division
    b. autonomic nervous system
    c. somatic nervous system
    d. none of the above
23. In most cases, adrenergic fibers release neurotransmitters that cause body processes to:
    a. conserve energy
    b. rest and repose
    c. use energy
    d. all of the above

ESSAY QUESTIONS

1. Describe the origin of a resting potential in a neuron, and describe what changes must occur for an action potential to take place.
2. Explain how an impulse is transmitted from one neuron to the next. Include a distinction between excitatory and inhibitory transmissions, and how they can lead to the complex processing that occurs in the brain.
3. Describe a typical reflex arc and its components. How is this advantageous for homeostasis?
4. Describe the division of labor that exists in the different parts of the brain.
5. Distinguish between cranial nerves and spinal nerves, and include a description of the branching that is typical of spinal nerves.
6. Compare and contrast the somatic and autonomic nervous systems, in terms of general pathways and functions.
7. How do the sympathetic and parasympathetic divisions of the autonomic nervous system differ?

Figure 9–1
What are the organs of the central nervous system?   The brain and spinal cord are the organs of the CNS.

Figure 9–2
Which neuroglia cells are found in the CNS?   Astrocytes, oligodendrocytes, microglia, and oligodendrocytes are located in the CNS.

Figure 9–3
What three structural features do all neurons have in common?   Every neuron contains a cell body, dendrites, and an axon.

Figure 9–4
In what direction do impulses travel as they pass through the axon?   Impulses always travel from the cell body to the axon.

Figure 9–5
What are the three functional distinctions of neurons?   Neurons may be sensory (carry impulses toward the CNS), association (carry impulses to other neurons within the CNS), or motor (carry impulses away from the CNS).

Figure 9–6
How is the high sodium ion concentration maintained on the outside of the membrane?   Sodium ions are continually pumped out of the cell by the sodium-potassium pump.

Figure 9–7
The rapid movement of what ions into the cell leads to depolarization?   The rapid movement of sodium ions into the cell causes depolarization.

Figure 9–8
What two changes in the plasma membrane lead to repolarization?   The opening of potassium-ion channels in the membrane, which permits the flow of potassium ions outward, and the closing of the sodium-ion channels lead to repolarization.

Figure 9–9
What is the benefit of saltatory conduction?   Saltatory conduction permits a more rapid propagation of the nerve impulse.

Figure 9–10
What is the chemical that carries an impulse across a synapse by diffusing from the presynaptic membrane to the postsynaptic membrane?   The chemical is called a *neurotransmitter*.

Figure 9–11
What two components of the nervous system does the spinal cord connect?   The spinal cord connects peripheral nerves with the brain.

Figure 9–12
What is the outermost meningeal layer?   The outermost meningeal layer is the dura mater.

Figure 9–13
How does the posterior gray horn of the spinal cord differ functionally from the anterior gray horn?   The posterior gray horn contains nerve fibers from sensory neurons, whereas the anterior gray horn contains nerve fibers from motor neurons.

Figure 9–14
Why does a reflex arc bring about a more rapid response than a coordinated, skilled movement?   A reflex arc produces a more rapid response because it travels the shortest distance possible from receptor to effector.

Figure 9–15
The cerebrum is a part of what region of the brain?   The cerebrum is part of the embryonic forebrain.

Figure 9–16
Trace the pathway of cerebrospinal fluid from its origin to its reabsorption.   CSF is initially produced through the choroid plexus of the lateral ventricles. It then passes through the foramen of Monro into the third ventricle, where more CSF is added in the choroid plexus there. CSF then flows through the cerebral aqueduct and into the fourth ventricle, where more CSF is added. From here, it enters the subarachnoid space around the spinal cord and brain, and the central canal of the cord, to eventually return to the top of the brain, where it is reabsorbed into arachnoid villi.

Figure 9–17
What ventricles are connected by the cerebral aqueduct?   The third and fourth ventricles are connected by the cerebral aqueduct.

Figure 9–18
What are the four lobes of each cerebral hemisphere?   A frontal lobe, a temporal lobe, a parietal lobe, and an occipital lobe are located in each hemisphere.

Figure 9–19
With what part of the diencephalon does the pituitary gland form a connection?   The pituitary gland forms a connection with the hypothalamus.

Figure 9–20
How do the connective tissue wrappings of a nerve parallel those of a skeletal muscle?   A nerve contains three layers, as does a muscle: the innermost layer surrounds individual cells (or cell processes), the middle layer surrounds groups of

cells (or processes), and the outer layer envelopes the organ itself. Also, both nerve and muscle connective tissue layers carry blood vessels.

Figure 9–21
Which cranial nerves are sensory, which are motor, and which are mixed in function? The olfactory, optic, and vestibulocochlear nerves are sensory, the hypoglossal and accessory nerves are motor, and the remaining are mixed in function.

Figure 9–22
Are the spinal nerves sensory, motor, or mixed in function? All spinal nerves are mixed nerves.

Figure 9–23
What is the direction of impulses that are traveling along the dorsal root of a spinal nerve? Impulses traveling along the dorsal root of a spinal nerve are entering the spinal cord.

Figure 9–24
What types of effectors receive impulses from postganglionic neurons? Smooth muscle, cardiac muscle, and glands receive impulses from postganglionic neurons.

Figure 9–25
How does the distribution of sympathetic preganglionic fibers differ from that of parasympathetic preganglionic fibers? Sympathetic preganglionic fibers extend from the spinal cord to sympathetic trunk ganglia located nearby, whereas parasympathetic preganglionic fibers extend to prevertebral ganglia located near effectors.

Figure 9–26
What are the general effects of sympathetic stimulation? Sympathetic stimulation usually causes an activity that prepares the body for an emergency situation.

CHAPTER 10

THE SPECIAL SENSES AND FUNCTIONAL ASPECTS OF THE NERVOUS SYSTEM

CHAPTER OUTLINE

SENSORY FUNCTIONS
Sensory Pathways
General Senses
Special Senses

INTEGRATIVE FUNCTIONS
Functional Regions of the Cerebral Cortex
Thought and Memory
Emotions: The Limbic System

MOTOR FUNCTIONS
Motor Origins
Motor Pathways

LEARNING OBJECTIVES

After studying this chapter, you should be able to:

1. Define the term *sensation*, and describe the features that characterize receptors.

2. Identify the components in a general and in a special sensory pathway.

3. Describe the structure and function of the general sensory receptors.

4. Identify the four special senses.

5. Describe the special sensory organ of smell and the olfactory pathway.

6. Describe the special sensory organs of taste and the gustatory pathway.

7. Identify the accessory structures and components of the eye.

8. Explain how the eye accommodates to near and far vision, and how light is converted to a nerve impulse.

9. Describe the visual pathway.

10. Identify the components of the ear.

11. Trace the path of sound waves through the ear to the generation of nerve impulses that reach the brain.

12. Identify the functional areas of the cerebral cortex.

13. Describe how thought, memory, and emotions are formed in the brain.

14. Identify the origins of motor impulses.

15. Distinguish between the major motor pathways.

As we saw in the previous chapter, the overall function of the nervous system is the maintenance of homeostasis. This is achieved by three main activities: monitoring the body and the external environment for changing conditions is the nervous system's **sensory** function; processing this sensory information to coordinate the proper response is its **integrative** function; and stimulating the response is its **motor** function. These three functions are made possible by the rapid conduction of impulses along nerve pathways in the substance of the brain, the spinal cord, and the peripheral nerves.

In this chapter we will examine the sensory, integrative, and motor functions of the nervous system. The special structures that are associated with these functions will also be studied, such as special sensory organs, functional parts of the brain, and motor pathways. It is important to have read Chapter 9 beforehand so that you will understand the structural relationships of the nervous system when these topics are discussed.

SENSORY FUNCTIONS

The sensory functions provide the brain with important information on conditions of the body in order to maintain homeostasis. They are performed by general and special sensory structures, which respond to changes in the environment.

Each time you open your eyes, hear a sound, feel pressure or warmth, or feel your stomach ache from hunger, you are sensing the world around and within you. These experiences are called **sensations**; a sensation is a state of awareness of the external or internal conditions of the body. Sensations are vital for your survival, for they provide the brain with information that is necessary to maintain the homeostatic balance. Without them, you would not see or hear that automobile rushing toward you; you would not know you are hungry and must eat to live another day; you would not feel the snow against your bare skin and know you must keep warm or otherwise freeze.

Sensory Pathways

The receptor is the origin of all sensory pathways. A sensory pathway extending from a receptor all the way to the brain consists of at least three neurons.

All sensory pathways begin with a **stimulus**, or change in the environment that is great enough to initiate a nerve impulse. The stimulus is converted to a nerve impulse by a **receptor**. From the receptor, the impulse is conducted along a sensory neuron to the central nervous system.

Receptors

As we discovered in the previous chapter, sensory receptors are variable in structure. For example, a receptor may be the simple dendrites of a sensory neuron that end bare in the skin, or it may be a complex organ such as the eye or ear. Despite the variation, however, all receptors share in common the capability to *excite*. That is, they are capable of generating an action potential (nerve impulse). This event occurs when a stimulus is received that is greater than threshold (the minimum stimulus able to generate an action potential).

The threshold level for each type of receptor is typically very low for one type of stimulus yet quite high for all other types of stimuli. In other words, receptors are sensitive to a particular environmental change but insensitive to all others. For example, the receptors in the eye have a low threshold to light and will initiate a nerve impulse in response to it, but have a very high threshold to slight changes in temperature or chemicals and will not respond to them. Thus, receptors are *stimulus-specific*.

In many receptors, the threshold level for a particular stimulus may rise after continuous stimulation. This phenomenon is called **sensory adaptation**. In these receptors the impulses are generated at decreasing rates until they stop completely and sensation ends. Sensory adaptation is experienced when you put on clothes; as you dress you can feel the clothes against your skin, but this sensation soon disappears as your touch receptors adapt to the stimuli.

Receptors are classified according to their types of sensitivity. **Mechanoreceptors** detect a mechanical or physical change in the receptor or nearby cells. They are sensitive to touch, pressure, muscle tension, hearing, equilibrium, and blood pressure. **Thermoreceptors** detect temperature changes, and **nociceptors** (NŌ-sē-sep-ters) detect pain that usually results from chemical or physical damage to nearby cells. **Photoreceptors** are sensitive to changes in the amount of light and are present only in the retina of the eye. **Chemoreceptors** detect chemicals dissolved in fluid, providing the senses of smell and taste. They also provide the detection of oxygen and carbon dioxide levels in the blood.

General Sensory Pathways

Conduction pathways that carry impulses from a simple receptor—such as those found in the skin, visceral organs, and muscles—to the brain are known as **general sensory pathways**. From the receptor, the impulse is conducted along three sensory neurons before it reaches its destination in the brain (Fig. 10–1a).

A single sensory neuron connects the receptor to the spinal cord. This is the **first-order neuron**, and it may terminate at the spinal cord or continue upward to terminate at the medulla oblongata. In either case, it synapses with a **second-order neuron** that conducts the impulse upward to the thalamus. From the thalamus, a **third-order neuron** conducts the impulse to the cerebral cortex for processing.

Special Sensory Pathways

Conduction pathways that carry impulses from complex receptors, such as those found within special sensory organs like the eye and ear, are called **special sensory pathways**. They exhibit more variation in the number of neurons than general sensory pathways.

Basically, a special sensory pathway includes *at least* three sensory neurons that connect a receptor with a specific region of the cerebral cortex (Fig. 10–1b). Since the special senses are located in the head (eye, ear, tongue, and nasal epithelium), the impulses travel

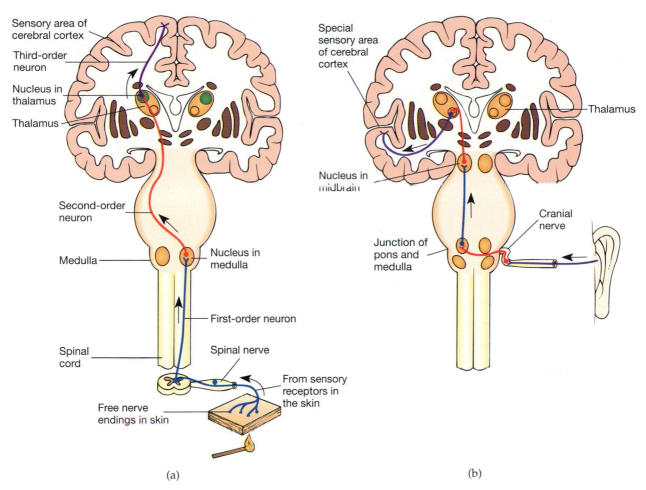

Figure 10–1

Sensory pathways. (a) An example of a general sensory pathway from a skin receptor to the cerebral cortex of the brain. This pathway contains three sensory neurons that carry the signal. (b) A special sensory pathway from a special sense organ to the cerebral cortex. The special sense organ in this example is the ear (the auditory pathway).

How many sensory neurons are involved in a general sensory pathway?

along cranial nerve routes to the brain. In most cases, impulses entering the brain pass through the thalamus before continuing to the cerebral cortex.

CONCEPTS CHECK

1. Why are sensations important tools for survival?

2. What are the different types of receptors, based on their sensitivity?

3. How many neurons are in a general sensory pathway to the brain?

General Senses

The general senses are sensations detected by simple receptors. They are touch and pressure, temperature, pain, and body position.

The simple receptors located in the skin, muscles, and visceral organs detect the general senses. These senses include those of touch and pressure, temperature, pain, and body position. Impulses travel from these receptors along general sensory pathways to the cerebral cortex, where the sensations are interpreted and processed.

Touch and Pressure

The sensations of touch and pressure are often called *cutaneous sensations*, because many of their receptors are located within the skin. They are detected by mechanoreceptors, which are widely distributed.

The sense of touch is primarily detected by specialized receptors in the skin, called **Meissner's corpuscles** (Fig. 10–2a). They are small, oval capsules of connective tissue that contain dendrites of two or more sensory neurons. When the connective tissue is moved even slightly by a mechanical disturbance, it contacts the dendrites, which respond by generating a nerve impulse. Meissner's corpuscles are abundant in the skin of the fingers, palms, soles, lips, and external genitals. A second, less common set of touch receptors in the skin are **Merkel's discs** (Fig. 10–2b), whose dendrites end in the skin's epidermis.

The sense of pressure requires a greater mechanical disturbance than touch and is longer-lasting and felt over a larger area. Its receptors are located within the deep regions of the skin, around joints and tendons, and in certain visceral organs, and they respond only to heavy pressure. These receptors are known as **Pacinian corpuscles** (Fig. 10–2d). They consist of a knoblike ending of a single sensory neuron surrounded by layers of connective tissue that resemble the layers of an onion.

Temperature

The sensation of temperature enables us to distinguish between heat and cold. Although it is poorly understood, it is thought to be detected by free nerve endings in the skin that are stimulated by temperature changes (Fig. 10–2c). As extreme temperatures are reached (less than 10°C and more than 45°C), pain receptors are also triggered, producing a burning sensation.

Pain

The sensation of pain is a "necessary evil," for it provides an important protective function. It is our early warning system that informs the brain of a homeostatic imbalance that must be dealt with. Although pain is unpleasant, it provides an opportunity to remove the source that is disturbing homeostasis by making us aware of it.

Pain is detected by branching dendrites of sensory neurons that end freely throughout the skin, muscles, and most visceral organs (Fig. 10–2c). It is currently thought that these dendrites are sensitive to chemicals that are released during an event involving the damage or destruction of cells. As greater numbers of cells are damaged or destroyed, more chemicals are released to stimulate more intense pain. A restricted blood flow to a body part may also stimulate pain receptors.

Figure 10–2
Receptors of the general senses.
Where can Meissner's corpuscles and Pacinian corpuscles be found?

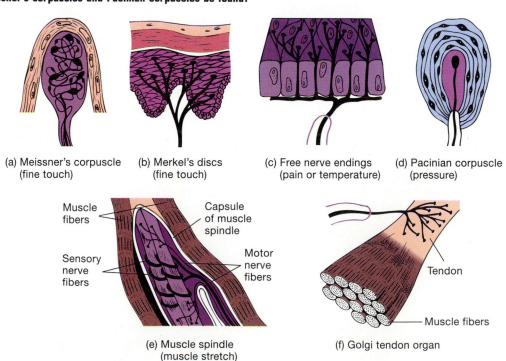

(a) Meissner's corpuscle (fine touch)

(b) Merkel's discs (fine touch)

(c) Free nerve endings (pain or temperature)

(d) Pacinian corpuscle (pressure)

Muscle fibers

Capsule of muscle spindle

Sensory nerve fibers

Motor nerve fibers

(e) Muscle spindle (muscle stretch)

Tendon

Muscle fibers

(f) Golgi tendon organ

Pain may be generated by receptors other than the specialized pain receptors. In this case, the source of pain is the excessive stimulation of any type of receptor. For example, staring into a bright light will hurt your eyes: the pain is generated by the photoreceptors in the eyes, not by pain receptors.

Pain that is generated by pain receptors located in the walls of visceral organs is known as **visceral pain**. These receptors respond differently from pain receptors closer to the body's surface. They tend to respond only to widespread disturbances, giving you sensations such as stomach or intestinal cramps, heartburn, headaches, and uterine cramps.

Visceral pain is also more difficult to trace to its source. For example, pain originating from the heart is felt along the left arm or shoulder. This phenomenon is known as **referred pain** (Fig. 10–3). This occurs because major nerve pathways that share impulses from different parts of the body are not easily distinguished by the brain. In our example, heart pain is referred to the left shoulder and arm because their nerve pathway to the brain is shared by the heart.

The conduction pathway of an impulse traveling from the source of pain extends along sensory neurons to the cerebral cortex, where recognition of the intensity of pain occurs. Fortunately, it is possible to reduce the intensity of pain by interfering with the transmission of impulses along this pathway. This may be done by a variety of techniques, including analgesic (pain-reducing) drugs and anesthetic drugs (to produce an absence of sensation, including pain), surgery, controlled breathing, acupuncture, hypnosis, massage, and biofeedback.

Body Position

You are made aware of the position of your body by the activities of muscles, tendons, joints, and equilibrium centers in the inner ear. The receptors that detect these activities are called **proprioceptors** (PRŌ-prē-ō-sep-ters). Specifically, proprioceptors provide information on the degree of muscle contraction, the amount of tension in tendons, the position of a joint, and the position of the head relative to the ground. They enable you to control your body movements without using your sense of vision, hearing, touch, or any other senses.

The primary receptors of body position are **muscle spindles** in skeletal muscle and **tendon organs** between a tendon and a skeletal muscle (see Fig. 10–2e and f). Muscle spindles are specialized muscle fibers

Figure 10–3
Areas of referred pain. The colored areas indicate locations on the skin surface where pain from internal organs can be felt. (a) Anterior. (b) Posterior.
Where would you experience pain sensations that result from a punctured lung?

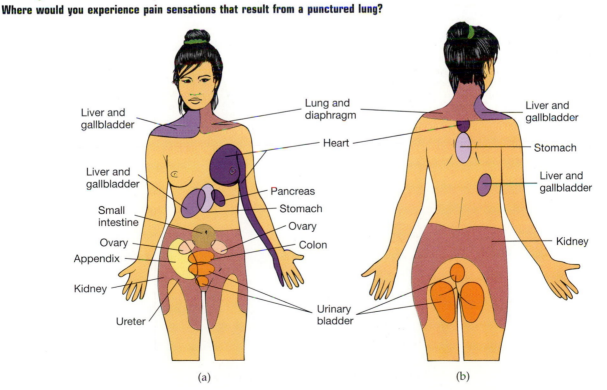

(a)   (b)

that are associated with sensory neurons. They are stimulated by muscle stretch and can monitor the actual length of a skeletal muscle during its contraction cycle. Tendon organs consist of two or more sensory neurons whose dendrites terminate within a connective tissue capsule. They are stimulated by tension in the surrounding tendon. Other proprioceptors are located in and around joints and within the inner ear.

CONCEPTS CHECK

1. What are the receptors that are sensitive to touch and pressure called, and where are they located?

2. How is the sensation of pain stimulated?

3. What do you suppose would happen if you did not have pain receptors?

4. How are muscle spindles and tendon organs stimulated?

Special Senses

Each special sense is detected by receptors located within supportive tissue. These receptors are highly specialized to respond to one type of stimulus.

The four special senses are smell, taste, sight, and hearing. Each special sense is detected by sensory receptors that are assisted and supported by other tissues. These specialized, complex structures are true organs since they are composed of two or more tissue types. They are, therefore, properly termed *special sensory organs*.

Smell

The sense of smell is also known as **olfaction**. It is detected by thousands of chemoreceptors located in the upper wall of the nasal cavities (Fig. 10–4). These receptors are bunched into a small space that measures about $\frac{1}{2}$ square inch, and are associated with other cells to form the **olfactory organs**. Because olfactory

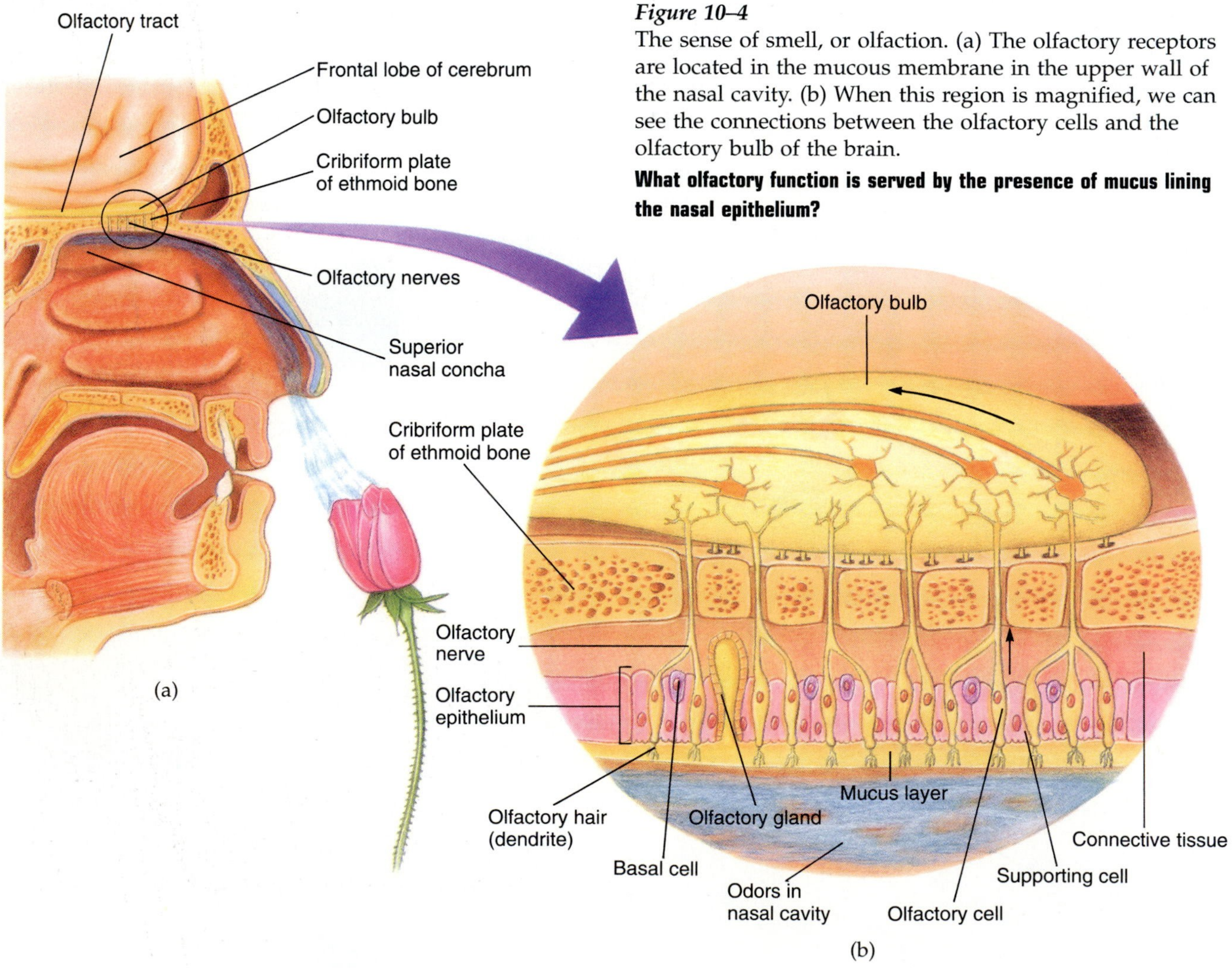

Figure 10–4
The sense of smell, or olfaction. (a) The olfactory receptors are located in the mucous membrane in the upper wall of the nasal cavity. (b) When this region is magnified, we can see the connections between the olfactory cells and the olfactory bulb of the brain.
What olfactory function is served by the presence of mucus lining the nasal epithelium?

Olfactory tract

Frontal lobe of cerebrum

Olfactory bulb

Cribriform plate of ethmoid bone

Olfactory nerves

Superior nasal concha

Cribriform plate of ethmoid bone

Olfactory nerve

Olfactory epithelium

Olfactory hair (dendrite)

Basal cell

Odors in nasal cavity

Olfactory gland

Olfactory cell

Mucus layer

Supporting cell

Connective tissue

Olfactory bulb

(a)

(b)

organs are located high in the nasal cavities out of the usual pathway of inhaled air, they are in a poor position to collect scents. This is why forcing more air through the nose by "sniffing" maximizes your sensitivity to smells. The olfactory organs provide many individual sensations of smell, perhaps 50 or more. The actual number varies among individuals, and declines with increasing age.

The olfactory receptor cells are neurons embedded within the mucous membrane of the nasal epithelium (Fig. 10–4). Each neuron is surrounded by columnar epithelial cells that provide support. At the free end of the olfactory cells are dendrites, or **olfactory hairs**, which extend beyond the epithelial layer and into a layer of mucus. At their basal end, the axon from each neuron extends through the cribriform plate of the ethmoid bone. The axon terminates at the olfactory bulb of the brain.

Olfactory Pathway

The conduction pathway for the sense of smell begins with the olfactory hairs of the receptor cells. When you inhale, gas molecules enter the nasal cavity and become dissolved in a layer of mucus. If these dissolved molecules reach the olfactory organs, the olfactory hairs generate an action potential that is conducted along the receptor cell through the nasal epithelium and cribriform plate into the cranial cavity. The impulse is transmitted to sensory neurons in the **olfactory bulbs**, which are a pair of swellings at the origin of the olfactory nerves (cranial nerve I) at the base of the brain. From the olfactory bulb, the impulse continues along an olfactory nerve to the frontal lobes of the cerebral cortex, where interpretation occurs.

Olfactory pathways are closely linked to the limbic system of the brain, which is the center for emotional expression. Olfactory sensations are therefore long-lasting and form a very important part of our memories and emotions. For example, a particular smell sensation sometimes has enough impact to sweep us back to a memory of an experience thought to be long-forgotten.

Taste

The sense of taste is called **gustation**. It is closely associated with the sense of smell, for both depend on chemoreceptors, and the two senses often operate simultaneously (such as when you are eating). The special organs of taste, known as **taste buds**, are scattered throughout the oral cavity. Most of the 10,000 or so taste buds are located on the surface of the tongue within small elevations called **papillae** (pa-PIL-ē). Others may be found in the roof of the mouth and the walls of the pharynx.

Each taste bud consists of a group of taste receptor cells, known as **gustatory cells**, that are surrounded by a capsule of supportive epithelial cells (Fig. 10–5).

Figure 10–5

The sense of taste, or gustation. (a) One of thousands of taste buds located on the tongue's surface. (b) A photomicrograph of a single taste bud.

What are the taste bud cells that are sensitive to dissolved chemicals called?

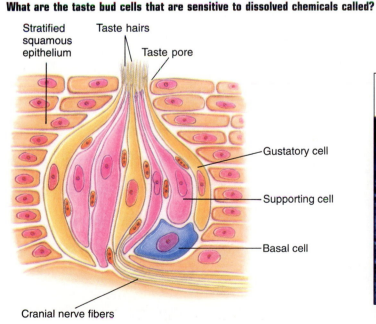

Stratified squamous epithelium
Taste hairs
Taste pore
Gustatory cell
Supporting cell
Basal cell
Cranial nerve fibers

(a)

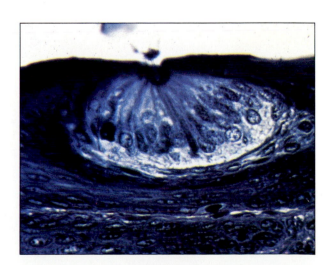

(b)

The free ends of the gustatory cells have microvilli, called **taste hairs**, that project through an opening in the taste bud known as a **taste pore**. The taste pore opens onto the surface of the tongue. At the base of the gustatory cells is a network of sensory nerve fiber endings.

There are four primary types of taste sensations, each of which is associated with a particular type of taste bud. The four taste sensations and the chemicals that stimulate them include:

1. Sweet
2. Sour
3. Bitter
4. Salty

All other taste sensations, such as coffee, cinnamon, garlic, and pepper, are combinations of the four primary tastes that may be modified by olfactory sensations. Each of the four types of taste buds is concentrated in a region on the surface of the tongue. These regions can be seen in Figure 10–6, which shows a taste "map" of the taste receptor distribution that is common to all of us.

Gustatory Pathway

When chemicals in your mouth are dissolved by saliva, they combine with taste hairs in the gustatory cells and cause a change. If this change is great enough, it leads to an action potential in the sensory nerve fibers at the bases of the cells. The actual mechanism of this exchange is poorly understood. From here the impulses travel on fibers of the facial, glossopharyngeal, and vagus nerves (cranial nerves VII, IX, and X) to the medulla oblongata. The impulses then pass to the thalamus, where they are directed to the gustatory center in the cerebral cortex.

Sight

The sense of sight, or **vision**, is the most complex sensory system of the body. In fact, of all sensory receptors in the body, nearly 70% are photoreceptors that detect light. These photoreceptors lie within the organs of sight, the **eyes** or **eyeballs.** In addition to the eyes are accessory structures that aid in vision.

Accessory Structures

The accessory structures to the eye include the eyelids, lacrimal apparatus, and extrinsic muscles of the eyeball. The lacrimal apparatus, the extrinsic muscles, and the eyeball itself are located within the orbit of the skull.

The **eyelids** protect the anterior surfaces of the eyes (Fig. 10–7). They meet at the medial and lateral corners of each eye and contain hairs extending from

Figure 10–7
Accessory structures of the eye, anterior view. The lacrimal gland, lacrimal canal, and nasolacrimal duct are shown superimposed through the skin and bone of the face. The arrows identify the direction of tear flow.
Why are tears continually secreted?

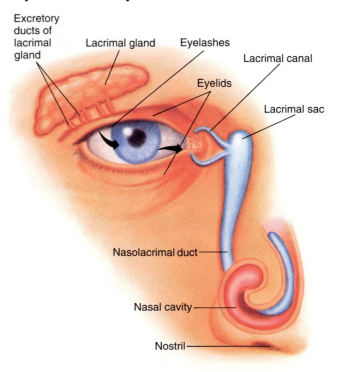

Figure 10–6
A "map" showing the locations of the four types of taste buds on the tongue: bitter, sour, salt, and sweet.
Taste buds that are sensitive to what primary sensation are located on the tip of the tongue?

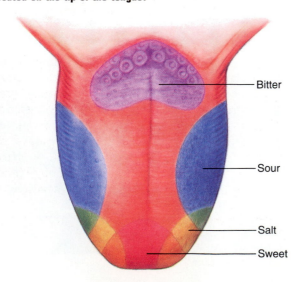

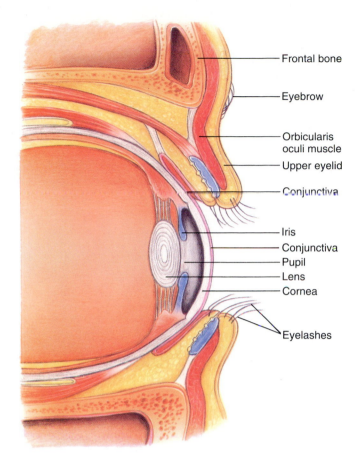

Figure 10–8
Accessory structures of the eye, lateral view. The eye and its orbit have been sagittally sectioned to reveal internal structures. Note the location of the conjunctiva in this drawing.

Does the conjunctiva cover any part of the eyeball?

each margin, known as **eyelashes**. Each eyelid consists of four layers: a thin outer skin layer, a skeletal muscle layer, a connective tissue layer, and an inner mucous membrane called the **conjunctiva**[1] (kon'-junk-TĒ-va) (Fig. 10–8). The conjunctiva folds back to cover much of the anterior surface of the eye as well. It secretes mucus to help moisten and lubricate the eyeball.

The **lacrimal** (LAK-ri-mal) **apparatus** associated with each eye consists of the lacrimal gland, lacrimal sac, and nasolacrimal ducts (Fig. 10–7). The **lacrimal gland** secretes tears, which are a dilute salt solution that is released continuously. Tears also contain an antibacterial enzyme, which fights against infection as the eye is moistened and lubricated. As the tears flush across the surface of the eye, they are collected by **lacrimal canals** that channel into the **lacrimal sac** located within the lacrimal bone. From here they pass into the nasal cavity by way of the **nasolacrimal duct**.

Table 10–1 EXTRINSIC MUSCLES OF THE EYE

Muscle	Controlling Cranial Nerve	Action
Lateral rectus	Abducens (VI)	Moves eye laterally
Medial rectus	Oculomotor (III)	Moves eye medially
Superior rectus	Oculomotor (III)	Elevates or rolls eye upward
Inferior rectus	Oculomotor (III)	Depresses or rolls eye downward
Inferior oblique	Oculomotor (III)	Elevates and turns eye laterally
Superior oblique	Trochlear (IV)	Depresses and turns eye laterally

The **extrinsic muscles** of the eye are skeletal muscles that move the eyeball. They originate from the walls of the orbit and insert upon the tough outer surface of the eyeball. There are six extrinsic muscles, which are listed in Table 10–1 with their sources of innervation and action; they are shown in Figure 10–9.

Structure of the Eye

The eye is a spherical structure about 2.5 cm (1 inch) in diameter. Its wall consists of three distinct layers, or tunics. These are the outer fibrous tunic, the middle vascular tunic, and the inner nervous tunic. Inside the eye are structures that divide the eye into fluid-filled compartments.

FIBROUS TUNIC   The **fibrous tunic** is the thick outermost layer of the eyeball. It contains two regions: the posterior sclera, and the anterior cornea (Fig. 10–10).

The **sclera**[2] (SKLE-ra) forms most of the fibrous tunic. It is composed of white fibrous connective tissue and is often called the "white of the eye." It forms a thick, tough protective layer that provides shape to the eyeball and protects its inner parts. It contains an abundant supply of blood vessels, some of which may come into view when the eyes are irritated. The posterior surface of the sclera is penetrated by the optic nerve.

The **cornea**[3] is the anterior, transparent part of the fibrous tunic that bulges outward slightly. It is the "window" of the eye, as light must pass through it before entering the internal structures and cavities. Its

---

[1]Conjunctiva: Latin *conjungere* = to bind, or join together.

[2]Sclera: Greek *sclera* = hard or tough.

[3]Cornea: Latin *cornea* = horny coat.

Labels (Figure 10–8): Frontal bone, Eyebrow, Orbicularis oculi muscle, Upper eyelid, Conjunctiva, Iris, Conjunctiva, Pupil, Lens, Cornea, Eyelashes

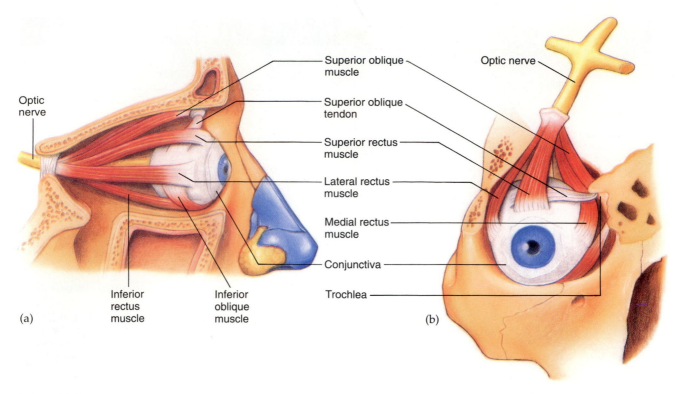

Figure 10–9
Extrinsic muscles of the eye. (a) Lateral view. (b) Anterior superior view.

The contraction of which extrinsic muscle rotates the eye upward?

transparency is due to its lack of blood vessels and its regular arrangement of protein fibers.

VASCULAR TUNIC   The **vascular tunic** is so called because of its abundance of blood vessels, which supply nourishment to numerous structures of the eye. Its most prominent components include the choroid, the ciliary body, and the iris (Fig. 10–10). Associated with the vascular tunic is the lens.

The **choroid**[4] (KŌ-royd) is a thin, dark-brown membrane that lines most of the internal surface of the sclera. The brown pigment it contains absorbs light rays and thereby minimizes the amount of reflection from the eye. This function is important, because reflection tends to distort images and reduce the eye's sensitivity to light. The blood vessels contained within the choroid serve to nourish the retina, which lies internally.

Anterior to the choroid, the vascular tunic is modified to form two structures composed of smooth muscle: the ciliary body and the iris. The **ciliary body** continues from the choroid to become the thickest part of the vascular tunic. It consists of smooth muscle fibers, which connect to the lens by way of suspensory liga-

ments. Attached to the inner margin of the ciliary body is the doughnut-shaped **iris**,[5] which you may recognize as the beautifully colored part of the eye that can be seen from the exterior. It is suspended between the cornea and the lens. The contraction of its muscle fibers changes the diameter of the opening in its center to regulate the amount of light entering the inner eye cavity. This opening is the **pupil**, which appears as a black spot in the center of your eye.

The **lens** is located immediately behind the pupil and iris. In young, healthy eyes it is perfectly transparent, although this transparency is gradually lost with advancing age (a condition called a *cataract*). Internally, it is composed of interlocking strands of transparent cytoplasm that originate from epithelial cells. Surrounding this clear, jellylike interior is an elastic capsule that is also transparent. The lens is held in position by suspensory ligaments, which extend from the muscles of the ciliary body to the capsule. These ligaments are held under constant tension by the ciliary muscles, although the amount of tension varies as your eye changes focus. When tension is relaxed, the elastic capsule expands and the lens becomes more convex. This occurs when the lens focuses on a close object.

[4]Choroid: *choroid* = membranelike.

[5]Iris: Greek *iris* = rainbow.

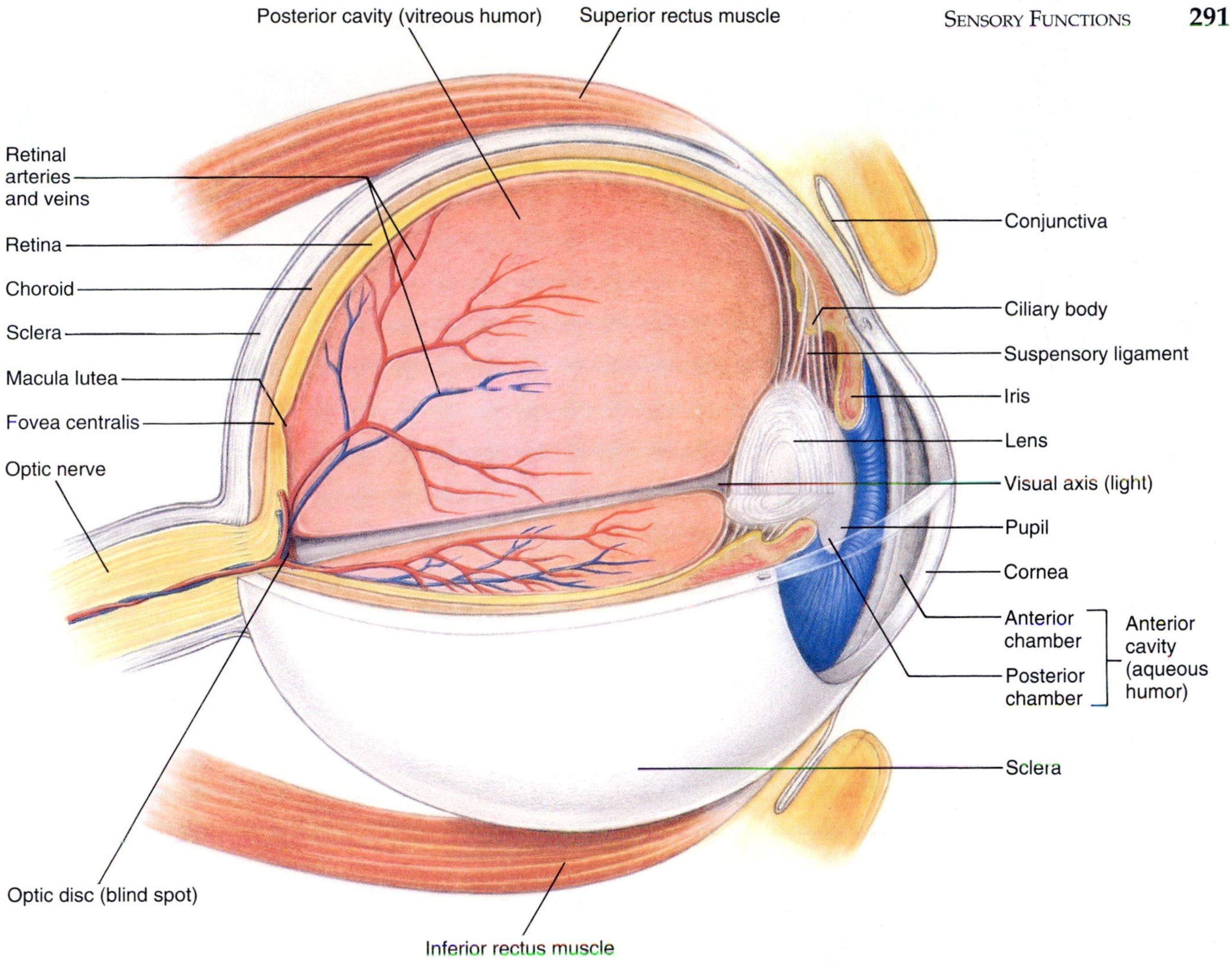

Posterior cavity (vitreous humor)   Superior rectus muscle

Retinal
arteries
and veins

Retina

Choroid

Sclera

Macula lutea

Fovea centralis

Optic nerve

Conjunctiva

Ciliary body

Suspensory ligament

Iris

Lens

Visual axis (light)

Pupil

Cornea

Anterior chamber | Anterior cavity (aqueous humor)

Posterior chamber

Sclera

Optic disc (blind spot)

Inferior rectus muscle

Figure 10–10
Structure of the eye. In this lateral view, the eye is shown partially sectioned along
the sagittal plane to show its internal cavities and components.
What structures do the suspensory ligaments attach to?

When tension is increased, the capsule flattens and the lens becomes less convex. This occurs when the lens focuses on a distant object. The alteration of lens shape for near and far vision is called **accommodation**.

The lens provides a physical separation between the two main compartments of the eye (Fig. 10–10). The anterior compartment is divided by the iris into two chambers: the **anterior chamber**, between the cornea and the iris; and the **posterior chamber**, between the iris and the lens. Within these two chambers a clear, watery fluid called the **aqueous humor**[6] circulates and is continuously recycled through the bloodstream. The posterior compartment behind the lens is known as the **posterior cavity**. It contains a thickened,

gel-like fluid that helps support the structure of the eyeball, known as the **vitreous humor**.[7] Unlike the aqueous humor, the vitreous humor is not continuously recycled.

NERVOUS TUNIC   The **nervous tunic** consists of the **retina**,[8] which is a thin, fragile layer of neurons that forms the inner lining of the eyeball's posterior wall. The retina contains its own supply of blood vessels, although most of its nourishment is provided by the choroid located external to it. The function of the retina is the detection of light and subsequent transport of nerve impulses to the optic nerve, which penetrates it.

[6]Aqueous humor: *aqueous* = watery + *humor* = fluid.

[7]Vitreous humor: *vitreous* = glassy, or glasslike + *humor*.

[8]Retina: *retina* = net.

Glaucoma and Intraocular Pressure

Aqueous humor is continually produced in the ciliary body in the posterior chamber and flows into the anterior chamber. In healthy eyes, it is drained through an expanded vein called the canal of Schlemm, where it reenters the bloodstream. Poor drainage of aqueous humor causes a painful elevation of pressure within the eye, called *chronic glaucoma*, which can damage the optic nerve if the pressure is not relieved. Eyedrops are available to relieve intraocular pressure by reducing the production rate of aqueous humor, or increasing its drainage. A total block of aqueous humor drainage may also occur if the anterior chamber is narrow: when the pupil dilates, the iris blocks the canal of Schlemm. This painful condition is called *acute glaucoma* and requires immediate treatment by surgical incision, which creates a drainage hole through the base of the iris. This procedure is known as an *iridectomy*.

The retina consists of three distinct layers of neurons (Fig. 10–11). The layer nearest the choroid contains the neurons that are specialized to respond to light, the **photoreceptor cells**. They are of two types: **rod cells**, each of which contains an elongated, cylindrical dendrite; and **cone cells**, whose dendrites are tapered to a point like cones. Rod cells are not color-sensitive but are sensitive to very small levels of light, while cone cells are color-sensitive and require more light but provide a sharper image. Both cells respond

Figure 10–11

The retina. This thin sheet of nerve tissue forms the inner lining of the eye's posterior wall. It is composed of three layers of neurons: photoreceptor cells (nearest the choroid), bipolar cells, and ganglion cells. The axons from the ganglion cells converge to exit from the retina by way of the optic nerve (at the optic disc).

What sensory neurons in the eyeball generate an action potential and conduct it through the optic nerve?

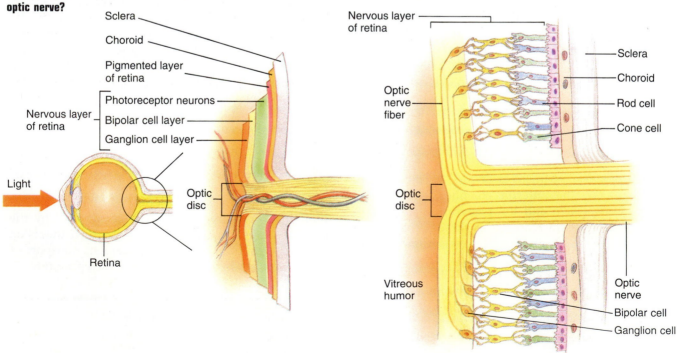

to light by initiating a signal, which is transmitted to the cells forming the middle layer of the retina, the **bipolar neurons**. Bipolar neurons transmit the signal to the inner layer, which is composed of multipolar neurons with large cell bodies, called **ganglion cells**. These cells contain long axons that converge to form the optic nerve, which emerges from the back of the eyeball and extends to the brain.

At a point near the center of the retina is the area where the axons from the ganglion cells converge to form the optic nerve. Because there are no photoreceptor cells here, images focused upon the area cannot be seen. This is your "blind spot," which is also known as the **optic disc** (Figs. 10–11 and 10–12). Slightly lateral to the optic disc, nearer to the center of the retina, is a yellowish spot called the **macula lutea**[9] (MAK-yoo-la LOO-tē-a) (Fig. 10–12). In its center is a depression that contains a concentration of cone cells (only) known as the **fovea centralis**[10] (FŌ-vē-a sen-TRA-lis). An image focused upon this depression is visualized with optimal sharpness, or *visual acuity*.

Pathway of Light Through the Eye

Light rays passing through the eye change as they leave the medium of air and enter the media of eye structures and fluids. Air, solids, and fluids have different densities, causing light to change in speed as it passes from one medium to the next. As light rays change in speed, their direction changes, causing them

to bend. The bending of light rays is called **refraction**. Thus, light rays are refracted as they pass through the cornea, the aqueous humor, the lens, and the vitreous humor of the eye (Fig. 10–13).

The amount of refraction that occurs at each surface of the eye is very precise. It is constant for all structures but the lens, which, as we have seen, has the ability to change its shape in order to move the visual focus—a feature called **accommodation**. A normal, or *emmetropic*, eye is an eye that is able to refract light rays from an object to focus a clear image onto the retina. However, visual problems occur when a lens is too strong or too weak, or when structural problems of the eyeball are present (Fig. 10–13). For example, when the amount of refraction is too great (or the eyeball is anatomically too long) and an image is focused in front of the retina, distant objects become blurred. This condition is called **myopia**[11] (mī-Ō-pē-a), and the individual is said to be *nearsighted* because close objects can still be focused on normally. If the image focuses behind the retina, the condition is called **hyperopia**[12] and the individual is *farsighted*, because distant objects can still be focused on properly. Hyperopia results from a weak or "lazy" lens or an eyeball that is too short. Unequal curvatures of the cornea or lens lead to **astigmatism**,[13] which causes blurred vision for near and far objects.

The combined refractory effects of the different parts of the eye form an image on the retina that is quite different from what your brain interprets (Fig. 10–13). The image is inverted (upside down), reduced in size, and reversed from left to right. We do not see this image, because the brain learns early in life to coordinate visual images with the locations of objects in the real world. As infants, we touch and pick up objects repeatedly until the brain learns to recognize the difference between retinal images and reality. The brain, therefore, rearranges the images very quickly for us so that we do not perceive an upside-down world.

Physiology of Vision

Once an image is formed on the retina, it must be converted from a pattern of light rays to nerve impulses. This is initiated by the rods and cones, which are activated by light to increase their rate of neurotransmitter release.

Figure 10–12
Photograph of a retina taken with an instrument called an ophthalmoscope.

Which region in this photomicrograph contains a concentration of cone cells?

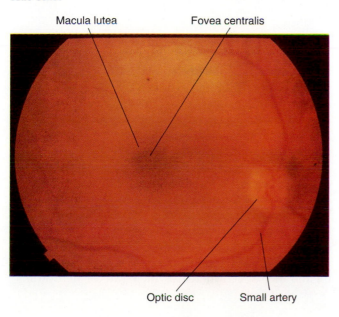

Macula lutea   Fovea centralis

Optic disc   Small artery

---

[9]Macula lutea: Latin *macula* = spot or point + *lutea* = yellow.

[10]Fovea centralis: Latin *fovea* = pit + *centralis* = central.

[11]Myopia: Greek *myein* = to close + *-opia* = condition of the eye.

[12]Hyperopia: Greek *hyper* = above + *-opia*.

[13]Astigmatism: Greek *a-* = without + *stigma* = point + *-ism* = a condition of.

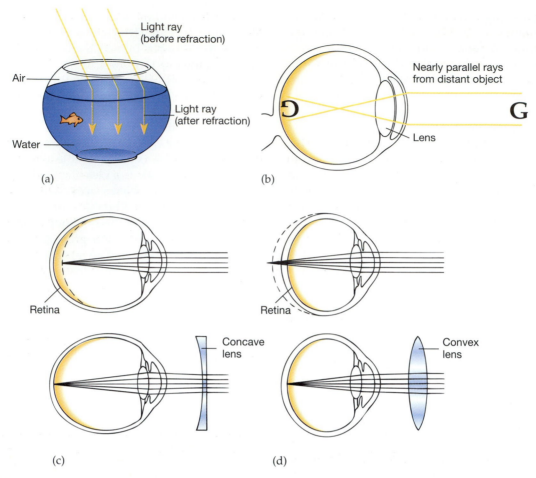

Figure 10–13

The refraction of light and visual accommodation. (a) Light rays are bent (refracted) as they pass from air into a liquid medium such as water. (b) When you view an object, light rays are refracted as they pass through the thick medium of the eyeball. As a result, the image is completely inverted on the retina. (c) A condition of myopia occurs when light rays from distant objects become focused in front of the retina. Myopia may be corrected by a concave lens that diverges light rays before they enter the eye. (d) Hyperopia occurs when light rays from distant objects are focused behind the retina. This condition may be corrected by a convex lens that converges the light rays as they enter the eye.

What structures of the eyeball bend light as it passes through?

Rods and cones play different roles in the perception of light. Rods are extremely sensitive to very small amounts of light. They are, therefore, useful in night or dim vision, when little light is available, and are not sensitive to colors. They limit your vision to general outlines of figures, for their route to the brain is usually shared by a common nerve fiber so that the brain's ability to interpret the image sharply is reduced. Cone cells, on the other hand, do not share common routes as often as rod cells and thus enable the brain to interpret images more sharply. They are, therefore, important for visual acuity, and are sensitive to color.

Both rods and cones employ light-sensitive pigments in the perception of light. These pigments break apart when they absorb energy from light, and it is their decomposition that leads to the generation of the action potential. In rods, the pigment is called **rhodopsin**, or visual purple. Rhodopsin is composed of a protein and a substance called *retinene*, which is synthesized from vitamin A. A deficiency of vitamin A in the diet leads to a condition called *night blindness*, due to a decline in sensitivity of rod cells and their subsequent inability to respond to dim light.

In the presence of even a very small amount of light, rhodopsin molecules break apart to activate an enzyme that triggers a series of reactions. These reactions lead to a change in permeability of the rod cell membrane, which increases their rate of neurotransmitter release, leading to the generation of a nerve impulse (in ganglion cells). Soon after its degradation, rhodopsin is resynthesized in preparation for the next light ray. The time during which rhodopsin is regenerated represents a period of insensitivity, explaining why you are unable to see for several seconds when entering a dark room from a well-lit room.

Cone cells contain pigments that are very similar to rhodopsin. These pigments also contain retinene, but the nature of the protein component differs from that of rod cells. In fact, there are three different types of cone cells, each of which contains a different protein component in the pigment. It is understood that each of these types of cone cells is sensitive mainly to one of three colors: red, green, or blue. Regardless of the protein component, the mechanism of cone cell activation is the same as in rod cells: decomposition of the pigment activates an enzyme, triggering a series of reactions that lead to a change in cone cell membrane permeability and—eventually, in ganglion cells—the action potential.

Visual Pathway

The nerve pathway for the perception of sight begins with the rods and cones in the retina (Fig. 10–11). From these photoreceptors, the stimulus passes to bipolar neurons and then to ganglion cells, both still within the retina. The ganglion cells generate an action potential if the stimulus is great enough. If so, the action potential passes along the axons of ganglion cells, which exit the eyeball by way of the optic nerve (cranial nerve II). The nerves from portions of each eyeball cross just anterior to the pituitary gland at the base of the brain, forming the X-shaped **optic chiasma** (Fig. 10–14).[14] Within the chiasma, some of the nerve fibers cross: those originating from the medial side of the retina cross, while those from the lateral side do not. The optic tract extending from the chiasma on the right side therefore contains fibers from the medial half of the left eye and lateral half of the right; the left optic tract contains fibers from the lateral half of the left eye and medial half of the right. Both optic tracts extend to the thalamus, with some of their fibers terminating in nuclei that control eye reflexes. From the thalamus, the impulses enter visual pathways that lead to the occipital lobes of the cerebral cortex. These centers, called the **visual cortex**, are where interpretation of visual signals takes place.

Hearing

The sense of hearing, or **audation**, is a very specialized sense. It is combined with the sense of **equilibrium**, which we will examine later in this section. Both senses are detected by mechanoreceptors located deep within the ear.

---

[14]Optic chiasma: *optic* = of the eye, referring to sight + *chiasma* = cross.

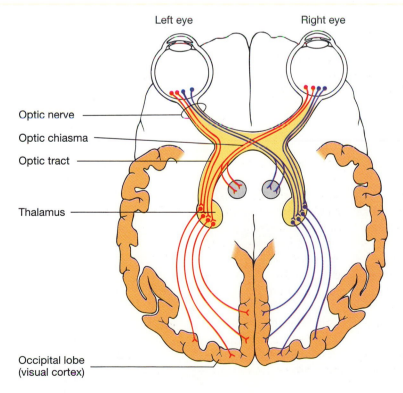

Left eye   Right eye

Optic nerve

Optic chiasma

Optic tract

Thalamus

Occipital lobe
(visual cortex)

Figure 10–14
The visual pathway. From this inferior view of the brain, notice that the right and left optic tracts contain fibers originating from both eyes.

Describe the pathway of a nerve impulse from the optic nerve to the visual cortex.

The Ear and the Sense of Hearing

The **ear** is the organ of hearing. It provides us with the perception of sound. Sound is produced by a disturbance in the air in the form of moving waves of energy, known as *sound waves*. We can detect sound when sound waves travel through the air to the ear and eventually reach the receptor cells, causing them to produce an action potential that races to the brain.

The ear consists of three portions: the outer ear, the middle ear, and the inner ear (Fig. 10–15). Each of these portions contains structures that help direct sound waves from the exterior to the receptors in the inner ear.

OUTER EAR   The **outer ear** consists of the external appendage on each side of your head, called the **auricle**,[15] and the tube that extends into the temporal bone, known as the **external auditory canal** (Fig. 10–15). The canal occupies the space in the bone formed by the external auditory meatus and contains glands in its lining of skin that secrete a waxy substance called **cerumen** (ear wax).

The auricle collects sound waves in the air and directs them into the external auditory canal. The canal channels the sound waves to the eardrum, which is the entrance to the middle ear.

MIDDLE EAR   The **middle ear** contains an air-filled space within the temporal bone, called the tympanic cavity. It also includes the eardrum, or tympanic membrane, and three small bones called the auditory ossicles. The middle ear is shown in Figure 10–15.

The **tympanic cavity**[16] lies between the inner surface of the tympanic membrane and the bone that forms the outer wall of the inner ear. It is lined with epithelium and contains the three auditory ossicles. It communicates with the throat by way of a narrow tube called the **auditory (eustachian) tube**. This tube allows air to pass between the throat and tympanic cavity. Its function is to equalize air pressure on both sides of the tympanic membrane. You demonstrate this function yourself when you are at higher altitudes and must yawn or chew gum to relieve the pressure in your head: yawning and swallowing bring atmospheric air

---

[15]Auricle: Latin *auri-* = ear + *-cle* = belonging to.

[16]Tympanic: *tympanic* = drumlike.

Figure 10–15
The structure of the ear. Ear components are not to scale.
What is the function of the three auditory ossicles?

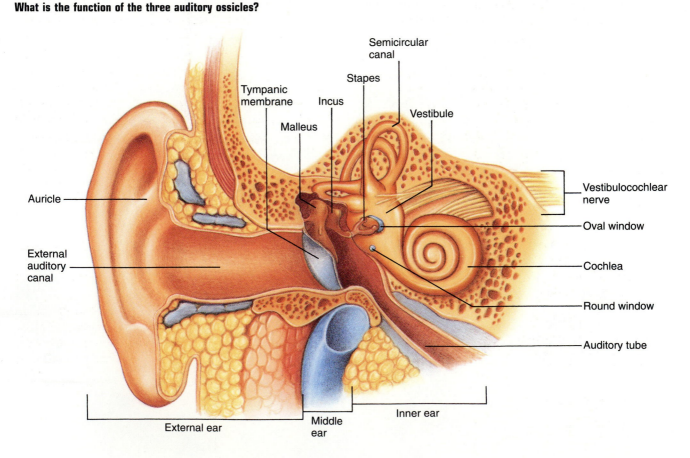

into the tympanic cavity through the auditory tube until the internal pressure of the tympanic cavity equals the atmospheric pressure. If this pressure is not relieved, pain, hearing impairment, and dizziness may follow. One unfortunate aspect of the auditory tube is the tendency for infections to travel from the mouth and throat to the middle ear by way of its mucous membrane lining. When these infections spread further from the middle ear into the tiny air cells of the temporal bone's mastoid process, the resulting condition called *mastoiditis* can be quite serious.

The **tympanic membrane** is a thin, semitransparent barrier separating the outer and middle ears. It consists of a mucous membrane that is covered with a thin layer of skin on its outer surface. It is slightly cone-shaped, with its apex pointing into the tympanic cavity, where it connects to one of the auditory ossicles (the malleus). The tympanic membrane receives sound waves from the external auditory canal. It vibrates in response to these sound waves. Its vibrations are then transmitted to the auditory ossicles.

The three **auditory ossicles** are the smallest bones of the body. They include the **malleus** (hammer), the **incus** (anvil), and the **stapes** (stirrup). They are held in place by tiny muscles and ligaments that attach to the walls of the tympanic cavity. The three bones are interconnected, forming a bridge from the tympanic membrane to the inner ear. As sound waves are transmitted across this bony bridge, they are amplified. When the tympanic membrane vibrates in response to sound waves, the vibration is transmitted first to the malleus, which responds by vibrating in unison. From the malleus the vibration is passed to the incus, and then to the stapes. The stapes transmits the vibration to an opening in the wall of the tympanic cavity, known as the **oval window**. The oval window opens into the cochlea of the inner ear. Within the inner ear, the vibrations from the stapes cause fluid to move, stimulating the receptors for hearing.

INNER EAR The **inner ear** is also known as the **labyrinth**,[17] for it consists of a winding, complicated series of passageways or canals. It contains two portions: an outer bony labyrinth, and an inner membranous labyrinth. The inner ear is shown in Figures 10–15, 10–16, and 10–17.

The **bony labyrinth** is a series of canals within the temporal bone. Its canals form the three regions of the inner ear: the **semicircular canals**, the **vestibule**,[18] and the **cochlea**[19] (KŌ-klē-a). Within its outer, bony walls

---

[17]Labyrinth: Greek *labyrinth* = maze.

[18]Vestibule: *vestibule* = entrance chamber.

[19]Cochlea: Greek *kochlias* = snail.

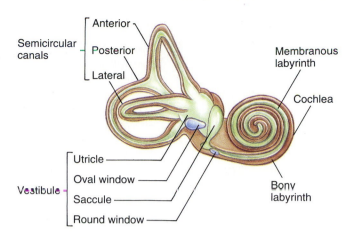

Figure 10–16
The inner ear, or labyrinth.
Which region of the labyrinth contains the organs of hearing?

lies the **membranous labyrinth**, which is an internal series of sacs and tubes. The membranous labyrinth conforms to the shape of the bony labyrinth, so it, too, helps form the semicircular canals, the vestibule, and the cochlea. Between the outer walls of the bony labyrinth and the walls of the membranous labyrinth is a fluid called **perilymph**.[20] Inside the membranous labyrinth is another fluid, the **endolymph**.[21]

The semicircular canals consist of three loops that lie at right angles to each other. The vestibule is a chamber between the semicircular canals and the cochlea. Both the canals and the vestibule function in the sensation of equilibrium, which we will discuss later.

The cochlea is the region of the inner ear whose shape resembles that of a snail shell, for the canals that form it are spiral and coiled. Its bony labyrinth is actually divided in half by a thin shelf of bone lined with membrane, forming upper and lower compartments. These are best seen in a cross section through a cochlear canal (Fig. 10–17). The upper compartment is the **scala vestibuli**, which extends from the oval window to the end of the cochlea. The lower compartment is called the **scala tympani**. It extends in the opposite direction from the end of the cochlea to a membrane-covered opening in the wall of the inner ear, called the **round window**.

Notice in the cross section shown in Figure 10–17 that there is a third channel through the cochlea. This is a portion of the membranous labyrinth and is called the **cochlear duct**. It passes between the two tubular compartments to end as a closed sac. The cochlear duct is separated from the scala vestibuli by the **vestibular**

---

[20]Perilymph: *peri-* = around + *lymph* = clear fluid.

[21]Endolymph: *endo-* = within + *lymph*.

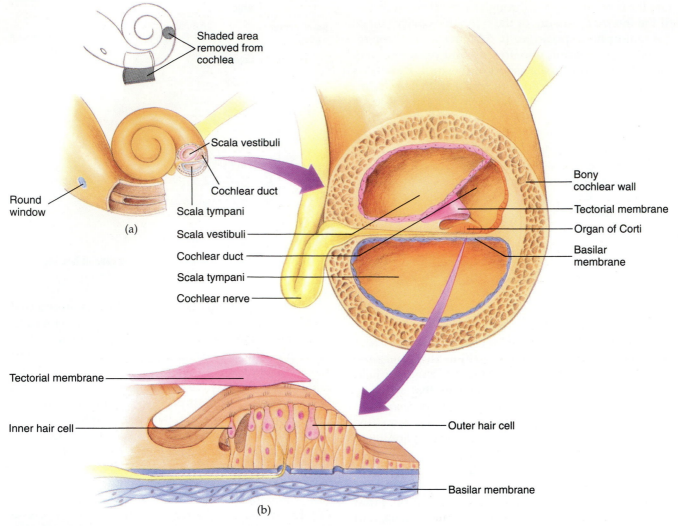

Figure 10–17
The organ of hearing. (a) The cochlea houses the cells that respond to sound vibrations. They are located in a central duct, called the *cochlear duct*. Perilymph circulates through the scala tympani and scala vestibuli, while endolymph is located in the cochlear duct. (b) The receptor cells lie within the organ of hearing, or organ of Corti, located between the basilar membrane and the tectorial membrane.

What fluid may be found within the cochlear duct?

membrane, and from the scala tympani by the **basilar membrane**. On top of the basilar membrane is the organ of hearing, called the **organ of Corti**. It consists of hair cells, which are the receptors supported by epithelial cells. The hair cells have long cilia at their free ends that extend into the endolymph of the cochlear duct. They contact a delicate, gelatinous membrane that lies over them called the **tectorial membrane**. The basal ends of the cells are in contact with nerve fibers of the cochlear nerve (a branch from the vestibulocochlear (VIII) nerve).

A vibration that is transmitted from the auditory ossicles to the oval window must be converted to fluid

motion of the endolymph before a sound can be heard. This conversion is summarized as follows (Fig. 10–18):

1. The movement of the stapes back and forth pushes the oval window in and out, producing waves in the perilymph of the inner ear.

2. Pressure waves in the perilymph pass through the scala vestibuli. The waves push the vestibular membrane inward, increasing the pressure of the endolymph within the cochlear duct. As perilymph waves reach the end of the cochlea, they pass to the scala tympani and are dissipated by passing through the round window and into the tympanic cavity.

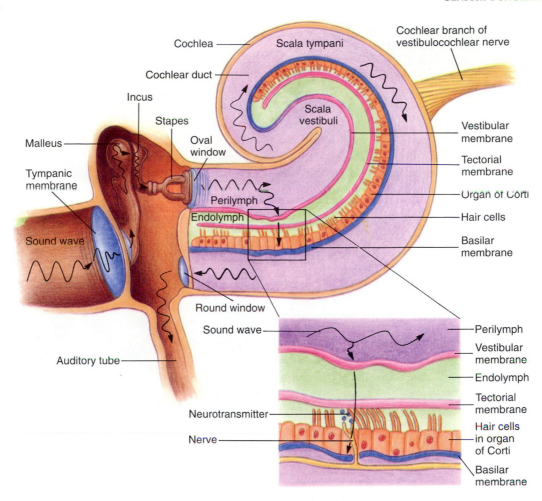

Figure 10–18
From vibration to nerve impulse transmission. This diagrammatic view of the middle ear and cochlea shows how a vibration on the tympanic membrane is transmitted through the ossicles to the oval window, and the subsequent movement of perilymph in the cochlea that results in contact between the hair cells and the tectorial membrane. The arrows indicate the progress a vibration makes through the ear.

How is a vibration dissipated quickly, readying the inner ear for the next vibration?

3. Pressure waves transmitted to the endolymph cause the basilar membrane to move slightly.

4. Movement of the basilar membrane causes the hair cells of the organ of Corti to bend against the tectorial membrane. The movement of the hairs causes the hair cell membranes to release neurotransmitters. If the movement is great enough, sufficient neurotransmitters will be released to stimulate sensory neurons in the cochlear nerve to generate an action potential.

Auditory Nerve Pathway

The hair cells within the organ of Corti convert the motion of endolymph into the release of neurotransmitters, which may stimulate a nerve impulse in the cochlear nerve by way of a sensory nerve fiber at the base of the hair cell. The sensory nerve fiber passes the impulse along the cochlear nerve, which unites with the vestibular nerve to form the vestibulocochlear nerve (cranial nerve VIII) soon after exiting the cochlea. The vestibulocochlear nerve carries the impulse to the medulla oblongata in the brain stem. Here impulses cross to the opposite side before they continue to the midbrain, then to the thalamus, and finally to the auditory area in the temporal lobe of the cerebral cortex.

The Sense of Equilibrium

Equilibrium is detected by receptor cells in the inner ear. The sensation of equilibrium is of two types: static equilibrium and dynamic equilibrium.

STATIC EQUILIBRIUM   **Static equilibrium** refers to the sensation of body position. Information regarding the exact position of the head that is needed to maintain posture is provided. Its receptors are located within the vestibule of the inner ear.

The membranous labyrinth of the vestibule contains two sacs called the **utricle** (YOO-trik-el) and **saccule** (SAK-yool), which are connected to each other by a small duct (see Fig. 10–16). Within their walls is a small, flat region of special cells known as the **macula** (MAK-yoo-la), which is shown in Figure 10–19. Two types of cells form the macula: supporting epithelial cells, and hair cells. The hair cells are the receptors of static equilibrium. They each contain long processes (microvilli and a single cilium) that extend into a thick, jellylike mass. Lying in this mass is a layer of calcium carbonate crystals, called **otoliths**.[22]

The position of the head is monitored by the movement of otoliths in the macula. When you tilt your head to one side, the otoliths shift their position in response to the inertia (the force of gravity). Their movement causes the jellylike mass to move and to pull on the processes of the hair cells, causing them to bend. This bending stimulates the release of neurotransmitters by the hair cells, which generates an action potential in the vestibular nerve (a branch of the vestibulocochlear nerve, cranial nerve VIII).

DYNAMIC EQUILIBRIUM   **Dynamic equilibrium** is the sensation of rapid movements, mostly of the head. The receptors that detect this sense aid in balancing the head and body when they are moved suddenly. They are located within the fluid-filled semicircular canals of the inner ear.

Within the membranous labyrinth of each of the three semicircular canals is an expanded region, called an **ampulla** (Fig. 10–20). It contains the sensory organs of dynamic equilibrium, called the **cristae**. Each crista contains a group of supporting cells and a group of hair cells. As in the macula, the hair cells are the receptors and their hairlike processes extend into a jellylike mass. In the crista, the mass is known as the **cupula**.

When the head shifts position rapidly, the cupula moves in response to the movement that results. Movement of the cupula causes the processes of the hair cells embedded within it to bend. This stimulates the hair cells to release neurotransmitters, which generate an action potential in sensory neurons of the vestibular nerve (a branch of the vestibulocochlear nerve). The part of the brain that receives most of the impulses is the cerebellum, which thus has the key role in maintaining both static and dynamic equilibrium. The cerebellum works in concert with the cerebral cortex, pro-

[22]Otolith: Greek *oto-* = ear + *lithos* = stone.

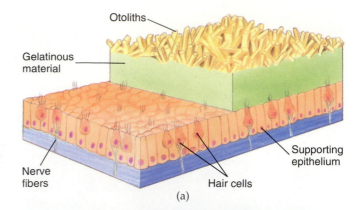

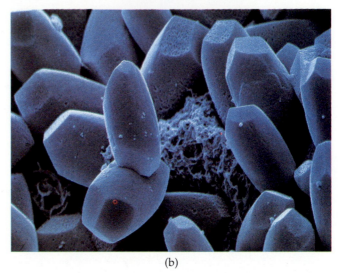

Figure 10–19
The components that detect static equilibrium. (a) Located within the utricle and saccule of the vestibule is the macula. Tilting the head causes the otoliths to shift position, causing the jellylike mass to move and stimulating the hair cells embedded within it. (b) Electron micrograph of otoliths.

What body activity is detected by the hair cells within the macula?

viding the body with the ability to respond to the sensation of equilibrium by sending out motor impulses to specific groups of skeletal muscles.

CONCEPTS CHECK

1. What are the organs for each of the four special senses called, and where are they located?

2. What part of the eye accommodates for near and far vision, and how does it do so?

3. How do rod cells and cone cells differ?

4. How is a sound wave converted to a nerve impulse within the inner ear?

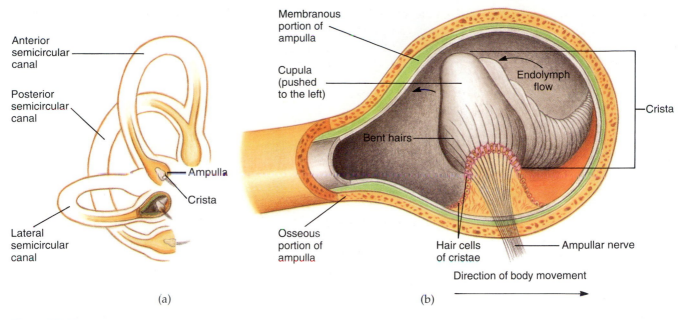

Figure 10–20
The components that detect dynamic equilibrium. (a) At the base of the semicircular canals are the ampullae, which contain the organs of dynamic equilibrium, the cristae. (b) Close-up of an ampulla. Sudden movement of the head causes the cupula to shift, stimulating the hair cells.

INTEGRATIVE FUNCTIONS

Most integrative activities of the nervous system are performed within functional regions of the cerebral cortex. They include interpretation of sensations, thought, memory, and emotions.

The ability of the nervous system to process and interpret sensory information before sending out a motor response is called its **integrative function**. Integration is performed primarily by the cerebral cortex of the brain. It includes activities such as thought, memory, and emotions.

Functional Regions of the Cerebral Cortex

The cerebral cortex is the final destination for most sensory impulses entering the brain. Once these impulses arrive at the cortex, they are interpreted and processed. If a motor response is appropriate, the cortex will initiate it and direct it along the proper route. These important functions of the cerebral cortex occur in specific regions, which have been identified and mapped by investigators studying the brain. They are grouped into sensory, association, and motor areas.

Sensory Areas

The sensory areas of the cerebral cortex contain large numbers of neurons that receive and interpret sensory impulses. They are located in several lobes of the cerebrum (Fig. 10–21).

The **general sensory area** is located in the parietal lobe behind the central sulcus. It receives sensations mainly from receptors in the skin. It provides you with the ability to pinpoint precisely where these sensations occur. One interesting aspect of this area is that the interpreted sensitivity of a body region is determined by the number of sensory receptors that region contains and not by its relative size. For example, the lips, tongue, and fingertips contain more receptors than the entire body below the shoulders, and are thus more sensitive despite their small size. This relationship is illustrated in Figure 10–22. Sensory impulses arriving at the general sensory area are often forwarded to other sensory areas for further interpretation.

The **somesthetic association area** is located behind the general sensory area. It receives impulses from the thalamus and general sensory area and interprets the nature of the sensation. It also stores memories of past sensory experiences, enabling you to relate the present experience to experiences from the past.

The **primary visual area** receives sensory impulses from the eyes. It is located in the occipital lobe and is associated with the **visual association area** nearby. Together these visual areas interpret images and analyze what is seen.

The **primary auditory area** receives sensory impulses from the ears and interprets the nature of the sound that is heard. It is located in the temporal lobe

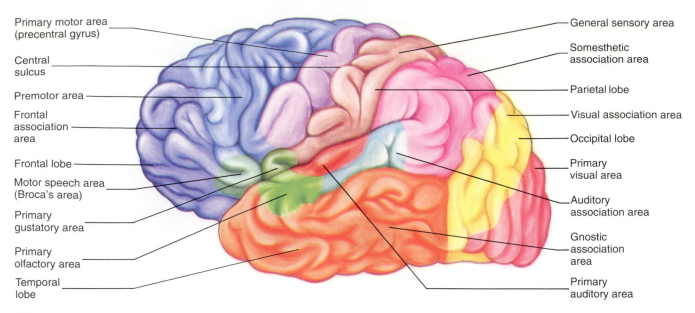

Figure 10–21
Functional regions of the cerebral cortex. The various regions are indicated in color on the surface of the cerebrum in this lateral view.

What is the function of the somesthetic association area?

of the cerebral cortex. Located nearby is the **auditory association area**, which distinguishes between recognizable sounds and translates the sound of speech into thought.

Other sensory areas include the **primary gustatory area** in the parietal lobe, which interprets taste, and the **primary olfactory area** in the temporal lobe, which interprets smell.

Association Areas

The **association areas** of the cerebral cortex receive impulses from each of the sensory areas. They provide a connection between sensory and motor areas. Together with sensory and motor components, the association areas recognize, analyze, and initiate a response to the sensory information. In other words, they *integrate* information. Association functions include learning and reasoning, memory storage and recall, language abilities, and even consciousness. In general, any area of the cerebral cortex not pinpointed as a primary sensory or motor area is considered to have an associative function.

One region of the cortex that has been identified as providing integrative functions is the **gnostic** (NOStik) **area**. It is located in the temporal lobe and is limited to one hemisphere only, usually the left. It integrates all sensory incoming signals into a conscious thought or understanding, and activates other parts of the cortex to cause the proper motor response. An example of a gnostic function would occur if you were

to drop a cup of hot tea on your lap while talking to a friend: you see the cup tip over and drop; you feel the burning sensation on your lap; you smell the steamy aroma of the tea; and you hear the crash as the cup hits the floor. Because of your conversation, these sensations do not dominate your thoughts, but their integration in the gnostic area signals an emergency response and you rise to your feet to quickly begin removing the spill from your clothes.

Motor Areas

The **motor areas** of the cerebral cortex are located primarily in the frontal lobes in front of the central sulcus (see Fig. 10–21). They receive impulses from the association areas for the initiation of the motor response. For the most part, individual motor neurons that arise from these areas pinpoint the origin of a particular motor pathway. In general, the different muscle groups are represented by the degree of precise muscle control and not by the size of the body part. For example, the lips, tongue, and vocal cords occupy a larger part of the motor areas than the entire trunk. This relationship is shown in Figure 10–22.

The **primary motor area** is located in front of the central sulcus in the frontal lobe. It consists of groups of motor neurons that control specific muscles or groups of muscles. The locations of neuronal groups are consistent among most individuals, and maps have been constructed that show these locations (Fig. 10–22). These are helpful in identifying neural dam-

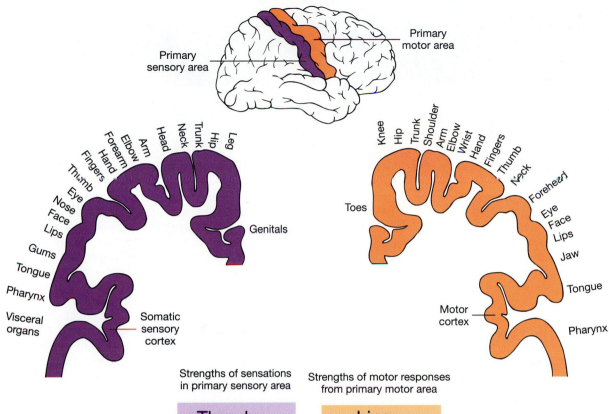

Strengths of sensations
in primary sensory area

Strengths of motor responses
from primary motor area

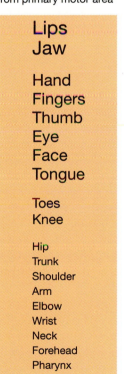

Strengths of sensations in primary sensory area	Strengths of motor responses from primary motor area
Thumb	Lips
Lips	Jaw
Genitals	Hand
Hand	Fingers
Fingers	Thumb
	Eye
Eye	Face
Nose	Tongue
Face	
Gums	Toes
Tongue	Knee
Leg	Hip
Hip	Trunk
Trunk	Shoulder
Neck	Arm
Head	Elbow
Arm	Wrist
Forearm	Neck
Elbow	Forehead
Pharynx	Pharynx
Visceral organs	

Figure 10–22
The relative strengths of sensations and motor functions of the cerebral cortex.

Are all sensations equally represented by the cerebral cortex?

age, and are used during open brain surgery where the surgeon may artificially stimulate a single group to test for the contraction of a particular muscle.

The **premotor area** is located in front of the primary motor area. It receives impulses from various areas of the brain, such as the primary motor area, the cerebellum, and the basal ganglia. The impulses it gen-

erates control coordinated, precise movements of skeletal muscles that are usually learned. Examples include skills that you develop, such as writing, typing, drawing, and sewing.

Other motor areas include a **visual motor area** and a **motor speech area**. The visual motor area, also called the *frontal eye field area*, controls scanning movements

Consciousness and Coma

The term *consciousness* refers to the ability of the brain to respond to stimuli. It is the process of mental awareness that involves the physical activity of large sets of interacting neurons. There are four levels of consciousness: (1) alertness, (2) drowsiness or lethargy, (3) stupor, and (4) coma. The levels of consciousness may be measured by use of an electroencephalogram (EEG), which measures electrical activity in the brain. Alertness is the highest state of consciousness, in which cortical activity is the greatest. The rapid processing in various areas of the brain enables you to respond to stimuli in a complex manner when you are alert. On the opposite side of consciousness is the most depressed state of the mind, **coma**. It is characterized by a lack of responsiveness to stimuli for an extended period of time. The EEG tracing of a patient in the coma state is typically reduced, indicating the depressed level of cortical activity. The human brain normally functions along the upper three levels of consciousness, and usually undergoes a cycling of behavior through the course of a 24-hour period. This type of cyclic activity is called a **circadian rhythm**. The coma state represents a diversion from the normal circadian rhythm.

Physiologically, coma differs from the higher levels of consciousness in the brain's decline in oxygen use. It is not a state of deep sleep, for during sleep the brain is active and its oxygen use resembles that of the waking state. Rather, coma is characterized by oxygen levels in the brain that lie below a resting baseline.

Coma may be induced by widespread trauma to the brain, which may arise from a blow to the head, hemorrhage (bleeding), or inflammation. It may also be caused by tumors, excessively low blood sugar levels (hypoglycemia), drug overdose, or severe blood loss elsewhere in the body. In cases where damage to the brain is irreparable, the coma becomes irreversible, although life support systems may be used in a hospital setting to keep vital organs alive. Once the EEG records a flat baseline (indicating a complete absence of cortical activity), **brain death** has occurred and the patient is determined to be deceased.

of the eye. The motor speech area, also known as **Broca's area**, controls the speech muscles, providing you with the ability to perform coordinated speech.

Thought and Memory

The functions of thought and memory are probably the most complex functions of the human body. They are performed within the association areas of the cerebral cortex. Thought and memory are the results of processing among the billions of neurons within these centers.

The term **thought** refers to the conscious understanding or development of an idea. The idea may be an image that is projected in the brain, similar to a visual image, or it may be a series of symbols in the form of a language. A thought is a very real experience within the brain, for it is produced as a result of billions of exchanges of neurotransmitters across billions of synapses, and the conduction of millions of impulses through millions of neurons. A thought usually originates from sensory impulses arriving at the association centers and develops within the association centers. It is the product of complex processing at synapses. The frontal and temporal lobes appear to be most active in the generation of thought, although many areas of the cortex may be active simultaneously.

Memory is the ability to recall past experiences, including thoughts. It is a neural event that is stored within the cortex for retrieval at a later time. It is a tremendous survival advantage to have a memory, for it enables **learning** to occur. Learning is the ability to acquire new knowledge or skills through experience; it provides an avenue to avoid repeating past mistakes.

The centers in the brain that are responsible for memory and learning include areas where actual memories reside and areas where memory information is integrated. These areas are interconnected by a network of fibers that relay information back and forth in a loop circuit. Memories are currently thought to reside within parts of the brain that must recall them to enable new information to be associated with the old. For example, visual memories are stored in the occipital cortex, auditory memories are stored in the temporal cortex, and taste memories lie in the parietal cortex. The memory integration centers have been identified through the efforts of brain research; they include the ventromedial prefrontal cortex (a region of the cortex beneath the front of the brain), parts of the diencephalon (thalamus and hypothalamus), and parts of the limbic system (the amygdala, which is part of the basal ganglia in the cerebrum; and the hippocampus, a sea horse–shaped region of the cerebrum in the temporal lobes). The limbic system will be discussed in the following section. Areas of the brain that are involved in memory are shown in Figure 10–23.

To recall a memory, a sensory perception is first formed in the cerebral cortex where the memory is stored. Connecting fibers carry this information, by way of impulses, first to the limbic system and then to the diencephalon and prefrontal cortex. Connecting fibers then route the impulse from the prefrontal cortex back to the sensory cortex where the perception was initially formed. This feedback loop through integration centers enables the memory to change, to become more durable, to influence the emotional state, or any combination of the three.

There are two types of learning processes, **short-term memory** and **long-term memory**. Short-term memory lasts for brief periods, usually from a few seconds to several hours, and enables you to recall small bits of information. For example, memorizing the answers to a quiz minutes before an exam utilizes this process. Short-term memory information is usually lost quickly and does not return. Investigations into the process of short-term memory have concluded that it involves changes in the strength of existing synaptic connections. Long-term memory, however, lasts for long periods of time (sometimes many years). Additional processing is required for long-term memory, which may be stimulated by repetitive thoughts, by associating something with a thought that is already well preserved, or by focused concentration for an extended period of time. Recent studies have shown that the processing for long-term memory involves something entirely different from short-term memory: the activation of genes, the expression of new proteins, and, as a result, the actual growth of new synaptic connections. Thus, the development of long-term memory leads to anatomic changes in the brain. However, our ability to process thoughts into long-term memory and retrieve them declines with aging, providing us with a memory bank that changes with time.

Emotions: The Limbic System

The **limbic system**[23] is a functional region of the brain that occupies parts of the cerebral cortex and basal ganglia of the cerebrum, the hypothalamus, the thalamus, and the brain stem (Fig. 10–23). It is the center of the brain that is responsible for our emotions, or feelings

[23]Limbic: Latin *limbus* = border.

Figure 10–23

The areas of the brain associated with memory. They include the ventromedial prefrontal cortex, the diencephalon (thalamus and hypothalamus), and the limbic system (amygdala and hippocampus). The limbic system is also associated with emotions.

What roles are performed by the limbic system in memory recall and in emotions?

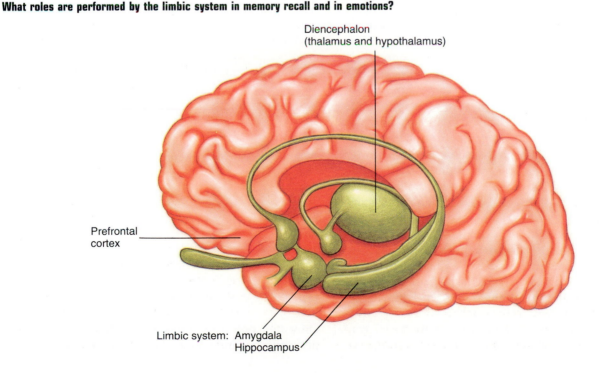

Diencephalon
(thalamus and hypothalamus)

Prefrontal cortex

Limbic system: Amygdala
Hippocampus

How Much of Your Mind Do You Use?

In past centuries, early thinkers tried to determine the location in the body of the mysterious entity we call the "mind"—that part of us that provides a sense of personal awareness and identity. Over the years, scientific discovery has shown us that the brain is the physical organ of the mind, in that it serves as a storehouse for thoughts, memories, and feelings. Science has also taught us that we actually use only a small percentage of our mind, or **total brain capacity**, during a particular period in our lifetime. However, although most investigators studying the brain agree that no one uses 100% of the brain, estimates of the volume that is actually used would vary with the individual and the activity being performed.

Although we may not know exactly how much of our brain capacity we are using, one fact is well established: the brain is a dynamic organ, capable of changing from one second to the next. For example, your brain is actually changing as you are reading these words. It is possible, in fact, to intentionally alter the way in which the brain is organized by changing the synaptic connections of the brain. This physical alteration of brain structure may occur when you develop certain kinds of abilities, ideas, or habits that differ from the ones you now have. In other words, by learning about and concentrating on activities that differ from those you know, you physically alter the organizational configuration of your brain. For example, if you wished to change your major in college from music to biology, new synaptic connections would be made as your long-term memory, thoughts, and activities transformed.

Increasing the volume of brain tissue that you actually use can, therefore, be done if you are in reasonably good health and are motivated to do it. The moment that you decide to learn a new lesson or activity marks the first step in expanding your neuronal network. This fact points to the incredible **plasticity** (ability to change form) of the physical entity of the mind, the brain. Some ancient peoples may have understood this concept, for according to the *Dhammapada* (an ancient Sanskrit text from India), all we are is the sum of all we have thought.

about ourselves and our place in the world. As we learned in the previous section, the limbic system also plays a role in memory processing.

The limbic system contains networks of nerve fibers that interconnect many nuclei with the various parts of the brain. These fibers extend between the higher and lower brain centers. Their interconnections allow the system to receive and integrate information from a wide variety of stimuli, ranging from smells that reach our consciousness to visceral sensations of which we are not aware. They also allow a complex response, which can be initiated through the autonomic nervous system to cause what are termed *psychosomatic disorders*. These include tension; hypersecretion of acid in the stomach, which can lead to ulcers; headaches; and muscle spasms. The most severe form of response to an emotional state is cardiac arrest, which can be brought on by extreme fear or anxiety. On the more positive side, emotional responses may generate a feeling of warmth, security, and a sense of well-being. These responses are possible because of the connections between the limbic system and the higher centers of thought within the cerebral cortex. By virtue of these connections we have the ability to be consciously aware of the emotional state of our lives.

CONCEPTS CHECK

1. Which sensory area of the cerebral cortex receives impulses from touch and pressure receptors?

2. What cerebral lobe contains most of the motor areas?

3. The integrative functions of thought and memory are performed within which areas of the cerebral cortex?

4. What parts of the brain are associated with emotions?

MOTOR FUNCTIONS

The motor system provides a means for responding to changes in our environment. Its pathways arise from integration centers throughout the central nervous system.

Once the central nervous system has received sensations and has integrated the information, it may gen-

Lobotomy

The frontal lobes of the cerebral cortex anterior to the pre-motor areas are associated with higher intellectual functions. Their role in brain function was largely recognized through work with mental patients during the first half of the twentieth century. The work sometimes included an operation used on uncontrollable patients, called the *prefrontal lobotomy*. This procedure included severing the connections between the frontal lobes, the limbic system, and the rest of the brain. The operation proved to be an effective method of changing moral and social attitudes, but it also caused severe intellectual and emotional deterioration, turning patients into unresponsive, docile beings void of emotions and thought. Today, the prefrontal lobotomy is no longer practiced; it has been replaced by psychoactive drugs that can target specific behavior patterns and do not cause intellectual or emotional damage.

erate motor impulses to bring about a response. These motor impulses flow along pathways through the nervous system until they reach their destination—the effectors. The origin and pathways of the motor impulses constitute the motor system, which enables us to respond to the world around us. In this section we will look at the motor functions of the nervous system by examining how they originate and the general pathways they follow.

Motor Origins

Sensory signals that inform us about our environment arrive at the central nervous system, where they are integrated. From integration centers in the CNS, motor impulses arise. These integration centers are located throughout the CNS: in the spinal cord, in the brain stem, in the cerebellum, in the basal ganglia, in the hypothalamus, and in the cerebral cortex.

The integration centers in the spinal cord are located within its gray matter along the entire length of the cord. The impulses they generate are reflexes, such as the reflex arc we examined earlier in Chapter 9. These reflexes serve as a quick-response system, since the motor impulses are generated soon after a sensory impulse is received, with very little integration. Most reflexes from the cord serve to stimulate skeletal muscle contraction.

The integration centers in the brain stem are located within nuclei in the medulla oblongata, pons, and midbrain. The motor impulses generated here are also regarded as reflexes, since the impulses require very little integration. Motor reflexes that arise from the brain stem are the visceral reflexes, which cause responses such as coughing, sneezing, breathing, change in heart rate, and vomiting.

The cerebellum contains integration centers within its gray matter. These centers are connected with centers in the cerebral cortex, basal ganglia, and brain stem. They receive sensory information regarding equilibrium from the inner ear, and coordinate this information with the cerebral cortex to generate motor impulses that control body posture, balance, and coordination. The effectors stimulated by the cerebellum by way of the cerebral cortex are skeletal muscles.

The basal ganglia are clusters of gray matter within the cerebrum. Their integration centers control semivoluntary movements, such as laughing, walking, and jumping. The basal ganglia stimulate skeletal muscle contraction also, but by way of the thalamus.

The hypothalamus is the center for involuntary integration and control. It is an origin for motor impulses that serve the autonomic nervous system and follow along their special pathways. The pathways lead to smooth muscle, cardiac muscle, and glands.

The cerebral cortex represents the highest level of integration. The region that is dedicated to the generation and control of motor impulses is called the **motor cortex**, which we have identified as the area in front of the central sulcus (see Fig. 10–21). The motor cortex is the primary site of origin for motor impulses requiring thought, memory, skillful body coordination, and precise muscular movements.

Motor Pathways

Motor impulses generated by the integration centers may pass along simple reflex arcs, autonomic pathways, or somatic pathways. Reflex arcs and autonomic pathways were discussed in Chapter 9. A discussion of somatic pathways, which provide impulses to skeletal muscles for voluntary body movement, follows.

Somatic Pathways

The **somatic pathways** conduct voluntary motor impulses from the motor areas of the brain to skeletal muscles (Fig. 10–24). For the most part, they originate from the motor cortex, descend through the cerebrum, and cross to the opposite side of the brain as they extend through the brain stem. From here they continue to a particular level of the spinal cord or to nuclei in a cranial nerve. Throughout this entire distance the impulse is conducted by a single neuron, simply called the **upper motor neuron**.

From the upper motor neuron the impulse is transmitted to the second neuron in the pathway, the **lower motor neuron**. In the case of a cranial nerve outflow, the upper motor neuron transmits the impulse to the lower motor neuron within a nucleus of a cranial nerve. From here the lower motor neuron carries the impulse to a skeletal muscle effector in the head or neck. In the case of a spinal nerve outflow, the upper motor neuron passes down a descending tract of the spinal cord until it synapses with a lower motor neuron within the anterior gray horn of the spinal cord. From here the lower motor neuron conducts the impulse to the skeletal muscle effector.

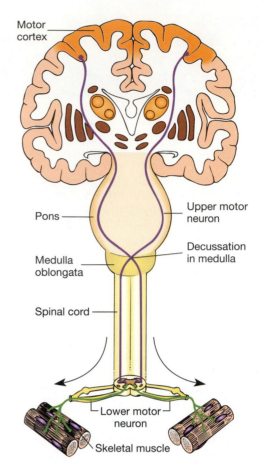

Figure 10–24
An example of motor pathways from the cerebral cortex to the peripheral nervous system by way of a spinal nerve outflow. The synapse between an upper motor neuron and a lower motor neuron may occur along any segment of the spinal cord.

How many motor neurons are usually involved in a somatic motor pathway?

CONCEPTS CHECK

1. Which part of the CNS contains integration centers that originate motor impulses passing to autonomic effectors?

2. How many motor neurons carry a single impulse from the cerebral cortex to a skeletal muscle?

CLINICAL TERMS OF THE FUNCTIONAL NERVOUS SYSTEM

Acataphasia   An inability to express thoughts in a connected manner. It is frequently caused by a lesion in the brain.

Achromatopsia   Total color blindness. This is usually caused by a genetic defect resulting in the lack of cone cells in the retina.

Alzheimer's disease   A progressive degeneration of brain tissue with a high incidence among individuals of both sexes over 50 years of age. It results in a gradual loss of memory of recent events (retrograde amnesia). Current studies suggest that the cause of this "disease of old age" may be

the gradual loss of cells from the cerebral cortex and the limbic system. There are several types of Alzheimer's disease; it is now known that at least one type is genetically linked.

Ametropia   A defect in the refractive powers of the eye in which images cannot be focused properly onto the retina.

Aphasia   A loss of language function in which the comprehension or expression of words becomes impaired. It is caused by trauma to the language centers of the cerebral cortex, such as a blow to the head.

Astereognosis An inability to recognize familiar objects by feeling their shapes with the eyes closed.

Cataracts An abnormal loss of transparency of the lens. It is a progressive disorder and is characterized by a gray-white opacity that can be seen in the lens through the pupil. It is caused by degenerative changes in the lens that most often occur after 50 years of age.

Cerebrovascular accident (CVA) Commonly called a **stroke**. It results in the destruction of brain tissue, and may be caused by blood deficiency due to a blocked vessel (ischemia) or sudden blood loss within the brain (hemorrhage).

Cochlear implant A surgical procedure involving the implantation of an artificial device in the cochlea. The implant receives impulses from an external receiver that converts the signals to nerve impulses, which are relayed to the vestibulocochlear nerve. It provides rudimentary sound perception.

Dementia A general term for mental deterioration. It is a symptom of Alzheimer's disease, AIDS, and other progressive degenerative diseases.

Epilepsy A group of neurological disorders characterized by repeated episodes of convulsive seizures, abnormal behavior, sensory disturbances, loss of consciousness, or all of these. These conditions are caused by uncontrolled electrical discharges from neurons of the cerebral cortex. Most forms of epilepsy are of an unknown cause, although some may be traced to brain trauma, intracranial infection, tumors, vascular disturbances, intoxication, or chemical imbalance.

Eustachitis An infection or inflammation of the eustachian tube.

Glaucoma An abnormal condition of elevated pressure in the eye. It is due to obstruction of the outflow of aqueous humor. Minor cases are usually treated with eyedrops that control the flow of aqueous humor, and serious cases may require the placement of a surgical hole through the iris (called an *iridectomy*).

Radial keratotomy A surgical procedure to improve myopia. It includes making small incisions in the cornea, allowing it to stretch and become flatter.

Sclerostomy A surgical procedure to relieve the pressure caused by glaucoma. It involves the creation of an opening through the sclera.

Vertigo A condition of dizziness, or spatial disorientation, especially due to heights.

CHAPTER SUMMARY

Homeostasis is achieved by the nervous system via three activities: sensory, integrative, and motor.

SENSORY FUNCTIONS

A sensation is a state of awareness of a particular stimulus.

Sensory Pathways All sensory pathways begin with a receptor and travel toward the central nervous system.

Receptors Receptors vary in structure. They may end bare as dendrites, or be a complex organ. Receptors are stimulus-specific, and many show sensory adaptation. Receptors are classified according to their types of sensitivity and include mechanoreceptors, thermoreceptors, nociceptors, photoreceptors, and chemoreceptors.

General Sensory Pathways A conduction pathway passing from a simple receptor to the brain consists of three sensory neurons: a first-order neuron, a second-order neuron, and a third-order neuron.

Special Sensory Pathways Conduction pathways passing from complex, special sensory organs to the brain. Each contains *at least* three sensory neurons.

General Senses Touch and pressure, temperature, pain, and body position are the general senses. They are detected by simple receptors and follow general pathways to the brain.

Touch and Pressure Also called *cutaneous sensations*. The receptors are located in the skin. They include Meissner's corpuscles, sensitive to fine touch; and Pacinian corpuscles, sensitive to deep pressure.

Temperature Changes in temperature are detected by receptors probably located in the skin.

Pain Pain is detected by receptors that are free nerve endings. Pain receptors make us aware of danger to the body.

Body Position Position of the body is detected by proprioceptors, which provide information on the degree of muscle contraction, the tension in tendons, and head position. They include muscle spindles in skeletal muscle, tendon organs between tendons and muscles, and inner ear receptors.

Special Senses Smell, taste, sight, and hearing are each detected by complex organs.

Smell

Location The sense of smell, or olfaction, functions by means of sensory receptors within olfactory organs located in the upper wall of the nasal cavity. The olfactory organs contain olfactory hairs, which respond to dissolved chemicals.

Olfactory Pathway Olfactory hairs generate the action potential, which passes to the olfactory bulbs and on to the frontal lobes of the cerebral cortex.

Taste

Location The sense of taste, or gustation, depends on receptors in organs known as *taste buds* within papillae on the tongue and in other areas of the mouth. The receptors are called *gustatory cells*, and they respond to dissolved chemicals.

Gustatory Pathway Gustatory cells stimulate an action potential in sensory neurons at their bases. The impulse travels along fibers of the facial, glossopharyngeal, and vagus nerves to the medulla and thalamus, and on to the cerebral cortex.

Sight The sense of sight, or vision, depends on photoreceptors in the eye.

Accessory Structures Structures associated with the eye include the eyelids, lacrimal apparatus, and extrinsic muscles of the eyeball.

Structure of the Eye The walls of the eye consist of three layers, or tunics:

Fibrous Tunic The thick outermost layer of the eyeball. It contains the posterior, white sclera and the anterior, transparent cornea.

Vascular Tunic The middle layer. It contains the thin posterior choroid, the thick anterior ciliary body, and the anterior, colored iris. The pupil is the opening in the center of the iris. The lens is held in position behind the iris and pupil by suspensory ligaments attached to the ciliary body. Tension on the lens can be varied by ciliary muscles, stretching or relaxing the lens to accommodate for near and far vision. Between the cornea and the lens is a cavity divided in half by the iris: between the cornea and iris it is called the *anterior chamber*; between the iris and lens it is the *posterior chamber*. The space behind the lens is the posterior cavity. The anterior and posterior chambers contain aqueous humor, and the posterior cavity contains vitreous humor.

Nervous Tunic The inner layer of the eyeball, consisting of the retina. The retina lines the posterior cavity and contains the photoreceptor cells (the rods and cones), bipolar neurons, and ganglion cells. Axons from ganglion cells converge to form the optic nerve, forming a blind spot called the *optic disc* where the optic nerve exits. The

fovea centralis contains many cone cells and is the area of sharpest vision.

Pathway of Light Through the Eye Light rays bend when they pass from air to the fluids and solids of the eyeball. Their bending is called *refraction*. The lens changes in shape to accommodate refraction and other factors.

Physiology of Vision Rods and cones convert light waves to a series of signals that result in the generation of an action potential in ganglion cells. Rods are sensitive to dim vision, and cones are sensitive to sharp vision and color. Both cells contain pigments that decompose when exposed to light. Pigment decomposition leads to nerve impulse generation.

Visual Pathway From rods and cones, the series of signals passes to bipolar neurons and then to ganglion cells, where the action potential is generated. Axons from ganglion cells extend out of the eye through the optic nerve and cross at the optic chiasma, and may terminate at the thalamus. From the thalamus the impulses pass to the occipital lobes of the cerebral cortex.

Hearing and Equilibrium The sense of hearing, or audation, depends on receptors within special organs in the ear. It is associated with the sense of equilibrium.

The Ear and the Sense of Hearing The ear consists of outer ear, middle ear, and inner ear portions.

Outer Ear The outer ear consists of the outer appendage, or auricle, and the auditory canal. These structures direct sound waves to the eardrum.

Middle Ear The middle ear consists of the tympanic membrane (eardrum) and three ear ossicles: the malleus (hammer), incus (anvil), and stapes (stirrup). These structures convert the sound wave to a vibration and transmit it to the oval window. The ossicles lie within a space, called the tympanic cavity, that communicates with the throat via the auditory (eustachian) tube.

Inner Ear Also called the labyrinth; it consists of the outer, bony labyrinth, which contains a fluid called perilymph; and the inner, membranous labyrinth, which contains endolymph. Together they constitute three regions of the inner ear: the semicircular canals, the vestibule, and the cochlea. The organ of hearing, called the organ of Corti, is located in the cochlea. It contains receptor cells that move in response to waves of endolymph and thereby release neurotransmitters that stimulate the nerve impulse.

Auditory Nerve Pathway The hair cells in the organ of Corti convert the motion of endolymph to the release of neurotransmitters. If sufficient, their release initiates a nerve impulse in a sensory neuron of the cochlear nerve (a branch of the vestibulocochlear nerve), which carries it to the medulla oblongata. The impulse then continues to the midbrain and then to the thalamus, and terminates in the temporal lobe of the cerebral cortex.

The Sense of Equilibrium There are two types of equilibrium: static and dynamic.

Static Equilibrium The sensation of body position, which is detected by receptors in the vestibule of the inner ear. Within the macula are two chambers, the saccule and utricle, which contain the hair cells sensitive to the position of the head.

Dynamic Equilibrium The sensation of rapid movements, which is detected by receptors in the semicircular canals of the inner ear. In each loop is an expanded region, called an ampulla, which contains the organs of dynamic equilibrium, the cristae. Hair cells in the cristae release neurotransmitters in response to rapid movement of the head, stimulating an action potential in the vestibular nerve (a branch of the vestibulo-cochlear nerve).

INTEGRATIVE FUNCTIONS

The central nervous system processes, or integrates, sensory information before initiating a motor response.

Functional Regions of the Cerebral Cortex

Sensory Areas Sensory areas of the cortex receive and interpret sensory information. They include the general sensory area and the somesthetic association area in the parietal lobes, the primary and associative visual areas in the occipital lobes, the primary and associative auditory areas in the temporal lobes, the primary gustatory area in the parietal lobe, and the primary olfactory area in the temporal lobe.

Association Areas Association areas of the cortex connect sensory areas to motor areas. One example is the gnostic area.

Motor Areas Motor areas of the cortex receive commands from association areas to initiate motor impulses. The primary and premotor areas are in the frontal lobes.

Thought and Memory Thought and memory are higher brain functions performed by association areas of the cerebral cortex. Thought is the conscious understanding or development of an idea. Memory is the ability to recall past experiences. Memory is stored in sensory areas of the cerebral cortex, which are connected to integrative centers. Short-term memory provides the recall of small bits of information for a short period of time. Long-term memory provides recall of information over a long period of time and requires more complex processing.

Emotions: The Limbic System The limbic system occupies parts of the cerebral cortex, basal ganglia, hypothalamus, thalamus, and brain stem. Its network interconnects aspects of brain function to create emotions, which are feelings about ourselves and our place in the world.

MOTOR FUNCTIONS

Motor functions arise from the stimulation of motor pathways in the cerebral cortex. They provide a means of responding to the world around us.

Motor Origins Motor impulses may arise from integration centers in the spinal cord, brain stem, cerebellum, basal ganglia, hypothalamus, or cerebral cortex. The site of origin determines the particular motor pathway and type of response.

Motor Pathways Motor pathways may follow simple reflex arcs, autonomic pathways, or somatic pathways.

KEY TERMS

accommodation (p. 294)
association (p. 303)
audation (p. 296)
choroid (p. 291)
cochlea (p. 298)

cornea (p. 290)
gustation (p. 288)
integration (p. 302)
iris (p. 291)

labyrinth (p. 298)
olfaction (p. 287)
receptor (p. 283)
retina (p. 292)

sclera (p. 290)
sensation (p. 283)
vestibule (p. 298)
vision (p. 289)

QUESTIONS FOR REVIEW

OBJECTIVE QUESTIONS

1. The receptors in the eye respond to:
   a. sound waves   c. light
   b. chemicals   d. all of the above
2. Mechanoreceptors generate a nerve impulse in response to:
   a. temperature changes
   b. physical changes in the receptor or nearby cells
   c. chemicals dissolved in fluid
   d. all of the above

3. The number of neurons in a general sensory pathway from the receptor to the brain is always:
   a. three   c. a variable number
   b. two   d. eight
4. Pacinian corpuscles in the skin detect:
   a. fine touch   c. deep pressure
   b. muscle stretch   d. temperature changes
5. Body position is detected by proprioceptors located in the inner ear and in:
   a. the skin   c. the cerebellum
   b. the muscles and tendons   d. visceral organs

6. When chemicals dissolve in the nasal cavity, they stimulate:
   a. gustatory cells
   b. rod cells
   c. olfactory hairs
   d. all of the above

7. The anterior, transparent part of the fibrous tunic is called the:
   a. sclera
   b. cornea
   c. iris
   d. ciliary body

8. The lens is attached to suspensory ligaments of the:
   a. pupil
   b. sclera
   c. cornea
   d. ciliary body

9. The fluid that is continuously produced in the anterior and posterior chambers of the eye is:
   a. aqueous humor
   b. perilymph
   c. vitreous humor
   d. plasma

10. Cone cells are concentrated in a depression in the retina to provide your best visual acuity. This area is called the:
    a. optic disc
    b. utricle
    c. fovea centralis
    d. choroid

11. The decomposition of rhodopsin in response to light exposure leads to the generation of an action potential in a:
    a. cone cell membrane
    b. hair cell
    c. ganglion cell
    d. cornea cell

12. From a rod or cone cell, the signal passes to a bipolar neuron and then to a ganglion cell, whose axon emerges out of the eyeball as the:
    a. optic chiasma
    b. optic nerve
    c. vestibulocochlear nerve
    d. vagus nerve

13. The organ of Corti is located within the _____ of the inner ear.
    a. vestibule
    b. cochlea
    c. bony labyrinth
    d. utricle

14. The stapes transmits a mechanical vibration from the incus to the:
    a. oval window
    b. round window
    c. tympanic cavity
    d. malleus

15. Hair cells in the organ of Corti convert the motion of _____ into the release of neurotransmitters.
    a. perilymph
    b. the round window
    c. endolymph
    d. the tympanic membrane

16. The movement of otoliths within the utricle and saccule stimulates hair cells to detect:
    a. sound waves
    b. rapid head movements
    c. static equilibrium
    d. dynamic equilibrium

17. The general sensory and somesthetic association areas are located in the:
    a. temporal lobe
    b. frontal lobe
    c. hypothalamus
    d. parietal lobe

18. The region of the cerebral cortex that generates impulses leading to learned, coordinated muscular activities such as writing and drawing is:
    a. the premotor area
    b. the somesthetic association area
    c. the parietal lobe
    d. Broca's area

19. Emotions are produced within complex pathways in the:
    a. cerebral cortex
    b. basal ganglia
    c. limbic system
    d. hypothalamus

20. Somatic motor pathways leading from the cerebral cortex follow along _____ until they reach the skeletal muscle effectors.
    a. pyramidal pathways
    b. extrapyramidal pathways
    c. second order neurons
    d. autonomic pathways

ESSAY QUESTIONS

1. Trace the pathway an impulse must follow from the initial stimulation of a touch receptor in the skin to its final destination in the brain.

2. How do pain receptors help in maintaining homeostasis?

3. Trace the events that result in vision, from a light ray that penetrates the cornea to the interpretation of sight by the brain. Include the visual pathway from the retina to the cerebral cortex.

4. How does the lens accommodate for near and far vision? Include in your answer why nearsightedness or farsightedness sometimes occurs.

5. Trace the pathway of sound waves through the outer, middle, and inner ears. Include an explanation of how sound wave energy is converted to a nerve impulse.

6. How does the brain produce the intangible functions of thought and memory? Where in the brain do these processes occur?

7. In what areas of the nervous system may motor impulses arise?

ANSWERS TO ART LEGEND QUESTIONS

Figure 10-1

How many sensory neurons are involved in a general sensory pathway?   Three neurons are involved in a general sensory pathway.

Figure 10-2

Where can Meissner's corpuscles and Pacinian corpuscles be found?   Meissner's and Pacinian corpuscles can be found in the dermis of the skin.

Figure 10-3

Where would you experience pain sensations that result from a punctured lung?   Pain sensations from a punctured lung would be felt along the neck, upper shoulders, upper chest, and upper back.

Figure 10-4

What olfactory function is served by the presence of mucus lining the nasal epithelium?   Mucus dissolves the chemicals that are inhaled, preparing them for contact with the olfactory hairs.

Figure 10–5

What are the taste bud cells that are sensitive to dissolved chemicals called?   The cells in taste buds that are sensitive to dissolved chemicals are the gustatory cells.

Figure 10–6

Taste buds that are sensitive to what primary sensation are located on the tip of the tongue?   Taste buds that are sensitive to the sweet sensation are located on the tip of the tongue.

Figure 10–7

Why are tears continually secreted?   Tears moisten and lubricate the eye and protect against infection.

Figure 10–8

Does the conjunctiva cover any part of the eyeball?   Yes, the conjunctiva covers the anterior surface of the eyeball (and the inner surface of the eyelid).

Figure 10–9

The contraction of which extrinsic muscle rotates the eye upward?   The superior rectus muscle rotates the eyeball upward.

Figure 10–10

What structures do the suspensory ligaments attach to?   The suspensory ligaments connect muscles of the ciliary body with the lens.

Figure 10–11

What sensory neurons in the eyeball generate an action potential and conduct it through the optic nerve?   Ganglion cells generate an action potential and conduct it through the optic nerve.

Figure 10–12

Which region in this photomicrograph contains a concentration of cone cells?   The fovea centralis contains a concentration of cone cells.

Figure 10–13

What structures of the eyeball bend light as it passes through?   The cornea and lens bend light as it passes through the eye. Also, aqueous humor (in the anterior chamber) and vitreous humor (in the posterior cavity) bend light as well.

Figure 10–14

Describe the pathway of a nerve impulse from the optic nerve to the visual cortex.   A nerve impulse from the eyeball passes along the optic nerve, portions of which cross those of the other eye at the optic chiasma. Thus, the impulse may cross at the chiasma to the opposite side of the brain (along the optic tract). The impulse may terminate at nuclei that control eye reflexes, or continue to the thalamus. Upon reaching the thalamus, the impulse is transmitted to another neuron, which conducts it to the occipital lobe of the cerebral cortex.

Figure 10–15

What is the function of the three auditory ossicles?   The auditory ossicles transmit a mechanical vibration from the tympanic membrane to the oval window. As it is transmitted, the vibration is amplified.

Figure 10–16

Which region of the labyrinth contains the organs of hearing?   The cochlea contains the organs of Corti, which are the organs of hearing.

Figure 10–17

What fluid may be found within the cochlear duct?   Endolymph is the fluid circulating within the cochlear duct.

Figure 10–18

How is a vibration dissipated quickly, readying the inner ear for the next vibration?   As perilymph waves reach the end of the cochlea, they pass to the scala tympani and are dissipated by passing through the round window and into the tympanic cavity.

Figure 10–19

What body activity is detected by the hair cells within the macula?   The action of tilting the head to one side is detected by the hair cells within the macula.

Figure 10–20

What function is served by the jellylike mass, the cupula, in detecting sudden head movements?   Movement of the cupula provides information to the brain regarding body position, particularly that of the head, in order to maintain equilibrium.

Figure 10–21

What is the function of the somesthetic association area?   The somesthetic association area receives impulses from the thalamus and general sensory area and interprets the nature of the sensation. It also stores memories of past sensory experiences, enabling you to relate the present experience with experiences in the past.

Figure 10–22

Are all sensations equally represented by the cerebral cortex?   No, sensations are unequally represented.

Figure 10–23

What roles are performed by the limbic system in memory recall and in emotions?   The limbic system plays a central role in both memory recall and emotions, for it consists of functional aspects of several regions of the brain that are active in these functions.

Figure 10–24

How many motor neurons are usually involved in a somatic motor pathway?   Two motor neurons are usually involved in a somatic motor pathway.

THE ENDOCRINE SYSTEM

LEARNING OBJECTIVES

After studying this chapter, you should be able to:

1. Distinguish between exocrine glands and endocrine glands.

2. Define a *hormone*, and describe why it affects only target cells.

3. Describe the second-messenger system used by water-soluble hormones.

4. Describe how genes are activated by lipid-soluble hormones.

5. Explain how prostaglandins affect the body and how they differ from hormones.

6. Distinguish between negative and positive feedback control mechanisms.

7. Describe the location and structure of each of the primary endocrine glands.

8. Identify the hormones produced by each primary endocrine gland and describe their effects.

The endocrine system works in concert with the nervous system in controlling body activities. Like the nervous system, it, too, provides control in order to keep the body functioning despite changing conditions in the internal or external environment. Thus, its primary role is in helping to maintain homeostasis.

In the previous two chapters, we discovered that the nervous system is capable of responding to changing conditions in an extremely rapid manner. The generation and conduction of nerve impulses occurs very rapidly along special nerve pathways. The endocrine system communicates to the body in a very different manner. It does so by the release of chemicals, known as **hormones**,[1] that circulate through the bloodstream until they reach the body structure they are to affect. Because the bloodstream flows much slower than the conduction of a nerve impulse along a nerve, the hormone takes much longer to reach its destination. Thus, the control provided by the endocrine system occurs more slowly.

Despite its slow response time, the endocrine system performs important functions the nervous system cannot. For example, the effects of hormones are much more long-lasting and widespread, enabling the endocrine system to regulate continuing processes that go on for long periods of time and affect large areas of the body. These processes include metabolism, altering the chemical composition of cells and surrounding fluids, growth and development, reproduction, and regulating the body's responses to stress. The proper control of these processes is essential for our survival.

---

[1]Hormone: Greek *hormon* = arousing, setting into motion.

COMPOSITION OF THE ENDOCRINE SYSTEM

The endocrine system is composed of the numerous glands that secrete their products into the extracellular space. These products are the hormones.

The endocrine system is composed of organs that produce and secrete hormones. Because these organs perform mainly a secretory function, they are also referred to as *glands*. As you should recall from Chapter 4, there are two types of glands in the body: **exocrine**[2] and **endocrine**[3] (Fig. 11–1). Exocrine glands secrete their products into ducts. The ducts transport the products into body cavities, into the space within organs, or onto the body surface. Exocrine glands include oil glands, sweat glands, mucus glands, and salivary glands. In contrast, endocrine glands secrete their products into the extracellular space surrounding the secretory cells. For this reason they are sometimes called *ductless glands*. Their products are the hormones, which diffuse from here to enter the bloodstream.

[2]Exocrine: Greek *exo-* = external + *crinein* = to release, or secrete.
[3]Endocrine: Greek *endo-* = internal + *crinein*.

The primary endocrine glands of the body are the pituitary gland, the thyroid gland, the parathyroid glands, the adrenal glands, the pancreas, and the sex glands, or gonads (the testes in the male and ovaries in the female). Endocrine glands that provide a minor role in body maintenance include the pineal gland and the thymus gland. In addition, there are certain organs of the body that contain a relatively small number of endocrine cells. Their role in body maintenance through hormone action is secondary to their main function. These organs include the kidneys, the stomach, the small intestine, and the placenta (associated with the developing fetus). These secondary endocrine structures will be examined in later chapters. The locations of the endocrine glands are shown in Figure 11–2.

CONCEPTS CHECK

1. How do exocrine and endocrine glands differ?

2. What are the primary endocrine glands in the body?

3. Are the primary endocrine glands the only structures that secrete hormones?

Figure 11–1
Exocrine and endocrine glands. (a) Exocrine glands secrete their products into ducts, which carry them to the body's surface or to the inside of a body cavity. (b) Endocrine glands release their products into surrounding tissues, where they find their way by diffusion into the bloodstream.

Identify several examples of exocrine glands we have studied previously.

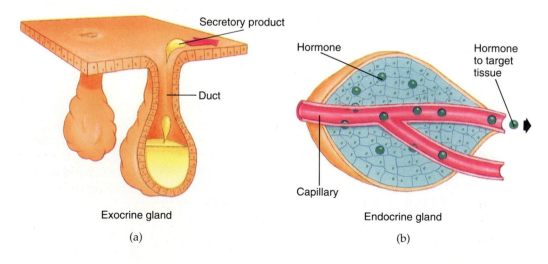

Exocrine gland
(a)

Endocrine gland
(b)

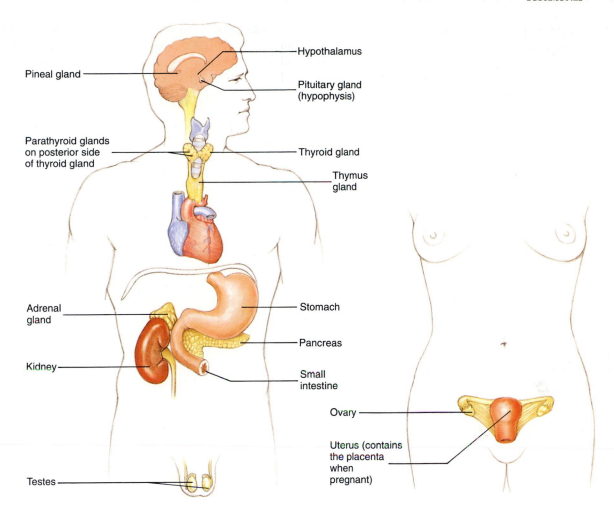

Figure 11–2
The endocrine glands.
Which endocrine glands play a role in maintaining homeostasis?

HORMONES

Hormones are chemicals that act only on target cells. Their effect is an alteration of target cell activity, and their release is regulated mainly by feedback control.

Hormones are the chemical units produced by endocrine glands. They are the means by which endocrine glands provide control of body activities to maintain homeostasis. How do they accomplish this control? Let's turn now to the nature of hormonal action so that we may answer this question.

Hormonal Action

Hormones act only upon cells to which they selectively bind. The two major classes of hormones induce changes in cellular activity in two different ways.

Hormones are released in very small quantities, for they are extremely potent compounds. Once released by secretory cells into the extracellular space, they find their way by diffusion into the bloodstream. Some may travel short distances, although most circulate with the blood throughout the body. In either case, a given hormone will have an effect only on a particular type of cell. This cell is called the **target cell.** Basically, all cells in the body are target cells for one or more hormones, although a select number of cells respond to a particular hormone. The effect is limited to the target cells because only they contain special protein molecules in their plasma membrane that serve as *receptors*, which recognize and bind to specific hormones while rejecting others. Cells other than target cells are not affected by a hormone, because they lack the appropriate receptors.

Once a hormone has united with the receptor on a target cell plasma membrane, it begins to exert its effect. In general, its effect is to alter the cell's metabolic processes. For example, a hormone may change the rates of enzyme activities, the rate of protein synthesis, the rates of secretion, or the rates at which materials are transported across the plasma membrane. These effects and others are largely determined by the chemical nature of the hormone.

Although there are many types of hormones that differ chemically, hormones may be grouped into two broad categories on the basis of their solubility characteristics: those that dissolve in water, or are *water-soluble*; and those that dissolve in lipids, or are *lipid-*soluble. A group of substances closely related to hormones, called *prostaglandins*, will also be considered here.

Water-Soluble Hormones

Hormones that are soluble in water include molecules that are composed of amino acids. These are proteins and proteinlike molecules. Because these hormones are soluble only in water, they cannot pass through the lipid plasma membrane. How can they produce an effect on the cell if they cannot penetrate the membrane? This is done by passing the signal to a **second-messenger system** located within the cell. The most common and best-understood second-messenger system is one that uses a compound called **cyclic AMP** (adenosine monophosphate), so it is this system we will now discuss (Fig. 11–3).

The initial contact between a water-soluble hormone and a target cell is made when the hormone chemically binds to a receptor that is linked to a membrane-bound enzyme. In our cyclic AMP system, the enzyme is called *adenylate cyclase*. The hormone is re-

Figure 11–3
Action of water-soluble hormones: the second-messenger system. In this example, cyclic AMP is the second messenger within the cell. (a) The binding of the hormone to the receptor molecule in the plasma membrane initiates the conversion of ATP to cyclic AMP by adenylate cyclase. (b) The presence of cyclic AMP causes the activation of enzymes in the cytoplasm, known as protein kinases. (c) Protein kinases initiate other reactions in the cell, producing a number of activated substrates as a result. (d) The activated substrates induce changes in the cell, which may include altering metabolic pathways, altering membrane permeabilities, or secreting cellular products such as hormones.

Why must water-soluble hormones rely on the action of second messenger chemicals to initiate an effect on a cell?

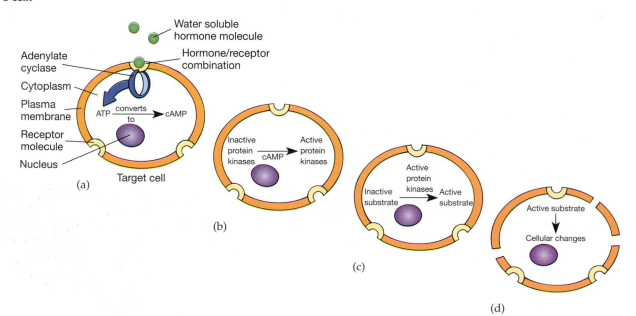

ferred to as the first messenger, and its contact with adenylate cyclase activates the conversion of ATP to cyclic AMP (cAMP). Cyclic AMP acts as the second messenger, diffusing throughout the cell to initiate a cascade of chemical reactions that result in the activation of enzymes, called *protein kinases.* The activated protein kinases proceed to activate other proteins within the cell, which react with other molecules to induce changes in the cell.

The enzymatic cascade activated by the second-messenger system has an enormous amplification effect within the cell. A single hormone molecule triggers a single enzyme, which catalyzes literally hundreds of reactions. These reactions are capable of producing profound changes in the cell, which include stimulating or inhibiting various metabolic pathways, promoting the synthesis of proteins, altering membrane permeabilities, and stimulating the secretion of hormones and other cellular products. Water-soluble hormones that serve as first messengers in this system include epinephrine, norepinephrine (NE), antidiuretic hormone (ADH), oxytocin (OT), calcitonin (CT), and parathyroid hormone (PTH).

Lipid-Soluble Hormones

Hormones that dissolve in lipids include mainly steroid hormones. Because the plasma membrane is composed of a bilayer of lipid molecules, steroid hormones can pass directly through it by diffusion to enter the target cell quite easily (recall that steroids are a type of lipid also; lipids dissolve in other lipids). Once inside, they bind to a protein receptor located within the cytoplasm (Fig. 11–4). The hormone-receptor complex that is formed moves into the nucleus of the cell, where it binds to a region of the DNA specific for the protein receptor component. This reaction "turns on" the synthesis of specific protein molecules by DNA through the processes of transcription and translation (see Chapter 3 to review these processes). Thus, lipid-soluble hormones activate genes to synthesize new proteins and enzymes.

The protein products that are newly formed include enzymes that promote the metabolic activities specified by the hormone. These enzymes may stimulate or inhibit metabolic pathways, or alter the rates of other cellular processes. The protein products may also

Figure 11–4
Action of lipid-soluble hormones. (a) Steroid hormones are able to diffuse directly through the plasma membrane. (b) Once inside the cell, a steroid hormone binds with a protein receptor. The newly formed complex migrates to the nucleus and enters it. (c) Once within the nucleus, the steroid-protein complex binds to a specific region of DNA. (d) The reaction between the steroid-protein complex and DNA activates genes to synthesize new proteins and enzymes.

The synthesis of what materials within a cell is stimulated by steroid hormones?

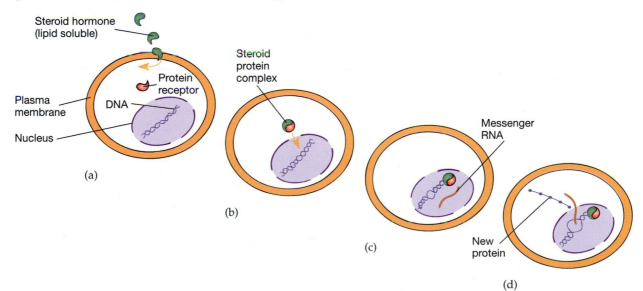

act as structural proteins in the plasma membrane. Lipid-soluble hormones that stimulate protein synthesis include aldosterone, cortisol, testosterone, estrogen, and thyroxine.

Prostaglandins

Prostaglandins (prō-sta-GLAN-dinz) are a group of chemicals that also have regulating effects on cells. They are lipids that are produced by many different parts of the body. These include the lungs, liver, digestive tract, kidneys, pancreas, brain, and various reproductive organs. Like hormones, they are extremely potent compounds and are released in very small quantities. They are often referred to as *local hormones* because they act on cells within close vicinity of their site of release.

Prostaglandins interact with hormones to regulate various metabolic activities. By virtue of their lipid structure, they can move quickly across plasma membranes by simple diffusion. This enables them to modify the effects of certain hormones directly within the cell. Specifically, prostaglandins stimulate or inhibit the formation of cyclic AMP, thereby modulating the effect of hormones that use cyclic AMP as a second messenger. Because they do not induce their own effect but instead modify the effect of a hormone, prostaglandins as a group are not considered true hormones.

There are many types of prostaglandins, each of which exerts a specific effect on target cells that lie nearby. For example, some prostaglandins reduce blood pressure and open airways by causing smooth muscles to relax, while others have the opposite effect by stimulating smooth muscle contraction. Other types inhibit the secretion of hydrochloric acid from the stomach wall, increase intestinal contractions, stimulate contraction of the uterus, regulate metabolism, cause inflammation, and even cause fever. Recent evidence has indicated that some prostaglandins have sleep-inducing and antineoplastic (tumor-destroying) activities as well.

CONCEPTS CHECK

1. Why doesn't a particular hormone generally induce changes in all body cells?

2. How does the second-messenger system pass information to the inside of a cell?

3. How do lipid-soluble hormones induce changes in the cell?

4. How does a prostaglandin differ from a hormone?

Hormonal Control

The control of hormone secretion rates is provided by chemical signals that relay information about the output levels, and in some cases, by nervous control.

We have seen that hormones are very potent substances. Any uncontrolled variation in their concentration can have devastating effects on the body. Therefore, the production and release of hormones must be regulated very carefully. As we shall see, the most important way of controlling endocrine gland activity is by a mechanism called *feedback control.*

Feedback Control

How does an endocrine gland "know" how much hormone to produce and release? This information, or **feedback,** is provided by way of chemical signals that are sent to the endocrine gland. If conditions change, the endocrine gland can act on the feedback and adjust its rate of secretion. There are two systems that operate in this manner: negative feedback systems and positive feedback systems.

Negative feedback systems control the amount of hormone released by providing a response in the opposite direction to that of the stimulus. In these systems, the secretion of a hormone that accelerates a body activity is inhibited by the negative feedback signal, and the secretion of a hormone that slows the same body activity is stimulated yet further. The negative feedback signal is provided by the secreted hormone or its products, which exert their negative effects on endocrine gland activity when they reach a certain level in the blood.

Parathyroid hormone provides an excellent example of a negative feedback system (Fig. 11–5). In this system, parathyroid hormone secreted by the parathyroid glands helps regulate calcium in the blood by stimulating the release of calcium from bones. When blood calcium levels rise above normal levels, the parathyroid glands are inhibited and their hormone production declines. Thus, a high blood calcium level provides a stimulus to the parathyroid glands to stop their production of parathyroid hormone. In this example, the effect of high product levels in the blood *inhibits* further hormone production. Negative feedback systems like this one are the most common method of hormone regulation in the body.

Positive feedback systems regulate hormone secretion by providing a response in the same direction as the stimulus. For example, when the desired response stimulated by hormone action occurs, a chemical feedback signal causes the endocrine gland to increase its rate of hormone release and more responses are stimulated (Fig. 11–6).

duction, and uterine contractions respond by gradually increasing in strength until birth is accomplished.

Nervous Control

A second way of controlling hormone release is by the nervous system. Nervous control is responsible for regulating only some endocrine glands, such as the adrenal medulla and secretory cells in the hypothalamus of the brain. These glands secrete hormones when they receive nerve impulses. When the impulses stop, hormone secretion stops also.

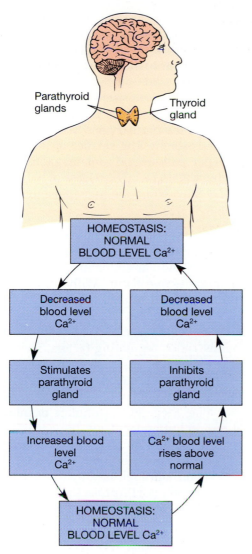

Figure 11–5
Negative feedback. The parathyroid gland produces a hormone that causes an increase in calcium ions (Ca^{2+}) in the blood. When Ca^{2+} levels rise above a normal level, the continued secretion of this hormone is inhibited. The inhibitory effect of rising levels is called *negative feedback*.
In a negative feedback system, what provides the signal to the endocrine gland, causing it to reverse its activity?

> **CONCEPTS CHECK**
>
> 1. What information is provided by negative feedback?
>
> 2. How does positive feedback differ from negative feedback control?
>
> 3. What is an example of a negative feedback system?

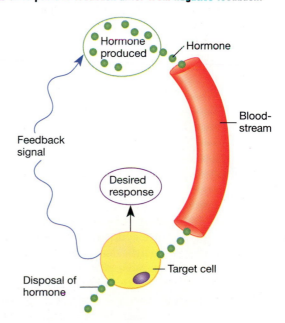

Figure 11–6
Positive feedback. As the desired response stimulated by hormone action is produced, a feedback signal stimulates continued secretion of the hormone. As a result, the desired response also continues.
How does positive feedback differ from negative feedback?

Positive feedback systems tend to cause extreme changes in conditions in the body and are therefore quite unstable and uncommon. One example, however, is in the production of oxytocin by the pituitary gland. Oxytocin stimulates uterine contractions during childbirth. Its rising levels in the blood cause the formation of products that stimulate further oxytocin pro-

THE ENDOCRINE GLANDS

The endocrine glands are hormone-secreting organs that are widely distributed.

The glands of the endocrine system do not form an interconnected series of parts, like what we've seen in the organs of the skeletal, muscular, and nervous systems. Instead, they are scattered in various areas throughout the body: for example, some are in the head, some in the neck, and some in the abdominal cavity. In this section we will examine the location and structure of each endocrine gland, as well as the hormones each produces and their primary effects on the body.

Pituitary Gland

The tiny pituitary gland produces many hormones, some of which control the activities of other glands. Both of its lobes receive their control from the hypothalamus.

The **pituitary** (pi-TOO-i-tar-ē) **gland**, or **hypophysis,** is located at the base of the brain (see Fig. 11–7). It is a remarkable little gland about the size of a pea and weighs only 0.5 gram (0.02 ounce). It is attached to the hypothalamus by a narrow stalk, called the **infundibulum,** and lies within a bony cavity formed by the sella turcica of the sphenoid bone. The pituitary gland produces many hormones, some of which control the activities of several other endocrine glands. It thereby influences a wide range of body functions.

The pituitary gland consists of two portions: an anterior lobe and a posterior lobe. The anterior lobe is the glandular part of the pituitary and accounts for about 75% of its total weight. The smaller posterior lobe is composed of nerve tissue and contains axons that originate in the hypothalamus. The release of hormones by both lobes is regulated by the hypothalamus.

Anterior Lobe

The **anterior lobe** (adenohypophysis) of the pituitary is composed of glandular epithelium enclosed within a capsule of connective tissue. Within its epithelium are five different types of secretory cells that release seven types of hormones. The release of these hormones is controlled by chemical secretions from the hypothalamus, called **regulating factors.** These chemicals are carried a short distance in the bloodstream, from the hypothalamus directly to the anterior lobe (Fig. 11–8). Once the regulating factors are received in the anterior lobe, they either stimulate or inhibit the release of certain hormones. Thus, by way of regulating factors, the hypothalamus controls the activity of the pituitary gland's anterior lobe.

The seven hormones released by the anterior lobe are growth hormone (GH), prolactin (PRL), thyroid-

Figure 11–7
The pituitary gland. (a) Lateral view of the brain, sagittal section. (b) Close-up view of the pituitary gland attached to the base of the brain.

What part of the brain forms a direct connection with the pituitary gland?

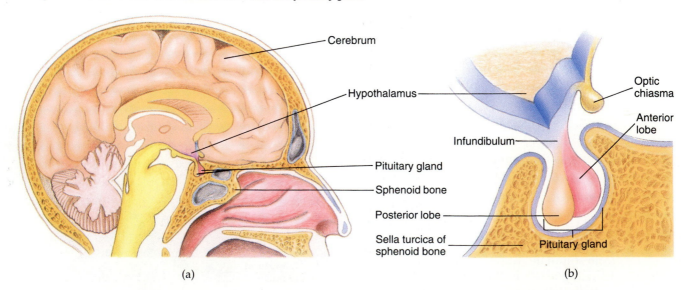

(a) (b)

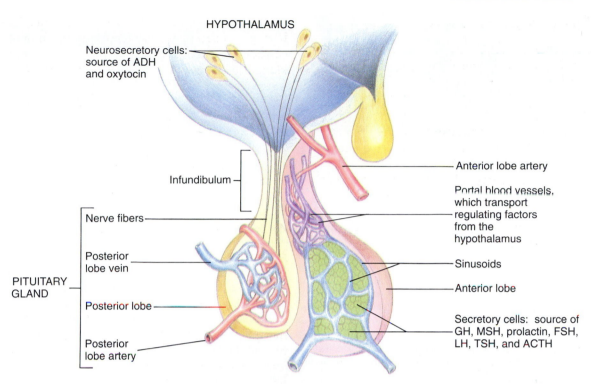

HYPOTHALAMUS

Neurosecretory cells: source of ADH and oxytocin

Infundibulum

Nerve fibers

PITUITARY GLAND

Posterior lobe vein

Posterior lobe

Posterior lobe artery

Anterior lobe artery

Portal blood vessels, which transport regulating factors from the hypothalamus

Sinusoids

Anterior lobe

Secretory cells: source of GH, MSH, prolactin, FSH, LH, TSH, and ACTH

Figure 11–8
Communication between the hypothalamus and the pituitary gland. The anterior lobe receives regulating factors from the hypothalamus by way of portal blood vessels. The posterior lobe receives neurons (neurosecretory cells) that originate in the hypothalamus. The neurosecretory cells produce the hormones ADH and oxytocin, which are transported to the posterior lobe prior to their release into the bloodstream.
Which lobe is composed of glandular epithelium, and which is composed of nerve tissue?

stimulating hormone (TSH), adrenocorticotropic hormone (ACTH), melanocyte-stimulating hormone (MSH), follicle-stimulating hormone (FSH), and luteinizing hormone (LH). These hormones and a summary of their effects are shown in Figure 11–9. Four of the anterior lobe hormones are called **tropic hormones,**[4] for they stimulate other endocrine glands. They are TSH, ACTH, FSH, and LH. Two of these are also called **gonadotropic hormones,** for they affect the gonads, or sex organs. They are FSH and LH.

Growth Hormone

Growth hormone (GH) stimulates body cells to grow and divide. It causes cells to increase in size by enhancing the movement of amino acids into the cells and their incorporation into newly synthesized proteins. It also increases the rate at which carbohydrates and fats are catabolized (that is, broken down to release energy). On a more short-term basis, the nutri-

tional status of your body affects the release of GH in order to maintain a relatively constant blood sugar level. When sugar levels are low, a condition called *hy-*

Figure 11–9
The pituitary hormones and the locations of their target cells.
What controls the release of posterior lobe hormones?

Pituitary	Hormone	Location of target cells
Anterior lobe	ACTH	Adrenal cortex
	TSH	Thyroid
	FSH, LH	Ovary, Testis
	Prolactin	Mammary glands
	MSH	Skin pigment
	GH	Mainly bone
Posterior lobe	ADH	Kidney tubules
	Oxytocin	Mammary glands
		Uterine muscles

---

[4]Tropic: *tropic* = turning or changing; acting on.

Growth Abnormalities

Excessively high or insufficient production of growth hormone (GH) by the anterior lobe of the pituitary gland causes abnormal growth patterns. Excessive production leads to gigantism and acromegaly. *Gigantism* is usually caused by the presence of a tumor of the pituitary gland and is the result of excessive bone growth before onset of puberty. The long bones of the appendages are affected the most, sometimes resulting in the afflicted person's reaching a height of 8 feet or more (the tallest individual on record was Robert Wadlow of England, who reached a height of 9'8"). Internal organs are also proportionately larger than those of normal people. If GH production suddenly increases after an individual has reached adult age (after the epiphyseal plates in bones have ossified), the bones cannot lengthen, but they can thicken. This results in a massive enlargement of bones of the hands, feet, jaw, and forehead. This condition is called *acromegaly.*

Insufficient production of GH during the normal growth years before puberty results in *pituitary dwarfism.* This condition is a direct effect of a lack of bone deposition and lengthwise bone growth at the epiphyseal plates. Although pituitary dwarfs are small, they have normal intelligence and their bodies are usually well proportioned. If pituitary dwarfism is detected early, its effects can be minimized by growth hormone therapy. Other causes of dwarfism include a genetic, incurable bone disease called *achondroplasia;* insufficient production of sex hormones during adolescence; insufficient production of thyroid growth hormones (*cretinism*); and severe childhood stress (caused by insufficient diet and rest, and emotional deprivation).

poglycemia exists and the hypothalamus is stimulated to release regulating factors. Once these factors reach the anterior lobe, GH is released into the bloodstream. The presence of GH on target cells causes the conversion of glycogen into glucose and its release into the blood. As a result, blood sugar levels rise. High levels of sugar in the blood, or *hyperglycemia,* cause the opposite effect (GH is inhibited). Thus, your blood sugar levels are kept relatively constant by negative feedback mechanisms involving GH. As we will learn a bit later, other hormones also interact to maintain the levels of sugar in the blood.

Melanocyte-Stimulating Hormone

Melanocyte-stimulating hormone (MSH) stimulates the production of melanin in the skin, causing the skin to increase in pigmentation. Its secretion is controlled by regulating factors released by the hypothalamus, which may stimulate or inhibit its production.

Prolactin

In combination with other hormones, **prolactin (PRL)** stimulates and maintains milk secretion by the mammary glands in females. The actual ejection of milk is controlled by a hormone released by the posterior lobe, called *oxytocin.* The combined secretion and ejection of milk from the mammary glands is an activity referred to as *lactation.* PRL levels are controlled by the hypothalamus through regulating factors that may stimulate or inhibit its secretion. Its levels are influenced by the hormonal changes that occur during menstruation, pregnancy, and delivery, which will be discussed in Chapters 19 and 20.

Thyroid-Stimulating Hormone

The production and secretion of hormones by the thyroid gland are stimulated by **thyroid-stimulating hormone (TSH).** Its control is provided by regulating factors released from the hypothalamus. It is influenced by the body's metabolic rate, levels in the blood of a thyroid hormone called *thyroxine,* and other factors.

Adrenocorticotropic Hormone

The production and secretion of certain hormones released by the outer region, or *cortex,* of the adrenal gland are controlled by the **adrenocorticotropic** (a-drē′-nō-kor′-tik-ō-TRŌ-pik) **hormone (ACTH).** The release of ACTH is controlled by regulating factors produced by the hypothalamus in response to low concentrations of adrenal cortex hormones. Its release is also influenced by various forms of stress.

Follicle-Stimulating Hormone

Follicle-stimulating hormone (FSH) has a different effect upon the two sexes. In the female, FSH stimulates the development of eggs, or ova, each month within the ovaries. It also stimulates the cells in the ovaries to secrete estrogens, the female sex hormones. In the male, FSH stimulates the production of sperm by the testes. FSH production is controlled by regulating factors released from the hypothalamus in response to estrogens in the female and to testosterone in the male, in the manner of a negative feedback system.

Luteinizing Hormone

Luteinizing (LOO-tē-i-nī-zing) **hormone (LH)** also plays a different role in each of the two sexes. In females, it works together with estrogens to stimulate the ovary to release an ovum (a process called *ovulation*) and prepare the uterus for implantation of the fertilized ovum. It also plays other roles that will be discussed in Chapter 19. In males, LH stimulates cells within the testes to produce and secrete testosterone. LH secretion is controlled by regulating factors released from the hypothalamus by way of a negative feedback system.

Posterior Lobe

The **posterior lobe** (neurohypophysis) of the pituitary gland is actually an extension of the hypothalamus (see Fig. 11–8). It consists mainly of neuroglial cells that support the terminal ends of axons. The cell bodies of these axons are located within the hypothalamus. These cell bodies secrete two hormones, oxytocin (OT) and antidiuretic hormone (ADH), and are therefore referred to as **neurosecretory cells.** Once they are produced, OT and ADH are packaged within little vesicles and transported down the axons of the neurosecretory cells to the posterior lobe for storage. Thus, the posterior lobe does not produce any hormones; rather, it stores those produced by the hypothalamus until it is time for their release into the bloodstream.

Oxytocin

Oxytocin (OT)[5] stimulates contraction of smooth muscle in the wall of the uterus. It also stimulates cells around mammary ducts to contract, thereby causing milk to eject. As you may suspect, its levels rise toward the end of pregnancy to ready the uterus for the birth process. Continued rising levels signal the beginning of uterine contractions for birth, and levels will rise further still until the baby is expelled. OT levels rise because of the positive feedback mechanism that regulates it (that is, rising levels stimulate more OT to be secreted). An extracted or synthetic form of OT, called Pitocin, is sometimes administered to the mother to initiate or speed labor. It may also be administered soon after birth to control bleeding and recover uterine muscle tone.

After birth, oxytocin levels decline but may rise again when the infant begins sucking at the mother's breast. Sucking stimulates neurosecretory cells in the hypothalamus to produce more OT, and the posterior lobe releases it into the bloodstream. Its rising levels not only cause milk to eject, but also stimulate uterine contractions to promote rapid recovery of the uterus to its prepregnancy size.

Antidiuretic Hormone

Antidiuretic (an-tī-dī-yoo-RET-ik) **hormone (ADH)** regulates fluid balance in the body. It is appropriately named, for the term **antidiuretic**[6] refers to a substance that prevents urination, and basically this is what ADH does. In more precise words, ADH helps the body conserve water by increasing the amount of water reabsorption in the kidneys. During reabsorption, water moves from kidney ducts into the bloodstream. The more water that is reabsorbed and conserved, the less water will be excreted as urine. ADH therefore causes a decrease in urine output and an increase in body fluid volume.

ADH secretion is regulated by the hypothalamus. This part of the brain contains receptors that detect the amount of osmotic pressure in the bloodstream, known as *osmoreceptors.* As you may recall from Chapter 3, osmotic pressure increases as fluids in a compartment become less dilute. Thus, when you become dehydrated from lack of water, the osmotic pressure of your bloodstream rises because the blood is becoming less dilute. An increase in osmotic pressure stimulates the osmoreceptors to trigger ADH synthesis in the neurosecretory cells. Once synthesized, ADH is transported to the posterior lobe and released into the bloodstream. When it reaches the kidneys, it causes more water to be reabsorbed, and your blood becomes more dilute. When you drink an excessive amount of water, the blood becomes more dilute and ADH secretion is inhibited. As a result, the kidneys reabsorb less water and the excess water is eliminated with the urine until your fluid volume is restored to normal. The factors influencing ADH secretion are summarized in Figure 11–10.

ADH also helps to regulate blood pressure. It does so by stimulating the contraction of smooth muscles in the walls of arterioles, a process called *vasoconstriction.* For this reason, ADH is sometimes referred to as *vasopressin.*

---

[5]Oxytocin: Greek *oxy-* = sharp, swift + *tokes* = birth + = *-in* (chemistry) = a neutral substance.

[6]Antidiuretic: *anti-* = against + *diuretic* = referring to passing of urine.

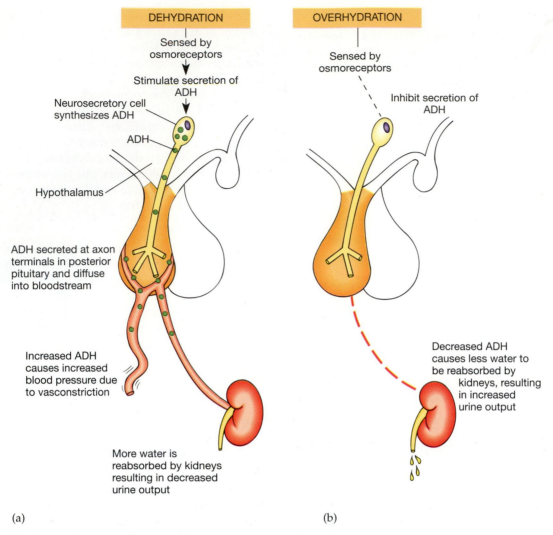

Figure 11–10
The control and action of ADH. (a) ADH secretion from neurosecretory cells in the hypothalamus is stimulated by dehydration (too little water in the blood). Increased levels of ADH stimulate the kidneys to reabsorb more water from the blood, resulting in a decrease of urine output. This reverses the state of dehydration. ADH also increases blood pressure by causing vasoconstriction. (b) A state of overhydration (too much water in the blood) inhibits the secretion of ADH. As a result, less water is reabsorbed by the kidneys and urine output increases, eliminating the excess water.

How does ADH increase blood pressure?

CONCEPTS CHECK

1. How is the secretion of anterior pituitary hormones controlled?

2. What is the function of growth hormone?

3. Where are posterior pituitary hormones actually produced?

4. What is the function of antidiuretic hormone?

Thyroid Gland

The hormones of the thyroid gland help regulate metabolism and growth, and reduce calcium and phosphate levels in the bloodstream.

The **thyroid gland**[7] is the prominent organ in the neck. It is located slightly below the larynx in front of the trachea (Fig. 11–11). It consists of two large *lateral lobes,*

[7]Thyroid: Greek *thyroid* = having the shape of a shield.

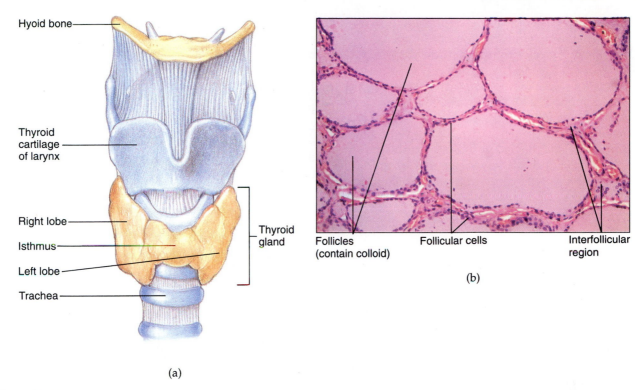

Figure 11–11
Thyroid gland structure. (a) The thyroid gland is located in the neck below the larynx. (b) This photograph of a magnified section through the thyroid shows the follicles, which store colloid.

What hormones produced by the thyroid gland regulate growth and metabolism?

right and left, that are interconnected by a constricted part called the *isthmus.*

The thyroid gland is composed of glandular epithelium within a thin outer capsule of connective tissue. Its epithelium is divided into many small compartments called **follicles.** The walls of each follicle are lined with simple cuboidal epithelium and enclose a cavity that contains a clear, viscous fluid known as *colloid.* The follicle cells produce and secrete the thyroid hormones, which may be stored temporarily in the colloid or released immediately into the bloodstream by way of capillaries.

The three primary hormones produced by the thyroid gland are thyroxine, also known as T_4 (because it has four atoms of iodine); triiodothyronine, also called T_3 (because it has three atoms of iodine); and calcitonin. Because thyroxine and triiodothyronine have similar effects on the body, they are often referred to collectively as the "thyroid hormones."

Thyroxine and Triiodothyronine

Thyroxine (T_4) and **triiodothyronine** (trī-ī-ō-dō-THĪ-rō-nin) **(T_3)** play important roles in metabolism and growth. They both contain iodine, and they both stimulate the rate of metabolism, promote protein synthe-

sis, increase the rate of glucose uptake, promote lipid metabolism, and accelerate actions of the nervous system. These effects are extremely widespread, since these important hormones influence nearly every cell in the body. It is therefore of vital concern that their levels be carefully regulated.

The concentrations of thyroid hormones in the blood are regulated by a negative feedback mechanism. It takes place between the hypothalamus, the anterior lobe of the pituitary, and the thyroid gland. When thyroid hormone levels in the blood drop, the anterior lobe is stimulated by the hypothalamus to secrete thyroid-stimulating hormone (TSH). Once TSH reaches the thyroid gland, it stimulates follicle cells to release thyroid hormones into the bloodstream. When levels are restored, the hypothalamus inhibits further secretion of TSH, and follicle cells stop releasing thyroid hormones.

Calcitonin

Calcitonin (kal-si-TŌ-nin) reduces the calcium and phosphate levels in the blood. It functions in concert with parathyroid hormone released by the parathyroid glands to regulate the concentrations of these ions. Calcium concentrations must be kept within narrow limits for normal nerve and muscle function, and both ions are essential mineral components of bone.

The release of calcitonin is controlled directly by the levels of calcium in the blood. As calcium levels rise, calcitonin secretion increases. The action of calcitonin is upon bone cells that dissolve the mineralized bone matrix, called *osteoclasts,* and upon kidney cells. Calcitonin inhibits osteoclast activity and stimulates the excretion of calcium and phosphate ions by the kidneys. The result of these activities is the lowering of calcium and phosphate concentrations in the blood.

CONCEPTS CHECK

1. Where is the thyroid gland located?

2. How are the secretions of thyroxine and triiodothyronine regulated?

3. What is the function of calcitonin?

Parathyroid Glands

The small parathyroid glands behind the thyroid in the neck secrete parathyroid hormone, which increases calcium and phosphate levels in the blood.

The **parathyroid glands** are four or five pea-shaped masses of glandular epithelium. They lie on the posterior side of the thyroid gland, usually two or three to each lobe (Fig. 11–12). Each gland consists of tightly packed secretory cells surrounded by a thin capsule of connective tissue. They secrete one hormone, called parathyroid hormone (PTH), which diffuses into nearby capillaries for its circulation throughout the body.

Parathyroid Hormone (PTH)

Parathyroid hormone (PTH) plays an important role in maintaining the calcium and phosphate levels in the blood. It increases blood calcium levels, thus opposing the effects of calcitonin released by the thyroid gland. It also decreases blood phosphate levels. Unlike most other hormones, its rate of secretion is not controlled by the hypothalamus or pituitary gland.

PTH is released by the parathyroid glands in response to low calcium levels in the blood (Fig. 11–13). Once released, PTH increases the numbers and activities of osteoclasts in bone tissue. The osteoclasts break down bone matrix to release calcium ions, which diffuse into the bloodstream. As a result, calcium levels in the blood rise. This effect is enhanced by the stimulation of calcium reabsorption in the kidneys, which reduces the amount of calcium normally lost by excretion. Also, PTH activates vitamin D, which increases the amount of calcium absorbed from the digestive tract into the bloodstream. Blood phosphate levels, on the other hand, decline in the presence of PTH. This is due to an increased rate of phosphate excretion by the kidneys.

CONCEPTS CHECK

1. Where are the parathyroid glands located?

2. How does parathyroid hormone increase blood calcium levels?

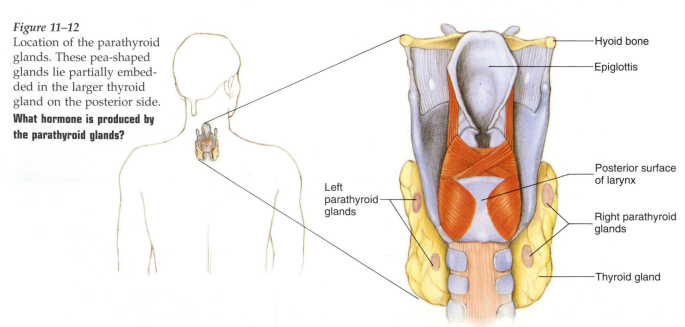

Figure 11–12
Location of the parathyroid glands. These pea-shaped glands lie partially embedded in the larger thyroid gland on the posterior side.
What hormone is produced by the parathyroid glands?

Hyoid bone

Epiglottis

Posterior surface of larynx

Right parathyroid glands

Thyroid gland

Left parathyroid glands

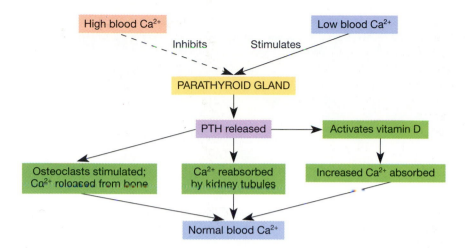

High blood Ca²⁺ ---- Inhibits ---- Stimulates ← Low blood Ca²⁺

PARATHYROID GLAND

PTH released → Activates vitamin D

Osteoclasts stimulated; Ca²⁺ released from bone | Ca²⁺ reabsorbed by kidney tubules | Increased Ca²⁺ absorbed

Normal blood Ca²⁺

Figure 11–13
Control and action of PTH.

What hormone counteracts the effects of PTH to maintain proper calcium ion levels?

Adrenal Glands

Each adrenal gland contains two portions. The hormones released by the inner medulla perform functions similar to functions of the sympathetic nervous system, and the outer cortex secretions perform many roles including the regulation of fluid balance, glycogen synthesis, and sex characteristics.

The **adrenal glands**[8] are paired, triangular masses that lie atop each kidney (Fig. 11–14). Like the kidneys, they are located behind the membrane that encloses the abdominal cavity, which is called the peritoneum. Their internal structure reveals the two parts they are di-

[8]Adrenal: Latin *ad-* = toward + *renal* = pertaining to the kidney.

Figure 11–14
Adrenal gland structure. (a) Each adrenal gland is located on top of a kidney. (b) Sectioning through an adrenal gland reveals its two parts, the outer cortex and inner medulla. (c) This photograph of a magnified section shows the cellular organization of the adrenal medulla.

How do the adrenal cortex and adrenal medulla differ in structure?

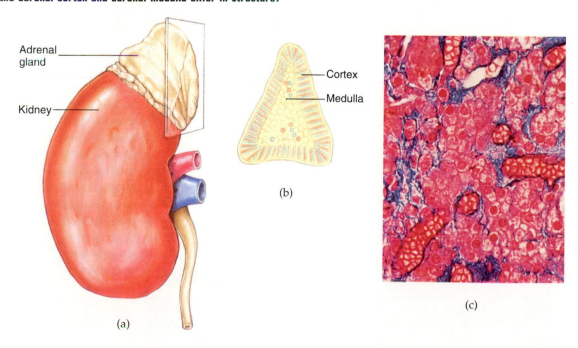

Adrenal gland

Kidney

Cortex

Medulla

(a)

(b)

(c)

The Calcium Flux

The body's need for calcium changes with body activity, diet, and health. For example, during times of strenuous exercise, calcium needs are high due to rates of muscle contraction, while during restful times calcium needs are low. Despite these fluctuating needs, calcium levels in the blood must be kept relatively constant. This regulation is provided by calcitonin and parathyroid hormone, and is illustrated in the diagram.

Although calcitonin and parathyroid hormone exert opposite effects, they work in concert to minimize calcium fluctuations. For example, when calcium levels rise more than 20% above normal, calcitonin is released from the thyroid gland and quickly acts to inhibit further bone decomposition. Conversely, when calcium levels drop more than 10% below normal, PTH is released to restore the blood calcium concentration. These upper and lower limits must not be exceeded, or serious consequences will follow. As an example, if calcium levels fall 40% below the average blood calcium level of 10 milligrams per 100 milliliters, nerve fibers become uncontrollably excitable. This may cause convulsions that are similar to epileptic seizures. They also cause muscles to undergo spasm and to go into tetany, or fixed contraction. If the muscles of the larynx or pharynx go into tetany, respiratory passages become blocked and death may follow if immediate treatment is not available. If calcium levels exceed the upper limits of health, dementia and permanent brain damage can result.

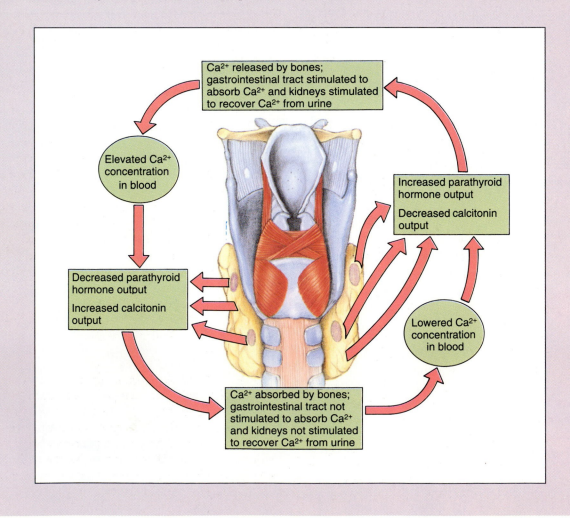

vided into: an inner medulla and an outer cortex. Although they are connected anatomically, these parts develop from different types of embryonic tissue and function as distinct glands.

Adrenal Medulla

The **adrenal medulla** is composed of modified nerve tissue. Its cells originate from autonomic neurons that change during the late embryonic stage of life. These cells retain some of their former characteristics, as they remain connected to preganglionic fibers from the sympathetic division of the autonomic nervous system. The adrenal medulla is therefore closely associated with the sympathetic division—so closely, in fact, that its cells may be regarded as postganglionic neurons. It secretes two hormones, epinephrine and norepinephrine.

Epinephrine and Norepinephrine

Epinephrine and **norepinephrine** are two closely related substances produced by the adrenal medulla. They are very similar in their composition and function. In fact, epinephrine is produced from norepinephrine by enzymes within secretory cells. Also, you may recognize the term *norepinephrine*: this is the same substance that is released by sympathetic neurons as a neurotransmitter in the "fight-or-flight" response. Together these two hormones have the same effect on the body as norepinephrine in its neurotransmitter role, although their effects last about ten times longer. This prolonged effect occurs because the hormones are removed from blood more slowly than the neurotransmitter is removed from synapses.

Epinephrine and norepinephrine are *sympathomimetic* (sim′-pa-thō-mī-MET-ik), which means that they cause changes in the body that mimic those produced by the sympathetic nervous system. These changes include a rise in metabolic rate; a routing of blood flow to vital organs such as the brain, heart, and skeletal muscles; a faster heart rate; increased alertness; and enlargement of the airways. Thus, the body is prepared physiologically to cope with emergency or threatening situations. We can think quickly, run faster, and fight harder for longer periods of time as a result of secretions of these hormones. The adrenal medulla therefore functions as the emergency gland of the body.

The secretion of epinephrine and norepinephrine is under nervous control. Although the hormones are released continuously under normal conditions, their rate of secretion is increased when anxiety is aroused. Anxiety is perceived by the hypothalamus, which responds by stimulating the adrenal medulla to increase its output. The signal between the hypothalamus and adrenal medulla is conducted along sympathetic preganglionic neurons.

Adrenal Cortex

The **adrenal cortex** is structurally similar to a more typical endocrine gland, since it is composed of glandular epithelium. It occupies the larger portion of the adrenal gland, and its epithelial cells are arranged in several distinct layers to form outer, middle, and inner zones. The secretory cells in each zone of the adrenal cortex secrete steroid hormones. These hormones are synthesized from cholesterol. They include three classes of compounds: mineralocorticoids (min′-er-al-ō-KOR-ti-koydz), glucocorticoids (gloo′-kō-KOR-ti-koydz), and sex hormones.

Mineralocorticoids

The primary mineralocorticoid is **aldosterone.** This steroid hormone maintains body fluid balance. It acts by regulating the concentrations of sodium and potassium ions. Its main target cells are in the kidneys, which are stimulated to conserve sodium and to excrete potassium. Other targets include sweat glands, salivary glands, and the digestive tract. In each case, the cells in these organs are stimulated by aldosterone to retain sodium and to remove excess potassium. The retention of sodium by aldosterone enables the body to hold an appropriate volume of water; should sodium be lost, water would follow it by osmosis, and the blood volume (and blood pressure) would drop dangerously low. Therefore, an indirect effect of aldosterone secretion is an increase in blood volume, which is accompanied by an increase in blood pressure.

The secretion of aldosterone is activated by a decrease in blood pressure arising from a drop in fluid volume, or by a decrease in sodium ion concentration in the blood. These changes are sensed by special cells in the kidneys, which respond by activating a series of reactions resulting in the stimulation of the adrenal cortex to produce more aldosterone. This regulatory mechanism is called the **renin-angiotensin pathway** and will be discussed further in Chapter 18. Aldosterone production may alternatively be stimulated directly by rising levels of potassium ions in extracellular fluids. As a result of increased aldosterone, the kidney is stimulated to increase its rate of potassium-ion removal from body fluids, and the excess potassium is excreted with the urine.

Glucocorticoids

Glucocorticoids include **cortisol** (hydrocortisone), **corticosterone,** and **cortisone,** of which cortisol is by far the most abundant. The primary role of cortisol (and other glucocorticoids) is to promote glucose and glycogen synthesis in liver cells, a process known as *gluconeogenesis.*[9] This effect tends to increase the levels of

---

[9]Gluconeogenesis: *gluco-* = pertaining to glucose + *neo-* = new + *genesis* = beginning.

sugar in the blood, thereby supporting body cells with adequate fuel supplies. By ensuring that cells receive sufficient energy resources, the adrenal cortex provides a backup support system to the stress responses provided by the adrenal medulla. Other effects of glucocorticoids include inhibition of allergic responses and reduction of inflammation. The secretion of glucocorticoids is regulated by ACTH from the pituitary gland.

Sex Hormones

The two classes of sex hormones released by the adrenal cortex are **androgens,** which have a masculinizing effect, and **estrogens,** which have feminizing effects. Both types of hormones are secreted by the cortex in both sexes, although in different amounts. Estrogens are produced in very small amounts in males, and androgens are produced in small amounts in females. Under certain abnormal conditions, however, such as the development of a tumor or a congenital enzyme deficiency, a sex hormone may be secreted at an accelerated rate. In males, for example, excess estrogens may cause breasts to develop and the testes to underdevelop. In females, excess androgens may cause the growth of facial and body hair, small breasts, and an enlarged clitoris.

CONCEPTS CHECK

1. What is the adrenal medulla structurally composed of?

2. What are the effects of epinephrine and norepinephrine?

3. What is the function of aldosterone?

Pancreas

The islets of Langerhans of the pancreas produce two hormones that have opposing effects. Together they help maintain proper glucose levels in the blood.

The **pancreas**[10] is a soft, oblong organ located in the abdominal cavity behind the stomach (Fig. 11–15). It is actually part of two body systems, since it performs two distinct functions. It is an endocrine gland, since it secretes two important hormones into the bloodstream; and it is a digestive organ, because of its secretion of digestive enzymes into ducts that empty into the small intestine. Thus, it is part of the endocrine system as well as the digestive system. In this chapter we will focus on the endocrine part of the pancreas.

The endocrine cells of the pancreas form clusters that are randomly distributed throughout it. They are called the **islets of Langerhans,** and they are closely associated with blood vessels. There are mainly two types of cells within the islets, each of which secretes a different hormone. About one-fourth of the cells are alpha cells, which secrete the hormone glucagon. Most of the remaining cells are beta cells, which secrete insulin. Both hormones play important roles in providing body cells with sufficient amounts of energy. They do this by regulating the amount of sugar (glucose) in the blood.

Glucagon

Glucagon[11] stimulates the conversion of glycogen into the simple sugar glucose. Since most of this activity occurs within liver cells, its main target is the liver. The effect of this conversion is the release of glucose into the bloodstream, which carries it to all cells in the body.

The release of glucagon is regulated by a negative feedback mechanism sensitive to glucose levels in the blood (Fig. 11–16). When levels become low (i.e., under conditions of hypoglycemia), alpha cells are stimulated to release glucagon. Conversely, its release is inhibited by high blood glucose levels (the condition of hyperglycemia). This seesaw type of regulation helps prevent glucose levels from reaching extreme high or low levels, but only in combination with the other pancreatic hormone, insulin.

Insulin

Insulin[12] has the opposite effect to that of glucagon on liver cells: it stimulates the formation of glycogen from glucose. It also inhibits the conversion of nutrient molecules other than carbohydrates into glucose. By performing these activities, insulin effectively removes glucose from the bloodstream and ties it up in the stored form of glycogen for later use. Its immediate effect, therefore, is to reduce the levels of glucose in the blood. In addition, insulin promotes the movement (by facilitated diffusion) of glucose into cells, especially skeletal muscle cells, cardiac muscle cells, and adipose cells. This further reduces blood glucose levels.

The secretion of insulin is regulated by a negative feedback mechanism that is similar to the one controlling glucagon release. As you can see in Figure 11–16, the rate of insulin secretion is also determined

---

[10]Pancreas: Greek *pan-* = all + *creas* = flesh.

[11]Glucagon: *glucagon* = agent producing glucose or sweetness.
[12]Insulin: Latin *insula* = islet + *-in* = neutral substance.

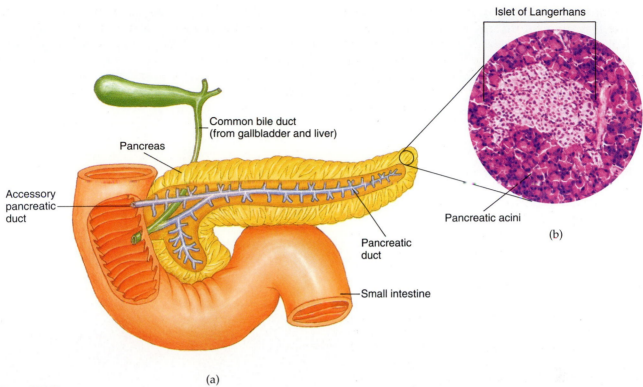

Islet of Langerhans

Common bile duct
(from gallbladder and liver)

Pancreas

Accessory
pancreatic
duct

Pancreatic acini

(b)

Pancreatic
duct

Small intestine

(a)

Figure 11–15
Pancreas structure. (a) The pancreas is located behind the
stomach in the upper abdominal cavity. Its central
pancreatic duct unites with the common bile duct from the
liver and gallbladder soon before entering the small
intestine. In this view, pancreatic tissue is removed to
reveal the pancreatic duct. (b) The endocrine cells of the
pancreas are located in clusters, called *islets of Langerhans*.
The cells that surround the islets are called *acini* and
secrete digestive enzymes into ducts (and are therefore
exocrine cells).

**What are the two types of cells in the pancreatic islets of
Langerhans, and what hormones do they produce?**

by glucose levels in the blood. High blood glucose levels stimulate insulin secretion, and low levels inhibit
it. As an example, soon after you eat a meal your blood
glucose levels are quite high because of the absorption
of nutrients into the bloodstream by the small intestine. Insulin is soon released by beta cells; it stimulates
the liver to draw glucose out of the bloodstream and
to bond it with other glucose molecules to form glycogen. Glucose molecules that pass through the liver
enter other body cells to provide them with energy.
Hours after the meal, when blood glucose levels decline, further insulin secretion is inhibited. However,
to prevent blood glucose levels from dropping too low,
glucagon is released and some glucose is returned to
the bloodstream. Thus, the two hormones function together to maintain a relatively stable blood glucose
concentration.

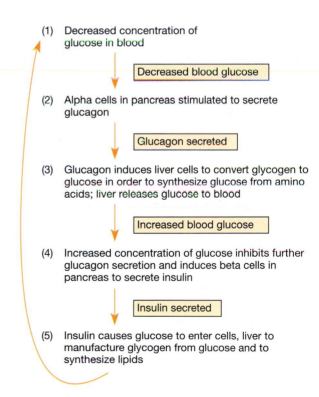

(1) Decreased concentration of
glucose in blood

Decreased blood glucose

(2) Alpha cells in pancreas stimulated to secrete
glucagon

Glucagon secreted

(3) Glucagon induces liver cells to convert glycogen to
glucose in order to synthesize glucose from amino
acids; liver releases glucose to blood

Increased blood glucose

(4) Increased concentration of glucose inhibits further
glucagon secretion and induces beta cells in
pancreas to secrete insulin

Insulin secreted

(5) Insulin causes glucose to enter cells, liver to
manufacture glycogen from glucose and to
synthesize lipids

Figure 11–16
Control and action of pancreatic hormones.

What change in homeostasis stimulates cells to secrete insulin?

HEALTH CLINIC

Diabetes Mellitus

The primary disease associated with pancreatic islets of Langerhans is **diabetes mellitus** (dī-a-BĒ-tēz mel-Ī-tus). It is a highly prevalent disorder, with an estimated 10 million sufferers in the United States. If it is not carefully controlled, it can be a life-threatening disease, for its long-term complications can reduce life expectancy by as much as one-third. This is illustrated by the fact that diabetics have a rate of heart disease 2 times greater, kidney disease 17 times greater, and blindness 25 times greater than nondiabetics. Almost 40,000 persons die each year as a result of diabetes mellitus and its complications, making it the third most common cause of death in the United States.

Two distinct clinical varieties of diabetes mellitus have been recognized. These are type I diabetes (insulin-dependent diabetes) and type II diabetes (non-insulin-dependent diabetes). Type I diabetes usually develops in persons less than 40 years of age, most frequently during adolescence. It is marked by a dramatic decrease in the number of beta cells. The decline in beta cells results in a deficiency of insulin. Recent investigations clearly indicate that the beta cells are destroyed by antibodies produced by the body's own white blood cells. How this autoimmune condition arises has not yet been explained.

Type I diabetes is controlled by daily injections of insulin in an effort to restore deficient levels of this important hormone. Insulin must be injected because it is a protein molecule; it cannot be taken orally, or it would be digested by enzymes in the digestive tract.

Type II diabetes accounts for about 90% of all diabetes cases. In contrast to type I diabetes, it develops gradually throughout adulthood and is most common in overweight persons over 40 years of age. It is characterized by the inability of target cells to take up insulin and use it, although sufficient insulin levels may be present in the blood. This causes unusually high levels of glucose in the blood, or hyperglycemia, which leads to dehydration and salt loss due to the osmotic imbalance, and a condition of acidosis (excessively low blood pH). Investigators have recently shown the occurrence of type II diabetes to be related to a decrease in the number of insulin receptors on the target cells. The decrease in insulin receptors reduces the ability of the cell to take up insulin from the blood and use it. Type II diabetes is usually milder than type I, and can be treated by careful diet management and maintaining an appropriate body weight. Daily injections of insulin are required only in more severe cases.

CONCEPTS CHECK

1. Where in the pancreas are hormones produced?

2. What are the roles of glucagon and insulin?

3. How are glucagon and insulin secretion rates controlled?

Gonads

The **gonads** are the sex organs; that is, they are the organs that produce the sex cells and secrete the primary sex hormones. In females, the **ovaries** are paired, almond-sized organs within the pelvic cavity. They secrete **estrogens,** which are the primary female sex hor-

mone. They also secrete another group of steroid hormones known as **progesterone.** In males, the **testes** are paired, egg-shaped organs located within a sac, the scrotum, outside the pelvic cavity. They secrete **testosterone,** the male sex hormone. The locations of the gonads were shown in Figure 11–2.

The regulatory mechanisms and functions of the sex hormones will be further discussed in Chapter 19, when this information can be tied together with other aspects of the reproductive system.

Pineal Gland

The **pineal gland** is a small structure within the cranial cavity associated with the brain. It is part of the diencephalon, where it forms a part of the roof of the third ventricle and is attached to the upper margin of the thalamus (see Fig. 11–2). For this reason, it is sometimes called the *epithalamus*. It is composed of modified neurons and supportive neuroglia. Its functional cells are innervated by sympathetic neurons.

There is some evidence that the pineal gland secretes at least one hormone, although its function remains in doubt. This hormone is called **melatonin,** which is thought to inhibit the secretion of FSH and LH from the anterior pituitary gland. Interestingly, the pineal gland is an important structure in reptiles and birds, where it has been shown to regulate reproductive cycles, hibernation cycles, and migration patterns, and is actually sensitive to light. It is even more interesting, perhaps, that human melatonin secretion appears to fluctuate with day-night and seasonal cycles, possibly providing us with our own system of rhythmic behavior.

Thymus Gland

The **thymus gland** is a prominent structure in infants and young children but diminishes in size with advancing age. It is a soft, irregularly shaped structure that lies in the mediastinum on top of the heart (see Fig. 11–2). During early life, it secretes a hormone known as **thymosin,** which stimulates the production of certain white blood cells (called T lymphocytes). Investigators also believe that the thymus acts as an incubator for these special cells once they are produced. Thus, the thymus gland plays an important role in immunity, which will be discussed in Chapter 14.

The endocrine glands, the hormones they secrete, and their primary effects on the body are summarized in Table 11–1.

Table 11–1 SUMMARY OF ENDOCRINE GLANDS AND THEIR EFFECTS

Gland	Hormone Secreted	Primary Effect
Pituitary Gland		
Anterior lobe	Growth hormone (GH)	Controls growth and development
	Melanocyte-stimulating hormone (MSH)	Stimulates melanocytes to increase skin pigmentation
	Prolactin (PRL)	Stimulates milk secretion by the mammary glands
	Adrenocorticotropic hormone (ACTH)	Stimulates the cortex of the adrenal glands
	Thyroid-stimulating hormone (TSH)	Stimulates the thyroid gland
	Follicle-stimulating hormone (FSH)	Stimulates development of ova in the ovaries and sperm in the testes
	Luteinizing hormone (LH)	Stimulates the secretion of sex hormones by the gonads
Posterior lobe	Oxytocin (OT)	Stimulates contractions of the uterus and the release of milk by the mammary glands
	Antidiuretic hormone (ADH)	Stimulates water reabsorption in the kidneys
Thyroid Gland	Thyroxine and triiodothyronine (T_3)	Control catabolic metabolism and protein synthesis in most body cells
	Calcitonin (CT)	Reduces calcium and phosphate levels in the blood
Parathyroid Glands	Parathyroid hormone (PTH)	Increases calcium levels in the blood, and reduces phosphate levels
Adrenal Glands		
Medulla	Epinephrine and norepinephrine	Prolong the conditions responsible for the "fight-or-flight" response, such as increase in metabolism, heart rate, and blood pressure

(Continued)

Table 11–1   (*Continued*)

Gland	Hormone Secreted	Primary Effect
Cortex	Mineralocorticoids (aldosterone)	Maintain body fluid balance by stimulating the retention of sodium and excreting potassium; help regulate blood pressure
	Glucocorticoids (cortisol, corticosterone, cortisone)	Stimulate glycogen formation and storage, increase body resistance to stress, and reduce inflammation
	Sex hormones (androgens and estrogens)	Stimulate sex characteristics, stimulate sex-cell development
Pancreas (Islets of Langerhans)	Glucagon	Stimulates the conversion of glycogen to glucose
	Insulin	Stimulates the conversion of glucose to glycogen, and glucose uptake in all body cells
Gonads		
Ovaries	Estrogens	Stimulate development of female sex characteristics, ovarian cycle, and menstrual cycle
Testes	Testosterone	Stimulates development of male sex characteristics, and sperm cell production
Pineal Gland	Melatonin	May inhibit gonad activities and regulate body rhythms
Thymus Gland	Thymosin	Stimulates the maturation of T lymphocytes

HOMEOSTASIS

The regulatory mechanisms that control endocrine gland secretion serve an overall function that has a widespread impact on the body: the maintenance of homeostasis. While the nervous system maintains homeostasis by sensing changes in the environment and responding rapidly by the transmission of nerve impulses, the glands of the endocrine system maintain stability by the release of hormones into the bloodstream. Although hormones are slower to act on body activities than are rapid nerve impulses, we have learned in this chapter that they affect a larger area of the body (or even the entire body) and their effects are longer-lasting. For example, while nerve impulses can provide you with a quick withdrawal reflex from a po-

tentially painful experience, hormones can alter your metabolic rate to enable your body to adjust to periods of time when food may be in short supply.

The positive effects of endocrine influence on homeostasis are best demonstrated when we consider the negative effects of its failure. A disorder of an endocrine gland may consist of either **hypersecretion** by the gland, reflected in an abnormally high blood level of the hormone; or **hyposecretion,** resulting from the underactivity of the gland. Although the effects of a disturbance are extremely variable—according to the degree of gland dysfunction, the age of the afflicted person, and even the person's sex—most glandular disorders profoundly influence body function, structure, or both. For example, hypersecretion by the thyroid gland produces an increased metabolic rate in all cells, resulting in weight loss, restlessness, and irritability. The more abnormal the output of the growth hormone, the greater these adverse effects. Other examples of the widespread effects of hypersecretion and

hyposecretion of endocrine glands are shown in Table 11–2. These examples demonstrate the consequences of improper endocrine output for the body and point to the overall importance of a balanced endocrine system in maintaining homeostasis in the body.

Table 11–2 HOMEOSTATIC DISTURBANCES CAUSED BY IMPROPER ENDOCRINE OUTPUT

Hormone	Hyposecretion Syndrome	Main Symptoms	Hypersecretion Syndrome	Main Symptoms
Growth hormone (GH)	Pituitary dwarfism	Retarded growth, abnormal fat distribution	Gigantism, acromegaly	Excessive growth
Thyroxine	Myxedema, cretinism	Low metabolic rate, impaired body development	Graves' disease	High metabolic rate
Parathyroid hormone (PTH)	Hypoparathyroidism	Low blood calcium, resulting in muscle weakness, tetany, neurological problems	Hyperparathyroidism	High blood calcium, resulting in muscle, brain, and nerve problems
Antidiuretic hormone (ADH)	Diabetes insipidus	Polyuria (high urine output), glycosuria	Insufficient ADH secretion syndrome	Increased body weight
Insulin	Diabetes mellitus	High blood glucose levels, leading to possible coma	Excess insulin	Low blood glucose, leading to possible coma
Mineralocorticoids	Hypoaldosteronism	Polyuria, low blood volume	Aldosteronism	Water retention, leading to weight gain and high blood pressure
Glucocorticoids	Addison's disease	Dehydration and inability to use fat reserves, tolerate stress, maintain blood glucose levels	Cushing's disease	Impaired use of glucose, excessive use of proteins
Androgens	Hypogonadism, eunuchoidism in males; no effect in females	Sterility, lack of body hair and other male features	No effect in males; androgenital syndrome in females	Development of male features
Estrogens	Hypogonadism, menopause in females; no effect in males	Sterility, lack of female features, no menstruation	Increased incidence of breast and ovarian cancer in females; gynecomastia in males	Tumor development in breasts and ovaries in females; development of breasts in males

CLINICAL TERMS OF THE ENDOCRINE SYSTEM

Addison's disease Caused by decreased production of aldosterone and cortisol by the adrenal cortex. It causes too much sodium to be excreted, leading to water loss and dehydration. Also, insufficient amounts of protein are converted to carbohydrate, and metabolism is disturbed. These metabolic effects can be deadly.

Aldosteronism Caused by an increased production of aldosterone by the adrenal cortex. It results in the excessive reabsorption of sodium by the kidneys. Because sodium is excessively concentrated in the body, large amounts of water are retained and blood pressure increases above normal levels.

Amenorrhea The cessation of menstrual periods due to hypersecretion of prolactin by the anterior lobe of the pituitary gland.

Cretinism Caused by a deficiency in the thyroid hormones (thyroxine and triiodothyronine), or hypothyroidism, in children. It results in a decrease in metabolic rate, causing reduced growth and development.

Cushing's syndrome Excessive production of cortisol by the adrenal cortex that leads to excessive accumulation of fat in the neck, face, and trunk. Muscles generally weaken and are reduced in size, and the skin bruises easily. It is usually caused by a tumor of the adrenal gland (Fig. a)

Diabetes insipidus Caused by a deficiency in antidiuretic hormone (ADH). It is characterized by the excretion of large quantities of dilute urine, accompanied by continued thirst.

Diabetes mellitus A disorder associated with a deficient availability to target cells of insulin secreted by pancreatic islets of Langerhans. It affects the metabolism of carbohydrates, fats, and protein, and its effects vary according to how much insulin, if any, is available to target cells. The effects of a lack of insulin are hyperglycemia (increased blood sugar levels), acidosis, and a lowered entry of glucose into body cells. The disease can be inherited, although it can also develop from trauma or disease to the pancreas, prolonged stress, and obesity. Two types exist: type I, or insulin-dependent diabetes, and type II, or non-insulin-dependent diabetes.

Exophthalmos Protrusion of the eyes due to hyperthyroidism.

Goiter An enlargement of the thyroid gland, causing the neck to swell (Fig. b). It is associated with either hypo- or hypersecretion by the thyroid gland and can be caused by an insufficient amount of iodine in the diet (when it is called an *endemic* goiter), which is easily corrected by adding iodine to the diet.

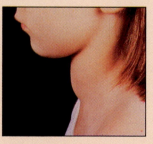

(b)

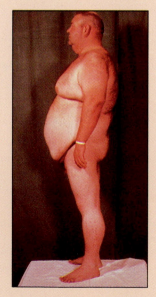

(a)

Graves' disease: Caused by excessive production of the thyroid hormones, or hyperthyroidism. Typically, the metabolic rate increases, causing loss of weight, nervousness, and hypertension.

Myxedema Caused by hyperthyroidism in adults. The skin thickens, causing a characteristic puffy appearance in the face and hands. The tongue thickens, nervous reflexes are slowed, and a lack of energy is the common complaint.

The endocrine system helps to maintain homeostasis by way of hormones that circulate throughout the bloodstream.

COMPOSITION OF THE ENDOCRINE SYSTEM

The endocrine system is composed of glands that release their secretions into the extracellular environment. It includes the pituitary gland, thyroid gland, parathyroid glands, adrenal glands, pancreas, and gonads. The pineal and thymus glands play minor roles, and certain organs from other systems contain endocrine cells that secrete hormones also.

HORMONES

Hormones are the chemical messenger units secreted by the endocrine glands.

Hormonal Action   Hormones have an effect only on cells to which they can selectively bind, called *target cells.* Their effect is to alter the cells' metabolic processes.

Water-Soluble Hormones   These types of hormones cannot penetrate the plasma membrane of a cell. They operate by way of a second-messenger system, the best-studied of which uses cyclic AMP as the second messenger. This system is a series of exchanges: from the hormone to plasma membrane receptors, to the second messenger in the cell, and finally to proteins and enzymes within the cell.

Lipid-Soluble Hormones   These types of hormones are usually steroid hormones, which can easily pass through the plasma membrane of the cell. They activate genes within the cell nucleus to synthesize new proteins and enzymes.

Prostaglandins   Although prostaglandins are not true hormones, because they do not produce their own changes in a cell but rather modify hormonal changes, their effects can be significant. They are secreted by numerous organs and affect only local areas.

Hormonal Control   Because hormones are very potent substances, their proper regulation is vitally important.

Feedback Control   The concentrations of hormones or their products can be gauged by the endocrine gland, and its rate of secretion can then be adjusted.

Negative feedback systems control the amount of hormone released by providing a response in the opposite direction from the stimulus.

Positive feedback systems control the amount of hormone released by providing a response in the same direction as the stimulus.

Nervous Control   Direct nerve connections may inhibit or stimulate certain glands that receive them.

THE ENDOCRINE GLANDS

Pituitary Gland   Located at the base of the brain and connected to the hypothalamus by the infundibulum. It contains two lobes.

Anterior Lobe   Composed of glandular epithelium. It produces seven hormones whose release is controlled by the hypothalamus by way of regulating factors in a negative feedback mechanism.

Growth Hormone   GH stimulates body cells to grow and divide. It does so by promoting protein synthesis and carbohydrate and fat catabolism.

Melanocyte-Stimulating Hormone MSH stimulates the production of melanin in the skin, increasing pigmentation.

Prolactin   PL stimulates and maintains milk secretion in females.

Thyroid-Stimulating Hormone   TSH regulates the secretion of hormones by the thyroid gland.

Adrenocorticotropic Hormone   ACTH regulates the secretion of hormones by the adrenal cortex.

Follicle-Stimulating Hormone   FSH in females stimulates development of ova and estrogen secretion in the ovaries. In males, it stimulates sperm production by the testes.

Luteinizing Hormone   LH in females stimulates ovulation and prepares the uterus for implantation of a fertilized ovum. In males, it stimulates the production of testosterone by cells in the testes.

Posterior Lobe   Composed of nerve tissue, it contains the axons of neurons originating in the hypothalamus called *neurosecretory cells.* The neurosecretory cells produce both posterior lobe hormones, which are passed to the lobe prior to secretion into the general circulation.

Oxytocin   OT stimulates contraction of smooth muscle in the wall of the uterus, and of cells around the mammary gland to eject milk.

Antidiuretic Hormone   ADH regulates fluid balance by increasing water reabsorption in the kidneys. It also helps regulate blood pressure.

Thyroid Gland   The large gland located in the neck, composed of glandular epithelium that secretes hormones into storage vesicles or directly into the bloodstream.

Thyroxine and Triiodothyronine   Both hormones stimulate metabolism, promote protein synthesis, increase glucose uptake, promote lipid metabolism, and accelerate nervous system activity. Their secretion rates are regulated by TSH through a negative feedback system.

Calcitonin   Reduces the calcium and phosphate levels in the blood by inhibiting osteoclast activity and by stimulating the excretion of ions by the kidneys.

Parathyroid Glands   Four or five pea-shaped masses of glandular epithelium behind the thyroid in the neck.

Parathyroid Hormone   PTH increases calcium levels in the blood, thus opposing calcitonin, and lowers phosphate levels.

Adrenal Glands   Paired glands located superior to each kidney. Each is divided into two portions: an inner medulla and an outer cortex.

Adrenal Medulla   Composed of modified nerve tissue. It receives impulses directly from the sympathetic nervous system.

Epinephrine and norepinephrine   Released by the adrenal medulla. These two hormones mimic the effects of the sympathetic nervous system. They are under nervous control.

Adrenal Cortex   Composed of glandular epithelium arranged in three distinct zones. It secretes three types of steroid hormones:

Mineralocorticoids   The most important is aldosterone, which maintains fluid balance by increasing the conservation of sodium and excreting potassium. It thereby causes a retention of water and an increase in blood pressure.

Glucocorticoids   The most important is cortisol, which promotes gluconeogenesis in liver cells. It also inhibits allergic responses and reduces inflammation.

Sex Hormones   These include androgens, which have a masculinizing effect, and estrogens, which have a feminizing effect.

Pancreas   This gland consists of endocrine and digestive portions. The endocrine portions are clusters of cells known as *islets of Langerhans*, where two hormones are produced. The secretion of these hormones is regulated by a negative feedback mechanism sensitive to blood sugar levels.

Glucagon   Stimulates the conversion of glycogen into glucose, thus elevating blood sugar levels.

Insulin   Stimulates the formation of glycogen from glucose, lowering blood sugar levels.

Gonads   The ovaries in the female, which secrete estrogens; and the testes in the male, which secrete testosterone.

Pineal Gland   A small gland in the diencephalon of the brain. It secretes melatonin, which may play a role in rhythmic behavior.

Thymus Gland   Large in infants and young children and small in adults. It lies in the mediastinum above the heart. It secretes thymosin in early life, which stimulates the production of T lymphocytes.

HOMEOSTASIS

The endocrine system plays a major role in the maintenance of homeostasis. It does so by the release of hormones, which provide long-lasting and widespread effects on body activities. Endocrine glands that become dysfunctional may secrete too much hormone, a condition called *hypersecretion*, or too little, a condition called *hyposecretion*.

KEY TERMS

aldosterone (p. 331)
androgens (p. 332)
antidiuretic (p. 325)
calcitonin (p. 327)
cortisol (p. 331)
estrogens (p. 332)
endocrine (p. 316)
exocrine (p. 316)
glucagon (p. 332)
hormone (p. 315)
insulin (p. 332)

QUESTIONS FOR REVIEW

OBJECTIVE QUESTIONS

1. Which of the following glands secrete substances into the extracellular space?
   a. salivary glands
   b. parathyroid glands
   c. exocrine glands
   d. sweat glands
2. A cell that contains membrane proteins enabling a hormone to selectively bind to its plasma membrane is called a(n):
   a. endocrine cell
   b. secretory cell
   c. target cell
   d. regulatory cell
3. The first messenger in a second-messenger system used by water-soluble hormones is a:
   a. hormone
   b. protein kinase
   c. cyclic AMP molecule
   d. cytoplasmic enzyme
4. In general, _____ hormones can pass directly through cell plasma membranes to activate genes.
   a. all
   b. water-soluble
   c. lipid-soluble
   d. prostaglandin
5. Which of the following is a lipid-soluble hormone?
   a. growth hormone
   b. aldosterone
   c. oxytocin
   d. prostaglandin
6. A hormone that inhibits its own continued production is regulated by:
   a. positive feedback
   b. nerve impulses
   c. negative feedback
   d. none of the above
7. The portion of the pituitary that is composed of glandular epithelium is known as the:
   a. posterior lobe
   b. cortex
   c. anterior lobe
   d. pineal
8. The anterior lobe of the pituitary secretes which tropic hormone that regulates the adrenal cortex?
   a. growth hormone
   b. ACTH
   c. ADH
   d. FSH
9. Neurosecretory cells of the hypothalamus manufacture oxytocin and:
   a. ACTH
   b. thyroxine
   c. antidiuretic hormone
   d. melatonin

10. Thyroid hormones are regulated by:
    a. the hypothalamus   c. TSH
    b. a negative feedback system   d. all of the above

11. The two hormones involved in maintaining calcium levels in the blood are:
    a. calcitonin and PTH   c. calcitonin and ACTH
    b. parathyroid hormone and insulin   d. PTH and glucagon

12. The adrenal medulla secretes epinephrine and norepinephrine, which mimic the response of the:
    a. adrenal cortex   c. sympathetic nervous system
    b. pituitary gland   d. central nervous system

13. A disease in which an abundance of water is retained by the body because of retention of sodium ions may be traced to the:
    a. posterior lobe of the pituitary gland
    b. pancreas
    c. adrenal cortex
    d. thyroid gland

14. Glucagon is released by the pancreas to stimulate the conversion of glycogen into glucose. The hormone whose effects oppose glucagon is called:
    a. cortisol   c. insulin
    b. growth hormone   d. thymosin

ESSAY QUESTIONS

1. Distinguish between negative feedback systems and positive feedback systems, and provide one example of each.
2. Describe how a hormone can induce a change in a target cell if it is water-soluble.
3. How does a steroid hormone induce changes in cells?
4. List the tropic hormones secreted by the pituitary gland, and indicate which gland each regulates. How are tropic hormone secretions themselves regulated?
5. How do calcitonin and parathyroid hormone work together in order to maintain proper calcium levels in the blood? Why are these levels important to maintain?
6. Describe how glucagon and insulin balance blood glucose levels following a meal. How may growth hormone influence this balance?

A N S W E R S   T O   A R T   L E G E N D   Q U E S T I O N S

Figure 11–1
Identify several examples of exocrine glands we have studied previously. Examples of exocrine glands include sweat glands, sebaceous glands, and salivary glands.

Figure 11–2
Which endocrine glands play a role in maintaining homeostasis? All endocrine glands play a role in maintaining homeostasis.

Figure 11–3
Why must water-soluble hormones rely on the action of second-messenger chemicals to initiate an effect on a cell? Water-soluble hormones must rely on second messengers because they are not capable of entering the cell, on account of their insolubility with the plasma membrane.

Figure 11–4
The synthesis of what materials within a cell is stimulated by steroid hormones? Synthesis of proteins (and enzymes) is stimulated by steroid hormones.

Figure 11–5
In a negative feedback system, what provides the signal to the endocrine gland, causing it to reverse its activity? The hormone produced by the endocrine gland, or its product, provides the signal to reverse glandular activity.

Figure 11–6
How does positive feedback differ from negative feedback? Positive feedback causes a response in the same direction as the stimulus, whereas negative feedback causes a response in the opposite direction from the stimulus.

Figure 11–7
What part of the brain forms a direct connection with the pituitary gland? The hypothalamus is directly connected to the pituitary gland.

Figure 11–8
Which lobe is composed of glandular epithelium, and which is composed of nerve tissue? The anterior lobe is composed of glandular epithelium, and the posterior lobe, nerve tissue.

Figure 11–9
What controls the release of posterior lobe hormones? The hypothalamus controls the release of posterior lobe hormones (these hormones are actually produced by secretory cells within the hypothalamus and transported to the posterior lobe).

Figure 11–10
How does ADH increase blood pressure? ADH increases blood pressure directly by stimulating the contraction of the walls of arterioles. Its role in decreasing urine output (thus increasing blood volume) also increases blood pressure indirectly.

Figure 11–11
What hormones produced by the thyroid gland regulate growth and metabolism? Thyroxine and triiodothyronine both regulate growth and metabolism.

Figure 11–12
What hormone is produced by the parathyroid glands? Parathyroid hormone (PTH), which increases the rate of calcium absorption.

Figure 11–13
What hormone counteracts the effects of PTH to maintain proper calcium ion levels? Calcitonin, produced by the thyroid gland, counteracts the effects of PTH.

Figure 11–14
How do the adrenal cortex and adrenal medulla differ in structure? The adrenal cortex is the outer layer of tissue and is composed of several zones of glandular epithelium (each zone produces a different steroid hormone). The adrenal medulla is the inner region and is composed of modified nerve tissue.

Figure 11–15
What are the two types of cells in the pancreatic islets of Langerhans, and what hormones do they produce? The two types of cells within the islets of Langerhans are alpha cells, which secrete glucagon, and beta cells, which secrete insulin.

Figure 11–16
What change in homeostasis stimulates beta cells to secrete insulin? High levels of glucose in the blood stimulate beta cells to secrete insulin.

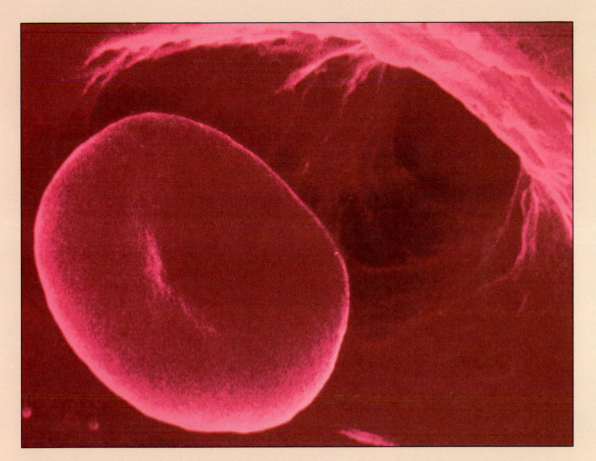

SYSTEMS THAT TRANSPORT AND PROTECT

Every cell of the human body must receive a continuous supply of oxygen and nutrients to survive. Cells must also have a means of disposing waste materials to avoid their potentially toxic accumulation. For a single-celled organism like an ameba, meeting these needs is much less complex than it is for multicellular organisms like ourselves. A human cell is not bathed in pond water, where nutrients and oxygen are within easy reach and waste materials are quickly diluted. On the contrary, a human cell is bathed in interstitial fluid—in most cases, a great distance from the external environment. Materials must be transported to and from the interstitial fluid for the needs of the cell to be met. As we shall see in this unit, this transportation function is accomplished by the movement of blood through the cardiovascular system. The important function of protection is closely associated with transportation, as elements in the blood work with another system, the lymphatic system, to defend the body from many forms of disease.

In Unit 4 we will examine the generalized system that provides transportation and protection, the circulatory system. The subject is covered in three chapters: the blood in Chapter 12, the cardiovascular system in Chapter 13, and the lymphatic system and immunity in Chapter 14.

THE BLOOD

LEARNING OBJECTIVES

After studying this chapter, you should be able to:

1. Identify the three primary functions of blood.

2. Describe the properties of blood.

3. Distinguish between plasma and formed elements.

4. Identify the dissolved substances in plasma.

5. Describe two ways in which blood can be analyzed.

6. Distinguish between the formed elements on the basis of their concentrations in blood, their structure, and their major function.

7. Indicate the function of hemoglobin.

8. Distinguish between the types of white blood cells on the basis of their structural and functional differences.

9. Identify the role of platelets in blood clot formation.

10. Describe the processes of blood vessel spasm, platelet plug formation, and coagulation, and relate each to the prevention of excessive blood loss.

11. Identify the importance of blood typing in performing blood transfusions.

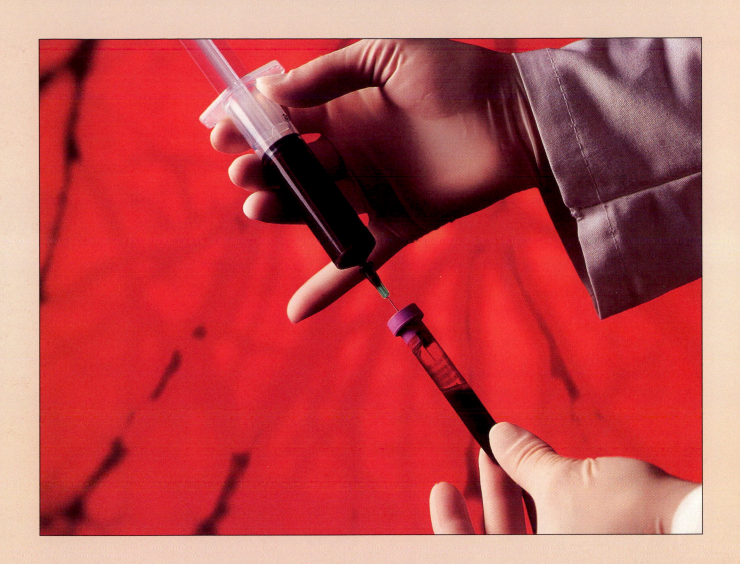

We classify blood as a part of the *circulatory system,* which also includes the *cardio-vascular system* (the heart and blood vessels) and the *lymphatic system* (the lymphatic organs, lymphatic vessels, and lymph). Blood may also be classified as part of the *hematologic system,* which includes red bone marrow, the lymph nodes, and the spleen.

Blood is a specialized type of connective tissue, as we learned in Chapter 4. It is quite unique, for it is the only liquid tissue in the body. Like other connective tissues, it contains cellular elements suspended in a large matrix. In blood, the cellular elements are known as **formed elements,** and the matrix consists of protein and other molecules suspended in a watery fluid called **plasma.** Both the formed elements and plasma play important roles in blood function and are vital in maintaining homeostasis.

CHARACTERISTICS OF BLOOD

Blood performs three main functions that are each important to maintaining homeostasis. They are transportation, protection, and regulation. The properties of blood are universal among all people, and are often used as indicators of the state of health.

Functions of Blood

Blood is a vital tissue with a set of characteristics that distinguish it from all other tissues in the body. These characteristics include the three main functions of blood, and a number of unique properties.

The greatest single benefit we obtain from our blood is that of homeostasis. The continuous flow of blood through the 60,000 miles of blood vessels helps to maintain the homeostasis of all body organs and tissues. It does so in several different ways:

1. **By transportation.** As blood moves through the body, cells receive from it vital substances, including nutrients that are brought from digestive organs, oxygen that is carried from the lungs, and hormones that are secreted from the endocrine glands. Cells add to the flowing blood their waste materials, such as carbon dioxide and nonprotein nitrogen substances (urea and uric acid), and their secretions.

2. **By protection.** The blood protects the body from harmful substances such as microorganisms and their toxins through the action of phagocytic white blood cells and specialized proteins called *antibodies.* Blood also protects against fluid loss following an injury by the clotting mechanism.

3. **By regulation.** The blood regulates the acid-base balance of body fluids by way of buffers, which neutralize the potentially harmful effects of too much carbon dioxide, lactic acid, and other compounds. The blood also helps regulate body temperature by cooling or heating parts of the body. This mechanism is controlled by the hypothalamus, which controls the volume of blood flow to different areas of the body.

Properties of Blood

Human blood has certain characteristics common to all people that distinguish it from all other types of tissue. These characteristics, or properties, may be used as indicators of the state of an individual's health. They include color, volume, viscosity, and pH.

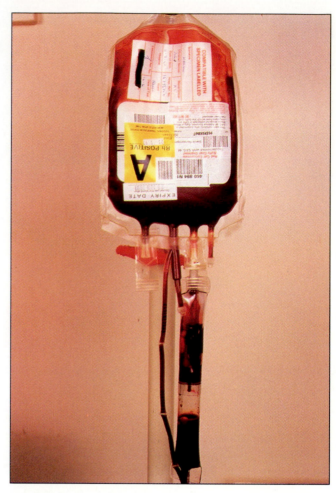

Figure 12–1
Human blood. This photograph is of a bag containing whole blood (type A positive) that is being given to a patient during a procedure known as a *blood transfusion.*

The red color of blood is due to a pigment carried by red blood cells (see Fig. 12–1). This pigment is a protein called **hemoglobin.**[1] In arterial blood, the abundant oxygen molecules are chemically bound to hemoglobin, giving the blood a crimson-red color. In venous blood, the oxygen molecules are not as prevalent, and the blood takes on a dark-red color with a slightly bluish tinge. When this dark color is seen through a layer of skin, our veins appear greenish blue, although we do not have green or blue blood!

The blood volume in a healthy adult does not vary much from day to day or year to year. It represents about 8% of total body weight, with most of this within the blood vessels (the remainder is mainly in the heart). An average male has 5 to 6 liters of blood, and an av-

---

[1]Hemoglobin: *heme* = blood + *globin* = globe-shaped protein.

erage female 4 to 5 liters. This difference is due to average body weight differences, not sexual differences.

Blood is thicker, denser, and more adhesive (that is, stickier) than water. This is mainly due to the presence of formed elements, particularly red blood cells. It causes blood to flow about five times more slowly than water. This resistance to flow is called *viscosity.* Thus, we call blood a *viscous* substance because it resists flow much more than water.

Blood is slightly alkaline (or basic), with a pH that usually ranges between 7.35 and 7.45. This range is maintained within its narrow limits despite changes in diet, cell secretions, and metabolic rate by a buffering system that removes hydrogen ions (H^+). Should the buffers fail and the blood become too acidic (below pH 6.0), body cells would stop functioning and the body would quickly lose homeostasis. This dangerous condition is called *acidosis* (or *acidemia*). A much less common condition may also occur in which there is too little acid in the blood. This is called *alkalosis* (or *alkalemia*), which also disrupts homeostasis.

CONCEPTS CHECK

1. What are the three functions of blood?

2. Why is the color of healthy blood red?

3. How is the narrow pH range of blood maintained?

PLASMA

Plasma consists mostly of water. It provides a transport medium for solutes and formed elements.

Plasma[2] is the liquid portion of blood. It is a clear, yellowish fluid that suspends the formed elements and is one of the three major extracellular fluids found in the body (the other two are interstitial fluid and lymph). The fluid nature of plasma makes it a valuable transport medium for a variety of substances carried by the bloodstream, including the formed elements, gases, metabolic waste materials, nutrients, and hormones.

Plasma is mostly composed of water (92%). Its remaining components (8%) are dissolved substances, or solutes. The solutes are mainly proteins, but also include breakdown products of metabolism, gases, nutrients, electrolytes, and hormones.

[2]Plasma: Greek *plasma* = something molded.

Plasma Proteins

The most abundant solutes in plasma are the **plasma proteins.** These proteins are, for the most part, synthesized in the liver and released into the bloodstream. About 50 different types of plasma proteins have been discovered so far, most of which fall into one of three main groups: albumins, globulins, and fibrinogen.

The **albumins** constitute about 55% of the plasma proteins. Because of their abundance they help to thicken the blood; that is, they increase its viscosity. The viscosity of plasma determines how easily it can flow through blood vessels and across blood vessel walls into the interstitial fluid. A measure of a fluid's ability to flow across membranes is termed *osmotic pressure.* Thus, albumins in the plasma help to maintain the osmotic pressure of blood.

The **globulins** are a group of proteins that occupy about 38% of the solute concentration in plasma. There are several types of globulins, the most important of which are the *gamma globulins.* These proteins serve as antibodies in the immune response, which we will discuss in Chapter 14, and thereby provide protection from various forms of disease.

The remaining 7% of the plasma proteins is **fibrinogen.** This protein is a precursor to the protein **fibrin,** which plays a major role in the blood-clotting mechanism (to be discussed later in this chapter in the section Hemostasis). When fibrinogen and other blood-clotting factors are selectively removed from a blood sample, the remaining sample is called *blood serum.* The removal of fibrinogen is a common procedure used to inactivate the blood-clotting mechanism, making the storage and reuse of blood much more convenient.

Other Plasma Solutes

In addition to the plasma proteins, plasma transports many other dissolved substances. One category of these solutes is a group of molecules that contain nitrogen atoms but are not proteins, often called *nonprotein nitrogenous (NPN) substances.* These molecules are by-products of the catabolic phase of metabolism. They include amino acids, urea, and uric acid. Amino acids and urea result from protein digestion, and uric acid from nucleic acid breakdown. Both urea and uric acid are removed from the blood by the kidneys and are excreted with the urine.

The gases that are dissolved in plasma include the two respiratory gases, oxygen and carbon dioxide, and any other gases inhaled by the lungs. While nitrogen is a large component of the gas volume, it plays no significant physiological role. As you will learn in the next section, most of the oxygen and some of the carbon dioxide carried by the blood are bound to the prominent protein molecule (hemoglobin) in red blood cells.

Nutrient molecules also form part of the solutes in plasma. They are produced by the digestion and absorption of food in the digestive tract, and enter the blood in order to be transported to all body cells. The plasma nutrients include simple sugars, fats, and amino acids.

The electrolytes dissolved in plasma have also arrived from absorption by the digestive tract or have been released by cells as a by-product of metabolism. In the order of their usual abundance, they include sodium, chloride, bicarbonate, potassium, calcium, phosphate, sulfate, and magnesium ions. These electrolytes are important in maintaining the acid-base balance (pH) and osmotic pressure of blood. Their concentrations are regulated mainly by the kidneys to keep their levels stable despite fluctuations in diet and cell secretions.

CONCEPTS CHECK

1. What is the major component of plasma?

2. What are the three main groups of plasma proteins?

3. Why are nutrients transported in the plasma?

FORMED ELEMENTS

The formed elements include red blood cells, white blood cells, and platelets. They perform most of the functions of blood.

The formed elements of blood include cells and cell fragments. They perform most of the functions provided by the blood and make up 45% of the total blood

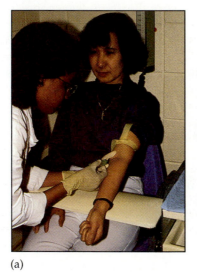
(a)

(b)

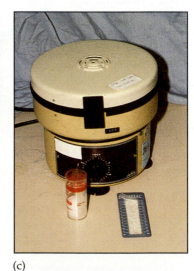
(c)

Figure 12–2
A method of determining blood cell composition is the hematocrit. (a) A blood sample is first drawn from the subject. (b) The whole blood sample is transferred from the syringe to a glass capillary tube. (c) It is then spun within a microcentrifuge, which separates blood components on the basis of their varying densities. (d) Red blood cells occupy the bottom of the sample tube, white blood cells and other materials form a layer (called the buffy coat) above the red cells, and plasma occupies the top portion. The hematocrit on the left is from a heathy patient, and that on the right is from a patient with an abnormally low number of red blood cells (a condition called anemia).
(e) A photograph comparing a normal hematocrit with one having an abnormally low number of red blood cells.

What is the purpose of performing a hematocrit?

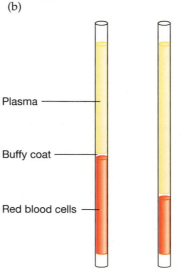
(d)

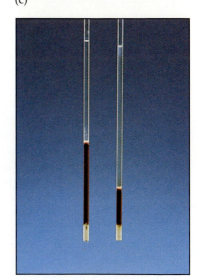
(e)

volume. The various types of elements are grouped into three categories: red blood cells, white blood cells, and platelets. Because these components each have different densities, they can be separated by spinning at high speeds in a centrifuge. This procedure is performed routinely to determine the percentage of blood cells in a sample of blood and is called **hematocrit** (Fig. 12–2). Another method of analyzing blood is called a **blood smear,** in which a blood sample is smeared across a microscope slide and is stained with dye. When a normal blood smear is viewed under a microscope, it looks much like Figure 12–3.

Formed elements are produced by a process called **hematopoiesis** (hem'-a-tō-poy-Ē-sis). Before birth, they are produced in the yolk sac, the liver, and the spleen. After birth and throughout the rest of life, hematopoiesis is confined primarily to red bone marrow, although some white blood cells are produced in lymphatic tissue. During the process of hematopoiesis, all the formed elements originate from a single population of cells called **stem cells** (or hemocytoblasts). These special cells differentiate to form the different cell lines, each of which leads to a particular type of formed element (Fig. 12–4). The major formed elements and their features are summarized in Table 12–1.

Red Blood Cells (Erythrocytes)

Red blood cells (RBCs), or **erythrocytes**[3] (ē-RITH-rō-sīts), are by far the most abundant type of formed element. They represent more than 95% of the elements in the blood and compose over 40% of the total blood

[3]Erythrocytes: Greek *erythros* = red + *-cyte* = cell.

volume. Although their numbers vary from time to time, the normal red blood cell concentration is 4,600,000 to 6,200,000 per cubic millimeter for adult males and 4,200,000 to 5,400,000 for adult females. Their primary function is the transport of the two respiratory gases, oxygen and carbon dioxide. Any reduction in the number of red blood cells or in their ability to transport gases is a disease condition known as *anemia*.

Structure

Red blood cells are small, flexible cells that are shaped like discs (Figs. 12–3, 12–4, and 12–5). They are thin in the center and thicker along the edges, giving them a biconcave appearance. This unusual structure is the result of the loss of nuclei during their early stages of development (Fig. 12–6). Thus, the mature, biconcave red blood cells circulating in our blood do not have nuclei. They also lack other types of organelles. It is generally agreed among investigators that the loss of nuclei occurs for two reasons, both of which provide important health benefits: the biconcave structure provides a larger surface area for the diffusion of gases; and the space previously occupied by the nucleus can be used to hold a greater number of pigment molecules.

Function

The primary function of red blood cells is the transport of oxygen from the lungs and the transport of carbon dioxide to the lungs. This function is provided by the large pigment molecules they contain, called **hemoglobin.** This protein occupies about one-third of the volume of the red blood cell; the molecule is illustrated

(*Text continues on p. 352*)

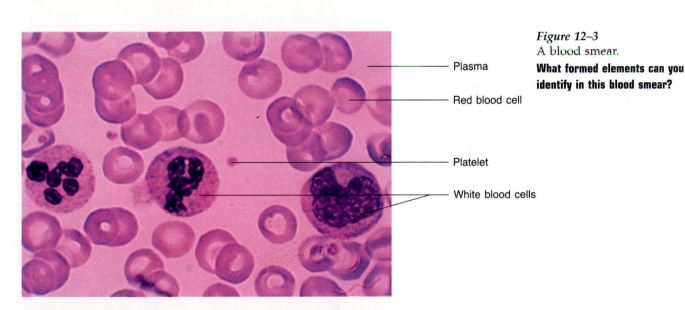

Figure 12–3
A blood smear.

What formed elements can you identify in this blood smear?

Plasma

Red blood cell

Platelet

White blood cells

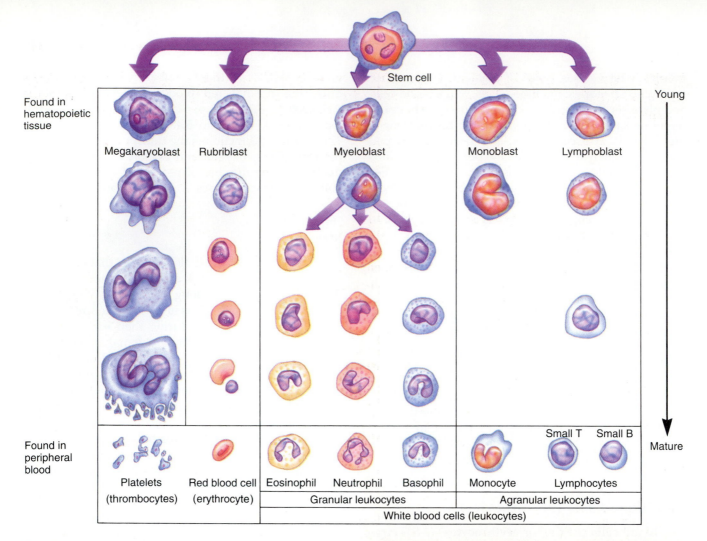

			Stem cell			Young
Found in hematopoietic tissue	Megakaryoblast	Rubriblast	Myeloblast	Monoblast	Lymphoblast	
					Small T / Small B	Mature
Found in peripheral blood	Platelets (thrombocytes)	Red blood cell (erythrocyte)	Eosinophil · Neutrophil · Basophil	Monocyte	Lymphocytes	
			Granular leukocytes	Agranular leukocytes		
			White blood cells (leukocytes)			

Figure 12–4
The development of blood cells. Notice that all cell lines originate from a common stem cell.

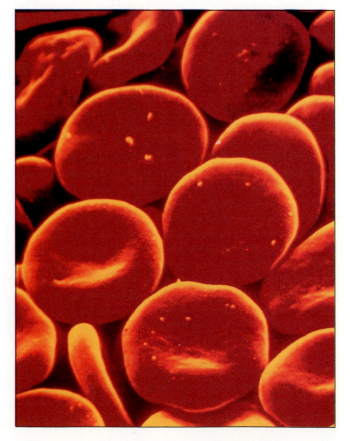

Figure 12–5
Red blood cells. This high-resolution photograph was taken with the use of an electron microscope (approximately 11,000x).

How would you describe the shape of a red blood cell?

350

Table 12–1   THE FORMED ELEMENTS OF BLOOD: A SUMMARY

Formed Element	Abundance (per cubic mm)	Structure	Function
Red blood cells (erythrocytes)	4.2 to 6.2 million	Biconcave discs; no nucleus	Transport oxygen from lungs to body cells, and carbon dioxide from body cells to lungs
White blood cells (leukocytes)	5000 to 10,000	Larger than red blood cells with nucleus	Protect from various forms of disease
Granulocytes		Highly visible granules in cytoplasm	
Neutrophils	55% to 60% of white blood cells	Cytoplasm stains pink in neutral stain; nucleus has two to five lobes	Phagocytic removal of foreign particles and damaged cells
Eosinophils	1% to 4% of white blood cells	Cytoplasm stains red in acid stain; nucleus has two lobes	Phagocytic removal of allergens
Basophils	0.5% or less of white blood cells	Cytoplasm stains blue in basic stain; nucleus is often S-shaped	Promote inflammation by secreting histamine
Agranulocytes		Cytoplasmic granules are not highly visible	
Monocytes	3% to 8% of white blood cells	Largest in size, with large round, oval, or lobed nucleus	Active phagocytic removal of large foreign particles and damaged cells
Lymphocytes	25% to 33% of white blood cells	About the same size as red blood cells, with large, round nucleus	Produce antibodies for the removal of toxins and viruses
Platelets (thrombocytes)	150,000 to 360,000	Round or oval disc about one-tenth the size of red blood cells without nucleus	Form platelet plugs and release factors in hemostasis

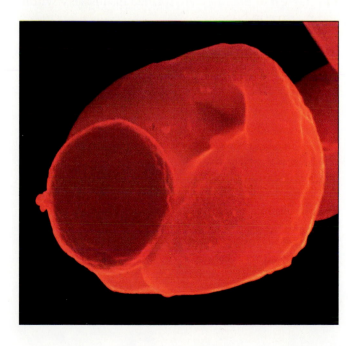

Figure 12–6
A red blood cell in the process of losing its nucleus.
Can a mature red blood cell divide (by mitosis)?

Anemia

Anemia is any one of a large number of disease conditions that involve a reduction in the number of red blood cells or in their ability to transport gases. The different types of anemias are classified on the basis of how they are caused. For example, *pernicious anemia* is caused by a shortage of dietary sources of folic acid and vitamin B_{12}, or of a substance produced by gastric glands of the stomach known as intrinsic factor. Folic acid and vitamin B_{12} are required for the synthesis of new red blood cells, and intrinsic factor is necessary for vitamin B_{12} to be absorbed across the small intestinal lining. Pernicious anemia causes a persistent feeling of drowsiness and lack of energy, due to the shortage of oxygen delivered to body tissues.

Another type of anemia is *hemorrhagic anemia*, which results from a loss of blood due to severe bleeding, or hemorrhage. If the volume of lost blood is not fatal, the reduced oxygen levels in the tissues will trigger the release of a hormone (erythropoietin) that stimulates the production of new red blood cells.

In yet another class of anemias known as *hemolytic anemia*, red blood cells are destroyed by the body's own phagocytic cells at a rate faster than they can be replaced. In this disease, the red blood cells are attacked because of genetic deficiencies, or because of damage by bacterial toxins or viruses. **Sickle cell anemia** is an example of a chronic hemolytic anemia; it is observed almost exclusively in black Africans (and, in the United States, Amer-

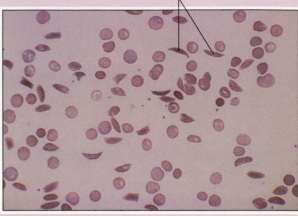

Sickle-shaped red blood cells

A blood smear from a patient suffering from sickle cell anemia. Compare this blood smear with the normal one shown in Figure 12–3.

icans of black African descent). It is characterized by sickle-shaped red blood cells, which are caused by an inherited inability to synthesize normal hemoglobin molecules (see the photograph). About 0.3% of African-Americans are affected, but 8% to 13% are carriers of the disease. The effects of this inherited disease include prolonged drowsiness, leg ulcers, fever, joint and abdominal pain, and clotting (thrombosis). There is currently no effective treatment, and few patients with the disease live beyond the age of 40 years.

in Figure 12–7. Hemoglobin selectively binds to oxygen molecules, much as a magnet binds to iron. In fact, hemoglobin contains iron atoms that actually provide the force that attracts and binds to oxygen. This causes the red blood cell to become bright red. When the blood circulates through a tissue that is low in oxygen, hemoglobin releases its oxygen, and it diffuses out of the red blood cell to the tissue. At the same time, a small amount of carbon dioxide diffuses into the red blood cell and binds to a different section on the hemoglobin molecule (most of the carbon dioxide remains in the plasma). Carbon dioxide is released when the blood reaches the lungs. Thus, hemoglobin performs the function of an important carrier molecule in red blood cells.

Life Cycle

Before birth, red blood cells are produced in the yolk sac, the liver, and the spleen. After birth and through-

out the rest of life, they are formed exclusively within red marrow in bone.

The life expectancy of a red blood cell is very short—only about 120 days. Most other body cells survive for years, and some survive an entire lifetime (such as skeletal muscle cells, cardiac cells, and neurons). The life expectancy of a red blood cell is limited because it lacks a nucleus and other organelles, making it impossible to perform repairs that are routinely conducted by other cells. In order to maintain a relatively constant red blood cell supply to replace this high rate of loss, their production rate must be carefully regulated.

Red blood cell production is regulated by special cells in the kidneys and, to a lesser extent, in the liver (Fig. 12–8). These cells can detect a reduction in oxygen levels in the blood. When they do so, they release a hormone called **erythropoietin**[4] (ē-rith'-rō-POY-ē-tin)

---

[4]Erythropoietin: Greek *erythros* = red + *poiein* = to make.

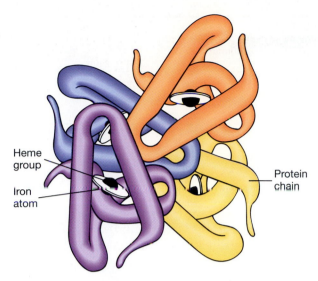

Figure 12–7
The hemoglobin molecule. It is a complex molecule composed of four distinct protein chains, each containing a heme group, which bears an iron atom. The iron atom attracts and binds to a molecule of oxygen.
What cells contain large amounts of this complex protein?

into the bloodstream. Erythropoietin travels to the red marrow, where it stimulates more red blood cells to be produced. When red blood cell volume is restored, high oxygen levels inhibit the further release of erythropoietin and the rate of production is reduced. As you may recall from your study of the previous chapter, this is a *negative feedback system,* which is a very common method of hormone regulation.

Approximately 10 billion red blood cells perish each hour in your body. With this enormous number of cells, the body must operate an efficient "recycling center" to reuse cell parts and remove toxic by-products. This is performed largely by white blood cells that wander throughout the body, called **macrophages,**[5] which phagocytize damaged and aging cells. The hemoglobin molecules are broken down by enzymes into a greenish pigment, called **biliverdin,**[6] which may be further reduced to an orange pigment called **biliru-**

[5]Macrophage: Greek *macro-* = large + *phagein* = to eat.

[6]Biliverdin: *bili-* = bile + *verde* = green + *-in* = chemical suffix.

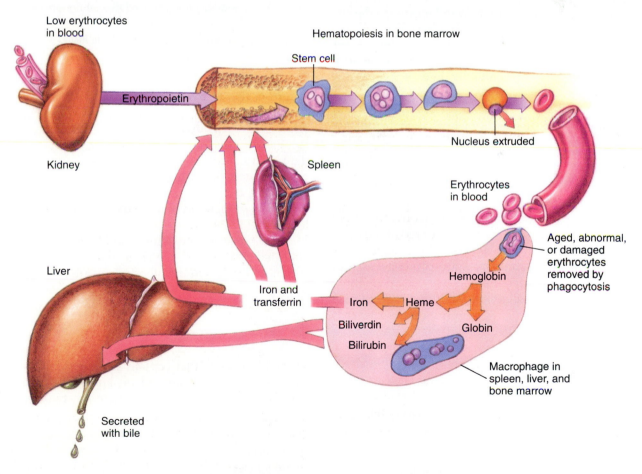

Figure 12–8
The life cycle of erythrocytes (red blood cells). The sequence of events is described in the text.
What is the role of the liver in recycling materials from red blood cells?

bin.[7] Biliverdin and bilirubin are released into the liver, which excretes them as bile pigment into the bile. Bile pigments are the main source of color in the stools and urine (see Chapter 16). The iron from hemoglobin is recycled by the red marrow for use in the formation of new hemoglobin molecules, or is stored in the liver or spleen for later use.

White Blood Cells (Leukocytes)

White blood cells (WBCs) or **leukocytes**[8] (LOO-kō-sīts), are much fewer in number than erythrocytes. They represent less than 1% of the total blood volume and number about 5000 to 10,000 in a cubic millimeter of blood. Despite this small number, their role in controlling diseases is very important. This is evidenced by the fact that any variation in their numbers, whether high or low, is an indication of a diseased condition.

Types

Unlike red blood cells, all white blood cells contain a nucleus and have the ability to wander outside the circulatory system. However, this is where the similarities among different types of white blood cells end, for they differ in the nature of their cytoplasm, their size, the shape of the nucleus, and how they respond to different staining techniques. We can begin classifying white blood cells into two groups, granulocytes and argranulocytes, on the basis of how their cytoplasm differs.

Granulocytes

White blood cells that have a cytoplasm containing highly visible pebblelike objects, known as *granules*, are classified as **granulocytes** (GRA-nyoo-lō-sīts). They are usually about twice the size of red blood cells, and each contains a nucleus that is partitioned into sections called *lobes*. Granulocytes are produced within red marrow along with red blood cells. There are three types of granulocytes: neutrophils, eosinophils, and basophils. Their names originate from the stains that are commonly used to bring out their distinguishing features (neutral, eosin, and basic).

The most abundant granulocytes in the blood are **neutrophils**[9] (NOO-trō-filz) (Fig. 12–9). They are characterized by the presence of cytoplasmic granules that stain pink in a neutral stain. Their nucleus may con-

PLATELETS (thrombocytes) 150,000 - 360,000/mm^3	
1 to 2 µm	
RED BLOOD CELLS (erythrocytes) 4,600,000 - 6,200,000/mm^3	
7 µm	

WHITE BLOOD CELLS (leukocytes) 5000 - 10,000/mm^3
Granulocytes (granular leukocytes)

Neutrophil 10 to 14 µm 50–70% Eosinophil 10 to 14 µm 1–4% Basophil 10 to 14 µm 0–0.5%

Agranulocytes (agranular leukocytes)

Lymphocyte 8 to10 µm 25–33% Monocyte 15 to 20 µm 3–8%

Figure 12–9
The formed elements and their relative abundances in circulating blood.
Which is the least common formed element in blood?

tain from two to five lobes that are interconnected by thin bridges. Neutrophils make up about 60% of all white blood cells in a normal blood sample.

Eosinophils[10] (ē′-ō-SIN-ō-filz) are much less abundant than neutrophils, making up only about 1 to 4% of the white blood cells in a blood sample. Their granules stain red in an acid stain that contains a dye known as *eosin*, and the nucleus usually has only two lobes (Fig. 12–9).

The rarest of the white blood cells are the **basophils**[11] (BĀS-ō-filz), which account for only 0.5% or less of white cells in the blood. They are characterized by large granules that stain blue in basic stain, and the nucleus is often bent into an S shape with two lobes (Fig. 12–9).

Agranulocytes

White blood cells that contain relatively few or very small cytoplasmic granules are known as **agranulocytes**[12] (ā-GRA-nyoo-lō-sīts). There are two types of

---

[7]Bilirubin: *bili-* + *ruber* = red + *-in*.

[8]Leukocytes: Greek *leuko-* = white + *-cyte* = cell.

[9]Neutrophil: Latin *neuter* = neither or neutral + Greek *-phil* = to love.

[10]Eosinophil: *eos* = dawn or rose-colored + *-in* = chemical suffix + *-phil*.

[11]Basophil: *baso-* = basic or alkaline + *-phil*.

[12]Agranulocyte: *a-* = without + *granule* = small particle + *-cyte* = cell.

Red Blood Cells and Your Diet

The vast number of erythrocytes that must be produced to replace damaged, aged, or lost cells is enormous. In order to maintain the demanding production rate, raw materials must be continuously available in large quantities. The materials that are most likely to be limited are two of the B-complex vitamins and iron. The vitamins, vitamin B_{12} and folic acid, are required for the synthesis of new DNA molecules. Although mature red blood cells do not contain a nucleus and, therefore, do not have DNA, their precursor cells have a nucleus and DNA like other body cells. In order to synthesize the DNA in these precursor cells, the basic building blocks of B-complex vitamins are needed. Because of the large numbers of red blood cells required continuously, the precursor cells are particularly sensitive to variations in the availability of these vitamins—more so than other body cells. B-complex vitamins can be obtained from leafy green vegetables, yellow vegetables, legumes (peas and beans), and whole grains.

Folic acid is particularly important during early pregnancy. In recent studies, it has become clear that a deficit of folic acid in the diet during the first two weeks of pregnancy causes certain birth defects, including spina bifida (incomplete development of the vertebral column). These findings have motivated the U.S. surgeon general to require all physicians with pregnant patients to prescribe folic acid supplements during pregnancy, especially during the first trimester.

Iron is required for hemoglobin synthesis. It is obtained from two sources: from the recycling of red blood cells, in which the iron is conserved and stored in the liver; and from diet. Through the diet, iron can be obtained from leafy green vegetables and red meat. In most people, however, much of the iron that is needed daily is obtained from body storage. Iron supplements are usually needed only following a traumatic loss of red blood cells, such as occurs during an injury. For many women, iron supplements may also be required to replace iron lost each month during menstruation. However, too much iron taken as supplements to the diet can have toxic side effects, the symptoms of which include gastrointestinal distress and headaches.

---

cells in this group: monocytes and lymphocytes. They are both produced in the red bone marrow like other blood cells. Agranulocytes are also produced by organs of the lymphatic system (lymph nodes, the spleen, and the thymus).

Monocytes are the largest cells found in the blood (Fig. 12–9). They are nearly three times larger than red blood cells and often twice the size of granulocytes. The nucleus may be round, oval, or lobed and often occupies most of the cell volume. Monocytes usually account for about 3% to 8% of the white blood cells in a blood sample.

Lymphocytes are typically about the same size as a red blood cell and thus are the smallest type of white blood cell (Fig. 12–9). The round nucleus is quite large, occupying nearly all of the cell volume. Lymphocytes make up about 25% to 33% of the white cells in a blood sample.

Function

All white blood cells provide the body with protection from disease. The defensive network they provide is aided by their ability to actively move about. Most white blood cells can actually squeeze between cells that form the walls of blood vessels, and thereby move freely between the cardiovascular system and most other areas of the body (Fig. 12–10). This process is called **diapedesis**[13] (dī-a-ped-Ē-sis) and is an important weapon in the battle against infection. Once outside the bloodstream and in the interstitial fluid, these active cells move about much like an ameba, extending streams of cytoplasmic arms (called *pseudopodia*) in the direction they are traveling. They are attracted to an area of infection by chemicals released by invading microorganisms and damaged cells. When they make contact with the microorganism or damaged cell, the white blood cells trap it and engulf it by **phagocytosis.** As more white blood cells arrive at the site of infection, they form a collection of living, dead, and broken cells and plasma called *pus*.

Phagocytosis is not the only weapon our white blood cells have to use against infection and other

---

[13]Diapedesis: Greek *dia-* = through + *pedan* = to leap.

HEALTH CLINIC

Inflammation: Friend and Foe

The inflammatory response is a reaction by the body's defenses to trauma caused by infectious microorganisms or physical injury. It involves the release by basophils of histamine, which is a chemical that causes capillary walls to dilate and increase in permeability. This increase in permeability leads to the mass movement of plasma into the extracellular environment. The accumulation of fluid in these areas is called *edema* and produces swelling, redness, heat, and pain. The purpose of producing edema in tissues is to promote the movement of phagocytic white blood cells to the site of infection. In this way, inflammation provides benefits to the defense system of the body.

However, there are times when inflammation can prove contrary to body health. In joint injuries, for example, the amount of swelling may become excessive, causing so much pressure and pain that the healing process becomes impaired. In these cases, fluids must be extracted from the injury site to control the amount of swelling. The edema that develops from allergies can also have adverse effects, which individuals with chronic hay fever experience all too frequently. Drugs called *antihistamines* are often taken to reduce inflammation in these cases. They counteract histamine by reducing capillary permeability.

forms of disease. In fact, the various types of white blood cells have specializations that enable them to combat disease in different ways. For example, certain lymphocytes produce highly specific proteins called **antibodies** that act against foreign particles and toxins entering the body. Antibody production forms the basis for *immunity*, which we will discuss further in Chapter 14. Basophils are also specialized, for they produce a substance that causes swelling (inflammation), called *histamine*. The accumulation of fluids during inflammation helps direct the movement of other white blood cells to the area of infection.

The primary cells specialized for phagocytosis are the neutrophils and monocytes. Neutrophils are very mobile and are often the first cells to reach a site of injury. Once there, they phagocytize bacteria and damaged cells. Monocytes are also very active, and their large size enables them to phagocytize whole cells and large numbers of bacteria. Eosinophils are not as mobile and active, but are effective in phagocytizing certain foreign particles that produce an allergic reaction, such as invading parasites, pollen grains, and mold spores.

Figure 12–10
The migration of white blood cells through capillary walls, or diapedesis.
Why do white blood cells migrate out of the bloodstream?

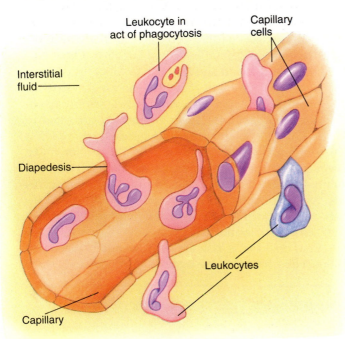

Leukocyte in act of phagocytosis

Capillary cells

Interstitial fluid

Diapedesis

Leukocytes

Capillary

Platelets (Thrombocytes)

Platelets, or **thrombocytes**[14] (THROM-bō-sīts), are formed elements that are fragments of complete cells (Figs. 12–3 and 12–9). During their developmental stage within red marrow, they are formed when a large precursor cell breaks apart. Its smaller fragments are the platelets that are released into the bloodstream for circulation. The larger fragments are further broken down in the bloodstream to form additional platelets.

[14]Thrombocyte: Greek *thrombo-* = clot + *-cyte* = cell.

Each platelet consists of cytoplasm surrounded by a plasma membrane. Platelets have no nucleus but contain most types of cytoplasmic organelles. A platelet is about one-tenth the size of a red blood cell and is in the shape of a round or oval disc. Platelets are less numerous than red blood cells, numbering between 150,000 and 360,000 per cubic millimeter in a normal blood sample.

Platelets play an important role in the prevention of fluid loss. They initiate the formation of blood clots, which plug up the breaks in blood vessel walls following an injury. The mechanism of blood clot formation is further discussed in the following section.

CONCEPTS CHECK

1. What is the most abundant formed element in blood?

2. Why must hemoglobin concentrations in the blood be maintained for health?

3. How do white blood cells move out of the bloodstream?

4. What are the general functions of each of the formed elements?

HEMOSTASIS

The stoppage of bleeding following a blood vessel accident is called hemostasis. It is performed in a three-step process.

Most of our blood is contained within blood vessels. When a blood vessel ruptures because of a defect or an injury, large amounts of plasma and formed elements may escape. This loss of vital fluids and cells quickly threatens homeostasis. To prevent a critical loss of homeostasis, the body has a special mechanism to stop the bleeding. This mechanism is called **hemostasis**[15] (note how it is spelled differently from homeostasis—don't confuse their meanings!). Hemostasis is a three-step process that includes: (1) blood vessel spasm; (2) platelet plug formation; and (3) coagulation. It is effective in stopping bleeding in smaller vessels, but in major vessel damage medical intervention is usually necessary.

Blood Vessel Spasm

The wall of a blood vessel contains smooth muscle that is controlled by involuntary responses. When the wall

---

[15]Hemostasis: Greek *hemo-* = blood + Latin *stasis* = stopping or standing still.

---

is cut or torn, the smooth muscle is stimulated to contract. Its contraction draws the open ends of the vessel wall together, minimizing blood loss. In some cases, the severed vessel is closed completely. Although this spasm lasts for less than 30 minutes, it provides time for the other hemostatic mechanisms to take effect. In some cases, the spasms may be prolonged by the release of a substance called *serotonin* by platelets. This chemical causes a sustained contraction of smooth muscle in the walls of blood vessels.

Platelet Plug Formation

When platelets arrive at the site of the vessel breakage, their characteristics change. They increase in size and take on irregular shapes, and their surfaces become very sticky. Their sticky nature causes them to adhere to collagen fibers from the blood vessel wall—which are often frayed at the breakage site—and to one another. As platelets continue to arrive at the scene and remain, they eventually form a large clump with the collagen known as a **platelet plug** (Fig. 12–11). This process can be very effective in plugging up an opening in a vessel wall if the damage is only minor. It is normally accompanied by coagulation, which is described next.

Coagulation

The process of **coagulation** is the most effective, and the most complex, of the blood-clotting mechanisms. It involves many substances that are normally present in the plasma, substances secreted by the platelets at the site of the injury, and substances released by the injured tissues themselves. It consists of a series of events that result in the formation of a **blood clot.**

The main event in coagulation is the conversion of the plasma protein **fibrinogen** into long threads of protein known as **fibrin.** This process begins when the injured blood vessel walls and nearby platelets release a substance called **thromboplastin.** Thromboplastin interacts with calcium ions and other substances to convert **prothrombin,** a protein present in the plasma in small amounts, into **thrombin.** Once produced, thrombin acts as an enzyme by combining soluble fibrinogen elements together, forming long, hairlike molecules of the insoluble protein fibrin.

The fibrin molecules that are newly formed tend to stick to the surfaces of the blood vessels. They soon form a netting, or meshwork, that traps the formed elements (Fig. 12–12). This mass of fibrin threads and blood cells forms the blood clot, which forms an effective barrier that prevents the further loss of cells and plasma through the injured vessel wall.

The process of coagulation normally takes about 2 to 8 minutes to complete. However, a lack of *any* of the

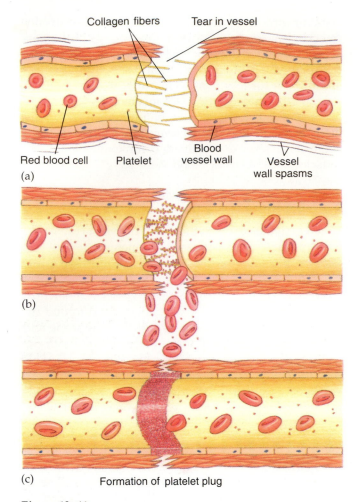

Collagen fibers   Tear in vessel

Red blood cell   Platelet   Blood vessel wall   Vessel wall spasms

(a)

(b)

(c)   Formation of platelet plug

Figure 12–11

The stages of platelet plug formation. (a) An injury occurs, causing an opening in the vessel wall. The blood vessel walls soon undergo spasm, which draws the torn ends of the vessel closer together. (b) Blood escapes through the injured vessel wall, as platelets begin to accumulate at the site of injury. The platelets adhere to each other and to the broken strands of collagen. (c) A platelet plug is formed, blocking further blood loss.

What feature of platelets promotes their ability to accumulate at an injury site?

factors required for coagulation can lead to a decreased ability of the blood to clot. For example, dietary deficiencies (especially vitamin K), liver diseases, and an inherited deficiency known as *hemophilia*, or bleeder's disease, each result in an inability to form clots due to a lack of one or more of the clotting factors. Following an injury, individuals suffering from these diseases must be given blood transfusions or receive injections of the factors they lack.

Once the blood clot has formed, permanent repair of the injury may begin. This involves the activity of fibroblasts from neighboring connective tissue, which further strengthen the clot and seal the vessel tear. During the early stages of repair, the blood clot is dissolved, in a process called *fibrinolysis*. In this process, enzymes are activated to digest the fibrin threads, returning fibrinogen into the plasma for its later reuse.

In some cases, blood may clot when it is not needed. This may occur if the internal wall of the vessel is roughened or if it contains an accumulation of fatty deposits, such as occurs in the disease *atherosclerosis*. The undesirable clot that may form is called a **thrombus,** which can cause death if it blocks circulation to a vital organ such as the heart or brain. If the thrombus detaches and becomes a free-floating clot, it is called an **embolus.** An embolus can also be life-threatening, since it may lodge in a blood vessel supplying a vital organ and cut off its blood supply.

The processes of blood vessel spasm, platelet plug formation, and coagulation in hemostasis are summarized in Figure 12–13.

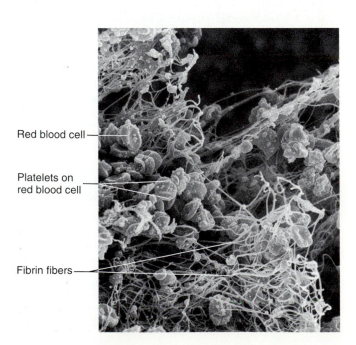

Red blood cell

Platelets on red blood cell

Fibrin fibers

Figure 12–12

Coagulation of blood. In this high-resolution electron micrograph, the threads of fibrin fibers are clearly visible. The tiny "bumps" on the red blood cells are platelets. (Magnified 3848x.)

What produces the fibrin threads that are visible in this electron micrograph?

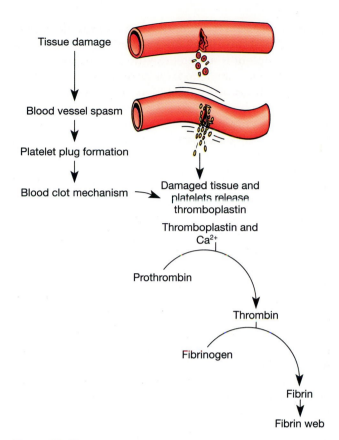

Figure 12–13
A summary of the events in hemostasis.
What is the role of fibrinogen in hemostasis?

When one type of blood was combined with another type, the cells clumped together, or **agglutinated**.[16] When this occurred in a living patient during a blood transfusion, the unfortunate patient's life was usually lost because of the massive destruction of red blood cells. On the other hand, blood of the same type did not agglutinate, and it became possible to give blood to a patient from a compatible donor. Eventually, investigators were able to determine why these reactions occurred, and to identify the different groups of blood in the human population.

The basis for blood grouping is the presence of surface proteins on the plasma membrane of red blood cells. These proteins are genetically determined, and are called **antigens** (or agglutinogens). In addition, there are other proteins within the plasma, known as **antibodies** (or agglutinins). The nature of the reactions of these two types of proteins with each other determines whether or not the blood will agglutinate. If the reactions between the antigens and antibodies cause the blood to clump, the two blood samples are not compatible. Conversely, if the proteins do not cause agglutination, the samples are compatible and can safely be mixed.

The process of **blood typing** involves identifying the antigens present on the red blood cell plasma membrane. It is a common procedure that enables one to determine if blood from two individuals is compatible and can safely be mixed in a blood transfusion. There are many systems of blood typing, but we will study only the two most common: the ABO system and the Rh system.

BLOOD GROUPS

Blood grouping is based upon the reaction between surface proteins on red blood cell plasma membranes and special plasma proteins. The two major systems are ABO and Rh.

It was first discovered during the latter half of the nineteenth century that, among certain groups of individuals, red blood cells have different characteristics.

ABO System

There are only two antigens in the ABO system. These are designated "A" and "B." An individual may have both antigens on his or her red blood cell membranes (and thus be described as type AB), have one of the two antigens (and thus be either type A or type B), or lack both of them (type O). Thus, everyone has one of four possible blood types: AB, A, B, or O.

In individuals who have type A, type B, or type O blood, the plasma contains antibodies that have formed during early life. These antibodies react only with antigens the individuals do *not* have. For example, whenever antigen A is absent in red blood cells (as it is in individuals with type B or type O blood), an antibody that reacts only with A is present. In this example, the antibody is called anti-A. Similarly, when-

[16]Agglutinate: Latin *ad-* = to, toward + *glutin-* = glue + *-ate* = to act or become.

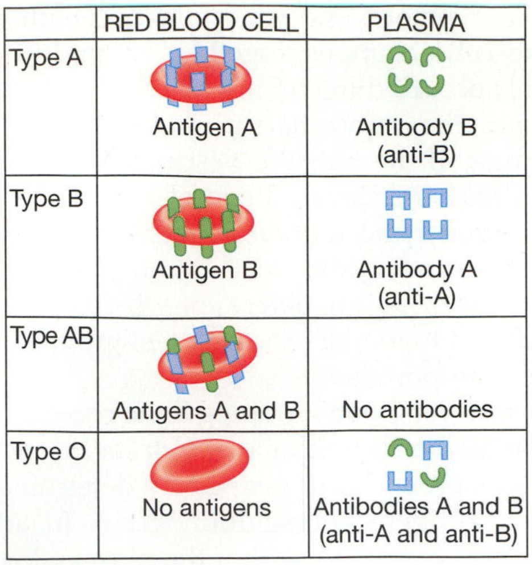

	RED BLOOD CELL	PLASMA
Type A	Antigen A	Antibody B (anti-B)
Type B	Antigen B	Antibody A (anti-A)
Type AB	Antigens A and B	No antibodies
Type O	No antigens	Antibodies A and B (anti-A and anti-B)

Figure 12–14
The ABO blood grouping system.
What recipient blood types can accept type A donor blood?

ever antigen B is absent, the antibody that reacts with B is present; it is called anti-B. Thus, persons with type B blood have anti-A antibodies, those with type A blood have anti-B antibodies, those with type AB blood have neither antibody, and those with type O blood have both anti-A and anti-B antibodies. These relationships are diagramed in Figure 12–14 and summarized in Table 12–2.

Blood Transfusions

A successful blood transfusion is one that replaces lost blood in a patient with blood obtained from a different source, usually a different individual, without developing an agglutination reaction. If two blood types are not compatible and agglutination occurs, then a transfusion reaction develops. This dangerous reaction is to be prevented, since it leads to destruction of red blood cells and the consequent liberation of hemoglobin and bilirubin into the body, which can eventually cause kidney failure and death. Therefore, the combination of incompatible types of blood must be avoided.

Before a blood transfusion is attempted, the blood types of the donor (the person who gives the blood) and the recipient (the person who receives the blood) must be carefully matched. Of primary concern is that the cells in the transfused blood be compatible with the recipient's blood. For example, a person with type A blood must not be given blood of type B or AB, because type A blood contains anti-B antibodies, which would react and agglutinate with the antigens in the type B or AB blood. Similarly, a person with type B blood must not be given type A blood, because of the anti-A antibodies; and a person with type O blood must not be given A, B, or AB blood, because both anti-B and anti-A antibodies are present.

In general, it is safest practice to use donor blood that is of the same type as recipient blood. The exceptions to this rule are in cases of emergency in which the proper blood type may not be available. In these cases, persons with type O blood may donate small amounts of blood to any recipient, since their red blood cells have no ABO surface antigens and, therefore, cannot react with a recipient's A or B antibodies (in the past, a person with type O blood was called a *universal donor*). Likewise, a person with type AB blood may receive small amounts of blood of any type, since the agglutination reaction that results is usually minimal. However, the mixing of type O blood or type AB blood must be made with caution and in small amounts, since some agglutination does occur. Why is this so? Because there are blood groups other than the ABO group that must be matched for compatibility. Also, the antibodies in the blood of any donor (including type O) can react with antigens in the blood of any recipient. Therefore, it is important to correctly match all blood groups before a transfusion to reduce the likelihood of agglutination, whenever it is possible to do so.

Rh System

The Rh blood-typing system is named after the *rh*esus monkey, in which it was originally discovered. It was later found that the Rh antigens are also present on the red blood cell membranes of humans. In this system,

Table 12–2 THE ABO BLOOD GROUP SYSTEM

Blood Type	Antigens on Erythrocytes	Antibodies in Plasma	Can Donate Blood to	Can Receive Blood from
A	A	Anti-B	Type A, AB	Type A, O
B	B	Anti-A	Type B, AB	Type B, O
AB	A and B	Neither	Type AB	Type A, B, AB, O
O	Neither	Anti-A and anti-B	Type A, B, AB, O	Type O

a person having Rh antigens on his or her red blood cells is said to be *Rh-positive*. Conversely, a person lacking Rh antigens is *Rh-negative.*

Like the ABO system, the presence of Rh antigens on red blood cell membranes is an inherited trait. Most individuals are born Rh-positive—a trait they inherit from their parents. However, the antibodies for Rh antigens (called anti-Rh) are not automatically formed as the anti-A and anti-B antibodies are in the ABO system. Instead, they are formed in Rh-negative persons only when they become sensitized.

Rh sensitization occurs when a person with Rh-negative blood receives a transfusion of Rh-positive blood, or the blood is otherwise mixed with Rh-positive blood. When this occurs, the recipient's immune system produces anti-Rh antibodies to protect against the foreign blood type. An agglutination reaction does not usually occur with the first combining of unlike blood types, since it takes time for the body to react and start making the anti-Rh antibodies. However, the second time and every time thereafter, the donor's blood will agglutinate when it comes into contact with the anti-Rh antibodies in the recipient's blood.

An example of Rh sensitization occurs when a pregnant Rh-negative woman is carrying an Rh-positive fetus for the first time (Fig. 12–15). This first preg-

Figure 12–15
Development of hemolytic disease of the newborn. (a) An Rh-negative mother pregnant for the first time with an Rh-positive fetus. In some cases, Rh-positive antigens (*green*) may diffuse through the placenta to the mother's bloodstream. (b) In time, the mother will develop anti-Rh antibodies in response. The first child will have been born before it could be affected by the antibodies. (c) A second Rh-positive fetus may receive anti-Rh antibodies from the mother, which will destroy the child's red blood cells if appropriate countermeasures are not taken.

If the second fetus has Rh-negative blood, would it respond unfavorably to the mother's anti-Rh antibodies?

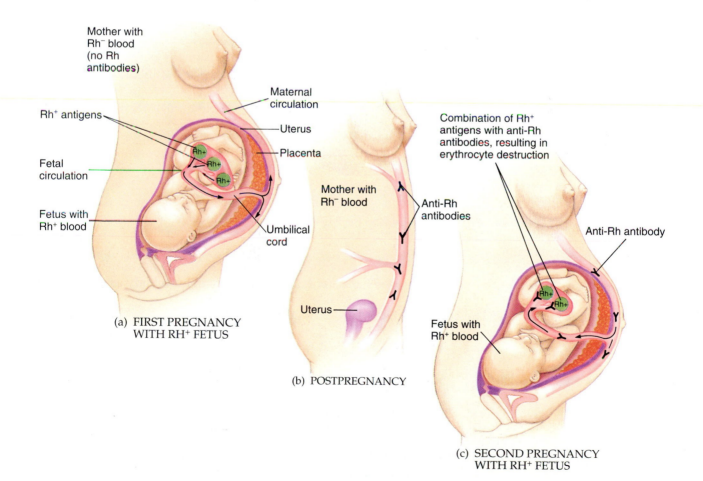

Mother with Rh⁻ blood (no Rh antibodies)

Maternal circulation

Rh⁺ antigens

Uterus

Fetal circulation

Placenta

Fetus with Rh⁺ blood

Umbilical cord

Mother with Rh⁻ blood

Anti-Rh antibodies

Combination of Rh⁺ antigens with anti-Rh antibodies, resulting in erythrocyte destruction

Anti-Rh antibody

Uterus

Fetus with Rh⁺ blood

(a) FIRST PREGNANCY WITH RH⁺ FETUS

(b) POSTPREGNANCY

(c) SECOND PREGNANCY WITH RH⁺ FETUS

nancy usually presents no major problems and results in the delivery of a healthy infant. However, during delivery the placental membrane separating the mother's blood from the fetal blood may tear, allowing the two to mix. The exposure to the infant's Rh-positive blood may sensitize the mother's immune system to produce anti-Rh antibodies. If the mother becomes pregnant again with an Rh-positive infant, her anti-Rh antibodies will cross the placenta and cause an agglutination reaction with the fetal red blood cells. This causes a condition in the fetus known as *erythroblastosis fetalis* or *hemolytic disease,* in which the unborn child suffers from anemia and hypoxia (lack of oxygen). Brain damage and even death may result unless blood transfusions are performed before birth to provide more red blood cells for oxygen transport. Fortunately, if the first-time pregnant mother knows she is Rh-negative and the fetus is Rh-positive, she can

avoid sensitization by receiving medical treatment with a substance called RhoGAM.

CONCEPTS CHECK

1. What is the difference between an antigen and an antibody?

2. If type A blood is mixed with an unknown blood type and it agglutinates, what type(s) may the unknown be?

3. If a mother has Rh-negative blood and her first child is Rh-positive, why would the child be at possible risk if it were to receive a blood transfusion from the mother?

HOMEOSTASIS AND THE BLOOD

As was pointed out earlier in this chapter, the blood performs an important role in maintaining homeostasis of the body. Each of its three major functions (transportation, protection, and regulation) is directly involved in this role. Consequently, any disruption of the normal functions of blood creates an imbalance of homeostasis to the body as a whole.

The function of transportation provided by the blood is essential for all cells. Every cell in your body relies on the blood to provide it with vital oxygen and nutrients and to carry away its waste materials. Without this transportation function, the cells would perish within minutes.

A sustained homeostasis also requires the protective function of immunity, provided by the white blood cells. What happens to homeostasis when immunity fails? The failure of white blood cell protection results in the rapid deterioration of health, which is caused by the rapid spread of infectious diseases. This occurs in the disease AIDS, which we will explore in Chapter 14.

The protective function of blood also includes the clotting mechanism, or hemostasis. The ability of the

blood to arrest the loss of fluids following an injury is vital. In patients whose clotting mechanism is incomplete, as we have seen in the disease called *hemophilia,* a loss of blood can upset homeostasis to the point of becoming life-threatening.

The regulatory mechanisms of maintaining the acid-base balance and of temperature control also play a direct role in maintaining homeostasis. Buffer systems in the circulating blood maintain the pH of all body fluids within the narrow range required for life. Any alteration of the buffers can upset this balance. To support the buffer concentration in the blood, the lungs constantly remove excess carbon dioxide (the source of carbonic acid) and the kidneys constantly remove excess hydrogen ions (the source of all acids). As a temperature regulator, the blood serves as a reservoir of fluid that can carry heat or cooling temperatures to various parts of the body. In this way, the blood helps to keep body temperature within a narrow range despite fluctuations in body activity or external temperature.

CLINICAL TERMS OF THE BLOOD

Anemia   A condition in which there is a reduction in the number of red blood cells or in the amount of hemoglobin per unit of blood. There

are several types, which include hemorrhagic anemia (due to loss of blood through profuse bleeding), pernicious anemia (due to shortage of

B vitamins), and hemolytic anemia (due to accelerated destruction of red blood cells by macrophages).

Anoxia   An oxygen deficiency in the blood.

Apheresis   A medical technique for cleansing the blood, in which a portion (such as plasma) suspected of containing harmful substances is removed and replaced with fresh ingredients.

Bacteremia   An acute infection of the bloodstream by bacteria. It sometimes follows certain surgical or dental procedures or insertion of venous or urethral catheters into patients' bodies. Symptoms include fever, chills, and skin rashes. It can be treated effectively with antibiotics.

Cyanosis   A bluish skin coloration due to oxygen deficiency in blood.

Direct transfusion   The transfer of blood directly from one person to another, without exposure of the blood to air.

Exchange transfusion   The direct transfer of blood from a donor to replace blood as it is removed from the recipient.

Hemophilia   A hereditary disorder of the coagulation process in hemostasis. In most cases one of two factors in the coagulation cascade is missing, resulting in the partial or complete inability to stop bleeding. Treatment involves supplying the patient with the missing factor through blood transfusions of fresh plasma from normal donors.

Hemorrhage   Severe blood loss due to an injury. It may occur externally when the skin and underlying blood vessels are torn, or internally. Massive hemorrhage results in cardiovascular collapse, or shock.

Heparinized whole blood   Whole blood placed in a solution of the anticoagulant heparin to prevent coagulation.

Indirect transfusion   Transfer of blood in which whole or fractioned donor blood is stored for later delivery to a recipient.

Leukemia   Cancer of the blood, which may involve any of the various types of white blood cells or their precursors. It is identified by the observation of abnormal cells in the bloodstream or in the bone marrow. It results in the proliferation of dysfunctional leukocytes, which leads to a failure of the immune system. The abnormal production of dysfunctional white blood cells also crowds out the production of erythrocytes, causing anemia. Treatment of this disease has improved considerably since 1980, because of the combined uses of antileukemic drugs, total body X-irradiation, and bone marrow transplants.

Leukocytosis   An elevated leukocyte count. This condition usually suggests an infection, an inflammatory process, or a malignancy (such as leukemia).

Leukopenia   A lowered leukocyte count. This condition may result from a viral infection, a failure in cell production, or a rapid destruction of cells. Some drugs can cause leukopenia by destroying white blood cells.

Malaria   An infection of the bloodstream by protozoan parasites (*Plasmodium*). Next to the common cold, it is the most common infectious disease in the world (60 million cases are reported each year). The parasite enters the bloodstream by way of a bite by a female *Anopheles* mosquito and proceeds to destroy erythrocytes on a cyclic basis. Malaria causes recurrent episodes of chills, fever, anemia, and spleen enlargement. Without treatment, it can cause death.

Polycythemia   An abnormal increase in the number of erythrocytes. It is accompanied by an increase in blood viscosity, leading to a rise in blood pressure. If left untreated, it can cause blood clots (thrombosis) and hemorrhage.

Septicemia   A persistent bacterial infection of the bloodstream. It is a more serious condition than bacteremia, with similar symptoms and treatment. It is sometimes called "blood poisoning."

Shock   A disturbance of blood circulation resulting in a critical reduction of blood to vital tissues. There are several types of shock, which include hypovolemic shock (a decreased blood volume due to a loss of formed elements and/or plasma), cardiogenic shock (failure of the heart), and vascular shock (lowered ability of blood vessels to transport blood). It is generally treated in an emergency situation by volume replacement of fluids followed by whole blood.

Sickle cell anemia   A chronic type of hemolytic anemia in which the red blood cells are sickle-shaped, containing an abnormal hemoglobin. It is an inherited trait and occurs in about 0.3% of the blacks in the United States. About 8% to 13% of American blacks are carriers of this disease, which has no effective treatment. Those stricken with it rarely live beyond 40 years.

Thalassemia   A chronic type of hemolytic anemia that results from the synthesis of dysfunctional hemoglobin molecules. It is an inherited trait that results in jaundice, spleen enlargement, and leg ulcers.

Blood is a type of connective tissue and is classified as part of the circulatory system. It consists of formed elements and plasma.

CHARACTERISTICS OF BLOOD

Functions of Blood Blood helps maintain homeostasis, by performing the following three functions:

Transportation Of respiratory gases, nutrients, metabolic wastes, and hormones.

Protection Against harmful microorganisms and toxins, and against fluid loss.

Regulation Helps maintain the acid-base balance by the presence of buffers, and helps regulate body temperature.

Properties of blood Color is red, owing to hemoglobin in red blood cells.

Male volume is 5 to 6 liters; female volume is 4 to 5 liters.

Blood is five times more viscous than water.

Blood pH is maintained within a limited range of 7.35 to 7.45 by buffers.

PLASMA

Plasma is the liquid portion of blood. It consists mostly of water, with dissolved substances (or solutes).

Plasma Proteins The most abundant solutes in plasma. They include albumins, globulins, and fibrinogen.

Other Plasma Solutes Include nonprotein nitrogenous substances, gases, nutrient molecules, and electrolytes.

FORMED ELEMENTS

The formed elements include red blood cells, white blood cells, and platelets.

Red Blood Cells (erythrocytes) The most abundant of the formed elements. The normal density range is 4.6 to 6.2 million per cubic millimeter of blood in males, and 4.2 to 5.4 million per cubic millimeter of blood in females.

Structure Shaped like biconcave discs, a form that allows storage of as much hemoglobin as possible.

Function Hemoglobin binds to oxygen and delivers it to body cells and carries away carbon dioxide.

Life Cycle Red blood cells last only 120 days. They are produced in the red bone marrow; production is regulated by the hormone erythropoietin secreted by cells in the kidneys in response to blood oxygen levels. Old and dead cells are removed by macrophages, and their materials are excreted or recycled.

White Blood Cells (leukocytes) Normally, the density ranges between 5000 and 10,000 per cubic millimeter of blood. Each contains a nucleus and can actively move about. Leukocytes are produced in the red bone marrow, although one type is also produced in lymphatic tissue.

Types:

Granulocytes Contain highly visible granules in their cytoplasm. They include neutrophils, eosinophils, and basophils.

Agranulocytes Contain granules that are not highly visible in the cytoplasm. They include monocytes and lymphocytes.

Function White blood cells protect the body from disease. The various types do this in different ways: lymphocytes produce antibodies; basophils cause inflammation; neutrophils and monocytes are phagocytic; and eosinophils battle allergens.

Platelets (thrombocytes) Incomplete cells that lack a nucleus, which have formed from the fragmentation of a large precursor cell in the bone marrow. They are small in size and number between 150,000 and 360,000 per cubic millimeter of blood. They play an important role in the prevention of fluid loss (hemostasis).

HEMOSTASIS

Hemostasis is the stoppage of bleeding. It involves three mechanisms:

Blood Vessel Spasm The contraction of smooth muscle tissue in the walls of blood vessels. Spasm closes the opening left by a cut or tear, but only temporarily.

Platelet Plug Formation At the site of a vessel breakage, platelets stick to one another and to collagen fibers in the vessel wall. This forms a clump that can seal a small break, called the *platelet plug*.

Coagulation A series of events that result in the formation of a blood clot. It involves the conversion of fibrinogen elements into a long protein thread called *fibrin*. Fibrin forms a meshwork that traps formed elements, creating a barrier over the breakage site.

BLOOD GROUPS

The basis for blood grouping is the presence of proteins on the plasma membrane of red blood cells (antigens) and other proteins in the plasma (antibodies). When foreign blood is introduced, as during a blood transfusion, a reaction of antibodies with the antigens may result in blood clumping, or agglutination, or in no clumping. Blood grouping identifies the types of antigens present in blood.

ABO System This system contains only two antigens: A and B. Type AB blood has both antigens, type A blood has A, type B blood has B, and type O blood has neither. The antibodies that react may be either anti-A or anti-B.

Rh System This system identifies a class of antigens called Rh. If it is present, the blood type is Rh-positive; if it is not present, the blood type is Rh-negative. Incompatible mixing may occur if Rh-negative blood is exposed to Rh-positive blood more than once. For example, in consecutive pregnancies of an Rh-negative mother and Rh-positive offspring, agglutination does not occur the first time, because the Rh-negative blood has not

Blood Transfusions The goal in a transfusion is no agglutination. Therefore, blood of different ABO types should not be mixed, although certain combinations are possible if low volumes of donor blood are used.

Rh System This system identifies a class of antigens called Rh. If it is present, the blood type is Rh-positive; if it is not present, the blood type is Rh-negative. Incompatible mixing may occur if Rh-negative blood is exposed to Rh-positive blood more than once. For example, in consecutive pregnancies of an Rh-negative mother and Rh-positive offspring, agglutination does not occur the first time, because the Rh-negative blood has not

been stimulated, or sensitized, to produce anti-Rh antibodies.

HOMEOSTASIS AND THE BLOOD

The blood is vital to body cells' continual supply of oxygen and nutrients and the removal of waste materials. It also plays an important role in pH maintenance, temperature regulation, and the transport of hormones and enzymes. Its importance to body homeostasis is made apparent when blood is lost from the body, producing the life-threatening condition of shock.

KEY TERMS

agglutination (p. 359)
agranulocyte (p. 354)
antibody (p. 359)
antigen (p. 359)
basophil (p. 354)

embolus (p. 358)
eosinophil (p. 354)
erythrocyte (p. 349)
erythropoietin (p. 352)
fibrinogen (p. 357)

granulocyte (p. 354)
hemoglobin (p. 349)
hemostasis (p. 357)
lymphocyte (p. 355)
macrophage (p. 353)

monocyte (p. 355)
neutrophil (p. 354)
plasma (p. 347)
platelet (p. 356)
thrombus (p. 358)

QUESTIONS FOR REVIEW

OBJECTIVE QUESTIONS

1. Which of the following is *not* an important function of blood?
   a. transports nutrients
   b. protect from injury
   c. transports enzymes and hormones
   d. protects against fluid loss
2. Blood is red in color because it contains the protein pigment:
   a. fibrinogen
   b. albumin
   c. hemoglobin
   d. erythrocyte
3. The pH of blood is kept within a limited range primarily by:
   a. buffers
   b. leukocytes
   c. hemoglobin
   d. hemostasis
4. The most abundant solute in plasma is a group of proteins called:
   a. fibrinogen
   b. albumins
   c. hemoglobin
   d. platelets
5. Solutes in the plasma include proteins, nonprotein nitrogenous substances, gases, nutrients, and:
   a. formed elements
   b. electrolytes
   c. platelets
   d. water
6. A procedure that is used to determine the composition of blood by spinning a sample at high speed in a centrifuge is called a:

   a. blood smear
   b. total blood makeup
   c. hematocrit
   d. none of the above
7. The most abundant of the formed elements in blood are:
   a. red blood cells
   b. platelets
   c. white blood cells
   d. lymphocytes
8. The primary function of red blood cells is:
   a. transport of hormones
   b. transport of oxygen and carbon dioxide
   c. production of albumin
   d. protection from fluid loss
9. Red blood cell production is regulated by erythropoietin, which is secreted by cells in the:
   a. red bone marrow
   b. liver
   c. kidneys
   d. heart
10. White blood cells that contain highly visible granules in their cytoplasm are called:
    a. erythrocytes
    b. granulocytes
    c. agranulocytes
    d. monocytes
11. A type of white blood cell that is in large numbers at a site of infection and actively phagocytizes foreign particles is a(n):
    a. lymphocyte
    b. basophil
    c. neutrophil
    d. eosinophil
12. Clumping of platelets at a blood vessel breakage site is called a(n):
    a. platelet plug
    b. embolus
    c. coagulation plug
    d. spasm

13. The inability of the body to synthesize prothrombin would result in:
   a. lack of oxygen to cells
   b. no platelet plug formation
   c. inability of blood to coagulate
   d. recurring vessel spasms
14. Type A blood contains which antibodies?
   a. anti-A
   b. anti-Rh
   c. anti-B
   d. both anti-A and anti-B
15. Anti-B antibodies form an agglutination reaction with which antigen?
   a. A
   b. both A and B
   c. B
   d. none
16. What causes Rh sensitization in a person with Rh-negative blood?
   a. exposure to more Rh-negative blood
   c. exposure to anti-Rh antibodies
   b. exposure to Rh-positive blood
   d. both (b) and (c)

ESSAY QUESTIONS

1. Describe the role of hemoglobin in the blood. Indicate where it originates, where it is found, its structure, its function, and what happens to it when a red blood cell dies.
2. Compare the functions of the various types of white blood cells in protecting the body from disease.
3. Discuss the process of hemostasis. Include each of the three mechanisms (spasm, platelet plug, and coagulation).
4. What is the basis for blood grouping? Include in your answer the concept of blood transfusion.
5. Compare and contrast the antigens and antibodies in both blood group systems, ABO and Rh.

ANSWERS TO ART LEGEND QUESTIONS

Figure 12–2
What is the purpose of performing a hematocrit?   A hematocrit is performed in order to determine the relative concentrations of red blood cells, white blood cells, and plasma in a blood sample.

Figure 12–3
What formed elements can you identify in this blood smear?   Red blood cells and white blood cells are clearly visible in the blood smear.

Figure 12–4
Which cell type loses its nucleus during development?   Red blood cells lose their nucleus during development.

Figure 12–5
How would you describe the shape of a red blood cell?   A red blood cell is a flattened, oval disc with a depression in the center.

Figure 12–6
Can a mature red blood cell divide (by mitosis)?   No, a mature red blood cell cannot divide, because it has no nucleus (and no DNA).

Figure 12–7
What cells contain large amounts of this complex protein?   Red blood cells are packed with hemoglobin molecules.

Figure 12–8
What is the role of the liver in recycling materials from red blood cells?   The liver stores iron released by macrophages that have phagocytized red blood cells. The liver also channels bilirubin to the bile for its release into the digestive tract.

Figure 12–9
Which is the least common formed element in blood?   The least common formed element in blood is the white blood cell (the least common white blood cell is the basophil).

Figure 12–10
Why do white blood cells migrate out of the bloodstream?   White blood cells migrate out of the bloodstream during pursuit of foreign particles or diseased or dying cells.

Figure 12–11
What feature of platelets promotes their ability to accumulate at an injury site?   Platelets change their structural characteristics: they increase in size, they take on irregular shapes, and their surfaces become very sticky.

Figure 12–12
What produces the fibrin threads that are visible in this electron micrograph?   A series of events that begins when prothrombin, released by platelets and the walls of damaged vessels, is converted to thrombin, which combines with plasma protein fibrinogen.

Figure 12–13
What is the role of fibrinogen in hemostasis? Fibrinogen is converted to fibrin when it combines with thrombin: fibrin consists of sticky threads that form the network of the blood clot.

Figure 12–14
What recipient blood types can receive type A donor blood? Type A and type AB recipients can safely receive blood from a type A donor.

Figure 12–15
If the second fetus has Rh-negative blood, would it respond unfavorably to the mother's anti-Rh antibodies? No, there would be no response, since the blood of the mother and fetus is compatible.

The CARDIOVASCULAR SYSTEM

LEARNING OBJECTIVES

After studying this chapter, you should be able to:

1. Identify the location and general features of the heart.

2. Describe the layers of the pericardium.

3. Identify the layers of the heart wall.

4. Describe the structure of the atria and ventricles.

5. Describe the structure and function of the heart valves.

6. Trace the flow of blood through the chambers of the heart.

7. Identify the sounds of the heart, and relate them to the cardiac cycle.

8. Describe the components and function of the heart conduction system.

9. Identify the electrical events measured in a normal electrocardiogram.

10. Define cardiac output, and describe how it is regulated.

11. Distinguish between the types of blood vessels on the basis of their structure.

12. Describe the factors influencing blood pressure and how it is regulated.

13. Distinguish between the pulmonary and systemic pathways in terms of function.

14. Identify the major arteries and veins in the pulmonary and systemic pathways.

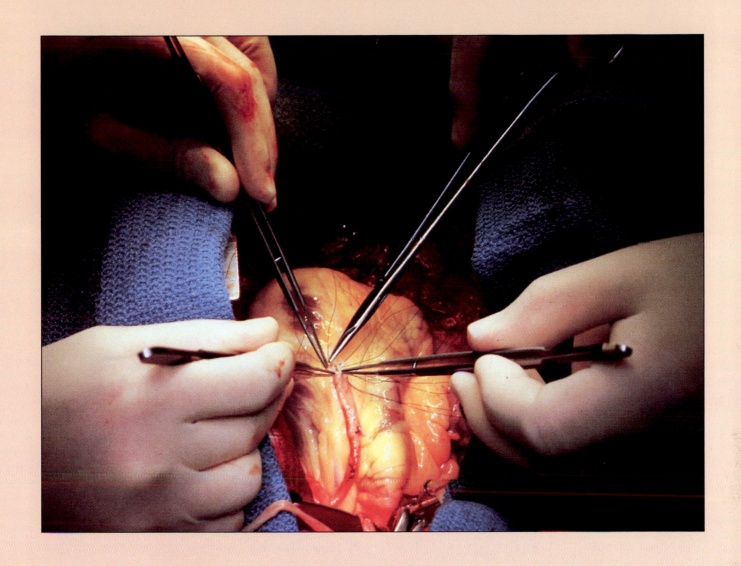

The cardiovascular system consists of the great muscle within our chest, the heart, and the many blood vessels that course throughout the body. Amazingly, the heart tirelessly beats over 100,000 times each day to push nearly 4000 liters (about 1000 gallons) of blood through nearly 60,000 miles of blood vessels. As these numbers suggest, the heart and blood vessels play a vital role in our lives. They move the blood throughout the body so that every living cell can obtain the life-sustaining molecules of oxygen and nutrients, and so that waste materials can be removed before their accumulation causes damage. The importance of moving the blood is made evident when we consider what happens when the heart stops beating, or when blood spills uncontrollably out of vessels: death follows within minutes.

In this chapter we will discuss the structure and function of the heart, and the tubes that transport the blood, the vessels. We will also examine the circulatory networks of the body.

THE HEART

The heart is the center of the cardiovascular system. It is a hollow, muscular pump that propels blood through the blood vessels.

General Characteristics

The heart is characterized by its location in the center of the thoracic cavity and by its structural features.

The heart is located in the thoracic cavity, where it lies between the lungs (Fig. 13–1). If you were to draw a vertical line down the center of the chest, you would find about two-thirds of the heart on the left of this midline and one-third on the right. Look carefully at Figure 13–1 and notice the triangular shape of the heart. Its pointed end, or **apex,**[1] is on the inferior margin contacting the diaphragm and points to the left. Its flattened superior margin is called the **base.** In most people, the heart is about the same size as the clenched fist, averaging about 14 centimeters long and 9 centimeters wide.

[1]Apex: Latin *apex* = pointed tip.

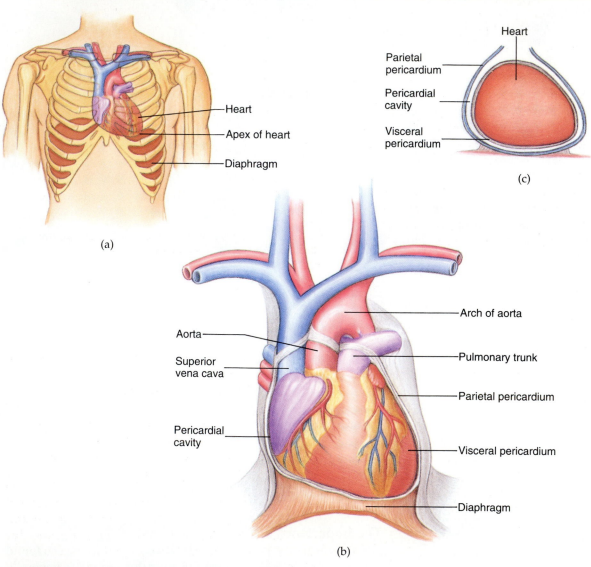

Figure 13–1
Location of the heart. (a) In this anterior view of the chest, the heart can be seen superimposed through the thoracic cage. (b) The sternum is removed and the ribs are pushed laterally in this anterior view. The outer covering of the heart, the parietal pericardium, has been cut away to reveal the heart surface. (c) The double-layered pericardium surrounds most of the heart, as this diagrammatic view shows.

Can you identify the location of the apex of your own heart on your chest surface?

Coverings of the Heart

The pericardium, which covers the heart, consists of two main protective layers.

The heart and the major blood vessels that attach to it are enclosed within a double-layered serous membrane called the **pericardium**[2] (par'-i-KAR-dē-um), which is shown in Figures 13–1 and 13–2. The outer layer is a loosely fitting sac called the **parietal pericardium,** or **pericardial sac.** It consists of two parts, a thick outer *fibrous layer* and a thin inner *serous layer.* The fibrous layer is composed of tough dense connective tissue that protects the heart and anchors it to the diaphragm, the sternum, and the major blood vessels attached to the heart. The serous layer consists of squamous epithelium underlaid by connective tissue. It turns downward at the base of the heart to continue over the heart surface as the inner layer of the pericardium, the **visceral pericardium.** The visceral pericardium, or **epicardium,**[3] is firmly attached to the surface of the heart; so much, in fact, that it is considered to be the outer layer of the heart wall as well as the inner layer of the pericardium.

Between the epicardium and the serous layer of the parietal pericardium is a potential space called the **pericardial cavity.** It contains a small amount of fluid that is secreted by serous cells of the pericardium. The fluid acts as a lubricant that reduces friction between the membranes as they glide against each other during heart activity. If the pericardium becomes swollen, such as in the disease *pericarditis,* fluid production is inhibited. This decline in lubrication causes the membranes to stick together, thereby leading to severe chest pain and possibly impeding heart activity.

Heart Wall

The heart wall consists of three layers, the most important of which is the middle layer, the myocardium.

The wall of the heart is composed of three layers: the outermost epicardium (already described), the middle myocardium, and the inner endocardium (Fig. 13–2). All three layers are well supplied with blood vessels.

The **epicardium** (visceral pericardium) serves as a thin protective barrier for the heart. As we learned, it is a serous membrane that is firmly attached to the heart. It often contains fat deposits, especially along the channels of blood vessels that course through the heart wall.

The **myocardium**[4] makes up the bulk of the heart wall. It is composed of cardiac muscle tissue and is the

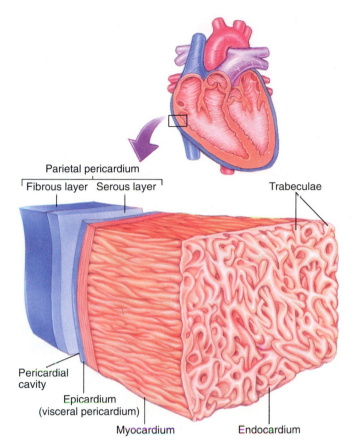

Parietal pericardium
Fibrous layer Serous layer

Trabeculae

Pericardial cavity

Epicardium (visceral pericardium)

Myocardium

Endocardium

Figure 13–2
Structure of the heart wall. In the lower diagram, the interior of the heart is on the extreme right side.
What layer of the heart wall is the thickest?

layer that actually contracts to provide the propulsion of blood. The cardiac muscle cells are arranged in spiral bundles that are supported and interwoven by connective tissue fibers. These interlacing bundles link all parts of the heart together. The connective tissue fibers form a network called the *fibrous skeleton,* which reinforces the structural integrity of the myocardium.

The **endocardium**[5] is a smooth, white membrane forming the inner layer of the heart wall. It lines the internal spaces of the heart (the heart chambers) and covers the heart valves. It is composed of squamous epithelial cells underlaid by a thin layer of connective tissue. This inner heart membrane, which is also found lining the inner walls of the blood vessels, is generally

[2]Pericardium: Greek *peri-* = around + *kardia* = heart.

[3]Epicardium: Greek *epi-* = upon + *kardia.*

[4]Myocardium: Greek *myo-* = muscle + *kardia.*

[5]Endocardium: Greek *endo-* = within + *kardia.*

known as **endothelium.** The endocardium is generally continuous with the endothelium of the major blood vessels entering and leaving the heart.

CONCEPTS CHECK

1. What is the outer layer of the pericardium called?

2. Which layer of the heart wall consists of cardiac muscle tissue?

3. What is the endothelium, and where is it found?

Heart Chambers

The heart chambers are the two thin-walled atria, which serve as receiving chambers, and two thick-walled ventricles, which are the main pumps.

The heart has four internal spaces, called **chambers.** The two superior chambers are the atria (right and left), and the two inferior chambers are the ventricles (right and left).

Atria

The right and left **atria**[6] (singular form is *atrium*) function as receiving chambers for blood entering the heart. Their role in pumping blood is minimal: they simply push blood "next door" into the ventricles. Their walls, therefore, are quite thin and have little myocardium (Fig. 13–3). Attached to the atria are small, earlike hollow appendages, known as **auricles,**[7] which increase their internal volume only slightly as they receive blood (Fig. 13–4).

The endocardium lining the atrial walls is smooth in texture, except for ridges formed by parallel bundles of underlying muscle. These ridges are called **pectinate muscles,** after their similarity in appearance to the teeth of a comb (Latin *pecten* = comb). These ridges are located in the auricles and the anterior atrial walls.

The right and left atria are internally separated by a partition, called the **interatrial septum.** This partition is complete in healthy hearts. On the posterior wall of the septum of the right atrium is an oval depression, the **fossa ovalis.** This is what remains of an

[6]Atria: Latin *atrium* = entrance room.
[7]Auricles: *auricle* (from Latin) = little ear.

Figure 13–3
Interior structure of the heart, anterior view.
Why does the left ventricle contain the thickest myocardium?

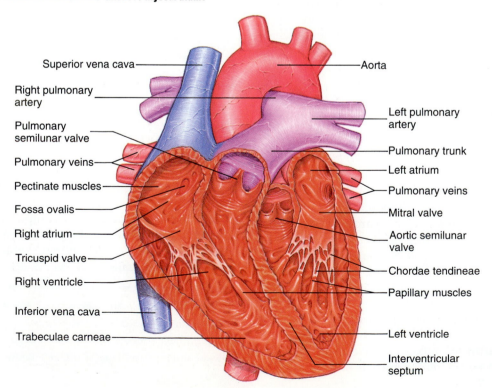

Superior vena cava — Aorta

Right pulmonary artery

Pulmonary semilunar valve — Left pulmonary artery

Pulmonary veins — Pulmonary trunk

Pectinate muscles — Left atrium

Fossa ovalis — Pulmonary veins

Right atrium — Mitral valve

Tricuspid valve — Aortic semilunar valve

Right ventricle — Chordae tendineae

Inferior vena cava — Papillary muscles

Trabeculae carneae — Left ventricle

Interventricular septum

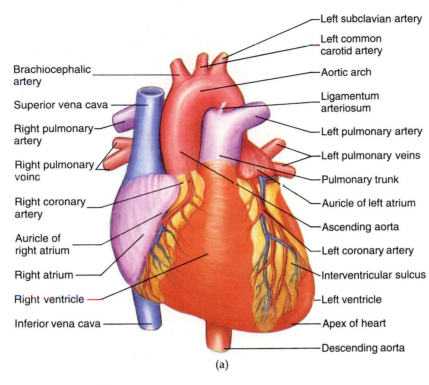

Left subclavian artery

Left common carotid artery

Aortic arch

Ligamentum arteriosum

Left pulmonary artery

Left pulmonary veins

Pulmonary trunk

Auricle of left atrium

Ascending aorta

Left coronary artery

Interventricular sulcus

Left ventricle

Apex of heart

Descending aorta

Brachiocephalic artery

Superior vena cava

Right pulmonary artery

Right pulmonary veins

Right coronary artery

Auricle of right atrium

Right atrium

Right ventricle

Inferior vena cava

(a)

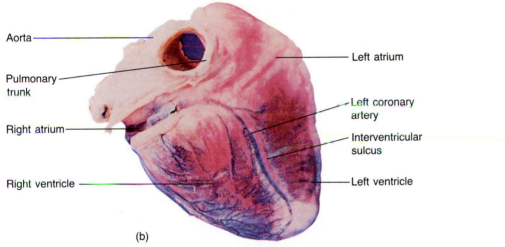

Aorta

Pulmonary trunk

Right atrium

Right ventricle

Left atrium

Left coronary artery

Interventricular sulcus

Left ventricle

(b)

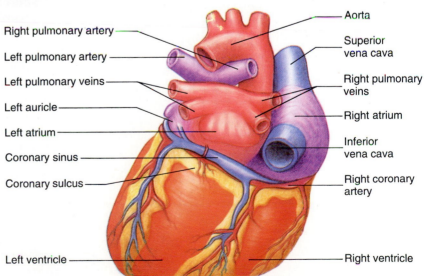

Right pulmonary artery

Left pulmonary artery

Left pulmonary veins

Left auricle

Left atrium

Coronary sinus

Coronary sulcus

Left ventricle

Aorta

Superior vena cava

Right pulmonary veins

Right atrium

Inferior vena cava

Right coronary artery

Right ventricle

(c)

Figure 13–4
Exterior structure of the heart. (a) Anterior view. (b) Photograph of the heart, anterior view. (c) Posterior view.
Identify the four chambers from the exterior surface of the heart.

opening that was once present in the fetal heart, in which blood was shunted from the right atrium to the left atrium in order to bypass the lungs. In the fetus this opening is called the **foramen ovale.** In some cases, children are born with the foramen ovale still open. This condition could result in brain damage due to lack of oxygen if not corrected surgically.

The atria receive blood from all areas of the body before pushing it through to the ventricles. The **right atrium** collects incoming blood from the *superior vena cava,* which drains blood from body regions above the heart; from the *inferior vena cava,* which drains blood from regions below the heart; and from the *coronary sinus,* which drains blood from the heart wall (see Fig. 13–4). The **left atrium** collects blood from four *pulmonary veins,* which drain blood from the lungs.

Ventricles

The right and left **ventricles**[8] provide the force necessary to push blood out of the heart and into the body's circulatory network (Fig. 13–3). Blood is pushed by the contraction of their thick layer of myocardium, which is many times thicker than that of the atria. Of the two ventricles, the left contains the thickest wall. This is required for the action of the left ventricle: it must generate enough force to push blood through all vessels of the body, except those supplying the lungs (the right ventricle pushes blood to the lungs).

The inner wall of each ventricle is marked by irregular folds of muscle lined with endocardium, called **trabeculae carneae**[9] (tra-BEK-yoo-lē KAR-nē-ē). Some of these folds form slender projections known as **papillary muscles,** which attach to certain heart valves and play a role in their function (to be discussed in the section on the valves).

The right and left ventricles are internally separated by a thick muscular partition called the **interventricular septum.** Its position can be determined externally, for it parallels a groove in the outer surface of the heart known as the **interventricular sulcus** (Figs. 13–3 and 13–4a). Another major groove in the heart surface, the **coronary sulcus,** separates the atria from the ventricles externally (Fig. 13–4c). Both sulci contain large blood vessels supplying the heart wall and a layer of fat.

Once the ventricles receive blood from the atria, they push it into the circulatory network of the body with great force. The right ventricle pumps blood into

the *pulmonary trunk,* which carries it to the lungs. The left ventricle pumps blood into the *aorta,* whose branches deliver blood to all remaining areas of the body.

Heart Valves

Both types of heart valves direct the flow of blood through the heart in one direction.

The **valves** of the heart permit blood to flow in one direction only. They operate like spring-loaded trap doors, snapping open or closed in response to the flow of blood. There are two types of valves in the heart: atrioventricular and semilunar.

Atrioventricular (AV) Valves

The **atrioventricular (AV) valves** are located between the atria and the ventricles. They consist of two or three triangular flaps, or cusps, that point downward into the ventricle (Figs. 13–3 and 13–5). The AV valve between the right atrium and right ventricle contains three cusps. For this reason it is commonly referred to as the **tricuspid valve.** The AV valve between the left atrium and left ventricle has only two cusps; it is often called the **bicuspid valve.** The bicuspid is also called the **mitral valve** after its similarity in shape to a bishop's miter, or headdress.

The cusps of the AV valves are composed of dense connective tissue covered by endocardium. Their pointed ends are attached to thin strands of connective tissue, called **chordae tendineae**[10] (KOR-dē tenDIN-ē-ē). These strands anchor the cusps to the papillary muscles in the wall of the ventricles.

The AV valves operate in response to the flow of blood in the heart (Fig. 13–6). They permit the one-way movement of blood from the atria to the ventricles. When an atrium contracts, it pushes blood downward through the opening between the cusps, permitting its entrance into the ventricles. When a ventricle contracts, the blood is pushed upward against the cusps, causing them to close off the opening. The cusps are prevented from everting (reversing their direction) into the atrium by their attachments to chordae tendineae, which anchor their pointed margins firmly to the wall of the ventricle by connecting to papillary muscles. Blood is thus prevented from entering the atrium during the ventricular contraction. Should the valves have

---

[8]Ventricles: *ventricle* (from Latin) = little belly.

[9]Trabeculae carneae: Latin *trabeculae* = little beams + *carneae* = fleshy.

[10]Chordae tendineae: Latin *chordae* = cords, narrow strands + *tendineae* = tendinous.

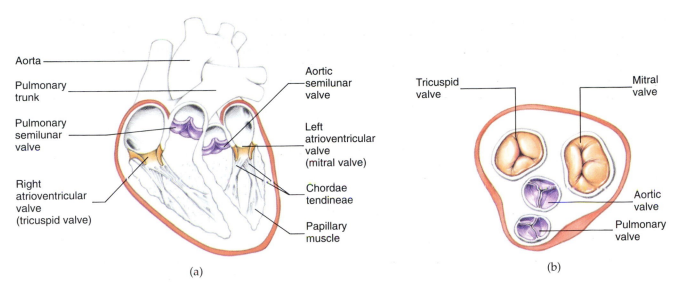

Figure 13–5
Location of the heart valves. (a) The valves are highlighted in color in this anterior view of a sectioned heart. (b) Top view of heart that has been sectioned along the horizontal plane.
Which heart valve is located between the left ventricle and the aorta?

a defect that prevents complete closure, some blood may enter the atrium. This condition is called a *heart murmur,* and in some extreme cases it may reduce blood flow to vital organs.

Figure 13–6
Operation of the atrioventricular valves. (a) The valve is open as blood is pushed from the atrium into the ventricle. (b) Soon after the ventricle has filled with blood, it contracts. As a result of the pressure of blood against the AV valve during ventricular contraction, the valve closes. Blood is thus prevented from passing from ventricle to atrium.
What do you suppose would happen to the flow of blood if the chordae tendineae were suddenly torn from their attachments to the papillary muscles?

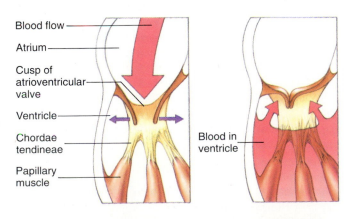

(a) Atrioventricular valve opened

(b) Atrioventricular valve closed

Semilunar (SL) Valves

The two **semilunar (SL) valves**[11] are located between the ventricles and the two major vessels carrying blood away from the heart—the pulmonary trunk and the aorta (Figs. 13–3 and 13–5). The SL valve between the right ventricle and the pulmonary trunk is called the **pulmonary valve,** and the one located between the left ventricle and the aorta is called the **aortic valve.**

Like the AV valves, the SL valves are composed of dense connective tissue covered with endocardium. However, this is where their structural similarities with AV valves end. Both SL valves consist of three half-moon (semilunar) shaped cusps that resemble shallow bowls (Fig. 13–5), with their convex surfaces facing the ventricle. Along one half of their convex margin (the rim of the bowl) they are attached to the vessel wall. The opposite half of their margin curves outward to project into the vessel opening. When an SL valve is closed, the free margins of the three cusps contact one another, closing off the opening into the vessel.

The function of the SL valves is to direct one-way flow of blood from the ventricles to the pulmonary trunk and aorta. They operate in response to blood flow and pressure changes that accompany ventricular contraction, as shown in Figure 13–7. When the ventricles contract, blood is pushed against the AV valves, causing them to close, thereby diverting blood flow toward the SL valves. The force of blood pushing against

[11]Semilunar: *semi-* = partial + *lunar* = of the moon.

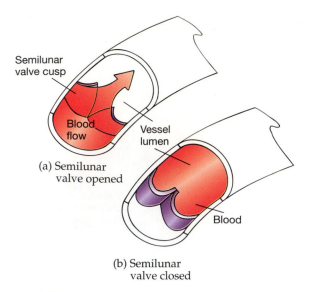

Semilunar
valve cusp

Blood
flow

Vessel
lumen

(a) Semilunar
valve opened

Blood

(b) Semilunar
valve closed

Figure 13–7
Operation of the semilunar valves. (a) A section through a major vessel (the pulmonary trunk or aorta), showing an SL valve in the open position as blood is pushed against it by the force of ventricular contraction. (b) Soon after contraction, blood within the major vessel is drawn back against the SL valve, closing it. Blood is thus prevented from returning to the ventricle as it relaxes.
When the SL valves close, what chambers do they prevent blood from entering?

the SL valves opens them, and blood is channeled through the SL valves into the vessels. When the ventricles relax, the drop in pressure within them forces the SL valves to close. This prevents blood from returning to the ventricles. Heart murmurs may result from incomplete closure of SL valves, although SL valve murmurs are not as frequent as AV valve murmurs.

Blood Flow Through the Heart

The flow of blood through the heart is directed by the pumping action of the chambers and the action of the valves.

Blood flow through the heart chambers and valves begins as blood enters the right atrium from the superior vena cava, the inferior vena cava, and the coronary sinus (Fig. 13–8). This blood is low in oxygen and high in carbon dioxide, for it has circulated throughout the body (except the lungs). When the right atrium contracts, blood is forced through the tricuspid valve into the right ventricle. The right ventricle then contracts, pushing blood against the tricuspid valve (forcing it

closed) and through the pulmonary valve. The blood enters the pulmonary trunk, which divides into many branches to carry the blood to the lungs. At the lungs, gas exchange occurs, restoring the oxygen level and reducing the carbon dioxide level in the blood. The refreshed blood reenters the heart at the left atrium through four pulmonary veins. Contraction of the left atrium pushes blood through the mitral valve into the left ventricle. When the left ventricle contracts, blood is pushed against the mitral valve (forcing it closed) and on through the aortic valve. The aorta carries blood by way of its many subdivisions throughout the body (except the lungs).

Supply of Blood to the Heart

The heart wall receives nourishment from its own circulatory network.

The heart is an active organ that requires a continual source of oxygen and nutrients and a means of removing wastes, just as any organ does. This is particularly important for the myocardium, whose constant rhythmic contractions keep blood flowing through our bodies. Although the chambers of the heart contain a large supply of blood with the necessary ingredients of oxygen and nutrients, there is no way this blood source can penetrate the chamber lining to nourish cells in the heart wall. Therefore, the wall of the heart has its own supply of blood vessels to meet its vital needs. The flow of blood through these vessels is called the **coronary circulation.**

The coronary circulatory network begins at the base of the aorta, where the *right* and *left coronary arteries* originate (Fig. 13–4a). These two vessels carry freshly oxygenated blood to the atria and ventricles of the heart by way of their smaller divisions. As blood passes through these smaller vessels, it delivers oxygen and nutrients to cells and collects carbon dioxide and other wastes. Most of the blood leaving the heart wall is collected by a large vein, the *coronary sinus*, which empties into the right atrium (Fig. 13–4c).

CONCEPTS CHECK

1. What are the structural differences between atria and ventricles?

2. How do the atrioventricular valves operate?

3. Where does blood flow once it is pushed out of the right atrium?

Capillary beds of head, neck, and upper appendages

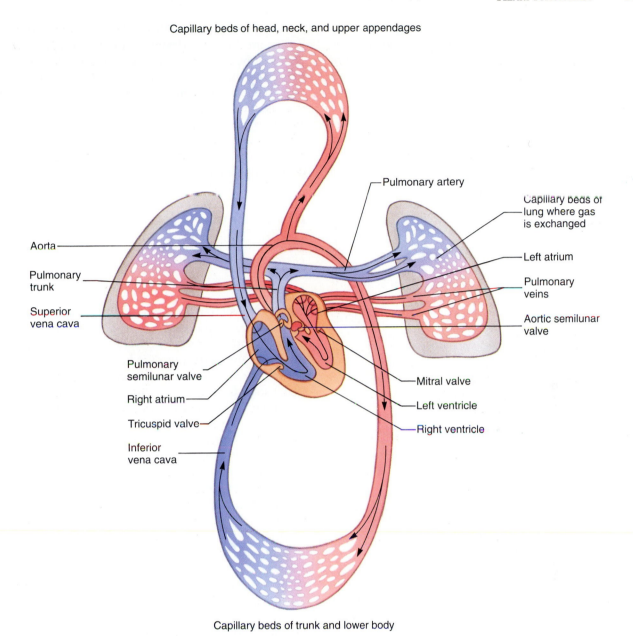

Figure 13–8
Blood flow through the heart. The arrows show the direction of blood flow. Red vessels are those that carry oxygen-rich blood, and blue vessels carry oxygen-poor blood.

Oxygen-poor blood enters the heart into which chamber?

HEART PHYSIOLOGY

The function of the heart is the propulsion of blood through the body. As you know, it performs this function by the rhythmic contractions of the heart wall. How does this rhythm operate, and how is it maintained? These questions and others will be answered as we examine the physiology of the heart.

Cardiac Cycle

The contraction of both atria, followed by both ventricles, constitutes a single cardiac cycle.

The heart beats in a rhythmic, cyclic fashion throughout your lifetime. This rhythm is produced by the coordinated, simultaneous contraction of both atria, followed by contraction of both ventricles. Once the ventricles relax, the atria contract again, followed again by the ventricles, and so on. The events required to complete a single heartbeat—the contraction of both atria followed by both ventricles—are collectively called a **cardiac cycle** (Fig. 13–9).

Accompanying every beat of the heart is a change of pressure in the chambers. When the atria contract, their internal volume is squeezed smaller, so that their pressure increases and blood is forced out. At the same time, the ventricles expand their volume by relaxing, causing their pressure to drop and blood to be drawn

Figure 13–9
Cardiac cycle. (a) Relaxation of the ventricles (ventricular diastole) and contraction of the atria (atrial systole) draws blood from the atria into the ventricles. (b) Contraction of the ventricles (ventricular systole) forces blood through the SL valves into the major vessels, and relaxation of the atria (atrial diastole) collects incoming blood.

Blood enters the aorta during which contraction cycle?

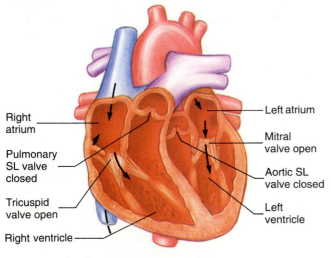

Right atrium

Pulmonary SL valve closed

Tricuspid valve open

Right ventricle

Left atrium

Mitral valve open

Aortic SL valve closed

Left ventricle

(a) Ventricular diastole

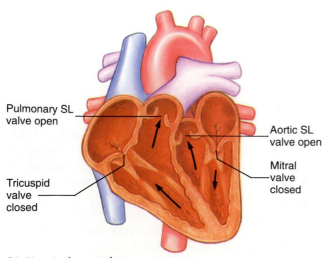

Pulmonary SL valve open

Tricuspid valve closed

Aortic SL valve open

Mitral valve closed

(b) Ventricular systole

in. Conversely, when the atria relax, their pressure drops and blood is drawn in from the major vessels. At the same time, the ventricles contract and their pressure rises, pushing blood outward. Thus, the rhythmic movement of blood in and out of the heart chambers is caused by pressure changes that result from the cycle of contraction and relaxation of the heart wall. In general, the state of contraction is referred to as **systole**[12] (SIS-tō-lē), and the state of relaxation is called **diastole**[13] (dī-AS-tō-lē).

Heart Sounds

The clicking shut of the heart valves produces the heart sounds.

When you listen to a heartbeat with a stethoscope, you hear only two sounds that repeat over and over again. These sounds are usually described as "lub-dup." The rhythm of the heart sounds is lub-dup, pause, lub-dup, pause, lub-dup, pause, and so on. They are the vibrational sounds produced by turbulence due to the closing of the heart valves, heard muffled through the chest wall. The first sound, the "lub," indicates closing of the AV valves, which occurs during the ventricular contraction (ventricular systole). The second sound, "dup," indicates closing of the SL valves during ventricular relaxation (ventricular diastole).

Heart sounds are used as a helpful tool in diagnosing problems associated with the heart. The most common group of dysfunctions that can be determined by listening to the heart is murmurs. Armed with only a stethoscope, a trained clinician can determine which of the four heart valves are faulty and to what extent.

Heart Conduction System

Each cardiac cycle is stimulated by special conducting cells in the heart.

The heart receives nerve impulses from the autonomic nervous system. When these nerves are cut, the heart continues to beat. How can this occur? The heart does not require external stimulation for contraction; that is, it has the ability to beat on its own. This is due to an intrinsic regulating system within the heart wall. This system consists of specialized cells that send electrical impulses to each cardiac muscle cell.

[12]Systole: *systole* = contraction.
[13]Diastole: *diastole* = expansion.

Heart Attack

More than 1 million people will suffer heart attacks in the United States this year, about one-third of which will prove fatal. Although this sounds grim, many medical advances have been made over the past decade, as evidenced by the fact that about 100,000 more people were dying from heart attacks ten years ago. The improvement is the result of important advances in the treatment and prevention of heart disease. For example, surgical procedures such as heart transplants and coronary bypass operations have become routine procedures for severe cases, and new classes of drugs can now stop a heart attack in progress. What is perhaps most significant is the new scenario that people can lessen the chance of a heart attack by reducing their risk factors.

Most of the problems associated with the heart are caused by poor blood circulation to the myocardium. A common warning sign of a problem with coronary circulation is **angina pectoris** (an-JI-na PEK-tor-is), in which the oxygen supply becomes reduced to a point of damaging cardiac muscle cells but not killing them. Some causes of angina pectoris include atherosclerosis (see the Clinical Terms box for this chapter), stress, heavy exercise following a meal, hypertension (see Clinical Terms box), anemia, and thyroid disease. Its symptoms include chest pain accompanied by a tightness in the chest, labored breathing, and often dizziness and weakness. If the oxygen supply is reduced to a point of actually killing cardiac muscle cells, a **myocardial infarction (MI)** occurs (*infarction* means the death of tissue due to an interrupted blood supply). This is commonly known as a **heart attack.** It is usually (about 90% of the time) caused by a thrombus—or clot—in one of the coronary arteries, an embolus (a drifting blood clot), or a spasm in one of the coronary arteries. Atherosclerosis dramatically increases the chances of an MI because it narrows the coronary vessel opening and encourages the formation of clots. The cardiac cells that die during an MI are not replaced by new functional cells, but by scar tissue that cannot contract. As a result, the heart muscle loses at least some of its strength. The degree of effect is dependent on the size and location of the infarcted, or dead, area. In extreme cases, an MI can cause complete heart stoppage, called **cardiac arrest.**

Recent evidence suggests that one's chances of avoiding a heart attack can be increased by preventive measures. These measures involve reducing the risk factors that can lead to heart disease, which include high blood cholesterol, high blood pressure, cigarette smoking, and obesity. The recommended preventive measures that should be followed include losing weight if obese, following a diet plan that includes foods low in cholesterol, not smoking, and regular exercise. For example, every 1% decline in serum cholesterol level is accompanied by a 2% to 3% decline in risk for a heart attack. Also, quitting smoking provides a 50% to 70% lowered risk within five years. Finally, people who exercise on a regular basis have been shown to have a 35% to 55% reduced chance of a heart attack. Other factors that show some evidence of reducing the risk of heart attack include taking low dosages of aspirin on a daily basis (a 33% lowered risk compared with nonusers), consuming one alcoholic drink per day (a 25% to 45% lowered risk compared with nondrinkers), and using estrogen replacement therapy after menopause (a 44% lowered risk compared with nonusers).

Cardiac muscle cells have an inherent ability to contract. If you were to carefully remove a cell from a living heart and place it in a culture dish, it would contract and relax in a rhythmic manner until it ran out of energy. When cardiac muscle cells are combined together, the cell with the fastest rhythm "sets the pace," stimulating other cells to contract at the same, faster rate. This is possible because the pace-setting cell sends out electrical signals to adjacent cells, stimulating their contraction. In a complete, healthy heart the rhythmic contraction of every cardiac muscle cell is initiated by pace-setting cells.

The conduction system of the heart consists of clusters of pace-setting cardiac cells that have become specialized to conduct electrical signals. The main site is a compact mass of cells located in the wall of the right atrium, called the **sinoatrial (SA) node** (Fig. 13–10). The SA node is also known as the *pacemaker*, for it initiates each cardiac cycle. It does so by generating an electrical impulse that quickly spreads out over both atria. This causes atrial contraction (atrial systole). It also stimulates the activity of a second cluster of cells known as the atrioventricular node, which relays the signal to the ventricles.

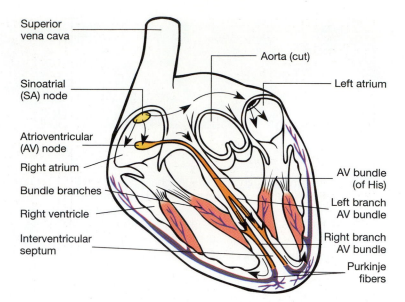

Figure 13–10
Heart conduction system. The arrows indicate the direction of impulse conduction from the sinoatrial node.

What cluster of cardiac cells initiates each cardiac cycle?

The **atrioventricular (AV) node** is located near the inferior portion of the interatrial septum on the right side. Once it is activated by the SA node, the AV node generates an impulse that is conducted down a bundle of conducting fibers, called the **AV bundle** (bundle of His). The AV bundle extends down the interventricular septum, sending branches right and left as it travels. At the apex of the heart, the AV bundle forms branches called **Purkinje** (per-KIN-jē) **fibers,** which pass further into the myocardium to reach cardiac muscle cells of the ventricles. Impulses passing down the AV bundle and through Purkinje fibers stimulate the ventricles to contract almost simultaneously (ventricular systole).

By the time the ventricles are stimulated to contract, the atria have had enough time to complete their contraction and return to a state of relaxation. This time delay is produced by the slowly conducting AV node. It allows time for the atria to empty their blood into the ventricles before the ventricles begin their contraction. Thus, we find a coordinated pumping action of first the atria, then the ventricles, during each cardiac cycle. If all four chambers were to contract at the same time, an insufficient amount of blood would be collected by the ventricles and cardiac arrest would soon follow because of a reduced blood supply to the heart wall.

CONCEPTS CHECK

1. What is a cardiac cycle?

2. What action of the heart produces the sounds heard in a beating heart?

3. Where are the pace-setting cells of the heart located?

Electrocardiogram (ECG)

The electrocardiogram measures the electrical events that occur during a cardiac cycle.

As we have just seen, the heart conduction system sets into motion a series of electrical events that stimulate each cardiac cycle. A recording of these events is called an **electrocardiogram (ECG or EKG).** It records the electrical changes that take place in the myocardium as it contracts and relaxes during a cardiac cycle.

An ECG is recorded by placing metal electrodes on the outer surface of the skin. Although it may not seem possible, the electrodes can actually detect electrical changes in the heart wall. This is possible because the electrical changes in the heart produce changes in the flow of ionic currents throughout the body—even in the skin. These changes in ionic current flow are detectable by the metal electrodes, since the electrodes are wired to an instrument that can convert very slight changes in ionic current to electrical charges, which are then amplified. When a change in electrical activity is detected by the instrument, the amplified current in the instrument surges slightly, causing pens on a recorder to move up and down. As the pens move, they draw on a laterally moving strip of paper, and a permanent written record is made. A normal ECG is shown in Figure 13–11.

Look carefully at Figure 13–11 and notice that there are several pen deflections that are easily distinguishable. This series of deflections represents a single cycle that is repeated with every beat of the heart. The deflections are called *waves;* they represent changes in electrical activity of the myocardium. The single series of waves in Figure 13–11 represents a single cardiac cycle.

Between cardiac cycles the pen draws a flat baseline, since no detectable electrical changes occur be-

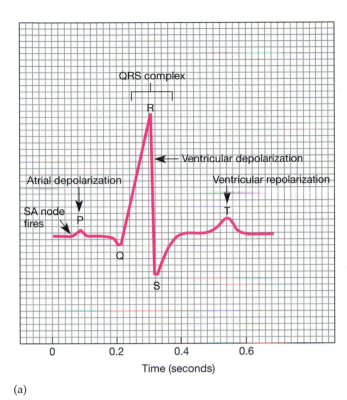

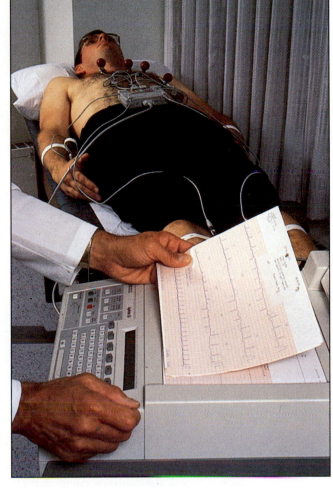

Figure 13–11
Electrocardiogram. (a) A single cardiac cycle is represented
by this series of deflection waves. (b) Obtaining an ECG
from a patient. The tracings shown are not normal.
What heart activity is represented by the QRS complex wave?

tween cardiac cycles. At this point, the myocardium is
polarized, or at rest. The first electrical activity occurs
when the SA node fires, sending a wave of electrical
activity (action potential) through both atria. This stimulates the atrial myocardium to depolarize. The skin
electrodes detect this current change, and it is recorded
as a small upward pen deflection called the **P wave.**
The pen recording returns to baseline as the atrial
fibers stop depolarizing.

Soon after the atrial myocardium has stopped depolarizing, the second event occurs: the impulse
reaches the ventricles, and they depolarize. Because of
the larger muscle mass of the ventricular myocardium,
the amount of electrical change is greater and the pen
deflection is more dramatic. This produces three
waves—Q, R, and S—which are combined to form the
prominent **QRS complex.**

As the ventricular myocardium repolarizes, its
electrical changes cause a third deflection of the pen.
This is the **T wave.** Repolarization of the atrial myocardium also produces electrical changes, but these
take place at the same time ventricular depolarization
occurs. Thus, they are masked by the QRS wave. Once

the ventricles have fully polarized, the pen returns to
baseline until the SA node fires again and a new cardiac cycle begins.

The ECG patterns we have discussed are valuable
tools used by cardiologists to detect problems associated with the heart. In a healthy heart, the size, duration, and timing of the waves tend to be consistent.
Thus, any changes in the pattern or timing of the ECG
may reveal a loss of cardiac function or problems with
the heart conduction system. For example, the time interval between the P wave and QRS complex represents the time it takes for the impulse to travel from
the SA node to the AV node and into the ventricle
walls. If the baseline between these waves is longer
than normal, it indicates damage to the conduction
pathways that may have been caused by a restriction
in blood flow. In severe disruptions of the conduction
system, normal heart rhythm can be restored and
maintained with an *artificial pacemaker.* This mechanical device is surgically inserted and sends out small
electrical charges to stimulate the heart at regular fre-
quencies. Several abnormal ECG recordings are shown
in Figure 13–12.

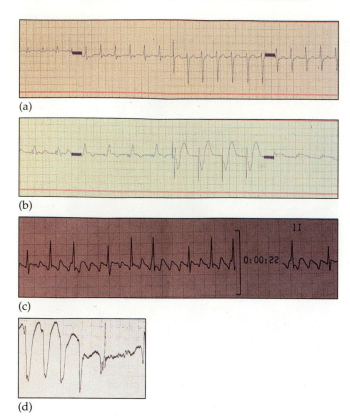

Figure 13–12
Abnormal ECG tracings. (a) In this example, part of the electrical signal passing to the AV node is blocked, resulting in a disrupted rhythm. This condition is called a *partial heart block.* (b) A *complete heart block,* resulting in independent atrial and ventricular contractions. (c) A P wave is lacking in this tracing, indicating a rapid, uncoordinated contraction of the atria. This condition is called *atrial fibrillation.* (d) An irregular heartbeat resulting from the unregulated contraction of all chambers. This condition is known as *ventricular fibrillation.*

Cardiac Output

Cardiac output is a measurable value that represents heart activity.

The rhythmic beating of the heart is the result of a series of cardiac cycles that repeat at regular intervals. At times, however, the body may require different volumes of blood to be delivered to different organs. For example, during strenuous exercise the skeletal muscles need lots of blood to be quickly sent, but at rest most body cells operate with a minimal blood flow. The heart must adjust to these changes in activity in order to meet the needs of the body at all times.

The heart adjusts the amount of blood it pumps in two ways: by altering the duration of each cardiac cycle, or **heart rate;** and by altering the volume of blood

ejected by the ventricles during their contraction, called **stroke volume.** In a resting adult, the average heart rate is 75 beats per minute, and the stroke volume averages about 70 milliliters (ml). A value that summarizes the changes in both heart rate and stroke volume is called the **cardiac output,** which is obtained by multiplying the two figures:

Heart rate × stroke volume = cardiac output

Using this equation, we find that the average cardiac output is 75/min × 70 ml = 5250 ml/min, or 5.25 liters per minute. Thus, in a resting adult, over 5 liters of blood is pumped through the heart each minute. If the heart rate or stroke volume were increased, such as during exercise, the cardiac output would also increase over this figure. Conversely, if the heart rate or stroke volume decreased, the cardiac output would drop to a value below this average.

An interesting application of the notion of cardiac output has been identified by investigators studying the heart. They have observed that the strength of a heart beat (or ventricular contraction) is determined by how far cardiac muscle cells are stretched by the filling of a chamber with blood. The more these cells are stretched by incoming blood, the more strongly the heart walls will contract to eject the blood. The response of the heart is, therefore, similar to that of a rubber band: the further it stretches, the greater its strength in contracting. This phenomenon is referred to as **Starling's law of the heart.** It provides an explanation of how cardiac output increases with increasing levels of exercise and blood flow.

Regulation of Heart Activity

The cardiac cycle is controlled by the medulla oblongata, which receives information from baroreceptors.

We have just seen that cardiac output is a measure of how active the heart is at a given time—that is, how fast it beats (heart rate) and how strong the ventricles contract (stroke volume). Changing the cardiac output adjusts the amount of blood flowing out of the heart and to tissues that are in need of it. However, we have also learned that each cardiac cycle is initiated by the SA node, which fires at regular, rhythmic intervals. How, then, may cardiac output be changed?

Heart activity may be altered by the cardiac center in the medulla oblongata of the brain. This reflex center is connected to the heart by way of autonomic fibers. Specifically, parasympathetic fibers extend from the medulla by way of the vagus nerve to the heart, where they branch to the SA and AV nodes.

Exercise and the Heart

The physical benefits of exercise are widespread, affecting not only muscles but other organs and systems as well. Exercise is especially beneficial to the heart. As we will see, these healthful benefits can be sustained by a regular program of exercise.

During exercise, the activity of skeletal muscles creates an increased demand for oxygen and nutrients. As a result, heart activity changes, sending an increased volume of blood to skeletal muscles. For example, blood flow to skeletal muscles at rest averages 3 to 4 ml/min per 100 grams of muscle tissue, but during exercise blood flow may increase to 80 ml/min or more. The factors influencing these changes include heart rate, stroke volume, and cardiac output.

Heart rate (the number of heart beats per minute) increases as exercise activity increases. During strenuous exercise, the heart rate may reach 200 beats per minute from its resting rate of about 75 beats per minute. However, as physical conditioning improves during an exercise program, the amount of increase in heart rate becomes progressively less. This occurs because of the increasing strength of the heart, which enables it to push a greater volume of blood with each beat.

Stroke volume (the volume of blood leaving the heart) increases during periods of exercise also. However, the influence of stroke volume is not great, and continues to diminish with increased athletic conditioning. At rest in an unconditioned individual, the stroke volume averages about 70 ml/min, but during moderate exercise this value quickly doubles. In a conditioned person, the stroke volume at rest may exceed 90 ml/min and during moderate exercise increase only slightly. In either case, the increase in stroke volume is due to an increased force of contraction of cardiac muscle and is caused by impulses from sympathetic nerves.

Cardiac output also rises during periods of exercise. Most of the increase in cardiac output is caused by the increase in heart rate, and not by an increase in stroke volume. Like heart rate and stroke volume, the better your body is conditioned by exercise, the less the cardiac output will increase. In other words, conditioning enables your heart to work less when propelling blood to your skeletal muscles, even during strenuous exercise. Thus, the heart activity of a well-conditioned person is more efficient than that of a poorly conditioned person.

In addition to physiological changes, prolonged periods of exercise performed during a regular program can also provide certain anatomical changes. For example, the mass of the heart increases permanently with an extended exercise program. This enlargement is caused by an increased thickness of the ventricular walls (called *cardiac hypertrophy*) or by a lengthening of the cardiac muscle fibers. The increased mass increases the strength of the heart, enabling it to increase stroke volume (and cardiac output). Also, increased exercise increases the density of capillaries supplying cardiac muscle, allowing more blood to flow to the strengthening heart walls.

The benefits of regular exercise on the heart are determined by the duration, frequency, and intensity of the program. Exercise physiologists agree that strenuous exercise, like jogging, bicycling, and aerobic dancing, is the most effective way to condition the cardiovascular system. However, milder forms of exercise (such as walking, hiking, and isometrics) may also provide conditioning, but to a lesser degree. In either case, exercise in any form contributes to a more efficient cardiovascular system and, consequently, to overall health.

Also, sympathetic fibers extend between the medulla and the heart, and similarly join the SA and AV nodes. Some sympathetic fibers terminate in the myocardium of the ventricles as well. The effect of parasympathetic impulses reaching the heart is to slow the heart rate by the release of acetylcholine (ACh), whereas sympathetic impulses increase both the heart rate and stroke volume by the release of norepinephrine (NE).

The balance that normally occurs between the two types of autonomic impulses is maintained by the medulla. At rest, the medulla favors the inhibitory effect of the parasympathetic impulses. During strenuous exercise, the stimulatory effect of sympathetic impulses is favored. How does the medulla know which type of impulse to favor? It is informed by way of nerve impulses that may originate in receptors sensitive to stretch, or pressure, located mainly in the aorta and carotid arteries. These receptors are called **baroreceptors;** they detect changes in blood pressure. During a rise in blood pressure, for example, the baroreceptors signal the cardiac center in the medulla to send parasympathetic impulses to the SA node of the heart. This causes the heart rate to slow and thus leads to a decline in blood pressure. The medulla may also receive signals from the cerebrum or hypothalamus, enabling the heart to alter its activity in response to emotions, sensory stimuli, and, in some cases, higher thought.

CONCEPTS CHECK

1. What heart action is represented by the QRS complex in a normal ECG?

2. How is the cardiac output determined?

3. What part of the brain contains a reflex center that regulates heart activity?

BLOOD VESSELS

The blood vessels of the cardiovascular system form a closed delivery system that transports blood to and from the heart. It consists of the arteries, arterioles, capillaries, venules, and veins, which are summarized in Table 13–1.

Arteries and Arterioles

Arteries and arterioles are three-layered tubes that carry blood away from the heart. Their main properties are contractility and elasticity.

Arteries are vessels that transport blood away from the heart. They are strong, elastic tubes that begin as the major arteries of the heart—the aorta and the pulmonary trunk. As the arteries extend away from the heart, they become progressively smaller and give off many branches. They eventually give rise to **arterioles,**[14] which have thinner walls and measure on average 0.5 mm in diameter.

An artery is composed of a three-layered wall surrounding a hollow interior, or **lumen**[15] (Fig. 13–13). The innermost layer is known as the **tunica intima;** it consists of an inner lining of endothelium enveloped by a layer of connective tissue (called the *basement membrane*) that is rich in elastic fibers. The middle layer is the **tunica media,** which consists of a thick layer of smooth muscle and numerous elastic fibers. The outer layer, called the **tunica adventitia,** is a thin layer of

[14]Arterioles: *arteriole* = tiny artery.
[15]Lumen: *lumen* = light space.

Table 13–1   THE STRUCTURE AND FUNCTION OF THE BLOOD VESSELS

Vessel	Structure	Function
Artery	Three-layered wall with thick tunica media, which gives it properties of contractility and elasticity.	Movement of blood away from the heart; maintenance of blood pressure.
Arteriole	Three-layered wall, with much thinner layers and smaller lumen than in arteries.	Movement of blood away from the heart; help control blood pressure by regulation of peripheral resistance (vasoconstriction and vasodilation).
Capillary	Microscopic size, with a single-layered wall of endothelium.	Thin walls permit the exchange of materials between the blood and interstitial fluid.
Venule	Three-layered wall with very thin layers, which gradually thicken as vessels become closer to the heart.	Movement of blood from capillary beds toward the heart.
Vein	Three-layered wall, with thinner tunica media and larger lumen than in arteries. Internal valves are present to aid the movement of blood in one direction.	Movement of blood from venules toward the heart.

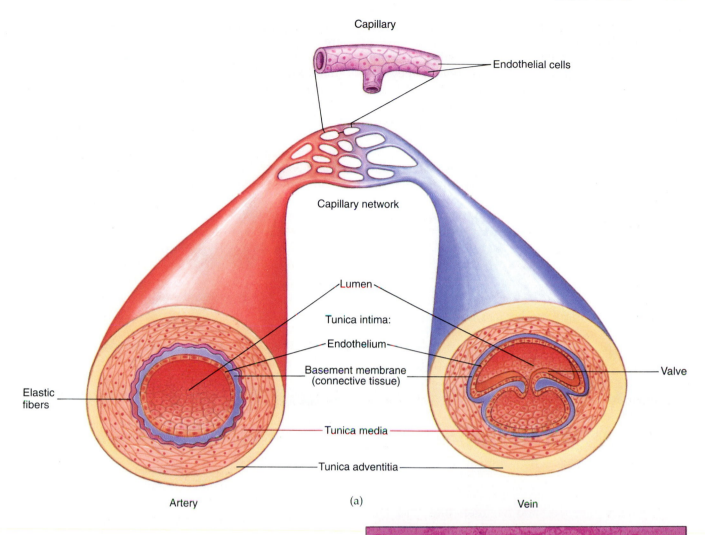

Capillary

Endothelial cells

Capillary network

Lumen

Tunica intima:

Endothelium

Basement membrane
(connective tissue)

Valve

Elastic
fibers

Tunica media

Tunica adventitia

Artery

(a)

Vein

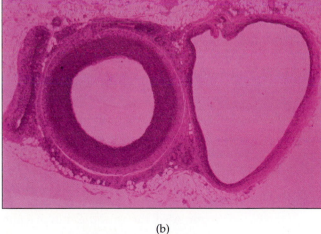

Figure 13–13
Artery, capillary, and vein structure. (a) A major artery
(*red*) and vein (*blue*) are shown in partial cross section in
this diagram. A capillary network is shown between them,
with a magnified view of a capillary section. (b) A
photograph of an artery and a vein in cross section.
What structural characteristics distinguish arteries from veins?

(b)

connective tissue that anchors the artery to neighboring structures.

The tunica media provides arteries with their two
major properties: *contractility*, or the ability to contract;
and *elasticity*, or the ability to stretch and return to the
original position. The smooth muscle in this layer contracts when it receives a stimulus from its nerve sup-

ply, which is provided by sympathetic fibers of the
autonomic nervous system. These fibers are called **vasomotor fibers.** The contraction they stimulate decreases the diameter of the vessel lumen. When this
occurs, the vessel is in a state of **vasoconstriction.** If
the nerve impulse is inhibited, the smooth muscle relaxes and the recoil effect of elastic fibers restores the

size of the lumen to its larger resting size. This state is called **vasodilation.**

The walls of arterioles are composed of the same three layers as those of arteries, although they are much thinner. As they continue their course away from the heart, they become yet smaller in diameter and their walls become thinner. Eventually, their walls will consist only of endothelium, a thin layer of smooth muscle, and a thin outer sheet of connective tissue. At this point, they branch into tiny capillaries. Throughout their length, their tunica media is of particular importance, for the vasoconstriction and vasodilation of the arterioles play major roles in the maintenance of blood flow and blood pressure.

Capillaries

Capillaries are thin-walled tubes that permit the exchange of materials between the blood and interstitial fluid.

The smallest vessels in the body are the **capillaries** (Figs. 13–13 and 13–14). They are microscopic, with an average lumen diameter of only 0.01 mm. This is large enough to allow only one red blood cell to pass through at a time. Their walls are extremely thin, for they are composed entirely of endothelium (a single layer of flattened cells with a thin sheet of connective tissue).

Capillaries permit the exchange of nutrients and gases between the blood and the interstitial fluid. This is possible because of the very thin wall of the capillary; the relatively thick walls of larger vessels pose too great a physical barrier for this to occur. As shown in Figure 13–14c, nutrients and oxygen move across the capillary walls into the extracellular space, and carbon dioxide and other waste materials move in the opposite direction.

The endothelium that forms the capillary walls provides a semipermeable membrane barrier, which permits the regulated movement of materials by diffusion, osmosis, facilitated diffusion, and active transport (discussed in Chapter 3). Diffusion is of particular importance; it is the means by which gas molecules, ions, and certain waste materials move into and out of the bloodstream. Also, the endothelium of some capillaries contains large pores that permit the bulk transport of molecules (as in the small intestine and kidneys), whereas other capillaries have very tight junctions with small pores or none at all (as in the brain, forming the blood-brain barrier). Most capillaries have pores of intermediate size, which permit the leakage of fluid from the bloodstream to the area surrounding cells. Once outside the capillary, this fluid is referred to as *interstitial fluid.*

Capillaries form networks within body tissues that are variable in arrangement. In general, they connect arterioles with venules. In some tissues a capillary may unite an arteriole directly with a venule without branching. However, in most tissues the capillaries form vast branching networks, called **capillary beds** (Fig. 13–14). These beds consist of **thoroughfare channels,** which directly connect to the arteriole or venule, and **true capillaries** in between. The true capillaries typically number between 10 and 100 per bed, depending on the tissue that is being supplied. In many cases a tiny band of smooth muscle surrounds the origin of a branching group of true capillaries; it is called a **precapillary sphincter.** It functions as a valve to regulate the flow of blood into those capillaries and is innervated by vasomotor fibers. Thus, a capillary bed may be filled with blood or nearly empty, depending on the nutritional requirements of the tissue or organ it is supplying.

Venules and Veins

Venules and veins are three-layered tubes that carry blood toward the heart.

Once the exchange of materials has taken place between the capillaries and the interstitial fluid, blood that is low in oxygen and nutrients and high in carbon dioxide and other waste materials moves from the capillaries into the larger vessels called **venules.**[16] Venules increase in size as they travel toward the heart and merge with other venules to form the larger **veins.** Veins complete the return trip of blood as they transport it into the heart for recirculation.

By definition, veins (and venules) carry blood toward the heart. Although they usually carry blood that is low in oxygen (called *deoxygenated* blood), certain veins also carry oxygenated blood like most arteries. For example, the pulmonary veins carry oxygenated blood toward the heart (from the lungs). Because of these exceptions, it is best to distinguish veins from arteries on the basis of direction of blood flow and not blood oxygen level.

Venules are formed when capillaries unite. Their walls consist of only a thin tunica intima, a very sparse tunica media, and a thicker tunica adventitia. These layers thicken and the lumen enlarges as the venules extend to the heart and form veins. Thus, the larger walls of veins consist of all three layers also. When compared with arteries of similar size, the walls of

---

[16]Venules: *venule* = tiny vein.

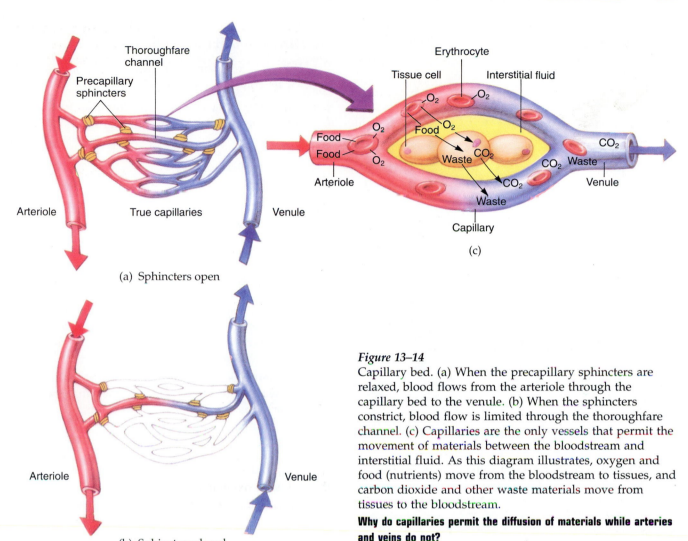

Thoroughfare channel

Precapillary sphincters

Arteriole

True capillaries

Venule

(a) Sphincters open

Arteriole

Venule

(b) Sphincters closed

Erythrocyte

Tissue cell

Interstitial fluid

O_2

O_2

Food

Food

O_2

Food

O_2

O_2

Waste

CO_2

CO_2

CO_2

CO_2

Waste

Arteriole

Waste

Venule

Capillary

(c)

Figure 13–14
Capillary bed. (a) When the precapillary sphincters are relaxed, blood flows from the arteriole through the capillary bed to the venule. (b) When the sphincters constrict, blood flow is limited through the thoroughfare channel. (c) Capillaries are the only vessels that permit the movement of materials between the bloodstream and interstitial fluid. As this diagram illustrates, oxygen and food (nutrients) move from the bloodstream to tissues, and carbon dioxide and other waste materials move from tissues to the bloodstream.
Why do capillaries permit the diffusion of materials while arteries and veins do not?

veins are always much thinner and their lumina much larger (Fig. 13–13). In particular, the tunica media is quite a bit thinner than in arteries because it contains smaller amounts of smooth muscle and elastic fibers. Its properties of contractility and elasticity are therefore very limited and much reduced compared with those of arteries.

Veins can hold large amounts of blood, owing to their large diameters. In fact, about 65% of the body's total blood supply is within the veins at a given time. Despite this large volume, they are usually only partially filled with blood and tend to be slightly collapsed. During certain times, however, they may fill completely with blood, causing the walls to expand or distend. This property of *distensibility* allows for variations in blood volume and blood pressure. We do not find this property in arteries, because the elastic tissue in arterial walls prevents them from maintaining a stretched, expanded state for an extended time period.

By the time the flow of blood reaches the veins from the venules, the blood pressure has dropped considerably. In fact, it has dropped below a level needed to push blood upward to the heart. Certain specializations of the body and of the veins themselves aid in the one-way movement of blood. For one, veins contain one-way valves that are formed from folds of the tunica intima. These valves resemble the semilunar valves of the heart in both structure and function (Fig. 13–13). They are especially numerous in the veins of the lower limbs, which must return blood to the heart against the force of gravity. When these valves become weakened and fail, blood forms pools below them, causing the vein walls to become overstretched. Veins in which this has occurred are called *varicose veins*. Varicosities in the veins surrounding the anal canal are called *hemorrhoids* (HEM-ō-roydz).

Other specializations for moving blood through veins include respiratory and skeletal muscle

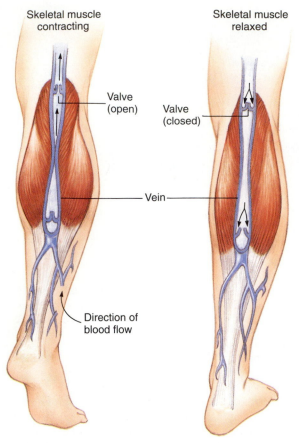

Skeletal muscle
contracting

Skeletal muscle
relaxed

Valve
(open)

Valve
(closed)

Vein

Direction of
blood flow

Figure 13–15
Skeletal muscle pump in venous blood flow. As skeletal muscles contract, they push against the vein walls to provide a pumping action for blood propulsion.
What role in venous blood flow is provided by valves within the veins?

Blood Pressure

Blood pressure is the primary force that pushes blood through arteries and arterioles. It is influenced by cardiac output, peripheral resistance, and blood volume, which are regulated by nervous, hormonal, and kidney factors.

Blood pressure is the force exerted by blood against the inner walls of the vessels. It provides an important driving force that keeps blood moving along. The standard measuring instrument for blood pressure utilizes a column of mercury, which moves in relation to changes in pressure that are measured in millimeters of mercury (abbreviated mm Hg). For example, if the blood pressure is 80 mm Hg, the pressure is great enough to lift a column of mercury a distance of 80 mm.

Blood pressure plays its primary role in the arteries and arterioles, where the pressure is highest (Fig. 13–16). In fact, in clinical settings the term *blood pressure* is used to describe only the arterial pressure in the major branches of the aorta. In capillaries and venules, the pressure is considerably less and the movement of blood is therefore slower. In veins, blood pressure plays a minor role, since the influence of valves, skeletal muscle activity, and respiratory activity provides most of what is needed to move the blood to the heart. Since blood pressure plays a minor role in blood flow among capillaries, venules, and veins, we will focus our following discussion on arterial blood pressure.

Arterial Blood Pressure

Blood behaves like any other flowing material in that it moves from a region of high pressure to a region of low pressure. The region of highest pressure is within the aorta. This is due to the pumping action of the heart, which pushes a large volume of blood at each beat into the constricted space of the aorta. The elastic walls of the aorta stretch to accommodate the surge of blood, which creates a pressure that reaches a maxi-

"pumps." The respiratory pump is activated each time you inhale: inhaling increases pressure within the abdomen, squeezing the local veins. Because the valves prevent backflow of blood, the blood is pushed upward to the heart. The skeletal muscle pump is active whenever your muscles move your limbs. Skeletal muscles that surround veins contract and relax during body movement, pushing blood toward the heart (Fig. 13–15). Again, the presence of valves prevents backflow, so the blood is pushed along in one direction.

Some veins are highly specialized for storing blood. These veins are called **venous sinuses** and are flattened veins with very thin walls composed only of endothelium. They are supported by tissues surrounding them rather than by the tunica media and tunica adventitia. Examples of venous sinuses include the coronary sinus of the heart, which collects blood draining from the heart wall, and the superior sagittal sinus above the brain, which drains blood and cerebrospinal fluid.

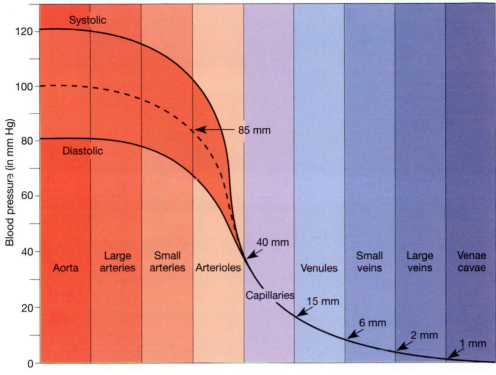

Blood pressure (in mm Hg)

120

100

80

60

40

20

0

Systolic

Diastolic

85 mm

40 mm

15 mm

6 mm

2 mm

1 mm

Aorta | Large arteries | Small arteries | Arterioles | Capillaries | Venules | Small veins | Large veins | Venae cavae

Route of blood

Figure 13–16
A blood pressure "hill." This graph illustrates the changes in blood pressure as blood passes from the heart and back.
At which point is blood pressure the greatest?

mum value. This pressure peak, called the **systolic pressure**, averages about 120 mm Hg in healthy adults. Beyond the junction between the heart and the aorta the pressure is lower, so blood moves down the pressure gradient into smaller vessels. Once the blood has left the aorta, the walls of the aorta recoil to their original shape and the aortic pressure drops to its lowest value (70 to 80 mm Hg). The elasticity of the aortic wall prevents the pressure from dropping further and thus keeps blood flowing along. The drop in pressure occurs during ventricular diastole (relaxation), and is called the **diastolic pressure.** Thus, the blood pressure within the aorta oscillates up and down, corresponding to the rhythmic beat of the heart.

The most common method of measuring arterial blood pressure involves the use of an apparatus called a **sphygmomanometer** (sfig'-mō-ma-NO-me-ter), which is shown in Figure 13–17. The cuff is first wrapped around the arm over the brachial artery. As air is pumped into the cuff, the brachial artery becomes compressed so that blood cannot pass through it. A release valve is then opened slightly to cause the pressure in the cuff to drop. Using a stethoscope, a tapping sound may soon be heard as blood begins to squeeze through the artery. The pressure at this moment is the systolic pressure. As the pressure continues to drop in the cuff, more blood is allowed to spurt through the artery, and the sound it makes becomes louder and changes in quality briefly before becoming inaudible. The pressure at the moment the sounds become inaudible is

the diastolic pressure. The tapping sounds that are heard through the stethoscope are known as Korotkoff (kō-RŌT-kof) sounds and represent turbulent blood flow through the artery. They continue as long as the cuff pressure is greater than the diastolic pressure.

Pulse

As blood moves from the aorta to small vessels, the arteries in which it flows behave the same way as the aorta. When blood enters, their elastic walls expand and stretch. Once the surge of blood moves on, the walls recoil to their original shape. This rhythmic expanding and recoiling of an arterial wall can be felt as the **pulse** where it nears the skin surface. The most common place to feel the pulse is on the radial artery in the wrist, but the common carotid artery in the neck or any other superficial artery lying against a bone or other firm surface may also be used.

The number of pulses counted each minute represents the number of heartbeats per minute. This is because each pulse wave is generated by a contraction of the heart (ventricular systole). The pulse is actually felt after the heartbeat has occurred because it takes time for the pulse wave to pass from the ventricle to the artery.

The average resting pulse rate ranges from 70 to 90 beats per minute in adults and from 80 to 140 beats per minute in children. The higher pulse rate in children is due to their higher metabolic rate. When the

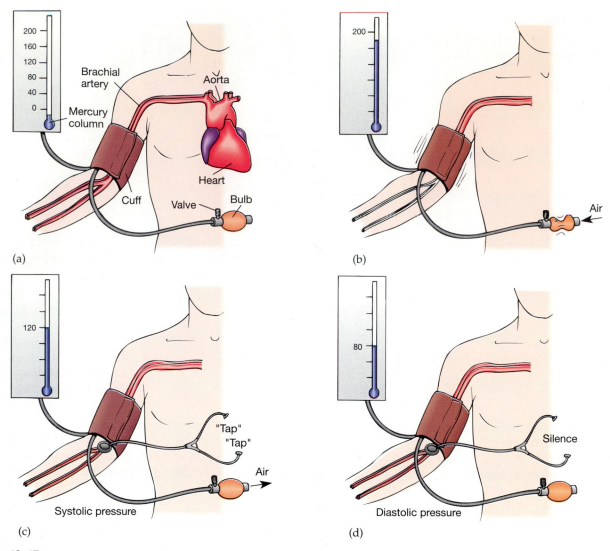

Figure 13–17

The use of a sphygmomanometer in measuring blood pressure. (a) A cuff is wrapped around the arm. (b) Pressure increases in the cuff as the bulb is pumped, causing the brachial artery to compress. (c) As cuff pressure is slowly released, the brachial artery opens slightly to permit blood to flow, producing tapping sounds that are heard through a stethoscope. The pressure at this point is the *systolic pressure.* (d) As the brachial artery opens completely, blood flows easily through it and the tapping sounds cease. This pressure point is called the *diastolic pressure.*

resting adult pulse rate exceeds 100 beats per minute, the condition is called *tachycardia*[17]; when it drops below 60 beats per minute, the condition is called *bradycardia.*[18]

Factors Affecting Blood Pressure

Arterial blood pressure is influenced by a number of factors. The main ones are cardiac output, peripheral resistance, and blood volume.

---

[17]Tachycardia: Greek *tachys* = swift + *kardia* = heart.

[18]Bradycardia: *bradys* = slow + *kardia.*

Cardiac Output

Recall that cardiac output represents the amount of blood that is pushed into the arterial system in a given time; its value is obtained by multiplying heart rate and stroke volume. Thus, if a change in heart rate or stroke volume occurs, a change in cardiac output will also occur. In other words, a change in how fast the heart beats or how strongly the ventricles contract will result in a change in the amount of blood being pushed into the arteries. This, in turn, will directly affect arterial blood pressure. In general, an increase in cardiac output increases blood pressure, and conversely, a decline in cardiac output will decrease blood pressure.

Peripheral Resistance

As blood flows through vessels, it makes physical contact with the vessel walls. This contact produces friction and drag, or **peripheral resistance,** which slightly slows the forward advance of blood. Peripheral resistance is largely determined by the diameter of the blood vessel lumen. In general, the smaller the diameter of the lumen, the more contact blood will make with it and, consequently, the greater will be the peripheral resistance. Peripheral resistance is also affected by the viscosity of blood. Thicker, more viscous blood increases peripheral resistance, whereas a less viscous blood decreases it. Regardless of the cause, an increase in peripheral resistance leads to an increase in blood pressure, and a decrease in peripheral resistance results in a decrease in blood pressure.

Any changes that alter peripheral resistance in a blood vessel also influence blood pressure. For example, the vasoconstriction of a vessel reduces its lumen diameter, thus increasing its peripheral resistance and causing an increase in blood pressure. Vasodilation of a vessel causes the reverse to occur. These changes affect blood pressure most profoundly when they occur in arterioles. In fact, arteriole diameter changes account for the single most important influence on blood pressure in the body.

Blood Volume

Blood volume is a measure of the amount of blood plasma and formed elements present in the cardiovascular system, with an average value of about 5 liters in a healthy adult. A change in this amount directly influences blood pressure. For example, a loss of blood volume due to hemorrhage causes an immediate drop in blood pressure. If the normal blood volume is restored by a blood transfusion, the blood pressure returns to its previous higher value.

The factors affecting blood pressure are summarized in Figure 13–18.

Regulation of Blood Pressure

Maintaining a steady pressure in our vessels is of primary importance, for without it a sufficient amount of blood may fail to reach vital organs and tissues. The factors that control blood pressure include nervous, hormonal, and kidney factors. They respond continuously to adjust cardiac output, peripheral resistance, and/or blood volume to achieve this control.

Nervous System Controls

Control of blood pressure by the nervous system is achieved mainly by adjusting cardiac output and peripheral resistance. Cardiac output may be increased or decreased by the activity of autonomic fibers that extend between reflex centers in the medulla oblon-

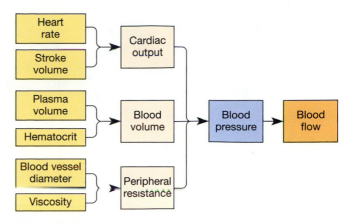

Figure 13–18
The factors that influence blood pressure and blood flow.
What is the effect on blood pressure if cardiac output, blood volume, or peripheral resistance is increased?

gata and the SA node in the heart. As we learned earlier, the nerve impulses that travel along these fibers may be parasympathetic impulses, which inhibit the heart rate, or sympathetic impulses, which stimulate the heart rate.

Peripheral resistance is regulated by the activity of vasomotor nerve fibers that extend between the medulla and smooth muscle in the tunica media of arterioles. The vasomotor fibers carry impulses that originate from a reflex center in the medulla, called the *vasomotor center.* An increase in the rate of impulse transmission stimulates vasoconstriction of the arterioles, whereas a reduced rate results in vasodilation.

As we have just seen, the medulla oblongata originates impulses that control both cardiac output and peripheral resistance. How does it determine when to stimulate and when to inhibit these factors? The medulla receives information from *baroreceptors* located in the major vessels above the heart (such as the aorta and carotid arteries), which we discussed earlier. Recall that these special receptors detect changes in blood pressure. Receptors sensitive to oxygen content or hydrogen-ion levels in the blood are also present in the large arteries above the heart, including the aorta and carotid arteries. These are called *chemoreceptors* and send impulses to the medulla to stimulate blood pressure as well. Chemoreceptors are stimulated by a sudden drop in blood oxygen levels or a rise in hydrogen-ion levels (a drop in pH). Higher brain centers may also influence the reflex centers in the medulla, which explains why certain emotional responses can have a profound effect on blood pressure. For example, fainting is caused by a sudden drop in blood pressure and is often brought on by an emotional disturbance. *(Text continues on page 394)*

HEALTH CLINIC

ATHEROSCLEROSIS

Atherosclerosis (a'-ther-ō-skle-RŌ-sis) is the second leading cause of death in the United States (cancer is number one). It is estimated that 50% of the deaths among the elderly are caused by atherosclerosis, through either coronary artery disease or strokes. This disease is due to the deposition of *plaques* in major arteries. A plaque, or atheroma, is a collection of fats and cholesterol. It tends to form along the inner wall of an artery when the inner surface becomes roughened and irregular. The roughening is a degeneration of the artery wall surface thought to be caused by a conversion of smooth muscle in the tunica media of the artery into a fatty material. The cause of the smooth muscle conversion is currently under investigation, and evidence is mounting that suggests the body's own white blood cells (macrophages) may damage this inner lining when blood cholesterol levels are high. Regardless of the actual cause, it is well established that degeneration of the smooth inner lining of the artery wall permits molecules of fat and cholesterol floating in the bloodstream to adhere to the roughened surface and to accumulate with time. Eventually, the plaque expands to a point of reducing the flow of blood through the vessel lumen. Blockage of the vessel lumen raises peripheral resistance, and consequently, blood pressure is increased.

As a result of blockage, the tissue or organ that is receiving the blood supply experiences a reduced oxygen availability. If the afflicted artery serves the myocardium of the heart wall, coronary artery disease results, which leads to angina pectoris and, if not treated, myocardial infarction. If the clogged artery serves the brain, a stroke will result.

Treatment of atherosclerosis includes the application of one or more techniques. *Angioplasty* has shown success in cases in which the atherosclerosis is not advanced. It involves the insertion of a balloon into the occluded (blocked) vessel. Once inserted, the balloon is expanded with air in an effort to widen the lumen. Another technique, which is used in more advanced cases, is *bypass grafting.* In this surgical procedure, a vein is taken from another area of the body (usually the saphenous vein in the leg) and inserted in the coronary circulation to reroute blood around the clogged vessel. The photograph on the opening page of this chapter, shows this procedure in progress. A third technique is the surgical removal of the plaque. This technique is used only if the occlusion is limited to one area of the affected vessel; it involves surgically opening the vessel and physically scraping out the plaque. Although these techniques improve the flow of blood to vital organs, the most desirable alternative lies in prevention of atherosclerosis.

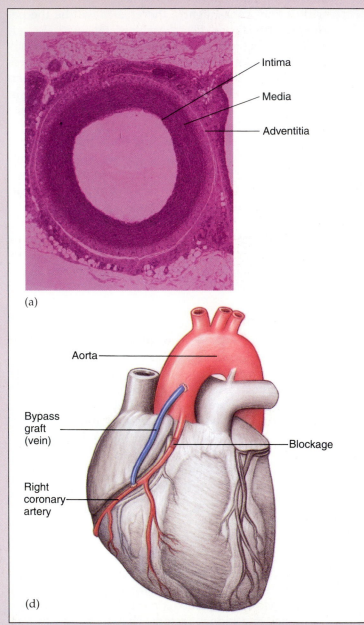

(a)

(d)

Atherosclerosis may be preventable in most individuals. Prevention may be aided by avoiding the recognized high risk factors, including tobacco smoke, foods that have high levels of fat and cholesterol, and lack of exercise. Avoiding these risk factors has been shown to lower a person's chances of developing atherosclerosis, even in cases in which family history reveals frequent occurrences of this disease.

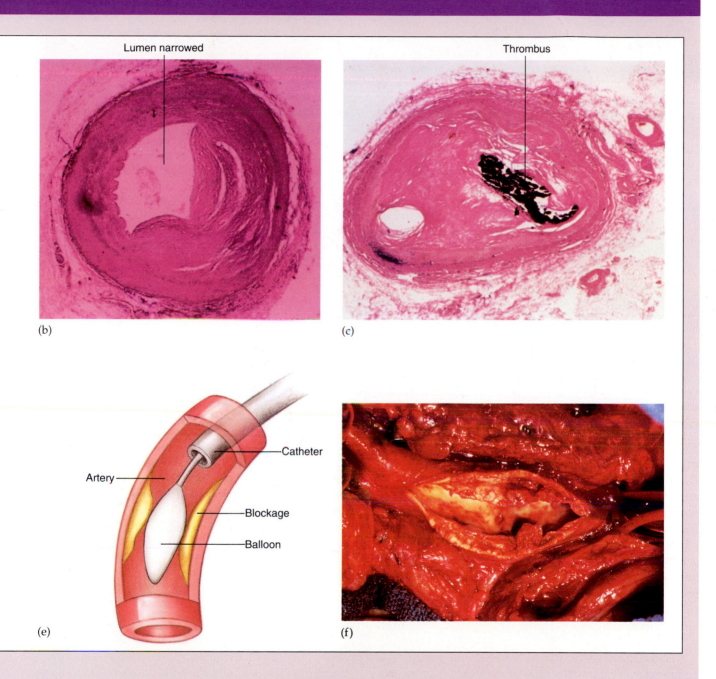

Figures a–f: The progression of atherosclerosis. (a) Cross section through a normal artery. (b) As the arterial wall degenerates, it thickens, narrowing the lumen. The thickened region is called a *plaque.* (c) A narrowed lumen has a greater tendency to develop a thrombus, which severely restricts blood flow. (d) The heart after bypass surgery, with the graft in place. (e) The technique of angioplasty. Once the balloon is in place at the occlusion site, it is filled with air and expands, compressing the blockage and widening the lumen. (f) The surgical procedure of endarterectomy, which involves opening the atherosclerotic artery and scraping away the plaque. The opened vessel in this photograph is the aorta.

Hypertension

Hypertension, or high blood pressure, affects over 15% of the U.S. population at some time in their lives. It is about twice as common among African-Americans as among the rest of the population. Hypertension is a chronic (long-lasting) condition that is characterized by a systolic pressure greater than 150 mm Hg and a diastolic pressure greater than 90 mm Hg. Because it requires the heart to work harder and puts added stress on arteries, it has an adverse effect on health. In fact, a strong correlation exists between increased sickness and early death and the level of hypertension a person may suffer from. Documented secondary effects include cerebral hemorrhage, renal disease, stroke, and myocardial infarction. Hypertension also increases the rate at which *arteriosclerosis*, a condition in which blood vessels lose their elasticity, may develop. Arteriosclerosis weakens the walls of the blood vessels, often leading to blood clot formation and vessel rupture.

Hypertension may be a secondary effect of other diseases, such as tumors of the adrenal gland, renal disease, or aldosteronism. However, the most common form of hypertension is essential (primary) hypertension, which cannot be linked to any other condition. Essential hypertension is thought to be an inherited disease.

Hypertension currently has no cure, but it can be controlled by not smoking, decreasing the salt content of the diet, weight control, restricting alcohol intake, and taking antihypertensive drugs. Recent studies also suggest that a diet high in potassium-rich foods, including legumes (beans and peas) and fresh fruit, may help considerably in reducing hypertension.

Hormonal Controls

There are several hormones that play a role in controlling blood pressure. They include norepinephrine (NE), epinephrine, atrial natriuretic factor (ANF), and antidiuretic hormone (ADH).

Epinephrine and **norepinephrine** are secreted by the adrenal medulla during times of stress. They enhance and prolong the fight-or-flight response that is activated by the sympathetic division of the autonomic nervous system. Epinephrine increases cardiac output by stimulating heart rate, and increases peripheral resistance by promoting arteriole vasoconstriction. Norepinephrine increases peripheral resistance in the same way.

Atrial natriuretic (nā-trē-yoo-RET-ik) **factor** is a hormone secreted by the atria of the heart. It promotes a reduction in blood volume, which leads to a decline in blood pressure. It does so by stimulating the kidneys to excrete more sodium and water from the body, thus causing a condition of dehydration that results in a drop in blood volume.

Antidiuretic hormone is produced by the hypothalamus and stimulates the kidneys to conserve water. This increases blood volume, causing a rise in blood pressure.

Kidney Controls

The kidneys provide the main long-term mechanism of blood pressure control. One mechanism they use affects blood volume. When the blood volume or blood pressure rises, the kidneys respond by allowing more water to pass from the body as urine. As blood volume drops, blood pressure decreases as well. Conversely, the kidneys may reduce the amount of water leaving the body as urine when blood volume or pressure is low, restoring it to a more favorable level.

A second mechanism the kidneys use involves the release of special chemicals. When arterial blood pressure drops, special cells in the kidneys release an enzyme called **renin** into the blood. Renin triggers the formation of another chemical, **angiotensin** (an-jē-ō-TEN-sin) **II,** which is a strong vasoconstrictor. Angiotensin II also stimulates the release of another hormone, **aldosterone,** which is produced by the adrenal gland and stimulates the reabsorption of sodium by the kidneys. As sodium moves into the bloodstream, water follows. Thus, a rise in both peripheral resistance and blood volume follows the release of renin, which in turn causes an increase in blood pressure.

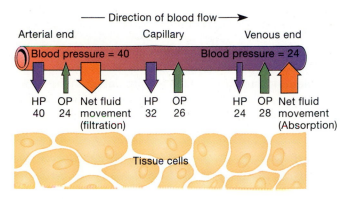

Figure 13–19
Capillary dynamics and fluid movement. At the arterial end of the capillary, net fluid movement is in an outward direction by filtration. At the venous end of the capillary, net fluid movement is inward by absorption. HP = hydrostatic pressure; OP = osmotic pressure.

At the capillary's arterial end, what factor forces fluid outward? At the capillary's venous end, what factor draws fluid inward?

Capillary Exchange

Materials move across capillary walls by diffusion, whereas fluid movement follows hydrostatic and osmotic pressure differences.

The 10 billion or so capillaries in the body are the only sites of exchange between the blood and interstitial fluid. Nutrients and oxygen move across the capillary walls into the extracellular space, and carbon dioxide and other waste materials move in the opposite direction. These materials pass through the thin barrier formed by the capillary endothelium by the processes of simple diffusion, osmosis, facilitated diffusion, and active transport.

A small amount of fluid also moves across the capillary walls. Its movement is governed by two forces: **hydrostatic pressure** and **osmotic pressure** (Fig. 13–19). Hydrostatic pressure is the blood pressure within capillaries that tends to push fluid from the capillary into the surrounding extracellular space. Osmotic pressure is generated by the movement of fluid by osmosis from the extracellular space into the capillary. Osmosis draws fluid into the capillaries because blood has a greater solute concentration than interstitial fluid (remember: fluid moves from a region of lower solute concentration to higher solute concentration in osmosis; see Chapter 3). The blood's high solute concentration is caused by the presence of large numbers of blood proteins, which are unable to cross the capillary wall because of their large size. Interstitial fluid, by contrast, contains far fewer proteins.

As indicated in Figure 13–19, blood pressure in the capillaries is greater where blood enters at their arteriolar ends than it is at their venous ends. This differ-

ence in blood pressure results in a hydrostatic pressure that exceeds osmotic pressure at the arteriolar end. As a result, more fluid is pushed out of the capillary (by filtration) than enters here. At the venous ends of capillaries, the drop in blood pressure results in a reduction of fluid movement out of the capillary. Consequently, more fluid enters the capillary by (absorption) at the venous end than exits.

The net movement of fluid out of the capillary at its arteriolar end results in an accumulation of fluid, now called *interstitial fluid,* in the extracellular environment. About 90% of the fluid that exits returns to the capillary bloodstream (mainly at the venule end of the capillary). The remaining 10% finds its way into lymphatic vessels, which carry the fluid through the body to eventually return it to the cardiovascular bloodstream. Exactly how this remaining amount is returned will be discussed in the next chapter, The Lymphatic System.

CIRCULATORY PATHWAYS

The blood vessels of the circulatory system are arranged into two distinct pathways: the pulmonary circulation and the systemic circulation. The pulmonary circulation is the route blood travels from the heart to the lungs and back. The systemic circulation is by far the more extensive, traveling from the heart throughout the remaining areas of the body, and back again. The basic organization of the circulatory networks is shown in Figure 13–20. In the following section we will study the circulation of the adult. Fetal circulation, which differs somewhat, will be discussed in Chapter 20.

Figure 13–20
The circulatory networks of the body.
What major artery initiates the systemic circulation?

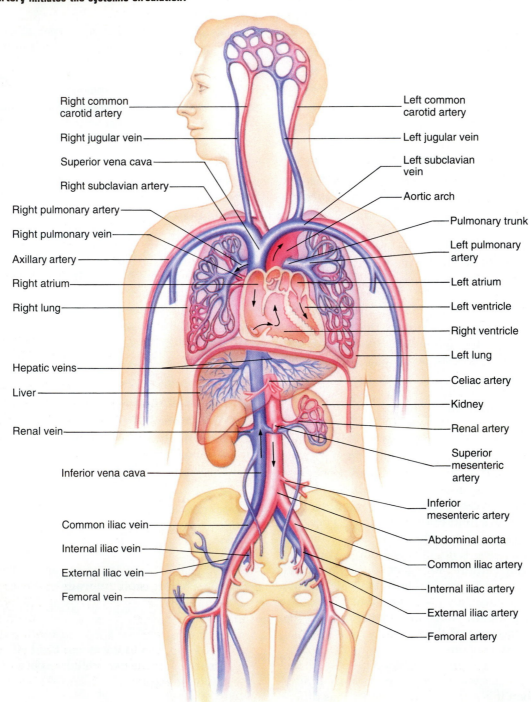

Right common carotid artery

Right jugular vein

Superior vena cava

Right subclavian artery

Right pulmonary artery

Right pulmonary vein

Axillary artery

Right atrium

Right lung

Hepatic veins

Liver

Renal vein

Inferior vena cava

Common iliac vein

Internal iliac vein

External iliac vein

Femoral vein

Left common carotid artery

Left jugular vein

Left subclavian vein

Aortic arch

Pulmonary trunk

Left pulmonary artery

Left atrium

Left ventricle

Right ventricle

Left lung

Celiac artery

Kidney

Renal artery

Superior mesenteric artery

Inferior mesenteric artery

Abdominal aorta

Common iliac artery

Internal iliac artery

External iliac artery

Femoral artery

Pulmonary Circulation

The pulmonary circulation consists of vessels that extend between the heart and lungs, and a vast lung capillary network where the exchange of gases occurs.

The **pulmonary circulation**[19] is the route blood travels for the purpose of oxygenation. In this route, blood that is low in oxygen and high in carbon dioxide is transported to the lungs, which contain inhaled air that has higher oxygen and lower carbon dioxide levels. As a result of this concentration gradient, oxygen diffuses into the blood and carbon dioxide diffuses out. Blood that has been refreshed in this way is then returned to the heart for its circulation throughout the rest of the body.

The pulmonary pathway begins with the **pulmonary trunk,** which carries blood from the right ventricle of the heart (Fig. 13–21). After extending diagonally upward for about 8 cm, the pulmonary trunk divides into a **right pulmonary artery** and a **left pulmonary artery,** which pass to their respective lungs. Within the lungs, the right and left pulmonary arteries subdivide into smaller arteries, whose branches eventually form many arterioles. The arterioles give rise to the **pulmonary capillaries,** which form dense networks surrounding the delicate air sacs of the lungs. Between the pulmonary capillaries and the air sacs the exchange of gases occurs. As oxygen is loaded into the red blood cells, the blood takes on a crimson-red color. The pulmonary capillary beds drain into venules, which unite to eventually form the two **pulmonary veins** exiting from each lung. The four pulmonary veins extend to the left atrium of the heart, completing the pulmonary circulation.

Systemic Circulation

The systemic circulation consists of the vessels that extend from the left ventricle to the right atrium of the heart, providing fresh blood to all body tissues except the lungs.

The **systemic circulation** provides the transport route for all body organs and tissues (except the lungs). It brings oxygenated blood to tissues and carries away carbon dioxide. It also transports nutrients from the digestive tract, hormones from the endocrine glands, enzymes and proteins from various cells, white blood cells to areas of infection or other forms of disease, and

[19]Pulmonary: *pulmonary* = of the lung.

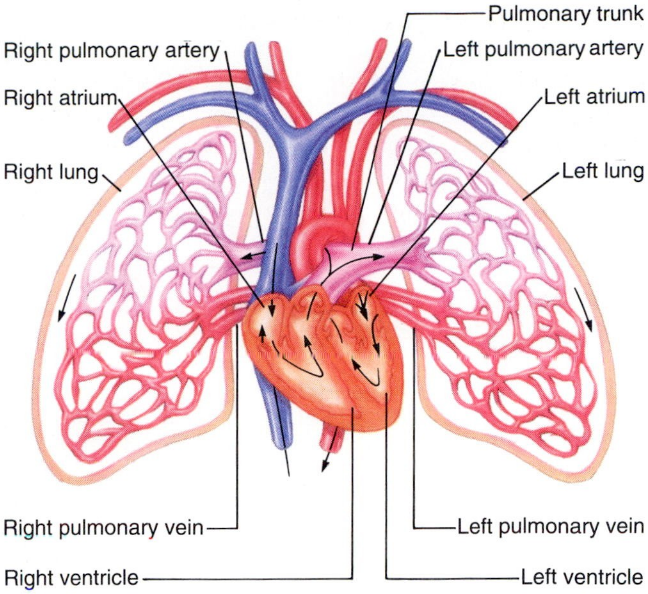

Figure 13–21
Pulmonary circulation.
What is the overall purpose of the pulmonary circulation?

nonprotein nitrogenous substances away from cells to the kidneys. Blood beginning this route is crimson red because of its high oxygen content; blood returning to the heart is a dark-red or muddy color because of its low oxygen content.

In the following discussion, only the major arteries and veins of the systemic pathway are described. As you read, keep in mind that the blood does not flow in only these vessels, but also flows through smaller arteries, arterioles, capillaries, venules, and smaller veins to provide a complete, closed circulatory network. Also, the arteries and veins are presented separately for your convenience. It may help you in understanding the terminology to know that they are often named according to the region of the body they supply. Because arteries and veins usually travel in parallel, the ones in the same region share similar names. For example, the *brachial artery* supplies the upper arm, or *brachium*, and travels in parallel with the *brachial vein*, which drains the brachium.

Systemic Arteries

The systemic arteries are branches of the largest artery of the body, the aorta. It begins at the left ventricle of the heart and extends upward a short distance before curving toward the back and turning downward to pass through the thoracic and abdominal cavities. Along its length it gives off numerous branches, which are shown in Figure 13–22 and detailed in Figure 13–23.

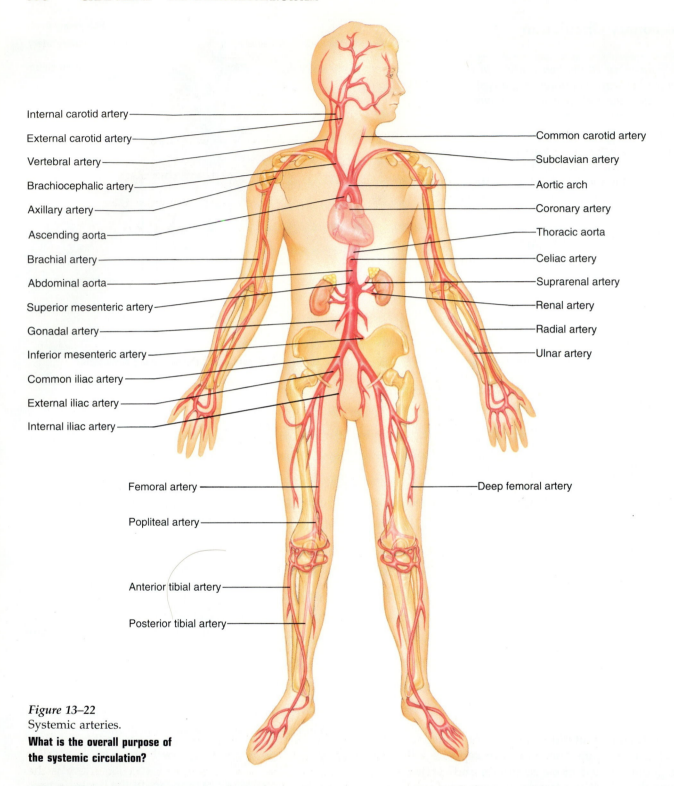

Internal carotid artery
External carotid artery
Vertebral artery
Brachiocephalic artery
Axillary artery
Ascending aorta
Brachial artery
Abdominal aorta
Superior mesenteric artery
Gonadal artery
Inferior mesenteric artery
Common iliac artery
External iliac artery
Internal iliac artery

Common carotid artery
Subclavian artery
Aortic arch
Coronary artery
Thoracic aorta
Celiac artery
Suprarenal artery
Renal artery
Radial artery
Ulnar artery

Femoral artery
Popliteal artery

Deep femoral artery

Anterior tibial artery
Posterior tibial artery

Figure 13–22
Systemic arteries.
What is the overall purpose of the systemic circulation?

Branches of the Aorta

The aorta originates at the left ventricle of the heart. From here it extends upward a short distance; this segment is called the **ascending aorta.** From the base of the ascending aorta arise the right and left **coronary arteries,** which supply the heart wall.

The segment of the aorta that curves to the back is the **aortic arch.** Along its short length it gives off three major branches: the **brachiocephalic** (brāk'-ē-ō-se-FAL-ik) **artery,** the left **common carotid** (ka-ROT-id) **artery,** and the left **subclavian** (sub-KLĀ-vē-an) **artery.**

As the aorta completes its arch and nears the vertebral column, it makes a downward turn and contin-

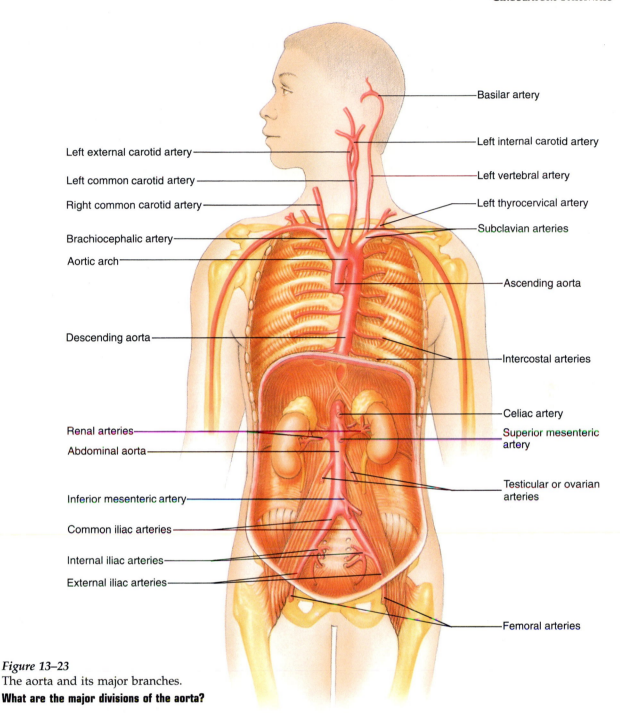

Basilar artery

Left internal carotid artery

Left vertebral artery

Left thyrocervical artery

Subclavian arteries

Ascending aorta

Intercostal arteries

Celiac artery

Superior mesenteric artery

Testicular or ovarian arteries

Femoral arteries

Left external carotid artery

Left common carotid artery

Right common carotid artery

Brachiocephalic artery

Aortic arch

Descending aorta

Renal arteries

Abdominal aorta

Inferior mesenteric artery

Common iliac arteries

Internal iliac arteries

External iliac arteries

Figure 13–23
The aorta and its major branches.
What are the major divisions of the aorta?

ues in this direction until it ultimately divides into two vessels. Throughout most of its length it lies directly in front of the vertebral column. This large segment of the aorta is called the **descending aorta** (Fig. 13–23).

The part of the descending aorta above the diaphragm is known as the **thoracic aorta.** Along its length it gives off numerous branches that extend between the ribs to supply the thoracic wall. These vessels are called **intercostal arteries.**

Below the diaphragm the descending aorta is referred to as the **abdominal aorta.** Its first major branch

is the **celiac** (SĒ-lē-ak) **artery,** which supplies the liver, spleen, stomach, and pancreas by way of its smaller branches. Below the origin of the celiac artery is the **superior mesenteric** (mes-en-TAR-ik) **artery,** which supplies the small intestine and part of the large intestine. A pair of arteries called the right and left **suprarenal** (soo-pra-RĒ-nal) **arteries** extend to the adrenal glands on top of each kidney, and below their origin are the right and left **renal arteries,** which pass to the kidneys. Below the renal arteries, the right and left **gonadal arteries** (**testicular arteries** in the male

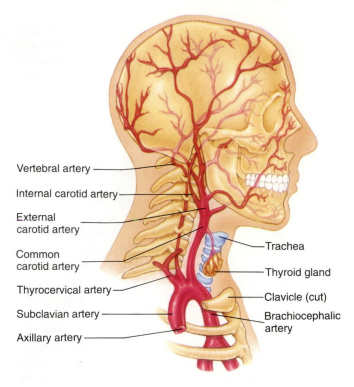

Figure 13–24
Arteries of the head and neck.

What is the primary artery supplying the brain with oxygen-rich blood?

and scalp by way of smaller branches. The internal carotid artery courses deep to the base of the skull. It enters the cranial cavity to supply the brain.

The right **subclavian artery** originates from the brachiocephalic artery, and as we have seen, the left **subclavian artery** arises from the aortic arch. Soon after their origin the subclavian arteries each give off two branches that extend upward to supply the head and neck. These are the **vertebral arteries** and **thyrocervical** (thī-rō-SER-vi-kal) **arteries.** The vertebral arteries pass upward through the transverse foramina of the cervical vertebrae. They enter the cranial cavity through the foramen magnum to supply the brain. The thyrocervical arteries extend a short distance into the tissues of the neck, giving off branches that supply the thyroid glands, parathyroid glands, larynx, trachea, esophagus, pharynx, and various muscles of the region.

Arteries of the Shoulder and Upper Limbs

The shoulder and upper limbs are supplied entirely by arteries arising from the subclavian arteries (Fig. 13–25). The distribution of the right and left subclavian is essentially the same on both sides of the body.

Once a subclavian artery reaches the lower shoulder near the armpit, or **axilla,** it continues downward as the **axillary** (AKS-il-ar-ē) **artery.** Branches from the axillary artery supply muscles and other structures in

and **ovarian arteries** in the female) extend downward to supply the gonads. The large **inferior mesenteric artery** is the last major branch of the aorta. It supplies most of the large intestine. The descending aorta ends at the level of the second lumbar vertebra, where it divides into two large vessels that supply the lower limbs: the right and left **common iliac** (IL-ē-ak) **arteries.**

Arteries of the Head and Neck

The head and neck regions are supplied with blood by arteries that originate from the common carotid and subclavian arteries (Figs. 13–23 and 13–24). The common carotid artery is the major supplier of the head region.

The right **common carotid artery** arises from the brachiocephalic artery. The left **common carotid artery,** as we have seen, originates from the aortic arch. These prominent vessels extend upward through the neck until they each branch into the **external** and **internal carotid arteries.** This occurs at the superior border of the larynx. At this junction is a slight dilation, called the *carotid sinus.* It contains baroreceptors that detect changes in blood pressure and aid in its control. The external carotid artery continues upward along the side of the head to supply parts of the neck, face, jaw,

Figure 13–25
Arteries of the shoulder and upper limb.
What are the branches of the subclavian arteries?

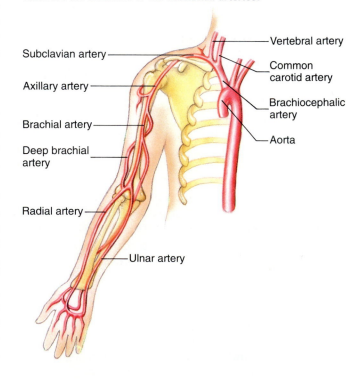

the axilla and thoracic wall. Once reaching the upper arm, or *brachium*, the axillary artery continues downward as the **brachial** (BRĀ-kē-al) **artery.** A major branch of the brachial artery that passes deep into muscles is the **deep brachial artery.** The brachial artery follows a course alongside the humerus to the elbow, where it divides into the **ulnar artery** and the **radial artery.** The ulnar artery extends downward along the ulnar side of the forearm to the wrist, and the radial artery courses along the radial side of the forearm to the wrist. These arteries supply the muscles of the forearm. At the wrist, they branch into smaller arteries that supply the wrist, hand, and fingers.

Arteries of the Pelvis and Lower Limbs

The pelvis and lower limbs are supplied with blood by branches of the common iliac arteries, which arise from the base of the abdominal aorta (Fig. 13–26). Like the distribution for the upper limbs, the distribution of the common iliac arteries is about the same on both sides of the body.

Figure 13–26
Arteries of the pelvis and lower limb.
What are the major branches of the common iliac arteries?

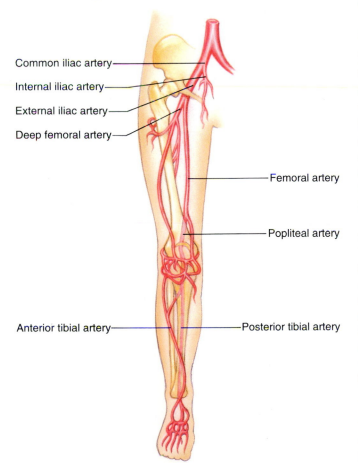

Common iliac artery
Internal iliac artery
External iliac artery
Deep femoral artery
Femoral artery
Popliteal artery
Anterior tibial artery
Posterior tibial artery

Each common iliac artery divides into an **internal iliac artery** and an **external iliac artery.** The internal iliac artery extends a short distance downward into the pelvic wall. Its branches supply visceral structures within the pelvic cavity, muscles of the pelvis, and external genitals.

The external iliac artery carries the main supply of blood to the lower limbs. It extends downward along the pelvis and under the inguinal ligament to enter the thigh, where it becomes the **femoral artery.** The femoral artery extends down the thigh along its anterior side. Along its length it gives off branches that supply the muscles and skin of the thigh, as well as the skin of the groin and lower abdomen. Its most important branch is the **deep femoral artery,** which supplies the flexor muscles of the thigh. As the femoral artery approaches the knee, it courses posteriorly to pass into the popliteal fossa. Here it becomes known as the **popliteal artery** and extends downward a short distance until dividing into the **anterior** and **posterior tibial arteries.** The anterior tibial artery descends between the tibia and fibula, giving off branches to the skin and muscles of the anterior and lateral portions of the lower leg. The posterior tibial artery extends downward behind the calf muscles. It gives off branches that supply the skin and muscles of the posterior and medial lower leg.

Systemic Veins

The systemic veins are large vessels that are formed by the convergence of smaller veins and venules, which drain blood from the body's extensive capillary networks (except those of the lungs). The blood is carried in a direction leading toward the heart, with the right atrium as the final destination. There are two major systemic veins that unite directly with the heart (in addition to the coronary sinus, which drains the heart wall): the superior vena cava and the inferior vena cava.

In our following discussion of systemic veins, we will trace the branches of the superior vena cava first. The branches of the inferior vena cava will be traced next. Our study will, for the most part, follow the true course of blood through the veins; that is, from the distal veins to those closest to the heart. Please note that this is opposite to our analysis of arteries, in which we began with those arteries closest to the heart and continued distally. This way of organizing the discussion serves to emphasize the difference in the direction of blood flow between arteries and veins.

Veins Draining into the Superior Vena Cava

The superior vena cava receives blood draining from all areas superior to the diaphragm, except the lungs. It receives blood from numerous smaller veins that drain specific areas of the head, neck, thoracic and abdominal walls, shoulder, and upper limbs. The most important of these veins are described in the following paragraphs and are shown in Figure 13–27.

Blood from the face, scalp, and superficial regions of the neck is drained by way of the right and left **external jugular veins,** which are shown in Figures 13–27 and 13–28. These vessels extend downward along ei-

Figure 13–27
Systemic veins.
What major veins empty into the right atrium?

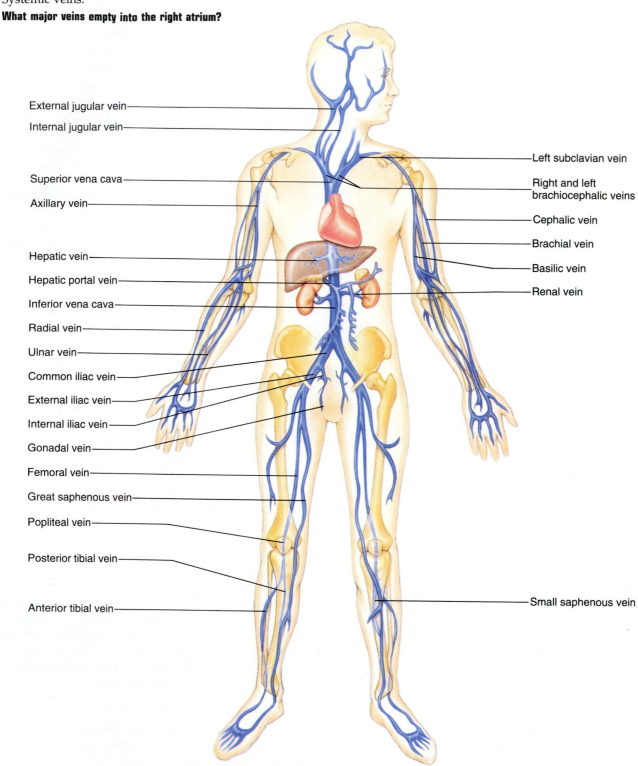

External jugular vein

Internal jugular vein

Superior vena cava

Axillary vein

Hepatic vein

Hepatic portal vein

Inferior vena cava

Radial vein

Ulnar vein

Common iliac vein

External iliac vein

Internal iliac vein

Gonadal vein

Femoral vein

Great saphenous vein

Popliteal vein

Posterior tibial vein

Anterior tibial vein

Left subclavian vein

Right and left brachiocephalic veins

Cephalic vein

Brachial vein

Basilic vein

Renal vein

Small saphenous vein

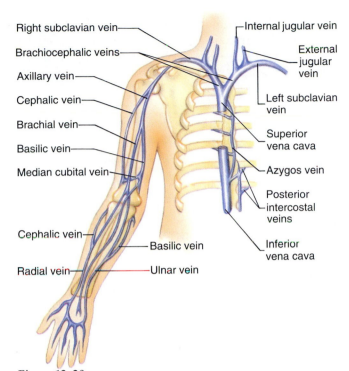

Figure 13–28
Systemic veins of the trunk wall, shoulder, and upper limb.

The subclavian vein is formed by the merging of what vessels?

ther side of the neck until they merge with the **subclavian veins.** Blood from the brain and deep areas of the head and neck is drained by the right and left **internal jugular veins,** which are slightly larger in size. They course down the neck in a medial position relative to the external jugular veins. At their proximal ends, the internal jugular veins unite with the subclavian veins, forming the large **brachiocephalic veins** (right and left). These two prominent vessels merge to form the **superior vena cava,** which delivers blood to the right atrium of the heart. Shortly before uniting with the heart, the superior vena cava receives a single vessel that drains the thoracic and abdominal walls. This is the **azygos (ā-ZĪ-gos) vein.**

The upper limbs are drained by two venous pathways: one courses deep through the muscles of the arm, while the other travels superficially just below the skin (Figs. 13–27 and 13–28). The deep veins parallel the pathway of the arteries and are named similarly: the **ulnar vein,** the **radial vein,** the **brachial vein,** and the **axillary vein.** The superficial veins are larger and can be easily seen through the skin. They form interconnecting networks, which also communicate with the deep veins just named. This provides blood many alternative routes to follow. The major superficial veins are the **basilic (ba-SIL-ik) vein,** which ascends from the forearm to the middle of the upper arm; the **cephalic (se-FAL-ik) vein,** which extends upward from

the wrist to the shoulder; and the **median cubital vein,** which interconnects the basilic and cephalic veins. The median cubital vein is a common site for drawing blood. At the shoulder, the cephalic vein unites with the axillary vein. Beyond the axilla, the axillary vein continues as the **subclavian vein,** which extends to its union with the internal jugular vein to form the brachiocephalic vein.

Veins Draining into the Inferior Vena Cava

The inferior vena cava drains blood from all areas of the body below the heart. It is a very large vessel, extending from its origin at the level of the second lumbar vertebra to its union with the heart at the right atrium (Fig. 13–27). Along its length it receives blood from the pelvis, the lower limbs, and the abdominal viscera by way of smaller veins. Like the arteries, the veins are essentially the same on both sides of the body.

Like the veins in the upper limbs, the veins of the lower limbs are arranged into two pathways: a deep pathway and a superficial pathway (Fig. 13–27). The deep veins parallel the course of the arteries serving the leg and share their names. They are the **anterior tibial vein,** the **posterior tibial vein,** the **popliteal vein,** the **femoral vein,** and the **external iliac vein.** The superficial veins interconnect with one another and with the deep veins to form a complex network that extends from the foot to the upper thigh. The main vessel in this network is the **great saphenous vein,** which is the longest vein in the body. It extends along the medial side of the leg from the ankle to the thigh, where it empties into the femoral vein just before it enters the pelvis. A second superficial vessel, the **small saphenous vein,** runs along the lateral side of the foot and up through the calf muscles, which it drains. It empties into the popliteal vein at the knee.

Several of the large veins draining the abdomen into the inferior vena cava parallel the course of the arteries, and therefore share similar names (Fig. 13–27). From inferior to superior, they are the **right gonadal vein** (right testicular vein or right ovarian vein), which drains the gonads; the **renal veins,** which drain the kidneys; and the **suprarenal veins,** which drain the adrenal glands. The **left gonadal vein** does not enter the inferior vena cava, but rather the left renal vein. In addition are the **hepatic veins,** which drain the liver and which unite with the inferior vena cava a short distance below the diaphragm.

Some prominent veins draining the abdominal viscera do not unite directly with the inferior vena cava, but rather travel in a direction leading toward the liver. They originate from the digestive tract and are collectively called the **hepatic portal system** (Fig. 13–29). In general terms, a *portal system* is one in which blood flows from one capillary network directly to another

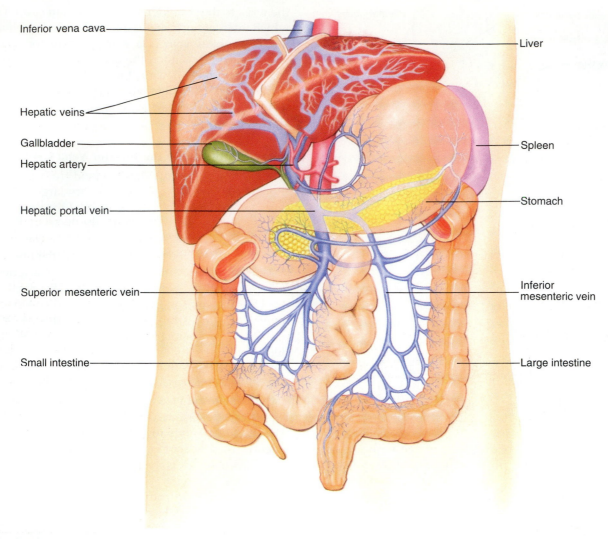

Inferior vena cava

Hepatic veins

Gallbladder

Hepatic artery

Hepatic portal vein

Superior mesenteric vein

Small intestine

Liver

Spleen

Stomach

Inferior mesenteric vein

Large intestine

Figure 13–29
The hepatic portal system of circulation.
What prominent vein carries blood from the digestive tract to the liver?

(instead of toward the heart). The hepatic portal system shunts blood from the capillaries of the digestive tract to capillaries of the liver.

The smaller veins of the hepatic portal system empty blood into the **superior** and **inferior mesenteric veins**, which converge to form the **hepatic portal vein** that continues to the liver. Thus, the liver receives blood from two sources: the hepatic portal vein and the hepatic artery. Blood that is high in oxygen enters through the hepatic artery, and blood that is low in oxygen enters through the hepatic portal vein. Although it is low in oxygen, venous blood from the digestive organs may be filled with nutrients and other substances absorbed in the intestine. As the blood passes slowly through special capillaries in the liver, hepatic cells remove materials required for the metabolic functions, and phagocytic cells eliminate bacte-

ria and other foreign matter that have penetrated the digestive lining. The roles of the liver are further described in Chapter 16. Once having passed through the liver capillaries, blood is collected by small veins that lead into the hepatic veins and is emptied into the inferior vena cava.

CONCEPTS CHECK

1. Which artery carries blood that is low in oxygen and high in carbon dioxide?

2. What are the three large branches of the aortic arch?

3. What major vein drains the body below the level of the heart?

HOMEOSTASIS

As we have seen in this chapter, the cardiovascular system supports the functions of blood by transporting it throughout the body. The pumping action of the heart and the specialized tubular structure of blood vessels keep blood flowing throughout the body. The purpose of this transport cycle is to carry refreshed blood to within close proximity of all cells, by passing it slowly through vast capillary networks. As blood moves through capillaries, vital molecules of oxygen and nutrients diffuse to surrounding cells and waste materials are carried away.

The function of transporting blood is a vitally important aspect of maintaining body homeostasis. Its importance is made evident when blood flow is reduced by disease or injury. A general condition in which cardiac output or blood volume is reduced to a point where body tissues fail to receive an adequate blood supply is referred to as *circulatory shock*. It is usually caused by the loss of blood by hemorrhage. The symptoms that typify circulatory shock include a weak, rapid pulse; shallow and rapid breathing; lowered body temperature; pale, clammy skin; blue color of the lips and fingertips (cyanosis); and mental confusion or unconsciousness.

During the early onset of circulatory shock, cardiovascular mechanisms quickly react in an effort to restore and maintain homeostasis. For example, the drop in blood pressure stimulates secretion of renin by the kidneys and epinephrine by the adrenal gland, which stimulate the constriction of most blood vessels. In an effort to maintain proper blood flow to the heart and brain during this emergency, the arterioles supplying these two organs are not constricted. Also, aldosterone is released by the adrenal gland and antidiuretic hormone (ADH) is secreted by the posterior pituitary gland, causing water to be retained. All of these responses combine in an effort to restore blood pressure to normal, but without compensating heart and brain activity.

In most victims, if blood loss exceeds one liter, the mechanisms attempting to restore homeostasis usually fail. This occurs because the supply of blood becomes insufficient to sustain the heart and the heart muscle thereby weakens. The decline in heart activity leads to cardiac arrest, and vital organs such as the brain, kidneys, and liver are damaged by the lack of available oxygen. This deteriorating state leads to a complete disruption of homeostasis, and death, unless body fluids are restored immediately (by transfusion).

CLINICAL TERMS OF THE CARDIOVASCULAR SYSTEM

Aneurysm   The formation of a sac within the wall of a vessel or heart chamber caused by local stretching or dilation. In extreme cases, it may produce a tear in the vessel wall leading to profuse hemorrhage.

Angina pectoris   A pain or tightness in the chest produced by a reduced blood flow to the heart. Angina does not cause cell death.

Arteriosclerosis   A loss of elasticity in arterial walls, resulting in a lack of responsiveness to properly regulate vessel diameter and blood pressure.

Atherosclerosis   A narrowing of arteries supplying organs or tissues, caused by the accumulation of plaques containing fats and cholesterol among the artery linings. It leads to a reduced blood flow.

Bacterial endocarditis   A bacterial infection of the inner lining of the heart and valves. If not treated it can cause inflammation of the heart and the production of emboli (floating clots).

Cardiac arrhythmias   Irregularity or loss of the heartbeat. A slowed heartbeat (less than 60 beats per minute) is called *bradycardia*, and an accelerated heartbeat (greater than 100 beats per minute) is called *tachycardia*.

Coarctation of the aorta   A congenital heart disease in which the aorta is narrowed at the arch just before it descends, forcing more blood to the upper parts of the body than the lower parts.

Congestive heart failure (CHF)   A failure of the heart to pump blood efficiently to body tissues. It is characterized by inadequate blood circulation (circulatory congestion) to other organs, which leads to increased pressure and fluid leakage in the lungs, liver, kidneys, and lower limbs. CHF is caused by damage inflicted on cardiac muscle as a result of cardiac disorders, such as myocardial infarction or damaged heart valves; increased blood volume, due to retention of sodium and water by the kidneys; or decreased oxygen-carrying capacity of the blood, due to anemia. Symptoms of CHF include shortness of breath, swelling of the lower limbs (edema), high venous pressure, and sometimes irritability and shortened attention span (due to reduced circulation to the brain). Treatment includes administration of the drug digitalis, which strengthens heart contraction and causes the kidneys to excrete more fluid (reducing blood pressure). Prolonged rest, administration of oxygen, and diuretics are also part of the treatment.

Heart block A failure of either the SA (SA block) or AV (AV block) node to generate impulses. Because the heart muscle has an intrinsic rhythm, the chambers will continue to beat but at a reduced rate and without proper coordination.

Heart fibrillation A type of cardiac arrhythmia in which the heart beats at a grossly irregular rhythm as a result of continuous, chaotic flurry of impulses from the heart conduction system.

Heart flutter A type of cardiac arrhythmia that produces a heart rate of as much as 300 beats per minute, stimulated by an accelerated cyclic wave of electrical activity from the heart conduction system.

Hypertension The elevation of either systolic or diastolic blood pressure, or both. It afflicts over 15% of the U.S. population and can lead to heart failure. It has few symptoms. It is controlled by lower salt intake, not smoking, weight control, and antihypertensive drug therapy.

Ischemia A local and temporary decline in blood supply due to a blockage in circulation. It causes the condition angina pectoris.

Murmur The leakage of blood through a closed valve. The leakage results from a defect in the valve, a somewhat common condition. Severe murmurs may cause a mixing of blood in the chambers, reducing blood flow to tissues and organs, and must be surgically corrected.

Myocarditis An infection of the heart muscle, often as a result of an infection originating elsewhere in the body.

Patent ductus arteriosus A congenital disease in which there remains an abnormal open connection between the pulmonary artery and the aorta, causing blood to shunt away from the lungs.

Pericarditis An infection of the pericardial sac. If not treated, it may result in thickening of the pericardium due to the replacement of normal membrane tissue by scar tissue. In severe cases, pericarditis can cause cardiac arrest and death.

Phlebitis Inflammation of the veins, sometimes caused by intravenous needles or catheters.

Septal defects Congenital defects resulting in the incomplete closure of the interatrial septum or the interventricular septum of the heart. These conditions cause insufficient blood flow and must be surgically corrected.

Stenosis A congenital defect of one or more of the heart valves, in which they are more narrow than normal. A stenosis causes incomplete valve closure during the cardiac cycle, which leads to development of heart murmur.

CHAPTER SUMMARY

The cardiovascular system consists of the heart and blood vessels.

THE HEART

General Characteristics The heart is located between the lungs in the thoracic cavity, mostly on the left side. Its pointed end is the apex, and the flattened superior end is the base.

Coverings of the Heart The serous membrane around the heart is the pericardium. It contains two layers: the outer, tough parietal pericardium (pericardial sac) and the thin, inner visceral pericardium (epicardium). Between them is a space, the pericardial cavity.

Heart Wall The wall of the heart is composed of three layers:

   Epicardium A thin outer serous membrane.

Myocardium The middle layer, composed of cardiac muscle, forming the bulk of the heart and the primary functional tissue.

Endocardium The thin inner layer, it is a membrane known as *endothelium.*

Heart Chambers There are four spaces, or chambers, in the heart.

Atria The two superior chambers. They are thin-walled and serve as receiving chambers for incoming blood. The right and left atria are separated by the interatrial septum.

Ventricles The two inferior chambers. They are thick-walled and serve as pumps for pushing blood out of the heart. The ventricles are separated from each other by the interventricular septum.

Heart Valves There are two types of valves:

Atrioventricular (AV) Valves Permit one-way flow of blood from the atria to the ventricles by closing during ventricular contraction. The right AV valve is also known as the *tricuspid* valve, and the left AV valve as the *bicuspid* or *mitral* valve.

Semilunar (SL) Valves Permit one-way flow of blood from the ventricles to the pulmonary trunk and aorta by closing during ventricular relaxation. The right SL valve is also known as the *pulmonary valve,* and the left is called the *aortic valve.*

Blood Flow Through the Heart Blood flows from the right atrium into the right ventricle through the tricuspid valve, then into the pulmonary trunk through the pulmonary valve. It reenters the heart by way of four pulmonary veins into the left atrium and then enters the left ventricle through the bicuspid (mitral) valve. It leaves the left ventricle through the aortic valve and enters the aorta.

Supply of Blood to the Heart The right and left coronary arteries branch from the base of the aorta to supply the heart wall. Blood is drained into the right atrium by way of the coronary sinus.

HEART PHYSIOLOGY

Cardiac Cycle The contraction of both atria followed by the contraction of both ventricles constitutes a single cardiac cycle. The pressure changes that follow the contraction and relaxation cycles provide the force to move blood.

Heart Sounds The heart sounds are produced by the closing of the heart valves ("lub-dup"): "lub" is the closing of the AV valves, and "dup" is the closing of the SL valves.

Heart Conduction System Each cardiac cycle is stimulated by the sinoatrial (SA) node in the wall of the right atrium. The AV node nearby transmits the signal to the ventricles by way of the AV bundle and Purkinje fibers.

Electrocardiogram (ECG) The ECG is a measurement of the electrical events during a cardiac cycle. It consists of the P wave, which represents atrial depolarization; the QRS complex, which represents ventricular depolariza-tion; and the T wave, which indicates ventricular repo-larization.

Cardiac Output Cardiac output is a measure of heart activity. It is determined by multiplying heart rate (beats per minute) by stroke volume (volume of blood ejected by the ventricles).

Regulation of Heart Activity Controlled by the cardiac reflex center in the medulla oblongata, which sends impulses down parasympathetic fibers to slow heart rate, or down sympathetic fibers to increase heart rate. The medulla receives sensory information on blood pressure from baroreceptors located in the aorta and carotid arteries.

BLOOD VESSELS

The tubes that carry blood, including the arteries, arterioles, capillaries, veins, and venules.

Arteries and Arterioles Three-walled tubes that carry blood away from the heart. Their middle layer (tunica media) contains smooth muscle and elastic fibers, giving them the properties of contractility and elasticity. These properties permit vasoconstriction and vasodilation of arterioles.

Capillaries Microscopic tubes with a single layer of cells making up their walls. Capillaries are the only site of exchange between the blood and interstitial fluid. They form capillary beds in most organs and tissues.

Venules and Veins Three-walled tubes that carry blood toward the heart. Veins have thinner walls than arteries and contain one-way valves that help in the movement of blood. Respiratory and skeletal muscle pumps also assist in blood movement.

Blood Pressure The force exerted by blood against the walls of vessels. It provides the main force for moving blood in arteries and arterioles.

Arterial Blood Pressure Blood pressure within arteries cycles with the beat of the heart. Systolic pressure is the high point caused by ventricular contraction. Diastolic pressure is the low, which parallels ventricular relaxation. Blood pressure is measured by use of a sphygmomanometer.

Pulse A measure of the number of beats per minute by the heart. It can be felt by pressing against a superficial artery, because of the presence of blood coursing through.

Factors Affecting Blood Pressure

Cardiac Output An increase in heart rate or stroke volume increases blood pressure, and a decrease causes a drop in blood pressure.

Peripheral Resistance Vasoconstriction causes an increase in blood pressure, and vasodilation causes a decrease in blood pressure. This is the main influence on blood pressure.

Blood Volume A drop in blood volume causes a drop in blood pressure, and a return of blood volume to normal restores blood pressure.

Regulation of Blood Pressure

Nervous System Controls The medulla oblongata controls heart activity by regulating heart rate, and peripheral resistance by regulating vasoconstriction or vasodilation of arterioles.

Hormonal Controls Epinephrine and norepinephrine increase cardiac output during times of stress. Atrial natriuretic factor promotes a reduction in blood volume. Antidiuretic hormone increases blood volume.

Kidney Controls Kidneys control blood volume, which directly affects blood pressure. They also release renin into the blood, triggering the formation of angiotensin II, which causes vasoconstriction, and release aldosterone, which increases blood volume.

Capillary Exchange In body tissues other than the lungs, oxygen and nutrients diffuse out of a capillary, and carbon dioxide and other waste materials diffuse in. Fluids also move across the capillary wall: net fluid movement is out of a capillary by filtration when the capillary hydrostatic pressure exceeds osmotic pressure, and net fluid movement is into a capillary by absorption when the capillary osmotic pressure exceeds hydrostatic pressure.

CIRCULATORY PATHWAYS

Pulmonary Circulation The pathway between the heart and lungs for the oxygenation of the blood. It consists of the pulmonary trunk, right and left pulmonary arteries, vast capillary beds in the lungs, and four pulmonary veins.

Systemic Circulation The pathway for blood between the heart and all body organs and tissues except the lungs.

Systemic Arteries

Branches of the Aorta

Ascending Aorta Gives rise to the coronary arteries.

Aortic Arch Gives rise to the brachiocephalic artery, left common carotic artery, and left subclavian artery.

Descending Aorta The thoracic aorta gives rise to intercostal arteries. The abdominal aorta gives rise to the celiac artery, the superior mesenteric artery, the suprarenal arteries, the renal arteries, the gonadal arteries, the inferior mesenteric artery, and the common iliac arteries.

Arteries of the Head and Neck The brachiocephalic artery gives rise to the right common carotid artery and the right subclavian artery. Both common carotid arteries give rise to external and internal carotid arteries. Both subclavian arteries give rise to the vertebral and thyrocervical arteries.

Arteries of the Shoulder and Upper Arm The subclavian artery continues as the axillary artery in the axilla, the brachial artery in the upper arm, and the radial or ulnar artery in the forearm.

Arteries of the Pelvis and Lower Limbs The common iliac artery divides into an internal iliac artery and an external iliac artery. The external iliac artery continues down the thigh as the femoral artery, as the popliteal artery behind the kneecap, and as the anterior and posterior tibial arteries in the lower leg.

Systemic Veins

Veins Draining into the Superior Vena Cava

Veins in the Head, Neck, and Trunk Walls The external jugular veins drain the brain. They merge with the subclavian veins—first the external jugular veins, and then the internal. The points of union of the internal jugular veins mark the brachiocephalic veins. The brachiocephalic veins empty into the superior vena cava. The azygos vein unites directly with the superior vena cava.

Veins of the Shoulder and Upper Limbs These empty into the subclavian vein. The deep veins include the radial vein, ulnar vein, brachial vein, and axillary vein. The superficial veins include the basilic vein and the cephalic vein.

Veins Draining into the Inferior Vena Cava

Veins of the Pelvis and Lower Limbs These drain into common iliac vein. The deep veins include the anterior tibial vein, posterior tibial vein, popliteal vein, femoral vein, and external iliac vein. The superficial veins include the great saphenous vein and the small saphenous vein.

Veins of the Abdominal Viscera Veins that drain directly into the inferior vena cava are the gonadal veins, renal veins, suprarenal veins, and hepatic veins. Veins of the hepatic portal system drain into the hepatic portal vein.

HOMEOSTASIS

The cardiovascular system performs the vital function of transporting blood throughout the body. It has regulatory mechanisms that ensure that the transport of blood is held relatively constant during environmental fluctuations, and even during body emergencies.

KEY TERMS

arteriole (p. 384)	diastole (p. 378)
atrium (p. 372)	electrocardiogram (p. 380)
blood pressure (p. 388)	endocardium (p. 371)
capillary (p. 386)	epicardium (p. 371)
cardiac output (p. 390)	myocardium (p. 371)

pericardium (p. 371)	systole (p. 378)
peripheral resistance (p. 391)	vasoconstriction (p. 385)
pulmonary (p. 397)	vasodilation (p. 386)
pulse (p. 389)	ventricle (p. 374)
systemic (p. 397)	venule (p. 386)

QUESTIONS FOR REVIEW

OBJECTIVE QUESTIONS

1. The apex of the heart is located:
   a. against the diaphragm
   b. at the superior margin
   c. at the top
   d. near the larynx

2. The thick outermost layer of the pericardium is called the:
   a. visceral pericardium
   b. myocardium
   c. epicardium
   d. pericardial sac

3. The layer of the heart wall that consists of cardiac muscle tissue is the:
   a. epicardium
   b. pericardial sac
   c. myocardium
   d. endocardium

4. Pectinate muscles and the fossa ovalis may be found in the walls of the:
   a. pericardial cavity
   b. ventricles
   c. right atrium
   d. coronary sinus

5. The chambers of the heart that have a thick myocardium for propelling blood through the body are called:
   a. ventricles
   b. venous sinuses
   c. atria
   d. arterioles

6. AV valves are prevented from everting into the atria during ventricular contraction because of the presence of:
   a. chordae tendineae and papillary muscles
   b. trabeculae carneae
   c. the interventricular septum
   d. semilunar valves

7. Semilunar valves are located between the ventricles and the:
   a. atria
   b. arteries leaving the heart
   c. veins leaving the heart
   d. veins entering the heart

8. Blood normally flows within the heart from the right ventricle through the:
   a. tricuspid valve
   b. pulmonary valve
   c. mitral valve
   d. aortic valve

9. The coronary arteries branch from the _____ to supply the heart wall.
   a. aorta
   b. inferior vena cava
   c. superior vena cava
   d. left ventricle

10. During a cardiac cycle, the state of simultaneous contraction by the ventricles is called:
    a. ventricular diastole
    b. atrial diastole
    c. ventricular systole
    d. atrial systole

11. The sound of the heartbeat is made by the:
    a. blood sloshing back and forth
    b. rhythmic breathing of the lungs
    c. valves closing
    d. heart hitting the thoracic wall

12. The specialized cardiac cells that "set the pace" by initiating each cardiac cycle are clustered in a region called the:
    a. AV bundle
    b. sinoatrial node
    c. AV node
    d. Purkinje fibers

13. The QRS complex in an electrocardiogram represents:
    a. atrial contraction
    b. ventricular relaxation
    c. ventricular contraction
    d. the SA node firing

14. Cardiac output is a measure of heart activity that includes:
    a. heart rate only
    b. stroke volume only
    c. heart rate and stroke volume
    d. none of the above

15. Cardiac output is controlled by an area of the brain called the:
    a. cerebral cortex
    b. midbrain
    c. medulla oblongata
    d. cerebellum

16. Blood vessels that carry blood away from the heart are called:
    a. veins
    b. arterioles
    c. arteries
    d. both (b) and (c)

17. The exchange of materials between the blood and interstitial fluid occurs across the walls of:
    a. the heart
    b. capillaries
    c. arterioles
    d. venules

18. Which layer of the arteriole wall provides for the properties of contractility and elasticity?
    a. tunica media
    b. tunica adventitia
    c. tunica intima
    d. none of the above

19. Which of the following vessels contain one-way valves?
    a. veins
    b. arterioles
    c. venules
    d. arteries

20. Systolic pressure in the aorta is created by the pumping action of the:
    a. arterioles
    b. left ventricle
    c. right ventricle
    d. right atrium

21. Blood pressure increases during:
    a. vasoconstriction
    b. an increase in cardiac output
    c. a rise in blood volume
    d. all of the above

22. The effect of epinephrine and norepinephrine on blood pressure is to:
    a. decrease it
    b. eliminate it
    c. increase it
    d. all of the above

23. The circulatory pathway that results in the oxygenation of blood is the:
    a. pulmonary circulation
    b. hepatic portal circulation
    c. systemic circulation
    d. coronary circulation

24. Arterial blood is supplied to the liver, spleen, stomach, and pancreas by which vessel?
    a. brachiocephalic artery
    b. celiac artery
    c. superior mesenteric artery
    d. hepatic portal system

25. A penetrating wound to the internal jugular vein would affect the return of blood to the heart from which body region?
    a. face and scalp
    b. right arm
    c. brain
    d. stomach and intestines

ESSAY QUESTIONS

1. Trace the pathway of blood through the heart chambers, valves, and major vessels attached to the heart.

2. Describe the mechanisms that operate the AV and SL valves during a normal cardiac cycle. Indicate the origin of the heart sounds.
3. Describe the cycle of contraction and the electrical events during a normal cardiac cycle, and relate them to the waves in an electrocardiogram.
4. Compare and contrast the structures of arteries, arterioles, capillaries, venules, and veins. Relate their structural characteristics to their properties and primary functions.
5. Distinguish between systolic and diastolic pressure.
6. Describe the factors that influence blood pressure and how they are regulated.
7. List the major branches of the aorta.

ANSWERS TO ART LEGEND QUESTIONS

Figure 13–1
Can you identify the location of the apex of your own heart on your chest surface?   To do this, you would place your fingers between the fifth and sixth ribs about four inches to the left of the midline.

Figure 13–2
What layer of the heart wall is the thickest?   The myocardium is the thickest layer of the heart wall.

Figure 13–3
Why does the left ventricle contain the thickest myocardium?   The left ventricle contains the thickest myocardium because it must push blood further than any other chamber.

Figure 13–4
Identify the four chambers from the exterior surface of the heart.   The four chambers are the right atrium, right ventricle, left atrium, and left ventricle. The atria are separated from the ventricles externally by the coronary sulcus, and the ventricles are separated from each other externally by the interventricular sulcus.

Figure 13–5
Which heart valve is located between the left ventricle and the aorta?   The aortic semilunar valve is located between the left ventricle and the aorta.

Figure 13–6
What do you suppose would happen to the flow of blood if the chordae tendineae were suddenly torn from their attachments to the papillary muscles?   If the chordae tendineae tore from the papillary muscles, the blood would be pushed directly into the atrium during ventricular contraction.

Figure 13–7
When the SL valves close, what chambers do they prevent blood from entering?   The pulmonary valve prevents blood from entering the right ventricle, and the aortic valve prevents blood from entering the left ventricle.

Figure 13–8
Oxygen-poor blood enters the heart into which chamber?   Oxygen-poor blood enters the heart through the right atrium.

Figure 13–9
Blood enters the aorta during which contraction cycle?   Blood enters the aorta during ventricular systole.

Figure 13–10
What cluster of cardiac cells initiates each cardiac cycle?   The sinoatrial node initiates each cardiac cycle.

Figure 13–11
What heart activity is represented by the QRS complex wave?   Depolarization of the ventricles is indicated by the QRS complex.

Figure 13–13
What structural characteristics distinguish arteries from veins?   The tunica media of arteries is thicker and contains abundant elastic fibers, the lumen is smaller, and there are no valves.

Figure 13–14
Why do capillaries permit the diffusion of materials while arteries and veins do not?   Diffusion occurs across capillary walls, and not across artery and vein walls, because only capillary walls are extremely thin (and in most cases porous).

Figure 13–15
What role in venous blood flow is provided by valves within the veins?   Valves prevent the backflow of blood (that is, the movement of blood away from the heart).

Figure 13–16
At which point is blood pressure the greatest?   Blood pressure reaches its maximum value in the aorta.

Figure 13–18
What is the effect on blood pressure if cardiac output, blood volume, or peripheral resistance is increased?   When any of these factors is increased, blood pressure rises.

Figure 13–19
At the capillary's arterial end, what factor forces fluid outward? At the capillary's venous end, what factor draws fluid inward?   At the arterial end, the capillary hydrostatic pressure exceeds the osmotic pressure, resulting in more fluid leaving than entering. At the venous end, the osmotic pressure exceeds hydrostatic pressure, causing more fluid to enter than exit.

Figure 13–20

What major artery initiates the systemic circulation? The aorta is the first vessel to receive blood from the left ventricle.

Figure 13–21

What is the overall purpose of the pulmonary circulation? The pulmonary circulation provides for oxygenation of blood.

Figure 13–22

What is the overall purpose of the systemic circulation? The systemic circulation carries oxygenated blood to all body tissues (except the lungs) and returns deoxygenated blood to the heart.

Figure 13–23

What are the major divisions of the aorta? The major divisions of the aorta are the ascending aorta, the aortic arch, and the descending aorta (which includes the thoracic aorta and the abdominal aorta).

Figure 13–24

What is the primary artery supplying the brain with oxygen-rich blood? The common carotid artery (right and left) is the primary artery supplying the brain.

Figure 13–25

What are the branches of the subclavian arteries? The subclavian arteries branch to the vertebral, thyrocervical, and axillary arteries.

Figure 13–26

What are the major branches of the common iliac arteries? The common iliac arteries branch to the internal iliac and external iliac arteries. The external iliac artery continues down the thigh as the femoral artery.

Figure 13–27

What major veins empty into the right atrium? The superior vena cava, inferior vena cava, and coronary sinus empty into the right atrium.

Figure 13–28

The subclavian vein is formed by the merging of what vessels? The subclavian vein is formed by the merging of the axillary and cephalic veins.

Figure 13–29

What prominent vein carries blood from the digestive tract to the liver? The hepatic portal vein transports blood from the digestive tract to the liver.

THE LYMPHATIC SYSTEM

LEARNING OBJECTIVES

After studying this chapter, you should be able to:

1. Identify the function of the lymphatic network.

2. Distinguish between plasma, interstitial fluid, and lymph.

3. Describe the pathway of lymph by identifying the structures it passes through.

4. Describe the structure and function of lymph nodes.

5. Identify the organs of the lymphatic system by their locations and structures.

6. Distinguish between nonspecific defense mechanisms and specific defense mechanisms.

7. Identify and describe the defense mechanisms that are nonspecific.

8. Distinguish between cell-mediated immunity and antibody-mediated immunity.

9. Define the terms *antibody* and *antigen*.

10. Explain the concept of self versus nonself.

11. Describe the types of T cells that arise following sensitization.

12. Describe the types of B cells that arise following sensitization.

13. Describe the processes of cell-mediated immunity and humoral immunity.

14. Describe the types of acquired immunity.

15. Explain the nature of AIDS transmission, and ways to prevent the spread of AIDS.

The lymphatic system is a part of the circulatory system. It plays two key roles in the body: (1) It carries fluid from the extracellular environment into the bloodstream, and thereby helps to maintain the fluid balance; and (2) it performs the primary role in defending the body against disease. The structures that provide these functions are a series of tubes called the lymphatic network; specialized organs, which include numerous lymph nodes, the spleen, the thymus, the tonsils, and Peyer's patches; and the components of immunity. In this chapter we will examine both roles of the lymphatic system, as well as the structures that perform these vital functions.

The Lymphatic Network

The lymphatic network consists of a series of tubes that transport lymph and return it to the bloodstream.

One major function of the lymphatic system is the transport of fluid from the extracellular environment to the body's cardiovascular system. In Chapter 13 we discovered that **interstitial fluid** is formed when plasma from the bloodstream is pushed through capillary walls. This watery fluid serves as a transport medium for gases, nutrients, and other molecules that travel between cells and the blood. Most of the interstitial fluid (about 90%) diffuses back into capillaries for the return trip to the heart, but some of it does not. However, if the fluid is not all returned to the bloodstream, the cardiovascular system will not have sufficient blood volume to operate properly. How is this remaining interstitial fluid returned? As we shall see, the accumulation of interstitial fluid in the extracellular

space generates a pressure gradient, which creates enough force to literally push the remaining interstitial fluid into the **lymphatic network.** This series of veinlike tubes drains interstitial fluid and returns it to the bloodstream in a one-way flow that moves slowly toward the heart. Once the interstitial fluid enters the lymphatic network, it is referred to as **lymph.**[1] The tubes of the lymphatic network include lymphatic capillaries, lymphatic vessels, lymphatic trunks, and collecting ducts.

Lymphatic Capillaries

The lymphatic network begins with microscopic, dead-end tubes known as **lymphatic capillaries** (Fig. 14–1a). These tiny vessels are found between cells and among capillary beds of nearly all tissues and organs in the

---

[1]Lymph: Latin *lympha* = clear spring water.

Figure 14–1
Lymphatic circulation begins with dead-end vessels known as lymphatic capillaries. (a) This diagram shows the flow of interstitial fluid *(arrows)* into blood capillaries or into lymphatic capillaries. (b) A sectional view of a lymphatic capillary. Gaps open between overlapping cells when interstitial fluid pressure rises, permitting the inflow of interstitial fluid.

What effect does the accumulation of small proteins in the extracellular space have on the flow of fluid?

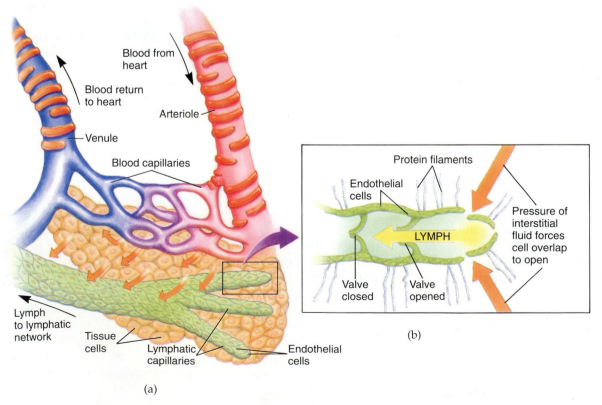

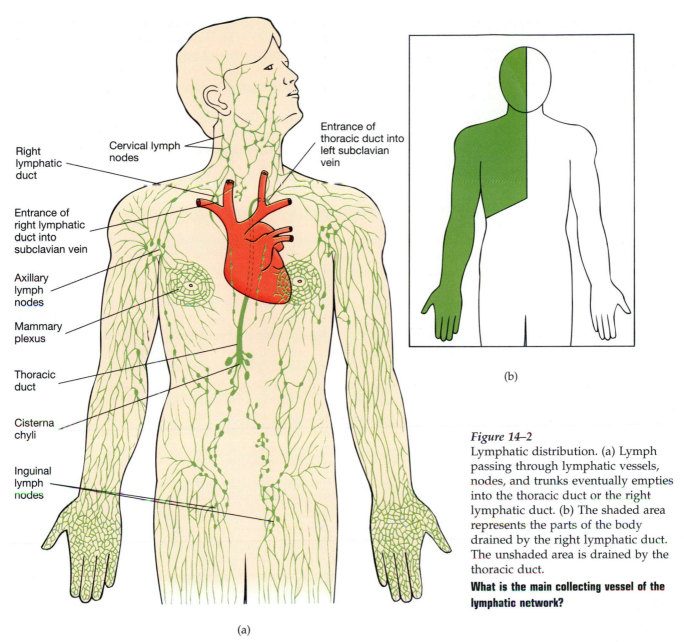

Right lymphatic duct

Cervical lymph nodes

Entrance of thoracic duct into left subclavian vein

Entrance of right lymphatic duct into subclavian vein

Axillary lymph nodes

Mammary plexus

Thoracic duct

Cisterna chyli

Inguinal lymph nodes

(a)

(b)

Figure 14–2
Lymphatic distribution. (a) Lymph passing through lymphatic vessels, nodes, and trunks eventually empties into the thoracic duct or the right lymphatic duct. (b) The shaded area represents the parts of the body drained by the right lymphatic duct. The unshaded area is drained by the thoracic duct.

What is the main collecting vessel of the lymphatic network?

body. Areas where they are not found include the brain and spinal cord, bone tissue, and epidermis.

Lymphatic capillaries are similar in structure to blood capillaries, as their walls consist of a single layer of endothelium. However, the endothelial cells of lymphatic capillaries are more loosely joined and tend to overlap each other (Fig. 14–1b). Also, bundles of protein filaments connect the cells to the extracellular space; the filaments pull on the overlapping cells to create gaps in the wall whenever the pressure in this space rises. These gaps allow the flow of interstitial fluid into the lymphatic capillaries. When the pressure is greater within the capillaries, the protein filaments permit the cells to overlap one another again, preventing lymph from leaking back into the extracellular space.

Lymphatic Vessels

Lymphatic vessels are formed as lymphatic capillaries merge with one another. They contain thicker walls and a larger lumen than lymphatic capillaries and are located closer to the heart. Their structure closely resembles veins, as their thin walls consist of the same three layers of tissue with the inner endothelium forming one-way valves. The valves of lymphatic vessels help prevent the backflow of lymph.

As lymphatic vessels extend toward the heart, they gradually increase in size and merge to form the larger lymphatic trunks and collecting ducts (Fig. 14–2). Along their course they lead to small, oval organs called **lymph nodes** and exit from them to continue on their way.

Lymphatic Trunks and Collecting Ducts

The **lymphatic trunks** are formed from the merging of numerous lymphatic vessels. They drain lymph from a large region of the body, such as the right upper limb or the pelvic region. They extend a short distance toward the heart until they converge to form one of the two **collecting ducts** in the lymphatic network: the thoracic duct or the right lymphatic duct. The distribution of major lymphatic vessels and lymph nodes, lymphatic trunks, and collecting ducts is shown in Figure 14–2.

The **thoracic** (thō-RAS-ik) **duct** is the main collecting vessel for the lymphatic network. It drains lymph from the left side of the head, neck, and thorax; the left upper limb; and the entire body below the diaphragm. It originates in the lower abdomen as an expanded vessel (called *cysterna chyli*) and travels upward against the back wall to the upper chest. The thoracic duct empties lymph into the bloodstream where it unites with the left subclavian vein. A small valve at this union prevents the entrance of blood into the thoracic duct.

The **right lymphatic duct** drains lymph from the right side of the head, neck, and thorax and the right upper limb. It extends a short distance in the neck to empty lymph into the right subclavian vein, with which it unites.

Movement of Lymph

Lymph is propelled through the lymphatic network by pressure gradients. These gradients are established by a number of factors, including the accumulation of protein in the interstitial fluid, skeletal muscle contraction, and breathing movements.

Lymph Formation

As we have already seen, interstitial fluid is formed by the movement of blood plasma out of blood capillaries. Like plasma, interstitial fluid is composed of water and dissolved substances, but it lacks larger substances such as cells and large proteins. However, some smaller proteins are able to leak through the capillary wall at the capillary's arterial end. These proteins thereby join the interstitial fluid. Many of these smaller proteins do not reenter the venous end of capillaries when much of the interstitial fluid is reabsorbed. As a result, the concentration of protein in the remaining interstitial fluid tends to rise, causing the *osmotic pressure* of the fluid to rise as well (recall that osmotic pressure rises when solute concentrations rise; in this case, the proteins are the solutes).

As the osmotic pressure rises in the interstitial fluid, the reabsorption of fluid back into the blood capillaries slows. This occurs because the difference in pressure between the two sides of the capillary wall is reduced. As a result, interstitial fluid begins to accumulate in the extracellular space, and thus its pressure begins to rise. The increase in pressure is great enough to force open the lymphatic capillary pores. Consequently, interstitial fluid flows into the lymphatic capillaries. Once within, the fluid is called *lymph.*

Flow of Lymph

The flow of lymph begins as interstitial fluid moves into the lymphatic capillaries. This movement occurs in response to the pressure gradient that exists between the extracellular space, where it is initially higher, and the inside of the lymphatic capillaries. As lymph begins to accumulate in the blind-ended capillaries, the pressure in these tiny tubes rises. This provides the force to move the lymph into the larger lymphatic vessels, where the pressure is lower. Movement of lymph beyond this point is accomplished by other forces.

Lymph moves through the lymphatic vessels, lymphatic trunks, and collecting ducts in much the same way blood moves through veins. Many of these vessels parallel the course of superficial veins, especially in the upper and lower limbs. They are thus surrounded by skeletal muscles. When these muscles contract and relax, they provide a skeletal muscle "pump" that squeezes and opens the vessels, pushing lymph along. Other vessels are located within the thoracic and abdominal cavities alongside deep veins. Lymph is propelled through them by the rhythmic expansion of the thorax during breathing—the respiratory pump—which squeezes and opens the vessels as the diaphragm rhythmically pushes visceral organs against them. In both cases, the presence of one-way valves directs the movement of lymph in the proper direction: toward the heart.

CONCEPTS CHECK

1. What are the two primary functions of the lymphatic system?

2. How does fluid flow from the extracellular space into lymphatic capillaries?

3. What is the main collecting vessel of the lymphatic network?

OTHER LYMPHATIC ORGANS

Lymphatic organs other than the vessels consist mostly of packed lymphocytes that perform defensive functions.

The vessels of the lymphatic network are true organs, for they contain tissue of more than one type that are combined to perform a general function. Their function, as we have seen, is the transport of lymph. Other organs of the lymphatic system are quite different from all other organs of the body. They are characterized by containing a special type of connective tissue, known as **lymphoid tissue.** This tissue contains collections of **lymphocytes,** a type of white blood cell that was described earlier, in Chapter 12. The lymphocytes play a role in mounting a defensive immune response to foreign materials such as pathogenic (disease-causing) microorganisms. The specialized organs of the lymphatic system include lymph nodes, the spleen, the thymus gland, the tonsils, and Peyer's patches.

Lymph Nodes

Lymph nodes are small, oval masses of lymphoid tissue located along the lymphatic pathways. They receive lymph from lymphatic vessels that drain distal regions and send it along its way toward the heart.

There are hundreds of nodes in the body: nodes are usually found in clusters or chains in the neck (cervical), armpit (axillary), groin (inguinal), and deep within the abdominal cavity. Their distribution is shown in Figure 14–2.

Lymph Node Structure

Lymph nodes vary in shape and size, but most of them are shaped like kidney beans and are less than 2.5 cm (1 inch) in length. An example of a lymph node is shown in Figure 14–3a.

A lymph node is a receiving and sending station for lymph. Along its convex margin it receives incoming lymph by way of lymphatic vessels that merge with it. These vessels are called **afferent lymphatic vessels.** At its concave margin, known as the **hilus,** the node sends lymph on its way toward the heart by way of **efferent lymphatic vessels.**

Each lymph node is kept structurally intact by an outer shell of dense connective tissue, known as the **fibrous capsule.** The protein fibers of this shell extend inward to divide the node into many compartments. Within these compartments are compact clusters of lymphocytes, which form the main structure of the node. They are called **lymph nodules.**

When a lymph node is sectioned, we can see that its internal structure consists of two portions: an outer area, called the **cortex,** and an inner part, known as the

Figure 14–3

Structure of a lymph node. (a) A cross section through a node shows its internal organization. Lymphocytes and other white blood cells are located within its internal chambers, or sinuses. (b) A photomicrograph of a sectioned lymph node (approximately 5x).

What happens to lymph as it is channeled through each lymph node?

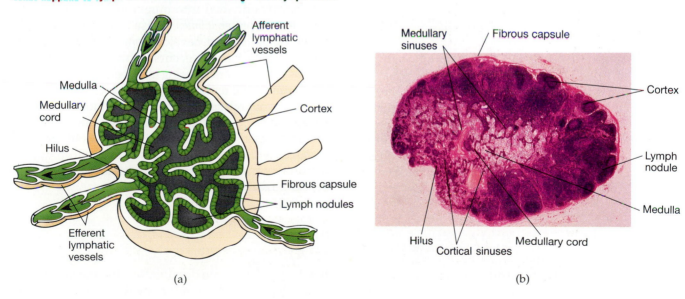

(a)   (b)

medulla (Fig. 14–3). The cortex consists of many lymph nodules, where lymphocytes and macrophages are located. It is also the site of lymphocyte maturation. In the medulla, the lymphocytes are arranged in strands called **medullary cords.**

Lymph Node Function

As lymph is channeled into and through lymph nodes on its way toward the heart, microorganisms and other foreign particles carried by the lymph are filtered and removed. These health-threatening substances—which include bacteria, toxins released by bacteria, and viruses—are inactivated by the large numbers of lymphocytes that are packed within the lymph nodes. Macrophages are also present; they engulf damaged cells and cell debris. These two types of white blood cells thus provide an important filtering mechanism for cleansing lymph as it flows through the node. The activities of lymphocytes and macrophages will be described later this chapter.

Spleen

The **spleen** is the largest of the lymphatic organs. It is located in the left side of the abdominal cavity just below the diaphragm and curls slightly around to the anterior side of the stomach (Fig. 14–4).

In many ways the spleen resembles an oversized lymph node. It is enveloped by a **fibrous capsule** whose protein fibers divide it internally into compartments, many of which contain lymphocytes and macrophages. However, this is where the similarities end. The spleen also contains venous sinuses that are normally filled with large numbers of red blood cells. These areas are referred to as **red pulp.** The regions of the spleen that contain white blood cells are called the **white pulp.**

The spleen functions as a large filter for removing foreign particles, old and defective red blood cells, and platelets from the bloodstream. These functions are accomplished by the lymphocytes and macrophages within the white pulp. It also provides a blood reservoir, as it stores a large volume of blood that can be called upon during emergency situations involving blood loss. This secondary function is performed by the red pulp.

Thymus Gland

The **thymus gland** is an organ we have studied before, since it plays a role in endocrine function (Chapter 11). In infants, it is a large, bilobed gland located in the thoracic cavity, where it lies behind the sternum (Fig.

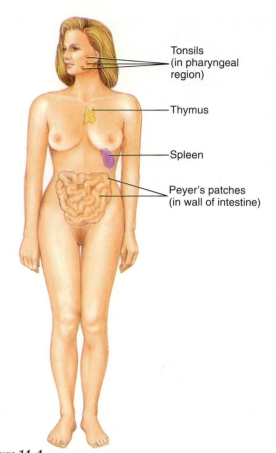

Figure 14–4
Locations of major lymphatic organs.
What type of white blood cell makes up much of the substance of the lymphatic organs?

Tonsils (in pharyngeal region)
Thymus
Spleen
Peyer's patches (in wall of intestine)

14–4). In adults it is very much reduced in size and function.

The thymus gland is composed of lymphoid tissue, which is divided into numerous clusters of lymphocytes known as *lobules*. These cells are active only during times of rapid development of the immune response, which usually occurs between the ages of 6 months and 5 years. During this period, the thymus cells promote the maturity of lymphocytes into T lymphocytes and release them into the bloodstream. T lymphocytes, or T cells, play an important role in the immune response, which we will learn about soon. By adolescence the thymus gland begins to atrophy, and by adult age it has been largely replaced by fibrous and fatty tissue.

Tonsils

The **tonsils** are located in the mouth and throat, where they lie partially embedded within the mucous membrane (Fig. 14–4). They are named according to their location: the two **palatine** (PAL-a-tīn) **tonsils** are on the back end of the palate; the two **pharyngeal** (fa-RIN-gē-al) **tonsils** lie against the wall of the upper throat (nasopharynx); and the two **lingual** (LĒNG-wal) **tonsils** are positioned at the base of the tongue.

The tonsils are composed of lymphoid tissue with lightly scattered lymphocytes within. These cells gather and engulf disease-causing microorganisms in the mouth and throat regions. During an infection, the palatine tonsils in particular may enlarge because of the proliferation of bacteria and white blood cells. This produces the sore throat symptoms with which you may be well acquainted. Sometimes bacterial invasion into channels leading inside a tonsil becomes a chronic problem, and the tonsils may be surgically removed by a procedure called *tonsillectomy*.

Peyer's Patches

Peyer's patches are isolated clusters of lymphoid tissue in the wall of the small intestine near its distal end. They contain lymphocytes and macrophages in large numbers. It is believed that these cells identify microorganisms and other foreign particles that reside in the intestine. It has also been proposed that the macrophages may provide a defensive barrier, preventing the movement of bacteria across the intestinal wall and into the abdominal cavity.

CONCEPTS CHECK

1. What is the function of lymph nodes?

2. In what ways are the lymphatic organs (other than vessels) similar?

3. Where are the tonsils and the spleen located?

THE DEFENSE MECHANISMS OF THE BODY

The defense mechanisms of the body guard against the disruption of homeostasis by nonself cells and their harmful products. They include nonspecific mechanisms and specific mechanisms.

Our bodies are constantly exposed to substances that can harm us. These substances include pathogenic microorganisms, toxic molecules, foreign particles, and even our own cells that are dying or diseased. Pathogenic microorganisms, or **pathogens,** are disease-causing agents that include viruses and living bacteria, fungi, and protozoans. Pathogens may cause harm by physically destroying cells, or by releasing poisonous substances that interfere with cell function. These substances are harmful molecules called **toxins.** Pathogens, toxins, and other harmful cells and particles can severely disrupt homeostasis if left unchecked by the body's defense strategies, leading to any one of a variety of immunological diseases.

The defensive measures that are used to ward off or control these threats to homeostasis are numerous and, in fact, are still under careful investigation through research efforts in the field of *immunology*. Current work in this growing field indicates that many of the defensive strategies used by the body rely on an ability to distinguish cells that belong to the body, known as **self cells,** from those that do not, called **nonself cells.** Amazingly, white blood cells are able to make this distinction by detecting the distribution and type of surface membrane proteins (receptors) that characterize cells. That is, all self cells contain a similar distribution of membrane receptors which can be read by white blood cells as "code" and which differs from the distribution of membrane receptors found in nonself cells. The specific distribution of membrane receptors is determined by a group of genes in the chromosomes of all self cells, called **major histocompatibility complex (MHC).** The proteins that are coded by these genes attach to the plasma membrane to provide the cell with its self identification; these proteins are called **MHC proteins.**

Immunologists distinguish between two general types of defensive strategies that the body normally employs. They are called **nonspecific mechanisms** and **specific mechanisms.** As we shall see, the lymphocytes and macrophages that compose lymphoid tissue play the key roles in both of these defense mechanisms.

Nonspecific defense mechanisms help prevent the entrance of foreign materials into the body. If the invaders are successful in gaining entrance, other nonspecific mechanisms seek to destroy them without concern for the type of foreign substance they may be. An example is the unselective destruction of all bacteria by phagocytosis.

Specific defenses key into the particular type of pathogen or toxin that is infecting the body. Once the invaders have been identified, they may be inactivated by a variety of defensive mechanisms. The selective production of special proteins, called **antibodies,**[2] is one way of inactivating the invaders. The specific defense mechanisms are collectively referred to as the **immune response.**[3]

---

[2]Antibodies: *anti-* = against + *body.*

[3]Immune: *immune* = safe, or exempt.

Nonspecific Mechanisms

Nonspecific mechanisms protect the body by defending against all types of invaders unselectively in a variety of ways.

There are many defense mechanisms that are nonspecific. The most important include physical barriers, phagocytosis, natural killer cells, proteins, and inflammation. The components that play major roles in these mechanisms are summarized in Table 14–1.

Physical Barriers

Physical barriers prevent the entrance of foreign substances into the body. The most important barrier is the **skin,** which is regarded as the body's first line of defense. As long as the skin remains unmarked by injury, the thick layer of keratin that forms its outer layer provides a surface barrier that is impenetrable to bacteria, viruses, and most toxic substances.

The **mucous membranes** provide a second important physical barrier. These membranes line all body cavities that open to the exterior. In addition to serving as barriers, mucous membranes also produce chemicals that help in the defensive strategy. For example, mucus produced in the intestinal and respiratory tracts traps microorganisms that gain entrance.

In general, an invasion by a pathogen that successfully penetrates the body's physical barriers is called an **infection.**

Phagocytosis

As we learned earlier, **phagocytosis** is the ingestion and destruction of particles by specialized cells. These cells contain flowing cytoplasmic extensions that bind to the particle and pull it within (Fig. 14–5). Once collected, the particle is digested by vacuoles inside the cell. A cell that performs this function is called a **phagocyte.**

Phagocytosis is a type of nonspecific defense because the white blood cells performing this function do not distinguish between different *types* of foreign cells. However, their phagocytic activity must be controlled to some extent to avoid the destruction of normal, healthy body cells. This is done by the selective

Table 14–1 COMPONENTS OF NONSPECIFIC MECHANISMS

Component	Description	Function
Skin and mucous membranes	Skin is an organ that covers the body surface. Mucous membranes are membranes covering openings into the body.	Provide a protective, impenetrable covering against the external environment
Monocytes	Large, active white blood cells, which are immature forms of macrophages.	Phagocytosis of foreign particles and cell debris
Macrophages	Large, usually stationary white blood cells attached to the linings of vessels, spleen, and lymph nodes. They originate as monocytes.	Phagocytosis of foreign particles and cell debris
Neutrophils	Active white blood cells.	Phagocytosis of foreign particles and cell debris
Natural killer cells	Active white blood cells, considered to be a form of lymphocyte.	Destruction of nonself cells by lysis
Complement	Protein molecules in the plasma.	Labels unwanted cells, and attracts white blood cells to a site of infection
Interferons	Protein molecules released by virus-infected cells.	Bind to plasma membranes to prevent further virus infection
Inflammation	Vasodilation and increased permeability of blood vessels.	Improves movement of white blood cells and other factors to a site of injury

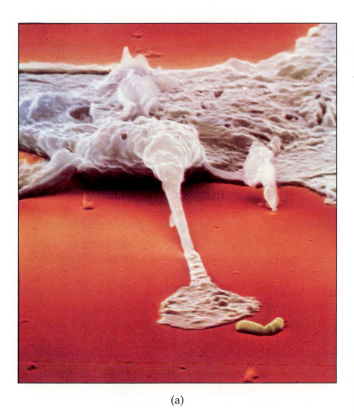

(a)

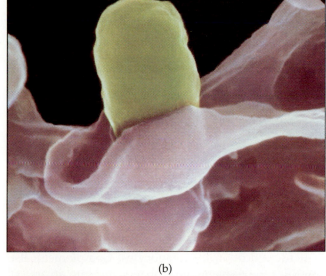

(b)

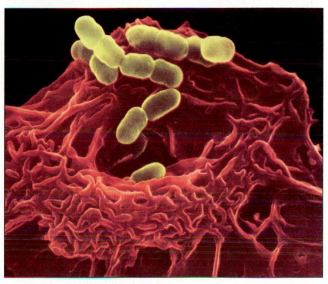

(c)

Figure 14–5
The macrophage is very active in phagocytosis. (a) A macrophage (*gray*) extends a pseudopod toward an *Escherichia coli* bacterium (*green*), which is beginning to divide. (b) The bacterium is trapped by the pseudopod. (c) The macrophage draws bacteria through a channel in its plasma membrane. Its membrane will seal over the bacteria, and lysosomal enzymes will destroy them.

How does a phagocyte distinguish between healthy cells and unwanted cells and particles?

ability of phagocytes to ingest only those cells or particles that are labeled with special proteins (other than MHC proteins). The nature of these proteins will be discussed later, but they include antibodies and complement. These proteins serve to identify the foreign cell or particle as being unwanted. Once labeled and identified, the unwanted cells or particles are ingested by the phagocyte.

Phagocytosis is performed by most white blood cells. The most active in this defensive activity are **monocytes** and **neutrophils,** which squeeze through blood vessel walls to reach sites of infection and inflammation. Other phagocytes may not be as active as these wandering cells. For example, monocytes give rise to their more mature form, known as **macrophages,** which often become attached to the walls of lymphatic vessels and other lymphatic organs.

Macrophages may also wander freely throughout the circulatory system, or they may fix themselves to the walls of blood vessels, the liver, and the lungs. Attached macrophages constitute an important part of the body's defense network, known as the **mononuclear phagocytic system.** As lymphatic fluid or blood moves by them, the cells of the mononuclear phagocytic system phagocytize foreign particles, thus cleansing the passing fluid.

Natural Killer Cells

Natural killer (NK) cells are a type of white blood cell that kills invading foreign cells nonspecifically by a method other than phagocytosis. Natural killer cells are unique in that they kill cancer cells and virus-infected cells by puncturing a hole in their plasma membrane (a process called *cell lysis*). These actively wandering cells are able to detect unwanted body cells by the disrupted sequence of membrane (MHC) proteins, which is caused by their diseased state. It is currently thought that natural killer cells are a type of lymphocyte.

Proteins

Certain groups of proteins that are normally found in the bloodstream provide an important role in nonspecific defense. One group is called **complement;** it includes more than 20 types of plasma proteins that are normally in an inactive state. Complement becomes activated by the onset of an infection. Once activated, it serves to mediate the destruction of foreign substances in the body by labeling them as unwanted, enabling their identification by phagocytes. Complement also amplifies the movement of phagocytes into an area of infection and stimulates their destructive capabilities, and may, in some cases, cause cell lysis directly.

A second group of proteins important in nonspecific defense are the **interferons** (in-ter-FĒR-onz). They are the body's main defense against viruses. Interferons are secreted by cells that have become infected by viruses. They diffuse to nearby cells and bind to their membranes. This causes an interference in the ability of viruses to proliferate in these cells. At one time interferons represented hope of a possible cancer cure, because of their antiviral activity and mounting evidence pointing to the viral cause of certain cancers.

However, clinical trials using a synthetic form of interferon have demonstrated only a limited effect on some tumors, such as melanoma and kidney cancer, and no effect on others, while producing side effects that can seriously disturb homeostasis. The single exception that responds to interferon therapy is a particular type of bone marrow cancer called *hairy-cell leukemia;* interferon helps to destroy cells of this cancer.

Inflammation

Inflammation is a response to body *stress,* or disruption of homeostasis, which often follows an infection or physical injury. It is a nonspecific mechanism because it may occur in any tissue of the body. It aids the defensive process by preventing the spread of the infectious agents to nearby tissues, removing dead and damaged cells, and setting the stage for tissue repair. The responses that characterize inflammation represent an attempt by the body to restore homeostasis.

The inflammatory responses are initiated when a damaged cell releases substances into the bloodstream. These substances include histamine and serotonin and produce two main responses by local tissues: vasodilation of blood vessels (widening of the blood vessel diameter) and an increased permeability of blood vessels (widening of pores in capillary walls). These responses produce four symptoms: redness, swelling, heat, and pain (Fig. 14–6). Immediately following an injury, redness appears on the skin's surface as a result of the vasodilation of local blood vessels, which leads

Figure 14–6
The process of inflammation. Its responses serve to flag immune mechanisms to the site of an injury.
What changes during inflammation promote the movement of phagocytes to the injury site?

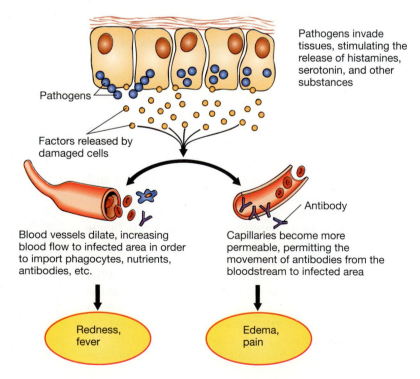

Pathogens invade tissues, stimulating the release of histamines, serotonin, and other substances

Pathogens

Factors released by damaged cells

Antibody

Blood vessels dilate, increasing blood flow to infected area in order to import phagocytes, nutrients, antibodies, etc.

Capillaries become more permeable, permitting the movement of antibodies from the bloodstream to infected area

Redness, fever

Edema, pain

to an increased flow of blood to the site. The added blood flow brings defensive substances, such as white blood cells, antibodies, and clotting proteins, to the injury site. It also improves the removal of dead cells, microorganisms, and toxins. The swelling is the result of fluids leaking across the capillary walls into the extracellular space, filling the space up like a water balloon. The leaking of fluids into the extracellular space is also known as **edema.** Heat is produced by the increase in blood flow from deeper areas of the body where the blood temperature is warmer. The pain is caused by pressure on local pain receptors by the increased fluid volume.

The vasodilation of local blood vessels serving the site of infection or injury, as we have seen, brings in more blood. This blood flow carries with it white blood cells, which remain at the site and accumulate to fight off invading microorganisms. Their tendency to accumulate in a local site produces a thick fluid known as *pus*, which contains living as well as nonliving white blood cells. Additional platelets and fibrinogen are also carried with the blood; they provide the clotting mechanism and begin the process of tissue repair.

CONCEPTS CHECK

1. How does a phagocyte distinguish between unwanted cells or particles and healthy cells?

2. What is the body's defense against viruses?

3. How does inflammation help defend the body from infection?

Specific Mechanisms: The Immune Response

The immune response relies upon the ability of lymphocytes to recognize specific antigens. Once they are recognized, a variety of defense strategies can be used against them.

Specific defense mechanisms involve responses that are directed against a particular pathogen or toxin. They involve the activity of a specialized cell, or the production of a certain antibody, that is capable of reacting against only a specific microbe or toxin. For example, a specific antibody that is distinct from all other types is produced for every particular type of toxin the body identifies.

The defense mechanisms that are specific constitute the **immune response** of the body, or **immunity.** They provide us with a complex line of defense against disease-causing microorganisms and their toxins that have gained entrance into the body. Two types of immune responses have been recognized, although they overlap in some ways. They are **cell-mediated immunity,** in which cells provide the main defensive strategy; and **humoral immunity,** which utilizes antibodies as the main weapon against invaders. The two immune responses will be examined further, but first we will consider the various components of immunity.

Components of Immunity

Components of immunity include antigens, antibodies, and the various types of lymphocytes. Both types of immune responses involve at least some of these components.

Antigens

Antigens[4] are substances that provoke an immune response when they enter the body. They are recognized by white blood cells as foreign. They are usually large molecules, such as proteins, polysaccharides, or nucleic acids. Antigens include toxic molecules released by microorganisms, or the whole microorganisms themselves. A whole microorganism is identified as an antigen because it contains antigenic (nonself) molecules on its surface membrane. Other substances also act as antigens, including chemicals injected in insect bites and stings, molds, certain foods, and cells and tissues that have been transplanted or transfused from another person. When the immune response mistakenly recognizes a person's own cells as being antigens, it causes a chain of events leading to the destruction of healthy cells and tissues. This type of immune failure is known as an **autoimmune disease** (see page 430 for a further discussion of autoimmune diseases).

Antibodies

An antibody is a protein molecule that is produced in response to a specific antigen. It binds to the antigen to form an **antigen-antibody complex,** which inactivates the toxic effects of the antigen. Antibodies are produced by a class of lymphocytes, as you will see shortly.

Antibodies belong to a family of proteins known as **immunoglobulins** (abbreviated **Ig**). There are five recognized classes of Ig, each of which has a distinctive structure and is involved in a different type of response. They are listed and described in Table 14–2. Within each class, additional types of antibodies can be synthesized in order to form a specific chemical

---

[4]**Antigens:** Greek *anti-* = against + *gennan* = to produce.

Table 14–2   THE FIVE CLASSES OF ANTIBODIES (IMMUNOGLOBULINS)

Antibody Class	Location	Response
IgG	In the blood, lymph, and intestines	Stimulate phagocytosis of bacteria and viruses, neutralize toxins, and trigger release of complement
IgA	In tears, saliva, mucus, milk, blood, and lymph	Provide localized protection of mucous membranes
IgM	In blood and lymph, and attached to surfaces of B cells	Stimulate lysis of bacteria in blood and lymph
IgD	In blood and lymph, and attached to surfaces of B cells	May stimulate the production of more antibodies
IgE	Attached to surfaces of basophil cells, mast cells, and B cells	Involved in the allergic response

match with the antigens they must bind. The structure of every antibody can therefore be modified slightly by the cell that produces it. As the body encounters new forms of antigens, new types of antibodies that can bind to them are synthesized and released.

Lymphocytes

The most important type of white blood cell in immunity is the **lymphocyte.** Lymphocytes originate from the red bone marrow before birth, where they may be found in their immature, nonfunctional forms (Fig. 14–7). Soon after their production, about half of them migrate to the thymus gland. Once within the tissues of the thymus gland, these cells mature into a form of lymphocyte called **T cells** (the *T* is from *thymus*). The remaining immature cells may remain in the bone marrow, or they may migrate to the spleen or to Peyer's patches in the wall of the small intestine, where they mature into a second form of lymphocyte known as **B cells** (the *B* is from *bursa* of Fabricius, a small pouch where B cells develop in certain birds).

Soon after this maturation process, T cells and B cells leave their site of development and migrate to lymphoid tissue (usually lymph nodes, but also the spleen, Peyer's patches, and bone marrow). Here they become "programmed" to recognize self cells and distinguish them from nonself cells. They are able to make this distinction by detecting the distribution and type of membrane receptors (MHC proteins) that characterize cells. Once a lymphocyte has developed this ability, it is called **immunocompetent.** Thus, an immunocompetent lymphocyte will attack a bacterium but not an epithelial cell belonging to the body.

T CELLS   T cells become available to defend the body (that is, become immunocompetent) shortly before or

after birth, when they migrate from the thymus to lymphoid tissue in the lymph nodes, spleen, bone marrow, and other areas, where they remain established for the life of the individual. Once established, the mature T cells continuously undergo mitosis to maintain the

Figure 14–7
Development of lymphocytes. T cells arise from red bone marrow early in life and migrate to the thymus gland, where they develop. After birth they migrate to lymphoid tissue, where they become immunocompetent. B cells also arise from red bone marrow; they then mature within it or in the spleen or Peyer's patches in the wall of the small intestine. Like T cells, they become immunocompetent in lymphoid tissue (mainly lymph nodes).

Where do both types of lymphocytes originate?

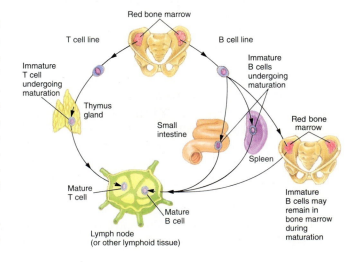

HEALTH CLINIC

Transplantation and Tissue Rejection

When a tissue or an organ in a patient is injured or diseased beyond repair, one alternative available to restore health is **transplantation.** In this procedure, some body part such as the heart, kidney, or liver may be surgically transferred from one person to another. Unfortunately, the recipient's body usually recognizes the proteins in the transplanted material as foreign (antigens). As a result, immunologic response is launched against the transplant. This reaction is called **tissue rejection.**

Tissue rejection can result in the complete destruction of transplanted tissue, and, in extreme cases, death of the patient. The rejection response can be minimized, however, by matching donor and recipient MHC proteins (a procedure similar to matching blood types prior to a blood transfusion). The use of immunosuppressive drugs also minimizes tissue rejection. These drugs suppress the recipient's immune response, usually by the destruction of lymphocytes. Unfortunately, the large-scale destruction of lymphocytes makes the recipient susceptible to infectious diseases. To offset this side effect, recipients are usually given antibiotic therapy until the lymphocytes have been replaced. However, a recently developed drug derived from a fungus, called *cyclosporin A,* has immunosuppressive capabilities without the large-scale destruction of lymphocytes. To some extent, this new drug has enabled heart, liver, and kidney transplants to become a routine event. Thanks largely to this immunosuppressive agent and improved surgical techniques, in 1989 (the most recent year for available statistics) there were 8890 kidney, 2160 liver, 1673 heart, 413 pancreas, and 67 heart-lung transplants in the United States alone.

colony and send newly produced T cells throughout the body. As a general rule, T cells are very active cells that can destroy nonself cells in a variety of ways.

Many types of T cells are normally present. These types have arisen from the activation, or *sensitization,* of mature immunocompetent T cells within lymphoid tissue. Sensitization begins when a macrophage circulating in the bloodstream identifies an antigen, phagocytizes it, and processes it within its cytoplasm (Fig. 14–8). The processed antigen is expressed onto the macrophage plasma membrane, along with MHC proteins. The macrophage then presents the processed antigen to a T cell, which responds by enlarging and dividing.

A T cell that has become sensitized develops special properties that enable it to react against the particular antigen. These properties are passed on to its descendent cells, producing a cell line with identical properties. Because the cells within a particular cell line are identical, they are known as **clones.** The most important T cell lines include killer T cells, helper T cells, suppressor T cells, and memory T cells which are shown in Figure 14–8.

Killer T cells are specialized in destroying virus-infected cells, cancer cells, and foreign cells (that may

Figure 14–8

Sensitization and differentiation of T cells in lymphoid tissue.

What are the four types of differentiated T cells?

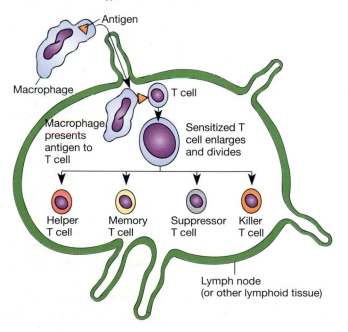

have entered the body by transplantation or grafting). They bind to these cells once they are recognized and release toxic chemicals called *lymphotoxins*.[5] The lymphotoxins cause the nonself cell to rupture. Killer T cells are also called *cytolytic T cells*.

Helper T cells stimulate the defensive activities of other cells. When they identify an antigen, they release proteins into the bloodstream that signal other cells to attack the invader. These proteins, called *lymphokines*,[6] stimulate killer T cells to grow and divide, attract neutrophils and monocytes to the site of intrusion, and enhance the ability of macrophages to ingest and destroy nonself cells. Helper T cells also stimulate the growth and division of B cells, and their production of antibodies. Thus, in a sense, the helper T cells orchestrate the defensive symphony of the body.

Suppressor T cells are vital for slowing and eventually stopping the defensive mechanisms that are called to action by the helper T cells. They release proteins that inhibit the helper T cell activity, indirectly suppressing killer T-cell, neutrophil, monocyte, and macrophage responses and reducing antibody production by B cells. This helps prevent uncontrollable or unnecessary activity that can lead to the destruction of healthy self cells.

Memory T cells provide the immune response with a "memory" of the specific antigen that caused the initial sensitization. When a second exposure to the same antigen occurs later, the time required for sensitization of the T cells and their subsequent division into the various cell lines is very much reduced.

B CELLS Mature, immunocompetent B cells arise following a period of development within red bone marrow, the spleen, or Peyer's patches in the intestinal wall (see Fig. 14–7). Like T cells, they migrate to lymph nodes and other lymphoid tissue, where they accumulate. They are not as numerous in circulating blood as T cells, accounting for only about 20% of the lymphocyte population in a normal blood sample. However, their role in immunity is very important.

Sensitization of a B cell begins when a macrophage circulating in the blood binds to an antigen, processes the antigen within its cytoplasm, and expresses the processed antigen on its plasma membrane. The processed antigen on the macrophage membrane is then presented to a B cell within lymphoid tissue (Fig. 14–9). If the B cell has received a chemical signal from a helper T cell, the binding of a B cell to the

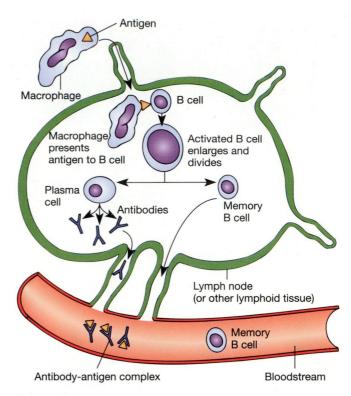

Figure 14–9
Sensitization and differentiation of B cells in lymphoid tissue.
What cells produce antibodies?

processed antigen sensitizes it, causing it to enlarge in a manner similar to T cells. In B cells, however, most of the descendents of the sensitized cell become **plasma cells.** These special cells experience a period of rest following their production and then swing into action to synthesize vast amounts of antibodies at an incredible rate (2000 antibody molecules per second!). The plasma cells are the only cells of the body that produce antibodies.

Descendents of the original sensitized B cells that do not develop into plasma cells form into **memory B cells.** These cells are found mainly in the bone marrow, with smaller numbers in the lymph nodes and spleen. They provide a "memory bank" of sensitized cells that can respond to the same antigen when it is encountered again. When this occurs, the immune response is much faster because the antibodies can be immediately synthesized. As a result, vast numbers of antibodies appear in the bloodstream hours after the initial contact with the recognized antigen.

The various components involved in specific mechanisms of immunity are summarized in Table 14–3.

---

[5]Lymphotoxins: *lympho-* = pertaining to a lymphocyte + *toxin* = poison.

[6]Lymphokines: *lympho-* = pertaining to a lymphocyte + Greek *kinein* = to move.

Table 14–3 COMPONENTS OF SPECIFIC MECHANISMS OF DEFENSE

Component	Description	Function
Monocytes/macrophages	Monocytes are large, active white blood cells that are immature macrophages	Macrophages are the first cells to identify an antigen and present it to a T cell or B cell. They are also phagocytic.
T Cells	Immunocompetent lymphocytes that have matured within the thymus gland	Active in cell-mediated immunity.
Killer T cells	A cell line of T cells following sensitization	Release lymphotoxins that destroy nonself cells.
Helper T cells	A cell line of T cells following sensitization	Release lymphokines that stimulate killer T cells, attract monocytes and neutrophils, and stimulate macrophages. They also stimulate B cells and humoral immunity.
Suppressor T cells	A cell line of T cells following sensitization	Inhibit the activity of helper T cells.
Memory T cells	A cell line of T cells following sensitization	Store information on the antigen structure.
B Cells	Immunocompetent lymphocytes that have matured in the red bone marrow or in Peyer's patches in the intestinal wall	Active in humoral immunity.
Plasma cells	A cell line of B cells following sensitization	Produce antibodies.
Memory B cells	A cell line of B cells following sensitization	Store information on the antigen structure.
Antibodies	Protein molecules called *immunoglobulins*, produced only by plasma cells	Chemically bind to specific antigens.

Cell-Mediated Immunity

Cell-mediated immunity is provided by the various types of T cells. It is summarized in Figure 14–10. This immune response is initiated when a macrophage identifies an antigen in the body, phagocytizes it, and processes it. The processed antigen is soon expressed on the macrophage plasma membrane. The macrophage then presents the antigen to a T cell, resulting in T-cell sensitization. The sensitized T cell then proliferates into the various T-cell lines (killer, helper, suppressor, and memory).

The killer T cells that result are capable of killing only one specific type of nonself invader cell. The lymphotoxins they release are specialized to rupture whole cells, such as bacteria and virus-infected human cells, but they cannot complete the task by themselves. They are assisted by helper T cells, which release lymphokines that call into action phagocytes (neutrophils, monocytes, and macrophages), stimulate the production of more killer cells, and activate humoral immunity. The suppressor T cells modulate the events by inhibiting cell activity, preventing these destructive cells from damaging healthy tissue. Memory T cells serve as a rapid recall mechanism to activate cell-mediated immunity during subsequent invasions by the same antigens.

Humoral Immunity

B cells provide the body with the humoral immune response, which is diagramed in Figure 14–11. This process begins when a macrophage identifies an antigen in the body, phagocytizes it, and processes it. The

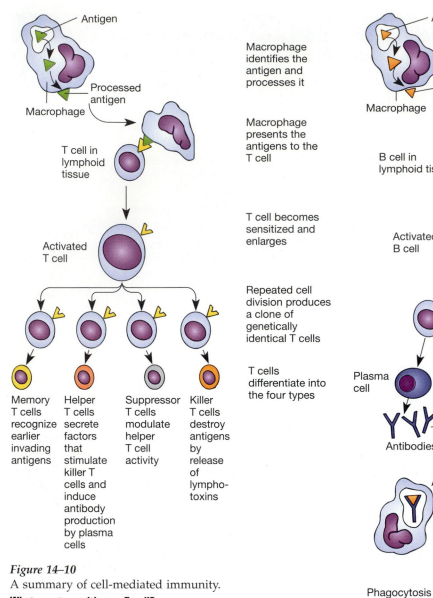

Figure 14–10
A summary of cell-mediated immunity.
What event sensitizes a T cell?

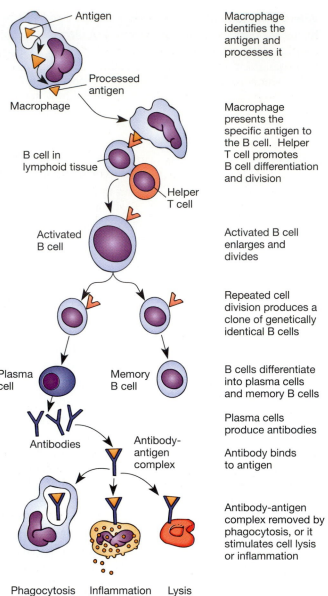

Figure 14–11
A summary of humoral immunity.
What type of T cell is needed to induce B cell sensitization?

macrophage soon expresses the processed antigen on its plasma membrane and presents this nonself substance to a B cell. Following contact with the antigen on the macrophage membrane, the B cell becomes sensitized to begin its production of the B cell lines. B cells also require a helper signal from helper T cells in order to be activated. B cells then proliferate to become antibody-secreting plasma cells and memory B cells.

The plasma cells produce and release vast numbers of antibodies following a lag period that lasts about two weeks. This productive flurry of activity lasts for only three to five days; then the plasma cells die. Antibody levels in the blood peak in about three weeks following the initial sensitization and then slowly decline to normal levels.

The antibodies that are produced by the plasma cells selectively bind to the specific antigens found on the nonself invader cell. The antibody-antigen complex that is formed inactivates the antigen and the nonself cell to which they are attached. However, the antibodies alone do not usually kill the nonself cell, although in some cases they may do so by direct lysis. The actual destruction of the nonself cell is usually accomplished by complement, a monocyte, a macrophage, or a natural killer cell. The antibodies that have bound to the antigens on the invader cell serve as a "flag," as they signal to these nonspecific proteins and cells to kill the cell they have attached to. Once activated, complement may destroy the unwanted cell by lysing it, monocytes and macrophages may destroy it by phagocytosis, or natural killer cells may destroy it by poking a hole through its membrane.

Memory B cells are also produced following sensitization of the B cells. These cells store the information needed to reproduce the specific type of antibody that worked against the antigen. Whereas plasma cells disappear in the bloodstream shortly after the immune response has ended, memory B cells remain for several years. The process of antigen identification, B cell sensitization, plasma cell proliferation, antibody production, and, finally, nonself cell destruction normally takes about three weeks. When an antigen is encountered that was previously identified, memory B cells recognize it. They rapidly produce plasma cells that synthesize and release the proper antibodies. As a result, the nonself invader cells are destroyed in only one to three days. The ability of the immune response to activate itself quickly when exposed to secondary invasions is called **immunization.**

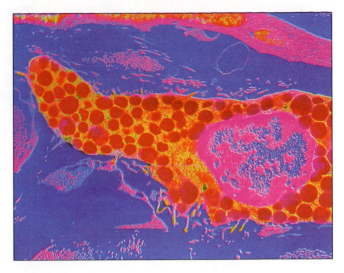

Figure 14–12
Electron micrograph of a mast cell releasing histamines in response to antigen hypersensitivity.

The Allergic Response

The **allergic response** is a type of humoral response, for it involves the release of antibodies by plasma cells. In most people, it provides a tolerance for many nonself molecules that come into contact with the skin and mucous membranes. About 10 percent of individuals are intolerant of many of these surface antigens, however. These individuals are said to be **allergic,** and the antigens they form a body reaction to are called **allergens.**

A person who is allergic experiences an immunological reaction to an allergen that differs from that of someone who is not. The plasma cells of an allergic person secrete a different type of antibody (called IgE) from those secreted by nonallergic individuals (called IgG or IgM). The reason for this defect in immunological regulation is not known. The IgE antibodies are secreted into the bloodstream in response to exposure to a particular allergen. They eventually come in contact with a type of cell called a **mast cell.**

Mast cells are found throughout the body but are most concentrated in the skin and mucous membranes. These cells contain numerous membrane receptors for IgE and store chemicals that cause inflammation, including histamine, serotonin, and prostaglandins. When IgE binds to a mast cell, it releases its inflammatory chemicals, which act on surrounding blood vessels and other tissues to cause inflammation and irritation (see Figure 14–12). The mucous membranes also increase their rate of mucus production, and the skin develops itchy skin lesions known as *hives* or *urticaria.*

Certain areas of the body may respond more than others when in contact with the allergen. For example, contact of an allergen with the eyes of certain allergic individuals causes the conjunctival lining over the eyes to become reddened, itchy, and watery. This condition is known as *allergic conjunctivitis.* When the allergen comes in contact with the nose, certain allergic individuals experience sneezing, increased mucus production, and itchiness. This is called *allergic rhinitis,* or hay fever. If the allergen lodges in airways in some individuals, *asthma* may result. A common symptom of asthma is an increased mucus production in the bronchial airway in response to the inflammation caused by sensitized mast cells, producing wheezing and shortness of breath. In some severe cases, treatment by bronchial dilators is necessary to prevent death from constriction of the airways.

The allergic response that can be the most serious threat to health is *systemic anaphylaxis.* It occurs when an allergic individual who is particularly sensitive to an allergen ingests it or is inoculated with it. Because of the method of transfer, the allergen can travel quickly throughout the body and cause mast cell sensitization everywhere. Common symptoms include hives, asthma, gastrointestinal distress, runny nose, irritated eyes, and, in some cases, a severe lowering of blood pressure leading to shock and possibly death. It often occurs within minutes following an allergen exposure, such as penicillin injection or an insect sting, and is frequently fatal without appropriate treatment.

Autoimmune Diseases

The concept of *autoimmune diseases,* or *autoimmunity,* was introduced on page 423. Here, we will consider this homeostatic disturbance in context with the proper functions of the immune response.

When the immune response is functioning properly, it is able to distinguish self from nonself and to eliminate nonself invaders from the body. The ability of lymphocytes and phagocytes to recognize body cells as self, as well as chemicals that are produced by these cells, prevents them from switching on the immune response in a self-destructive manner. The recognition of self is known as *immunologic tolerance.* The presence of MHC proteins on plasma membranes provides the opportunity for this recognition.

Current research in the field of immunology suggests that, in some way in certain persons, some white blood cells lose the ability to detect MHC proteins and thereby lose their immunologic tolerance. As a result, immunologic mechanisms are activated against self molecules. In another scenario, the T cell populations that modulate the events of immunity, called suppressor T cells, may be lacking or dysfunctional. As a result, phagocytes may attack healthy cells once the unwanted cells have been removed, since they continue to receive mobilization signals but are not receiving the stop signals. In either case, the actual cause of these immunologic dysfunctions appears to have a genetic basis. The group of diseases that often result are known as the autoimmune diseases.

The direct effect of autoimmunity is a degeneration of tissue, followed by the destruction of its normal function. Damage is often caused by activation of complement, and macrophages have also been shown to be recruited into the area of attack. In some cases, activated T cell clones may also play a role in tissue degeneration. In each autoimmune disease, a reversal of the immune dysfunction is required to stop the progression of tissue destruction. However, precisely how to produce this reversal has yet to be discovered.

The diseases that have been identified as autoimmune are extremely variable in their focus of injury, duration, and nature of progression. They may involve a single organ such as the brain (as in encephalomyelitis and multiple sclerosis), or a type of tissue that is widespread throughout the body (as in rheumatoid arthritis and systemic lupus erythematosus, or SLE). They may occur over a short duration (as in rheumatic fever and glomerulonephritis) or extend through the remainder of a lifetime (as in type I diabetes mellitus, some types of pernicious anemias, and SLE). Finally, they may be medically treatable (as in Addison's disease and thyroiditis), or be terminal with no known treatment (as in multiple sclerosis). About the only common denominator of autoimmune diseases appears to be the cause: the immune response gone awry.

Acquired Immunity

Acquired immunity refers to a person's ability to mount a defensive immune response as a result of a previous contact with an antigen. It may be obtained by natural contact with an antigen, or by artificially introducing an antigen or antibody into the body. Also, your body may actively respond to an antigen by mounting the immune response, or the immunity may be provided passively by transfer from another person or animal. On this basis, there are four types of acquired immunity, whose descriptions follow.

Naturally Acquired Active Immunity

Naturally acquired active immunity develops from the exposure to pathogens or toxins during the course of daily living, which stimulates activity of the im-

mune response. Examples include immune responses against measles, chickenpox, and influenza (the flu). Immunity resulting from the first exposure may last a lifetime, or for only several years. During the first exposure, however, the symptoms of the disease are expressed, since immunity has not yet been acquired.

Naturally Acquired Passive Immunity

In **naturally acquired passive immunity,** the transfer of antibodies occurs from a person with immunity to a person without immunity. This situation arises during pregnancy, during which antibodies (IgG) from the mother pass across the placenta to her developing fetus. These antibodies protect the child against diseases, such as polio, diphtheria, and rubella, during pregnancy and the first few months after birth. After several months, the short-lived antibodies are broken down and the infant must rely on its own immune response. However, certain antibodies (IgA) are also present in breast milk, which can provide extended immune protection if the mother nurses her child. Specifically, these antibodies protect the digestive tract from infection and are believed to assist in the colonization of the tract by normal populations of microorganisms (known as *intestinal flora*).

Artificially Acquired Active Immunity

Artificially acquired active immunity results from the deliberate, artificial introduction of an antigen to stimulate the immune response. The antigen is called a **vaccine,** which usually consists of killed or weakened living pathogens, or inactivated toxins. The process of introducing a vaccine is referred to as a **vaccination.** Examples of vaccines include those used to protect against diphtheria, tetanus, and pertussis (DPT) and against mumps, measles, and rubella (MMR). A vaccine may be introduced by injection, or orally if it can be safely absorbed through the digestive tract.

Artificially acquired active immunity produces long-lasting protection without the symptoms of the disease. For example, it provides lifelong immunity against polio and measles. Therefore, it is often the preferred method of stimulating the immune response.

Artificially Acquired Passive Immunity

Human or animal antibodies may be transferred to an individual by injection, providing him or her with **ar-** **tificially acquired passive immunity.** These foreign antibodies are injected in an active state and are thereby capable of reacting against the disease-causing antigens immediately. The immunity provided is short-lived, since the foreign antibodies are soon eliminated by the recipient's own immune response.

The foreign antibodies used in artificially acquired passive immunity are called *immune serum globulins* (formerly called *antisera*), because they are usually stored in blood serum before use (plasma less the blood-clotting factors). Immune serum globulins are used when the patient's condition does not enable the natural production of immunity. They are not used in prevention against diseases, but rather in response to diagnosed diseases. Specific immune serum globulins are available against rabies, hepatitis, tetanus, and many poisons such as snake venoms.

CONCEPTS CHECK

1. What is required for a lymphocyte to be activated (sensitized)?

2. What cell lines result from T cell sensitization?

3. What is the role of antibodies during antibody-mediated immunity?

4. How does a vaccination provide immunity?

HOMEOSTASIS

The lymphatic system supports homeostasis of the body. In instances in which its function becomes impaired, such as in the disease AIDS, homeostasis is critically disrupted.

As we have previously learned in this chapter, the lymphatic system provides two functions: it carries fluid from the extracellular environment into the bloodstream and it defends the body from disease. The first of these functions supports homeostasis of the body, since it recycles valuable body fluids. The importance of fluid recycling is made apparent when we consider the consequences of a sudden loss of these fluids: a critical body emergency follows shortly (called *circulatory shock*), which results in death if fluids are not restored quickly. The second function of the lymphatic system, that of defense against disease, also supports homeostasis, as we are about to discover.

Immunodeficiency

On a daily basis, we are exposed to pathogens and toxins that could produce disease if they are not challenged by our immune responses. The diseases they produce are the result of the attack on—and, in many cases, the destruction of—cells and tissues. Damage to tissues disrupts homeostasis in the areas affected. This disruption of homeostasis can spread to other tissues if left unchecked. The ability of white blood cells in the bloodstream, in the lymphatic network, and in the extracellular space to identify and destroy nonself cells and toxins largely prevents foreign materials from inflicting homeostatic damage.

However, in some cases, the white blood cells themselves are attacked by invading pathogens. In other cases, white blood cell production or development may be blocked. In either case, if the numbers of white blood cells become significantly reduced, the body's ability to defend against all pathogens and toxins is eventually impaired. This condition is called **immunodeficiency.** It may develop as a result of malnutrition, age, cancer, or other diseases that affect white blood cell production. Immunodeficiency may also arise from a genetic defect that results in the inability to synthesize one or more types of lymphocytes. For example, a rare disease known as *severe combined immunodeficiency (SCID)* results when a person is born with the inability to produce active B cells and T cells. Treatment may be provided by transplantation of bone marrow containing healthy stem cells. Without treatment, there is little hope of avoiding a life-threatening infection. For example, the longest-surviving untreated patient lived in a plastic bubble to protect him from infection for nearly every day of his 12-year lifetime (Fig. 14–13).

Acquired Immunodeficiency Syndrome (AIDS)

Far and away the most publicized immunodeficiency is one that is caused by a virus. Known as **acquired immunodeficiency syndrome (AIDS),** it is one of the most devastating diseases known, and has reached global epidemic proportions. AIDS is a lethal disease that acts by lowering the victim's immunity, allowing a second, unrelated infection to produce fatal symptoms.

AIDS was first identified in the United States in June 1981 when a very rare form of pneumonia began occurring among homosexual males. At about the same time, a type of skin cancer that normally occurs only in persons over the age of 70 years was being seen in homosexual males under 70 as well. Called Kaposi's sarcoma, it causes large pigmented spots on the skin and can spread to other organs. The males afflicted by these conditions soon died, despite efforts to save them. This led investigators to suspect that the pneumonia and sarcoma were secondary to a primary infectious disease that was suppressing immunity. About two years later (in 1983), a group of researchers in the United States and France isolated the virus that causes the immunodeficiency of AIDS. This virus is now called **human immunodeficiency virus,** or **HIV.**

A person infected with HIV can receive confirmation of its presence about six months after infection, by a simple blood test. The blood test analyzes for the presence of antibodies specific to HIV, which are produced by healthy B cell lymphocytes in response to HIV exposure. Although symptoms, such as skin lesions, lung congestion, swollen lymph nodes, and frequent infections, may not be expressed for years, the person identified as HIV-positive is considered a carrier of AIDS. Such a carrier is said to have AIDS-related complex, or ARC. Infection is a lifelong condition, and those infected are capable of transmitting the disease to others. The incubation period for the virus (the time interval from infection to development of AIDS symptoms) may last eight to ten years, although the risk of developing AIDS symptoms increases each year following infection. Once symptoms begin to show, death usually follows within two to three years.

Currently, there is no cure for AIDS. Treatment includes expensive drug therapies directed against viral

Figure 14–13
David was a victim of severe combined immunodeficiency (SCID). He had to spend his life within the controlled, pathogen-free environment of this plastic bubble, because of the absence of a functioning immune response.

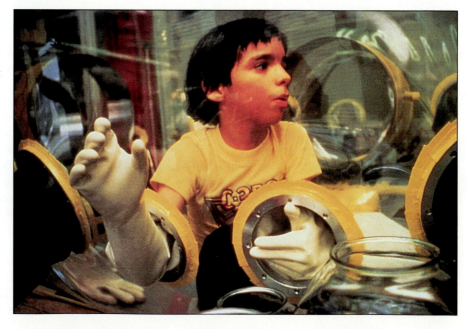

activity, which have limited successes, and therapies targeted against secondary infections as they arise. Despite this dismal news there is growing hope, as therapeutic vaccines containing proteins (called gp120 and gp160) cloned from HIV's outer shell are currently being tested, and preliminary tests suggest that they might slow or stop the progression of the disease among people already infected. However, much work is required before these vaccines can be determined to be effective against HIV.

The most frequent victims of AIDS are homosexual males, intravenous drug users, and hemophilia patients. These are so-called high-risk groups, since their incidence of infection is much greater than that of other groups within the general population. Infection among high-risk groups is more frequent because of their direct exposure to body fluids, such as blood and semen. However, all persons are capable of developing AIDS, but only if they are exposed to infected body fluids. In fact, there is an increasing trend toward the spread of AIDS among the heterosexual population in countries worldwide. All sources agree that this trend is due to heterosexual conduct without protection against possible infection.

HIV: Mode of Infection

The structure of a virus typically consists of a protein coat that envelops DNA or RNA in the center. In addition, certain viruses, such as HIV, contain an added layer of lipid around the protein (Fig. 14–14). Virus structure is amazingly simple; too simple, in fact, to perform vital life functions such as metabolism and growth. The only function a virus is capable of is combining with a host cell and injecting its nucleic acid inside. Once within the host cell, the viral nucleic acid (DNA or RNA) redirects cell functions to make copies of the virus. Soon after large numbers of new viruses have been produced, they leave the host cell to infect other cells. The host cell is damaged or killed by the mass exodus of virus particles, or it is killed by the body's own antiviral immune response. In the case of HIV, viral particles may lie dormant for years. Once activated, they are capable of killing many types of cells, among them helper T cells.

HIV appears to prefer helper T cells over other cells in the body to infest and kill, although it has been shown to be capable of attacking other T cells and monocytes as well. Once it enters the body, its usual pattern is to initially infest helper T cells, causing flu-like symptoms (fever, headache, fatigue, and swollen glands) that last for several weeks. The virus then enters a period of dormancy in which there is no active reproduction for about six months, although this time period is highly variable (dormancy has been known to last as long as ten years in some people). Following this dormant phase, HIV begins its reproductive phase, invading primarily helper T cells and destroying them to produce new viruses (Fig. 14–15). It has been estimated that this reproductive phase destroys 60% to 90% of the helper T cells in the body within several months. The result of helper T-cell destruction is a suppression of cell-mediated immunity.

Development of AIDS

The progression of AIDS into a debilitating disease occurs with the decline of helper T cells. Once cell-mediated immunity has become suppressed by HIV activation, infectious agents take advantage of this

Figure 14–14
The AIDS virus, HIV.
How may this virus be transmitted from an infected person to one who is not infected?

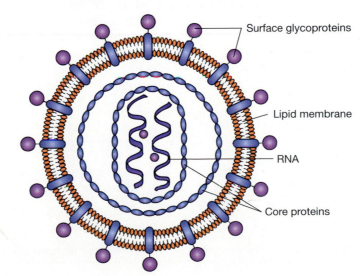

Surface glycoproteins

Lipid membrane

RNA

Core proteins

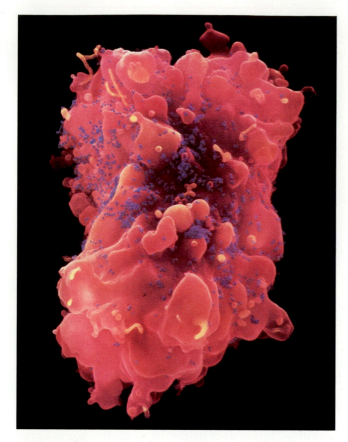

Figure 14–15
A helper T cell under attack by HIV.

opportunity to cause secondary diseases. These pathogens, which are normally destroyed by a healthy immune defense, are appropriately called *opportunistic pathogens*. The diseases they cause are at present used to diagnose AIDS; they include *pneumonia* caused by the microorganism *Pneumocystis carinii*, a type of skin cancer called *Kaposi's sarcoma*, a liver disease called *hepatitis B*, a systemic bacterial infection known as *toxoplasmosis*, a loss of weight caused by chronic diarrhea known as *HIV wasting syndrome*, and a degeneration of functional brain tissue called *AIDS dementia*. These secondary diseases are the usual causes of death among AIDS sufferers.

Transmission and Prevention

The method of HIV transmission is by way of blood transfer from an infected individual to a noninfected individual. HIV is also transmittable by way of reproductive fluids, such as semen and vaginal secretions. The virus is found in these fluids within lymphocytes and, in some cases, within macrophages. As a result of body fluid transmission, the most common victims of AIDS are intravenous drug users who share needles, individuals who fail to practice safe sex with infected (homosexual or heterosexual) partners, and, in increasingly rare circumstances, individuals receiving blood transfusions from blood donated from infected individuals. The virus is also transmitted from infected mothers to their infants before or during birth, or through breast feeding.

Current scientific research indicates that AIDS is not transmitted through casual contact. Once outside the body, HIV is a fragile virus that can be easily controlled and eliminated. For example, it cannot exist in the 20% oxygen atmosphere we live and breathe in, so it cannot be transmitted through the air. Dish washing and clothes washing at 135° for 10 minutes will destroy HIV, and chemicals such as 30% hydrogen peroxide, rubbing alcohol, household bleach, and germicidal soap are also effective killing agents. Chlorinating pools and spas properly will also prevent any possible HIV survival. Because of the fragile nature of HIV, casual contact with infected persons is considered safe. For example, family members, coworkers, and friends can safely coexist with AIDS patients, as long as a body fluid exchange is prevented.

At the present time, there is no vaccine or drug program to prevent the spread of AIDS. The only effective prevention is to stop the transmission of HIV. Sexual transmission, which is the most common mode of infection, will be prevented if infected persons do not have sexual intercourse (vaginal, oral, or anal) with noninfected persons. If intercourse is practiced, effective barriers, such as latex condoms and spermicides, should be used at all times. Careful, proper use of barriers during intercourse eliminates the considerable risk that comes from not knowing if a partner is an HIV carrier. Transmission through contaminated needles could be prevented if intravenous drug users were to use sterile needles only, or modify their drug dependency. Transmission through contaminated blood can be prevented if donated blood is screened carefully for the presence of HIV antibodies and HIV. Until an effective drug program or vaccine is discovered, AIDS prevention is limited to these few measures.

CONCEPTS CHECK

1. How may immunodeficiency result in a disruption of homeostasis?

2. What cells are primarily attacked by HIV during infection?

3. How may acquiring HIV be prevented?

CLINICAL TERMS OF THE LYMPHATIC SYSTEM

Acquired immunodeficiency diseases A class of disease characterized by the destruction of all or part of the immune response. It can be caused by the use of toxic drugs, exposure to radiation, or attack by infectious agents. AIDS is a type of acquired immunodeficiency disease.

Anaphylaxis An acute systemic allergic reaction that includes hives, respiratory distress, shock, and gastrointestinal distress as symptoms. It is a severe life-threatening emergency that can occur with exposure to drugs and insect stings or bites.

Atopic diseases Allergic disorders resulting from the release of inflammatory substances by mast cells that have been sensitized by the interaction of antigens with IgE antibodies. They include allergic rhinitis (hay fever) and bronchial asthma.

Autoimmunity A general condition in which the immune response produces autoantibodies to a natural antigen, causing injury and destruction to otherwise healthy tissues. Examples include glomerulonephritis, acquired hemolytic anemia, rheumatoid arthritis, and systemic lupus erythematosus.

DiGeorge's syndrome A congenital disease in which a child is born without the thymus and parathyroid glands. Death usually occurs by the age of 2 years because of uncontrolled infection.

Hodgkin's disease A form of cancer characterized by the presence of large, multinucleate cells in affected lymphoid tissue. Its symptoms usually include swelling of the lymph nodes and spleen, edema of the face and neck, jaundice, anemia, and an increased susceptibility to disease. Hodgkin's disease occurs most often in young adults between the ages of 15 and 38 years, and in adults over 50 years. Its cause is unknown, although treatment with radiation therapy and chemotherapy is often successful. Left untreated, Hodgkin's disease is fatal.

Immunization The ability of the immune response to reactivate quickly during repeated exposures to infectious disease.

Immunosuppression The administration of agents that interfere with the immune response's ability to respond. Drugs that are toxic to cells, corticosteroid hormones, and radiation provide immunosuppression. They may be administered to reduce the rejection of organ transplants.

Immunotherapy The use of tumor-specific antigens that may be obtained from transplanted tumors to destroy cancers. It has been used in conjunction with chemotherapy (the use of toxic chemicals) to treat several types of cancer.

Lymphoma A malignant tumor (cancer) of lymphoid tissue. Normally involving lymph nodes in a particular region, it is a highly metastatic form of cancer because of the ease with which tumor cells can be carried to distant sites through lymphatic vessels.

Monoclonal antibodies A pure antibody preparation that combines with only one specific type of antigen. They are produced by injecting laboratory animals with the antigen, removing the resulting B cell clones, and fusing the clones with tumor cells. The resulting cells are called *hybridomas* and divide to produce only one kind of antibody. Monoclonal antibodies are useful in diagnosing diseases (including syphilis, hepatitis, rabies, and some cancers), as therapeutic agents against cancer, in assisting cytotoxic chemicals used in cancer chemotherapy by locating cancer cells, and in determining pregnancy.

Neutropenia A defect in the nonspecific defense mechanism characterized by a decreased neutrophil count in the blood.

Splenomegaly An enlargement of the spleen. Enlargement, or hypertrophy, is caused by the proliferation of white blood cells. This condition usually follows infectious diseases such as typhus, scarlet fever, and syphilis.

Vaccine A suspension of parts of microorganisms, inactivated whole microorganisms, or inactivated toxins that may be administered to induce the immune response. The first vaccinations were given to immunize against smallpox. Vaccinations are now given routinely to immunize against typhoid, measles, mumps, polio, and other infectious diseases.

The lymphatic system performs two main functions: (1) It carries fluid from the extracellular space to the bloodstream and (2) it defends the body against disease.

THE LYMPHATIC NETWORK

The flow of interstitial fluid from the extracellular space is primarily into the lymphatic network and is caused by a pressure gradient. Once within the network, the fluid is called lymph.

Lymphatic Capillaries Blind-end vessels where the flow of lymph begins, similar in structure to blood capillaries.

Lymphatic Vessels Larger continuations of the capillaries that carry lymph toward the heart. They are similar in structure to veins, and likewise contain one-way valves.

Lymphatic Trunks and Collecting Ducts The trunks are larger vessels that empty lymph into one of two collecting ducts: the larger thoracic duct and the smaller right lymphatic duct. The collecting ducts drain lymph into the bloodstream.

Movement of Lymph Lymph flows by pressure gradients into lymphatic capillaries and into lymphatic vessels. It moves toward the heart through vessels by action of skeletal muscle (the skeletal muscle pump) and respiratory movements (the respiratory pump) and is aided by one-way valves.

OTHER LYMPHATIC ORGANS

Lymph Nodes Small, oval masses of lymphoid tissue that are composed mainly of lymphocytes. They are concentrated in the neck, armpit, groin, and abdominal cavity regions. As lymph flows through them, their lymphocytes and macrophages remove foreign particles.

Lymph Node Structure Nodes have the shape of a kidney bean, about 2.5 cm long. They receive afferent lymphatic vessels that bring lymph through the convex margin, and efferent lymphatic vessels that carry lymph away from the hilus, or concave margin. A lymph node is surrounded by a fibrous capsule. Internally it consists of clusters of lymphocytes, called *lymph nodules.* The internal structure is divided into an outer region, the cortex, and an inner region, the medulla.

Lymph Node Function As lymph flows through the nodes, it is filtered of foreign materials and thereby cleansed.

Spleen The largest organ of the lymphatic system, located in the left side of the abdominal cavity below the diaphragm. The spleen is surrounded by a fibrous capsule. Internally it contains a white pulp, composed of lymphocytes and macrophages, and a red pulp, made up of venous sinuses. The spleen functions as a large filter for removing foreign particles and old cells, and as a blood reservoir.

Thymus Gland A soft, bilobed structure located in the thoracic cavity above the heart in infants. It is very much reduced in adults. The thymus is composed of lymphoid tissue and in infants is the site where T lymphocytes mature.

Tonsils Located in the mouth and throat; they include the palatine tonsils, pharyngeal tonsils, and lingual tonsils. The tonsils are composed of lymphoid tissue. Its white blood cells destroy pathogens in the mouth and throat regions.

Peyer's patches Located in the wall of the small intestine. They are areas of lymphoid tissue containing white blood cells that may destroy pathogens trying to pass through the intestinal wall.

THE DEFENSE MECHANISMS OF THE BODY

Defense mechanisms require white blood cells to distinguish between self and nonself cells. This is done by the identification of self proteins on the plasma membrane of cells, called *major histocompatibility complex (MHC)* proteins. There are two ways the body defends itself from disease: by nonspecific means and by specific means.

Nonspecific Mechanisms Mechanisms that do not distinguish between the types of foreign particles and substances.

Physical Barriers The skin and mucous membranes prevent entrance of foreign particles into the body.

Phagocytosis If foreign particles penetrate the physical barriers, several types of white blood cells destroy them by engulfing and digesting them. Phagocytes must distinguish between self cells and nonself cells. This is done by recognition of protein labels, including antibodies and complement. Monocytes and neutrophils are mobile phagocytes. Macrophages are also phagocytic, but they are mainly stationary and make up much of the mononuclear phagocytic system.

Natural Killer Cells Leukocytes that kill cancer cells and virus-infected cells by puncturing a hole through the plasma membrane.

Proteins Include complement, which promotes the movement of white blood cells to an area of infection, and interferons, which disrupt the ability of viruses to infect cells.

Inflammation A nonspecific mechanism that causes redness, swelling, heat, and pain. It promotes the flow of blood, and, therefore, movement of white blood cells, to an area of infection or trauma.

Specific Mechanisms: The Immune Response A response in which the foreign substance or invader is specifically identified.

Components of Immunity

Antigens Foreign substances that stimulate an immune response.

Antibodies Proteins produced by cells that react with antigens by binding with them, forming an antigen-antibody complex. This inactivates the antigen.

Lymphocytes Originate from red bone marrow, and may mature there or in the thymus or Peyer's patches. During this process they are programmed to distinguish between self and nonself. Once programmed, they are called *immunocompetent*.

T Cells Undergo development in the thymus gland and then migrate to lymphoid tissue. Once they bind to an antigen, they develop into cell lines, of which several types have particular importance. Killer T cells produce lymphotoxins, which rupture nonself cells. Helper T cells stimulate the defensive activities of other lymphocytes. Suppressor T cells modulate the reaction of other lymphocytes by inhibiting their activity. Memory T cells store information about the specific antigen for the next encounter.

B Cells Undergo development in the bone marrow or in Peyer's patches and migrate to lymphoid tissue. Once they bind to an antigen, they develop into two cell lines: plasma cells, which synthesize and release antibodies, and memory B cells, which store information about the specific antigen for the next encounter.

Cell-mediated Immunity Involves T-cell function of all cell lines.

Humoral Immunity Involves B-cell function of plasma cell antibody production and memory B-cell storage of information.

The Allergic Response A type of humoral response in which a person forms antibodies to certain allergens, producing inflammatory responses of the skin and mucous membranes.

Acquired Immunity The ability to develop immunity after an initial exposure to a particular type of antigen.

Naturally Acquired Active Immunity Immunity as a result of previous exposure to pathogens under natural conditions.

Naturally Acquired Passive Immunity Immunity caused by the transfer of antibodies from one person to another, such as occurs during pregnancy.

Artificially Acquired Active Immunity Immunity caused by the artificial introduction of a vaccine.

Artificially Acquired Passive Immunity Immunity induced by the artificial introduction of antibodies from an animal or a person.

HOMEOSTASIS

The lymphatic system supports body homeostasis by recycling fluids into the bloodstream and defending the body from disease.

Immunodeficiency The lack of components of immunity causes an inability to defend the body against disease.

Acquired Immunodeficiency Syndrome (AIDS) A disease caused by a virus, called HIV, which destroys mainly helper T cells. It results in severe immunodeficiency that is fatal.

HIV: Mode of Infection The virus that causes AIDS, HIV, selectively destroys helper T cells (but also other leukocytes) after an initial dormant period. This suppresses cell-mediated immunity.

Development of AIDS Following a period of dormancy that may last six months or more, secondary infections begin to appear. They include pneumonia, Kaposi's sarcoma, dementia, and diarrhea. Secondary infections are opportunistic and are the actual cause of death.

Transmission and Prevention HIV is spread through the transfer of blood, semen, or vaginal fluids from an infected person to one who is not infected. It occurs through unprotected sexual intercourse, sharing intravenous needles, and receiving infected blood during transfusion; and it can be passed from an infected mother to her child through the placental barrier or in breast feeding. Since there is no vaccine or drug treatment yet, prevention is by avoiding these circumstances.

KEY TERMS

allergen (p. 429)
antibody (p. 423)
antigen (p. 423)
complement (p. 422)

immunity (p. 423)
immunization (p. 429)
immunocompetent (p. 424)
immunodeficient (p. 431)

infection (p. 420)
inflammation (p. 422)
interferon (p. 422)
lymph (p. 414)

lymphoid tissue (p. 417)
pathogen (p. 419)
toxin (p. 419)
vaccine (p. 431)

OBJECTIVE QUESTIONS

1. The primary function of the lymphatic system is to:
   a. return fluid to the bloodstream
   b. defend the body from disease
   c. transport nutrients to the liver
   d. both (a) and (b)

2. Fluid enters lymphatic capillaries from the:
   a. extracellular space
   b. thoracic duct
   c. lymphatic vessels
   d. bloodstream

3. The fluid within lymphatic vessels is called:
   a. plasma
   b. lymphatic fluid
   c. lymph
   d. interstitial fluid

4. The major lymphatic collecting duct of the body drains lymph into the:
   a. right lymphatic duct
   b. right atrium
   c. left subclavian vein
   d. right jugular vein

5. The movement of lymph against gravity is made possible by the action of the respiratory pump, the skeletal muscle pump, and:
   a. the heart
   b. pulsation of the thoracic duct
   c. one-way valves
   d. osmotic pressure

6. The primary type of cell within lymphoid tissue is the:
   a. neutrophil
   b. epithelial cell
   c. lymphocyte
   d. monocyte

7. As lymph flows through a lymph node, the fluid is _____ by lymphocytes and macrophages.
   a. redirected to the spleen
   b. converted to interstitial fluid
   c. filtered and cleansed
   d. used as waste

8. The functional compartments within lymph nodes that contain clusters of lymphocytes are known as:
   a. afferent lymphatic vessels
   b. lymph nodules
   c. cortex
   d. white pulp

9. The largest lymphatic organ in an adult is the:
   a. thymus gland
   b. pharyngeal tonsils
   c. spleen
   d. liver

10. Phagocytosis of bacteria by monocytes and neutrophils is a type of:
    a. nonspecific defense mechanism
    b. humoral response
    c. cell-mediated immune response
    d. all of the above

11. A protein substance that amplifies the movement of leukocytes into an area of infection and stimulates their destructive capabilities is characteristic of:
    a. interferon
    b. complement
    c. antibodies
    d. antigens

12. Inflammation can be beneficial to the immune response because it:
    a. carries leukocytes to the site of infection or injury
    b. causes pain
    c. reduces the rate of tissue repair
    d. reduces the amount of blood flowing to the site of infection or injury

13. A foreign substance that stimulates an immune response is called a(n):
    a. antibody
    b. complement
    c. antigen
    d. plasma cell

14. T cells reach their maturity within the:
    a. lymph nodes
    b. liver
    c. thymus gland
    d. spleen

15. The T-cell line that releases lymphokines during an infection and stimulates the activity of B cells is that of the:
    a. helper T cells
    b. memory T cells
    c. killer T cells
    d. suppressor T cells

16. The cells that produce and release large quantities of antibodies are called:
    a. memory B cells
    b. plasma cells
    c. helper T cells
    d. killer T cells

17. The specific defense mechanism that involves the use of B-cell lines and antibodies is called:
    a. humoral immunity
    b. B-cell sensitization
    c. cell-mediated immunity
    d. allergic immunity

18. People who are not tolerant of various surface antigens and respond to them by the release of inflammatory chemicals by mast cells are called:
    a. sensitive
    b. allergic
    c. autoimmune
    d. immunosuppressive

19. The use of vaccines to induce immunity is called:
    a. naturally acquired active immunity
    b. naturally acquired passive immunity
    c. artificially acquired active immunity
    d. artificially acquired passive immunity

20. HIV is transmitted through:
    a. blood
    b. the air
    c. saliva
    d. contact with skin

ESSAY QUESTIONS

1. Describe the movement of lymph from its origin in the extracellular space to its drainage into the bloodstream. In your description, include the mechanisms that push lymph along.

2. Discuss the flow of lymph through a lymph node and how it is filtered and cleansed.

3. Describe the nonspecific defense mechanisms that operate in a healthy body.

4. Compare and contrast the two mechanisms of specific defense, that is, cell-mediated immunity and humoral immunity.

5. Indicate the activities of helper T cells in both cell-mediated immunity and humoral immunity.

6. Describe the mechanisms of the allergic response. Can you think of a way to reduce the effect of an allergen on an allergic individual?

7. Explain how a vaccine can induce immunity against a particular antigen.

Figure 14–1
What effect does the accumulation of small proteins in the extracellular space have on the flow of fluid? The accumulation of small proteins causes an increase in osmotic pressure, which leads to the movement of fluid into the lymphatic capillaries by osmosis.

Figure 14–2
What is the main collecting vessel of the lymphatic network? The thoracic duct is the main collecting vessel.

Figure 14–3
What happens to lymph as it is channeled through each lymph node? Lymph is filtered of foreign cells and particles as it passes through lymph nodes.

Figure 14–4
What type of white blood cell makes up much of the substance of the lymphatic organs? Lymphocytes make up much of the substance of lymphatic organs (called *lymphoid tissue*).

Figure 14–5
How does a phagocyte distinguish between healthy cells and unwanted cells and particles? A phagocyte distinguishes on the basis of protein labels present (or absent) on the plasma membranes or particle surfaces.

Figure 14–6
What changes during inflammation promote the movement of phagocytes to the injury site? Changes that promote phagocyte migration include an increased blood flow to the injured area (caused by dilation of blood vessels), an increase in capillary permeability, and the release of attractive substances by the injured cells.

Figure 14–7
Where do both types of lymphocytes originate? Both T cells and B cells originate from red bone marrow.

Figure 14–8
What are the four types of differentiated T cells? The differentiated T cells include helper T cells, memory T cells, suppressor T cells, and killer T cells.

Figure 14–9
What cells produce antibodies? Only plasma cells produce antibodies.

Figure 14–10
What event sensitizes a T cell? Sensitization of a T cell occurs when a macrophage presents a processed antigen to the T cell and it is accepted.

Figure 14–11
What type of T cell is needed to induce B cell sensitization? A helper T cell must provide a signal for the B cell to undergo sensitization.

Figure 14–14
How may this virus be transmitted from an infected person to one who is not infected? HIV may be transmitted only by physical contact with body fluids, which may include unprotected sex, use of a contaminated needle, or transfusion of contaminated blood.

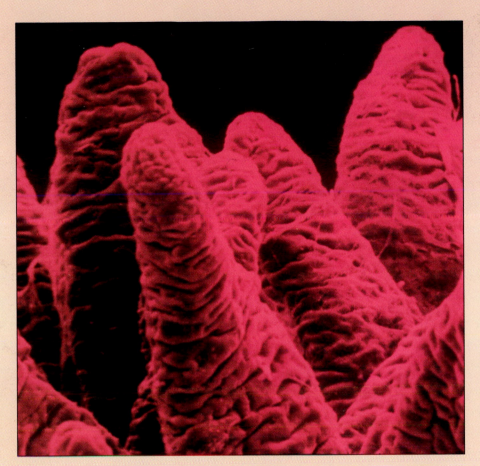

Villi of the small intestine

METABOLIC PROCESSING SYSTEMS

Metabolism includes all the processes by which the body synthesizes new materials for growth and repair and produces the energy it needs to perform all of its functions. It is performed by virtually all cells, and is vital for the support of our bodies as a whole.

Numerous systems of the human body support metabolic processes by providing raw materials and removing waste materials on a continuous, unending basis throughout your lifetime. The raw materials include oxygen, water, and nutrients. Oxygen is brought into the body by the respiratory system, and water and nutrients are made available to cells by the digestive system. As we have seen, the cardiovascular system transports these materials from the respiratory and digestive organs to all body cells. Waste materials resulting from diet are eliminated from the body by the digestive system. Metabolic waste products, as well as excess water or salts, are ultimately removed by the urinary system.

In this unit we will investigate the systems of the body that are involved in metabolic processing: the respiratory system, in Chapter 15; the digestive system and nutrition, in Chapters 16 and 17; and the urinary system, in Chapter 18.

THE RESPIRATORY SYSTEM

LEARNING OBJECTIVES

After studying this chapter, you should be able to:

1. Identify the main function of the respiratory system.

2. Distinguish between pulmonary ventilation, external respiration, and internal respiration.

3. Identify the organs of the respiratory system on the basis of their location, structure, and functions.

4. Describe the structure and function of alveoli and the respiratory membranes.

5. Describe the structural features of the lungs.

6. Describe the events involved in inspiration and expiration.

7. Identify the measures of respiratory volume.

8. Describe the events involved in external respiration.

9. Describe the events involved in internal respiration.

10. Identify the main source of respiratory control and describe how it works.

11. Describe the factors that affect breathing other than the respiratory center.

The respiratory system helps meet the metabolic needs of the body by bringing oxygen into the bloodstream, where it can be transported to all body cells. It also removes the waste product carbon dioxide from the blood and channels it outside the body. Several events are involved in this transfer of gases. The first occurs when you breathe in and out. It is called **pulmonary ventilation** and involves the movement of air between the external environment and the air sacs within the lungs. Once fresh air has filled the lungs, gas molecules diffuse between the air sacs and the tiny capillaries that surround them. This is the second event in respiration and is called **external respiration.** The third event is the movement of oxygen and carbon dioxide between the bloodstream and body cells. This exchange of gases is called **internal respiration.** The entire process of gas exchange between the external environment and body cells is referred to as **respiration.** Don't confuse these terms with **cellular respiration,** which you were introduced to in Chapter 3. Cellular respiration is a metabolic process that uses the oxygen molecules brought into cells and produces molecules of carbon dioxide as waste. Because cellular respiration does not involve gas exchange, it is not a function characteristic of the respiratory system. Rather, it is a function of all individual body cells.

ORGANS OF THE RESPIRATORY SYSTEM

The respiratory organs are a series of chambers and tubes that process inhaled air as it travels from the nasal cavity to the bronchial tree.

The respiratory system is structurally divided into two portions: the **upper respiratory tract,** in the head and neck, and the **lower respiratory tract,** mainly within the thoracic cavity (Fig. 15–1). The organs of the upper respiratory tract include the nose, the pharynx, and the larynx. The lower respiratory tract begins at the base of the neck and consists of the trachea, the bronchi, the smaller subdivisions of the bronchi in the lungs known as bronchioles, and the tiny air sacs of the lungs, the alveoli. The collective network of bronchial branches, alveoli, and supporting tissues makes up the lungs.

A second way of dividing the organs of the respiratory system is based on their general functions. As you will discover, most respiratory organs conduct air between the external atmosphere and the air sacs in the lungs. Because of this function, these organs are often collectively called the **conduction zone.** Organs of the conduction zone (nose, pharynx, trachea, bronchi, and bronchioles) also warm and humidify the air as it passes. Actual gas exchange occurs only between the alveoli and a vast capillary network. The alveoli and their associated structures are thus referred to as the **respiratory zone.**

Nose

The **nose** provides an initial receiving chamber for inhaled air (Fig. 15–1). It is formed by the two nasal bones and numerous cartilages. The two openings at its base are the nostrils, or **external nares**[1] (NAR-ēz), which are separated by a partition in the midline called the **nasal septum.** The nasal septum continues behind the external nares to divide the nasal cavity in half. Each external naris opens directly into a small chamber, known as the **vestibule**[2] (Fig. 15–2). Its walls contain coarse hairs that help filter particles from the inhaled

---

[1]Nares: Latin *naris* = nostril.
[2]Vestibule: *vestibule* = entrance chamber.

Figure 15–1
The respiratory system and surrounding structures.
What are the two functional divisions of the respiratory system?

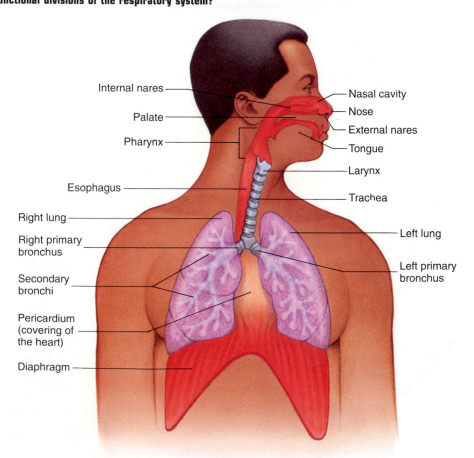

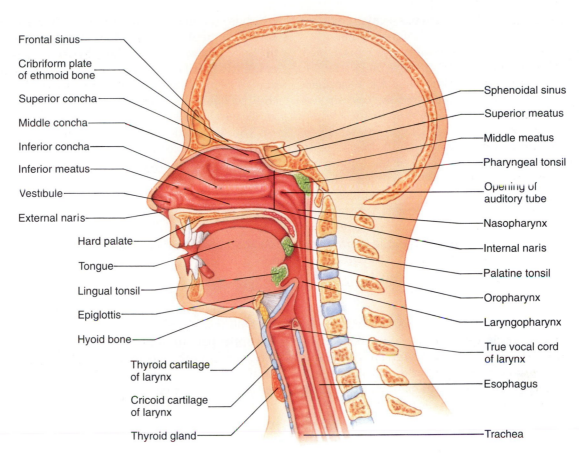

Figure 15–2
The upper respiratory tract. The head and neck are shown sagittally sectioned in this lateral view.

Through what opening is air channeled as it passes from the nasal cavity to the pharynx?

air. The vestibule opens into the much larger space behind it, the **nasal cavity.**

The nasal cavity is divided by the nasal septum into right and left chambers. Attached to the lateral walls of both chambers are three bony projections called **nasal conchae** (see Chapter 7). These scroll-like bones form shelves within each chamber of the nasal cavity, dividing the chambers into narrow passageways called **meati**[3] (mē-Ā-tī), which are illustrated in Figure 15–2. The nasal conchae increase the surface area of the nasal cavity, thereby maximizing physical contact between inhaled air and the nasal cavity walls.

The entire nasal cavity is lined with mucous membrane that has an abundant supply of blood vessels. As incoming air is channeled through the meati, it is warmed from the heat provided by the nearby bloodstream. Air is also moistened by the evaporation of water from the mucous membrane. The external layer of the mucous membrane contains many cilia, which beat in a rhythmic fashion to create a flow of mucus that moves toward the pharynx. The sticky mucus traps small particles such as bacteria and dust and carries them to the pharynx, where they are usually swallowed.

Associated with the nasal cavity are several other cavities in the skill, the **paranasal sinuses** (see Chapter 7). They communicate with the nasal cavity by way of small ducts and are also lined with mucous membrane. Because these ducts are narrow, they plug up quickly during inflammation and infection. Mucus and other fluids become trapped and build up within the blocked sinuses, producing pressure that can be very painful. This condition is called *sinusitis.*

Pharynx

The throat, or **pharynx**[4] (FAR-inks), is a chamber that extends from the back of the nasal cavity to the larynx (Fig. 15–1). Its walls are formed by skeletal muscle and are internally lined with mucous membrane. At its su-

---

[3]**Meati:** Latin *meatus* = opening or passage.

[4]**Pharynx:** Greek *pharynx* = throat.

perior end it receives air from the nasal cavity by way of two small openings, called the **internal nares.**

The pharynx is divided into three segments, which are continuous with one another (Fig. 15–2). The superior part of the pharynx is known as the **nasopharynx.** It receives the internal nares and the openings to the two auditory tubes, which extend to the middle ear. The portion of the pharynx that you can see when you look into a mirror with your mouth opened wide is the **oropharynx.** Below the level of the tongue begins the third segment of the pharynx, the **laryngopharynx** (la-rin′-gō-FAR-inks), which unites with the larynx in the neck. Both the oropharynx and the laryngopharynx provide a common passageway for air traveling between the nasal cavity and the larynx, and for food passing from the mouth to the esophagus.

Larynx

The voicebox, or **larynx**[5] (LAR-inks), is a short passageway that connects the pharynx with the trachea (Fig. 15–1). It provides passage to air traveling between these structures, and prevents solid material from entering the trachea. It houses the vocal cords, and thereby provides for the production of sound.

The walls of the larynx are composed mainly of cartilage and are lined internally with ciliated mucous membrane. The cartilage consists of nine pieces that are arranged to form a boxlike structure (Fig. 15–3). They include the large **thyroid cartilage**[6] in the front,

which is enlarged in males to form what is commonly known as the *Adam's apple;* the **cricoid** (KRĪ-koyd) **cartilage,**[7] located below the thyroid cartilage; the **epiglottic cartilage,** which is a tongue-shaped piece that is completely covered with mucous membrane; and three pairs of smaller cartilages that are attached to the vocal cords and muscles of the throat.

The epiglottic cartilage, or **epiglottis,**[8] is suspended by muscles and ligaments over the opening into the larynx, known as the **glottis.** It is supported in an elevated position to permit the unobstructed passage of air. During swallowing, however, the larynx moves upward and the epiglottis presses downward, closing off the glottis. This prevents the entrance of food or liquid into the lower respiratory tract and channels it instead into the nearby esophagus, which lies behind.

Inside the larynx, the mucous membrane lining contains two pairs of folds that extend inward from the walls in a horizontal direction (Figs. 15–3 and 15–4). The upper pair are supported by skeletal muscle fibers that help move the larynx upward during swallowing. Because these folds do not function in sound production, they are referred to as **false vocal cords.** The lower pair of folds contain elastic fibers. When air rushes between them, they vibrate back and forth. This generates sound waves, which can be converted to speech when the muscles of the pharynx, jaws, and tongue operate to modify the openings through which exhaled air passes. These lower folds are therefore called the

---

[5]Larynx: Greek *larynx* = upper windpipe.
[6]Thyroid: *thyroid* = shieldlike.

[7]Cricoid: Greek *krikos* = ring + *-oid* = having the shape of.
[8]Epiglottic: *epi-* = upon + *glottis* = tonguelike structure.

Figure 15–3
The larynx. (a) Anterior view. (b) Lateral view. The larynx has been sectioned along the sagittal plane to show its internal features.
What type of tissue provides the larynx with a firm but elastic structure?

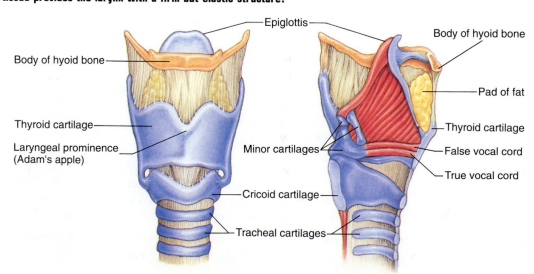

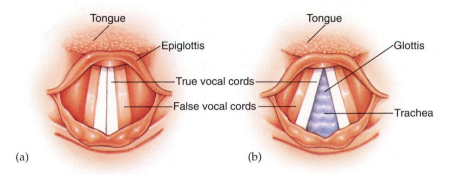

Figure 15–4
Larynx actions. (a) Top view looking down into the larynx. Notice that the glottis is closed in this drawing. (b) Glottis is opened.
Why can't you swallow solid food or liquid properly while you are talking?

true vocal cords (or, more simply, just **vocal cords**). It is interesting to note that there are no cilia covering the true vocal cords. Because cilia normally move mucus, the lack of cilia over the cords results in its accumulation. In order to remove the mucus, you must do so consciously by "clearing the throat."

Trachea

The windpipe, or **trachea**[9] (TRĀ-kē-a), is a tubular air passageway that is about 12 cm (about $4\frac{1}{2}$") long and 2.5 cm (about 1") wide. It is located in front of the esophagus and can be felt as a hard, ridged structure just deep to the anterior surface of the neck. It extends from the larynx downward into the thoracic cavity, where it divides into the right and left bronchi (Fig. 15–1).

The walls of the trachea are supported by rings of cartilage, smooth muscle, and elastic fibers. The cartilage rings contain open ends that face the back, and thus appear like a stack of letter C's. The open ends of the cartilage rings are shown in Figure 15–5, which is a posterior view of the trachea and adjacent structures. The open ends of the rings are connected by smooth muscle and connective tissue, which allows the nearby esophagus to expand as food is pushed down it. The cartilage rings provide the trachea with a rigid frame,

[9]Trachea: Greek *trachea* = rough artery.

Figure 15–5
A posterior view of the larynx, trachea, and bronchial branches.
What prevents the trachea from collapsing after you exhale?

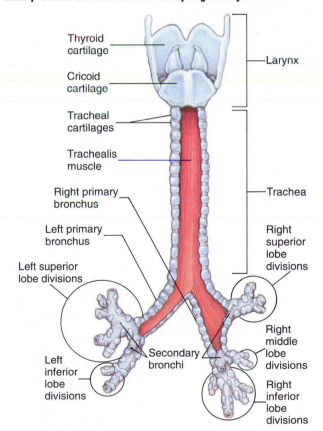

preventing it from collapsing and closing off the air passageway.

The trachea is internally lined with ciliated mucous membrane. It contains large numbers of mucus-secreting cells (this tissue is called PSCC, or pseudo-stratified ciliated columnar, epithelium; it was covered previously in Chapter 4), whose secretions form a dense carpet of mucus over the cilia. The cilia beat synchronously, moving microorganisms, dissolved gases, and other particles that have become trapped in the sticky mucus toward the pharynx like a conveyor belt. Once reaching the pharynx, the mucus is usually swallowed or expelled as sputum. This system of mucus transport extends throughout the conduction zone (except the true vocal cords) and is called the **mucociliary transport system** (Fig. 15–6). It helps keep microorganisms, dust particles, and noxious gases from reaching the alveoli in the lungs.

Bronchial Tree

The distal end of the trachea splits into the right and left **primary bronchi**[10] (BRONG-kī) within the thoracic cavity. Each primary bronchus extends a short distance from the trachea to a lung, although the two paths are not symmetrical: the left bronchus branches at a sharper angle than the right. As a result of this anatomical difference, materials accidentally inhaled usually end up in the right bronchus. Once the bronchi reach the lungs, they branch extensively into smaller and smaller tubes. Because this branching network resembles the branches of a tree, the divisions of the bronchi within the lungs are referred to as the **bronchial tree** (Fig. 15–7).

Each primary bronchus is very similar in structure to the trachea. Its walls are supported by incomplete rings of cartilage and smooth muscle, and it is internally lined with ciliated mucous membrane. As its branches divide into smaller tubes within the substance of the lung, the cartilage rings gradually diminish in size and number to form partial rings or plates, which eventually disappear altogether. The mucous membrane also becomes progressively thinner. The resulting tubes are supported only by a band of smooth muscle and elastic tissue and are lined with a thin mucous membrane layer. These small tubes are called **bronchioles** (BRONG-kē-ōlz) and are very numerous within each lung. The bronchioles continue to divide into yet smaller tubes, called **alveolar ducts,** which terminate as round, microscopic pouches known as **alveoli**[11] (al-VĒ-ō-lī).

---

[10]Bronchi: Greek *bronchos* = windpipe.

[11]Alveoli: Latin *alveolus* = tiny cavity.

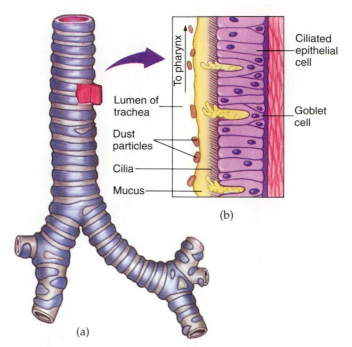

Figure 15–6
The mucociliary transport mechanism. (a) Anterior view of the trachea and primary bronchi. The cut section exposes the tracheal lining. (b) A magnified, sectioned view of the tracheal lining. The mucus secreted by goblet cells and the particles it has trapped are shown moving upward toward the pharynx like a conveyor belt. Movement is provided by the synchronous beating of the cilia.
What is the function of the mucociliary transport mechanism?

The function of the bronchial tree is to provide a passageway for air between the trachea and the alveoli. It is vitally important for this airway to be kept open and clear. Should it become blocked, suffocation may follow if the blockage is complete and not removed within minutes.

Alveoli and Respiratory Membranes

There are about 300 to 500 million alveoli in the lungs of an average adult. Their function is to provide the only site of gas exchange between the external environment and the bloodstream. The vast numbers of alveoli provide an enormous surface area for this exchange to occur; each lung has an internal surface area about 80 times greater than the external body surface area, or about the area of a tennis court (70 m[2]). Without this expansive surface area, gas exchange could not occur at a fast enough rate to support the body's metabolic needs.

The structure of alveoli is highly efficient for supporting the diffusion of gases. Each alveolus consists of a microscopic air space surrounded by a thin wall. This wall separates one alveolus from another, and from nearby capillaries (Fig. 15–8). It is composed of a

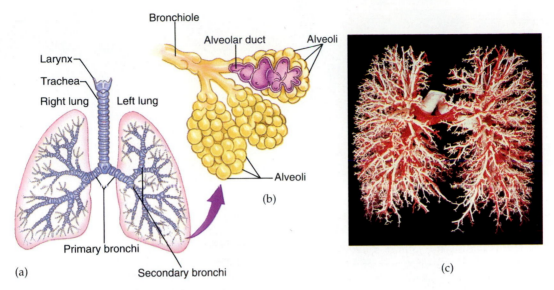

Figure 15–7
Bronchial tree and alveoli. (a) Major branches of the bronchi, which occur within the substance of each lung. (b) A magnified view of a bronchiole and its branches, with clusters of grapelike alveoli arising from them. (c) A photograph of a human bronchial tree (prepared by a process called plastination).

Why is the distribution of bronchi and their branches referred to as the bronchial tree?

Figure 15–8
Alveolar structure. (a) Electron micrograph of sectioned lung tissue. A single alveolus is shown in the center. (b) A magnified view of a cluster of alveoli, which have been sectioned and magnified yet further in the larger diagram. (c) A highly magnified view of a respiratory membrane, where the diffusion of gases actually occurs.

Across what layers must a molecule of oxygen pass as it diffuses from an alveolus and into a capillary?

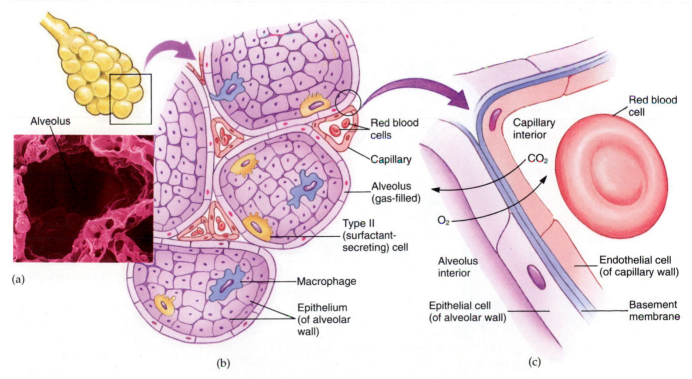

The Effects of Smoking

Certain substances that are inhaled into the respiratory passages can damage the mucociliary transport system and the lungs. One of these substances is tobacco smoke. When tobacco smoke is inhaled, it penetrates the mucus layer and comes into contact with cilia. As a result, the cilia become partially paralyzed and the action of the mucociliary transport system is slowed. Tobacco smoke also stimulates the production of additional mucus. The combination of these effects leads to a decreasing ability to clear the passageways of mucus. Consequently, potentially harmful substances that are inhaled dissolve in the excess mucus and remain trapped in the airways for extended periods of time. Their prolonged contact eventually causes a structural change in the epithelial layer lining the airways (converting PSCC epithelium toward stratified squamous epithelium), which can eventually lead to the development of cancer.

In the case of long-term smoking, the lungs become seriously affected. In time, microorganisms and excess mucus begin to accumulate in the lower airways within the lungs. The excess mucus plugs these passageways, stimulating a cough reflex to dislodge and remove it. When excess mucus accumulates over a long period of time, as occurs with chronic smoking, bacterial infection will often occur. A bacterial infection causes an immune response within the lungs, resulting in large-scale rupturing of white blood cells and the development of a thickened, green-colored mucus that causes increased coughing and irritation. The inhibition of the mucociliary transport system also leads to the accumulation of the by-products of tobacco smoke, such as tar, which literally coats the inner linings of the lungs in heavy smokers.

In addition to suppressing the mucociliary transport system, smoking tobacco has also been shown to inhibit the immune response. In persons who smoke on a daily basis, infections occur more frequently and generally last longer. There is also mounting evidence that smoking inhibits the response by white blood cells to attack and destroy cancer cells. In fact, scientists have known since 1972 that a direct correlation exists between the development of several types of cancers and smoking. The best-documented evidence is with lung cancer, which is the leading type of terminal cancer among both men and women (it has surpassed breast cancer in women as the number one killer every year since 1987, and is still on the rise). Although the mechanism is still not well understood, smoking tobacco on a regular basis has been shown to increase your chances of developing lung cancer by at least 80% (in both males and females). Recent evidence also indicates that the passive inhalation of tobacco smoke (inhaling smoke generated by others around you) may be almost as dangerous. Thus, the consensus of opinion among the medical community regarding smoking tobacco is very clear: smoking significantly lowers your life expectancy, and even that of people around you.

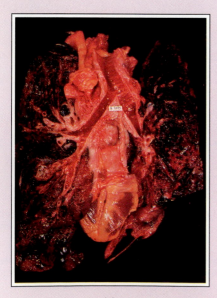

(a)

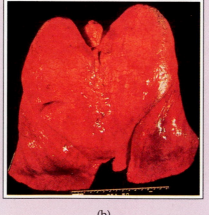

(b)

Smoking and the lungs. (a) The sectioned lungs in this photograph have been taken from a chronic smoker. They are blackened by the accumulation of smoke by-products, such as tar, and by the effects of prolonged inflammation. (b) A photograph of normal lungs, removed from a nonsmoker.

single layer of squamous epithelium. Interspersed between epithelial cells are specialized cells that secrete a layer of detergentlike lipid molecules called **surfactant.** The surfactant normally lines the inner surface of the alveolar wall, along with a thin layer of watery fluid. The fluid is required to keep the alveolar surface moist, which is necessary for the diffusion of gases to proceed through the alveolar wall. The water in this fluid, however, exerts a strong attractive force (called *surface tension*) that causes alveolar walls to be drawn together and collapse when air leaves the alveolar chamber during expiration. The surfactant counteracts this collapsing force, enabling the alveoli to inflate quickly after an expiration. Without surfactant, the surface tension would be so great that it would require an enormous amount of muscular effort to reopen the alveoli. For example, premature infants born before the seventh month of gestation have not yet developed the ability to produce surfactant. As a result, they risk death by suffocation because of the exhaustive effort required to reopen their alveoli during each breath. Although treatment is available by maintaining the infants in respiratory tents or incubators until their alveolar cells can begin producing surfactant, mortality from this condition, which is known as *respiratory distress syndrome (RDS)* or *hyaline membrane disease (HMD)*, remains the major cause of death among premature infants (about 25,000 deaths per year in the United States). However, new treatment techniques are being explored, including the direct administration of surfactant into alveoli, offering parents a renewed hope.

Closely associated with each alveolus are several capillaries. The wall of an alveolus is in very close proximity to the capillary wall. These walls are close together to promote the rapid diffusion of gases. They are collectively referred to as the **respiratory membrane.** As Figure 15–8c demonstrates, the respiratory membrane consists of the close arrangement of the epithelial wall of the alveolus, a basement membrane of connective tissue, and the endothelial wall of the capillary. The thickness of the respiratory membrane averages only 0.5 μm, providing a relatively thin, penetrable barrier for gas diffusion. The combination of the many millions of respiratory membranes within each lung provides an immense surface area, which allows for an efficient diffusion of gases in the lungs.

Lungs

The right and left **lungs** are composed of the many branches of tubes that constitute the bronchial tree, the millions of alveoli and their accompanying capillary networks, and supportive tissues. As organs, the lungs function as the site of gas exchange between the atmosphere and the bloodstream.

General Characteristics

The lungs are soft, spongy organs, each shaped much like a cone (Fig. 15–9). They occupy most of the thoracic cavity, extending from the diaphragm to just above the clavicles. The ribs border their surfaces to the front and back. The right lung is somewhat thicker and broader than the left, owing to the position of the liver, which presses the diaphragm slightly upward on the right side.

The narrow superior portion of each lung is called the **apex,** and the broad inferior portion is the **base.** The inferior surface of the base is concave, because of its fit over the convex diaphragm. The surface that lies against the ribs to the front and back is called the **costal surface,** and that which faces the midline toward the location of the heart is called the **medial surface.** Each

Figure 15–9
Structure of the lungs. The close-up reveals the locations of the pleurae relative to a lung and the pleural cavity.
What membrane is firmly attached to the outer wall of each lung?

HEALTH CLINIC

Asthma: The Manageable Disease

Asthma is a disease common in both children and adults. Recent studies have shown that up to 10% of all children have been diagnosed with asthma, and the condition may be present while remaining undiagnosed in up to 5% more. Diagnosed cases of asthma are slightly less common in adults (8%), because of a recent rise in diagnosed cases among children. In addition to these high rates of incidence, there is evidence that deaths due to asthma are increasing. While the reasons for this mortality are not yet known, several factors appear to be involved. They include improper diagnosis of the condition, inadequate treatment, and environmental factors such as increased exposure to dust mites and other indoor allergens.

Asthma researchers and diagnosticians generally recognize two types of asthma: extrinsic (allergic) asthma, in which the attack is a direct cause of exposure to pollens, dust, molds, or tobacco smoke; and intrinsic (idiopathic) asthma, in which the attack is triggered by nonallergenic factors such as infections, irritations, emotions, or stress. Extrinsic asthma is well documented to be caused by a genetic predisposition to one or more allergens. Upon exposure to an allergen, the immune system initiates a hypersensitivity response, triggering inflammation in the bronchial airways, which results in an accumulation of fluid in the lungs (a condition called *pulmonary edema*) and surrounding tissues, constriction of the smooth muscles lining the walls of the bronchioles, and increased discharge of mucus. The swelling, muscle spasms, and increased mucus cause widespread bronchial narrowing, which reduces the airflow to the lungs. Intrinsic asthma, on the other hand, has been suggested to be caused by an allergic reaction to a protein released by cells in response to stress, producing effects similar to those described for

extrinsic asthma. However, this protein has yet to be identified.

Symptoms of asthma include periodic episodes of wheezing (whistling sounds heard through the chest during breathing), coughing, and shortness of breath. These episodes may occur frequently over a period of one or two days and then subside, or they may occur frequently for a prolonged period of time. For most asthma sufferers, various stimuli, including inhaled allergens (pollen, dust, tobacco smoke, and others), certain ingested foods, acute viral respiratory tract infections, exposure to cold air, emotional reactions, and nonsteroidal anti-inflammatory drugs may initiate airway obstruction.

Once asthma has been diagnosed, the patient must learn immediately that asthma is not a curable disease. Rather, it is a chronic condition that must be monitored and managed with care. Only if a long-term treatment plan is properly followed will the asthma patient be able to control the uncomfortable symptoms.

The long-term treatment plan usually includes education of the patient concerning the disease and its management, avoidance of the factors that lead to attacks (tobacco smoke, dust, and other allergens), control of bronchial constriction with bronchodilating drugs (usually with inhalers), and control of inflammation with steroidal anti-inflammatory drugs. With serious attention paid daily to the treatment plan, asthma sufferers may be able to completely manage their condition. For example, even severe asthma patients are currently told by their physicians that there is no reason to restrict physical activities, including swimming, contact sports, or jogging, as long as the treatment plan is followed carefully.

---

lung contains its attachments to the trachea and heart on its medial surface only. This collection of attachments, which include a primary bronchus, large blood vessels, and nerves, is called the **root.**

Serous Membranes

Two layers of serous membrane surround each lung (Fig. 15–9). They are collectively referred to as the **pleurae**[12] (PLOOR–ē). The outer layer, the **parietal pleura,**

---

[12]**Pleurae: Greek *pleura* = rib or side.**

---

lines the thoracic wall and mediastinum. It continues around the heart and between the lungs, where it combines with the parietal pleura of the opposite lung to form a ligament that supports both lungs by surrounding the root. From the root of each lung, the parietal pleura folds inward to the lung surface to form the inner layer, called the **visceral pleura.** The visceral pleura surrounds the lung and is firmly attached to its outer surface.

Between the two pleural layers surrounding each lung is a potential space, called the **pleural cavity.** It contains a thin film of fluid produced by serous cells in the pleurae. The pleural fluid lubricates the surfaces of the two pleural membranes to reduce friction as the lungs expand and contract during breathing. If fluid production is reduced or the membranes become swollen, a painful condition called *pleurisy* results, in which the membranes scrape against each other with every breath.

Divisions

Each lung is divided into smaller compartments. The first division separates the substance of the lung into compartments called **lobes** (Fig. 15–9). The right lung contains three lobes and is larger than the left, which contains only two lobes. The lines of division between lobes extend completely through the lung and are called *fissures.* Each lobe is supplied by a major branch of the bronchial tree and is enclosed by connective tissue.

Lobes are further divided into smaller compartments known as **segments.** Each segment is supplied by a smaller branch of the bronchial tree and is also enclosed by connective tissue. Within each segment are numerous **lobules,** each of which receives a single bronchiole, an arteriole, a venule, and a lymphatic vessel. Lobules contain the branches of the bronchiole— the alveolar ducts—as well as many alveoli and their capillary networks.

CONCEPTS CHECK

1. What is the respiratory organ between the nasal cavity and larynx?

2. What is the function of the cartilage rings in the trachea and primary bronchi?

3. What are the components of the respiratory membrane?

4. What membranes surround each lung?

MECHANICS OF BREATHING

Inspiration requires the contraction of the respiratory muscles to expand the volume of the thorax and lungs. Expiration requires no muscle activity, but rather the recoil of elastic tissue, while forced expiration requires muscle contraction to push air outward.

Breathing, or *pulmonary ventilation,* provides for an exchange of air between the external environment and the spaces within the alveoli of each lung. It sets the stage for the vital process of gas exchange with the bloodstream. Breathing involves two events: **inspira-**tion (inhalation, or breathing in) and **expiration** (exhalation, or breathing out).

Inspiration

As you may recall from previous chapters, pressure is caused by the collisions of molecules within a space. In the external environment, the collisions of molecules in the air at sea level are capable of pushing a column of mercury 760 mm upward in a tube. This value represents the normal air (or atmospheric) pressure, and is demonstrated in Figure 15–10a.

Air pressure may be varied by changing certain physical parameters, such as raising or lowering the temperature, changing the number of molecules in the space, or changing the volume of the space. Each of these variables affects the number of collisions between molecules as they move randomly about (Fig. 15–10b). If any of these variables were used to decrease air pressure in a space that is open to the external environment, a pressure gradient would be formed that would cause air molecules to rush in. For example, if you increased the volume of a sealed container, you would decrease the number of collisions between the molecules in it and thereby decrease its pressure below normal air pressure. If you suddenly opened the lid of the container, what do you suppose would happen? In response to the pressure gradient, air molecules would rush into the container until its air pressure was restored to normal.

Air rushes into the lungs during inspiration in a similar way. The physical variable that provides the pressure gradient is *volume.* At rest, the air pressure inside the alveoli of the lungs is about the same as normal air pressure (760 mm of mercury). An increase in volume of the lungs causes the air pressure in the alveoli to drop about 1 to 3 mm of mercury below the external air pressure. Although this pressure gradient seems quite small, it is large enough for air to be drawn into the air passageways and into the alveoli.

The first step in inspiration is contraction of the respiratory muscles. These include the **diaphragm,** which presses downward into the abdominal cavity; and the **external intercostal muscles** between the ribs, which raise the ribs and elevate the sternum (Fig. 15–11). The result is the expansion of the thoracic cavity.

As the thoracic cavity expands, the parietal pleurae, which are attached to its inner wall, are pulled outward. This increases the volume of the pleural cavity, causing the pressure within this space to drop. The sudden decrease in pleural cavity pressure draws the visceral pleurae and the attached lung surfaces outward, much the way a household vacuum cleaner draws dust into its canister.

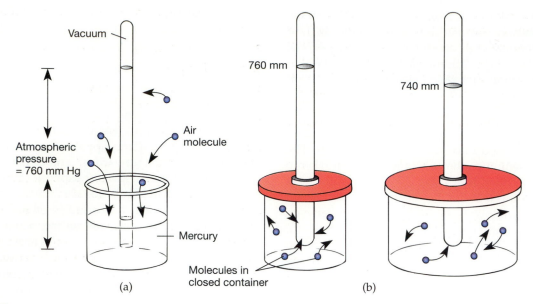

Figure 15–10
The influence of pressure gradients in breathing.
(a) Atmospheric pressure is formed by the random movement of molecules in the air; it can push a column of mercury in a vacuum up a distance of 760 mm (the arrows represent molecules in motion). (b) When the volume of a sealed container containing a fixed number of molecules is increased, the pressure within the container drops. This phenomenon occurs when the lungs expand during inspiration.

Figure 15–11
Mechanics of breathing. (a) Expansion of the thoracic cavity occurs when the diaphragm and external intercostal muscles contract. (b) During inspiration the diaphragm pushes downward and the ribs are elevated. (c) During expiration the diaphragm moves upward as it relaxes and the ribs are depressed.
What muscles enable you to inhale?

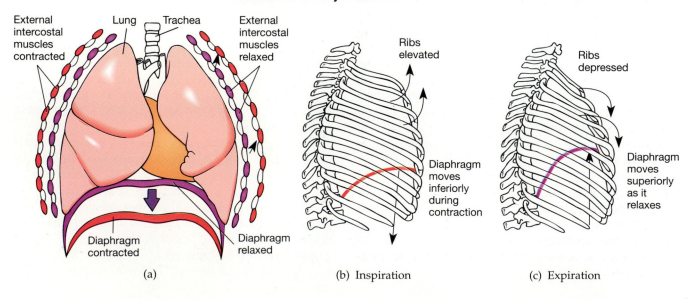

As the surface of the lungs is pulled outward, the internal volume of the alveoli within it expands. This decreases the alveolar pressure by about 1 to 3 mm of mercury, thereby establishing a pressure gradient between the alveoli of the lungs and the external atmosphere. As a result, air rushes into the alveoli, thereby restoring equilibrium.

The sequence of events in inspiration can be summarized as follows:

1. The diaphragm and external intercostal muscles contract.
2. The thoracic cavity expands.
3. Pleural cavity pressure decreases.
4. The lung surface is pulled outward, causing the lung volume to increase.
5. Alveolar pressure falls below atmospheric pressure.
6. Air rushes into the alveoli.

Expiration

As we have now seen, contraction of the respiratory muscles initiates the process of inspiration by expanding the thoracic cavity. Unlike inspiration, however, expiration is a passive process, since it does not rely upon muscle contraction. It relies instead on the ability of the lungs and thoracic wall to recoil, somewhat as a rubber band does after it has been stretched.

Once inspiration is complete, the diaphragm and external intercostal muscles relax. The thoracic cavity recoils to its original volume, forcing the attached parietal pleurae to return to their original size as well. As a result, the pleural cavity pressure rises quickly, allowing the visceral pleurae and lung surfaces to recoil to their original shape. As this occurs, the internal diameter of the alveoli gradually decreases as their elastic walls recoil, pushing air outward and into the air passageways. The elastic nature of the alveolar walls is so great, in fact, that the alveoli tend to recoil inward and collapse in on themselves like a rapidly deflating balloon. Their total collapse is prevented by a slightly lower pressure in the pleural cavity, which keeps the alveoli slightly inflated even after expiration. Also, the presence of surfactant in the inner walls of the alveoli prevents them from adhering to one another if they come into contact. Should these provisions fail because of pressure leaks (resulting, for instance, from an open wound in the thoracic wall; the presence of air in the pleural cavity is called *pneumothorax*) or other dysfunctions (such as tumors or compression from other organs), the collapse of the lung or a part of it would result. This dangerous condition, called *atelectasis* (at-e-LEK-ta-sis), obstructs the movement of air.

The process of normal expiration may be summarized as follows:

1. The diaphragm and external intercostal muscles relax.
2. The thoracic cavity decreases in size.
3. Pleural cavity pressure increases.
4. Alveolar pressure becomes greater than atmospheric pressure.
5. Air flows out of the alveoli.

Expiration may not always be a passive process. There may be times when you want to force most of the air out of your lungs. This is called **forced expiration** and follows the controlled contraction of the internal intercostal muscles and the muscles of the abdominal wall. These muscles compress the thoracic cavity, causing its pressure to increase beyond normal. The increase in pressure leads to a further increase in pressure within the alveoli, which forces more air out of the lungs.

Respiratory Volumes

Many factors influence the amount of air that moves into and out of the lungs. For example, it varies during different body activities such as sleep, light exercise, and heavy exercise. Physical differences among individuals, such as sex, height, age, and weight, also influence the volume of air movement. In addition, respiratory volumes may vary according to the state of health of the lungs and other respiratory organs. A common method of examining lung function is by measuring respiratory volumes under different conditions and comparing these figures with average values. The instrument that is used is called a *spirometer*;[13] one is shown in Figure 15–12. It is useful in evaluat-

---

[13]Spirometer: Latin *spirare* = to breathe + Greek *metron* = measure.

Figure 15–12
A spirometer, which is used in determining respiratory volumes.

The Return of Tuberculosis

Tuberculosis, or **TB,** was the leading cause of death in the United States at the beginning of the twentieth century. It was a serious, widespread killer that led physicians to isolate sufferers in special hospitals, called *sanatoriums,* that were often far from population centers. In these hospitals, the victims of TB were quarantined from the rest of society in the hope that their disease would not spread to others. During this time before effective drug therapy came into being, treatments were limited to drastic measures: since the bacteria that caused this disease require high levels of oxygen to survive, doctors would puncture and collapse a patient's infected lung in the hope of suffocating the microbes. Bright sunlight and fresh air were also erroneously thought to help toward recovery. In some cases, TB would go into a state of remission (for reasons other than those believed at the time), but in most cases it resulted in a gradual deterioration until death. With the introduction of sulfa drugs in the early part of the twentieth century, and antibiotics several decades later, TB declined in incidence. For a time, the medical community believed that complete eradication was possible.

Tuberculosis is a contagious infection caused by the bacterium *Mycobacterium tuberculosis* (or tubercle bacillus). It is spread through the air in tiny droplets of sputum that are expelled when an infected person coughs or speaks. The bacteria then enter their next victim by way of the respiratory tract and most often establish themselves in the lungs, although they can travel to almost any organ (they have been found in the brain, kidneys, heart, liver, and bones). Once established, they usually enter a dormant stage in the body, sometimes for decades, and are kept in check by the immune response. The disease is rarely detected early, because its initial symptoms are very mild and the bacterium is in low concentrations. TB becomes active when the immune system falters, as may occur because of an extended viral infection, increasing age, serious injury, or even substance abuse (alcohol or drugs). Once activated, it establishes large colonies in the lungs and other organs and destroys functional tissue, replacing it with fibrous plaques called *tubercles.* Even when activated, TB tends to waste its victims slowly. Gradually, the patient loses weight, energy, and state of mind, until death comes. Nineteenth-century doctors called this deadly disease "consumption" because of the way it ate away, or consumed, its victims.

Although TB was brought under control decades ago and its eradication seemed a possibility, a growing body of evidence indicates that it is back in full fury. Public health experts have noticed a reemergence of TB in recent years, particularly in inner cities. Beginning in 1980, the number of TB cases has risen each year in significant numbers. Some physicians are even suggesting that TB may develop into the plague of the 1990s, just as AIDS was the plague of the 1980s. This resurgence is due to several causes: the growth of immigration from areas where TB is still prevalent (Southeast Asia and Central America); the growing number of AIDS sufferers, who commonly serve as reservoirs for TB; the difficulty of maintaining drug treatment programs for the underprivileged, for the disease sometimes requires daily treatment for over a year; and the development of drug-resistant strains of TB. At the present time, the medical community is reeducating itself to the importance of early diagnosis and treatment of this deadly disease, and research is gearing up to the development of new drug treatments that are effective against all TB strains. For now, TB skin tests still provide reliable diagnoses, and most drug therapy programs (the most common of which includes taking a drug called *isoniazid* every day for six to nine months) are effective against early infections.

---

ing losses in respiratory function and in following the course of some respiratory diseases.

On the average in a healthy adult, normal quiet breathing moves about 500 ml of air into and out of the lungs with each breath. This is called the **tidal volume (TV).** The amount of air that can be inhaled forcibly over the tidal volume is the **inspiratory reserve volume (IRV).** The inspiratory reserve volume normally averages about 3100 ml. The maximum amount of air that can be forcibly exhaled after a tidal expiration is about 1200 ml and is called the **expiratory reserve volume (ERV).** The volume of air remaining in the lungs after a forced expiration averages 1200 ml. This is the **residual volume (RV).** The total amount of exchangeable air is found by adding the tidal volume, inspiratory reserve volume, and expira-

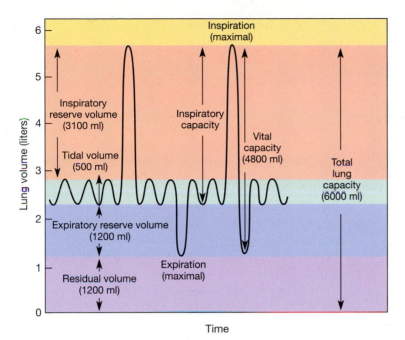

Figure 15–13
Respiratory volumes. This graph illustrates tracings taken from a spirometer. The values represent average figures.

Which respiratory volume represents the maximum volume of air you are capable of inhaling during a deeply forced breath?

tory reserve volume (TV + IRV + ERV). This sum averages about 4800 ml and is called the **vital capacity (VC).** Another value, known as the **total lung capacity (TLC),** is the sum of the vital capacity and the residual volume. It averages about 6000 ml of air. The average respiratory volumes are graphed in Figure 15–13 and summarized in Table 15–1.

During a normal tidal breath, a portion of the air that enters the respiratory tract never reaches the alveoli. The air remains in the air passageways. It amounts to about 150 ml and is called the **anatomic dead space**

volume. The volume of air that actually reaches the alveoli with each tidal breath averages 350 ml.

CONCEPTS CHECK

1. Why is contraction of the respiratory muscles required for inspiration?

2. How do the mechanics of expiration differ from those of inspiration?

3. What is the vital capacity of the lungs?

Table 15–1   AVERAGE RESPIRATORY VOLUMES

Type	Definition	Average Volume
Tidal volume	Amount of air moving into or out of the lungs during quiet breathing	500 ml
Inspiratory reserve volume	Maximum amount of air that can be inhaled forcibly over the tidal volume	3100 ml
Expiratory reserve volume	Maximum amount of air that can be exhaled forcibly over the tidal volume	1200 ml
Residual volume	Amount of air remaining in the lungs following a forced expiration	1200 ml
Vital capacity	The total amount of exchangeable air, determined by the sum of the tidal volume, inspiratory reserve volume, and expiratory reserve volume	4800 ml
Total lung capacity	The total amount of air contained in the fully inflated respiratory system, determined by the sum of the vital capacity and the residual volume	6000 ml

EXCHANGE OF GASES

Successful gas exchange requires the independent diffusion of oxygen and carbon dioxide between alveoli and the blood, and between the blood and body cells.

Gas exchange involves two independent processes, external respiration and internal respiration. Both processes involve the movement of gases by diffusion; that is, the gases move from regions of higher concentration to regions of lower concentration. The rate at which they move depends on how concentrated they may be, or the amount of pressure they exert.

The air we breathe is actually a mixture of several gases, including nitrogen (78%), oxygen (21%), and carbon dioxide (0.04%). The amount of pressure each gas creates is called its **partial pressure.** Partial pressure of a particular gas may be determined by multiplying its percentage of the mixture and the total pressure of all gases in the mixture. For example, if the total pressure of all gases in a mixture is 760 mm Hg (atmospheric pressure at sea level) and if 21% of the mixture is oxygen, then the partial pressure of oxygen in the mixture is calculated as follows:

$$760 \text{ mm Hg} \times 21\% = 160 \text{ mm Hg}$$

Thus, 160 mm Hg represents the partial pressure of oxygen, or P_{O_2}, in the atmosphere at sea level. When a mixture of gases dissolves in the blood, each gas diffuses in response to its own partial pressure. Its rate of diffusion is not affected by other gases in the mixture. As a result, each gas will diffuse independently until its partial pressure becomes equalized in the two regions.

Although gases may exist in several states as they are carried with the bloodstream (dissolved in plasma, bound to proteins, and combined with other chemicals to form ions), only those gases dissolved in plasma affect the partial pressure of a gas. For example, most of the oxygen carried in the blood is bound to hemoglobin, while a very small fraction is dissolved in the plasma. Only the plasma oxygen determines the partial pressure, although it is the smaller amount. This enables the oxygen bound to hemoglobin to increase enormously in volume without affecting the partial pressure of oxygen in the blood. This phenomenon occurs during oxygenation of the blood in the lungs.

The two gases that are most important physiologically are oxygen and carbon dioxide. They are present in both inhaled and exhaled air. Their partial pressures enable them to diffuse independently of one another, permitting the processes of external and internal respiration, as we shall see in the following sections.

External Respiration

External respiration is the exchange of gases between the alveoli and the bloodstream (Fig. 15–14). It occurs across the thin respiratory membrane, which was described earlier in this chapter.

External respiration relies on the fact that there is always a greater partial pressure of oxygen in the alveoli than there is in the blood plasma. This establishes a pressure gradient across the respiratory membrane, which forces the movement of oxygen molecules from the alveoli into the oxygen-poor blood of the capillaries. Most of the oxygen entering the bloodstream (about 98%) is bound by hemoglobin within the red blood cells. Oxygen-filled hemoglobin is referred to as **oxyhemoglobin.** From the alveolar capillaries, hemoglobin carries its bound oxygen through the bloodstream and onward to the tissue capillaries, setting the stage for internal respiration. A small amount of oxygen (about 2%) is carried in a dissolved state within the plasma throughout the journey.

The movement of oxygen from alveoli into the bloodstream is accompanied by the movement of carbon dioxide in the opposite direction. Normally, the blood plasma has a greater partial pressure of carbon dioxide than the alveoli. This establishes a pressure gradient, which leads to the movement of carbon dioxide from the blood into the alveoli.

Internal Respiration

Internal respiration is the exchange of gases between capillaries in the body (other than in the lungs) and body cells. It includes the movement of oxygen from the capillaries to the interstitial fluid, and on to cells. It also includes the diffusion of carbon dioxide from cells into the interstitial fluid, and onward into capillaries. Internal respiration is diagramed in Figure 15–14.

Like external respiration, internal respiration relies on the establishment of a pressure gradient. This gradient occurs between the capillaries and the interstitial fluid, and between the interstitial fluid and the cytoplasm of cells. For the diffusion of oxygen, the partial pressure of oxygen is always greatest in the capillaries, slightly less in the interstitial fluid, and least in the cytoplasm. In response to this gradient, oxygen initially diffuses from its bound state with hemoglobin to the blood plasma. From the plasma, oxygen diffuses across the capillary wall to the interstitial fluid, and onward into the cytoplasm of cells. For the diffusion of carbon dioxide, the partial pressure of this gas is normally highest in the cytoplasm, slightly less in the interstitial fluid, and least in the capillaries. Carbon dioxide therefore moves from the cytoplasm to the interstitial fluid and then into the capillaries.

EXTERNAL
RESPIRATION

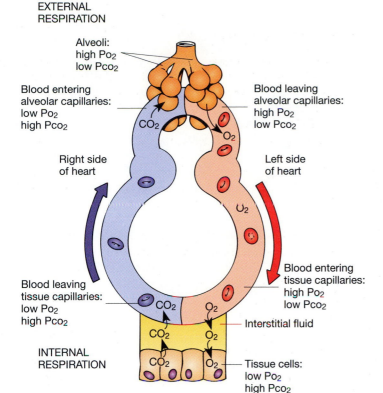

Figure 15–14
External and internal respiration. During external respiration, gases move in response to partial pressure gradients between alveoli and capillaries in the lungs. The net movement of carbon dioxide is into the alveoli, and the net movement of oxygen is into the capillaries. During internal respiration, gases move in response to partial pressure gradients between capillaries and other tissues. Oxygen diffuses into the tissues, and carbon dioxide diffuses into the capillaries.

Why does oxygen diffuse into capillaries from the alveoli?

Once carbon dioxide enters the bloodstream, it changes in form (Fig. 15–15). Most of the carbon dioxide in the bloodstream (about 60% to 70%) is transported as bicarbonate ions (HCO_3^-), mainly in the plasma (with a small amount within red blood cells). The remainder is transported either as carbon dioxide in combination with hemoglobin in red blood cells (25%), or as carbon dioxide dissolved in the plasma (about 10%).

The first step in the change of carbon dioxide to bicarbonate ions occurs shortly after it enters the red blood cell. In this reaction, carbon dioxide and water combine to form carbonic acid. This reaction occurs quickly, because of the help of an enzyme called *carbonic anhydrase*. In the next step, carbonic acid dissociates into a bicarbonate ion and a hydrogen ion. These reactions are shown as follows:

$$CO_2 + H_2O \xrightarrow{\text{carbonic anhydrase}} \underset{\text{carbonic acid}}{H_2CO_3} \longrightarrow \underset{\text{bicarbonate ion}}{HCO_3^-} + \underset{\text{hydrogen ion}}{H^+}$$

Once this reaction is complete, the bicarbonate ions move from the red blood cell into the plasma, and most of the hydrogen ions remain within the red blood cell to temporarily combine with hemoglobin molecules.

As circulating blood enters the capillaries supplying the alveoli, the bicarbonate ions are quickly converted back to carbon dioxide and water in the reverse action. This process begins when bicarbonate ions react with hydrogen ions, some of which were previ-

ously released from their combination with hemoglobin. The result is the formation of carbonic acid, which quickly splits apart to form water and carbon dioxide when carbonic anhydrase is present:

$$\underset{\text{bicarbonate ion}}{HCO_3^-} + \underset{\text{hydrogen ion}}{H^+} \xrightarrow{\text{carbonic anhydrase}} \underset{\text{carbonic acid}}{H_2CO_3} \longrightarrow H_2O + CO_2$$

The carbon dioxide thus formed diffuses from red blood cells into the plasma. It thereby becomes available to diffuse across the respiratory membrane into the alveoli.

The removal of hydrogen ions from solution provides an important buffering system for the blood. As we have just seen, hydrogen ions are initially removed when they combine with hemoglobin, and again later when they recombine with bicarbonate ions to form carbonic acid and then carbon dioxide and water. The removal of hydrogen ions by the bicarbonate ions is especially important, because of the large volume of bicarbonate in the blood (recall that most of the carbon dioxide in the blood is carried in the form of bicarbonate). The result of hydrogen-ion removal is the prevention of blood pH from decreasing to dangerously acidic levels. The condition that results from a drop in pH, known as *acidosis*, can become life-threatening if measures are not taken to remove the excess hydrogen ions. The most effective measure is an increased breathing rate and depth. For example, a sudden accumulation of carbon dioxide in the blood, such

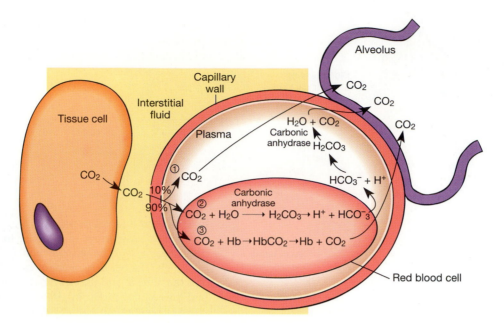

Figure 15–15
Carbon dioxide transport. Soon after carbon dioxide diffuses from tissue cells into the bloodstream (left side of diagram), it is carried in the blood in several different ways. (1) About 10% of the carbon dioxide may be dissolved in the plasma, while the remainder enters red blood cells to be (2) transported as bicarbonate ions (HCO_3^-, about 65%) following its conversion, or (3) combined with hemoglobin ($HbCO_2$, about 25%). Once the blood reaches the alveoli in the lung, the reactions are reversed to release carbon dioxide from the plasma, produce carbon dioxide from bicarbonate ions, and release carbon dioxide from hemoglobin. Much of the carbon dioxide thereby diffuses into the alveoli (right side of diagram).
What advantage to the body is provided by the conversion of carbon dioxide to bicarbonate ions?

as occurs during brief, strenuous activity, can potentially overwhelm the buffering system by producing hydrogen ions faster than they can be removed. This burden is quickly compensated by increased breathing, which removes the excess carbon dioxide, thereby restoring the balance normally achieved by the buffering system.

CONCEPTS CHECK

1. What is the pressure of a single gas in a mixture called?

2. External respiration occurs across what thin barrier?

3. What is the final destination of oxygen in internal respiration?

CONTROL OF BREATHING

Breathing rate and depth are controlled by the respiratory center in the brain. They are influenced by factors that act upon the respiratory center.

Breathing is influenced by several mechanisms in the body. The main source of control is through the respiratory center in the brain. It is also affected by chemical changes in the blood, the degree of stretch of the lungs, and a person's mental state.

Respiratory Center

Breathing is a rhythmic, involuntary process that is controlled by a group of neurons in the brain stem known as the **respiratory center.** These neurons may be found in the medulla oblongata and pons (Fig.

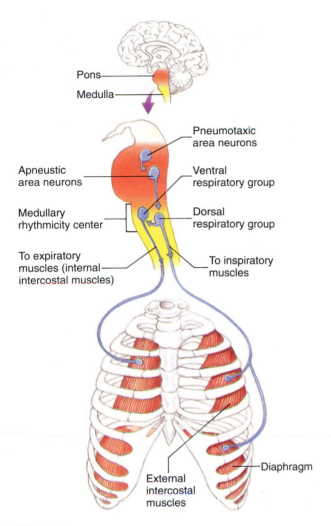

Figure 15–16
Respiratory center. The control centers for respiration are located in the brain stem. They transmit stimulatory or inhibitory signals to the respiratory muscles by way of nerves.

What component of the respiratory center maintains the respiratory rhythm?

the neurons cease abruptly. They remain inactive during expiration and then begin another burst of activity to stimulate another inspiration sequence. This cycle of activity and inactivity repeats itself over and over again to produce rhythmic, tidal breathing.

The neurons of the ventral respiratory group are inactive during normal tidal breathing. They are activated when the need arises to breathe more forcefully. The impulses they generate travel to the diaphragm and external intercostal muscles to stimulate forceful inspiration, and to other thoracic and abdominal muscles to stimulate forceful expiration.

The neurons of the **pneumotaxic area**[14] in the pons are associated with the dorsal respiratory group in the medulla. They transmit impulses to this group to regulate the duration of their inspiratory bursts; their effect is to inhibit the rate of breathing. When the pneumotaxic signals are strong, the inspiratory bursts have shorter durations. This decreases the breathing rate. Conversely, when the pneumotaxic signals are weak, the inspiratory bursts have longer durations, increasingly the breathing rate. By limiting inspiration in this manner, the pneumotaxic area permits expiration to proceed.

The neurons in the **apneustic area**[15] in the lower part of the pons inhibit expiration. They do this by sending stimulatory impulses to the dorsal respiratory group in the medulla, which activate it and prolong inspiration. The apneustic neurons are active only when the neurons of the pneumotaxic area are inactive.

15–16). The areas within these structures where the neurons are most concentrated are the rhythmicity center in the medulla, the pneumotaxic area in the pons, and the apneustic area in the pons.

The **medullary rhythmicity center** contains two groups of neurons extending the length of the medulla, known as the dorsal respiratory group and the ventral respiratory group. The dorsal group controls the basic rhythm of breathing. Its neurons produce bursts of impulses in a cyclic manner, which travel to the diaphragm and external intercostal muscles and stimulate their contraction. After several seconds of activity,

Factors That Affect Breathing

Although the rate and depth of breathing are controlled by the respiratory center, several factors influence it. The most important include chemical changes in the blood, the degree of stretch of the lungs, and a person's mental state.

The concentrations of certain chemicals in the blood vary when physical demands on the body change. These chemicals include oxygen, carbon dioxide, and hydrogen ions. There are areas in the body that are highly sensitive to these chemicals, particularly to carbon dioxide. One area is the **chemosensitive area** in the medulla, which monitors only carbon dioxide levels in the arteries circulating the brain. Out-

[14]Pneumotaxic: Greek *pneumo-* = lung + *taxis* = ordering or arrangement.
[15]Apneustic: Greek *a-* = not + *-pneustic* = pertaining to breathing.

Respiration and SCUBA Diving

When a diver submerges under water, a new world of exploration and pleasure is opened up. However, along with this new world comes new problems, such as surviving in an environment with little oxygen and changing pressures. Although oxygen can be bottled in tanks and brought into the water with you (by way of a Self Contained Underwater Breathing Apparatus, or SCUBA), the pressure changes cannot be as easily accommodated and must be well understood before diving is attempted.

For every 33 feet below the water surface, the pressure increases by 1 atmosphere (760 mm Hg), because of the increasing weight of the water column. As a diver descends to increasing depths, the air inside the body is compressed (squeezed into a smaller space) by the high pressure surrounding the body. This compression is felt mainly in the chest because of the air trapped in the lungs, although its main effect is on a larger scale: the compression of air molecules within the body forces nitrogen gas (N_2) to be absorbed by the blood and tissues. The movement of nitrogen into body tissues does not become noticeable, however, until the diver exceeds a depth of 100 feet. Beyond this depth, the number of nitrogen molecules in the brain begins to have a narcotic effect, which may lead to errors in decisions that can have disastrous consequences. Also,

the longer the time spent at these depths, the greater the amount of nitrogen absorbed, so divers must monitor their length of stay very closely. However, this so-called "rapture of the deep" is only one concern of diving. Another is the problem of ascending to the surface.

When the pressure under water becomes great, such as in depths of 100 feet, significant amounts of nitrogen in the compressed air within the body are absorbed into tissues. As a diver ascends to the surface, this nitrogen is released by the tissues. If the ascent is slow, the nitrogen has enough time to be transported by capillaries and delivered to the lungs to be exhaled. However, if the ascent is too rapid, the nitrogen forms into bubbles in the blood and tissues, causing tremendous pain and possibly death. This condition is called the "bends," because divers suffering from it were observed to bend over from the pain. Treatment for the bends involves time in a decompression chamber, in which the pressure is raised until the nitrogen bubbles dissolve. Pressure in the chamber is then gradually lowered to give nitrogen the time to move through capillaries and into the lungs.

In addition to a slow, careful ascent to the surface, a diver must also be cautious of the air in the lungs. During descent, the lung volume decreases because of the compression

of all air within the body. As the diver ascends, the compression on the lungs gradually lessens as the pressure decreases, resulting in an expansion of lung volume. Should the diver ascend with a lung full of inhaled air without exhaling to compensate for the increase in lung volume, the lungs may overinflate, causing severe tissue damage. Thus, it is important to exhale gradually as the ascent is made.

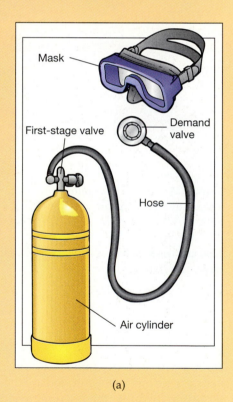

(a)

---

side the nervous system are **chemoreceptors** located within certain arteries (the aorta and the carotid arteries). The cells in these areas monitor levels of all three chemicals (oxygen, carbon dioxide, and hydrogen ions). They respond by sending impulses to the respiratory center that cause the respiratory rate to increase.

During emergency circumstances in which carbon dioxide levels have risen to levels beyond normal, the respiratory center responds by increasing the rate of ventilation, resulting in "blowing off" the excess carbon dioxide. This increased rate is known as **hyperventilation.**

(a) The basic equipment of SCUBA (Self Contained Underwater Breathing Apparatus) diving, which was developed shortly after World War II. Before using this gear, it is critically important to understand the techniques that are necessary to avoid injury or death when making deep dives. (b) The diver in the photograph makes a decompression stop during a dive of 120 feet. This stop is an important safety precaution that permits the gradual release of nitrogen from tissues. A rapid ascent from this depth can lead to a painful, and sometimes fatal, development of nitrogen bubbles in the blood and tissues known as the bends.

(b)

The depth of breathing is regulated by the degree of stretch in the lungs. Within the walls of bronchi and bronchioles are located **stretch receptors.** When these receptors become overstretched during inspiration, they transmit impulses to the respiratory center in the brain. As a result, further inspiration is inhibited and expiration follows. In this way, stretch receptors pro-

tect the lungs from overinflation. As air leaves the lungs, the lungs deflate and the stretch receptors are no longer stimulated, and the stage is set for the next inspiration.

A person's mental state can influence breathing rhythm both involuntarily and voluntarily. Emotional factors such as stress, fear, or pain can increase the rate of breathing automatically. Also, because the respiratory center has connections with the cerebral cortex, you can voluntarily alter breathing rate. The ability to stop breathing is limited, however, by the buildup of carbon dioxide in the blood. When this reaches a certain level, the respiratory center sends impulses to respiratory muscles and breathing resumes whether it is wished or not. Thus, it is impossible for a person to commit suicide by holding his or her breath.

CONCEPTS CHECK

1. How is inspiration stimulated?

2. What area of the respiratory center controls the rate of breathing?

3. What factor regulates the depth of breathing?

4. How does the respiratory system adjust during exercise?

HOMEOSTASIS

The respiratory system assists the body in maintaining homeostasis in two ways: it ensures that sufficient levels of oxygen are supplied to the bloodstream; and it eliminates carbon dioxide from the bloodstream to the external environment. As we have seen, the mechanisms that regulate its function (the reflex centers in the brain stem) can do so by altering the rate and depth of respiration. This is done in an effort to adjust for an increased or decreased need for oxygen by body tissues, and to prevent a potentially toxic accumulation of carbon dioxide.

An example of a homeostatic adjustment by the respiratory system occurs during strenuous exercise. During exercise, the depth of breathing (tidal volume) may increase sixfold, while the rate of breathing may experience a fourfold increase. Pulmonary ventilation, which is normally 5 to 6 liters per minute at rest, may rise 150 liters each minute during extremely active, short-term exercise. This pattern of breathing is called **hyperpnea** (hī-PERP-nē-a) to distinguish it from the emergency breathing reaction characteristic of hyperventilation. The deeper, more rapid breathing during exercise is sufficient to maintain constant oxygen levels in the blood. This is evidenced by the fact that, during exercise, oxygen and carbon dioxide partial pressure levels in the blood remain at about the same levels as they are during rest. Thus, the adjustment of the respiratory system to exercise is extremely efficient in its ability to maintain homeostasis during periods of activity.

CLINICAL TERMS OF THE RESPIRATORY SYSTEM

Anoxia   An absence or deficiency of oxygen within tissues.

Apnea   A temporary state in which there is an absence of breathing.

Asphyxia   A condition in which there is a deficiency of oxygen and an excess of carbon dioxide in the blood and other tissues.

Asthma   A respiratory disease characterized by pulmonary edema (fluid accumulation in lung tissues), inflammation, and repeated attacks of spasms of the smooth muscles in the walls of the smaller bronchi. The edema, inflammation, and spasms narrow the openings in the bronchi, restricting airflow. Symptoms include expiratory wheezing and shortness of breath.

Cystic fibrosis   An inherited disorder of the exocrine glands manifested by the secretion of very thick and sticky mucus and an unusual secretion of sweat and saliva. The thick mucus blocks the smaller airways within the lungs, often trapping air inside alveoli that causes either overexpansion of the lungs or lung collapse. This leads to progressive lung damage that is fatal. The thick mucus also disturbs digestion and absorption in the digestive tract.

Dyspnea   Difficulty in breathing.

Emphysema   A chronic disease of the alveoli and alveolar ducts that leads to the destruction of their walls, resulting in a decreased surface area for gas exchange and decreased elastic recoil of the lungs during expiration. Because of the growing need to expand the thoracic wall forcefully to breathe, the use of accessory muscles of inspiration can often cause the chest to develop a barrel shape, which is a characteristic of advanced emphysema. The destruction of the alveolar walls is thought to be caused by leukocytes introduced during inflammation. Thus, any cause of chronic inflammation of the lungs may lead to emphysema, including cigarette smoking and the prolonged inhalation of dust and pollutants.

Hemothorax   The presence of blood in the pleural cavity.

Hypoxia   A reduced availability of oxygen in the blood.

Lung cancer   The most fatal type of cancer in the United States. It metastasizes (spreads and infiltrates) quickly to other tissues and organs to form secondary tumors. There is a strong association between lung cancer and cigarette smoking: the occurrence of lung cancer is about 20 times higher in cigarette smokers than it is in nonsmokers. Also called *bronchogenic carcinoma,* it most commonly affects male smokers at or past middle age. The incidence in women is increasing, and it has surpassed breast cancer as the most common malignancy in women in the United States. Symptoms include coughing, lung and throat irritation, and blood-tinged sputum.

Pneumoconiosis   A chronic lung disease resulting in the reduction of respiratory membrane surface.

It is caused by excessive exposure to toxic dusts, such as coal or cigarette smoke.

Pneumonia   An infection of the alveoli caused by viruses, bacteria, or fungi. It is characterized by the accumulation of fluid within alveoli, which quickly fills with the infectious pathogens and attacking white blood cells. Oxygen exchange is usually diminished. If not treated, the infection can cause death.

Pneumothorax   A condition in which air enters the pleural cavity, disrupting cohesion between the pleurae and causing collapse of the lung.

Pulmonary embolism   A blockage of blood flow to the lungs by a moving blood clot, or embolus, that has originated elsewhere in the body. Its symptoms include a sudden shortness of breath, due to the loss of lung function in a region that was previously supplied blood. Approximately 10% of pulmonary emboli are fatal.

Quinsy   A bacterial infection of the tonsils that causes them to swell, often leading to obstruction of the pharynx.

Respiratory infections   Infections of the mucous membrane lining the respiratory tract, and often obstructing airflow. They include the common cold (acute coryza) and flu (influenza), pharyngitis, acute epiglottitis, laryngitis, tracheitis, and bronchitis. When the infection reaches the alveoli, it is known as *pneumonia.*

Rhinitis   Inflammation of the nasal cavity lining

Sinusitis   Inflammation of the paranasal sinus lining.

Sudden infant death syndrome   Also known as *SIDS* or *crib death;* a completely unexpected and unexplained death of an apparently healthy infant. It is the most common cause of death in infants between the ages of two weeks and one year in the United States, and strikes infants who were premature with greater frequency. It is characterized by sudden death during sleep when the infant simply stops breathing. Many investigators therefore believe that abnormal sleep patterns are responsible.

Tracheotomy   An incision through the tracheal wall to clear the air passageway or for exploration.

The movement of oxygen and carbon dioxide between the external environment and the cells is called *respiration*. It includes three events: pulmonary ventilation, external respiration, and internal respiration.

ORGANS OF THE RESPIRATORY SYSTEM

Nose The nose is bordered by the external nares and the vestibule. The space within the nose is the nasal cavity. Its mucous membrane lining warms and moistens inhaled air, and mucus streams carry trapped particles to the throat. The cavity is divided into right and left chambers by the nasal septum. It opens to the pharynx through the internal nares.

Pharynx The chamber between the internal nares and the larynx. Its walls are composed of skeletal muscle lined with mucous membrane. It is divided into the nasopharynx, the oropharynx, and the laryngopharynx.

Larynx The boxlike chamber between the pharynx and the trachea. Its walls are made up of nine pieces of cartilage: thyroid, cricoid, and epiglottic cartilage and three pairs of smaller cartilages. Mucous membrane folds extend across the glottis to form two layers: an upper layer, the false vocal cords; and a lower layer, the true vocal cords. Vibration of the true vocal cords as air rushes past produces sound.

Trachea Located between the larynx and the bronchi, it is supported by incomplete rings of cartilage and is internally lined with mucous membrane. Like other parts of the conduction zone, its lining forms the mucociliary transport system, which carries particles trapped in sticky mucus toward the pharynx.

Bronchial Tree The trachea divides in the chest into the right and left primary bronchi, which branch into smaller and smaller tubes to form the bronchial tree. Small bronchi that lack cartilage rings are called the *bronchioles;* they continue into alveolar ducts, which terminate in round alveoli.

Alveoli and Respiratory Membranes Alveoli are composed of a thin epithelial wall with interspersed cells that produce surfactant. A respiratory membrane consists of the alveolar wall, a basement membrane of connective tissue, and the endothelial wall of a capillary.

Lungs Both lungs consist of branches of the bronchial tree, alveoli, and supportive tissues.

General Characteristics The lungs are located in the thoracic cavity within the rib cage. The narrow superior portion is the apex, and the broad inferior portion is the base. The costal surface lies against the ribs, the medial surface faces the heart, and the root attaches to each lung at the medial surface.

Serous Membranes The outer layer of membrane surrounding each lung is the parietal pleura, which attaches to the thoracic wall. The inner layer is the visceral pleura, which is attached to the outer lung surface.

Divisions Each division is enclosed by connective tissue and is supplied air by a branch of the bronchial tree. Each lung is divided into two or three lobes. Each lobe is divided into segments. Each segment is divided into lobules.

MECHANICS OF BREATHING

Pulmonary ventilation is the movement of air between the external environment and the alveoli.

Inspiration Initiated by the contraction of the respiratory muscles, the diaphragm, and the external intercostal muscles. The muscles expand the volume of the thoracic cavity, increasing the volume of the pleural cavity and alveoli. Because of the lower pressure within the alveoli relative to atmospheric pressure, air rushes in.

Expiration Begins when the respiratory muscles relax, causing the thoracic cavity to return to its original volume. In response to this recoil, the alveolar walls also recoil to their original volume, forcing air out. Expiration does not require muscle contraction, although forced expiration requires contraction of thoracic and abdominal muscles.

Respiratory Volumes
Tidal Volumes (TV) Normal quiet breathing; averages about 500 ml.
Inspiratory Reserve Volume (IRV) Forcible inspiration over the tidal volume; averages about 3100 ml.
Expiratory Reserve Volume (ERV) Forcible expiration over tidal volume; averages about 1200 ml.
Residual Volume (RV) Air remaining in lungs following a forced expiration; averages about 1200 ml.
Vital Capacity (VC) Sum of tidal volume, inspiratory reserve volume, and expiratory reserve volume; averages about 4800 ml.
Total Lung Capacity (TLC) Sum of vital capacity and residual volume; averages about 6000 ml.
Dead Space Volume The portion of the tidal volume that never reaches the alveoli.

EXCHANGE OF GASES

Oxygen and carbon dioxide move independently of one another. Each diffuses in response to changes in its partial pressure.

External Respiration The diffusion of oxygen and carbon dioxide across the respiratory membranes between alveoli and the bloodstream. Oxygen diffuses from alveoli into the bloodstream because the partial pressure of oxygen is higher in the alveoli. Carbon dioxide diffuses from the bloodstream into the alveoli because the partial pressure of carbon dioxide is higher in the capillaries.

Internal Respiration   The diffusion of oxygen and carbon dioxide between the bloodstream and interstitial fluid, and between the interstitial fluid and the cytoplasm of cells. Oxygen diffuses from the blood into the interstitial fluid and into cells because the partial pressure of oxygen is highest in the blood and lowest in cells. Carbon dioxide diffuses from cells into interstitial fluid and into the blood because its partial pressure is highest in the cells and lowest in the blood.

CONTROL OF BREATHING

Breathing is mainly controlled by the respiratory center in the brain.

Respiratory Center   Located in the brain stem.
  Medullary Rhythmicity Center   Located in the medulla. It generates impulses that stimulate contraction of the respiratory muscles for inspiration.
  Pneumotaxic Area   Regulates the breathing rate by inhibiting inspiration.

  Apneustic Area   Inhibits expiration.
Factors That Affect Breathing
  Chemicals   The chemosensitive area in the medulla and chemoreceptors in certain arteries detect levels of oxygen, carbon dioxide, and hydrogen ions in the blood. They stimulate hyperventilation if carbon dioxide levels rise too high.
  Stretch Receptors   Located in the walls of the lungs. They prevent overextension of the lungs during inspiration.
  Mental States   Breathing rate and depth are under both voluntary and involuntary control. However, you cannot stop breathing completely using conscious thought. Certain emotions can increase breathing rate.

HOMEOSTASIS

The respiratory system helps maintain homeostasis by keeping the oxygen and carbon dioxide levels in the blood relatively constant, despite changing conditions.

KEY TERMS

alveoli (p. 448)
bronchi (p. 448)
expiration (p. 455)

inspiration (p. 453)
larynx (p. 446)
partial pressure (p. 458)

pharynx (p. 445)
pleura (p. 452)
respiration (p. 443)

surfactant (p. 451)
trachea (p. 447)
ventilation (p. 443)

QUESTIONS FOR REVIEW

OBJECTIVE QUESTIONS

1. Which of the following is the process by which oxygen and carbon dioxide diffuse between alveoli and the bloodstream?
   a. pulmonary ventilation
   b. external respiration
   c. internal respiration
   d. cellular respiration

2. Air initially enters the respiratory tract through the:
   a. external nares
   b. vestibule
   c. nasal cavity
   d. internal nares

3. The paranasal sinuses communicate by way of narrow ducts with the:
   a. nose
   b. pharynx
   c. nasal cavity
   d. larynx

4. The segment of the pharynx that borders the larynx is the:
   a. nasopharynx
   b. oropharynx
   c. laryngopharynx
   d. glottis

5. Sound is produced as air causes the _____ to vibrate.
   a. muscles of the pharynx
   b. false vocal cords
   c. true vocal cords
   d. glottis

6. The prominent cartilage of the larynx that is known as the *Adams apple* in males is the:
   a. thyroid cartilage
   b. cricoid cartilage
   c. epiglottic cartilage
   d. none of the above

7. The trachea is prevented from collapsing by:
   a. rigid muscles
   b. incomplete rings of cartilage
   c. the location of the esophagus
   d. the thick mucous membrane

8. A branch of the bronchial tree that is not supported by cartilage and terminates as round pouches called alveoli is a(n):
   a. alveolar duct
   b. right primary bronchus
   c. bronchiole
   d. pleura

9. The diffusion of gases between the alveolar walls and capillary walls occurs across a thin barrier known as the:
   a. visceral pleura c. surfactant
   b. respiratory membrane d. endothelium
10. Each lung is divided into smaller _____ by cleavage lines known as *fissures*.
    a. segments c. lobules
    b. lobes d. lungs
11. The first step in inspiration is:
    a. pleural cavity recoil
    b. outward pull on the visceral pleura
    c. respiratory muscle contraction
    d. alveolar expansion
12. Lung collapse is prevented during expiration by:
    a. holding your breath
    b. a lower pressure in the pleural cavity
    c. surfactant lining the alveolar walls
    d. both (b) and (c)
13. The total amount of air that moves into and out of the lungs during forced breathing is called the:
    a. tidal volume
    b. vital capacity
    c. expiratory reserve volume
    d. total lung capacity
14. The movement of oxygen from the bloodstream into body cells occurs because:
    a. the partial pressure of oxygen is always higher in the blood
    b. the partial pressure of oxygen is always lower in the blood
    c. the pressure of gases is usually higher in body cells
    d. there is more carbon dioxide in the cells
15. Most of the carbon dioxide in the blood plasma is in the chemical form of:
    a. oxyhemoglobin c. bicarbonate ion
    b. carbonic acid d. carbon dioxide
16. Rhythmic breathing is stimulated in a cyclic manner by the _____ located in the medulla oblongata.
    a. pneumotaxic area c. stretch receptors
    b. medullary rhythmicity center d. chemosensitive area

ESSAY QUESTIONS

1. Describe the path a molecule of air must follow from the atmosphere to an alveolus. Identify all structures along this path.
2. Describe a path a molecule of oxygen must follow from an alveolus to a body cell.
3. Describe the process of pulmonary ventilation. Include in your description an explanation of pressure changes.
4. Describe the role of partial pressure in external and internal respiration.
5. How is lung collapse prevented during forced expiration?
6. What are the controlling and regulating mechanisms for breathing, and where are they located?
7. Describe the path of an atom of carbon in carbon dioxide as it travels from a body cell, through the blood, and out of the body.

ANSWERS TO ART LEGEND QUESTIONS

Figure 15-1
What are the two functional divisions of the respiratory system? The two functional divisions are the conduction zone, containing the nose, pharynx, larynx, trachea, and bronchi; and the respiratory zone, containing the bronchioles and alveoli.

Figure 15-2
Through what opening is air channeled as it passes from the nasal cavity to the pharynx? Air passes through the internal nares in order to enter the pharynx from the nasal cavity.

Figure 15-3
What type of tissue provides the larynx with a firm but elastic structure? Hyaline cartilage provides the larynx with a firm, elastic structure.

Figure 15-4
Why can't you swallow solid food or liquid properly while you are talking? Swallowing properly is not possible while talking because talking requires air to be exhaled and the epiglottis to be in the open position. Swallowing with the epiglottis raised would result in dropping solid food or liquid into the glottis, which would stimulate the cough reflex.

Figure 15-5
What prevents the trachea from collapsing after you exhale? The trachea is prevented from collapsing by the presence of C-shaped rings of cartilage that make up most of its walls.

Figure 15-6
What is the function of the mucociliary transport mechanism? The mucociliary transport mechanism provides a method of trapping and removing inhaled particles.

Figure 15-7
Why is the distribution of bronchi and their branches referred to as the bronchial tree? The term *bronchial tree* is used because the branching of bronchi into smaller and smaller lines resembles the branching of a tree.

Figure 15-8
Across what layers must a molecule of oxygen pass as it diffuses from an alveolus and into a capillary?   A molecule of oxygen must pass through the alveolar epithelium, a thin basement membrane, and the capillary endothelium in order to enter a capillary lumen from an alveolus.

Figure 15-9
What membrane is firmly attached to the outer wall of each lung? The visceral pleura is attached to the outer wall of each lung.

Figure 15-10
Does pressure increase or decrease when the volume in a container increases?   Pressure in a sealed container decreases when the container volume increases (because of fewer collisions of molecules).

Figure 15-11
What muscles enable you to inhale?   The diaphragm and external intercostal muscles expand the chest cavity, leading to inspiration.

Figure 15-13
Which respiratory volume represents the maximum volume of air you are capable of inhaling during a deeply forced breath?   The vital capacity (IRV + ERV + TV) represents the maximum volume of air that can be inhaled, which averages about 4800 ml.

Figure 15-14
Why does oxygen diffuse into capillaries from the alveoli?   Oxygen diffuses into capillaries because the partial pressure of oxygen in alveoli is greater than the partial pressure of oxygen in capillaries.

Figure 15-15
What advantage to the body is provided by the conversion of carbon dioxide to bicarbonate ions?   Bicarbonate ions provide a source of pH buffer, which removes hydrogen ions from solution and thereby minimizes the impact of acids on body tissues.

Figure 15-16
What component of the respiratory center maintains the respiratory rhythm?   The dorsal respiratory group within the medullary rhythmicity center maintains the respiratory rhythm.

THE DIGESTIVE SYSTEM

LEARNING OBJECTIVES

After studying this chapter, you should be able to:

1. Identify the two divisions of the digestive system.

2. Define the six processes of digestion.

3. Describe the components of the peritoneum and the organs they are associated with.

4. Describe the structures associated with the mouth.

5. Identify the tongue and pharynx, and describe their role in swallowing.

6. Describe primary and permanent dentition, and the structure of a tooth.

7. Identify the salivary glands, and describe the function of saliva.

8. Describe the structural features of the stomach, and identify its functions.

9. Describe the structure of the pancreas, and identify the function of pancreatic juice.

10. Identify the regulatory mechanisms of pancreatic secretions.

11. Describe the structure and functions of the liver and gallbladder.

12. Describe the structure and functions of the small intestine and large intestine.

Every time we put food into our mouths we begin the remarkable process of digestion. This process converts the food we eat into a form that can be utilized by cells for growth, repair, and energy, and it rids the body of indigestible remains. Usable food may thus become a part of us, or it may be used as fuel to power our activities. Food that is not usable is removed from the body as waste. In this chapter we will investigate the process of digestion and the organs that perform it.

Introduction to Digestive Structure and Function

The organs of the digestive system are organized to maximize the efficiency of reducing foods to smaller particles so that they can be utilized by body cells.

The digestive system consists of a number of organs, each of which perform a function that is vitally important to the overall function of the system as a whole: the breakdown of foods into particles small enough to be absorbed into the bloodstream. In this section, we will investigate the general organization of the digestive system and the specific processes they perform.

Organization

The organs of the digestive system are divided into two main groups: the **alimentary canal**[1] and the **accessory organs.** The alimentary canal, which is also known as the **gastrointestinal (GI) tract,** is a long,

[1]**Alimentary:** *aliment* = nourishment (from Latin) + *-ary* = characterized by.

Figure 16–1
Organs of the digestive system.
What are the organs of the alimentary canal? What are the accessory organs of the digestive system?

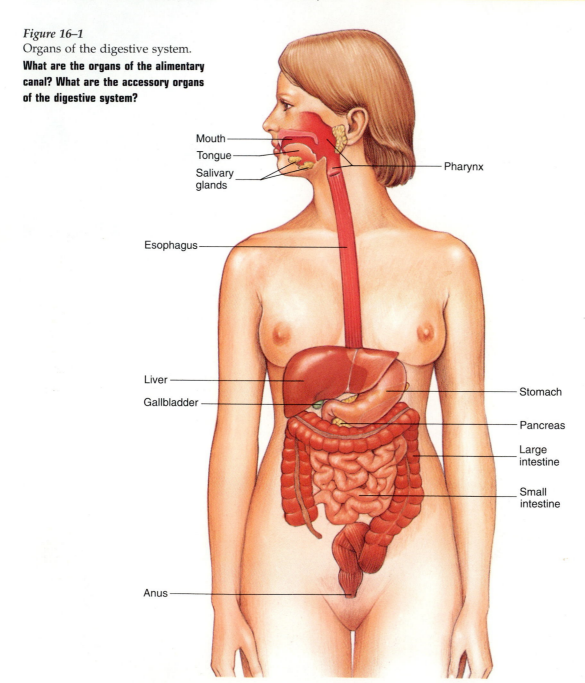

Mouth
Tongue
Salivary glands
Pharynx
Esophagus
Liver
Gallbladder
Stomach
Pancreas
Large intestine
Small intestine
Anus

winding, continuous tube that extends from the mouth to the anus. It averages about 9 meters (30 feet) in length, and consists of the organs that digest food, absorb nutrients, and form solid waste. They are the mouth, pharynx, esophagus, stomach, small intestine, and large intestine. The accessory organs are located either within the alimentary canal or outside it and communicate with it by way of ducts. The accessory organs assist the organs of the alimentary canal in their functions. They are the teeth, tongue, salivary glands, pancreas, liver, and gallbladder. The digestive organs are shown in Figure 16–1.

Digestive Processes

The digestive system operates much like a factory that packages goods. It receives raw materials, processes the materials into a usable form, transports usable products to a distribution network (the bloodstream), and disposes of waste materials. It utilizes an "assembly-line" process involving the interaction of all the digestive organs, each of which performs its own specific role. The most important roles include ingestion, propulsion, mechanical digestion, chemical digestion, absorption, and defecation. These six processes are defined here and are summarized in Figure 16–2. They will be discussed further as we study the digestive organs later in this chapter.

Ingestion—the process of bringing food into the digestive system. The usual point of entry is the mouth.

Mechanical digestion—the breakdown of food particles by mechanical means, which include chewing (**mastication**), mixing with the tongue, churning and mixing in the stomach, and mixing in the small intestine.

Propulsion—the movement of food through the alimentary canal. It involves **swallowing** and **peristalsis** (per-i-STAL-sis). Swallowing is a series of muscular events that push food through the mouth and throat. Peristalsis[2] is the movement of food through the remainder of the alimentary canal by a series of alternate waves of muscle contraction and relaxation. The process of peristalsis is diagramed in Figure 16–3.

Chemical digestion—the breakdown of large molecules of food into their basic chemical building blocks, which are small enough to be

---

[2]**Peristalsis**: Greek *peri-* = around + *stalsis* = placement; thus, a wrapping around, or constriction.

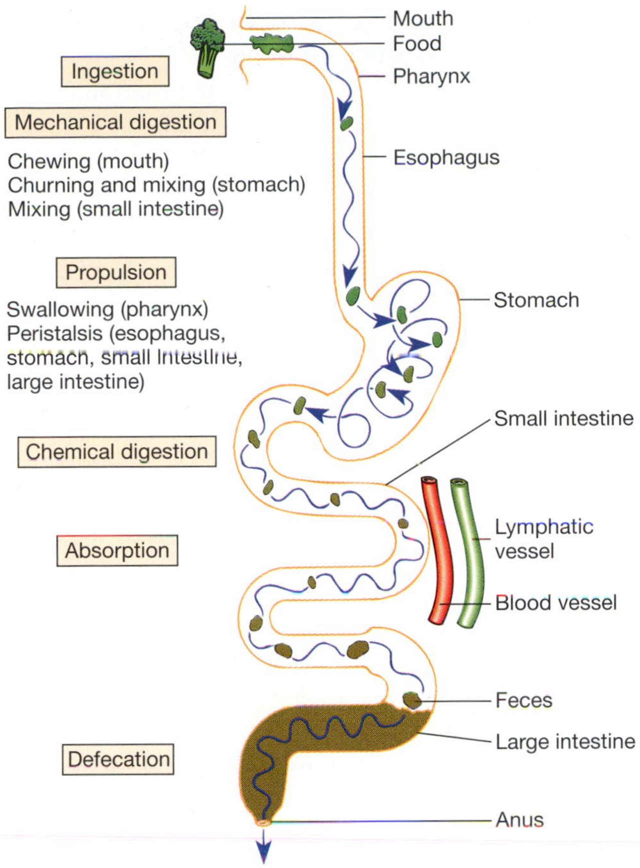

Figure 16–2
Digestive processes. The important processes of digestion are indicated by colored boxes in this diagram of the alimentary canal. The arrows represent the movement of materials through the canal.

In which organ does nutrient absorption occur?

Figure 16–3
Peristalsis. The solid object is food, which is propelled through much of the alimentary canal by the muscles that line its walls. The muscles perform a series of contraction and relaxation cycles, producing a wavelike effect that pushes food forward.

What digestive organs perform peristalsis?

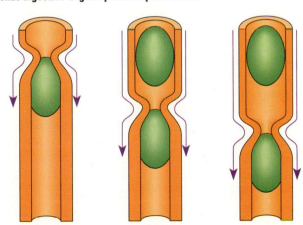

absorbed. It is largely performed by enzymes in the stomach and small intestine.

Absorption—the transport of digested food material from the cavity, or lumen, of the alimentary canal to the blood or lymph. It is primarily performed by the small intestine.

Defecation—the elimination of indigestible material from the body in the form of feces. It is performed by the large intestine.

CONCEPTS CHECK

1. What are the organs of the alimentary canal?

2. What is the definition of propulsion?

3. What is the difference between mechanical and chemical digestion?

SPECIAL FEATURES OF THE DIGESTIVE SYSTEM

The digestive system is, in part, characterized by the presence of the largest serous membrane of the body, the peritoneum, and a similarity of wall structure among most organs of the alimentary canal.

The digestive system contains features that distinguish it from other systems. One of these features is an extensive membrane that is closely associated with many of its organs, known as the peritoneum. Another feature is the similarity in structure among most organs of the alimentary canal.

Peritoneum

As you may recall from Chapter 4, a serous membrane is a thin, moistened sheet of lubricating tissue that covers the walls of the body cavities and the surfaces of organs. We observed it as the pericardium associated with the heart, and again as the pleurae associated with the lungs. The most extensive serous membrane in the body is the **peritoneum**[3] (per-i-tō-NĒ-um). It is located in the abdominopelvic cavity, where it lines the walls as the **parietal peritoneum** and covers the external sur-

faces of most digestive organs as the **visceral peritoneum,** or **serosa.** The potential space between these membranes is called the **peritoneal cavity;** it contains a small amount of serous fluid that reduces friction between organs as they move about.

The parietal peritoneum contains many folds that extend to numerous organs within the peritoneal cavity, serving to suspend or anchor them. The most important of these folds include the falciform ligament, the lesser omentum, the mesentery, and the greater omentum. These peritoneal folds are shown in Figure 16–4.

The **falciform ligament** connects the liver to the anterior abdominal wall and the diaphragm. The **lesser omentum**[4] extends between the medial margin of the stomach and the liver, and also attaches to the anterior abdominal wall. The **mesentery**[5] suspends the coils of the small intestine from the posterior abdominal wall. The extensive **greater omentum** is a double-layered fold of the visceral peritoneum of the stomach. It hangs from the lateral margin of the stomach over the coils of the small intestine. Its abundance of fat deposits gives it the appearance of a lacy apron draping over the lower abdomen. There are large numbers of lymph nodes within the greater omentum, providing it with an immunological role.

Wall Structure of the Alimentary Canal

The organs that form the major portion of the alimentary canal—the esophagus, stomach, small intestine, and large intestine—share a similar wall structure. Each contains four distinct tissue layers, which are shown in Figure 16–5. They are the mucosa, submucosa, muscularis, and serosa. The arrangement of cells within these layers differs among the organs, however. These differences reflect the different roles they play in the digestive process and will be described as we examine each organ later. But first, let's focus on the similarities of wall structure.

Mucosa

The **mucosa**[6] is a mucous membrane that lines the lumen of the alimentary canal. It consists of a thin inner surface layer of epithelium, a middle layer of loose (areolar) connective tissue (called the *lamina propria*), and a small amount of smooth muscle in the external

---

[3]**Peritoneum:** Greek *peri-* = around + *ton-* = stretch + *-eum* = characteristic thing or substance.

[4]**Omentum:** Latin *omentum* = a covering.

[5]**Mesentery:** Greek *mes-* = middle + *enteron* = intestine.

[6]**Mucosa:** Latin *muc-* = mucus + *-osa* = belonging to, relating to.

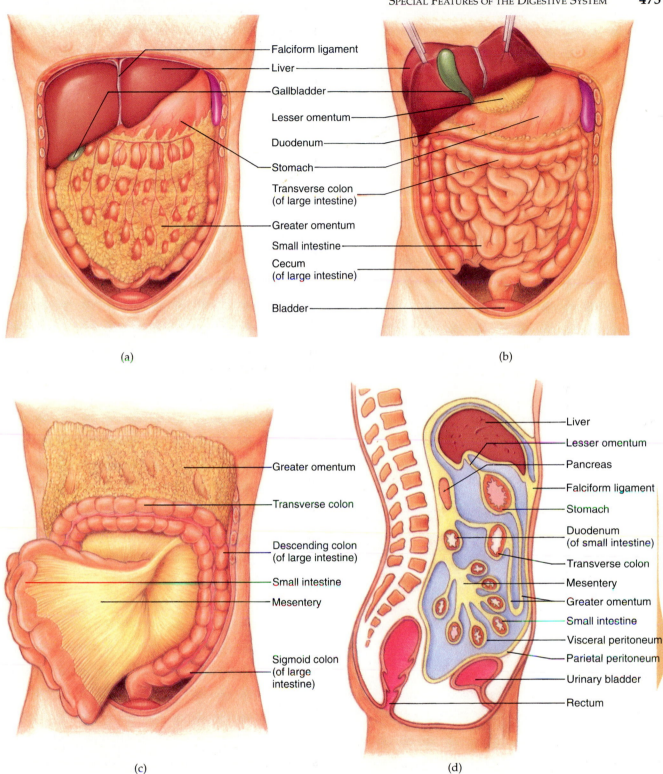

Figure 16–4

Digestive structures in the abdominopelvic cavity. (a) Anterior view of the trunk with the skin, muscle, and parietal peritoneum forming the trunk wall removed. (b) Same view of the trunk, but in this drawing the greater omentum is removed and the liver and gallbladder are pulled upward. (c) Same view as in part (a), but with the greater omentum intact and pulled upward (concealing the stomach and liver) and the small intestine pulled to the right side. (d) Sagittal section through the trunk, lateral view.

How is the peritoneum similar to the pericardium associated with the heart, and the pleurae associated with the lungs?

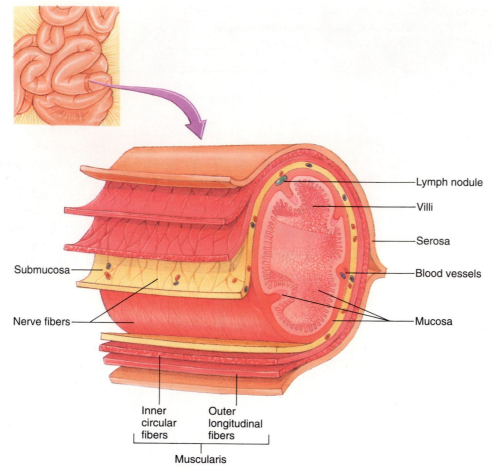

Figure 16–5
Structure of the alimentary canal wall. This drawing of a canal section shows the wall layers that are common to the esophagus, stomach, small intestine, and large intestine: namely, the mucosa, submucosa, muscularis, and serosa. Since this drawing is of a section of the small intestine, it does not illustrate the features that are unique to the esophagus, stomach, or large intestine (which are shown in subsequent figures).

What is the innermost lining of the alimentary canal called?

layer (called the *muscularis mucosa*). In some organs it forms folds, which serve to increase the functional surface area. Its major functions are protection from invading microorganisms, absorption of digested food materials, and secretion of mucus and digestive enzymes.

Submucosa

External to the mucosa lies the **submucosa.**[7] It is composed of loose (areolar) connective tissue rich in blood vessels, lymphatic vessels, nerve endings, and small glands. Its blood vessels nourish surrounding tissues and carry away absorbed materials.

[7]**Submucosa:** Latin *sub-* = below, beneath + *mucosa.*

Muscularis

The **muscularis** is a region of smooth muscle that externally encircles the submucosa layer. In most organs it contains two layers of fibers: an inner, circular layer and an outer, longitudinal layer. At several points along the alimentary canal the circular layer is thickened to form *sphincters,* which act as muscular valves to control food passage from one organ to another. The muscularis provides for the mixing of food during mechanical digestion, as well as the propulsion of food through the alimentary canal by peristalsis.

Serosa

The **serosa** is also known as the visceral peritoneum, which was introduced to you in the preceding section. It is the outer covering of the alimentary tube. It is composed of loose connective tissue with a single layer of flattened epithelial cells, and thus is similar to other serous membranes. The epithelial layer secretes serous fluid, which reduces friction between contacting surfaces.

CONCEPTS CHECK

1. What portion of the peritoneum covers the surfaces of many abdominal organs?

2. With what organ is the falciform ligament associated?

3. What are the functions commonly provided by the mucosa?

4. What tissue is the submucosa composed of?

DIGESTIVE ORGANS

The organs of the digestive system are presented in the normal sequence of food passage, beginning with the mouth.

Mouth

The mouth and its accessory organs, the teeth, tongue, and salivary glands, perform the first stages of mechanical and chemical digestion.

The **mouth** receives food and prepares it for swallowing by mechanically breaking it down into smaller particles (by the process of chewing, or mastication) and moistening it with liquid secretions collectively known as **saliva.** The saliva also begins the process of chemical digestion. The mouth is bordered by the lips, cheeks, and palate. Accessory organs associated with the mouth include the tongue, the teeth, and the salivary glands. The space within the mouth between the tongue and palate is known as the **oral cavity.** The mouth and its associated structures are shown in Figures 16–6 and 16–7.

Lips and Cheeks

The **lips** are usually the first structures that come in contact with food. They are filled with sensory recep-

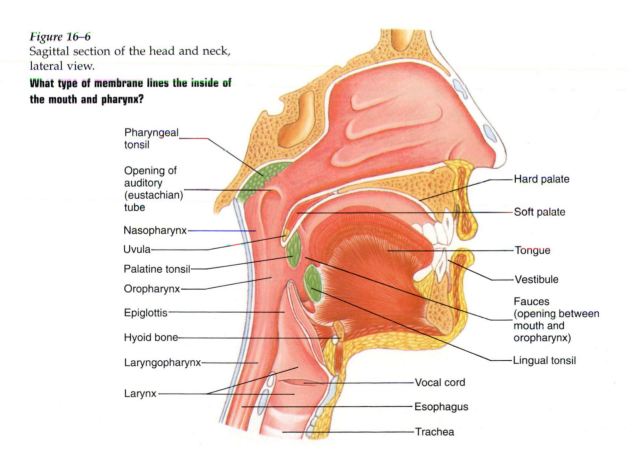

Figure 16–6
Sagittal section of the head and neck, lateral view.

What type of membrane lines the inside of the mouth and pharynx?

Pharyngeal tonsil

Opening of auditory (eustachian) tube

Nasopharynx

Uvula

Palatine tonsil

Oropharynx

Epiglottis

Hyoid bone

Laryngopharynx

Larynx

Hard palate

Soft palate

Tongue

Vestibule

Fauces (opening between mouth and oropharynx)

Lingual tonsil

Vocal cord

Esophagus

Trachea

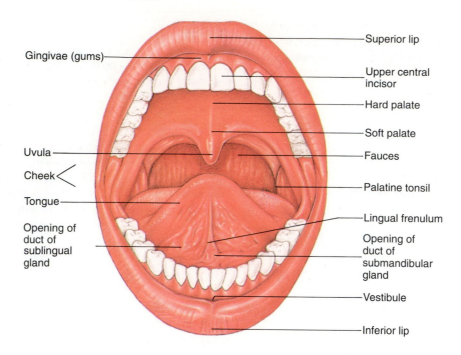

Figure 16–7
The mouth. The tongue is in a raised position in this drawing of the open mouth.
What bony platform forms the anterior roof of the mouth?

tors that enable us to detect fine touch, temperature, and pain. They are composed of skeletal muscle fibers overlaid by a thin layer of skin on the outside and mucous membrane on the inside. Their reddish color is due to the abundance of blood vessels just below the skin layer.

The **cheeks** form the lateral walls of the mouth. They consist of skeletal muscle and fat, and are lined externally with skin and internally with mucous membrane. Both the cheeks and the lips are separated from the teeth by a narrow space called the **vestibule.**[8]

Palate

The **palate** forms the roof of the oral cavity. It contains two sections, both of which are covered with mucous membrane: a hard, bony anterior part known as the **hard palate;** and a muscular, posterior part called the **soft palate.** The soft palate forms an archway that borders the opening of the oral cavity into the pharynx. This opening is called the **fauces.** Along both sides of the archway are the palatine tonsils, which lie partially embedded within small depressions. Extending from the archway in a downward direction is a fingerlike

projection called the **uvula** (YOO-vyoo-la). During swallowing, muscles of the throat draw the soft palate and uvula upward, closing the opening between the pharynx and nasal cavity (the internal nares). This prevents food from entering the nasal cavity.

Tongue

The **tongue** is an accessory organ that occupies most of the oral cavity when the mouth is closed (Figs. 16–6 and 16–7). It is anchored to the floor of the mouth by a membranous fold extending from its midline along its undersurface, called the **lingual frenulum**[9] (LIN-gwal FREN-yoo-lum). It is also firmly attached at its posterior margin to the pharynx and hyoid bone, where the lingual tonsils are located.

The tongue is internally composed mainly of skeletal muscle fibers, and its external surface is covered with mucous membrane. Other skeletal muscles connect it internally to the hyoid bone in the neck. These two sets of muscles change the shape of the tongue and move it about, providing for the mixing of food with saliva as it is chewed. Movement of the tongue

---

[8]**Vestibule:** *vestibule* = entry chamber.

[9]**Lingual frenulum:** *lingual* = pertaining to the tongue + Latin *frenulum* = little bridle.

also enables us to form speech. Its mucous membrane contains tiny projections, called **papillae,** that provide friction for moving food and contain the taste buds (see Chapter 10).

Teeth

The **teeth** provide us with the ability to chew food, a process called **mastication**. The hardness and resistance of teeth to breakage give them a special advantage over most foods we put into our mouths. During mastication, the teeth tear and grind food into smaller particles in preparation for swallowing. In addition to making swallowing considerably easier, mastication of food also increases the particle surface area. This improves the process of chemical digestion by enzymes, since enzymes can act only at surfaces.

Dentition

Two different sets of teeth form during normal development (Fig. 16–8). The first set begins to appear at about 6 months of age and continues to erupt until all 20 teeth are present by about 24 months. The teeth in this primary dentition are referred to as **deciduous teeth.**[10] As the second set of teeth begin to develop and grow, the roots of the deciduous teeth are absorbed, causing them to loosen and fall out between the ages of 6 and 12 years. By the end of adolescence all **permanent teeth** have erupted except for the third molars, or wisdom teeth, which usually emerge between the ages of 17 and 25 years. In a full set there are 32

[10]Deciduous: *deciduous* = falling off, sheddable; thus, temporary.

Figure 16–8
Dentition. (a) The deciduous teeth of the upper and lower jaws. (b) Photograph of a central incisor. (c) Photograph of a canine. (d) Photograph of a first premolar. (e) Photograph of a first molar. (f) The permanent teeth of the upper and lower jaws.
What are molar teeth mainly used for?

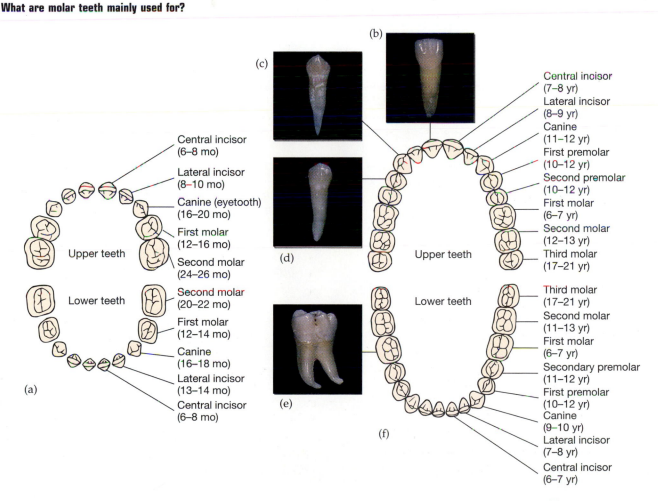

permanent teeth, although in many people some or all of the wisdom teeth fail to erupt and the number is therefore less.

There are several types of teeth in your mouth, which differ in shape according to the particular role they play in chewing food (Fig. 16–8). The **incisors** are shaped somewhat like a pointed chisel and are useful in cutting. The cone-shaped **canines** serve a tearing role, and are the pronounced teeth in predators like cats and dogs. The **premolars** (bicuspids) and **molars** have broad, flattened surfaces that are useful in grinding and crushing.

Tooth Structure

The structure of each tooth is most notable for its hard, corrosion-resistant qualities. There are two regions to a tooth: the **crown,** which is the visible part of the tooth above the gums (or gingivae); and the **root,** which is buried below the gum line. The constricted area where they meet is called the *neck* of the tooth (Fig. 16–9).

The crown is covered by a glossy, heavily mineralized substance called the **enamel.** It is the hardest substance in the body and thereby serves to protect the softer parts of the tooth that lie beneath. Unfortunately,

the cells that produce enamel die soon after the tooth erupts, making it impossible for the body to replace it once it cracks or becomes damaged by decay.

The bulk of the tooth lies beneath the thin layer of enamel and is composed of a substance slightly harder than bone, called **dentin.**[11] The dentin surrounds the central **pulp cavity,** which contains soft tissues including connective tissue, blood vessels, and nerves, collectively known as **pulp.** A small part of the pulp cavity extends through the roots, allowing for passage of the blood vessels and nerves. These narrow channels are called **root canals** and open at the base of the root by way of a small hole.

Each root extends through the gums to become embedded within the jawbone. The root's outer surface is covered by a hardened connective tissue coat called **cementum,** which attaches the tooth to the surrounding **periodontal** (per-ē-ō-DON-tal) **ligament.**[12] This ligament contains bundles of collagen fibers that anchor the tooth in place within the jaw.

[11]Dentin: Latin *dent-* = tooth + *-in* a substance.

[12]Periodontal: Greek *peri-* = around + *odont-* = tooth + *-al* = pertaining to.

Figure 16–9
Structure of a tooth. The molar in this drawing is sectioned to show its internal features.

Once tooth enamel has worn down, what layer remains to protect the tooth?

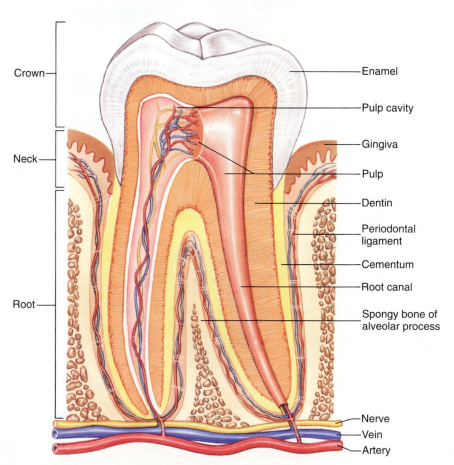

HEALTH CLINIC

Tooth and Gum Disorders

Tooth decay, or *dental caries,* is the main cause of tooth loss up to the age of 35. It is caused by the activities of bacteria, the most common of which is *Streptococcus mutans.* The bacteria convert sugars in the mouth to acid, which demineralizes the surface of the teeth. The enamel is the first to slowly dissolve, followed by the dentin beneath it. Once the enamel has been damaged, a condition of dental caries, or a cavity, has been created. If the decay is not prevented from further erosion, it will eventually reach the central pulp cavity, causing death of the pulp. The result is an abscess in the bone of the jaw surrounding the root of the tooth. Once decay has reached this advanced stage, root canal therapy and restorative procedures are required to save the tooth. If the condition is recognized before the pulp cavity is reached, a dentist may protect the tooth from further decay by inserting materials like silver amalgam, gold, and the more recent resins to fill the cavity.

After the age of 35, the most common cause of tooth loss is gum disease, or *periodontal disease.* Again, bacteria are the main cause. They reside within dental plaque, which is a nearly invisible film that forms over teeth. Unless it is regularly removed by proper mouth hygiene or professional attention, plaque becomes a refuge for bacteria that cause periodontal disease and dental caries. The by-products of bacterial activities first begin to irritate the gums, a condition known as *gingivitis,* which results in tender, bleeding gums. Gingivitis can be reversed by proper cleaning methods. However, if it is not, destruction of the attachment between gum and tooth tissue occurs, and more decay of gum tissue may follow if not professionally treated.

Saliva plays an important role in the prevention of tooth and gum disease. It has an ability to neutralize the cavity-causing acids on the surface of the teeth, and helps rinse away plaque during its early stages of formation. Saliva also supplies the teeth with calcium and phosphorus, which help restore cracks and scratches in the enamel.

Salivary Glands

The **salivary glands** are a collection of accessory organs surrounding the mouth that secrete a fluid called *saliva.* Saliva consists of 99.5% water and 0.5% solutes. The abundant water in saliva provides a liquid medium for dissolving soluble food molecules, which is necessary for the sense of taste. The solutes include mucus, which lubricates and binds food particles to ease the process of swallowing; and two enzymes. One enzyme (lysozyme) destroys bacteria to protect the mucous membrane and teeth from infection and decay. A second enzyme, known as **salivary amylase,** begins the chemical digestion of starchy foods.

The secretion of saliva is stimulated by involuntary nerve impulses originating from the brain when you see, smell, taste, or even think of food. Most of the saliva entering the mouth is produced by three pairs of salivary glands, which secrete their products into ducts that transport the saliva into the oral cavity. They are the parotid glands, the submandibular glands, and the sublingual glands (Fig. 16–10). A small amount of saliva is also provided by small glands that are part of the mucous membrane lining the mouth, called *buccal glands.*

The **parotid glands**[13] are the largest salivary glands. They are located in front of and slightly below each ear, between the skin of the cheek and the masseter muscle. Both parotid glands secrete a clear, watery fluid that is rich in salivary amylase.

The **submandibular glands**[14] are located along the inner surface of the jaw in the floor of the mouth. They secrete a more viscous fluid than the parotids, owing to the presence of mucus in their secretions.

The smaller **sublingual glands**[15] lie in front of the submandibular glands under the tongue. The fluid they secrete contains an abundance of mucus, and so is thick and stringy.

[13]Parotid: Greek *para-* = beside + *otid* = of the ear.

[14]Submandibular: *sub-* = below + *mandibular* = pertaining to the lower jaw.

[15]Sublingual: *sub-* = below + *lingual* = pertaining to the tongue.

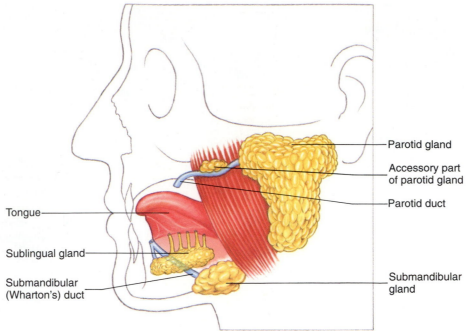

Figure 16–10
Salivary glands. In this lateral view of the head and neck, the skin and some
muscles are removed.
Which is the largest salivary gland?

Digestion in the Mouth

The digestion of food begins in the mouth. When food
first enters, it is subjected to mastication and mixing.
As we have learned, this is a form of mechanical digestion. It serves to reduce the size of food particles in
order to make swallowing easier, and also to increase
the food particle surface area for the improved exposure to enzymes during chemical digestion.

As food is being chewed, the salivary glands secrete saliva into the mouth. Once in contact with the
food particles, salivary amylase immediately begins
the process of chemical digestion by splitting starch
and glycogen molecules into simpler forms. Many of
the simpler molecules produced by this enzymatic activity include the disaccharide *maltose,* which is a combination of two glucose molecules.

The mastication and mixing of food particles with
saliva result in the formation of a compacted mass,
called a **bolus.** Mucus from the saliva coats the outer
surface of the bolus, allowing it to easily slide down
the throat during swallowing.

CONCEPTS CHECK

1. What is the space between the tongue and
   palate called?

2. What functions are provided by the tongue?

3. Where is enamel located on a tooth?

Pharynx

*The pharynx and esophagus transport food from the mouth
to the stomach without processing it further. Food is moved
along by swallowing and peristalsis.*

The pharynx is the chamber located behind the oral
cavity extending from the internal nares to the larynx (Fig. 16–6). As we learned in Chapter 15, its
walls are composed of skeletal muscle lined with
mucous membrane, and it is divided into three segments: the **nasopharynx,** the **oropharynx,** and the
laryngopharynx.

With regard to its digestive function, the pharynx
transports food from the oral cavity to the esophagus.
It therefore plays an important role in the act of swallowing, which is diagramed in Figure 16–11. Swallowing begins as the food bolus is pushed from the
oral cavity into the pharynx by the tongue. As this
occurs, the soft palate rises to prevent food from entering the nasal cavity. This is soon followed by the
contraction of muscles in the wall of the pharynx,
which move the larynx upward as the epiglottis
presses downward. This action closes off the opening
into the airway. Muscles in the lower part of the pharynx then relax, opening the passage into the esophagus. The food bolus is pushed into the esophagus and
onward to the stomach by peristalsis, which begins in
the pharynx and continues along the length of the
esophagus (shown in Figure 16–12).

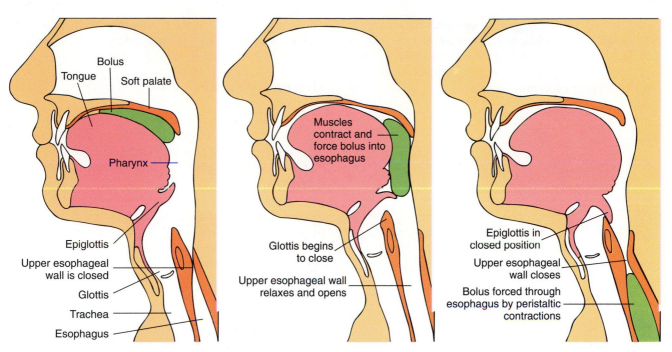

Figure 16–11

The swallowing reflex. (a) The food bolus is positioned by the tongue to the back of the mouth. (b) The soft palate rises, and the tongue and pharynx position the bolus to enter the esophagus. (c) The epiglottis presses downward over the glottis, and the bolus enters the esophagus as it relaxes to receive it.

What benefit is obtained when the epiglottis drops over the glottis during swallowing?

Esophagus

The **esophagus** is a muscular tube that extends from the pharynx to the stomach for about 25 cm (10 inches). It is located behind the trachea in the neck and upper thorax, and is collapsed when it is not propelling food to the stomach. In order to reach the stomach, it penetrates the diaphragm through an opening called the **esophageal hiatus.** The esophagus is shown in Figures 16–1 and 16–12.

The wall of the esophagus consists of an inner mucosa, a submucosa, and a muscularis. Instead of a serosa as its outer layer, the esophagus is externally covered with connective tissue that blends into surrounding structures. The mucosa of the esophagus contains protective stratified squamous epithelium, providing it with a degree of resistance to abrasion. The muscularis layer begins as skeletal muscle, and changes gradually into smooth muscle as the esophagus nears the stomach. Near its union with the stomach, the circular fibers of the muscularis are slightly thickened to form a sphincter, called the **lower esophageal sphincter.** It acts as a valve that is normally closed to prevent the upward movement of material from the stomach. The sphincter opens as peristaltic waves reach the stomach. In some people the lower

Figure 16–12
Peristalsis of the esophagus.

How does peristalsis push food along?

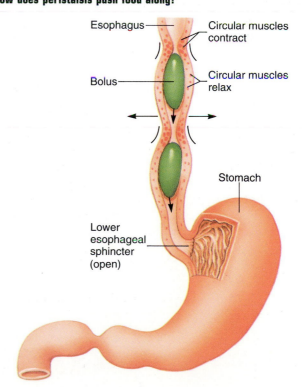

esophageal sphincter is quite weak and thereby permits the leakage of acidic gastric juices into the esophagus. The juices cause an irritation in the lining of the esophagus, which produces a condition of local pain called *heartburn,* especially after a meal.

CONCEPTS CHECK

1. What roles do pharyngeal muscles play in swallowing?

2. How is food moved through the esophagus?

3. What prevents the upward movement of material from the stomach?

Stomach

The stomach exhibits specialized structural features that enable it to perform mechanical digestion, begin the chemical digestion of proteins, and propel material into the small intestine.

Soon after the esophagus penetrates the diaphragm and enters the abdominal cavity, the alimentary canal expands suddenly to form the **stomach** (Fig. 16–12).

The stomach is a temporary storage container for food, and an important site for mechanical and chemical digestion.

Structure of the Stomach

The structural features of the stomach are shown in Figure 16–13. This large, pouchlike organ is about 25 cm (10 inches) long and in most people can hold over 2 liters of food. When it is empty, its inner linings form deep folds known as **rugae**[16] (ROO-jē), which disappear as its walls stretch to receive incoming food. The convex lateral margin of the stomach is called the **greater curvature,** and the concave medial margin is known as the **lesser curvature.** As we observed earlier, the greater omentum extends from the greater curvature and the lesser omentum from the lesser curvature.

The stomach is divided into four regions: the cardia, the fundus, the body, and the pylorus. The **cardia** is a small area surrounding the opening that receives food from the esophagus. The **fundus**[17] is the expanded region that bulges above the cardia; it serves as a temporary holding area for food. The **body** is the main part of the stomach, which lies between the fun-

[16]Rugae: Latin *rugae* = wrinkles.
[17]Fundus: Latin *fundus* = base.

Figure 16–13
Structure of the stomach.
How many layers of smooth muscle are in the stomach wall?

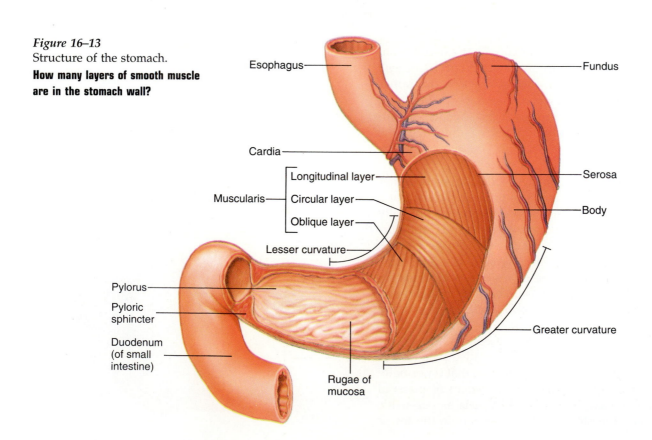

Esophagus — Fundus — Cardia — Longitudinal layer — Circular layer — Muscularis — Oblique layer — Lesser curvature — Serosa — Body — Pylorus — Pyloric sphincter — Duodenum (of small intestine) — Rugae of mucosa — Greater curvature

dus and the pylorus. The **pylorus**[18] is the narrowed, inferior region. At the terminal end of the pylorus the muscular wall thickens to form a powerful sphincter muscle, called the **pyloric sphincter.** This muscle serves as a valve to control the movement of material from the stomach to the small intestine.

Stomach Wall

The wall of the stomach contains the four basic layers that typify most organs of the alimentary canal: the mucosa, submucosa, muscularis, and serosa. However,

the mucosa and muscularis exhibit differences that relate to the specific roles of the stomach.

The mucosa of the stomach is characterized by millions of tiny openings that can be seen only with a magnifying lens or microscope (Fig. 16–14). They are called **gastric pits**[19] and are lined with the surface epithelium

---

[18]Pylorus: Greek *pylorus* = gatekeeper.
[19]Gastric: Greek *gaster* = stomach + *-ic* = pertaining to.

Figure 16–14
Microscopic structure of the stomach wall. (a) A sectioned stomach, exposing the stomach mucosa. (b) Electron micrograph of the mucosal surface (approximately 800x). The holes are the gastric pits. (c) A small portion of the stomach wall is drawn in this magnified section to show its four basic layers.

What three major secretions emerge from the gastric glands during a meal?

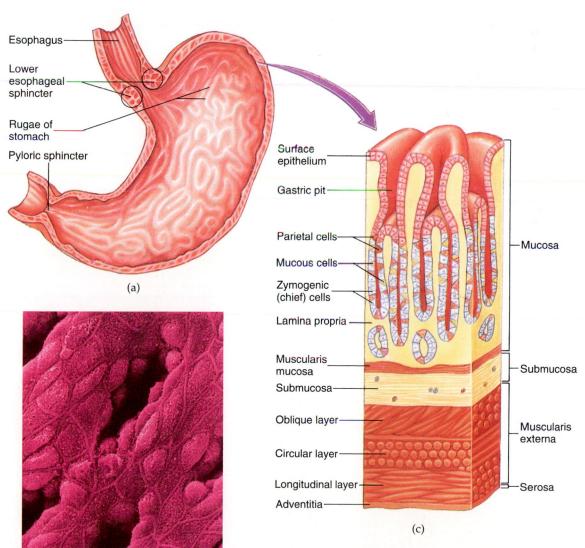

of the mucosa. The gastric pits lead into tubelike **gastric glands,** which secrete a collection of chemicals that are referred to as **gastric juice.** Under normal dietary conditions, about 2 to 3 liters of gastric juice pour out from the gastric glands per day. The gastric glands contain a variety of secretory cells, including **zymogenic (chief) cells,** which secrete digestive enzymes; **parietal cells,** which secrete hydrochloric acid (HCl); and **mucous cells,** which secrete mucus. Parietal cells also secrete a substance called *intrinsic factor.* The composition of gastric juice is summarized in Table 16–1.

The muscularis of the stomach differs from that of other organs in the alimentary canal because of the presence of an additional layer of muscle fibers. This third layer is innermost and consists of smooth muscle fibers that extend in an oblique direction. It provides the stomach with the ability to mix and churn its contents thoroughly, which aids in mechanical digestion as well as increases the chance that all food particles will come in contact with the gastric juice.

Functions of the Stomach

The stomach functions in mechanical and chemical digestion of food, a minor amount of absorption, the propulsion of material to the small intestine (the next organ of the alimentary canal), and the release of intrinsic factor.

Mechanical Digestion

Mechanical digestion in the stomach is accomplished by churning and mixing actions, which are demonstrated in Figure 16–15. These are provided by the muscularis of the stomach wall. Mechanical digestion is more thorough than in other organs of the alimentary canal because of the additional layer of muscle in the muscularis.

Chemical Digestion

Chemical digestion is provided by the gastric juice. Although there are several enzymes present, by far the most important is **pepsin.**[20] It is secreted by zymogenic cells in an inactive form called **pepsinogen** so that it will not digest the cells that produced it. Once entering the cavity of the gastric gland, the pepsinogen comes in contact with hydrochloric acid released by parietal cells, converting it rapidly into pepsin. Pepsin is a powerful protein-splitting enzyme capable of beginning the breakdown of nearly all proteins, and is most active in an acidic environment such as that provided by hydrochloric acid. Pepsin and its acidic environment are prevented from digesting the stomach wall by the protective layer of mucus that coats the mucosa. However, in some people this defensive layer is not enough, and holes in the stomach wall result, called *gastric ulcers.*

Absorption

A limited amount of absorption occurs across the stomach lining. The materials that can cross this barrier include small amounts of water, certain salts, glucose, alcohol, aspirin, and some lipid-soluble drugs.

Propulsion

The propulsion of food through the stomach and into the neighboring small intestine is provided by peristalsis of the stomach wall (Fig. 16–15). The material that is ready to enter the small intestine has been converted by mechanical and chemical digestion into a semifluid paste of small food particles and gastric juice, known as **chyme.** Peristaltic waves push chyme to the pylorus, where it accumulates. As chyme is

---

[20]Pepsin: Greek *pepsis* = digestion + *-in* = a substance.

Table 16–1 THE COMPOSITION OF GASTRIC JUICE

Secretion	Source	Function
Pepsinogen	Zymogenic cells	Inactive form of pepsin
Pepsin	Converted from pepsinogen in the presence of HCl	Chemical digestion of proteins
HCl	Parietal cells	Provides an acidic environment that is needed for pepsin activation
Mucus	Mucous cells	Provides a protective layer along the mucosal lining
Intrinsic factor	Parietal cells	Enables the absorption of the essential vitamin B_{12} across the small intestinal wall

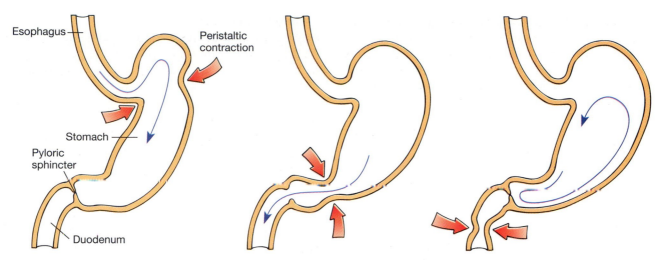

Figure 16–15
Mixing and propulsion in the stomach. The blue arrows indicate movement of
stomach contents, and the red arrows indicate peristaltic waves of the stomach wall.
The waves of peristalsis pass down the stomach, pushing chyme slowly through the
pyloric sphincter. With each wave, most of the chyme does not enter the duodenum,
but rebounds back into the stomach. As a result, chyme is mixed thoroughly before it
is pushed again toward the pyloric sphincter.
What benefit to chemical digestion is provided by mixing in the stomach?

pressed against the pyloric sphincter, the muscle begins to relax and the valve opens slightly. The walls of
the pylorus then pump small amounts of chyme
through the restricted opening and into the small intestine. As a general rule, a meal rich in carbohydrates
moves through the stomach more rapidly than a high-
protein meal, and much more rapidly than a meal high
in fats, which may take 6 hours or more.

Intrinsic Factor

An additional role of the stomach is provided by the
secretion of a substance, known as **intrinsic factor,** by
parietal cells of the gastric glands. This substance aids
the absorption of vitamin B_{12} by the small intestine. Vitamin B_{12} is vital for survival, for it is involved in the
production of mature red blood cells.

Regulation of Stomach Functions

The activities of the stomach are regulated primarily
by involuntary control centers of the brain, and by hormones.

The secretion of gastric juice is stimulated initially
by the hypothalamus, which sends impulses to gastric
glands to begin their activity as soon as you see, smell,
hear, or even think of food. Continued activity by the
gastric glands is initiated by a reflex mechanism, which
begins when food enters the stomach and stretches the
stomach walls. Stretching stimulates receptors in the
stomach wall to quickly route impulses to the medulla
and back to secretory cells in the stomach wall. Once

stimulated, these cells release a hormone known as
gastrin. This important hormone stimulates the continued secretion of gastric juice. Gastrin also stimulates
peristalsis of the stomach.

As chyme enters the small intestine, stomach
digestive activity must be slowed and eventually
stopped to prevent possible damage by its gastric juice.
Further gastric secretions are thereby inhibited by the
release of two intestinal hormones. These hormones
are produced by secretory cells in the intestinal mucosa and are called **secretin** and **cholecystokinin**
(KŌ-lē-sis-tō-KĪ-nin), or **CCK.** Both hormones also inhibit stomach peristalsis and stimulate the secretion of
pancreatic enzymes and bile into the small intestine.
In addition, the presence of acid in the upper part of
the small intestine, pressure in the walls of the small
intestine as it fills, and other factors trigger nerve reflexes that stimulate contraction of the pyloric valve,
causing it to close. This prevents the continued movement of caustic gastric juices into the small intestine.

CONCEPTS CHECK

1. What secretions are provided by the stomach
   mucosa?

2. What mechanisms protect the mucosa from
   the digestive actions of pepsin?

3. What is intrinsic factor?

Pancreas

Much of the pancreas consists of cells that secrete a wide spectrum of digestive enzymes, which are transported by ducts into the small intestine.

The **pancreas** is an accessory organ of the digestive system. It is closely associated with the small intestine, to which it provides a variety of digestive enzymes. As we learned in Chapter 11, the pancreas also plays an important role in endocrine activities.

Structure of the Pancreas

The pancreas is a soft, oblong organ that lies behind the stomach in the upper abdominal cavity (see Figure 16–1). It extends across the abdomen from its **head** near the duodenum, to its **tail**, which touches the spleen. Its middle portion is known as the **body.** A closer view of the pancreas is shown in Figure 16–16a.

Within the pancreas are groups of exocrine secretory cells called **acini** (A-si-nī). One acinus group is shown in Figure 16–16b. Notice that it consists of numerous cells that encircle the end of a duct. The secretory cells release a mixture of enzymes known as **pancreatic juice** into the duct. The ducts from numerous acini drain into the centrally located **pancreatic duct,** which in most individuals fuses with a tube originating from the liver known as the **common bile duct** just before entering the small intestine. The common bile duct carries fluid into the first segment of the small intestine, the duodenum.

Scattered among the acini are small clusters of endocrine cells. These are the islets of Langerhans, which contain the cells that produce the hormones insulin and glucagon. These cells are ductless, releasing their products into the bloodstream.

Functions of the Pancreas

The digestive function of the pancreas is the secretion of pancreatic juice, which finds its way into the duo-

Figure 16–16
Structure of the pancreas. (a) In this external view the ducts that lie within are shown exposed. (b) A magnified view of a part of the pancreas, showing a single acinus group. The cells that secrete digestive enzymes are the acinar cells, which encircle the duct.

What digestive organ receives pancreatic juice from the pancreas?

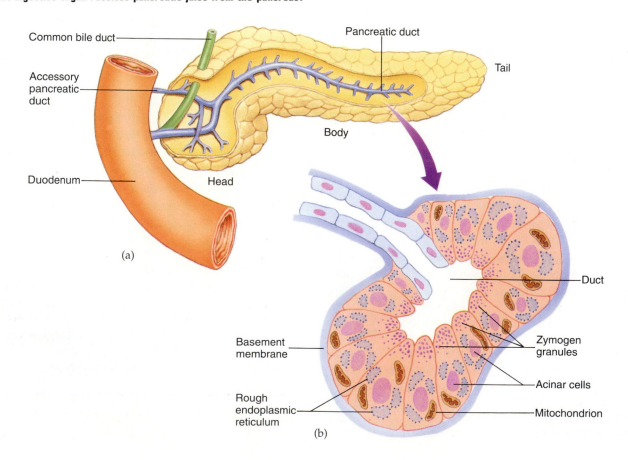

Common bile duct

Accessory pancreatic duct

Duodenum

Pancreatic duct

Tail

Body

Head

(a)

Basement membrane

Rough endoplasmic reticulum

Duct

Zymogen granules

Acinar cells

Mitochondrion

(b)

denum of the small intestine. In an average adult, about 1500 ml of this watery, clear fluid is produced each day. It provides the small intestine with the ability to complete the process of chemical digestion. Pancreatic juice is an alkaline fluid that contains enzymes capable of digesting carbohydrates, proteins, fats, and nucleic acids.

The pancreatic enzyme that digests carbohydrates is known as *pancreatic amylase.* It continues the splitting of carbohydrates that was begun in the mouth by salivary amylase. Similarly, pancreatic amylase splits starch and glycogen molecules into the much simpler compound containing only two glucose subunits, maltose.

Pancreatic juice contains several enzymes that digest proteins. They include *trypsin, chymotrypsin,* and *carboxypeptidase.* These enzymes are initially secreted in inactive forms, much like pepsin in the stomach, to prevent destruction of tissue. They are later activated by other enzymes after they reach the lumen of the small intestine. When the active forms come in contact with proteins, they split the molecules into their amino acid subunits.

The enzyme that aids in the digestion of fats is called *pancreatic lipase.* This enzyme splits triglycerides of fat into their subunits of fatty acids and glycerol.

Pancreatic juice also contains two enzymes that digest nucleic acids. These enzymes are called *nucleases;* they split nucleic acids into their nucleotide subunits.

Regulation of Pancreatic Secretions

The release of pancreatic juice is controlled by involuntary centers of the brain, and by hormones. As the hypothalamus stimulates stomach mucosal cells to secrete gastric juice, it simultaneously stimulates cells of the acini in the pancreas to release pancreatic juice. Hormonal control is provided by the release of two hormones, both of which are secreted by cells in the small intestine. They are secretin and cholecystokinin (CCK).

Secretin is released into the bloodstream in response to the presence of hydrochloric acid in the small intestine. Once it reaches the pancreas, it stimulates the release of a form of pancreatic juice rich in bicarbonate ions. These ions neutralize the acid in chyme and provide a favorable environment for intestinal enzymes. As we learned earlier, secretin also inhibits gastric secretions and gastric peristalsis.

Cholecystokinin is released into the bloodstream in response to the entry of fats and proteins into the small intestine. Upon reaching the pancreas, it stimulates the release of a form of pancreatic juice rich in digestive enzymes. Like secretin, cholecystokinin also inhibits gastric secretions and peristalsis.

CONCEPTS CHECK

1. What cells of the pancreas secrete digestive enzymes?
2. What food substances can be digested by pancreatic enzymes?
3. What two hormones regulate pancreatic secretions?

Liver

The cells that make up the substance of the liver secrete bile, which aids in the chemical digestion of fats in the small intestine. They also perform many other metabolic functions that are vitally important.

The **liver** is the largest visceral organ of the body. It occupies the upper right side of the abdominal cavity and presses against the diaphragm. Like the pancreas, the liver is associated with the small intestine by way of a communicating duct (the common bile duct). The liver is an accessory organ of the digestive system, for one of its primary functions is the production of bile that is necessary in fat digestion and absorption. It is one of the most vital of all body organs, for it has many metabolic and regulatory roles.

Structure of the Liver

The liver is enclosed by a thin layer of fibrous connective tissue, which extends through its interior to divide it into two main sections, called *lobes.* They are the large **right lobe** and the smaller **left lobe** (Fig. 16–17). The falciform ligament, a fold of the peritoneum mentioned earlier, separates the right lobe from the left lobe and suspends the liver from the diaphragm and anterior abdominal wall. The lesser omentum also anchors the liver to the lesser curvature of the stomach.

Each lobe is further divided into numerous **liver lobules** (Fig. 16–18), which are the structural and functional subunits of the organ. Liver cells, which are known as **hepatocytes**[21] (he-PA-tō-sītz), are arranged within each liver lobule into columns that converge toward a **central vein.** Columns of hepatocytes are separated from adjacent columns by channels lined with endothelial cells. These channels are filled with flow-

[21]Hepatocytes: Greek *hepato-* = pertaining to the liver + *-cytes* = cells.

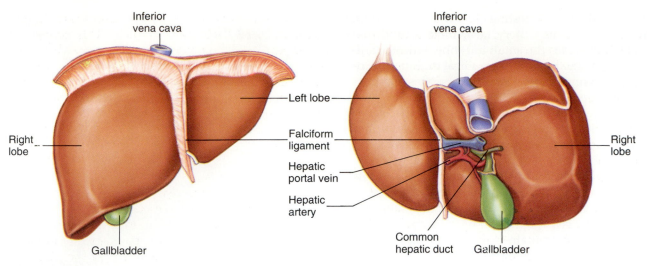

Figure 16–17
Liver structure. (a) Anterior view. (b) Posterior view.
What structure separates the two main lobes of the liver?

Figure 16–18
A liver lobule. The liver consists of hundreds of these microscopic subunits, which consist of packed hepatocytes that modify blood as it passes through. Vessels colored green in this drawing transport newly produced bile. Arrows show the direction of blood and bile flow.
What cells manufacture bile?

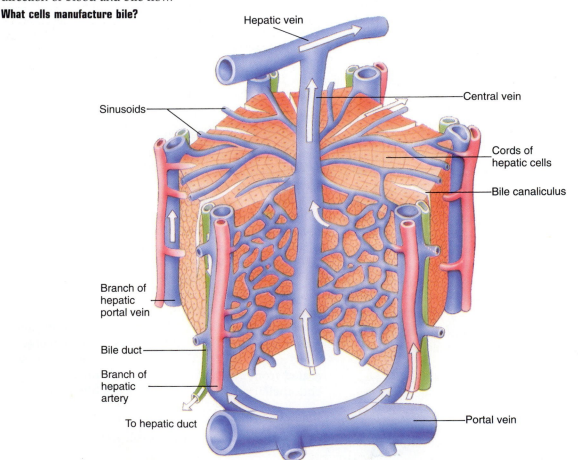

ing blood and are called **sinusoids.** Blood from both the hepatic portal vein and the hepatic artery (discussed in Chapter 13) passes through the sinusoids and drains into the central vein. Attached to the walls of the sinusoids are phagocytic cells, called **Kupffer cells,** which remove bacteria arriving with blood from the digestive tract.

In addition to the sinusoids are smaller channels, called **bile canaliculi** (kan'-a-LIK-yoo-lī). They carry a fluid secreted by the hepatocytes, known as **bile.** Bile is a yellowish-green fluid that contains water, bile salts, bile pigments (bilirubin and biliverdin), cholesterol, and electrolytes. The bile salts in this mixture are the only active participants in the preparation of fats for chemical digestion (to be discussed in the next section); the other chemicals in bile are waste materials. Bile is carried from the bile canaliculi to larger ducts, which converge to form the large **hepatic duct** emerging from the liver. Near the duodenum of the small intestine, the hepatic duct unites with the **cystic duct**[22] from the gallbladder to form the **common bile duct,** which joins the duodenum (Fig. 16–19).

Functions of the Liver

The digestive role of the liver is the secretion of bile. The bile salts in bile do not act as enzymes, but instead aid the actions of fat-digesting enzymes in the small

---

[22]Cystic: *cyst* = sac, bladder.

intestine. Bile salts do this by breaking apart clumps of fat molecules into tiny droplets. This process is called **emulsification** and acts in the same way soap or detergents break oils into smaller droplets. The result of emulsification is an increase in the surface area of the fat droplet, exposing it to the action of fat-splitting enzymes. The enzymes can thereby digest the fat molecules more effectively. In addition to enhancing digestion of fats, bile salts also promote the absorption of fatty acids and vitamins A, D, E, and K across the small intestinal lining.

The liver performs many other vital functions, most of which are related to metabolism. For example, the liver plays a prominent role in maintaining the proper levels of glucose in the bloodstream; in other words, blood sugar homeostasis. Under the direction of hormones released by the pancreas, the liver may store glucose as glycogen for later use or liberate stored glycogen as glucose.

The liver also plays an important role in the metabolism of fats (lipids). It packages fats into forms that can be stored or transported to distant cells. These forms are bound to small proteins and are called **lipoproteins.** The lipoproteins vary in their fat-protein composition, which considerably affects how easily they are transported. In general, the higher the percentage of fat in the lipoprotein, the lower its density, and vice versa. Lipoproteins are therefore classified as high-density lipoproteins (HDLs), low-density lipoproteins (LDLs), and very low-density lipoproteins (VLDLs). The liver is the primary source of VLDLs, which trans-

Figure 16–19
The gallbladder and associated structures.

In addition to bile, secretions from what other organ are carried through the common bile duct?

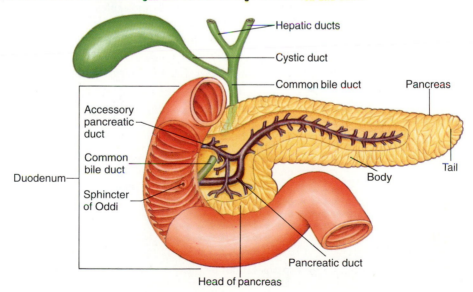

Hepatic ducts
Cystic duct
Common bile duct
Pancreas
Accessory pancreatic duct
Common bile duct
Duodenum
Sphincter of Oddi
Body
Tail
Pancreatic duct
Head of pancreas

port triglycerides synthesized in the liver to adipose tissue for storage. LDLs transport cholesterol to cells outside the liver for membrane or hormone synthesis. HDLs transport cholesterol from cells to the liver, where it becomes part of bile. As we will learn in the next chapter, HDLs are beneficial lipoproteins because they are used in cell repair and growth and do not, therefore, tend to accumulate in body spaces where they are not needed. On the other hand, LDLs can cause health problems, such as *atherosclerosis*, because they tend to accumulate in the walls of blood vessels when their levels in the blood are high.

Perhaps the most important metabolic role of the liver is in protein metabolism. The liver converts amino acids, which are potentially toxic to cells, into products that can be used or removed safely as waste when filtered through the kidneys. Usable by-products of amino acids are necessary for the synthesis of hemoglobin in red blood cells, plasma proteins, numerous other body proteins, and vitamins. Amino acids and other nitrogen-containing waste substances are converted by the liver to urea and uric acid, which are excreted with water and salts as liquid urine.

The liver provides a variety of other functions. For example, it stores glycogen; vitamins A, D, and B_{12}; and iron. The liver also detoxifies harmful substances that arrive in the bloodstream—such as alcohol—by changing their composition. If the liver is unable to change the substance's harmful form, the liver will store it. Toxic materials (such as heavy metals like lead and mercury), pesticides, and other environmental contaminants are examples. In addition, it destroys and removes worn-out red blood cells and other debris by phagocytosis.

Except for the removal of red blood cells, bacteria, and other cell debris, which are performed by Kupffer cells, the varied functions of the liver are performed by hepatocytes. The liver functions are summarized in Table 16–2.

Gallbladder

The **gallbladder** is a small, thin-walled sac that serves as an accessory organ associated with the liver. It is located immediately behind the liver, where it lies in a shallow depression tucked in against the lower margin of the right lobe (Fig. 16–17b). The wall of the gallbladder is internally lined with epithelium, which is supported by connective tissue and an external layer of smooth muscle.

The main functions of the gallbladder are to store and concentrate bile secreted by the liver. It receives bile by way of the **cystic duct,** which extends between the hepatic duct and the gallbladder (Fig. 16–19). The union of the cystic duct and the hepatic duct forms the **common bile duct,** which extends to the duodenum

Table 16–2 FUNCTIONS OF THE LIVER	
Function	**Specific Role**
Secretion of bile	Bile simplifies complex fat molecules by emulsification, which prepares the smaller particles for further digestion by enzymes.
Blood sugar homeostasis (carbohydrate metabolism)	Converts glucose to glycogen, glycogen to glucose, and other nutrient molecules to glucose.
Fat metabolism	Packages fats (triglycerides, cholesterol, and phospholipids) and proteins for transport in the blood as lipoproteins. Also converts excess supplies of carbohydrates and proteins into fat.
Protein metabolism	Catabolizes amino acids for the production of energy. Converts by-products, such as ammonia, into nontoxic forms, such as urea. Also synthesizes plasma proteins.
Storage	Stores glycogen; vitamins A, D, and B_{12}; and iron.
Detoxification	Removes toxic materials, such as worn-out blood cells and foreign particles, by phagocytosis. If toxins cannot be phagocytized, they are stored.

HEALTH CLINIC

Cirrhosis of the Liver

The importance of the liver to normal body functions is made especially apparent when the liver begins to fail. Liver failure may result from any one of several health problems. Accumulation of toxic materials in liver tissue, prolonged damage from viral infection, and traumatic injury are the most common causes. The nature of the damage that each of these problems can inflict on the liver is basically the same: the destruction of healthy hepatocytes. Once hepatocytes are destroyed, the space they occupied is filled by nonfunctional scar tissue composed of connective tissue and fat. Basically, the scarring of the liver from any cause is referred to as **cirrhosis of the liver.**

Cirrhosis of the liver is an irreparable condition. Once hepatocytes are killed in large numbers, they cannot be replaced. The harmful things that destroy them include long-term, heavy intake of alcohol (it is generally considered that one must drink the equivalent of over one pint of whiskey daily for at least ten years to be at risk), prolonged or heavy exposure to heavy metals (such as mercury and lead), heavy exposure to certain poisons (such as those common in pesticides), and viruses (such as the virus that causes hepatitis B). As already mentioned, traumatic injury can also destroy functional hepatocytes.

In its later stages, cirrhosis causes symptoms that include abdominal swelling (called *ascites*), *jaundice* (the accumulation of bile pigments in the skin), and abdominal pain. Once cirrhosis is in its later stages, death due to liver dysfunction usually follows. In recent years, however, liver transplants have been proving increasingly successful. Although the transplant is limited mostly to younger patients because of the unavailability of healthy livers, a growing number of clinics experienced in liver transplantation have been releasing patients with functioning livers and renewed spirits.

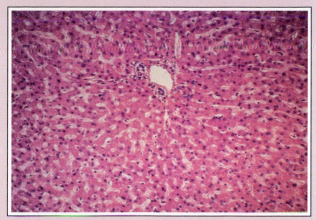

(a)

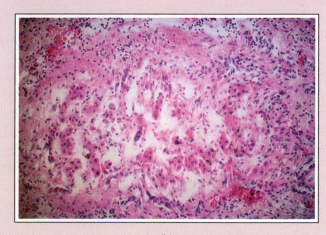

(b)

(a) Microscopic view of a normal liver lobe. (b) Microscopic view of a liver lobe with advanced cirrhosis.

of the small intestine. At the common bile duct's entrance into the duodenum is a small sphincter muscle, known as the **sphincter of Oddi.** This muscle is normally in a contracted state, causing bile to collect in the duct and gallbladder. Its contraction is inhibited by peristaltic waves in the small intestine, resulting in a small amount of bile squirting through into the duodenum with each wave of peristalsis.

CONCEPTS CHECK

1. How are hepatocytes arranged in the liver?

2. What is the role of bile in digestion?

3. What are the metabolic functions of the liver?

Small Intestine

The small intestine is structurally specialized to complete the process of chemical digestion, absorb nutrients, and propel indigestible material to the large intestine.

The **small intestine** is the body's most important digestive organ. It completes the processes of mechanical and chemical digestion, and is the main site of nutrient absorption.

Structure of the Small Intestine

The small intestine is the longest segment of the alimentary canal, extending from the stomach to the large intestine for about 6 m (20 feet). Because of its great length, it occupies a large part of the abdominal cavity even though it is only 2.5 cm (1 inch) in diameter. Its twisting, sausagelike coils are suspended in the abdominal cavity by the mesentery, which anchors it to the posterior abdominal wall. As you can see in Figure 16–1, it is framed by the large intestine from the sides and top.

The small intestine is divided into three segments: the duodenum, the jejunum, and the ileum. The **duodenum**[23] (doo-ō-DĒ-num) receives chyme from the stomach through the pyloric valve. It is only about 25 cm (10 inches) long and is relatively immovable. At the distal end of the duodenum the small intestine continues as the more mobile **jejunum**[24] (je-JOO-num). The jejunum extends for about 2.5 m (8 feet) to the third segment, the ileum. The **ileum**[25] (IL-ē-um) is the longest segment at about 3.6 m (12 feet). It unites with the large intestine at the **ileocecal** (il'-ē-ō-SĒ-kal) **valve**,[26] which helps regulate the flow of material from the ileum to the large intestine.

Small Intestinal Wall

The wall of the small intestine contains the four layers that we find in most other organs of the alimentary canal. However, the mucosa and submucosa have considerable differences from those of other organs. These differences provide for the special functions the small intestine is noted for: absorption and the completion of chemical digestion.

The mucosa is extremely specialized for the absorption of nutrients. If you were to view it without using a microscope, its inner lining would appear velvety—somewhat like a very fine shag carpet. This is due to the presence of tiny projections of the mucosa, known as **intestinal villi**.[27] Shown in Figure 16–20, they project about 1 mm into the lumen of the small intestine, where they come in contact with its contents. Covering each villus is the lining epithelium of the mucosa, whose cells are bristled with thousands of tiny microprojections called **microvilli.** The combined effect of microvilli and villi is to vastly increase the surface area of the intestinal lining, which improves the amount of physical contact between the lining and the contents traveling through the lumen. This makes possible the highly efficient absorption of digestive products. Within each villus are blood capillaries and a lymphatic vessel (called a **lacteal**), which carry absorbed nutrients away with the blood.

Between the bases of adjacent villi the mucosa contains numerous openings, or pits, that lead into tubular glands. These glands are called **intestinal glands** and are the source for a colorless secretion containing water and mucus with a neutral pH. The glands are particularly numerous in the duodenum.

The submucosa contains numerous lymphatic nodules, known as **Peyer's patches,** which protect the body against infectious microorganisms that may try to penetrate the intestinal wall. The submucosa also contains a collection of mucous glands, called **Brunner's glands.** Found only in the walls of the duodenum, Brunner's glands secrete an alkaline mucus that helps neutralize the acidic chyme entering from the stomach.

The mucosa and submucosa are together arranged in deep, permanent folds. These folds force the movement of chyme through the lumen in a spiral flow, causing it to mix with intestinal juice and increasing its contact with the villi. These folds are called **plicae circulares**[28] (PLI-kē sur'-kyoo-LAR-ēz).

Functions of the Small Intestine

The small intestine performs three important digestive functions: (1) it completes chemical digestion that began in the mouth and continued in the stomach; (2) it is the primary site of nutrient absorption; and (3) its peristaltic waves propel indigestible materials into the large intestine.

Chemical Digestion

The small intestine completes the chemical digestion of food by using a mixture of enzymes and bile. By way of the common bile duct, it receives a wide spec-

---

[23]Duodenum: Latin *duodenum* = 12 fingerbreadths in length.

[24]Jejunum: Latin *jejunum* = empty.

[25]Ileum: Latin *ileum* = flank, or groin.

[26]Ileocecal: *ileo-* = pertaining to the ileum + *cecal* = of the cecum, or blind pouch.

[27]Villi: Latin *villus* = tuft of hair.

[28]Plicae circulares: Latin *plicae* = folds + *circulares* = circular.

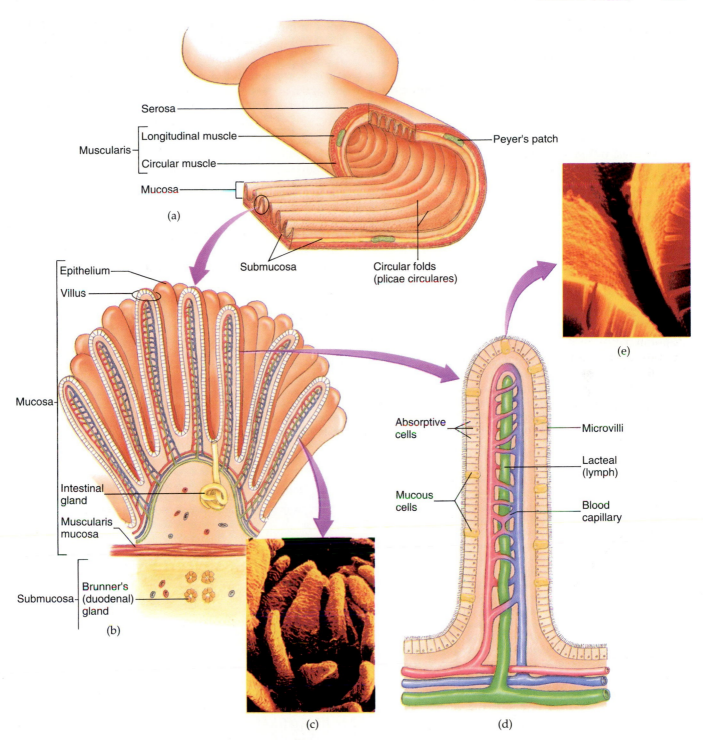

Serosa

Muscularis— Longitudinal muscle

Circular muscle

Mucosa

(a)

Peyer's patch

Submucosa

Circular folds
(plicae circulares)

Epithelium

Villus

Mucosa

Intestinal
gland

Muscularis
mucosa

Submucosa— Brunner's
(duodenal)
gland

(b)

(c)

(e)

Absorptive
cells

Mucous
cells

Microvilli

Lacteal
(lymph)

Blood
capillary

(d)

Figure 16–20

Structure of the small intestine. (a) All four layers of the
small intestinal wall are shown in this drawing. (b) Close-
up of the sectioned mucosa. Notice the numerous villi.
(c) Electron micrograph of the mucosal surface. This view
shows numerous villi projecting into the intestinal lumen.
(d) A single villus in a sectioned view to show its internal
features. (e) Electron micrograph of microvilli on the free
surface of cells lining intestinal villi. Each tiny rod-shaped
structure is a microvillus.

What benefit to absorption is provided by villi and microvilli?

trum of pancreatic enzymes from the pancreas, and bile from the liver. Although the intestinal glands lack digestive enzymes, the microvilli of the intestinal mucosa contain packets of intestinal enzymes. Like the pancreatic enzymes, these powerful molecules break down a wide spectrum of food molecules. They include *maltase, lactase, sucrase,* and *dextrinase,* which digest carbohydrates (simplify disaccharides to **monosaccharides,** or simple sugars); *amino peptidase,* which breaks down short protein chains into free amino acids; *intestinal lipase,* which breaks down fats; and *nucleases,* which cleave nucleic acids. Thus, by the time food has passed through the length of the small intestine, essentially all molecules that can be digested have been simplified to units small enough to be absorbed.

Absorption

The small intestine is the main site of nutrient absorption, which occurs from its lumen across the mu-

cosa lining and into the blood vessels and lacteals. It is effective in absorbing nutrients, including water and electrolytes. Its effectiveness is due mainly to the vast surface area of its mucosa, which is provided by the invaginations of the villi and microvilli. The role of the small intestine in absorption is diagramed in Figure 16–21.

Carbohydrate digestion begins in the mouth and is completed in the small intestine. The resulting simple sugars, or monosaccharides, are transported into the lining epithelium of the villi by facilitated diffusion or active transport. Once inside the epithelial cells, the monosaccharides move passively into blood capillaries by simple diffusion. The absorption of monosaccharides is extremely efficient, resulting in the movement of nearly 100% of usable molecules.

The digestion of protein by pepsin in the stomach, and later in the small intestine by enzymes from the pancreas and intestinal mucosa, results in the avail-

Figure 16–21
Chemical digestion and absorption in the small intestine. During chemical digestion, the three basic types of nutrients entering the small intestine are reduced to their subunits. During absorption, these smaller molecules are transported or diffuse across the epithelial cells lining the lumen. Monosaccharides and amino acids diffuse through the cell to enter a blood capillary, while fat subunits are packaged into chylomicrons, which diffuse through the cell to eventually enter a lacteal.

How does the absorption of digested proteins and fats differ?

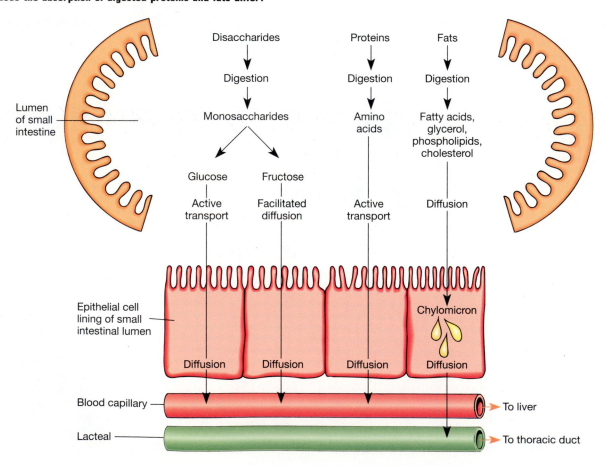

ability of amino acids. These small molecules are absorbed by active transport through the intestinal lining. Once across, amino acids move passively by diffusion through the cells and across blood capillary walls.

Fat digestion occurs by the action of bile salts and enzymes within the small intestine. Most of the fat consumed in an average diet is in the form of triglycerides, which are simplified into fatty acids and glycerol during chemical digestion. Other fats include phospholipids and cholesterol. The fatty acid, glycerol, phospholipid, and cholesterol molecules are readily absorbed into the lining epithelium. Once within the epithelial cells, the molecules of fat are resynthesized into fat droplets (called *chylomicrons*), which migrate to the lacteals of the villi. Lymph in the lacteals carries these droplets to the bloodstream.

Large quantities of water and electrolytes are also absorbed by small intestinal villi. Out of the 3 liters of water entering the small intestine each day in an average diet, only 1 liter passes into the large intestine. The 2 liters absorbed by the small intestine move into the lining epithelium by osmosis. The electrolytes, which include sodium, potassium, chloride, hydrogen, iron, calcium, and others, move into the lining epithelium by active transport and diffusion.

Propulsion

The muscles in the wall of the small intestine provide it with the ability to mix its contents and move them along toward the large intestine. The mixing actions help increase the contact between nutrients and intestinal villi, improving the efficiency of absorption. The contractions that mix the chyme are small, local movements that move it back and forth.

The propulsion of chyme through the small intestine is provided by peristalsis. The peristaltic waves are produced by weak, wavelike contractions of the wall that stop after pushing the chyme a short distance. The result is a slow propulsion that takes between 3 and 10 hours to move chyme all the way through. Should the intestinal wall become irritated because of a harsh diet or infection, however, the peristaltic contractions may dramatically increase, causing a very rapid movement of material through it. This prevents a normal absorption of water, electrolytes, and even nutrients, resulting in abdominal cramps, frequent defecation, and watery stools. This condition is called *diarrhea;* it can produce serious problems in water and electrolyte balance if it persists for a long time.

CONCEPTS CHECK

1. What structural features of the intestinal mucosa enhance absorption?

2. What secretions complete the chemical digestion of food particles?

3. What nutrients can be absorbed by intestinal villi?

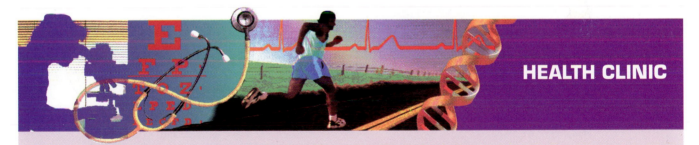

HEALTH CLINIC

Lactose Intolerance

The enzyme *lactase* is one of the intestinal enzymes present on the microvilli lining the small intestine in many people. In some individuals, however, the production of lactase may be incomplete throughout their lifetime, or it may stop around adolescence. This condition is called *lactose intolerance.* Because lactase breaks down the primary carbohydrate in milk, lactose, a lack of it reduces the ability to properly digest milk and its products. As a result, when milk products are ingested they pass quickly through the stomach and small intestine undigested. Once within the large intestine, the bacterial inhabitants proliferate because of the abundance of lactose, producing extreme discomfort by causing cramps, diarrhea, and increased gas generation. Special diets must be prepared that eliminate milk products or substitute them with milk-like derivatives. Alternatively, a person suffering from this condition may take lactase supplements orally before eating.

There appears to be an inherited basis for lactose intolerance. Only about 15% of Caucasians develop it, while 80% to 90% are estimated to suffer from it in the black and Asian populations. The reason for this distribution among populations remains unknown.

Large Intestine

The large intestine is specialized to absorb water and form the feces. It also provides the process of defecation.

The **large intestine** is the final segment of the alimentary canal. It extends about 1.5 m (5 feet) from its union with the small intestine at the ileocecal valve to the anus. Its large diameter, which measures about 7 cm (almost 3 inches), provides it with its name. The main functions of the large intestine are to dry out indigestible material by absorbing water, and to eliminate this unwanted material from the body.

Structure of the Large Intestine

The large intestine frames the small intestine on both sides and the top (Fig. 16–22). It is divided into four main segments: the cecum, the colon, the rectum, and the anal canal.

The **cecum**[29] (SĒ-kum) receives material from the ileum of the small intestine. It is a short, pouchlike seg-

ment that lies slightly below the ileocecal valve. Attached to its lower margin is a narrow, wormlike extension called the **vermiform appendix.**[30] Although it contains lymphatic tissue, the appendix is a potential trouble spot because it contains a narrow channel that ends blindly, providing an ideal location for the growth of bacteria. Infection of the appendix is a condition called *appendicitis*, which can become life-threatening if the appendix is allowed to rupture and spread infectious material throughout the peritoneal cavity, causing a condition called *peritonitis*.

The **colon**[31] (KŌ-lun) is the longest segment of the large intestine. It is divided into four regions. The first, the **ascending colon,** continues from its union with the cecum up the right side of the abdominal cavity to a point just below the liver. Here, it turns sharply to the left to become the **transverse colon,** which extends across the abdomen to a point just below the spleen at the left corner of the abdomen. It then makes a sharp turn downward, continuing as the **descend-**

---

[29]Cecum: Latin *cecum* = blind pouch.

[30]Vermiform appendix: Latin *vermiformis* = wormlike + *appendix* = attachment.

[31]Colon: Greek *kolon* = the large intestine.

Figure 16–22
Structure of the large intestine. Part of the ileum, cecum, and ascending colon is cut open to view their internal features.
What purpose is served by the unique structure of the large intestine's muscularis?

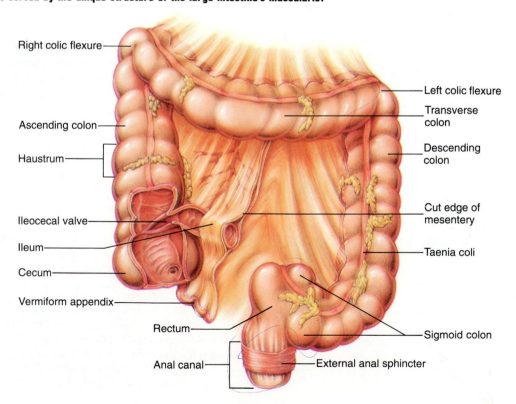

Right colic flexure

Ascending colon

Haustrum

Ileocecal valve

Ileum

Cecum

Vermiform appendix

Rectum

Anal canal

Left colic flexure

Transverse colon

Descending colon

Cut edge of mesentery

Taenia coli

Sigmoid colon

External anal sphincter

ing colon. Once within the pelvic cavity, it makes two more turns to form the S-shaped **sigmoid colon.** After a short distance, the sigmoid colon continues as the rectum.

The **rectum**[32] is located against the sacrum in the pelvic cavity. It extends downward as a straight tube until it converges with the last segment of the large intestine, the anal canal.

The **anal canal** opens to the exterior at the **anus.**[33] The anus has two sphincters: an internal, involuntary sphincter composed of smooth muscle; and an external, voluntary sphincter composed of skeletal muscle (Fig. 16–22). Both sphincters are ordinarily closed until defecation begins.

Wall of the Large Intestine

Like several other organs of the alimentary canal, the large intestinal wall consists of a mucosa, a submucosa, a muscularis, and a serosa. Its mucosa lacks the villi characteristic of the small intestine, although its mucosa and submucosa form deep folds that are similar to the plicae circulares. Between these folds and elsewhere in the mucosal lining are enormous numbers of mucus-secreting cells, whose mucus binds together the colonic contents and thereby eases the passage of feces through the large intestinal lumen. Also, the mucosa of the anal canal is organized into parallel ridges or folds, called **anal columns,** which serve to reduce friction with the feces during defecation (Fig. 16–23).

The muscularis of the large intestine is quite different from that of other organs of the alimentary canal. The outer longitudinal layer of smooth muscle does not cover the entire organ, but is arranged in three distinct bands. These bands, called **taenia coli**[34] (TĒ-nē-a, KŌ-lī), extend the entire length of the colon. One such band is visible in Figure 16–22. The contraction of these muscles gathers the colon into pouches, called **haustra**[35] (the singular form is *haustrum*).

Functions of the Large Intestine

The chyme entering the large intestine consists largely of indigestible food materials and water that was not absorbed by the small intestine. The functions of the large intestine are to process the indigestible waste further by absorbing water to form the feces, and to

---

[32]Rectum: Latin *rectum* = straight tube.

[33]Anus: Latin *anus* = ring.

[34]Taenia coli: Latin *taenia* = ribbons or bands + *coli* = of the large intestine.

[35]Haustra: Latin *haustra* = drawers.

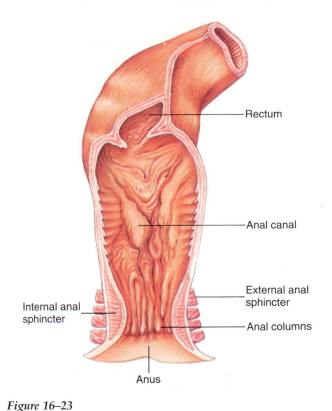

Figure 16–23
The rectum and anal canal.
Pressure exerted against the rectal wall initiates what response?

propel it to the exterior in the process known as defecation.

Feces Formation

The mucosa of the large intestinal wall absorbs much of the remaining water in the chyme as it passes through the lumen (most of the water in the alimentary canal was previously absorbed by the small intestinal mucosa). It also absorbs electrolytes that are dissolved in the water. As a result, the chyme gradually hardens as it moves toward the rectum, becoming what is called the **feces.** Feces are composed of about 75% water, with the remaining 25% including indigestible food materials, mucus from the intestinal wall, and bacteria. The color is usually due to the presence of bile pigments that have been altered by bacteria. The bacterial content is called the *intestinal flora,* and is more highly concentrated than the normal bacterial growths anywhere else in the body. The role of these microorganisms is unclear, but there is evidence that certain vitamins (such as vitamin K) and other nutrients are produced by intestinal bacteria. Metabolism by the intestinal flora produces a mixture of gases, including carbon dioxide, methane, hydrogen sulfide, and hydrogen, although most of the gas in the large intestine is swallowed air.

Defecation

The propulsion of feces through the large intestine and out through the anus is referred to as **defecation** or **elimination.** Movement through the large intestine is accomplished much as through the small intestine, except slower. It normally takes 18 to 24 hours for material to pass through the colon, whereas it takes 3 to 10 hours to pass through the 6 m of the small intestine. As in the small intestine, muscle contractions provide for mixing and peristalsis.

Peristaltic contractions of the large intestine travel only a short distance before stopping, then resume elsewhere along its length. These gentle waves of contraction occur only several times a day and serve to push the feces slowly into the rectum. The feces are stored in the rectum until elimination. Once the rectum becomes full, nerve receptors in the rectal wall are stimulated to initiate a defecation reflex. The result is the intensification of peristalsis, followed by relaxation of the internal sphincter muscles. If the external sphincters also relax (recall that they are voluntary), the peristaltic contractions force the feces out through the anus. If defecation is prevented voluntarily, the defecation reflex diminishes and the urge to defecate is reduced until pressure by the feces again stimulates the reflex.

The normal frequency of defecation is highly variable among individuals, ranging from several times a day to once every few days or so. An accumulation of feces in the large intestine beyond that which is normal for a person produces a condition known as *constipation.* This uncomfortable state causes abdominal discomfort, headache, nausea, and lack of appetite, which are the result of intestinal wall distension (overstretching). The opposite condition, in which defecation is frequent and the feces contain large quantities of water, is called *diarrhea.* It results from poor absorption of water or increased fluid secretion, and is commonly caused by bacterial and viral infections of the intestinal wall. Severe or prolonged diarrhea can result in a significant loss of water and electrolytes, which may lead to a more serious condition called *metabolic acidosis* (discussed in Chapter 18).

CONCEPTS CHECK

1. What are the divisions of the large intestine?

2. How does the outer muscle layer of the large intestine differ from that of other organs of the alimentary canal?

3. What are the steps involved in the process of defecation?

HOMEOSTASIS

The digestive system supports homeostasis of the body by supplying nourishment to all living cells. As we have seen, the digestive system converts food particles to their basic subunits, which are transported across digestive surfaces into the bloodstream. Once in the blood, these vital nutrients are carried to every cell and used to fuel the metabolic needs of cell repair, cell growth, cell division, and the synthesis of materials used by other areas of the body.

For many people, supplying their cells with enough raw materials in the form of food on a daily basis is not a serious problem. However, some people are not as fortunate, as the daily unavailability of food becomes a life-threatening health condition known as **malnutrition.** Malnutrition may also occur among people who eat plenty of food, but of low nutritious quality. In addition, it may result from the inability to properly absorb or assimilate food once it is ingested. For example, certain congenital defects of the stomach or intestine, or severe gastrointestinal infections, may impair the digestive process. Certain psychological disorders may also lead to malnutrition. Examples include *anorexia nervosa,* a condition in which the patient refuses to eat, and *bulimia,* a condition in which the patient induces vomiting or defecation after eating. These conditions are usually caused by distorted self image, which leads to an unhealthy concern over gaining weight.

Regardless of the cause, malnutrition always leads to the eventual utilization of body proteins for nourishment, which disrupts homeostasis. The resulting lack of plasma proteins lowers the osmotic pressure of the blood, leading to the accumulation of interstitial fluid in the extracellular spaces. As a result, fluid accumulates in body spaces (a symptom called *edema*). At the same time, skeletal muscle tissue is lost by the utilization of proteins, and the body loses significant amounts of weight (a symptom called *emaciation*). In extreme cases of malnutrition, the disruption of homeostasis is so complete that death results.

HEALTH CLINIC

How to Avoid Diverticulosis

Nearly half of all Americans beyond the age of 50 years have a disease of the alimentary canal called **diverticulosis**. In this common condition, small pouches called *diverticula* (which is Latin for "openings off the main path") develop on the wall of the large intestine. Each pouch is a protrusion through the inner lining, or a *hernia*, of the intestinal wall. Diverticula range in size from that of a small pea to slightly over one inch in diameter, and are most common on the sigmoid portion of the colon.

Diverticulosis is thought to be caused by excess pressure within the large intestine. As age increases, the outer wall of the large intestine thickens, narrowing the diameter of the lumen. This makes the passage of feces more difficult, leading to a buildup of pressure during defecation that pushes diverticula outward like a balloon. In some cases of diverticulosis, the pressure becomes so great that diverticula become irritated and inflamed, and may burst if not treated. This dangerous condition is called *diverticulitis*; it can lead to the spread of bacteria throughout the peritoneal cavity and peritonitis, which is life-threatening.

Recent evidence obtained from surveying hospital records (called *epidemiological evidence*) suggests that diverticulosis may be avoidable. This evidence indicates that the thickening of the intestinal wall, which precedes the formation of diverticula, may be a function of lifestyle rather than a natural part of the aging process. Among persons over 50 years of age who had maintained a diet high in fiber throughout much of their adult lives, the incidence of diverticulosis was less than 20%, but in those who maintained a diet low in fiber throughout much of their adult lives the incidence was much higher: over 50%. These values suggest that a diet high in fiber helps to reduce your chances of developing diverticulosis later in life. Dietary fiber, which consists mainly of indigestible vegetable material common in most fruits and grains, will be discussed further in Chapter 17.

CLINICAL TERMS OF THE DIGESTIVE SYSTEM

Achalasia Failure of the smooth muscle lining the alimentary canal to relax at some junction, such as that between the stomach and the small intestine.

Achlorhydria Lack of hydrochloric acid in the gastric juice, resulting in the insufficient digestion of protein.

Aphagia The inability to swallow.

Cancers of the digestive organs Include carcinomas that may arise from the epithelium of the lips, cheek, tongue, palate, or back of the throat. If treated early, they usually regress. Stomach cancer and pancreatic cancer are both highly metastatic, and usually fatal. Cancer of the small intestine is uncommon, but colon cancer is quite common and is fatal if not treated early.

Cholecystitis Inflammation of the gallbladder. This condition is often caused by the presence of bile (gall) stones, a condition known as *cholelithiasis*. The stones block the movement of bile from the gallbladder, causing the hardening of its contents and resulting in visceral pain.

Colitis Inflammation of the colon and rectum. Colitis may be chronic, usually as an inherited disorder; or acute, when it is caused by an irritation in the bowel due to diet or infectious organisms.

Diverticulitis Inflammation of the small pouches (diverticula) that sometimes form in the lining and wall of the large intestine. It may be caused by the accumulation of food debris in the pouches. Predisposing factors include chronic constipation and low-fiber diets.

Dysentery A severe infection of the intestinal wall characterized by diarrhea and cramps. It may be caused by bacteria, viruses, or protozoans that are inadvertently ingested.

Enteritis Inflammation of the small intestine. It may be caused by infectious microorganisms, injury or trauma, intestinal ulcers, or dietary factors. It may be acute or chronic; a chronic form is Crohn's disease, which is a relapsing form that is thought to be due to an infectious agent. Symptoms include nausea, vomiting, diarrhea, and abdominal pain.

Gastritis Inflammation of the stomach, generally caused by irritation of the most sensitive part, the mucosa. Gastritis may take the form of gastric ulcer formation; causes include infectious microorganisms, injury or trauma, and dietary factors. Symptoms include nausea, vomiting, diarrhea, and abdominal pain.

Hemorrhoids Enlarged veins in the lining of the anal canal.

Hepatitis Inflammation and death of liver tissue. It is usually caused by viruses, although it may be caused by chemical toxins, such as alcohol. *Type A* hepatitis (also known as *infectious* hepatitis) is caused by a virus that can be transmitted through contaminated foods, liquids, or body fluids. A second type, called *type B* hepatitis (or *serum* hepatitis), is virally transmitted by way of blood and blood products. A third type, known as *non-A, non-B* hepatitis (or *type C* hepatitis), is thought to be caused by at least two viruses, one resembling the type A virus and one resembling the type B virus, and is the usual type transmitted by blood or blood products. Yet another type of hepatitis may be caused by excessive alcohol consumption or other long-term exposure to chemicals. In all cases, hepatitis symptoms include anorexia (refusal to eat), malaise (lack of energy), nausea, vomiting, diarrhea, and often fever. Patients who exhibit these symptoms longer than six months are said to have *chronic hepatitis*, which can result in permanent liver damage that can lead to death.

Hiatal hernia A protrusion of the stomach through the diaphragm and into the thoracic cavity. Symptoms include extreme discomfort during eating and prolonged heartburn after eating.

Pancreatitis Inflammation of the pancreas due to escape of pancreatic juice from the ducts into the pancreatic tissues, causing digestion of the pancreas. Acute pancreatitis is usually fatal and is often caused by excessive alcohol consumption.

Peptic ulcer A disorder caused by increased acid secretions and digestive enzymes eroding the mucosal lining. If it occurs in the stomach, it is called a *gastric* ulcer, and if in the small intestine, it is called an *intestinal* ulcer. Symptoms include abdominal pain and blood in the sputum.

Thrush An infection of the mouth caused by the fungus *Candida albicans*. It is most common in newborn infants and individuals treated with high doses of antibiotics or immunosuppressive drugs.

CHAPTER SUMMARY

INTRODUCTION TO DIGESTIVE STRUCTURE AND FUNCTION

Organization The digestive organs are organized to form the alimentary canal—a tube extending from the mouth to the anus, and the accessory organs associated with the alimentary canal.

Digestive Processes Digestion consists of numerous processes, which include the following:

Ingestion Bringing food into the mouth.
Propulsion Movement of food through the alimentary canal.
Mechanical Digestion Breakdown of food particles by mechanical means.
Chemical Digestion Breakdown of food molecules by chemical means.
Absorption Movement of digested material from the lumen of the alimentary canal into the bloodstream.
Defecation Elimination of indigestible material from the body.

SPECIAL FEATURES OF THE DIGESTIVE SYSTEM

Peritoneum The largest serous membrane of the body. It consists of the outer, parietal peritoneum, lining the inner wall of the abdominal cavity; and the inner, visceral peritoneum (serosa), covering many abdominal organs.

The parietal peritoneum includes the following folds:

Falciform Ligament Connects the liver to the anterior abdominal wall and diaphragm.

Lesser Omentum Connects the upper margin of the stomach and part of the liver to the anterior abdominal wall.

Mesentery Connects the small intestine to the posterior abdominal wall.

Greater Omentum A double layer extending downward over the abdomen from the lower margin of the stomach.

Wall Structure of the Alimentary Canal The walls of the esophagus, stomach, small intestine, and large intestine share the following layers:

Mucosa Inner lining of mucous membrane. It provides protection, absorption, and secretion functions.

Submucosa External to the mucosa. It is composed of loose connective tissue. Its blood vessels nourish tissues and carry away absorbed materials.

Muscularis External to the submucosa. It is usually composed of an inner, circular layer and an outer, longitudinal layer. Its contractions propel food by peristalsis.

Serosa The external layer. It is composed of loose connective tissue and a single layer of flattened epithelial cells.

DIGESTIVE ORGANS

Mouth The organ that receives food and prepares it for swallowing. The cavity within it between the tongue and palate is the oral cavity.

Lips and Cheeks Forming the walls of the mouth, they are composed of skeletal muscle covered externally with skin and internally with mucous membrane. The space between the lips and cheeks and the teeth is the vestibule.

Palate Forming the roof of the mouth. Its anterior part is the hard palate, and the posterior part is the soft palate. The uvula hangs from the soft palate near the fauces, or opening of the oral cavity into the pharynx.

Tongue An accessory organ occupying most of the oral cavity when the mouth is closed. It is composed mainly of skeletal muscle, some of which extends to the hyoid bone. Movement of the tongue mixes food with saliva, helps propel food into the pharynx, and is essential for speech.

Teeth Provide for the mechanical digestive process of mastication.

Dentition Two sets of teeth are provided in a lifetime: 20 teeth in the deciduous set, and 32 in the permanent set if all wisdom teeth arrive.

Tooth Structure There are two regions to a tooth: the crown, above the gum line; and the roots, below. The crown is covered with enamel; its bulk is mostly made of dentin. Inside is the pulp cavity—containing connective tissue, blood vessels, and nerves—which penetrates through the roots as the root canals. The roots are embedded in the jawbone and covered by cementum, which attaches the root to the periodontal ligament.

Salivary Glands Accessory organs that secrete saliva into the mouth by way of ducts. There are three pairs of major salivary glands. The parotid glands are the largest salivary glands; they are located in front of and below each ear between the skin and the masseter muscle. Their secretion contains salivary amylase, which begins chemical digestion of carbohydrates. The submandibular glands are located along the inner surface of the jaw in the floor of the mouth. Their secretion contains both serous fluid and mucus. The sublingual glands are located in front of the submandibular glands under the tongue. Their secretion contains mucus.

Digestion in the Mouth Mechanical digestion is begun by chewing, or mastication, and mixing. Chemical digestion begins in the mouth by salivary amylase, which simplifies starch and glycogen to the disaccharide maltose.

Pharynx The chamber behind the oral cavity from the internal nares to the larynx. It is divided into the nasopharynx, oropharynx, and laryngopharynx. The muscles in its walls perform the act of swallowing.

Esophagus The muscular tube between the pharynx and the stomach. Its wall consists of a protective mucosa, a submucosa, and a muscularis that changes from skeletal to smooth muscle as it descends. The muscularis thickens at the union with the stomach to form the lower esophageal sphincter, which is normally closed to prevent the movement upward of contents of the stomach.

Stomach

Structure of the Stomach The convex lateral margin is the greater curvature, and the concave medial margin is the lesser curvature. The stomach is divided into four regions: the cardia, fundus, body, and pylorus. It unites with the small intestine at the pyloric sphincter, which serves as a valve to control the movement of material.

Stomach Wall The mucosa and the muscularis of the stomach wall exhibit several distinguishing features. The mucosa contains holes called *gastric pits*, which lead into tubelike gastric glands. Cells of the gastric glands secrete a mixture called gastric juice. The cells include zymogenic cells, which secrete digestive enzymes; parietal cells, which secrete hydrochloric acid and intrinsic factor; and mucous cells, which secrete mucus. The muscularis contains an additional layer, an innermost layer of oblique muscle fibers.

Functions of the Stomach

Mechanical Digestion Mixing and churning are provided by the muscularis.

Chemical Digestion The digestion of proteins is begun mainly by pepsin. It is secreted by zymogenic cells as inactive pepsinogen and converts to active pepsin in the acid environment of the stomach.

Absorption Limited to small amounts of water,

some salts, glucose, alcohol, aspirin, and some lipid-soluble drugs.

Propulsion Peristaltic contractions push chyme through the pyloric sphincter.

Intrinsic Factor Secreted by parietal cells, it aids the absorption of vitamin B_{12} across the small intestinal wall.

Regulation of Stomach Functions

Nervous Control Secretion of gastric juice is stimulated by the hypothalamus in the brain.

Hormonal Control Gastric juice secretion is maintained by the hormone gastrin, produced by cells in the stomach wall. Once chyme leaves the stomach, further gastric secretions are inhibited by two hormones secreted by cells in the intestinal wall, secretin and cholecystokinin.

Pancreas

Structure of the Pancreas Located behind the stomach in the upper abdominal cavity. It consists of head, body, and tail regions. Its acinar cells secrete pancreatic juice into ducts that converge to form the central pancreatic duct, which in most individuals unites with the common bile duct. Through the duct, pancreatic juice is emptied into the duodenum of the small intestine.

Functions of the Pancreas Secretion of pancreatic juice, which contains enzymes that simplify all food groups into more basic nutrients.

Regulation of Pancreatic Secretions Under control of hormones secreted by cells in the intestinal wall.

Secretin Released in response to acid in the small intestine. It stimulates the release of pancreatic juice rich in bicarbonate ions.

Cholecystokinin Released in response to fats and proteins in the small intestine. It stimulates the release of pancreatic juice rich in digestive enzymes.

Liver

Structure of the Liver Divided into two main lobes: right lobe and left lobe. Each lobe is further divided into smaller lobules. Lobules contain columns of hepatocytes that radiate from a central vein. Between hepatocyte columns are endothelium-lined channels, called *sinusoids,* which carry blood that drains into the central vein. Phagocytic Kupffer cells lie in wait in the sinusoids. Channels smaller than sinusoids, called *bile canaliculi,* are also present. They carry bile secreted by hepatocytes into larger ducts, and finally into the large hepatic duct, which unites with the cystic duct from the gallbladder to form the common bile duct.

Functions of the Liver Digestive function is the secretion of bile. Bile salts in bile aid in fat digestion within the small intestine, by emulsification. They also promote the absorption of fatty acids and certain vitamins. Other functions include maintaining blood sugar homeostasis; packaging fats into lipoproteins for their transport to other body tissues; metabolism of proteins and their conversion into urea; storage of glycogen, iron, and several vitamins; detoxification of poisons; and removal of worn-out red blood cells and other debris.

Gallbladder A small sac located between the two main lobes of the liver. It stores and concentrates bile, receiving it from the liver by way of the cystic duct. Bile enters the small intestine by way of the common bile duct.

Small Intestine

Structure of the Small Intestine The longest segment of the alimentary canal, it is divided into three regions: the duodenum, the jejunum, and the ileum.

Small Intestinal Wall The mucosa contains tiny projections, called *villi,* that increase the absorptive surface area. Cells that line the lumen also have microvilli, which aid in this regard. The submucosa is thrown into folds with the mucosa, called *plicae circulares,* which enhance contact between food and the villi. It contains Peyer's patches, which protect against infectious microorganisms, and Brunner's glands, which are mucous glands in the duodenum that help neutralize the acid in chyme entering from the stomach.

Functions of the Small Intestine

Chemical Digestion Pancreatic enzymes, bile, and intestinal enzymes in the microvilli combine to complete chemical digestion of all foods into absorbable nutrients.

Absorption The small intestine serves as the primary site for nutrient absorption, which is carried on by a variety of means, as well as water and electrolyte absorption.

Propulsion Contractions of the muscularis propel chyme through by peristalsis.

Large Intestine

Structure of the Large Intestine The last segment of the alimentary canal. It is divided into four segments: the cecum, colon, rectum, and anal canal. The colon is itself divided into the ascending colon, the transverse colon, the descending colon, and the sigmoid colon.

Wall of the Large Intestine The mucosa lacks villi and contains an abundance of mucus-secreting cells. In the anal canal, it forms parallel ridges known as *anal columns.* In the muscularis, the outer longitudinal layer is divided into three bands, called *taenia coli,* which gather the colon into pouches called *haustra* when they contract.

Functions of the Large Intestine

Feces Formation Feces are formed as water and electrolytes are absorbed into the large intestinal lumen from the chyme.

Defecation The process of feces elimination, provided by peristaltic contractions that push feces into the rectum. Rectal muscles contract when the rectum becomes full. Defecation can be controlled voluntarily by the external sphincters.

HOMEOSTASIS

The digestive system helps maintain homeostasis by simplifying foods into their basic subunits for utilization by cells. Malnutrition represents a disruption of the homeostatic balance and can be life-threatening.

KEY TERMS

absorption (p. 474)
bile (p. 491)
colon (p. 498)
defecation (p. 474)
digestion (p. 473)

duodenum (p. 494)
elimination (p. 500)
emulsification (p. 491)
gastric (p. 484)
ileum (p. 494)

ingestion (p. 473)
jejunum (p. 494)
malnutrition (p. 500)
mastication (p. 473)
peristalsis (p. 473)

propulsion (p. 473)
sphincter (p. 476)
villi (p. 494)

QUESTIONS FOR REVIEW

OBJECTIVE QUESTIONS

1. Which of the following digestive organs is an accessory organ?
   a. stomach
   b. liver
   c. small intestine
   d. mouth

2. The breakdown of large molecules of food into their basic building blocks by enzymes is called:
   a. mechanical digestion
   b. absorption
   c. chemical digestion
   d. propulsion

3. The outer layer of the peritoneum that lines the inner abdominal wall is the:
   a. parietal peritoneum
   b. greater omentum
   c. serosa
   d. visceral peritoneum

4. The peritoneal fold that supports the small intestine is called the:
   a. parietal peritoneum
   b. mesentery
   c. serosa
   d. lesser omentum

5. The inner layer of much of the alimentary canal is the:
   a. serosa
   b. submucosa
   c. mucosa
   d. muscularis

6. The roof of the mouth is formed by the:
   a. uvula
   b. palate
   c. gingivae
   d. cheeks

7. The hardened material that forms the outer covering of a tooth's crown is called the:
   a. enamel
   b. pulp
   c. dentin
   d. cementum

8. The digestive enzyme in saliva that begins the process of chemical digestion is:
   a. amylase
   b. protease
   c. lipase
   d. pepsin

9. The accessory organ located in front of and below each ear between the skin and masseter muscle layers is the:
   a. esophagus
   b. sublingual gland
   c. parotid gland
   d. tongue

10. Food travels from the pharynx to the stomach by way of the:
    a. laryngopharynx
    b. esophagus
    c. trachea
    d. mouth

11. The narrowed, inferior region of the stomach that contains a prominent sphincter muscle is the:
    a. cardia
    b. fundus
    c. pylorus
    d. body

12. The stomach functions in:
    a. mechanical digestion
    b. production of intrinsic factor
    c. chemical digestion of proteins
    d. all of the above

13. The hormone released by cells in the stomach wall that stimulates digestive secretions and peristalsis is:
    a. secretin
    b. gastrin
    c. cholecystokinin
    d. pylorin

14. Pancreatic juice is a mixture of digestive enzymes that are released into the:
    a. pancreas
    b. small intestine
    c. stomach
    d. large intestine

15. When acid enters the small intestine from the stomach, cells in the intestinal wall secrete a hormone called _____, which stimulates the production of pancreatic juice rich in bicarbonate ions.
    a. cholecystokinin
    b. secretin
    c. gastrin
    d. amylase

16. Bile is produced by cells in the:
    a. gallbladder
    b. small intestine
    c. liver
    d. pancreas

17. The digestive organ that performs many important metabolic functions is the:
    a. pancreas
    b. liver
    c. stomach
    d. duodenum

18. The segment of the small intestine that receives pancreatic juice and bile by way of the common bile duct is the:
    a. duodenum
    b. ascending colon
    c. ileum
    d. jejunum

19. Villi are fingerlike projections of the mucosa found only in the:
    a. duodenum
    b. colon
    c. small intestine
    d. stomach

20. Carbohydrates that are absorbed by the small intestine are in the form of basic nutrients called:
    a. fatty acids
    b. amino acids
    c. monosaccharides
    d. triglycerides

21. When a person undergoes surgery to remove the gallbladder, it is most likely that the patient will experience some difficulty in the digestion of:
    a. carbohydrates
    b. proteins
    c. lipids
    d. nucleic acids

22. The transverse colon receives undigestible materials from the:
a. ileum
b. ascending colon
c. small intestine
d. rectum

23. Defecation is performed by the:
a. large intestine
b. stomach
c. small intestine
d. all of the above

ESSAY QUESTIONS

1. Describe the path a molecule of food passes through as it is ingested, swallowed, and propelled through the alimentary canal. Include the secretions it encounters along the way.

2. Discuss the chemical digestion, absorption, and catabolism of carbohydrates. Include the areas of the body in which each of these processes occurs.

3. Explain the process of peristalsis, and describe how it propels food through the digestive tube.

4. Describe the actions of the mouth, tongue, and pharynx during the act of swallowing.

5. Describe the effects of gastric juice on swallowed food. Indicate why the mucosal lining normally avoids damage.

6. Identify the features of the small intestine that provide it with the efficient ability to absorb nutrients.

7. Describe the functions of the liver, including its digestive and its metabolic roles.

8. Describe the structural features of the alimentary canal that serve to increase its efficiency by either increasing the surface area of contact with food particles, or the time of contact between food particles and digestive chemicals and surfaces.

ANSWERS TO ART LEGEND QUESTIONS

Figure 16-1
What are the organs of the alimentary canal? What are the accessory organs of the digestive system? The mouth, pharynx, esophagus, stomach, small intestine, and large intestine are the organs of the alimentary canal. The teeth, tongue, salivary glands, pancreas, liver, and gallbladder are the accessory organs.

Figure 16-2
In which organ does nutrient absorption occur? Nutrient absorption occurs in the small intestine.

Figure 16-3
What digestive organs perform peristalsis? The esophagus, stomach, small intestine, and large intestine perform peristalsis.

Figure 16-4
How is the peritoneum similar to the pericardium associated with the heart, and the pleurae associated with the lungs? The peritoneum, pericardium, and pleurae are all serous membranes. Also, they each contain a parietal layer and a visceral layer.

Figure 16-5
What is the innermost lining of the alimentary canal called? The innermost lining is called the *mucosa*.

Figure 16-6
What type of membrane lines the inside of the mouth and pharynx? A mucous membrane lines the inside of the mouth and pharynx.

Figure 16-7
What bony platform forms the anterior roof of the mouth? The bony platform forming the roof of the mouth is the hard palate.

Figure 16-8
What are molar teeth mainly used for? Molar teeth are used for grinding food particles during chewing.

Figure 16-9
Once tooth enamel has worn down, what layer remains to protect the tooth? The layer of dentin will protect the tooth after the enamel has worn down. However, protection is limited, since dentin is composed of a softer material than enamel.

Figure 16-10
Which is the largest salivary gland? The largest salivary gland is the parotid gland, located on the angle of the jaw.

Figure 16-11
What benefit is obtained when the epiglottis drops over the glottis during swallowing? The positioning of the epiglottis over the glottis prevents food from entering the trachea during swallowing.

Figure 16-12
How does peristalsis push food along? Peristalsis pushes food by the sequential wave of contraction of circular muscles.

Figure 16-13
How many layers of smooth muscle are in the stomach wall? Three layers are in the stomach wall: the inner oblique layer, the middle circular layer, and the outer longitudinal layer.

Figure 16-14
What three major secretions emerge from the gastric glands during a meal? Pepsinogen, hydrochloric acid, and mucus are secreted by the gastric glands.

Figure 16-15
What benefit to chemical digestion is provided by mixing in the stom-

ach? Mixing increases the contact time between gastric juice and food materials in the stomach.

Figure 16-16
What digestive organ receives pancreatic juice from the pancreas? The small intestine (at the duodenum) receives pancreatic juice.

Figure 16-17
What structure separates the two main lobes of the liver? The falciform ligament separates the right and left lobes of the liver.

Figure 16-18
What cells manufacture bile? Hepatocytes (liver cells) manufacture and release bile.

Figure 16-19
In addition to bile, secretions from what other organ are carried through the common bile duct? Pancreatic juice from the pancreatic acini cells is transported through the common bile duct, in addition to bile.

Figure 16-20
What benefit to absorption is provided by villi and microvilli? Villi and microvilli increase the absorptive surface area of the small intestinal mucosa and thus vastly increase the amount of physical contact between nutrients and the absorptive surface.

Figure 16-21
How does the absorption of digested proteins and fats differ? Amino acids must be actively transported across the absorptive surface; then they can diffuse through the cell to a capillary. Fat subunits diffuse across the absorptive surface and then are converted to chylomicrons within the epithelial cell. Chylomicrons enter a lacteal and are carried by the lymphatic network.

Figure 16-22
What purpose is served by the unique structure of the large intestine's muscularis? Bands of the outer layer of the muscularis, called taenia coli, gather the intestine into pouches (called haustra), which collect the stools as they are formed.

Figure 16-23
Pressure exerted against the rectal wall initiates what response? Pressure against the rectal wall triggers the defecation reflex, stimulating relaxation of the internal sphincter and contraction of the rectal walls.

CHAPTER 17

NUTRITION AND METABOLISM

CHAPTER OUTLINE

NUTRIENTS
Carbohydrates
Fats
Proteins
Vitamins
Minerals

BIOAVAILABILITY

TRANSPORT
Carbohydrate Transport
Fat Transport
Protein Transport
Vitamin and Mineral Transport

METABOLISM
Cellular Respiration
Metabolic Rate and Body Temperature

HOMEOSTASIS AND TEMPERATURE REGULATION

LEARNING OBJECTIVES

After studying this chapter, you should be able to:

1. Describe carbohydrates, fats, and proteins and explain how the body utilizes them.

2. Identify the dietary requirements of carbohydrates, fats, and proteins.

3. Explain why vitamins and minerals are needed for a balanced diet.

4. Describe the benefits of eating a variety of foods.

5. Describe nutrient transport in the body.

6. Define metabolism, anabolism, and catabolism.

7. Describe the processes of glycolysis, anaerobic respiration, and aerobic respiration.

8. Discuss how metabolic rate is measured and how body temperature is regulated.

Energy refers to the ability to perform work. It is what drives chemical reactions to proceed in a direction they otherwise would not go. Since most functions of the human body are basically a series of chemical reactions, our bodies require energy in order to operate. Under normal circumstances, all of the energy needed to power our body's chemical reactions is obtained from the food we eat. As we learned in the previous chapter, food must be processed by the digestive system so that its basic subunits may be released for their absorption into the bloodstream. Once in the bloodstream, the food subunits are transported to the liver, and onward to body cells. Inside the cells the food subunits are processed further to produce energy, or used as building materials for the construction of cell parts or products.

The process by which food is taken into the body and changed into a form that can be utilized by cells is called **nutrition.** It involves ingestion, digestion, absorption, transport, and metabolism. In the previous chapter, we examined the first three of these processes. In this chapter, we continue our discussion of nutrition by looking at transport and metabolism. We will also explore the nature of the basic food subunits that our bodies depend on for energy: nutrients. It is this topic that we turn to first.

NUTRIENTS

Adequate nutrition is necessary to maintain health. It requires the ingestion of all nutrients on a regular basis, especially essential nutrients.

The modification, absorption, and utilization of food substances by the body is called *nutrition*. Food substances that the body uses are called **nutrients,** and include mainly carbohydrates, fats, protein, vitamins, minerals, and water. For the most part, nutrients must be brought into the body by ingestion. Once they are within, the body has a remarkable ability to change these nutrients into forms that can be utilized best.

Not all nutrients are ingested into the body. Some nutrients are synthesized from raw materials (obtained from other ingested nutrients) inside the metabolic machinery of cells. For example, the preferred nutrient of the body for energy, glucose, may be obtained from glycogen stored in the liver or skeletal muscle. Glycogen is a carbohydrate that is synthesized from raw materials in liver cells.

However, some nutrients cannot be synthesized by the body, but are nonetheless required for certain functions. These molecules are called **essential nutrients.** They include about 50 types of nutrients that must be obtained through ingestion. B-complex vitamins and vitamin C are examples of essential nutrients.

In this section we will examine the basic groups of nutrients that are in a normal, healthy diet: carbohydrates, fats, proteins, vitamins, minerals, and water. How these basic groups are used by the body, and why we need a constant supply of them to sustain our health, will be considered.

Carbohydrates

Carbohydrates are large molecules that are composed of simple sugar subunits. They are needed by the body as a primary source of energy to fuel cellular activities. We obtain carbohydrates from the ingestion of starch from vegetables and fruits, and glycogen from meats. Starch and glycogen are known as **polysaccharides,** or complex carbohydrates, since they contain many subunits of sugar molecules. We can also obtain glycogen by synthesizing it through the assembly of glucose molecules, which is performed by liver cells.

During early digestion, starch and glycogen molecules in food are broken down into simpler forms, the **disaccharides** (two-sugar compounds; the most abundant disaccharide formed is the complex sugar maltose). In the small intestine, disaccharides are simplified further into the simple sugar molecules, or **monosaccharides,** which can be absorbed into the bloodstream. The simpler forms of carbohydrates may also be ingested directly, such as disaccharides from table sugar (sucrose), and monosaccharides from honey and certain fruits (fructose and galactose). One form of carbohydrate that is ingested with vegetables and fruits that cannot be digested but is important for providing roughage, or *fiber,* to the diet is cellulose. As dietary fiber, cellulose facilitates the movement of ingested material through the digestive tract, increases the bulk of the stool, and facilitates defecation.

Dietary Requirements

Carbohydrates are not the only source of fuel to power cellular activities. Therefore, individuals can survive for long periods of time with very little carbohydrate intake. For example, the Eskimos live on a diet that is extremely low in carbohydrate. However, persons following an excessively low-carbohydrate diet run the risk of losing tissue proteins and fats, which are alternatively broken down to fuel cellular activities. Most nutritional sources agree that about 100 grams of carbohydrates per day is the minimum amount required to maintain adequate blood glucose levels. Current recommendations call for 125 to 175 grams per day with an emphasis on complex carbohydrates. Sources of carbohydrates, recommended levels, and the consequences of varying from these levels are summarized in Table 17–1.

Fats

The most abundant form of dietary fats is the **triglycerides,** or neutral fats. The other types of fat molecules—**phospholipids** and **cholesterol**—are also ingested but in smaller amounts. Two types of triglycerides are normally consumed: those that contain only single bonds between carbon atoms and are solid at room temperature, called **saturated fats;** and those that contain one or more double bonds between carbon atoms and are liquid at room temperature, known as **unsaturated fats.** Saturated fats are found in animal products such as meat and dairy foods, and in some plant products such as coconut. Unsaturated fats are found in most vegetable oils, nuts, and seeds.

As a fuel source, each gram of fat contains several times more potential energy than a gram of carbohydrate, and so provides an efficient backup to carbohydrate metabolism. Fats also help the body absorb fat-soluble vitamins and provide basic raw materials for the synthesis of plasma membranes in cells.

One type of fatty acid that the body is unable to synthesize, but that is nonetheless required, is linoleic acid. It is needed for the synthesis of certain phospholipids, which are used for the formation of plasma

Table 17-1 SUMMARY OF CARBOHYDRATE, FAT, AND PROTEIN NUTRIENTS

Food Sources for Adults (RDA)	Recommended Daily Amounts	Problems	
		Too Little	Too Much
Carbohydrates Starches, such as vegetables, seeds, and nuts; simple sugars, such as sugars in fruit and candy; glycogen from meat	125 to 175 g, or about 55% of total food intake	Loss of weight; acidosis due to use of fat for energy	Obesity; nutrient deficiency
Fats Animal fats, such as meat, eggs, and milk; plant fats such as chocolate, plant oils	80 to 100 g, or less than 30% of total food intake; 250 mg or less of cholesterol	Loss of weight; rapid loss of heat	Obesity; increased blood cholesterol levels with higher risk of heart disease
Proteins Complete proteins, such as in animal products including meat, eggs, and milk; incomplete proteins, such as legumes, nuts, and seeds	0.8 g per kilogram of body weight	Loss of weight and tissue wasting; anemia and edema	Obesity

membranes in cells. Linoleic acid must be ingested and is thus regarded as an **essential fatty acid.**

Cholesterol differs from the other fats in that it is not used for energy. Instead, it is important as a structural component of bile salts, steroid hormones, and plasma membranes of cells.

Dietary Requirements

Fats represent about 40% of the calories in an average American diet. This figure is higher than it should be for optimal health. The American Heart Association recommends that fats should represent 30% or less of total caloric intake. Also, saturated fats should be limited to 10% of total fat intake, and daily cholesterol intake should be no more than 250 mg (the amount of cholesterol in one egg yolk). A growing body of evidence indicates that levels higher than these recommended values may contribute to cardiovascular diseases, such as *atherosclerosis.* Sources of fats, recommended levels of intake, and problems associated with variations from recommended amounts are summarized in Table 17–1.

Proteins

Proteins are simplified into their subunits, the amino acids, by the process of chemical digestion. These amino acids are either used directly, or are modified by the liver into raw materials that can be used to synthesize new amino acids and proteins. Newly synthesized proteins are vital for cell structure and function.

Although cells can synthesize many amino acids from raw materials obtained through digestion, eight amino acids that are required for the construction of body proteins cannot be synthesized. Therefore they are termed **essential amino acids** (Fig. 17–1). The most complete diet includes protein sources that contain these eight essential amino acids. All eight of these amino acids may be obtained at once through any animal source (including meat, milk, cheese, and eggs). However, to obtain the eight essential amino acids in a vegetarian diet, a variety of vegetables must be consumed, since no single vegetable source contains all eight. Thus, animal proteins are nutritionally complete, but plant proteins are incomplete, since they lack one or more of the essential amino acids.

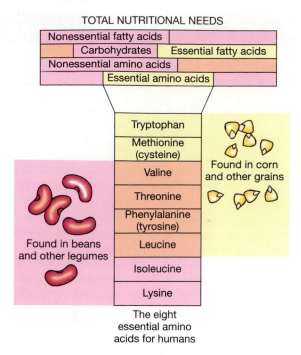

TOTAL NUTRITIONAL NEEDS

Nonessential fatty acids
Carbohydrates — Essential fatty acids
Nonessential amino acids
Essential amino acids

Tryptophan
Methionine (cysteine)
Valine
Threonine
Phenylalanine (tyrosine)
Leucine
Isoleucine
Lysine

Found in corn and other grains

Found in beans and other legumes

The eight essential amino acids for humans

Figure 17–1
Essential amino acids form one of the important building blocks for the body's nutritional needs. There are eight amino acids that cannot be synthesized by the body but are nonetheless required for nutrition. Vegetarian diets must be carefully managed to obtain all the essential amino acids. For example, corn has little isoleucine and lysine, and beans have little tryptophan and methionine. When they are combined in a meal, all amino acids may be obtained.

Why must essential nutrients be obtained from the food you eat?

Dietary Requirements

Proteins that are consumed provide a direct source for the essential amino acids, as well as raw materials for forming nonessential amino acids and other compounds that require nitrogen. The amount of protein a person requires depends on age, body weight, and metabolic rate. As a general rule, nutritionists recommend a daily protein intake of 0.8 gram per kilogram of body weight, or about 56 grams for a 154-pound man and 48 grams for a 128-pound woman. This converts to about 2 ounces of protein daily, which could be met by a single glass of milk and a small helping of meat. For pregnant and nursing women, an increase of 30 and 20 grams per day, respectively, is required to maintain fetal growth or milk production. Sources of proteins, recommended levels of consump-

tion, and problems associated with variations on recommended amounts are summarized in Table 17–1.

Vitamins

Vitamins are organic compounds that are required in small amounts for growth and good health. Unlike carbohydrates, fats, and proteins, they are not used for energy and they do not form structural building blocks. Instead, vitamins are essential for helping the body use those nutrients that do. They act as *coenzymes*, by assisting in the regulation of physiological processes.

Vitamins are classified as either fat-soluble or water-soluble. **Fat-soluble vitamins** that are ingested bind to fats and are absorbed along with the fat products of digestion. They include vitamins A, D, E, and K. Vitamins A and E must be ingested, for they cannot be synthesized in sufficient quantities by the body. However, these vitamins can be stored in significant amounts within tissues. Vitamin D is converted by sunlight (ultraviolet radiation) from cholesterol molecules in the skin, and vitamin K is synthesized by bacteria in the intestinal flora. Fat-soluble vitamins are not easily damaged by cooking and food processing.

Water-soluble vitamins include the B-complex vitamins and vitamin C. They are absorbed along with water across the intestinal wall (except for vitamin B_{12}, which must be attached to intrinsic factor to be absorbed). They must be obtained in the food you eat, for your body cannot synthesize sufficient amounts of them to maintain health. Water-soluble vitamins are not stored in large amounts in the body, and are excreted with the urine if they are not needed. Unlike fat-soluble vitamins, they are damaged during food processing and cooking. The vitamins and their characteristics are summarized in Table 17–2.

Minerals

Dietary **minerals** are inorganic molecules that are ingested along with the organic components of food. They make up about 4% of the body by weight, with most of this (about 75%) in the bones and teeth. Thus, the most important minerals are calcium and phosphorus, although potassium, sulfur, sodium, chloride, and magnesium are also essential. Trace amounts of other minerals are also needed by the body, including iron, manganese, copper, iodine, cobalt, and zinc. The major minerals and trace minerals are further described in Table 17–3.

Minerals in the body are usually incorporated into structures that need reinforcement for strength. For example, calcium, phosphorus, and magnesium are in

Your Diet and Cancer

Cancer experts estimate that dietary factors contribute to at least one-third of the 500,000 cancer deaths occurring each year in the United States. The cancers that cause these nutrition-related deaths include colon cancer, breast cancer, and prostate cancer—three of the most common types of cancer in the developed world. After more than a decade of painstaking research into the relationship between nutrition and the development of these cancers, two factors have emerged as the most important dietary influences. They are the role of fat as a promoter of cancer, and fiber as a protector against cancer.

According to evidence compiled from numerous population studies, being overweight is an established risk factor for colon cancer, breast cancer, and prostate cancer. For example, in a large-scale 12-year study conducted by the American Cancer Society, it was found that women who were 40% over their ideal weight had up to a 55% greater risk of dying from cancer than women of normal weight. Also, overweight men had up to 33% more deaths from cancer than normal-weight men. In this same study, exercise was found to reduce the risk of developing colon cancer in men. In another study, it was found that women who consumed large servings of red meat (beef, pork, or lamb) as a main dish every day had $2\frac{1}{2}$ times the risk of developing colon cancer than women who ate red meat less than once a month. In yet another study, excessive calorie intake of any kind, whether from carbohydrates, fats, or protein, increased the risk of colon cancer. From these studies and others, experts conclude that a high caloric intake of food that results in weight gain significantly increases the risk of dying from cancer.

On the opposite side of the coin are the benefits you can gain with fiber. Fiber is indigestible material that is present in most vegetables, fruits, and whole-grain cereals. Persons with a diet high in fiber have been shown to develop cancers of the breast and colon less frequently than those with a diet low in fiber. In addition, many of the foods high in fiber contain chemicals called *antioxidants*, such as vitamin A and beta carotene, which some investigators have suggested may be able to knock out carcinogens that can lead to breast and colon cancer. These antioxidants are effective only in moderate concentrations; when they are ingested in high doses, they often induce problems of toxicity that can lead to a serious disruption of homeostasis.

Although current experimental evidence in support of the anticancer benefits of antioxidants is not as conclusive as that in support of fiber, antioxidants are being studied further in an effort to understand their potential. For example, two synthetic antioxidants that have been used since the mid-1950s as food additives, called BHA and BHT, have recently been shown to inhibit cancer development (in the lungs, liver, stomach, skin, breast, and colon). Other antioxidants that are suspected to inhibit cancer include myristicin, a major component of the oil present in parsley; quercetin, a component in onions and garlic; ellagic acid, found in many fruits and nuts; tannic acid, a component of brewed tea leaves; and phytate, a component of soybeans and other legumes. However, more studies are needed before the medical community is ready to hail antioxidants as a valuable weapon in the fight against cancer—as valuable, at least, as our present knowledge of the health benefits of a diet high in fiber and low in calories.

bone and teeth to provide them with their strength and hardness. Many other minerals are in an ionized form in body fluids, such as sodium and chloride ions. Still other minerals are bound to organic compounds to form molecules such as hormones, enzymes, and functional proteins. The concentrations of these minerals are under homeostatic control by various regulating factors. This balance is required to ensure that the amount of minerals absorbed and retained by the body is roughly equal to the amount lost.

CONCEPTS CHECK

1. What are the six groups of nutrients needed to maintain health?

2. How important are carbohydrates to a healthy diet?

3. How may the essential amino acid requirement be satisfied?

Table 17–2 VITAMIN REQUIREMENTS

Vitamin	Sources	Utilization	Problems	
			Too Little	**Too Much**
Fat-Soluble Vitamins				
Vitamin A	Carotene found in yellow and leafy green vegetables; fish oils; egg yolk; and liver	Synthesis of pigments in rods and cones; skin pliability; bone remodeling	Night blindness; dry skin and hair; skin sores; clouding of cornea	Nausea; vomiting; headache; hair loss; bone pain
Vitamin D	Produced in skin by irradiation of cholesterol by ultraviolet light	Increases blood calcium by enhancing calcium absorption	Insufficient bone calcification; rickets in children; osteomalacia in adults; irritability; poor muscle tone	Vomiting; diarrhea; weight loss; kidney damage
Vitamin E	Vegetable oils; whole grains; leafy green vegetables	Helps maintain integrity of cell membranes; acts as an antioxidant	Extremely rare	Hypertension; slow wound healing
Vitamin K	Mostly synthesized by intestinal flora; some from certain vegetables	Component of clotting proteins, and enzymes in metabolic pathway	Slow clotting of blood	Not known
Water-Soluble Vitamins				
Vitamin C	Fruits and vegetables	Synthesis of connective tissues and of serotonin for vasoconstriction; conversion of cholesterol to bile salts	Collagen-related deficiency diseases; bone, joint, and tooth pain; poor wound healing; scurvy	Enhanced movement of bone minerals
Vitamin B_1 (thiamine)	Meats; eggs; whole grains; legumes; leafy green vegetables	Carbohydrate metabolism	Beriberi; reduced appetite; gastrointestinal upset; muscle cramps	Not known
Vitamin B_2 (riboflavin)	Milk; egg white; whole grains; meat	Important component of electron transport enzymes	Dermatitis; sensitivity to light; blurred vision	Not known
Vitamin B_6 (pyridoxine)	Meat; potatoes; tomatoes	Amino acid metabolism; formation of antibodies and hemoglobin	Skin lesions around mouth; anemia and nerve disorders in infants	Peripheral nerve numbness
Niacin	Meat primarily; also peanuts and potatoes	Important component in electron transport chain enzymes	Pellagra; headache; weight loss; dermatitis; diarrhea; nausea; vomiting	Hyperglycemia; flushing of skin

Table 17-2 *(continued)*

Vitamin	Sources	Utilization	Problems	
			Too Little	**Too Much**
Pantothenic acid	Meat; legumes; egg yolk	Oxidation of glucose and fatty acids during cellular respiration	Loss of appetite; abdominal pain; mental depression	Not known
Biotin	Liver; egg yolk; legumes; nuts; also some produced by intestinal flora	Amino acid metabolism	Dry, scaly skin; muscle pain; anorexia; nausea	Not known
Vitamin B$_{12}$	Liver; meat; dairy foods (other than butter); eggs	Assists a wide spectrum of enzymes; synthesis of DNA; formation of red blood cells	Anemia; anorexia	Not known
Folic acid	Liver; lean beef; green vegetables; eggs; whole grains; some produced by intestinal flora	Formation of red blood cells; assists enzymes	Anemia; diarrhea; gastrointestinal distress	Not known

Table 17-3   MINERALS REQUIRED FOR HEALTH

Mineral	Sources	Utilization	Problems	
			Too Little	**Too Much**
Calcium	Milk and milk products; green leafy vegetables; egg yolk; shellfish	Bone calcification; transmission of nerve impulses; blood clotting	Rickets in children; osteomalacia in adults	Depressed nerve function; kidney stones; calcium deposits
Phosphorus	Meat; milk products; fish; legumes; nuts; whole grains	Bone calcification; energy storage and transfer	Rickets in children	Not known
Magnesium	Milk products; nuts; whole grains; legumes; green leafy vegetables	Assists ATP synthesis; normal muscle and nerve irritability	Neuromuscular tremors	Diarrhea
Potassium	Most fruits and vegetables; meat; fish; cereals	Nerve impulse conduction; muscle contraction; protein synthesis; osmotic pressure in cells	Severe diarrhea; vomiting; muscle weakness; paralysis	Muscle weakness

(continued)

Table 17-3 *(continued)*

Mineral	Sources	Utilization	Problems	
			Too Little	**Too Much**
Sulfur	Meat; milk; eggs; legumes	Component of many proteins and some vitamins	Not known	Not known
Sodium	Table salt; cured meats; cheese	Abundant in interstitial fluid; important in maintaining water and pH balance and neuromuscular function	Nausea; abdominal cramping; convulsions	Hypertension; edema
Chlorine	Table salt	Abundant in interstitial fluid along with sodium; also important in maintaining water and pH balance	Muscle cramps	Vomiting
Trace Minerals				
Iron	Meat; liver; egg yolk; legumes	Component of hemoglobin for oxygen transport	Anemia; anorexia; gastrointestinal problems	Liver damage
Iodine	Fish oil; iodized salt; shellfish	Needed for synthesis of thyroid hormones	Hypothyroidism (cretinism in infants, myxedema in adults); goiter	Depressed synthesis of thyroid hormone
Manganese	Nuts; legumes; whole grains; green leafy vegetables; fruit	Synthesis of fatty acids, cholesterol, hemoglobin, and urea; also needed for normal nerve function	Not known	Not known
Copper	Liver; whole grains; legumes; meat	Synthesis of hemoglobin, myelin, and melanin	Possible anemia, although rare	Possible toxic effects to liver, although rare
Zinc	Seafood; meat; legumes; cereals; nuts	Component of numerous enzymes involved in growth; senses of taste and smell; healing; metabolism	Loss of sensations; growth retardation	Loss of motor functions
Cobalt	Liver; meat; fish; milk	Red blood cell development	Anemia; anorexia; weight loss	Goiter; heart disease
Fluorine	Fluoridated water	Tooth structure; bone calcification	Increase in dental caries	Mottling of teeth; possible toxicity
Selenium	Meat; cereals; seafood	Component of certain enzymes	Not known	Vomiting; irritability; fatigue
Chromium	Liver; meat; cheese; whole grains; wine	Component of enzymes in carbohydrate metabolism	Not known	Not known

BIOAVAILABILITY

Not all of the nutrients in the food we eat can be absorbed into the bloodstream. Often, their tendency to be absorbed is affected by other nutrients that may or may not be available.

The term **bioavailability** refers to the body's ability to absorb only a portion of the vitamins and minerals it takes in. The various vitamins and minerals vary considerably in this regard. For example, studies show that our bodies use only about 30% to 40% of the calcium we consume, 10% of the iron, and about 5% of the manganese. What determines the amount of a given nutrient your body can absorb and utilize? For one, the combination of foods that are consumed in the meal can influence bioavailability. For example, the combination of vitamin C and iron can improve absorption of both. Similarly, vitamin D helps with the absorption of calcium. Also, the source of the nutrients can affect bioavailability. Some substances in certain foods interfere with the ability to absorb nutrients. For example, oxalic acid found in dark-green leafy vegetables interferes with the absorption of calcium, zinc, and iron, as does phytic acid present in some grains. A third influence on bioavailability is the body's needs. For example, a healthy man absorbs less than 1% of the iron in a balanced diet, but a woman with an iron deficiency, such as in anemia, will absorb as much as 35% of the iron in her diet. A pregnant woman who needs large amounts of all nutrients will absorb even more. Other factors that influence bioavailability include drugs, such as aspirin, which can interfere with the absorption of vitamin C and folacin; and chronic diseases, especially of the digestive tract.

How do you keep track of the various nutrient interactions so that you get the nutrition you need? There are three commonsense rules to follow:

1. *Eat a variety of foods that include all groups: carbohydrates, fats, and proteins.* This is the best way to balance the positive and negative interactions of nutrients. Introduce variety at every meal by choosing among lean meats and fish, whole-grain cereal products, low-fat dairy products, and vegetables and fruits. Eating a variety of foods from these groups will provide you with what is called a "balanced diet" and is about the most beneficial thing (along with exercise) that you can do to promote good health.

2. *Get your nutrients from foods, not pills.* Foods contain nutrients that often enhance each other's bioavailability, whereas pills do not. Large doses of vitamin or mineral supplements can actually sabotage the absorption process by introducing imbalances and can produce toxic effects on their own.

3. *Avoid the fad diets.* These are diets that are short-range and advertise fast results. They should be avoided because they often fail to supply necessary nutrients. They may also eliminate beneficial interactions among nutrients and amplify negative ones.

CONCEPTS CHECK

1. Why are vitamins important in a healthy diet?

2. Why is eating a variety of foods important to maximize nutrient utilization?

TRANSPORT

The transport of nutrients from the alimentary canal to body cells occurs in a variety of ways, depending on the nature of the nutrient.

Once nutrients are absorbed across the intestinal wall, they are transported by the bloodstream (or, in the case of some fats, through the lymphatic network) to eventually reach body cells and enter them. Because each of the basic groups of nutrients have different characteristics, the ways in which they are transported differs. In this section, we will briefly examine how each basic group is transported from the alimentary canal to body cells.

Carbohydrate Transport

As we have seen, carbohydrates may be absorbed into the bloodstream in the form of simple sugars, or monosaccharides. Only three monosaccharides can be absorbed directly into the bloodstream: glucose, fructose, and galactose. Only glucose is ultimately delivered to and used by body cells; fructose and galactose are converted to glucose by the liver before they enter the general circulation.

Glucose is a very important nutrient molecule. It is the primary fuel source of the body, and is readily used to generate energy in the form of ATP. The generation of ATP is provided by the process of cellular respiration, which occurs within the mitochondria of all body cells.

After absorption, glucose is carried by the bloodstream in its basic form. It is first transported to the liver, where it may be combined with other glucose molecules to form glycogen and stored. If glucose levels in the blood are low, such as between meals, it may be transported immediately to body cells for processing to produce energy. If glucose levels in the blood are in excess, as they may be following a heavy meal, glucose may alternatively be converted to fat for long-term food storage. The interconversion of glucose and glycogen is regulated by the pancreatic hormones glucagon and insulin (described in Chapter 11).

Fat Transport

Fats (triglycerides, phospholipids, and cholesterol) are reduced in size by the action of bile salts within the lumen of the small intestine (during emulsification; see Chapter 16). Prior to absorption, the reduced droplets of triglycerides are chemically digested to their subunits of glycerol and fatty acids (Fig. 17–2). Together with cholesterol and phospholipids, the fat subunits combine with bile salts to form small packages called **micelles,** which are soluble in water. Once formed, the micelles "ferry" the fat subunits to the intestinal epithelium. The fat subunits are then released from the micelle and diffuse through the intestinal wall by simple diffusion.

The fat subunits pass through the cells lining the small intestine and enter the lacteals, which are lymphatic vessels within the villi. During their passage, the fatty acids and glycerols are reconverted to triglycerides for transport with the lymph. The fat-laden lymph, which is a milky-white fluid called **chyle** (kīl), is carried to the bloodstream by way of lymphatic vessels.

Once in the bloodstream, triglycerides, cholesterol, and phospholipids combine with proteins to form molecules called **lipoproteins.** It is in this form that fats are carried throughout the bloodstream. As we learned in the previous chapter, the higher the percentage of fat in the lipoprotein complex, the lower is its density, and vice versa. On this basis, there are three categories of lipoproteins: high-density lipoproteins (HDLs), low-density lipoproteins (LDLs), and very low-density lipoproteins (VLDLs).

HDLs transport fats from various tissues throughout the body, where they may be stored, to the liver. In the liver, HDLs are broken down to become part of the bile. The energy that is released during their breakdown is used to power cellular activities. Since HDLs

Figure 17–2
Absorption of fats across the small intestinal wall and their transport into the lymphatic system.

Why are high-density lipoproteins (HDLs) considered to be beneficial forms of lipoprotein?

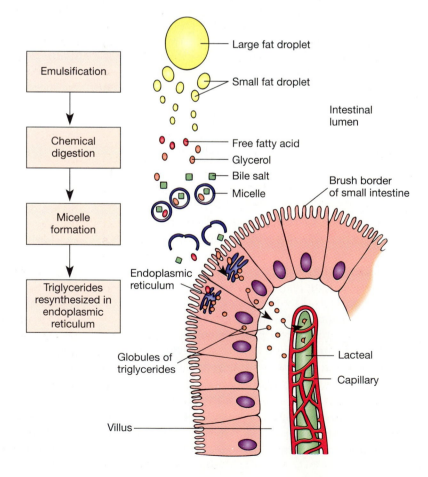

What Is Your Cholesterol Level?

Levels of cholesterol in the blood are measured by withdrawing a small amount of blood, which is analyzed in a lab, and determining the number of milligrams of cholesterol in a deciliter (about one-tenth of a quart) of blood. The average American has a cholesterol level of about 210 mg/dl. This was once considered to be an acceptable level, but no longer. The so-called normal levels help to produce "normal" levels of 1.5 million heart attacks each year. Researchers now believe that to be at low risk for heart disease, people under 30 should maintain blood cholesterol levels of 180 mg/dl or less, and those over 30, of less than 200 mg/dl. Although there is no magic number to make you automatically safe from heart disease, the risk rises with increasing levels of blood cholesterol—markedly once levels of 200 mg/dl are exceeded.

When your cholesterol is measured, the first test administered is usually concerned with the total amount of cholesterol in the blood. If you should test high, a second test is commonly administered to determine what form the cholesterol is in: is it part of the high-density lipoprotein (HDL) complex, or the low-density lipoprotein (LDL) complex? This distinction is important, for these lipoproteins serve two different body needs: the HDL carries less cholesterol, and it performs the beneficial function of pick-

ing up cholesterol and transporting it back to the liver for reprocessing or excretion; LDL brings cholesterol from the liver into the blood and releases it. Therefore, HDL is a beneficial, or "good," form, and LDL is a nonbeneficial, or "bad," form. Thus, if your HDL levels are low and your LDL levels are high, you are more likely to develop atherosclerosis. Average HDL levels in adult Americans are about 45 to 65 mg/dl, with women averaging higher than men. Studies suggest that levels above 70 mg/dl of HDL cholesterol may protect against heart disease, while those below 35 indicate greater risk.

How do you lower your blood cholesterol levels, or maintain the low levels you may have? Substitute unsaturated fats, such as canola oil, safflower oil, or corn oil, for saturated fats in your diet whenever possible. Saturated fats are high in cholesterol content, and come mainly from animal products. Unsaturated fats are low in cholesterol and originate from plant products. Also, reduce your total fat intake, and lose weight if you are currently overweight. Exercise can help too, because it lowers total cholesterol and raises HDL levels. Finally, don't smoke—smoking increases total cholesterol and reduces HDL. It also is an independent risk factor for heart disease as well.

---

actually *reduce* the amount of fat (including cholesterol) in the bloodstream, they are often referred to as "good" cholesterol. The function of LDLs is the transport of fat (including cholesterol) from the liver to various tissues. Once they reach the tissues, they are usually deposited into adipose cells for long-term energy storage. LDLs are recognized as "bad" cholesterol, since their presence in the blood indicates a buildup of cholesterol in body tissues. VLDLs are converted to LDLs, usually in the liver.

Protein Transport

Protein digestion begins in the stomach and continues in the small intestine until essentially all proteins ingested are simplified to their basic subunits, amino acids. Amino acids are generally transported across the intestinal epithelium by active transport and enter the bloodstream intact. Once in the blood, they are carried to the liver. Liver cells may metabolize amino acids for

the production of energy (as ATP), or they may use amino acids in the synthesis of plasma proteins (such as albumins and fibrinogen). Most of the amino acids, however, pass through the liver and are carried on by the bloodstream to other body cells, where they are used for protein synthesis.

Amino acids are used by the body in a variety of ways. For example, they are important for the synthesis of structural proteins. Proteins such as collagen in connective tissue, myosin and actin in muscle tissue, and keratin in the skin are vital for the support of body tissues. Amino acids are also necessary for the synthesis of important functional proteins and enzymes, such as hemoglobin in red blood cells, various neurotransmitters, the enzymes that participate in driving metabolic reactions within cells, and the numerous digestive enzymes. Amino acids may also be used as a source of energy, such as when their blood levels are in excess or when other, preferred energy sources (carbohydrates and fats) are in low supply.

Vitamin and Mineral Transport

Water-soluble vitamins, which include the B-complex vitamins and vitamin C, are absorbed along with water through the intestinal epithelium (except vitamin B_{12}, which must accompany intrinsic factor to be absorbed). Like most other absorbed nutrients, they are transported by the bloodstream first to the liver and then on to other body cells. If the body is in need of them, these vitamins enter body cells by active transport, where they function as coenzymes to assist in metabolic reactions. However, if the body does not need them, they are usually not stored but are excreted with the urine.

Fat-soluble vitamins, such as vitamins A, D, E, and K, bind to ingested fats in the small intestine and are absorbed along with them. In the bloodstream, these vitamins are carried to the liver and on to body cells, where they pass through plasma membranes easily. Like water-soluble vitamins, fat-soluble vitamins act as coenzymes by assisting metabolic reactions. Except for vitamin K, fat-soluble vitamins can be stored in body cells. As a result, poisoning can result from ingesting too much of a fat-soluble vitamin; this condition is called *hypervitaminosis.*

Most minerals are quickly absorbed across the intestinal epithelium and are carried in the bloodstream to body cells in need of them. Some minerals, such as calcium, must be combined with other substances to cross the epithelial border. In the case of calcium, vitamin D must be present to carry it across in combination with ingested fats.

CONCEPTS CHECK

1. The cells of what organ perform the interconversion of glucose and glycogen?

2. In what molecular form are fats carried through the bloodstream?

3. How may body cells utilize amino acids?

METABOLISM

The biochemical reactions of metabolism use nutrients to build new molecules, or to obtain energy to fuel body activities. They can be measured as the metabolic rate.

Once inside body cells, newly absorbed nutrients undergo a variety of biochemical reactions. They may be used as raw materials in the synthesis of new products, or they may be simplified further into smaller molecules for the production of usable energy. As we learned in Chapter 1, these reactions are collectively called **metabolism.** They include the reactions in which larger molecules are built from smaller ones, known as **anabolism,** and the processes that break down complex structures into simpler molecules, called **catabolism.** The digestion of foods into their more basic subunits for absorption involves many catabolic reactions. **Cellular respiration,** which is the breakdown of glucose, fatty acids, and amino acids within cells to release the energy locked within their chemical bonds, is also a form of catabolism. In both anabolism and catabolism, the high-energy molecule ATP plays a central role. The collective process of metabolism is diagramed and summarized in Figure 17–3.

Cellular Respiration

The chemical breakdown of nutrients within the machinery of the cell is a process known as **cellular respiration.** Its purpose is to produce energy in a form that can be used by cells to power their activities. During cellular respiration, adenosine diphosphate (ADP) is combined with an inorganic phosphate group (PO_4^{2-}) to form adenosine triphosphate, or ATP. The energy to perform work is thus stored in the chemical bond between ADP and PO_4^{2-}. When ATP is converted back to ADP and PO_4^{2-}, the energy in the bond is released to drive chemical reactions, such as those involved in the synthesis of molecules (anabolism), muscle contraction, and active transport.

Once food has been digested and absorbed into cells, the chemical bonds within the nutrient molecules are exposed to oxygen and thereby break apart. This process is called *oxidation.* The energy that is released in these bonds is used to synthesize ATP. Whenever a substance is oxidized, it loses electrons. During catabolism, these electrons are attracted to special carrier molecules, called coenzymes (derived from vitamins), which transport the electrons to other molecules. During their transport, the energy in the broken bonds of the nutrient molecules is transferred by way of these electrons to molecules of ADP, and thereby stored in the form of ATP. In this section we will investigate the oxidation reactions of the three major nutrient groups: carbohydrate, fats, and proteins.

Carbohydrate Metabolism

The products of carbohydrate digestion are monosaccharides. As we learned earlier, glucose is the most important monosaccharide, for it is the preferred energy source of cells. Glucose may also be used anabolically to store energy, in the form of glycogen. However, glycogen accounts for only about 1% of the body's energy storage (the remainder is in the form of fats),

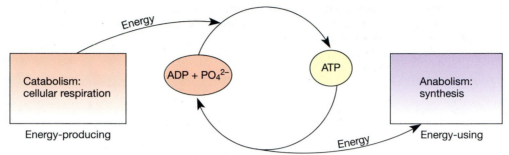

Figure 17–3
The cycling of energy between catabolism and anabolism. Cellular respiration (the primary form of catabolism) generates energy that is stored in ATP. This energy may be used in the anabolic synthesis of molecules.

What usually happens to ATP after it releases its energy in a cell?

so the most important use of glucose is in the catabolic production of energy. The catabolism of glucose is a multiple-step process, which begins with glycolysis.

Glycolysis

Glycolysis (glī-KOL-i-sis) is a series of biochemical reactions that occur in the cytosol of most cells. It results in the breakdown of glucose, which contains six carbon atoms, into two three-carbon molecules of **pyruvic acid** (Fig. 17–4). Two ATP molecules are required to start glycolysis, and four ATP molecules are produced for a net gain of two ATP molecules.

Glycolysis begins when glucose is modified by the addition of two phosphate groups from two ATP molecules. The modified glucose is then cleaved into two three-carbon molecules, each with a phosphate group. The three-carbon molecules are soon oxidized, and the released energy is available for use to power other reactions in the metabolic pathway. Electrons are also re-

leased in this process and are stored for later use by electron-carrier molecules.

Glycolysis proceeds whether or not oxygen is available. This capability is important in tissues such as skeletal muscle when the demand for ATP increases rapidly when oxygen supply may be temporarily low. This situation arises during strenuous exercise. If the cell is receiving a deficient supply of oxygen, the pyruvic acid and energy produced in glycolysis will be processed by anaerobic respiration. If, on the other hand, the cell is receiving a sufficient oxygen supply, the pyruvic acid and energy produced in glycolysis will be used in aerobic respiration to produce additional ATP.

Anaerobic Respiration

The catabolic process of **anaerobic respiration** is the continued breakdown of glucose molecules in the absence of oxygen. It results in the net production of two molecules of **lactic acid** and two molecules of ATP (Fig. 17–5). The ATP produced from anaerobic respiration may be used as energy by the body when insufficient amounts of oxygen are delivered to cells.

The first phase of anaerobic respiration is the conversion of glucose to two molecules of pyruvic acid by glycolysis. The second phase is the conversion of pyruvic acid to lactic acid, which requires an input of energy. This energy may be obtained from electrons on carrier molecules generated during glycolysis.

Lactic aid produced by anaerobic respiration is released by cells and transported by the bloodstream to the liver. When oxygen is again in good supply, the lactic acid in the liver can be converted back to glucose. This process requires additional energy, which is made available by aerobic respiration. The oxygen that is required to produce the ATP needed for the synthesis of glucose from lactic acid is called the **oxygen debt** (see Chapter 8).

Figure 17–4
Glycolysis.

What is the product of glycolysis?

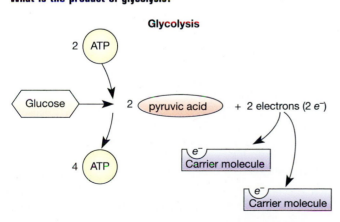

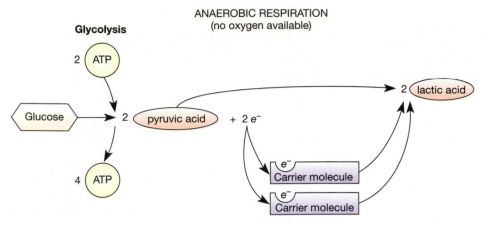

Figure 17–5
Anaerobic respiration. When oxygen is not available to the cell, pyruvic acid
produced during glycolysis is converted to lactic acid.
What is the product of anaerobic respiration?

Aerobic Respiration

Aerobic respiration is the breakdown of glucose in the
presence of oxygen. For the breakdown of a single glucose molecule, aerobic respiration produces 36 (or, in
some cases, 38) molecules of ATP, 6 molecules of water, and 6 molecules of carbon dioxide (Fig. 17–6). The
process of aerobic respiration takes more time to complete than anaerobic respiration, although it produces

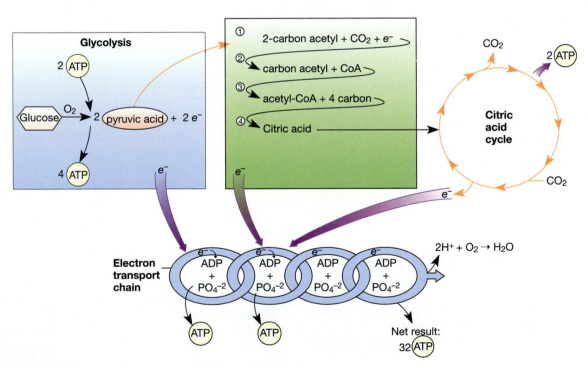

Figure 17–6
Aerobic respiration. The net gain of ATP generated by aerobic respiration includes
2 from glycolysis, 2 from the citric acid cycle, and 32 from the electron transport
chain, for a total of 36 ATP molecules.
What does oxygen do during aerobic respiration?

more ATP molecules from each molecule of glucose. To draw an analogy with a track meet, we could say that if aerobic respiration is like a 50-meter sprint in that it is over quickly but provides little benefit toward physical conditioning, aerobic respiration would be like a marathon run in that it takes more time to complete but its conditioning benefits are enormous.

The first phase of aerobic respiration is glycolysis, similar to anaerobic respiration. In the second phase, pyruvic acid moves from the cytoplasm into the mitochondrion, where enzymes remove a carbon atom from the three-carbon pyruvic acid molecule to form a two-carbon acetyl molecule and carbon dioxide. Energy is released in this reaction and can be used later to produce ATP. It is temporarily stored on an electron-carrier molecule. The acetyl group combines with a molecule obtained from vitamin B, called *coenzyme A* (abbreviated CoA), to form **acetyl-CoA.** In the third phase, acetyl-CoA combines with a four-carbon molecule to form a six-carbon molecule of citric acid.

The citric acid thus formed enters a cyclic series of reactions, known as the **citric acid cycle** (or Krebs cycle). Here, the citric acid molecule goes through a series of *oxidation-reduction reactions,* in which electrons and hydrogen ions are removed. The purpose for these reactions is to transfer energy from the bonds in citric acid into the electron transport chain, where ATP will be generated to store this energy. Citric acid is first converted back to a four-carbon molecule, which may then combine with another acetyl-CoA to begin the cycle again. During the cycle, two carbon atoms are lost as the waste product carbon dioxide, energy in the form of two molecules of ATP is released, and electrons are produced. The electrons are stored on electron-carrier molecules, which transport the electrons to the electron transport chain, where more ATP can be generated.

The **electron transport chain** consists of a chain of energy-relay molecules attached to the inner mitochondrial membrane. They transfer electrons that have originated from glycolysis, acetyl-CoA generation, and the citric acid cycle from one molecule to the next. As electrons are transferred, molecules of ADP and PO_4^{2-} are combined to form ATP. The net result is the generation of 32 (or 34) molecules of ATP for each transfer cycle.

The final phase of aerobic respiration is the combination of hydrogen ions released from the electron transport chain with oxygen to form water. A lack of oxygen to accept the hydrogen ions causes the electron transport chain to stop because of the drop in pH, which destroys the carrier molecules. Oxygen is therefore needed by the cell not only to oxidize molecules, but also to remove hydrogen ions that are continually produced.

Fat Metabolism

Triglycerides, cholesterol, and phospholipids are the body's main energy storage molecules and are important molecules in cell structure. Between meals, when the blood glucose levels begin to decline, some of the triglycerides stored within adipose tissue are reduced to fatty acids and glycerol (see Fig. 17–7). Once in the bloodstream, they are transported to other tissues, particularly skeletal muscles and the liver, where the fatty acids can be further metabolized for the release of energy by a process called *fatty acid oxidation.*

The catabolism of fatty acids occurs by a series of reactions in which two carbon atoms are removed from

Figure 17–7
Catabolism of proteins, carbohydrates, and fats. All three nutrients may undergo conversion for the production of energy in the form of ATP.

Notice the central role of acetyl CoA in this illustration. What preferred energy source is simplified to acetyl CoA by the process of glycolysis?

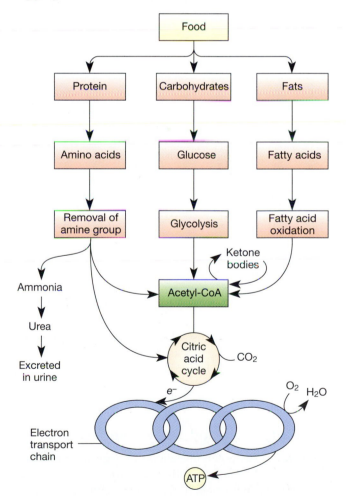

the fatty acid chain to form acetyl-CoA. This process continues, two carbon atoms at a time, until the entire fatty acid has been converted into molecules of acetyl-CoA. The acetyl-CoA may then enter the citric acid cycle, where it can be used to generate ATP. Alternatively, two molecules of acetyl-CoA may combine in the liver to form **ketone bodies.** These molecules are released into the bloodstream and travel to other tissues, particularly skeletal muscle, where they are converted back to acetyl-CoA for the generation of ATP.

Ketone bodies are generally produced in small quantities, since the body prefers to use glucose as an energy source. In some cases, however, ketone bodies may be produced in larger quantities to compensate for an inability to acquire glucose (such as occurs in starvation, and in diabetes mellitus), leading to an accumulation of ketones in the bloodstream. This abnormal condition, called *ketosis,* can be identified in a patient by the smell of acetone on the breath, since acetone is a type of ketone (acetone smells like nail polish remover, since it is its main component). Ketosis can lead to a drop in pH if the ketone concentration rises significantly, since ketones are acidic. If ketosis exceeds the capacity of the body's buffering system, it can lead to the more dangerous condition of *acidosis,* which can cause death.

Protein Metabolism

The products of protein digestion are amino acids. Once absorbed into cells, amino acids may be used to synthesize needed proteins, or serve as a secondary energy source. If amino acids are used for the production of energy, an amine group (NH_2) is removed from the amino acid, leaving ammonia (NH_3) and a keto acid. During this process energy is released, which can enter the electron transport chain to generate ATP. The ammonia produced is toxic to cells, and is converted by the liver to a less toxic molecule called **urea.** Urea is carried by the blood to the kidneys, where it is excreted in urine. The keto acid can enter the citric acid cycle, or it can be converted to pyruvic acid, acetyl-CoA, or even glucose, for the later generation of additional ATP (see Fig. 17–7). Although amino acids may be used to generate ATP, they are more important as structural building blocks for proteins and enzymes. Consequently, their breakdown usually involves the sacrifice of valuable tissue.

Metabolic Rate and Body Temperature

The process of metabolism uses energy to synthesize new materials, and to produce additional revenues of energy that can be used immediately or stored for later use. The use of energy through metabolism is expressed as a measure of heat, which is known as **calories.** A calorie is the amount of heat needed to raise the temperature of 1000 grams of water from 14°C to 15°C. Calories are used to indicate the amount of energy it takes to metabolize foods. For example, a double-decker hamburger has about 500 calories, which means that it will take 500 calories of heat for the body to metabolize it. The body's use of energy within a given time period is called the **metabolic rate.** It is the sum of all the chemical reactions and mechanical work of the body.

Factors Affecting Metabolic Rate

Metabolic rate may be increased or decreased by a variety of factors. One of these is nervous stimulation, which increases the metabolic rate of individual body cells when the sympathetic nervous system is triggered by a stressful situation. Hormones, such as norepinephrine, may also increase metabolic rate. A third factor is body temperature. Each 1°C rise in temperature increases the metabolic rate by about 10%. Other factors that can increase metabolic rate include exercise and food ingestion. Factors that can decrease metabolic rate include increased age, sleep, malnutrition, and hypothermia.

Basal Metabolic Rate

Since many factors affect metabolic rate, it is normally measured under controlled, standardized conditions that include fasting for 12 hours beforehand and lying in a reclined position at rest in a low-stress environment at a controlled room temperature. The measurement obtained is a measure of the energy the body needs to perform its most essential activities, such as breathing, maintaining the heartbeat, and maintaining kidney function. It is called the **basal metabolic rate (BMR).** In past years, measuring a person's BMR was very helpful in determining the amount of the hormone thyroxine that was being produced by the thyroid gland. The more thyroxine produced, the higher the metabolic rate. Today, thyroid activity is more easily determined by blood tests.

Total Metabolic Rate

The **total metabolic rate (TMR)** is the amount of energy consumed by the body to fuel all activities. It includes the BMR as well as voluntary muscle activity. In most people, the BMR accounts for a surprisingly large percentage of the TMR. For example, in a man whose total energy needs per day are about 3000 calories, about 2000 calories are used to support vital activities.

Estimating Caloric Intake

For most people, the most energy-efficient diet consists of consuming about the same number of calories as that lost during a normal day's activity. The approximate number of calories you use daily (the total metabolic rate, or TMR) can be estimated as follows. First, calculate the number of calories you burn in a resting state (the basal metabolic rate, or BMR). This may be approximated by multiplying your kilogram weight by 24 calories per day (one kilogram is equal to 2.2 pounds). Thus, if you weigh 50 kg (110 pounds), your approximate BMR is 1200 calories. Next, estimate the number of calories you burn off in physical activity each day. If you are inactive, you'll burn about 30% of your BMR in activity, bringing the TMR to 1560 calories. If you are moderately active (no strenuous exercise, but some walking), add 40% to your BMR. This would bring the TMR to 1680 calories. Finally, if you are active (some strenuous exercise), 50% can be added to the BMR, for a total of 1800 calories. Although these figures are rough estimates, they can provide you with an idea of the advantages of strenuous exercise. If your consumption of food is not excessive and you feel you would benefit from losing weight, most experts agree that raising your TMR by increasing your current level of exercise is preferable to going on a "crash diet" of low food consumption.

Heat Production and Loss

The oxidation of foods we eat is not completely efficient. Only a part of the energy released during catabolism (about 43%) is used to fuel biological activities, such as skeletal muscle contraction, active transport across plasma membranes, and anabolism. Most of the energy results in the production of heat. Because the amount of heat produced is so great, it must be removed to prevent body temperature from rising to dangerous levels. The average body temperature is about 37°C (98.6°F), which must be homeostatically controlled. The primary routes of heat loss are through radiation, conduction, convection, and evaporation, which are illustrated in Figure 17–8.

Radiation refers to the exchange of heat from one object to another without physical contact. It involves the flow of energy in the form of infrared conduction waves. An example of radiation occurs whenever you feel heat waves released by a household heater or fireplace. Radiation results in a loss of heat from the body if the objects around you are cooler than body temperature. Over 50% of body heat is lost by radiation when the temperature in a room is 21°C (70°F).

Conduction is the exchange of heat by direct contact. It occurs whenever your bare feet touch a cold floor, you put on chilled clothes, or you lean against a cold concrete wall. About 3% of body heat is normally lost through conduction.

Convection is the transfer of heat between the body and the air. For example, your body is cooled

Figure 17–8
Mechanisms of heat loss between the body and the environment.

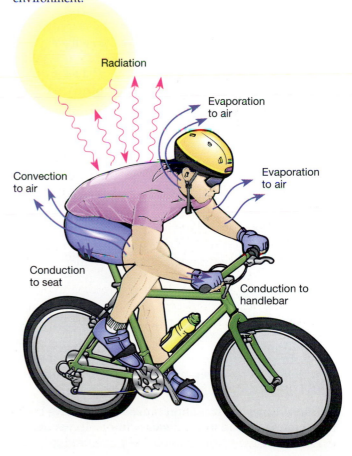

525

when a breeze of cold air makes contact with your skin. Normally, about 15% of body heat is lost by convection.

Evaporation refers to the conversion of a liquid to a vapor. It results in heat loss to the body because of the loss of water from the skin. Evaporation accounts for about 22% of body heat loss, although this figure increases considerably with increased air temperature. However, increasing relative humidity actually lowers the evaporation rate, which explains why hot, humid climates seem so much warmer than hot, dry climates.

CONCEPTS CHECK

1. What is the difference between catabolism and anabolism?

2. How many molecules of ATP are normally produced from the catabolism of a single molecule of glucose?

3. What factors affect metabolic rate?

4. How is basal metabolic rate measured?

HOMEOSTASIS AND TEMPERATURE REGULATION

In order to maintain a constant internal body temperature, the amount of heat that is released must equal the amount of heat produced. There are various methods the body employs to regulate this heat balance, many of which you will recognize from previous chapters. They include vasoconstriction of blood vessels in the skin, which conserves body heat by reducing blood flow and heat transfer from the internal organs to the skin; skeletal muscle contraction, which produces additional heat by the increased metabolic activity; and stimulation by the hormone thyroxine, which raises the body's coverall metabolic rate to produce additional heat.

Each of these homeostatic mechanisms is regulated by a single center in the body: the hypothalamus.

Often referred to as the body's thermostat, the hypothalamus receives sensory information regarding internal body heat, and relays impulses to the smooth muscles lining the skin's blood vessels, skeletal muscles throughout the body, and the thyroid gland in response to any change. The control center within the hypothalamus consists of a group of neurons, referred to as the *preoptic area*. These neurons respond to changes in blood temperature within nearby blood vessels: if blood temperature rises, the neurons fire impulses more rapidly; if blood temperature lowers, the neurons fire more slowly. By way of the responses of the preoptic area, the hypothalamus is able to regulate internal body temperature despite changes in the surrounding environment.

CLINICAL TERMS OF NUTRITION

Fever   An elevated body temperature resulting from a reset of the hypothalamus "thermostat." Chemicals called *pyrogens*, which are produced by white blood cells during an infection, stimulate the release of prostaglandins. The prostaglandins stimulate the hypothalamus to adjust body temperature to a setpoint higher than normal. Aspirin, acetaminophen, and ibuprofen are effective in reducing fever by interfering with the production or circulation of prostaglandins.

Heat stroke   A condition caused by excessive body temperature, usually over 106°F. It results in

dizziness, abdominal pain, delirium, and loss of consciousness. Because of loss of fluids, circulatory shock often occurs as well. If the body temperature is not reduced quickly, heat stroke can cause irreparable damage to brain tissue, the kidneys, and the liver.

Hyperglycemia   An elevated level of glucose in the blood. It may follow a meal rich in simple sugars, but if a meal has not been recent, it is one of the primary indications of diabetes mellitus.

Hypoglycemia   A depressed level of glucose in the blood. Hypoglycemia is a symptom of pancreatic cancer, malnutrition, or improper administration of insulin in a diabetic patient.

Obesity A state of excessive weight that exceeds by 20% the desirable standard. Because of the accumulation of fat, it is considered a serious health risk for cardiovascular disease, hypertension, pulmonary disease, arthritis, gallbladder disease, diabetes mellitus, colon cancer, and uterine cancer.

Phenylketonuria (PKU) A genetic disease of metabolism, in which the amino acid phenylalanine cannot be completely metabolized. It results in elevated levels of phenylketone in the blood and is frequently associated with other inherited disorders such as mental retardation.

CHAPTER SUMMARY

Nutrition is the process by which food is taken into the body and changed into a form that can be utilized by cells. It involves ingestion, digestion, absorption, transport, and metabolism.

NUTRIENTS

The food substances needed by the body for energy, repair, and growth are called *nutrients*. Essential nutrients are those that cannot be synthesized by the body; they must be obtained from ingestion of food.

Carbohydrates Molecules composed of simple sugar subunits. They are used as the primary energy source in the body. They are obtained as starch from fruits and vegetables, and as glycogen from meats.

  Dietary Requirements Current recommendations are that 125 to 175 grams of carbohydrates be ingested each day.

Fats Dietary fats consist mostly of triglycerides (glycerol and fatty acids), but also phospholipids and cholesterol. They are used by the body as its most efficient fuel source and also help in the absorption of vitamins and are needed for the formation of plasma membranes in cells.

  Dietary Requirements Fats should represent no more than 30% of total caloric intake, and cholesterol should not exceed 250 mg per day.

Proteins Proteins consist of amino acids. They are used for cell structure and function, and may provide an energy source as well. There are eight essential amino acids, or required amino acids that cannot be synthesized by body cells and so must be obtained by ingestion.

  Dietary Requirements About 0.8 gram of protein per kilogram of body weight.

Vitamins Organic compounds required in small amounts for good health. Most vitamins are not synthesized by the body. They are necessary for helping the body use nutrients that are used as energy and structural building blocks. Vitamins may be fat-soluble (A, D, E, and K), or water-soluble (B complex and C).

Minerals Dietary minerals provide strength to certain tissues and ions for body functions and are bound to organic compounds to form hormones, enzymes, and functional proteins.

BIOAVAILABILITY

Some substances that are ingested assist in the absorption of other substances. However, some substances that are ingested interfere with the absorption of important nutrients. Therefore, it is best to eat a variety of foods that include all groups: carbohydrates, fats, and proteins.

TRANSPORT

Once food is digested and absorbed, it must be transported throughout the body to where it is needed.

Carbohydrate Transport Carbohydrates are absorbed in the form of monosaccharides, mainly glucose. Glucose is the preferred energy source of the body and is used to generate ATP. Its interconversion with glycogen occurs in the liver and is regulated by pancreatic hormones (glucagon and insulin).

Fat Transport Triglycerides, phospholipids, and cholesterol are absorbed and carried through the lymphatic network in packages called *micelles*. Once in the bloodstream, they combine with protein to form lipoproteins: LDLs, HDLs, and VLDLs. LDLs carry fats from the liver to body tissues, HDLs carry fats from body tissues to the liver for their breakdown, and VLDLs are converted to LDLs. High LDL levels in the blood increase the risk of cardiovascular disease, whereas high HDL levels decrease the risk.

Protein Transport Amino acids absorbed across the intestinal lining are carried by the bloodstream to the liver, where they may be metabolized to release energy. Most amino acids, however, pass directly to body cells and are used for protein synthesis.

Vitamin and Mineral Transport Water-soluble vitamins absorbed through the intestinal lining are carried by the bloodstream to the liver, and then on to body cells in need of them. Here, they assist metabolic reactions. If not needed, they are excreted. Fat-soluble vitamins are absorbed along with fats, and are transported by the bloodstream to the liver. They pass to body cells, where they assist in metabolic reactions if they are needed. If not needed, they can be stored.

METABOLISM

Anabolism is the synthesis of new materials from raw nutrients. Catabolism is the breakdown of molecules for the release of energy.

Cellular Respiration   The catabolism of nutrients for the production of energy.

Carbohydrate Metabolism   Involves a series of steps that reduce glucose to yield 36 (or 38) molecules of ATP, 6 molecules of water, and 6 molecules of carbon dioxide.

Glycolysis   The breakdown of glucose into two molecules of pyruvic acid, and a net gain of two ATP molecules.

Anaerobic Respiration   The breakdown of glucose in the absence of oxygen, to yield two molecules of lactic acid that can later be reconverted.

Aerobic Respiration   The breakdown of glucose to acetyl-CoA. Acetyl-CoA then enters the citric acid cycle, where two ATP molecules are generated and electrons are transported to the electron transport chain, where in turn 32 (or 34) more ATP molecules are produced. Hydrogen ions are removed by combining with oxygen to form water.

Fat Metabolism   Fat is the most common form of energy storage. Triglycerides may be catabolized to produce molecules of acetyl-CoA, which can enter the citric acid cycle to generate ATP. Ketone bodies are also produced.

Protein Metabolism   Amino acids are important as building blocks for protein synthesis, but can be used as a secondary energy source. Amino acids are broken down to yield an amine group, ammonia, and a keto acid. Energy is produced, and more can be generated by the conversion of the keto acid. Ammonia is toxic, and is converted to urea and excreted.

Metabolic Rate and Body Temperature   Metabolic rate is the body's use of energy within a given time period.

Factors Affecting Metabolic Rate   Nervous stimulation, hormones, body temperature, exercise, food ingestion, sleep, malnutrition, hypothermia, and age affect metabolic rate.

Basal Metabolic Rate   The metabolic rate of vital activities within the body, such as breathing, maintaining the heartbeat, and maintaining kidney function.

Total Metabolic Rate   The amount of energy consumed by the body to fuel all activities, including basal metabolic rate and voluntary muscle activity.

Heat Production and Loss   Much of the energy produced during catabolism is lost as body heat. This heat must be disposed of to avoid excessive body temperatures. It is eliminated in the following ways:

Radiation   The transfer of heat from one object to the environment by waves moving through space. Radiation accounts for a large percentage of heat lost during normal circumstances.

Conduction   The exchange of heat by direct contact.

Convection   The drawing away of heat by the motion of air.

Evaporation   The transfer of heat by the vaporization of water from a surface, such as the skin.

HOMEOSTASIS AND TEMPERATURE REGULATION

Internal body temperature must be regulated carefully to avoid excessive extremes. It is done so by a variety of mechanisms, such as vasoconstriction, skeletal muscle contraction, and thyroxine release, all of which are regulated by the hypothalamus.

KEY TERMS

acetyl-CoA (p. 523)
aerobic (p. 522)
anabolism (p. 520)
anaerobic (p. 521)

calorie (p. 524)
catabolism (p. 520)
citric acid cycle (p. 523)
coenzyme (p. 512)

electron transport chain (p. 523)
essential nutrient (p. 510)
glycolysis (p. 521)
nutrient (p. 510)

nutrition (p. 509)
vitamin (p. 512)

QUESTIONS FOR REVIEW

OBJECTIVE QUESTIONS

1. A substance that is needed by the body but cannot be obtained for synthesis is called a(n):
   a. mineral
   b. essential nutrient
   c. vitamin
   d. coenzyme

2. Which of the following could be described as nutrients that assist metabolic reactions in cells but must be ingested regularly, since they cannot be stored?
   a. fat-soluble vitamins

b. water-soluble vitamins
c. vitamin D and vitamin E
d. amino acids

3. Nutrients that are important primarily as building blocks for cellular structure and function are:
   a. carbohydrates   c. fats
   b. proteins   d. minerals

4. The preferred energy source of the body is:
   a. glucose   c. fructose
   b. triglycerides   d. amino acids

5. The preferred percentage of fat in a healthy diet should be no more than:
   a. 50%   c. 90%
   b. 40%   d. 30%

6. An important group of nutrients that do not form structural building blocks, but rather are necessary for incorporating other nutrients, are the:
   a. carbohydrates   c. vitamins
   b. minerals   d. essential amino acids

7. Fat subunits are transported in the bloodstream in a large molecular form that incorporates them with proteins. The large molecule is called a:
   a. micelle   c. polysaccharide
   b. lipoprotein   d. coenzyme

8. The first phase in the catabolism of glucose results in the formation of pyruvic acid. This phase is called:
   a. electron transport chain   c. glycolysis
   b. citric acid cycle   d. anaerobic respiration

9. The complete breakdown of a glucose molecule during aerobic respiration usually produces how many molecules of ATP?
   a. 15   c. 30
   b. 36   d. 58

10. The metabolic rate of essential body activities only is called:
    a. total metabolic rate
    b. basal metabolic rate.
    c. essential metabolic rate
    d. cellular respiration

11. Body temperature is regulated by the:
    a. hypothalamus
    b. medulla
    c. skeletal muscles
    d. process of vasoconstriction

ESSAY QUESTIONS

1. Explain why cholesterol can be beneficial to the body in some forms but increase risk factors for cardiovascular disease when in other forms.

2. Describe the cellular respiration of glucose, and indicate the products that are formed.

3. Identify a balanced diet, and discuss the reasons why it is important to maintain it.

ANSWERS TO ART LEGEND QUESTIONS

Figure 17–1
Why must essential nutrients be obtained from the food you eat? Essential nutrients must be obtained from food because they cannot be manufactured by the body.

Figure 17–2
Why are high-density lipoproteins (HDLs) considered to be beneficial forms of lipoprotein? HDLs are beneficial because they reduce the amount of fat (including cholesterol) in the bloodstream.

Figure 17–3
What usually happens to ATP after it releases its energy in a cell? ATP is reduced to ADP ($+ PO_4^{2-}$), and is usually regenerated to ATP by catabolic processes.

Figure 17–4
What is the product of glycolysis? Two molecules of pyruvic acid and four molecules of ATP (two ATP are used, for a net gain of two ATP) result from glycolysis.

Figure 17–5
What is the product of anaerobic respiration? The product of anaerobic respiration is two molecules of lactic acid and two molecules of ATP.

Figure 17–6
What does oxygen do during aerobic respiration? Oxygen performs the role of an oxidizer, which causes molecules to lose electrons with the result of breaking bonds. The electrons are carried by electron-carrier molecules and are used to generate ATP. Oxygen also binds with hydrogen ions to form water.

Figure 17–7
Notice the central role of pyruvic acid in this illustration. What preferred energy source is simplified to pyruvic acid by the process of glycolysis? Glucose is the preferred energy source that is simplified to pyruvic acid.

THE URINARY SYSTEM

LEARNING OBJECTIVES

After studying this chapter, you should be able to:

1. Identify the functions of the urinary system.

2. Identify the structural features of the kidneys.

3. Describe the blood vessels associated with the kidneys.

4. Identify the components of a nephron.

5. Describe the flow of blood through a nephron.

6. Identify the juxtaglomerular apparatus.

7. Describe the processes of filtration, reabsorption, and secretion and state where they occur.

8. Describe how urea and uric acid are formed and how they are excreted.

9. Describe the processes that regulate kidney function.

10. Identify the role of the kidneys in maintaining body fluid composition and pH.

11. Identify the structure and function of the ureters, urinary bladder, and urethra.

12. Describe the process of micturition.

The urinary system is the "sanitary engineer" of the body, for it maintains the purity and homeostatic balance of the body's fluids by removing unwanted waste materials and recycling other materials. Its most important organs are the kidneys, which filter gallons of fluids from the bloodstream every day. They remove metabolic wastes, toxins, excess ions, and water that leave the body as urine while returning needed materials back to the blood. This function is called **excretion;** it is also performed by the skin, lungs, and liver, but to a far lesser extent. In addition to performing excretion, the kidneys also regulate fluid and electrolyte levels and help regulate blood pressure and control red blood cell production in the bone marrow. The other organs of the urinary system—the ureters, urinary bladder, and urethra—transport or store urine once it is formed.

KIDNEY STRUCTURE

Each kidney contains a large number of subunits known as nephrons, which receive large volumes of blood for processing.

The kidneys are a pair of organs vital to our survival. Their special structure enables them to perform functions no other organ can accomplish.

External Structure

The kidneys are each about the size of your fist and in the shape of a kidney bean, with the concave (indented) margin facing the body's midline (Fig. 18–1).

They lie against the posterior abdominal wall behind the peritoneum. This positioning is termed *retroperitoneal.* The right kidney is crowded by the liver above it and lies slightly lower than the left kidney.

Both kidneys are externally supported by several layers of connective tissue (Fig. 18–2). These include the outermost **renal fascia,**[1] which anchors each kidney and the nearby adrenal gland to surrounding tissues; the **adipose capsule** beneath it, providing the kidney with a cushioning layer of fat; and the inner **renal capsule,** a thin, transparent protective layer of fibrous connective tissue.

[1]**Renal: Latin** *ren-* = kidney + *-al* = belonging to.

Figure 18–1
Organs of the urinary system. Anterior view of the trunk with the organs of the digestive system removed. This drawing is of a female.

What major arteries supply the kidneys with fresh blood?

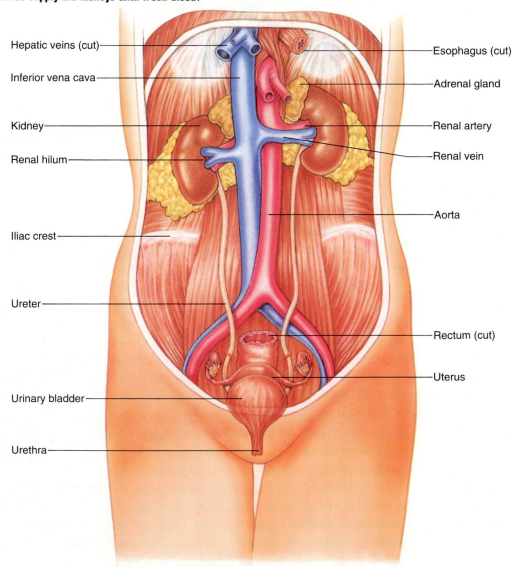

Hepatic veins (cut)
Inferior vena cava
Kidney
Renal hilum
Iliac crest
Ureter
Urinary bladder
Urethra

Esophagus (cut)
Adrenal gland
Renal artery
Renal vein
Aorta
Rectum (cut)
Uterus

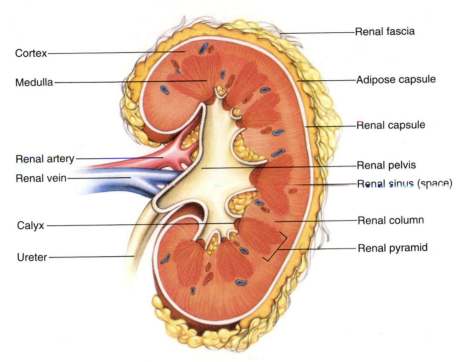

Figure 18–2
Structure of the kidney. This drawing shows the left kidney sectioned through the frontal plane. The blood vessels within the substance of the kidney are not shown.
What are the three internal regions of each kidney?

Internal Structure

The concave margin of each kidney is called the **hilum** (HĪ-lum) and is where the renal artery, renal vein, nerves, and ureter unite with the kidney (Fig. 18–1). The hilum opens into an internal space known as the **renal sinus** (Fig. 18–2). Within the renal sinus are fat, connective tissue, and a membrane-lined basin called the **renal pelvis.** From the renal pelvis extend several funnel-shaped channels. These are the **calyces**[2] (KAL-i-sēz; singular form is *calyx*), which collect newly formed urine from the kidney tissue and direct it into the renal pelvis.

Internally, the kidney is divided into three distinct regions: the innermost renal pelvis, just described; the **renal medulla;** and the outer **renal cortex** (Fig. 18–2). The medulla contains a number (usually eight to ten) of cone-shaped **renal pyramids.** The pyramids appear striped, or striated, because of the presence of parallel bundles of microscopic tubules. The narrow apex of each pyramid points toward the renal pelvis, where its tip is surrounded by a calyx, and the broad base borders the cortex. The cortex is granulated in appearance and usually stains lighter than the medulla. Portions of it extend between the renal pyramids and are called

renal columns. The substance of the renal medulla and cortex consists of a vast supply of blood vessels associated with many tiny, microscopic ducts and tubules. The tubules constitute the functional subunits of the kidney, and are called **nephrons.**[3]

Blood Supply

The circulatory network of the kidneys is extensive, for it must carry large volumes of blood that is cleansed and modified continuously. The kidneys receive blood from the large **renal arteries,** which branch from the abdominal aorta (Fig. 18–3). When the body is at rest, these vessels provide the kidneys with about one-fourth the total cardiac output (about 1200 ml) every minute. As each renal artery approaches a kidney, it branches into five **lobar arteries,** which enter the renal sinus. The lobar arteries quickly branch into numerous **interlobar arteries,** which extend between the renal pyramids. Once reaching the outer cortex, the interlobar arteries give rise to arteries that curve over the outer margins of the pyramids. These vessels extend between the cortex and the medulla and are called **arcuate arteries.** Small **interlobular arteries** extend from the arcuate arteries into the cortex. Branches of the in-

---

[2]Calyces: Latin or Greek *calyx* = cup.

[3]Nephrons: Greek *nephros* = kidney + *-on* = unit of.

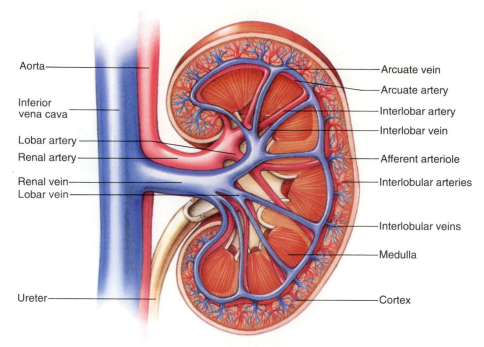

Figure 18–3
Blood circulation in the kidney. In this drawing, the left kidney is sectioned through the frontal plane and all blood vessels are included.
Which vessels do the interlobar arteries branch into?

terlobular arteries, called **afferent arterioles,** lead to the nephrons.

Blood is drained from the kidney by a series of veins that parallel the arteries and have similar names. The **renal vein** collects the draining blood and delivers it to the inferior vena cava.

Nephron Structure

About 1 million nephrons make up the functional tissue within each kidney. Although they are microscopic in size, a single nephron occupies space in both the cortex and medulla regions. One nephron unit is shown in Figure 18–4.

Each nephron consists of two parts: an expanded, bulblike end called the **renal corpuscle** and a thin, twisting duct called the **renal tubule.** The renal corpuscle is located in the cortex, whereas the renal tubule contains portions in the cortex and in the medulla.

The renal corpuscle consists of a bowl-shaped structure, called the **Bowman's capsule,** that partially surrounds a capillary network. The capillary network is called the **glomerulus**[4] (glō-MER-yoo-lus) and resembles a ball of yarn tucked into the bowl of the Bowman's capsule. Its endothelial walls contain unusually large pores, called **fenestrae.**[5] Notice from Figure 18–5 that the Bowman's capsule is hollow: its outer wall is composed of a single layer of squamous epithelium, and its inner wall follows closely the twisting curves of the glomerulus. The cells that form the inner wall

are highly specialized and are called **podocytes**[6] (PŌ-dō-sīts). Podocytes have numerous branches, which are separated from the branches of adjacent podocytes by narrow gaps known as **filtration slits.** The podocytes, an underlying basement membrane, and the glomerular wall provide a **filtration membrane** between the lumen of the glomerulus and the cavity of the Bowman's capsule.

The renal tubule contains several segments that are continuous with one another (Fig. 18–4b). Each segment is lined with a single layer of cells that may be flat or cube-shaped and that play important roles in absorption or secretion. The renal tubule begins with a coiled segment that leads from its union with the Bowman's capsule, makes several turns, and then begins a descent into the medulla (in a renal pyramid). This segment is the **proximal convoluted tubule.** The segment of the renal tubule that descends into the medulla toward the renal pelvis, makes a sharp U-turn, and ascends back to the cortex is the **loop of Henle.** The descending portion of the loop is termed the *descending limb,* and the portion that returns to the cortex is the *ascending limb.* Upon reaching the cortex, the tubule coils once again, forming the **distal convoluted tubule.** This segment enters a larger tubule, called the

[4]**Glomerulus:** Latin *glomer-* = ball of yarn + *-ulus* = tiny.
[5]**Fenestrae:** Latin *fenestra* = window.
[6]**Podocytes:** Greek *podo-* = foot + *-cyte* = cell.

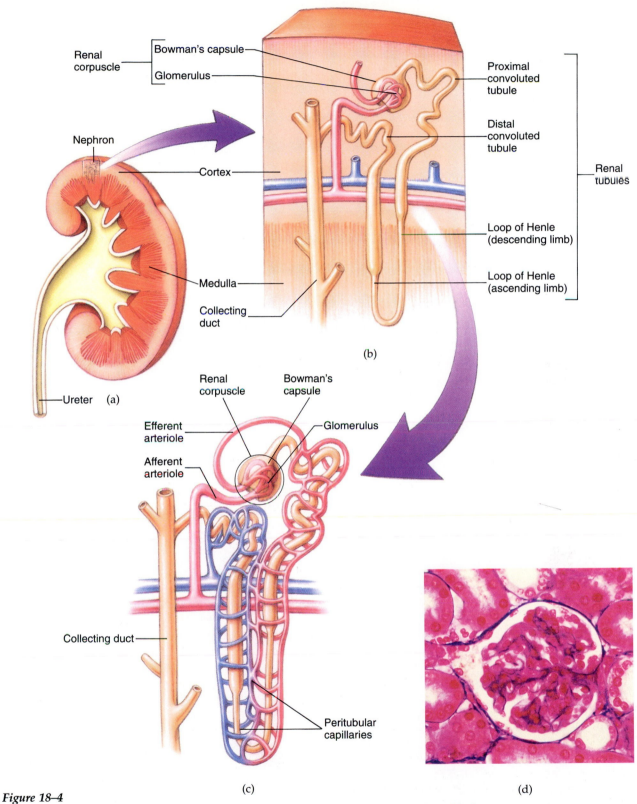

Figure 18–4
Nephron structure. (a) The location of a single nephron is indicated in the sectioned kidney. There are about 1 million of these units in each kidney. (b) In this simplified view of a nephron, the capillaries associated with it are not shown. Note how parts of the nephron are located in the medulla and in the cortex. (c) A single nephron with its blood supply. (d) A photomicrograph of a sectioned kidney, showing cross sections of a renal corpuscle in the center and of renal tubules surrounding it.

The loop of Henle extends into what region of the kidney?

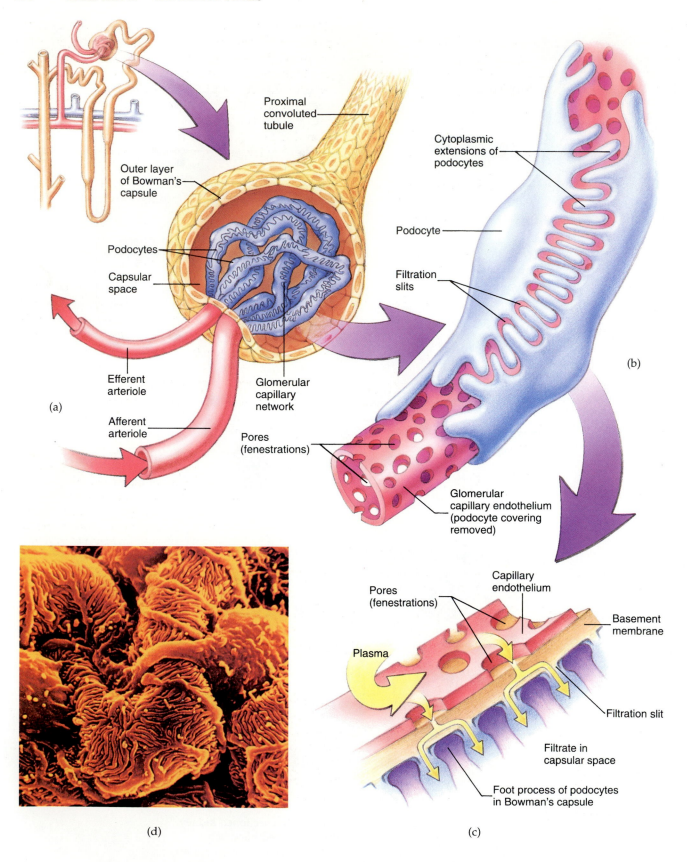

Proximal convoluted tubule

Outer layer of Bowman's capsule

Podocytes

Capsular space

Efferent arteriole

(a)

Afferent arteriole

Glomerular capillary network

Pores (fenestrations)

Cytoplasmic extensions of podocytes

Podocyte

Filtration slits

(b)

Glomerular capillary endothelium (podocyte covering removed)

Pores (fenestrations)

Capillary endothelium

Basement membrane

Plasma

Filtration slit

Filtrate in capsular space

Foot process of podocytes in Bowman's capsule

(c)

(d)

◄ *Figure 18–5*

The renal corpuscle. (a) A section through a renal corpuscle. In this drawing, the podocytes of the Bowman's capsule are shown in their proper position overlying the glomerulus. (b) A close-up view of a section of the glomerulus and two podocytes. Note the close association between the podocytes and the glomerular wall. (c) The filtration membrane. Net filtration pressure pushes fluid and small molecules through the pores in the capillary endothelium, through the basement membrane, and through filtration slits between podocytes. (d) In this electron micrograph the filtration slits between podocytes are clearly visible.

In what structure does filtrate collect once it is formed?

collecting duct, which collects the newly formed urine and channels it from the cortex through the medulla to empty it into a calyx.

Blood Flow Through the Nephron

The nephron receives its blood supply from the **afferent arteriole,** which unites with the glomerulus (Fig. 18–4c). Once blood has passed through the glomerulus, it is drained by another arteriole, called the **efferent arteriole.** Because the afferent arteriole is larger in

diameter than the efferent, blood flow slows as it begins to exit. This "backs up" the blood in the glomerulus, providing it with an extraordinarily high pressure. The high blood pressure in the glomerulus is necessary for forcing fluids and solutes out of the bloodstream and into the Bowman's capsule during filtration. Blood is carried from the efferent arteriole to a second capillary network, known as the **peritubular capillaries.** These tiny vessels are porous, low-pressure capillaries that permit the movement of materials across their walls. They travel a course that clings to the renal tubule before draining into the venous circulation.

Juxtaglomerular Apparatus

Each nephron contains a region known as the **juxtaglomerular** (juks'-ta-glō-MER-yoo-ler) **apparatus.**[7] It is located where the distal convoluted tubule contacts the afferent arteriole (Fig. 18–6). It consists mainly of densely packed epithelial cells from the distal tubule,

[7]**Juxtaglomerular:** Latin *juxta-* = near + *glomerul-* = little ball of yarn + *-ar* = pertaining to.

Figure 18–6

The juxtaglomerular apparatus. (a) A cutaway view of the renal corpuscle, showing the outer Bowman's capsule, the glomerulus within, and the location of the juxtaglomerular apparatus nearby (*within circle*). (b) A close-up view of a sectioned juxtaglomerular apparatus, revealing the juxtaglomerular cells of the afferent arteriole and the macula densa of the distal convoluted tubule.

What two structures as they come into physical contact house the cells of the juxtaglomerular apparatus?

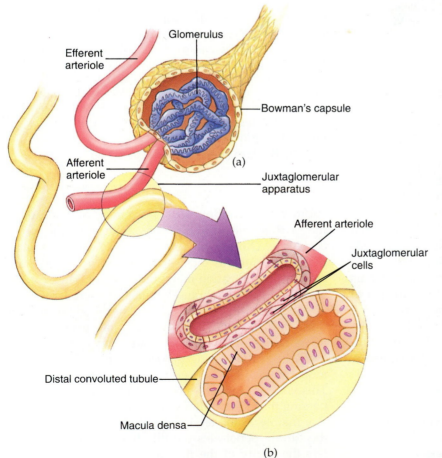

Glomerulus

Efferent arteriole

Bowman's capsule

Afferent arteriole

(a)

Juxtaglomerular apparatus

Afferent arteriole

Juxtaglomerular cells

Distal convoluted tubule

Macula densa

(b)

collectively known as the **macula densa,** and modified smooth muscle cells from the arteriole, called **juxtaglomerular cells.** These specialized cells play important roles in regulating the rate of urine formation and regulating blood pressure, which we will examine later in this chapter.

CONCEPTS CHECK

1. In what region of the kidney are the pyramids located?

2. What are the components of the filtration membrane?

3. Where in the nephron is the juxtaglomerular apparatus?

KIDNEY FUNCTIONS

The nephrons of the kidneys process blood by filtration, reabsorption, and secretion. The result is the excretion of urea and uric acid, and maintenance of body fluid volume, electrolyte concentration, and pH.

The kidneys remove unwanted substances from the bloodstream and maintain the fluid balance of the body. They accomplish these tasks by forming urine and regulating its rate of output and its content.

Urine Formation

Through the blood vessels in the kidneys pass incredible volumes of blood each day. As blood passes through, it is processed by the kidneys. Out of the 1200 milliliters of blood that pass into the glomeruli each minute, about 10% (120 ml) is pushed into the renal tubules as blood plasma. On a daily basis, this amounts to about 180 liters (45 gallons) of fluid, or about 35 times your total blood volume! It is fortunate for us that nearly all (about 99%) of this fluid is returned to the bloodstream, or we would spend most of our time urinating and replacing this tremendous volume. The remaining 1% (a volume of about 1 to 2 liters per day) finds its way out of the kidneys to be excreted as **urine.** As a result of the processing undertaken by the kidneys, urine normally contains mostly water, but also urea, uric acid, and numerous ions.

There are three processes that are involved in urine formation, all of which are performed by the nephrons of the kidneys. They are filtration, reabsorption, and secretion and are introduced in Figure 18–7.

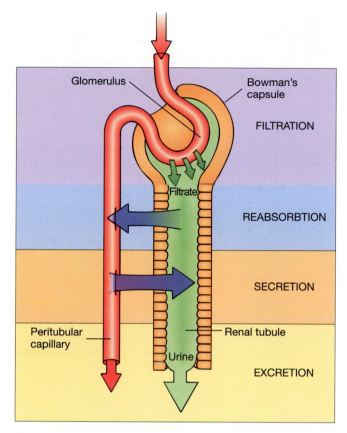

Figure 18–7
The processes involved in urine formation. Filtration involves the movement of fluid and dissolved molecules from the bloodstream (carried by the glomerulus) into the Bowman's capsule. Reabsorption is the movement of mostly water from the renal tubules back into the bloodstream, and secretion is the movement of excess ions and other molecules from the bloodstream into the renal tubules. The resulting fluid exits the kidney and is excreted as urine.

What are the three processes of urine formation?

Filtration

Glomerular **filtration** is the movement of blood plasma across the filtration membrane of the renal corpuscle. During this process, blood cells and large molecules, such as most proteins, are restricted from crossing the filtration membrane barrier. They cannot penetrate the membrane because of their large size, and therefore they remain within the bloodstream. The fluid and dissolved substances that penetrate the membrane constitute the **filtrate.** The volume of filtrate produced during glomerular filtration is enormous: about 180 liters per day, as indicated previously. This volume is much greater than the amount of fluid passing through capillaries in other tissues and organs due mainly to the larger size of glomerular pores. As the filtrate penetrates the filtration membrane, it enters the cavity of the Bowman's capsule to begin its journey through the renal tubule.

Table 18-1 COMPARATIVE AMOUNTS OF SUBSTANCES IN THE BLOOD PLASMA, GLOMERULAR FILTRATE, AND URINE

Substance	Plasma	Filtrate	Urine
Glucose (mg/100 ml)	100	100	0
Urea (mg/100 ml)	26	26	1850
Uric acid (mg/100 ml)	4	4	55
Sodium ions (mEq/liter)	142	142	128
Potassium ions (mEq/liter)	5	5	60
Bicarbonate ions (mEq/liter)	27	27	14
Phosphate ions (mEq/liter)	2	2	40

The force that pushes plasma through the filtration membrane is the high level of hydrostatic pressure within the glomerulus (exerted by the force of blood pushing against the glomerular walls). Glomerular hydrostatic pressure is unusually high because of the "backup" of blood that results from narrowing of the efferent arterioles. This pressure overcomes opposing forces (including osmotic pressure within the glomerulus and hydrostatic pressure within the Bowman's capsule) to maintain a final pressure level of about 10 mm of mercury. The resulting force is called the **net filtration pressure (NFP).** Although the NFP may appear to be a small value (it is about one-half of the hydrostatic pressure within the capillaries of most other tissues and organs), it is sufficient to push large quantities of fluid and small molecules through the large pores of the glomerular wall, through the basement membrane, and between the processes of podocytes. Once penetrating through these barriers, the fluid and solutes enter the Bowman's capsule.

The rate of filtration is determined by the amount of filtration pressure that is present. In general, when the filtration pressure increases, the volume of the filtrate increases, and more urine is formed. When the filtration pressure decreases, filtrate volume decreases as well, resulting in less urine formed. Filtration pressure is influenced primarily by the volume of blood in the glomerulus. As we shall learn in a later section of this chapter, glomerular blood volume is regulated by cells within the juxtaglomerular apparatus.

Reabsorption

Filtrate and urine composition are quite different. Filtrate is essentially the same as blood plasma, except for its lack of proteins, whereas urine contains a much greater concentration of most ions, urea, and uric acid (Table 18–1). Also, glucose is present in the filtrate but usually absent in the urine. Why are their compositions so different? As the filtrate passes through the renal tubules, about 99% of the water, essentially all of the nutrients, and many essential ions are removed and transported into the nearby peritubular capillaries. This reclamation process serves to return needed materials to the bloodstream, resulting in a concentration of wastes that remain in the tubules to gradually form urine. This process is called tubular **reabsorption.**

The movement of substances during reabsorption is always in one net direction: from the renal tubule into the bloodstream (Fig. 18–8). Movement may oc-

Figure 18–8

Movement of substances during reabsorption. The direction of movement is always from the renal tubule into the bloodstream.

What substance is normally reabsorbed in enormous volumes?

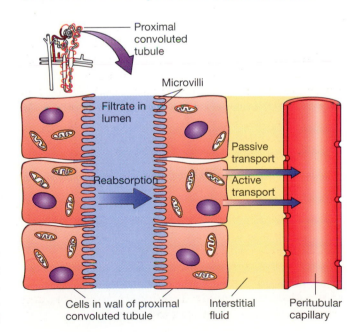

cur by active transport, or passively by osmosis, diffusion, or facilitated diffusion. The mode of movement depends on the nature of the substance being transported. For example, amino acids move across by active transport, sodium ions move by simple diffusion, and water moves in large quantities by osmosis. In any case, substances move through the single cell layer of the renal tubule wall, across a small space filled with interstitial fluid, and through the endothelial wall of the peritubular capillaries to be carried away with the bloodstream.

Reabsorption begins in the proximal convoluted tubule and continues throughout the length of the renal tubule (Fig. 18–9). The proximal convoluted tubule is the main site of water and solute reabsorption, since its walls are highly permeable to water. As proteins, glucose, and numerous ions are actively transported out of the tubule, water quickly follows by osmosis in the same direction. As a result, about 65% of the filtrate is initially reabsorbed.

The first segment of the loop of Henle, the descending limb, continues the reabsorption of water begun in the proximal convoluted tubule. The amount of solutes being actively transported declines, however, resulting in an increased concentration of solutes in the filtrate. By the time the filtrate has passed through the descending limb, the filtrate is as concentrated as the surrounding interstitial fluid, and another 15% of the filtrate has been reabsorbed.

As the filtrate continues along its path through the ascending limb of the loop of Henle, many solutes are reabsorbed with only a small amount of water. For example, chloride ions are carried out of the tubule by active transport, and sodium ions by diffusion. As a result of the removal of solutes, the filtrate has become more dilute by the time it enters the distal convoluted tubule.

The distal convoluted tubule and collecting duct further reabsorb more solutes (mainly by active transport) and water (by osmosis). The result is a reduction of the filtrate by another 19% of its original volume, leaving about 1% of the original volume of filtrate to flow out of the kidneys as urine.

Secretion

Some unwanted substances that are not removed from the blood by filtration may be removed by another process, called tubular **secretion.** It is essentially the opposite of reabsorption, in that the net movement of substances is from the peritubular capillaries into the renal tubule. The substances involved include ions that

Figure 18–9
The process of reabsorption changes from one segment of the renal tubule to the next.

What is the fluid called that passes through the collecting duct?

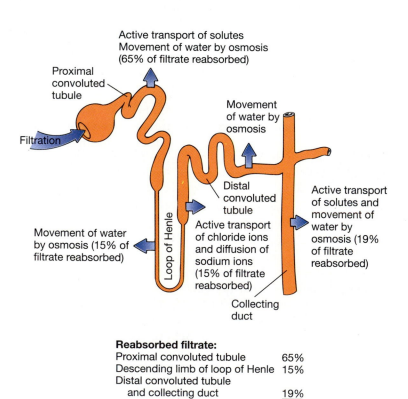

Active transport of solutes
Movement of water by osmosis
(65% of filtrate reabsorbed)

Proximal convoluted tubule

Filtration

Movement of water by osmosis

Distal convoluted tubule

Active transport of chloride ions and diffusion of sodium ions (15% of filtrate reabsorbed)

Loop of Henle

Movement of water by osmosis (15% of filtrate reabsorbed)

Active transport of solutes and movement of water by osmosis (19% of filtrate reabsorbed)

Collecting duct

Reabsorbed filtrate:

Proximal convoluted tubule	65%
Descending limb of loop of Henle	15%
Distal convoluted tubule and collecting duct	19%
Total reabsorbed filtrate	99%
Filtrate remaining as urine	1%

become toxic if they are allowed to concentrate in the blood, and certain drugs and other molecules not produced in the body. These substances may move into the tubule by active transport, diffusion, or facilitated diffusion. For example, hydrogen ions, potassium ions, and penicillin are actively transported into the renal tubule, but urea moves in passively by diffusion. The proximal convoluted tubules are the most active in secretion, although some secretion occurs in the distal convoluted tubules as well.

Secretion is important for removing harmful substances that were not removed from the blood during filtration (such as penicillin), eliminating unwanted substances that have been reabsorbed along with needed materials (such as urea and uric acid), removing excess potassium ions from the bloodstream, and controlling blood pH. Blood pH is controlled by adjusting the quantity of hydrogen ions secreted.

Urea and Uric Acid Excretion

Urea and **uric acid** are nitrogen-containing waste products resulting from metabolism. If they are allowed to accumulate in body tissues, they can cause serious damage and become life-threatening. An important role of the kidneys is in their removal on a continuous basis.

During amino acid metabolism in the liver, ammonia is released as a by-product. This toxic substance is quickly converted to urea by liver cells and released into the bloodstream. Urea enters the renal tubule by filtration, and about 50% of it is reabsorbed by diffusion along with other substances. The remaining 50%

is excreted in the urine. Because urea is a by-product of amino acid metabolism, its levels in the blood usually correspond to the amount of protein in the diet.

Uric acid results from the metabolism of certain nucleic acids. It enters the renal tubules by filtration, but it is reabsorbed into the bloodstream by active transport. The kidneys are finally able to remove about 10% of the uric acid in the blood by secretion.

Regulation of Urine Concentration and Volume

The composition and volume of urine fluctuate in response to changing conditions of the body. For example, after you eat a heavily salted meal, the urine volume is greater and has a high salt content, but after heavy exercise the urine volume is considerably less, with a different salt content. These fluctuations and others are produced by the kidneys as they work to maintain a constant blood composition and volume. In other words, the amount of water and solutes filtered, reabsorbed, or secreted varies with changing conditions of the body so that blood homeostasis can be maintained. Kidney function resulting in urine formation must therefore be regulated carefully. There are several mechanisms that perform kidney regulation, which are described in what follows and summarized in Table 18–2.

Renin and Angiotensin

As we learned earlier, the overall force within the glomerulus that pushes plasma through the filtration membrane is called *net filtration pressure*. This pressure

Table 18-2 FACTORS CONTROLLING URINE CONCENTRATION AND VOLUME

Factor	Source	Effect
Renin	Juxtaglomerular cells in kidney	Production of angiotensin II, which stimulates secretion of aldosterone; this ultimately leads to a more concentrated urine and an increase in blood pressure
Aldosterone	Adrenal gland	Increased sodium and chloride reabsorption, leading to more water reabsorption and a more concentrated urine
Antidiuretic hormone (ADH)	Posterior pituitary gland	Increased water reabsorption, leading to a more concentrated urine
Atrial natriuretic factor	Right atrium of the heart	Decreased water and solute reabsorption, leading to a dilute urine and reduced blood pressure
Sympathetic impulses	Hypothalamus	Vasoconstriction of afferent arterioles, which leads to a reduced urine volume

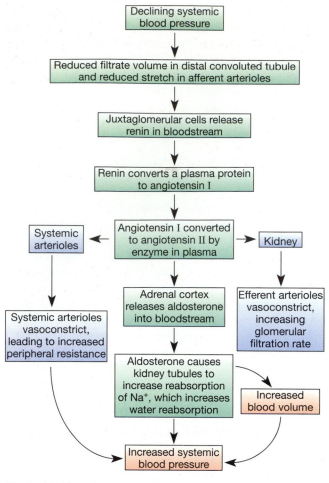

Figure 18–10
The sequence of events in blood pressure restoration by the renin-angiotensin system.

What enzyme released by juxtaglomerular cells is needed for the appearance of angiotensin II in the bloodstream?

is influenced mainly by a mechanism within the juxtaglomerular apparatus that is triggered when the filtration rate begins to decline (Fig. 18–10). As it declines, certain juxtaglomerular cells are stimulated to release an enzyme called **renin.** Renin acts on a plasma protein produced by the liver, converting it to another protein, **angiotensin I.** Angiotensin I is then converted in the bloodstream by another enzyme to an activated form known as **angiotensin II.** Angiotensin II acts on the adrenal gland by stimulating its secretion of aldosterone. The effect is the increased reabsorption of sodium, chloride, and water, which increases blood volume and, hence, blood pressure. The increase of blood pressure causes a corresponding increase in filtration pressure, and more filtrate is subsequently produced.

Aldosterone

Released by the adrenal gland, the hormone **aldosterone** regulates the rate of active transport in the distal convoluted tubules and collecting ducts. Recall that the release of aldosterone is stimulated by angiotensin II. Aldosterone's presence increases the reabsorption of sodium and chloride ions from the nephron to the peritubular capillaries. As sodium ions are transported out of the nephron, water follows because of the osmotic gradient that is formed. The net effect of aldosterone is the increased reabsorption of sodium, chloride, and water. As a result, the volume of urine produced is decreased. A decrease in urine volume leads to a greater volume of fluid in the bloodstream, which causes an increase in blood pressure. In the absence of aldosterone, large amounts of sodium and chloride remain in the renal tubule and are excreted as part of the urine. The relationships between aldosterone, renin and angiotensin II, urine volume, and blood pressure are diagramed in Figure 18–10.

Antidiuretic Hormone

The volume of water that is reabsorbed by the distal convoluted tubules and collecting ducts is regulated by **antidiuretic hormone (ADH),** which is released by the posterior pituitary gland. When the volume of blood begins to decline, the corresponding drop in blood pressure signals the pituitary gland to release more ADH, which increases the permeability of the distal tubules and collecting ducts to water. As a result, a greater volume of water is reabsorbed from the filtrate, and blood volume is thereby restored. An increase in ADH production therefore results in the formation of a more concentrated urine, and a decrease results in a more dilute urine.

Atrial Natriuretic Factor

Atrial natriuretic factor is released by cells in the right atrium of the heart when blood pressure inside that chamber increases. This hormone reduces the ability of the kidneys to reabsorb water and solutes, thus leading to the formation of large volumes of urine. As a result, blood volume (and blood pressure) decline.

Sympathetic Stimulation

Nerve impulses from the sympathetic nervous system may stimulate the contraction of smooth muscles in the walls of the afferent arterioles. As a result, the arterioles undergo vasoconstriction, which reduces the flow of blood passing through the glomeruli. This causes the net filtration pressure to decline, leading to a reduction in filtrate and urine volume. This chain of events commonly occurs during heavy exercise or excitement.

Renal Failure and Dialysis

Renal failure is the loss of the kidney's ability to respond to the changing conditions of the body. It results in the rapid loss of body fluid maintenance, electrolyte balance, and pH regulation, as well as the accumulation of toxic metabolic waste products in tissues. The loss of kidney function can be acute or chronic. **Acute renal failure** is the abrupt stoppage of kidney function. It is usually reversible, temporarily resulting in reduced urine output, an increase in urea in the blood, and sometimes local pain. It can be caused by a loss of body fluids, such as occurs in excessive bleeding, diarrhea, and heavy use of diuretics. It can also be caused by damage to the nephrons by injury, infectious microorganisms, hypertension, or toxic chemicals and drugs. This type of acute renal failure is also called *glomerulonephritis*. **Chronic renal failure** is the progressive loss of functioning nephrons; it is manifested by reduced glomerular filtration rates, accumulation of urea in the blood, electrolyte imbalances, and acid-base imbalances. If not treated by dialysis or kidney transplant,

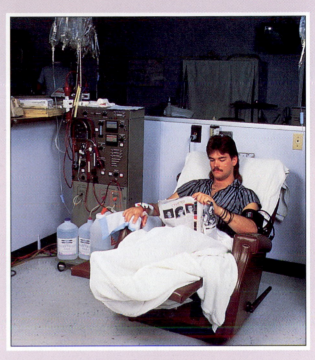

A patient undergoing dialysis.

chronic renal failure usually leads to death. It can be caused by metabolic disorders, accumulation of toxins, inherited connective tissue disorders, and hypertensive vascular diseases.

During total kidney failure, the formation of urine stops completely. This results in a rapid accumulation of toxic wastes and a progressively acidic pH of body fluids, which can lead to death in about 8 to 12 days. In order to remove the toxic wastes that build up in the blood as a result, dialysis may be performed if kidney transplantation is not possible. In dialysis (see the illustrations), the patient's blood is channeled through cellophanelike tubing whose walls are permeable only to certain substances, such as nitrogenous wastes and potassium, but not to blood cells and proteins. The tubing is immersed in a solution similar to normal, waste-free plasma. As the blood circulates through the tubing, nitrogenous wastes and excess ions diffuse from the blood to the dialyzing solution across the tubing walls. Patients are usually dialyzed three times a week, and they can often undergo this procedure at home. The main drawbacks to dialysis are the constant risk of infection and hemorrhage during dialysis, the development of *uremia* (accumulation of urea in the blood) between periods of dialysis, and its high cost. Dialysis is, however, the only way of keeping chronic kidney patients alive and functioning until the only present cure for this disease, a kidney transplant, becomes available.

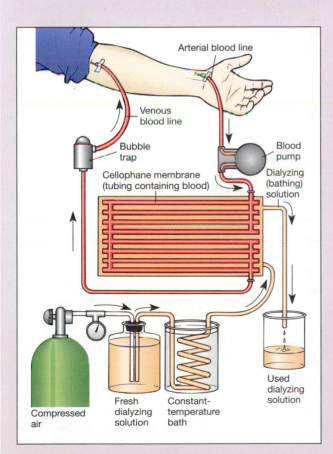

The process of dialysis, which provides an artificial means of replacing kidney function.

HEALTH CLINIC

Testing for Healthy Kidney Function: The Urinalysis

A standard test for urine content helps a physician determine the functional state of the kidneys. The test is called a **urinalysis** and involves the analysis of the chemical content of a urine sample. However, a urinalysis provides information regarding the function of more than just the kidneys, thanks to the fact that urine contains cellular products that originate from almost every organ in the body. It can also yield early information about infections that may occur anywhere. Presented here is a table of some standard and more specific tests performed in urinalysis, the expected quantities of substances normally present, and the clinical results arising from a deviation from the normal. In addition to these tests, a microscopic examination of urine may also reveal the presence of parasites, red blood cells, white blood cells, and other elements not usually present in urine.

Test	Normal Quantities	Clinical Implications
Acetone and acetoacetate	0	Amounts increase in diabetic acidosis
Albumin	0 to trace amounts	Amounts increase in hypertension, kidney disease, and heart failure
Ammonia	20 to 70 mEq/liter	Amounts increase in diabetes mellitus and liver disease
Bacteria	Under 10,000/ml	Numbers increase in urinary tract infections
Bile and bilirubin	0	Quantities increase in melanoma and obstructions of the gallbladder

Test	Normal Quantities	Clinical Implications
Calcium	Under 250 mg/24 h	Quantities increase in hyperparathyroidism and decrease in hypoparathyroidism
Corticosteroids	2 to 10 mg/24 h	Quantities increase in Cushing's syndrome and decrease in Addison's disease
Creatinine	1 to 2 g/24 h	Quantities increase in infections and decrease in kidney disease, anemia, leukemia, and muscle atrophy
Glucose	0	Quantities increase in diabetes mellitus and pituitary gland disorders
Urea	25 to 35 g/24 h	Quantities decrease with excessive protein break-down; decrease indicates impaired kidney function
Urea clearance	Over 40 ml of blood cleared of urea/per minute	Quantities increase in kidney disease
Uric acid	0.6 to 1.0 g/24 h	Quantities increase in gout and decrease in several diseases of the kidneys

Maintenance of Body Fluids

For the body to efficiently maintain homeostasis, the intake of substances such as water and electrolytes must equal their removal. The organs that play a role in controlling this balance include the skin, the liver, the organs of the alimentary canal, and the kidneys. Of these, the kidneys have the greatest immediate effect on fluid balance. As we have seen, they process large amounts of blood continually by filtration, reabsorption, and secretion. As blood that has passed through the kidneys is transported to capillary networks elsewhere, the blood plasma is exchanged with the interstitial fluid and the bathed cells receive the revitalized fluids. Thus, although the kidneys process blood only, the fluids present in other body environments are similarly affected. For example, after a heavily salted meal, the concentration of sodium and chloride ions in the blood is initially quite high. The high salt levels soon reach the interstitial fluid in the extracellular environment, and then the cytoplasm in the intracellular environment. If not corrected quickly, the high salt levels can lead to cell damage by the rapid movement of water into the cells (by osmosis). The kidneys prevent this from occurring by removing excess ions in the blood by the process of secretion. During times of low salt levels, the kidneys compensate by reducing the amount of water that is reabsorbed. The

kidneys are therefore endowed with the ability to regulate fluid volumes and electrolyte volumes in all body compartments.

The relationships between the fluid compartments of the body are summarized in Figure 18–11.

Regulation of pH

The concentration of hydrogen ions, or pH, in body fluids is controlled by buffers, the respiratory system, and the kidneys. It is maintained within a narrow margin, between 7.35 and 7.45, despite fluctuations that occur with diet, exercise, and stress. If the pH declines much below this range, a life-threatening condition called *acidosis* occurs. This increase in hydrogen ions can irreversibly damage enzymes and structural proteins. If the pH increases, a less frequent, but equally dangerous, condition called *alkalosis* occurs. Thus, it is vitally important for the body to maintain the pH in all fluids.

Buffers

A **buffer** is a chemical that resists a change in the pH of a solution when either acids or bases are added. The buffers in the body fluids contain weak acids or bases that combine with excess hydrogen ions in the pres-

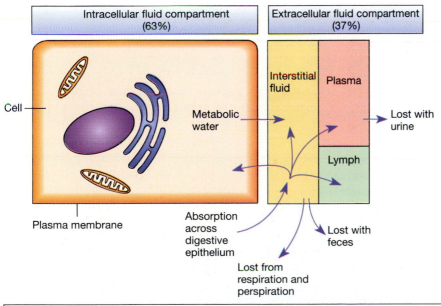

Figure 18–11
Major fluid compartments of the body. The arrows indicate the movement of water. Average daily volumes of water movement are summarized in the box.

What is the effect of increased levels of aldosterone on the volume of fluid in the extracellular compartment?

Intracellular fluid compartment (63%)	Extracellular fluid compartment (37%)

Cell — Plasma membrane

Metabolic water — Interstitial fluid — Plasma — Lost with urine

Lymph

Absorption across digestive epithelium — Lost with feces

Lost from respiration and perspiration

Average daily input of water:	200 ml (10%) from metabolism 1800 ml (90%) from digestive absorption
Average daily output of water:	850 ml (35%) from respiration (lungs) and perspiration (skin) 150 ml (8%) from feces 1000 ml (57%) from urine

ence of a strong acid, or release hydrogen ions in the presence of a strong base. These reactions tend to neutralize the solution. The three main buffer groups in body fluids are proteins, phosphates, and bicarbonates. Of these, the most important is the bicarbonate system, because it can be regulated by the respiratory and urinary systems. In this buffer system, carbon dioxide combines with water to form carbonic acid, which in turn forms hydrogen ions and bicarbonate ions:

$$H_2O \ + \ CO_2 \ \leftrightarrow \ H_2CO_3 \ \leftrightarrow \ H^+ \ + \ HCO_3^-$$

water + carbon dioxide   carbonic acid   hydrogen ion + bicarbonate ion

As you can see, this reaction can proceed in either direction. The reaction between carbon dioxide and water is catalyzed by an enzyme found in red blood cells and in kidney tubule cells, called *carbonic anhydrase.* It accelerates the reaction in either direction. For example, when carbon dioxide levels are high, more carbonic acid is formed. Consequently, more hydrogen ions are formed also, resulting in a decreased pH. During the reverse reaction, when carbon dioxide levels are low, hydrogen and bicarbonate ions combine to form carbonic acid, which dissociates to form carbon dioxide and water. As a result, the pH increases.

Respiratory System

The respiratory system has the capability to respond rapidly to a change in pH. For example, as carbon dioxide levels increase in the blood and pH levels decline, neurons in the respiratory center of the brain are stimulated to cause an increase in the rate and depth of ventilation. As a result, carbon dioxide is exhaled through the lungs at a greater rate. As carbon dioxide levels in body fluids decline, the concentration of hydrogen ions declines also, moving the pH toward its original value. If, on the other hand, carbon dioxide levels become too low and the pH of fluids rises, the rate and depth of ventilation are reduced. A smaller amount of carbon dioxide therefore leaves the body and begins to accumulate in body tissues. As carbon dioxide levels rise, so do hydrogen-ion levels, resulting in a drop in pH.

Kidneys

The kidneys play an important role in pH regulation also. Although they can remove large amounts of excess hydrogen ions, they respond to a change in pH more slowly than the respiratory system does. The kidneys respond to a drop in pH by increasing the rate of secretion of hydrogen ions. At the same time, they also increase the rate of reabsorption of bicarbonate ions. As a result, excess hydrogen ions are removed from the blood and bicarbonate ions are returned to the blood, restoring the pH to a higher, more normal value. The kidneys respond to a rise in blood pH by reducing the rate of hydrogen-ion secretion and bicarbonate-ion reabsorption, thus lowering the pH.

CONCEPTS CHECK

1. What materials are excluded from passing through the filtration membrane?

2. What factors regulate the reabsorption of water from the renal tubules?

3. Filtration pressure is regulated by what mechanism?

4. The secretion of hydrogen ions is important in what regard?

URETERS

The ureters carry newly formed urine from the kidneys to the urinary bladder.

The **ureters** are paired, tubular organs that transport urine from the kidneys to the urinary bladder (Fig. 18–1). They arise from the renal pelvis of each kidney and extend in a downward direction along each side of the vertebral column behind the peritoneum (their location is retroperitoneal, like that of the kidneys). Once reaching the pelvic cavity, they curve medially to unite with the urinary bladder from underneath.

The wall of each ureter consists of three layers. The inner layer is a mucous membrane and is continuous with the inner linings of the renal pelvis and the urinary bladder. The mucus that coats its inner surface protects the underlying cells from exposure to the passing urine. The middle layer is composed of smooth muscle fibers and elastic connective tissue, and its coordinated contraction results in peristaltic waves that propel urine through the ureter. The outer layer is composed of fibrous connective tissue, which protects the underlying layers.

Bordering the opening between each ureter and the urinary bladder is a flaplike fold of mucous membrane. This fold acts as a one-way valve by allowing urine to enter the bladder, but preventing it from backing up into the ureter.

Acidosis and Alkalosis

A failure of the buffering systems in the blood, the respiratory system, and the urinary system to maintain normal pH can result in the conditions of acidosis or alkalosis. If these conditions are not compensated by changes in the body to improve the buffering system, or by medical intervention, they can become life-threatening disturbances to homeostasis.

Acidosis, or acidemia, occurs when the pH of blood drops below 7.35. In slight cases, acidosis is usually limited to causing symptoms of disorientation, headache, and nausea. However, if the drop in pH is not corrected, a critical condition may develop that results in central nervous system collapse and death. There are two types of acidosis, which differ by the nature of their cause. **Respiratory acidosis** is an abnormal accumulation of carbon dioxide. It is usually caused by a failure of the respiratory system to eliminate ("blow off") carbon dioxide at an adequate rate. As a result, carbon dioxide (and carbonic acid) accumulate in the bloodstream, increasing the hydrogen-ion levels. **Metabolic acidosis** is caused by the excessive production of acidic compounds, or by a loss of bicarbonate buffers. Metabolic acidosis may result from many sources, including a failure of the kidneys to re-

move acid end products of metabolism, diabetes mellitus, heavy exercise (accumulation of lactic acid), the ingestion of acidifying agents (such as ammonium chloride), and prolonged diarrhea (loss of alkaline intestinal fluids).

Alkalosis, or alkalemia, occurs when the pH of blood rises above 7.45. A major clinical effect of this condition is hyperexcitability of the nervous system. It includes involuntary, spontaneous stimulation of muscles throughout the body, resulting in muscle spasms, tetanic (sustained) contractions, and convulsions. If the respiratory muscles become affected, death may follow. As we have seen in acidosis, alkalosis may occur in two different ways. **Respiratory alkalosis** is caused by an excessive loss of carbon dioxide, and is produced by hyperventilation. Hyperventilation can be caused by voluntary effort, anxiety, or stimulation of respiratory reflex centers in the brain. Fever, intoxication with alcohol, and high altitudes all stimulate these reflex centers to trigger hyperventilation. **Metabolic alkalosis** usually results from a rapid elimination of hydrogen ions from the body and may be caused by severe vomiting or by ingestion of sodium bicarbonate (used in treating gastric ulcers).

URINARY BLADDER

The urinary bladder is an expandable, saclike organ that receives urine from the ureters and stores it until released into the urethra.

The **urinary bladder** is a hollow, muscular organ that temporarily stores urine. It is located at the floor of the pelvic cavity behind the symphysis pubis, and its top surface only is covered by the peritoneum (Fig. 18–12). In many ways its features are similar to those of a balloon: its walls are extremely elastic, so that when it is empty it shrivels in shape and its inner wall contains many wrinkles. As it fills with urine its wall becomes smoother, and it expands dramatically upward to form a pear-shaped structure that often presses against other

organs. Its capacity when full averages about 500 ml (1 pint), although it can hold more than twice that amount if necessary.

The interior of the bladder has openings for the two ureters and the single urethra. The triangular region that is outlined between the three openings contains a smoother wall and is known as the **trigone**.[8] This region is clinically important, for the trigone is a frequent site of urinary tract infections.

The wall of the bladder consists of four layers. The innermost layer is a mucous membrane, with the epithelium composed of highly elastic transitional epithelium. The mucous coat that normally lines the bladder's interior wall protects underlying cells from

---

[8]Trigone: *trigone* = triangular area.

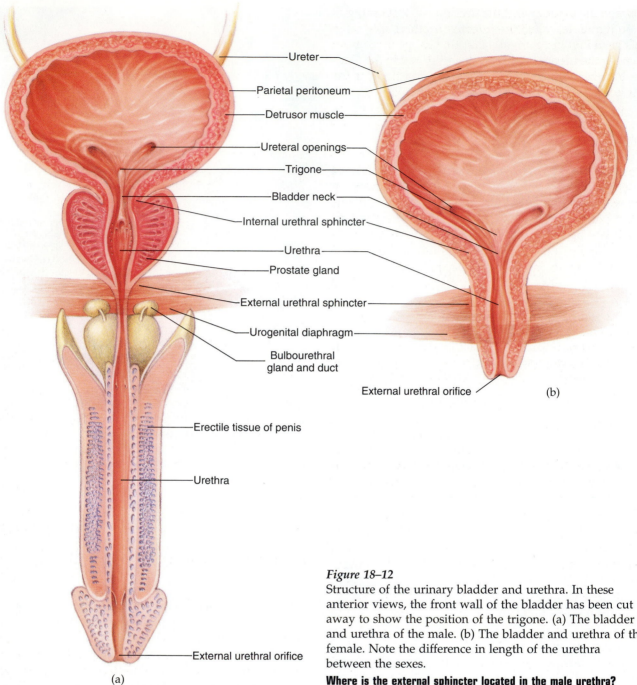

Ureter

Parietal peritoneum

Detrusor muscle

Ureteral openings

Trigone

Bladder neck

Internal urethral sphincter

Urethra

Prostate gland

External urethral sphincter

Urogenital diaphragm

Bulbourethral gland and duct

Erectile tissue of penis

Urethra

External urethral orifice

(a)

External urethral orifice

(b)

Figure 18–12
Structure of the urinary bladder and urethra. In these anterior views, the front wall of the bladder has been cut away to show the position of the trigone. (a) The bladder and urethra of the male. (b) The bladder and urethra of the female. Note the difference in length of the urethra between the sexes.
Where is the external sphincter located in the male urethra?

the harmful effects of urine exposure. The mucous membrane is supported by a second layer that is composed of connective tissue. External to this is a thick layer of smooth muscle, called the **detrusor** (dē-TROO-zer) **muscle,** which consists of longitudinal and circular layers of fibers. The detrusor muscle participates in the process of urination. The outermost layer is fibrous connective tissue, except on the superior surface, where the bladder is covered by the peritoneum.

URETHRA

The urethra transports urine from the urinary bladder to the exterior. Its structure differs between males and females.

The **urethra** is a muscular tube that drains urine from the floor of the urinary bladder and transports it out of the body to the exterior (Fig. 18–12). At the junction between the bladder and the urethra is a thickening of smooth muscle, called the **internal urethral sphincter.** This is an involuntary sphincter that keeps urine from entering the urethra while it is being stored in the bladder. As the urethra extends through the floor of the pelvic cavity, a second sphincter surrounds it. This is the **external urethral sphincter.** It is composed of skeletal muscle and is therefore under voluntary control.

The nature of the urethra differs considerably between males and females. In females it is about 3 to 4 cm (1.5 inches) long and is tightly bound to the anterior wall of the vagina by connective tissue. It opens to the exterior by way of the external urethral orifice, which lies between the vaginal opening and the clitoris. In males the urethra is about 20 cm (8 inches) long and extends from the urinary bladder to the end of the penis, where it opens as the external urethral orifice. Near its emergence from the bladder, the male urethra passes through a gland that surrounds it, known as the *prostate gland.* Later in life, it is quite common for the prostate gland to enlarge, resulting in difficulty with urinating. In addition to carrying urine, the male urethra also carries reproductive fluids.

MICTURITION

Micturition is the process of emptying the bladder, and usually involves both involuntary and voluntary responses.

The act of emptying the bladder is called **micturition**[9] (mik'-too-RISH-un). Also referred to as *voiding* or *urination,* it is a process that combines involuntary and voluntary activities. It begins when the urinary bladder has collected over 200 ml of urine. At this point the bladder wall has stretched to a point of activating stretch receptors, which send sensory impulses to the spinal cord (Fig. 18–13). Motor impulses are quickly sent from the spinal cord to the bladder by way of a reflex arc, causing the detrusor muscle to contract and the internal sphincter muscle to relax. As detrusor muscle contractions intensify, stored urine is pushed through the internal sphincter into the urethra. At this point, one experiences the urge to void. If the time is not appropriate, urine flow may be stopped voluntarily by the external sphincter muscle and the urine stored until a later time. Other skeletal muscles may also help stop urine flow, such as the muscles in the floor of the pelvic cavity and the abdominal wall. When the decision to urinate is made, the external

---

[9]**Micturition:** Latin *micturire* = to urinate + *-ition* = process of.

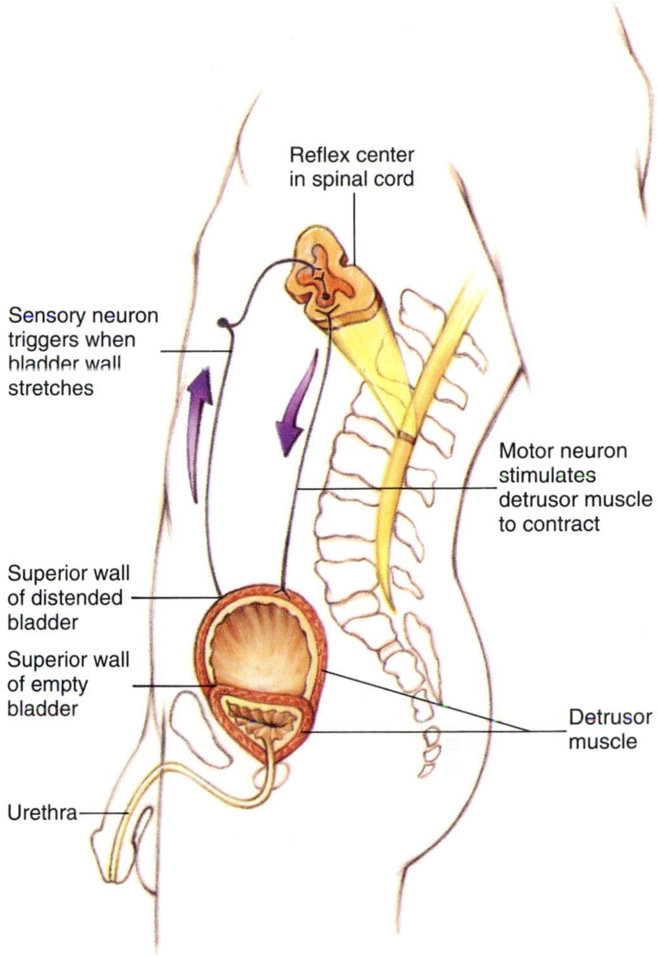

Figure 18–13
Micturition. In this lateral view of the male trunk, the urinary bladder is shown in both its distended and empty forms.
What is the function of the detrusor muscle?

sphincter muscle is allowed to relax, and urine begins to exit. Further contraction of the detrusor muscle forces urine remaining in the bladder out. After several moments the detrusor muscle relaxes, and the bladder begins to fill with urine once again.

CONCEPTS CHECK

1. What feature of the ureters protects their inner lining from damage by urine?

2. What is the detrusor muscle?

3. What is the structural basis for the voluntary control of micturition?

HOMEOSTASIS

The primary function of the urinary system is the homeostatic maintenance of the body's fluid and salt balance. As we have learned, there are several mechanisms that play a role in regulating this balance. They include the release of renin by kidney juxtaglomerular cells when blood pressure declines, the release of aldosterone from the adrenal gland when sodium and chloride ions need to be retained by the blood, the release of ADH from the posterior pituitary gland when blood volume declines, the release of atrial natriuretic factor by heart cells when blood pressure increases, and the vasoconstriction of arterioles by sympathetic neurons when blood pressure declines.

The combined effect of these regulatory mechanisms is the control of the composition and volume of body fluids, despite fluctuations that occur when conditions of the body change. For example, a strenuous bicycle ride in warm weather results in a large water and salt loss due to perspiration. To adjust for this water loss, the kidneys are directed (by increased levels of ADH, aldosterone, and renin) to reabsorb more water and salt, resulting in a lower urine output. At the same time, sympathetic stimulation of arterioles reduces blood flow to the kidneys, also contributing to a decreased urine output. Unless the water loss is extensive, this adjustment keeps the blood volume from changing much. Therefore, blood pressure remains relatively constant. Another example occurs when you eat a heavily salted meal. As we have seen, the presence of an abundance of sodium and chloride in the blood tends to draw water from other body compartments and would lead to dehydration of tissues. To avoid this, the kidneys are directed (by increased levels of ADH) to conserve water by increasing its reabsorption, while maintaining the rate of salt secretion. Consequently, urine output is reduced while the salt content of the urine remains about the same, with a net effect of excreting excess salts in a concentrated urine. Again, the blood volume and the salt content of the blood remain relatively unchanged, because of the kidney's rapid adjustment. Thus, as shown in both examples, the homeostasis of the blood and other body fluids is maintained despite changing conditions in the body.

Homeostatic functions of the urinary system are not limited to body fluid and salt balance. As we have seen in this chapter, the kidneys remove urea and uric acid from the bloodstream; these are potential hazards if allowed to accumulate in body tissues. The kidneys also help maintain the pH balance in fluids by harboring the enzyme carbonic anhydrase, which plays a role in the bicarbonate buffering system. As an example of acid-base homeostasis, let's consider a patient suffering from the serious lung disease *emphysema,* which results in a reduced alveolar surface area for the diffusion of gases. As a result of the reduced diffusion, carbon dioxide tends to be retained in the bloodstream, leading to a drop in pH. In an effort to remove excess hydrogen ions, the kidneys remove carbonic acid from the bloodstream to produce an acidic urine, which is excreted. Thus, the kidneys provide vital support of the body's internal stability, or homeostasis, even during many types of changes that would otherwise disrupt health.

CLINICAL TERMS OF THE URINARY SYSTEM

Anuria   The absence of urine. It may be due to kidney failure or to an obstruction in the pathway of urine flow.

Calculi   Kidney stones. This development is due to an increased concentration of salts in the urine. It may be in the form of uric acid crystals that precipitate out of the bloodstream and accumulate in the lower extremities (gout), or an excess of calcium salts resulting from hyperparathyroidism. Clinical symptoms include obstruction, pain, and possibly infection.

Congenital Polycystic kidney disease   An inherited disease that results in the abnormal development of renal tubules and collecting ducts. It causes the development of cysts, which enlarge and destroy kidney function. It is a relatively common cause of kidney failure.

Cystitis Inflammation of the urinary bladder. This relatively common affliction is often caused by the entry of bacteria into the urinary system through the urethra.

Diuresis An increased production of urine. A *diuretic* is a substance that causes diuresis.

Dysuria Pain during urination. Dysuria is usually caused by irritation of the urethra, which may result from bacterial infection or physical injury.

Enuresis A condition of uncontrolled urination.

Glomerulonephritis Inflammation of the glomeruli. It is caused by a hypersensitive immune reaction (humoral response) in the glomeruli as a result of bacterial or viral infections. A severe form of this condition, called *anti-GBM nephritis* or *chronic glomerulonephritis,* occurs when antibodies are directed against the glomerular basement membranes. It often results in kidney failure.

Hematuria The presence of blood in the urine. It is usually caused by lesions in the urinary tract. Blood originating from the upper portion of the tract may be due to glomerulonephritis, a malignant tumor, infection, trauma, or a stone. Lower tract bleeding may be due to infection, a malignant tumor, urethritis, or a stone.

Oliguria A reduced output of urine.

Polyuria An excessive output of urine.

Pyelonephritis An inflammation of the renal pelvis. It is often caused by bacteria that have ascended the urinary tract.

Uremia A condition characterized by toxic levels of urea in the blood. It is usually an indication of kidney failure.

Urethritis Inflammation of the urethra. It is frequently caused by bacterial infections and results in pain during urination.

Urinary tract tumors Most commonly, carcinomas arising from lining epithelium. They are usually of low-grade malignancy, and if treated quickly the prognosis is favorable.

CHAPTER SUMMARY

The most important organs of the urinary system are the kidneys, which provide the function of excretion.

KIDNEY STRUCTURE

The kidneys are bean-shaped organs located behind the peritoneum in the abdominal cavity. They are externally supported by the outer renal fascia, an adipose capsule, and the inner renal capsule. The concave margin is the hilum, which opens into a space called the renal sinus. The three regions of the kidney are the basinlike renal pelvis, the renal medulla, and the outer renal cortex. The medulla contains numerous renal pyramids.

Blood Supply Arterial blood supply to a kidney is by way of the renal artery to lobar arteries, to interlobar arteries, to arcuate arteries, to interlobular arteries, and then to afferent arterioles, which supply the glomeruli. Veins parallel these arteries and have similar names.

Nephron Structure The nephron is the functional subunit of the kidney. Each nephron consists of a renal corpuscle and a renal tubule. The renal corpuscle consists of a capillary network, called the glomerulus, enveloped by the bowl-shaped Bowman's capsule. The glomerulus contains large pores, and the Bowman's capsule has an inner layer of cells called podocytes. The renal tubule consists of the proximal convoluted tubule, the loop of Henle with descending limb and ascending limb, the distal convoluted tubule, and the collecting duct.

Blood Flow Through the Nephron Blood enters the glomerulus by the afferent arteriole and leaves by the efferent arteriole. The afferent arteriole has a larger diameter, which creates a high pressure within the glomerulus. Blood then flows through the peritubular capillaries surrounding the renal tubules before draining through veins.

Juxtaglomerular Apparatus A region of specialized cells located where the distal convoluted tubule contacts the afferent arteriole.

KIDNEY FUNCTIONS

The kidneys remove unwanted substances from the blood, maintain fluid balance, and regulate pH.

Urine Formation About 1 to 2 liters of urine is produced each day by three processes performed by the kidneys.

Filtration The movement of small substances through the filtration membrane barrier by filtration pressure. The material entering the Bowman's capsule is called *filtrate.*

Reabsorption The return of water and other substances from the filtrate within the renal tubules into the peritubular capillaries. The proximal convoluted tubules are the most active segment in reabsorption.

Secretion The movement of ions and other substances from the peritubular capillaries into the renal tubules.

Urea and Uric Acid Formation Urea and uric acid are formed as harmful by-products of amino acid and nucleic acid metabolism. They are removed from the bloodstream by filtration and secretion.

Regulation of Urine Concentration and Volume

Renin and Angiotensin A drop in filtration pressure signals the juxtaglomerular apparatus to produce renin, which converts a protein to angiotensin I. In the blood, angiotensin I is converted by another enzyme to angiotensin II. Angiotensin II acts on the adrenal gland to increase its rate of aldosterone secretion, leading to increased sodium, chloride, and water reabsorption in the kidneys.

Aldosterone Increases the rate of sodium and chloride reabsorption. The increase is followed by increased water reabsorption.

Antidiuretic Hormone A drop in blood pressure signals the production of more ADH, which increases the rate of water reabsorption by renal tubules and thereby restores blood pressure.

Atrial Natriuretic Factor An increase in blood pressure in the right atrium stimulates the release of atrial natriuretic factor, which reduces reabsorption in the kidneys. This lowers blood pressure.

Sympathetic Stimulation Sympathetic impulses constrict afferent arterioles, reducing filtration pressure and leading to a reduction in urine output.

Maintenance of Body Fluids The kidneys are the primary organs in the body that regulate the water-salt balance of all body fluids in all body compartments.

Regulation of pH Hydrogen-ion concentration (pH) is maintained within a narrow range by three factors:

Buffers Chemicals that accept hydrogen ions in the presence of an acid, or release hydrogen ions in the presence of a base. The most important buffer is the bicarbonate system.

Respiratory System A change in ventilation rate and depth can adjust carbon dioxide levels, and hence pH levels, in body fluids.

Kidneys The kidneys remove excess hydrogen ions in the blood by secretion.

URETERS

The ureters are paired tubes that carry urine from the renal pelvis of each kidney to the urinary bladder. The walls consist of three layers: an inner mucous membrane, a middle layer of smooth muscle, and an outer layer of connective tissue. Mucus protects the inner lining, and the muscle layer undergoes peristaltic contractions.

URINARY BLADDER

The urinary bladder is a hollow, muscular organ located at the floor of the pelvic cavity. It temporarily stores urine, which it receives from the ureters. The triangular region that receives the openings from the ureters and the urethra is the trigone. The wall of the bladder consists of four layers: an innermost mucous membrane of transitional epithelium, a connective tissue layer, a smooth muscle layer called the *detrusor muscle,* and an outer layer of mostly fibrous connective tissue.

URETHRA

The urethra is a muscular tube that carries urine from the urinary bladder to the outside. It contains an internal urethral sphincter of smooth muscle between the bladder and the urethra, and an external urethral sphincter of skeletal muscle.

MICTURITION

The act of emptying the bladder. It is initiated by the activation of stretch receptors in the wall of the bladder, which send an impulse to the spinal cord that quickly returns to cause the internal sphincter to relax and the detrusor muscle to contract. If the voluntary sphincter relaxes also, urine flows out.

HOMEOSTASIS

The main function of the kidneys (and urinary system) is the maintenance of homeostasis. Their roles in homeostasis include maintaining the fluid and salt balance, removal of urea and uric acid, and maintaining pH.

KEY TERMS

Bowman's capsule (p. 534)
buffer (p. 545)
electrolyte balance (p. 540)
excretion (p. 531)

filtration (p. 538)
glomerulus (p. 534)
micturition (p. 549)
nephron (p. 533)

peritubular capillaries (p. 537)
reabsorption (p. 539)
renal (p. 532)
secretion (p. 540)

urea (p. 541)
uric acid (p. 541)
urine (p. 538)

QUESTIONS FOR REVIEW

OBJECTIVE QUESTIONS

1. The primary organ of excretion is the:
   a. skin   c. kidney
   b. urethra   d. lung
2. The renal pyramids are located within the:
   a. renal pelvis   c. renal cortex
   b. renal medulla   d. ureter
3. The functional subunit of each kidney is called the:
   a. renal tubule   c. renal pelvis
   b. nephron   d. lobule
4. Interlobular arteries carry blood to the _____, which pass to the glomeruli.
   a. interlobar arteries
   b. efferent arterioles
   c. afferent arterioles
   d. peritubular capillaries
5. The glomerulus differs from other capillary networks in that it:
   a. carries blood under low pressure
   b. is lined with endothelium
   c. contains large pores
   d. all of the above.
6. The cluster of cells located where the distal convoluted tubule contacts the afferent arteriole is called the:
   a. juxtaglomerular apparatus   c. glomerulus
   b. Bowman's capsule   d. podocytes
7. Water and solutes move from the glomerulus into the Bowman's capsule by the process of:
   a. filtration   c. secretion
   b. reabsorption   d. active transport
8. Water is returned to the bloodstream during reabsorption by:
   a. active transport   c. osmosis
   b. secretion   d. micturition
9. Hydrogen ions are removed from the bloodstream primarily by the process of:
   a. filtration   c. reabsorption
   b. secretion   d. osmosis
10. Urea is removed from the bloodstream primarily by:
    a. secretion   c. reabsorption
    b. filtration   d. active transport

11. An increase in the amount of ADH released by the pituitary gland results in:
    a. an increase in water reabsorption
    b. a more concentrated urine
    c. an increase in blood pressure
    d. all of the above
12. Aldosterone promotes the increased reabsorption of water in the kidneys by:
    a. inhibiting sodium reabsorption
    b. inhibiting renin production
    c. stimulating sodium reabsorption
    d. promoting ADH secretion
13. What substance is released by juxtaglomerular cells in response to low glomerular filtration pressure?
    a. aldosterone   c. angiotensin II
    b. renin   d. hydrogen ions
14. The detrusor muscle contracts and the external urethral sphincter relaxes during:
    a. micturition   c. filtration
    b. urinary bladder filling   d. tubular reabsorption
15. The presence of which of the following molecules in the filtrate could signal that the filtration membrane of the glomerulus is damaged?
    a. glucose
    b. urea
    c. sodium and potassium
    d. proteins

ESSAY QUESTIONS

1. Describe the external and internal structural features of the kidney. Relate these structures to kidney function.
2. Describe the components of a nephron, and indicate their location relative to the overall kidney organization.
3. Describe the processes that result in urine formation (filtration, reabsorption, and secretion).
4. Explain reabsorption in the renal tubules by indicating the activities that are prominent in their various segments.
5. Why is hydrogen-ion secretion by the kidneys important to maintaining homeostasis? What other body activities remove excess hydrogen ions?
6. Describe the process of micturition.

ANSWERS TO ART LEGEND QUESTIONS

Figure 18–1
What major arteries provide the kidneys with fresh blood?   The renal arteries supply the kidneys with oxygenated blood.

Figure 18–2
What are the three internal regions of each kidney?   The cortex, medulla, and pelvis are the three internal regions of a kidney.

Figure 18–3
Which vessels do the interlobar arteries branch into?   The interlobar arteries branch into the arcuate arteries in the cortex of each kidney.

Figure 18–4
The loop of Henle extends into what region of the kidney?   The loop of Henle dips down into the medulla of the kidney.

Figure 18–5

In what structure does filtrate collect once it is formed?   Filtrate is collected in the capsular space of the Bowman's capsule as it is formed by filtration.

Figure 18–6

What two structures as they come into physical contact house the cells of the juxtaglomerular apparatus?   Juxtaglomerular cells are located within the distal convoluted tubule and afferent arteriole where these structures come into contact.

Figure 18–7

What are the three processes of urine formation?   The three processes of urine formation are filtration, reabsorption, and secretion.

Figure 18–8

What substance is normally reabsorbed in enormous volumes?   Water is normally reabsorbed in very large volumes.

Figure 18–9

What is the fluid called that passes through the collecting duct?   The fluid within the collecting duct is called *urine*.

Figure 18–10

What enzyme released by juxtaglomerular cells is needed for the appearance of angiotensin II in the bloodstream?   Renin is required to convert a protein released by the liver to angiotensin II.

Figure 18–11

What is the effect of increased levels of aldosterone on the volume of fluid in the extracellular compartment?   Increased levels of aldosterone cause an increased volume in the extracellular fluid compartment and a smaller volume of excreted urine.

Figure 18–12

Where is the external sphincter located in the male urethra?   The external sphincter is located at the base of the penis at the urogenital diaphragm.

Figure 18–13

What is the function of the detrusor muscle?   Contraction of the detrusor muscle pushes urine out of the bladder and into the urethra during micturition.

A human embryo, 6 weeks old.

THE CYCLE OF LIFE

The continuation of our species, *Homo sapiens*, is the primary concern of Unit 6. It includes an examination in Chapter 19 of the reproductive system, the only system that is not directly involved in maintaining internal homeostasis. As we are about to see, this system consists of groups of organs that differ markedly between males and females but whose overall goal is the same: the production of sex cells to create a fertilized egg capable of becoming a new individual. Chapter 20 continues our life story by exploring fertilization and early development and the changes the embryo must undergo before it is ready for birth. Human development is examined yet further in Chapter 20 as we consider the actual birth process and the changes that occur in the body throughout life. A brief study of human genetics and inheritance follows.

THE REPRODUCTIVE SYSTEM

LEARNING OBJECTIVES

After studying this chapter, you should be able to:

1. Identify the organs of the male system, on the basis of their structures and functions.

2. Describe the process of spermatogenesis and its results.

3. Describe the tubes that carry sperm through the male system.

4. Identify the male accessory glands, and discuss the nature of their secretions.

5. Describe the external genitalia of the male.

6. Explain the processes of erection, emission, and ejaculation and describe the neural mechanisms that influence male reproduction.

7. Describe the effects of GnRH, LH, and FSH on male reproduction.

8. Identify the effects of testosterone on male reproduction.

9. Identify the organs of the female system on the basis of their structure and functions.

10. Describe the processes of oogenesis, follicle development, and ovulation.

11. Describe the neural mechanisms that influence female reproduction.

12. Identify the roles of GnRH, LH, and FSH in the development of female puberty.

13. Describe the effects of estrogen on female reproduction.

14. Define the terms *menses* and *menopause*.

The primary function of both male and female reproductive systems is to produce offspring. The two systems perform this function jointly, although their structures differ. The structural differences between male and female organs reflect the different roles that are performed by the two sexes. For example, the male reproductive system consists of organs that produce sperm and deliver it to the female. On the other hand, the female system contains organs that produce eggs, support the developing embryo and fetus if the egg has been fertilized, and give birth to offspring. Both male and female reproductive organs that produce the sex cells (sperm and egg) are called **gonads**. Both types of sex cells are called **gametes**.

ORGANS OF MALE REPRODUCTION

The male reproductive organs include the testes, which produce testosterone and sperm; a system of ducts that transport the sperm to the outside; and several accessory organs that nourish and support the sperm along their journey. Sperm is conveyed into the female by the penis.

The primary organs of the male reproductive system are the testes (singular form is testis), which are the male gonads. They produce the male sex hormone, testosterone, and the male gametes, the sperm (spermatozoa). The remaining male organs include a number of ducts that store, support, and transport the sperm to the outside; accessory glands whose secretions support the sperm and provide it with a fluid medium; and external sex organs.

Testes

The **testes**[1] are paired, oval organs that lie suspended within a saclike pouch of skin called the **scrotum** outside the abdominopelvic cavity (Fig. 19–1). They are located outside the body to promote the production of healthy sperm, for sperm are heat-sensitive, and the scrotum maintains an internal temperature cooler than normal body temperature.

Structure

Each testis is about 4 cm (1½ inches) long and 2.5 cm (1 inch) wide. It is enclosed within a tough outer shell of fibrous connective tissue, which extends into the interior of the testis to subdivide it into about 250 small

Figure 19–1
Male reproductive organs. This sagittal section of the lower trunk region reveals the major organs of the male reproductive system.

What organ contains the testes?

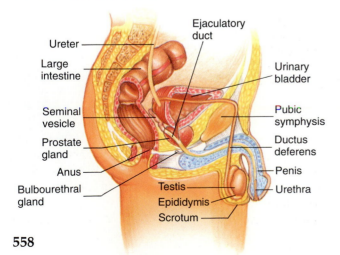

Ureter
Large intestine
Seminal vesicle
Prostate gland
Anus
Bulbourethral gland

Ejaculatory duct
Urinary bladder
Pubic symphysis
Ductus deferens
Penis
Urethra

Testis
Epididymis
Scrotum

compartments known as *lobules* (Fig. 19–2). Within each lobule are from one to four tightly packed **seminiferous** (sem'-ĭ-NIF-er-us) **tubules**.[2] Between the tubules is a delicate connective tissue membrane that contains clusters of endocrine cells. These cells are called **interstitial cells** (or cells of Leydig) and secrete the male sex hormone **testosterone**. The seminiferous tubules extend to the posterior edge of the testis, where they unite to form a network of small tubes known as the **rete** (RĒ-tē) **testis**. The rete testis gives rise to another set of tubes, called the **efferent ductules**, which exit the testis before uniting to form the single duct of the epididymis (whose description soon follows).

The walls of the seminiferous tubules are lined with a specialized layer of cells, called the **germinal epithelium**. There are two types of cells within this layer: supportive cells and germ cells. The supportive cells provide nourishment and support for the germ cells, and the germ cells produce new sperm cells by the process of *spermatogenesis*.

Spermatogenesis

Spermatogenesis is the production of sperm cells, which occurs within the seminiferous tubules of the testes. The sperm are produced from precursor cells that divide by meiosis. As you may recall from Chapter 3, meiosis is a type of cell division that results in the production of four daughter cells, each of which contains one-half the chromosome number of the original parent cell. In the case of spermatogenesis, each daughter cell matures into a sperm cell. The other form of cell division, mitosis, results in the formation of two daughter cells that are genetically identical to each other and to the parent cell. Recall also from Chapter 3 that mitosis is the means by which the body achieves growth, tissue repair, and cell replacement, and therefore it occurs in most body cells. Meiosis, on the other hand, occurs only in the male and female gonads and results in the formation of male and female gametes.

In a developing male fetus, the testes begin to develop within the abdominopelvic cavity. Soon before or after birth the testes descend into the scrotum, where they will remain for the rest of life. Before the onset of puberty, the testes remain about the same size as they were following their descent. Once puberty begins, the interstitial cells increase in size and in number, and the seminiferous tubules enlarge. At this point, the production of new sperm cells by spermatogenesis begins.

---

[1]**Testes:** Latin *testis* = testis, or testicle.

[2]**Seminiferous tubules:** *semin-* = seed + *fer* = to carry; *tubule* = tiny tube.

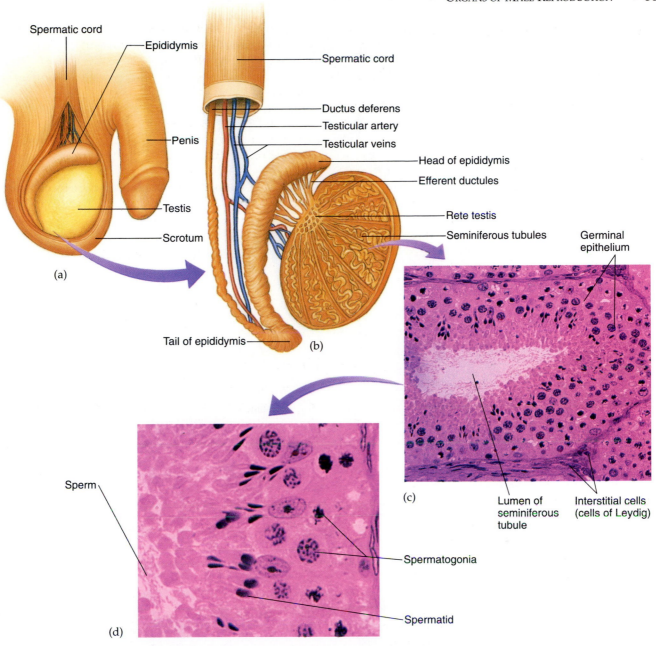

Figure 19–2
Structure of the testis. (a) Each testis is bordered by an epididymis, which lies with it within a compartment of the scrotum. (b) The internal organization of the testis, epididymis, and spermatic cord is revealed when the outer layers of connective tissue are removed. (c) A photomicrograph of a single seminiferous tubule in cross section (64x). (d) When the germinal epithelium of a seminiferous tubule is viewed under higher magnification, the stages in sperm development may be observed (160x).

What is produced by the interstitial cells (cells of Leydig)?

Among the germ cells of the germinal epithelium are stem cells known as **spermatogonia.**[3] They actively

---

[3]**Spermatogonia:** Greek *spermato-* = pertaining to seed + *gone* = generation.

---

divide by mitosis on an almost continuous basis. Some of their daughter cells form **primary spermatocytes,** which divide by meiosis, while other daughter cells remain as spermatogonia and continue to divide by mitosis (Fig. 19–3). In the course of meiosis, a single primary spermatocyte divides to form two secondary

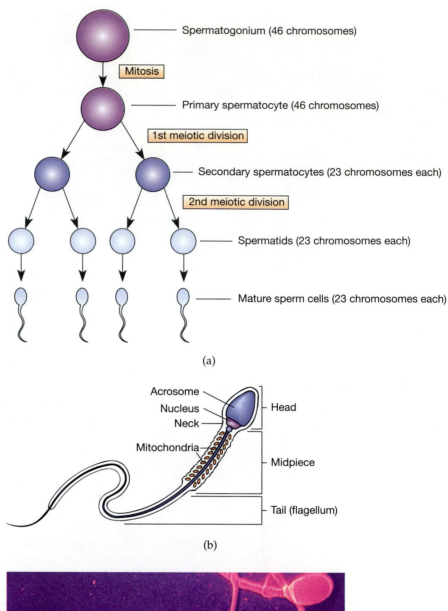

Spermatogonium (46 chromosomes)

Mitosis

Primary spermatocyte (46 chromosomes)

1st meiotic division

Secondary spermatocytes (23 chromosomes each)

2nd meiotic division

Spermatids (23 chromosomes each)

Mature sperm cells (23 chromosomes each)

(a)

Acrosome
Nucleus
Neck
Head

Mitochondria
Midpiece

Tail (flagellum)

(b)

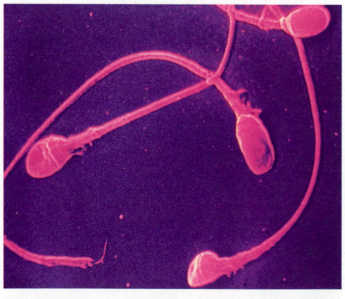

(c)

Figure 19–3
Sperm cell. (a) Spermatogenesis, which results in the production of four sperm cells from a single spermatogonium. This process takes place within the germinal epithelium of the seminiferous tubules.
(b) Structure of a mature sperm cell.
(c) A high-powered electron micrograph of sperm (3220x).

What is the function of the sperm cell's flagellum?

spermatocytes, each of which divides again to form two **spermatids.** Thus, four small spermatids form from each primary spermatocyte. Each spermatid contains 23 chromosomes (one-half the complete human chromosomal complement) and begins to develop into a tadpolelike structure while it is still part of the germinal epithelium. It will eventually form a head, a midpiece, and a flagellum, at which point we call it a **sperm cell** or **spermatozoon.** Once formed, the sperm cell breaks away from the germinal epithelium and continues its development as it drifts toward the epididymis.

Spermatogenesis occurs on a continuous basis throughout the life of a male. Regardless of his daily activities, a normal male produces millions of sperm cells each day. In some cases, however, sperm cell production may be insufficient to fertilize an egg. This condition is called *sterility* and may be caused by emotional anxiety, trauma from injury to the testes, an inherited disorder that causes insufficient numbers of sperm or disabled sperm to be produced, or tumors.

CONCEPTS CHECK

1. What are the male gonads?

2. Where are sperm cells produced?

3. What cells secrete testosterone?

Ducts

Sperm cells that have been produced in the testes soon find their way into a series of ducts. These ducts, beginning with the epididymis, store, support, and transport sperm from the testes to the urethra.

Epididymis

The **epididymis**[4] (ep'-i-DID-i-mis) is a comma-shaped structure on the posterior side of each testis (Fig. 19–1). It consists of a single, tightly coiled tube that receives sperm from the efferent ductules of the testis (Fig. 19–2b). If stretched out, this tube would measure about 6 m (20 feet) long, although its diameter is similar to that of a thin thread. After emerging from the testis, the tube courses downward before curving upward as the next duct, the ductus deferens.

Sperm cells that arrive in the epididymis are immature, and incapable of self-propulsion. They are moved through the epididymis slowly by its peristaltic contractions. As they move, they continue to mature. By the time sperm cells have exited, they can move independently by whipping their flagella and are capable of fertilizing an egg.

Ductus Deferens

The **ductus deferens**[5] (or **vas deferens**) carries sperm from the epididymis to the urethra (Figs. 19–1 and 19–4). Each ductus deferens is a single tube with a thick, muscular wall that extends upward from the epididymis and through the scrotum. The ductus deferens then passes through the abdominal wall by way of a small opening called the *inguinal canal*, crosses the lateral wall of the pelvic cavity, and loops over the back of the urinary bladder. Just outside the prostate gland it empties into the **ejaculatory duct**, which carries sperm through the prostate gland and into the urethra.

Along its route through the scrotum, the ductus deferens is associated with other structures to form the **spermatic cord.** The cord consists of the ductus deferens, the testicular artery and veins, lymphatic vessels, the testicular nerve, the cremaster muscle, and connective tissue. The cremaster muscle consists of bands of skeletal muscle that form the outer layer of the spermatic cord; it serves to elevate the testes closer to the body during periods of cold external temperature in order to maintain the narrow temperature limits of developing sperm cells. During the surgical sterilization procedure known as a *vasectomy*, the ductus deferens is separated from the other components of the spermatic cord, severed completely, and tied or cauterized.

Urethra

The male **urethra** extends from the urinary bladder to the distal end of the penis (Figs. 19–1 and 19–4). It is a passageway for both urine and male reproductive fluids. Its wall is lined with mucous membrane and contains a relatively thick layer of smooth muscle. It also contains numerous mucous glands, which empty their secretions directly into the lumen. The opening at the distal end of the urethra is called the **external urethral orifice.**

CONCEPTS CHECK

1. What type of muscle contraction helps transport sperm cells through the ducts?

2. Which duct is part of the spermatic cord?

3. What fluids does the male urethra transport?

---

[4]Epididymis: Greek *epi-* = upon + *didymos* = a twin, or one of a pair of things (the testes).

[5]Ductus deferens: Latin *ductus* = duct or tube + *deferens* = one that carries away from.

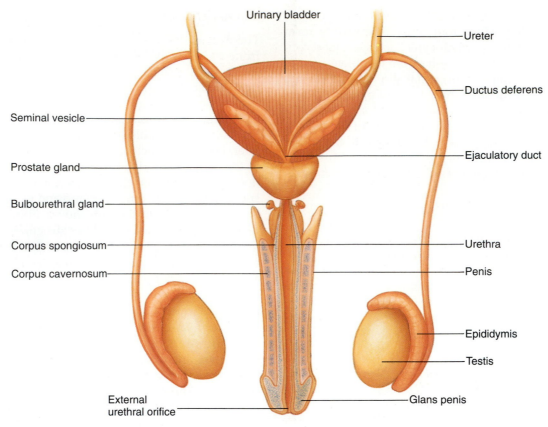

Figure 19–4
Male reproductive tract. This posterior view reveals the route traveled by the
ductus deferens and the urethra. The penis is shown sectioned in this drawing.
Can you trace the route of sperm travel from the testes to the external urethral orifice?

Accessory Glands

There are three types of organs that serve as accessory
glands to the male reproductive tract. They are the
seminal vesicles, the prostate gland, and the bulbourethral glands. The collective function of their secretions is to provide a liquid, nutritious medium for
the support of sperm cells. The mixture of sperm cells
and fluids that results is called **semen**, or seminal fluid.
The volume released at one time ranges from 2 to 7
milliliters, and the average number of sperm cells present in the fluid is about 120 million per milliliter.

Seminal Vesicles

The **seminal vesicles**[6] are a collection of glands located at the base of the urinary bladder (Figs. 19–1 and

19–4). Each gland consists of numerous saclike structures and empties its secretions by way of a short duct
into the ductus deferens. The union between the seminal vesicle ducts and the ductus deferens actually
forms the ejaculatory duct, which continues on to the
urethra. Secretions from the seminal vesicles are rich
in sugar (fructose), which nourishes and activates the
passing sperm. Seminal vesicle secretions account for
about 60% of the fluid volume of semen.

Prostate Gland

The single **prostate gland**[7] is about the size and shape
of a walnut. It is located against the lower margin of
the urinary bladder, where the gland encircles the urethra (Figs. 19–1 and 19–4). Because of its location, an
increase in size (*hypertrophy*) tends to constrict the urethra, making urination difficult and painful. Hyper-

---

[6]**Seminal vesicles:** *semin-* = seed + *-al* = pertaining to + *vesicles* = little bladders.

[7]**Prostate:** Latin *prostare* = to stand before.

The Troublesome Sex Gland: The Prostate

Throughout most of a man's life, the walnut-sized prostate gland produces a fluid that is an important component of semen, the fluid that carries sperm through the urethra during sexual climax. However, in later years this small structure is a primary cause of discomfort and, in some cases, can even develop a life-threatening cancer. The trouble usually begins soon after age 50.

For reasons not well understood, the prostate gland often increases in size as a man ages. In fact, over half of all men in their sixties suffer from an enlarged prostate, a condition known as *benign prostatic hyperplasia (BPH)*. In this condition, enlargement of the gland can result in compressing the urethra, which leads to a weakened or hesitant urine stream, leaking, a sensation that the bladder is not completely emptied, and a need to get up several times a night to urinate. Fortunately, men suffering the discomforts of BPH are not at increased risk of developing cancer, and treatment is usually quite effective. In past years, treatment was limited to surgical removal of part of the prostate, which involved insertion of a lighted instrument through the urethra to view the prostate and electrically cut part of it out. About 400,000 such procedures were performed annually in the United States after 1980. However, recent advances in drug therapy are showing increased hope, particularly through a drug called Proscar, whose early trials have shown an ability to stop further growth of the prostate and in about 24% of cases to cause actual shrinkage.

Although the symptoms of BPH can be annoying, they are not a serious threat to life itself, like cancer of the prostate, which shares identical symptoms. Prostate cancer is second only to lung cancer as a deadly form of cancer in men; in 1992 it caused an estimated 34,000 deaths in the United States. In past years, the high incidence of death was largely due to late diagnosis. Until recently, the only available technique for identifying this disease was by a digital rectal exam, in which a physician inserts a finger into the rectum and probes for any hard lumps that may indicate the presence of cancer. While this test can identify some malignancies, it is dependent on the presence of well-developed lumps, which are usually detectable only after metastasis (spreading to other tissues) has already begun. Within the last several years, a protein known as *prostate-specific antigen (PSA)* has been under careful study and is showing great hope as an early indicator of prostate cancer. It is identified by way of a simple blood test, and high concentrations of it are considered sufficient cause to evaluate the possibility of prostate cancer more carefully (usually by biopsy). PSA is normally produced by cells lining the prostate, and it serves to help liquefy semen to improve sperm cell motility during ejaculation. However, it is normally produced in low levels; levels greater than 4 microgams per liter are considered a possible risk factor that warrants further investigation.

trophied prostate is a common condition in men over 50 years of age. The prostate gland secretes a thick, milky fluid, which passes through numerous short ducts before reaching the urethra. This fluid is more alkaline than the secretions of the vagina. Since acidic environments tend to inhibit sperm cell motility, the neutralizing effect of prostate fluid on vaginal fluid may aid sperm motility during fertilization.

Bulbourethral Glands

The paired **bulbourethral** (bul'-bō-yoo-RĒ-thral) **glands**, or Cowper's glands, are tiny, pea-sized structures located below the prostate gland (Figs. 19–1 and 19–4). Each produces a thick, clear mucus that drains into the urethra by way of a single short duct. The fluid

is thought to provide lubrication for sexual intercourse, although most of the lubrication is provided by female glands.

CONCEPTS CHECK

1. What is the fluid mixture of sperm cells and glandular secretions called?

2. Which accessory gland is wrapped around the urethra?

3. What accessory gland delivers its secretions into the ductus deferens?

External Genitalia

The external organs, or external genitalia, of the male include the scrotum and the penis.

Scrotum

The **scrotum**[8] (Fig. 19–1) is a saclike pouch of skin that hangs below the lower abdominal wall. Internally, it is divided into right and left chambers by a septum. Each chamber contains a testis and is lined with a serous membrane that reduces friction between the testis and the scrotal wall. The inner wall of the scrotum is lined by the involuntary **dartos muscle,** which assists the cremaster muscle (of the spermatic cord) during cold weather in drawing the testes closer to the abdominal wall for warmth. Contraction of the dartos muscle causes the skin of the scrotum to wrinkle.

Penis

The **penis**[9] is an external cylindrical organ that contains the distal portion of the urethra, which carries semen and urine to the outside (Figs. 19–1, 19–4, and 19–5). During sexual arousal, it has the special ability to enlarge and become firm. This process is called *erec-*

---

[8]Scrotum: Latin *scrotum* = bag.
[9]Penis: Latin *penis* = tail.

---

tion; it enables the penis to be inserted into the female vagina during sexual intercourse.

The penis is externally covered with a layer of skin. Below the skin is a thin layer of superficial fascia, which separates the skin from three internal columns of erectile tissue (Fig. 19–5). The erectile tissue contains many spaces, which fill with blood during erection. Two of the columns form the dorsal and lateral portions of the penis and are called **corpora cavernosa.**[10] The third column, the **corpus spongiosum,**[11] is considerably smaller and forms the ventral portion of the penis. As you can see in the cross-sectional view (Fig. 19–5b), the urethra courses through the corpus spongiosum.

At its distal end, the corpus spongiosum is enlarged to form a cap over the penis, called the **glans penis.**[12] A loose fold of skin covers over the glans penis; it is called the foreskin or **prepuce.** The prepuce is frequently removed by surgical means in a procedure known as *circumcision* (usually performed soon after birth as an optional procedure). Circumcision provides

---

[10]Corpora cavernosa: Latin *corpora* = bodies + *cavernosa* = containing hollow areas.

[11]Corpus spongiosum: Latin *corpus* = body + *spongiosum* = spongy.

[12]Glans penis: Latin *glans* = acorn + *penis*.

Figure 19–5

Structure of the penis. (a) Anterior view, with the prepuce partially removed to reveal the glans penis and a portion of the shaft cut away. (b) Cross section of the penis.

What is the function of the penis?

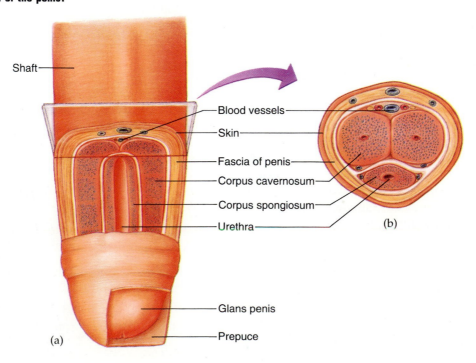

some advantage, since it prevents the buildup of bacteria that sometimes occurs in the cleft of the prepuce, particularly when hygiene is not regularly performed. The skin of the glans penis is very thin and hairless and is well supplied with sensory receptors.

CONCEPTS CHECK

1. What organ houses the testes outside the abdominopelvic cavity?

2. What column of erectile tissue in the penis contains the distal portion of the urethra?

PHYSIOLOGY OF MALE REPRODUCTION

The functions of the male system are the generation and release of sperm for fertilization of the female. They are accomplished by neural and hormonal mechanisms.

The male reproductive system is influenced by neural and hormonal mechanisms. Neural mechanisms are mainly involved in the processes of erection, orgasm, and ejaculation. Hormones play a profound role in the development of reproductive structures, the regulation of reproductive functions, the development of secondary sex characteristics, and sexual behavior.

Neural Mechanisms

Neural control of male reproductive functions involves the production of autonomic impulses, which originate from the hypothalamus or as a reflex from the spinal cord in response to sensory stimuli. When sensory stimuli are not present, the spaces within the erectile tissue of the penis are empty. When sensory stimuli begin, parasympathetic nerve impulses pass to the walls of the arteries supplying the penis, causing the arteries to dilate. As a result, arterial blood enters the erectile tissue spaces. As the erectile tissue fills with blood, it swells, compressing the veins that carry blood away from the penis. Thus, much more blood enters the erectile tissue than leaves, causing the penis to swell and become erect. This is the first step in sexual arousal, referred to as **erection**.

Soon after parasympathetic impulses begin traveling to the vessels of the penis, sympathetic impulses are conducted to the walls of the ducts. They stimulate peristaltic contractions of the testicular ducts, the epididymides, the ductus deferentes, and the ejaculatory ducts. Movement of sperm cells through the ducts, which is called **emission**, results.

As the urethra fills with incoming semen, sensory impulses are stimulated in its wall and are conducted to the spinal cord. Motor impulses are then routed from the cord to skeletal muscles at the base of the penis, causing them to rhythmically contract. This increases the pressure within the erectile tissue, and also aids in forcing semen through the urethra during the process of **ejaculation**. As the semen begins its rapid flow through the urethra, secretions from the bulbourethral gland flow in first, followed by secretions from the seminal vesicles and prostate gland that mix with sperm cells and a small amount of testicular fluid.

The combination of sensory impulses, erection, emission, and ejaculation invokes a pleasurable experience. It is a feeling of physical and mental tension that is followed by complete relaxation. The latter parts of the sexual response process are referred to as **orgasm**, or male climax. In some individuals, ejaculation is not a requirement for orgasm.

Once ejaculation is completed, sympathetic impulses travel to the artery walls supplying the penis, causing their constriction. Gradually, more blood exits the erectile tissue than enters. With the exit of blood, the penis returns to its original flaccid state.

Hormonal Mechanisms

The hormones that influence male reproductive functions arise from the hypothalamus, the anterior pituitary gland, and the testes. The hormones released by the hypothalamus and pituitary gland regulate the activity of the testes, which secrete the primary male sex hormone testosterone.

Hypothalamic and Pituitary Influences

As a male approaches the age of puberty, neurons in the hypothalamus begin to secrete **gonadotropin-releasing hormone (GnRH)**, which enters nearby blood vessels to target the anterior pituitary gland. The presence of GnRH stimulates cells in the anterior pituitary to secrete two hormones that are collectively called *gonadotropins:* **luteinizing hormone (LH)** and **follicle-stimulating hormone (FSH)**. LH and FSH are named after their effects on female reproduction, although their influence on male function is also important (for this reason, LH is also known as *interstitial cell–stimulating hormone,* or *ICSH*). In the male, LH promotes the development of interstitial cells in the testes. Once they reach maturity, the interstitial cells will secrete testosterone. FSH prepares the germinal epithelium of the testes to become responsive to the effects of testosterone. Once they are responsive, these cells stimulate the process of spermatogenesis when in the presence of FSH and testosterone.

Testosterone

The male sex hormones are collectively called **androgens**. By far the most abundant—and functionally significant—androgen is **testosterone**. As we have seen, it is secreted by interstitial cells in the testes once they have undergone maturity, as a result of the influence of LH. After puberty, its secretion continues throughout life.

The early presence of testosterone in the body of a young male causes many changes. These changes begin to occur during a phase of development known as **puberty**, which usually begins between the ages of 10 and 12 years and is completed by age 18. The changes include the functional development and enlargement of the testes and other reproductive organs, and the development of *secondary sexual characteristics*. The secondary sexual characteristics make the male body different from the female; they are described in Table 19–1.

The amounts of testosterone that are released by the testes determine the extent of the secondary sexual characteristics. For example, high levels of testosterone in the blood contribute to a large amount of body hair, a deep voice, and broad shoulders, whereas low levels result in little body hair, a high voice, and narrow shoulders. Thus, it is important that the testosterone blood levels be kept relatively constant. This is achieved by the influence of the hypothalamus in a negative feedback system (Fig. 19–6). In this system, an increase in testosterone levels inhibits the production of GnRH by the hypothalamus, which in turn reduces the production of LH and testosterone. Conversely, as

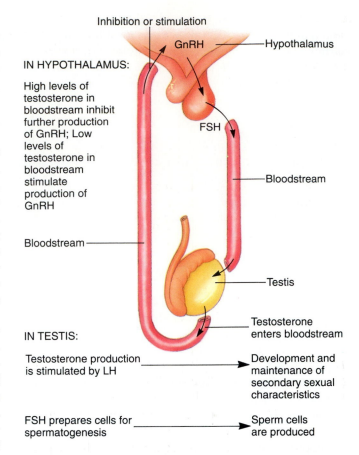

IN HYPOTHALAMUS:

High levels of testosterone in bloodstream inhibit further production of GnRH; Low levels of testosterone in bloodstream stimulate production of GnRH

Inhibition or stimulation

GnRH — Hypothalamus

FSH

Bloodstream

Bloodstream

Testis

Testosterone enters bloodstream

IN TESTIS:

Testosterone production is stimulated by LH → Development and maintenance of secondary sexual characteristics

FSH prepares cells for spermatogenesis → Sperm cells are produced

Figure 19–6
Influence of hormones on male reproduction. Regulation is provided by a negative feedback system.

What part of the brain regulates testosterone production in the testes?

Table 19-1 EFFECTS OF HORMONES ON MALE REPRODUCTION		
Hormone	**Primary Source**	**Effects**
GnRH	Hypothalamus	Stimulates anterior pituitary gland to secrete LH and FSH
LH (ICSH)	Anterior pituitary gland	Promotes development of interstitial cells in the testes
FSH	Anterior pituitary gland	Prepares germinal epithelium of seminiferous tubules to become sensitive to testosterone
Testosterone	Interstitial cells of the testes	Stimulates growth of reproductive organs, male sexual behavior, and, in the presence of FSH, spermatogenesis; also stimulates the following secondary sexual characteristics: 1. Enlargement of the larynx and vocal cords 2. Increase of body hair, especially in the face, chest, axilla, and pubic areas, with a possible decrease in scalp hair 3. Increase of muscle and bone growth, leading to broad shoulders and narrow waist 4. Thickening of the skin

testosterone levels in the blood decrease, the hypothalamus increases its rate of GnRH production. Consequently, more LH is produced by the pituitary gland, stimulating the production of more testosterone.

CONCEPTS CHECK

1. What are the neural events that trigger erection?

2. What is the effect of GnRH on testosterone production?

3. What are the secondary sexual characteristics in the male?

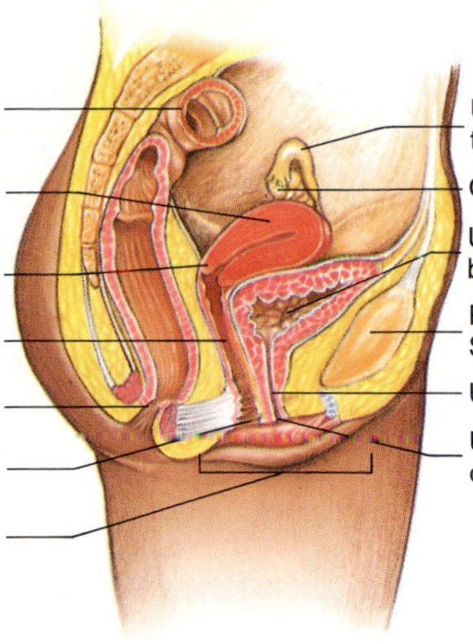

ORGANS OF FEMALE REPRODUCTION

The female reproductive organs produce and transport the eggs, promote the success of fertilization by receiving the penis, and support the offspring during its first nine months of development.

The primary female organs of reproduction are the ovaries. They produce the female gametes and sex hormones, and are thus referred to as the female gonads. Other female reproductive organs provide a site for the development of offspring, provide for the birth process, and receive the penis during sexual intercourse.

Ovaries

The **ovaries**[13] are paired, oval organs located opposite one another against the lateral walls of the pelvic cavity (Fig. 19–7). They are about 3.5 cm ($1\frac{1}{2}$–2 inches) long and 2 cm (less than 1 inch) wide in most females, and are supported by ligaments attached to the pelvic wall and the nearby uterus.

Structure

Each ovary is externally covered by a single layer of visceral peritoneum. Beneath this thin membrane is a layer of epithelial cells, called the **germinal epithelium**. Although the term *germinal* is used to identify this layer of cells, female sex cells are not present here.

---

[13]Ovaries: Latin *ovarium* = place for eggs.

Figure 19–7
Female reproductive organs. This sagittal section of the lower trunk region reveals the major organs of the female reproductive system.

Which organs are the female gonads?

The area deep to the germinal epithelium consists of tough, fibrous connective tissue that forms a protective shell around the ovary. Beneath this covering lies the interior, which is divided into two regions: an outer cortex and an inner medulla.

The **cortex** of the ovary consists of dense connective tissue. Within the cortex are numerous saclike **ovarian follicles**, which may be found in various stages of development. Each ovarian follicle contains a single **oocyte**[14] (ovum), the female gamete.

The ovarian **medulla** consists of connective tissue that is more loosely arranged. It receives blood vessels, lymph vessels, and nerves and is continuous with the ligament (called the *ovarian ligament*) that attaches the ovary to the uterus.

Oogenesis

The production and development of female gametes differ considerably from those of male sperm cells. As we observed earlier, sperm production begins at puberty and continues throughout life. In females, the total number of oocytes that are produced in a lifetime are present in the ovary at birth. By the time a female is born, she may have as many as a half million oocytes in her ovaries. Also, maturation and release occur between puberty and about the age of 50 on a monthly cycle. As we shall see shortly, this cycle is called the **menstrual cycle**.

---

[14]Oocyte: Greek *oo-* = egg + *-cyte* = cell.

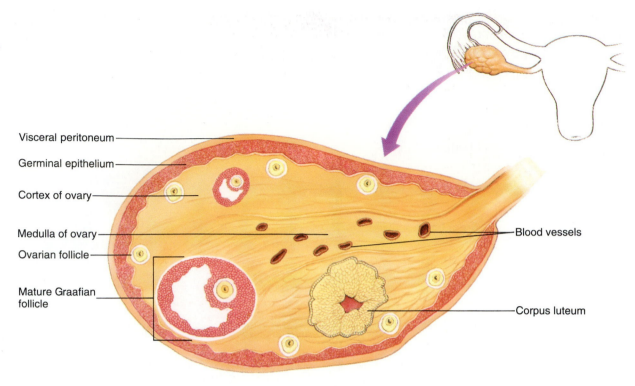

Figure 19–8
Structure of the ovary. The right ovary is sectioned across the frontal plane in this drawing. Notice the numerous follicles in their various stages of development.
What structures within the ovary contain the oocytes?

The oocytes that are present in the ovary at birth are called **primary oocytes** (Fig. 19–9). They represent the first stage in the process of egg development, or **oogenesis** (ō′-ō-JE-ne-sis). At this point oogenesis stops; it does not resume until puberty is about to begin. Each primary oocyte is surrounded by a single layer of flattened cells to form a structure called a **primordial follicle**.

At the onset of puberty, hormonal changes stimulate some of the primary oocytes to resume development. Although about 20 oocytes are stimulated to begin dividing by meiosis each month, only one com-

Figure 19–9
Oogenesis, which results in the development of a single oocyte capable of being fertilized and three nonfunctional polar bodies from a single ovarian cell.
What are the products of meiosis I in the female?

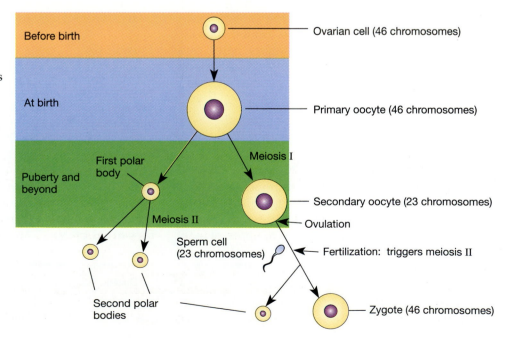

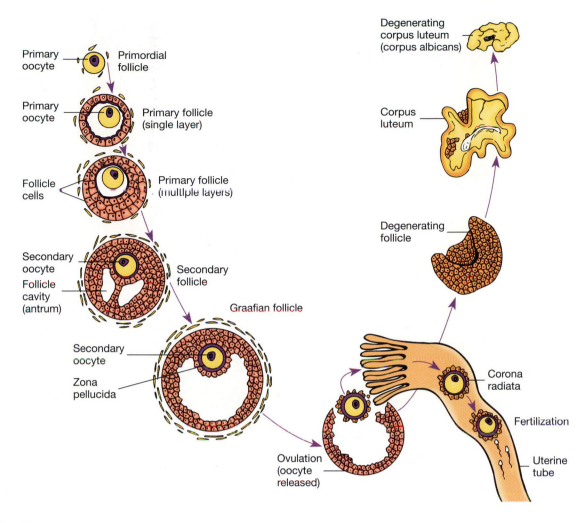

Figure 19–10
Follicle development, ovulation, and corpus luteum development. Follicle and corpus luteum development take place within the cortex of the ovary. Ovulation is the complete release of the secondary oocyte from the ovary. In most cases, the released oocyte is swept into the uterine tube, where fertilization may occur if sperm cells are present.

After ovulation, the Graafian follicle transforms into what structure?

pletes division. The result of the first meiotic division is two unequally sized daughter cells: a large **secondary oocyte** and a smaller **first polar body** (Fig. 19–9). The first polar body usually proceeds immediately into the second division of meiosis to produce two very small cells (called **second polar bodies**). These cells soon die and are reabsorbed. On the other hand, the secondary oocyte does not yet begin another division. Its further development is arrested, for at this point it is ready to be fertilized by a sperm cell. Should this occur, the fertilized secondary oocyte will quickly undergo its second meiotic division to produce an additional **second polar body** and a **zygote**. The zygote represents the fertilized egg, and the second polar body will degenerate. In time, the zygote will give rise to an embryo, which will be described in the next chapter.

Follicle Development

Throughout childhood, primordial follicles are in a state of suspended animation. As hormone levels (FSH from the pituitary gland) begin to rise and puberty begins, a number of changes occur (Fig. 19–10). First the flattened cells of the follicle begin to enlarge. At this point, the follicle is called a **primary follicle**. This is quickly followed by enlargement of the primary oocyte as it prepares for its first meiotic division. The follicle cells then begin to rapidly divide by mitosis, forming a multiple-layered arrangement. As a result, the primary follicle increases in size also. A cavity called the

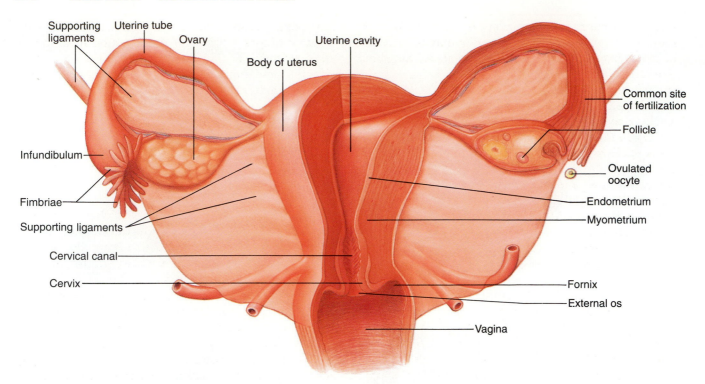

Figure 19–11
Female reproductive organs, frontal view. The ovary, uterine tube, uterus, and vagina are cut on the left side to show their internal anatomy.

What tube transports the oocyte to the uterus after ovulation?

antrum between the oocyte and follicle cells soon appears; it fills with a clear fluid as it expands. Once this cavity has formed, the same follicle is called a **secondary follicle**. As time proceeds, the secondary follicle continues to enlarge until it reaches a diameter of about 10 mm (a little less than $\frac{1}{2}$ inch). Its follicle cells undergo further change and begin to secrete the primary female sex hormone, **estrogen**. These follicle cells also secrete a substance that forms a transparent membrane around the oocyte, called the **zona pellucida**.[15] By this time, the size of the follicle causes it to bulge out from the surface of the ovary like a blister, and it will soon be ready for ovulation. The process of development thus far has taken about ten days, and the mature follicle that has formed is now called the **Graafian follicle**.

This entire sequence of events is repeated every month, on an approximately 28-day cycle beginning with puberty and ending at about the age of 50. Because only one follicle is permitted to mature at a time,

only about 400 to 500 of the 400,000 or more follicles in the two ovaries are prepared for ovulation in a lifetime.

Ovulation

After about 14 days of development, the Graafian follicle has matured to the point of being ready for expulsion out of the ovary. This process, known as **ovulation**, begins when the wall of the ovary at the site of the ballooning follicle ruptures (Figs. 19–10 and 19–11). The secondary oocyte is expelled into the peritoneal cavity as a result of this eruptive event, which some women can feel as a twinge of pain in the lower abdomen. This sensation is thought to be caused by stretching of the ovarian wall as it bursts. The expelled oocyte remains surrounded by some of its protective follicle cells, which now form an outer mantle called the **corona radiata**.[16] Normally, the expelled oocyte

---

[15]Zona pellucida: Latin *zona* = band or layer + *pel- (per)* = through + *lucida* = shining; thus, transparent or clear layer.

[16]Corona radiata: Latin *corona* = crown + *radiata* = spreading out in a circle.

and its corona radiata are swept by currents of fluid into a uterine tube where fertilization may occur. If the oocyte is not fertilized, it will degenerate within about five days.

After ovulation, most of the follicle cells remain within the ovary. They soon become transformed into a yellowish, glandular mass called the **corpus luteum**[17] (LOO-tē-um), which is shown in Figures 19–8 and 19–10. Cells of the corpus luteum begin to secrete two hormones: progesterone and estrogen. If the oocyte becomes fertilized and pregnancy occurs, the corpus luteum will enlarge. Its output of hormones maintains the state of pregnancy for the first three months, until the placenta takes over. If pregnancy does not occur, the corpus luteum persists for about 10 to 12 days before it degenerates.

CONCEPTS CHECK

1. At the onset of puberty, which meiotic division occurs in oogenesis?

2. At which point does a follicle begin secreting estrogen?

3. What happens to follicle cells in the ovary after ovulation?

Female Accessory Organs

The accessory organs of the female reproductive system include structures that play a role in fertilization, development of offspring, and the birth process. They are the uterine tubes, the uterus, and the vagina.

Uterine Tubes

The paired **uterine tubes** are also called the **fallopian tubes**, or oviducts. Each is located along one side of the pelvic cavity wall, extending between an ovary and the centrally located uterus (Fig. 19–7). They are about 10 cm (4 inches) long and about 1 cm in diameter, and are internally lined with a ciliated mucous membrane. At the end near the ovary, they widen to form a funnel-shaped part that opens directly into the peritoneal cavity (Fig. 19–11). This is the **infundibulum**[18] (in-fun-DIB-yoo-lum), which is fringed along its outer margin with fingerlike projections called **fimbriae**[19] (FIM-brē-ē). The fimbriae nearly touch the surface of the ovary as they wave about in the peritoneal fluid. They serve to collect the oocyte during ovulation, and

their cilia sweep it into the uterine tube. Fertilization usually occurs within the uterine tube near the infundibulum. If fertilization is successful, the actions of the cilia combined with contractions of the tube's muscular wall transport the developing embryo to the uterus.

Uterus

The **uterus** is a pear-shaped organ about the size of a woman's clenched fist. It is located near the floor of the pelvic cavity between the urinary bladder and the rectum (Fig. 19–7). It is a hollow, thick-walled structure that provides support for the developing embryo and fetus. Its powerful contractions provide the force necessary to push the fetus out during birth, and it is also the site of menstruation.

The uterus does not rest upon the floor of the pelvis, but is suspended by ligaments that tilt it downward into the vagina. It consists of two main regions: the upper, dome-shaped **body**, which receives the uterine tubes; and the lower, narrow **cervix**[20] (Fig. 19–11). Internally, the space within the body is the **uterine cavity**, and that within the cervix is the **cervical canal**. The cervical canal opens into the vagina by way of the **external os**.

The wall of the uterus consists of three distinct layers: a serous layer, a muscular layer, and a vascular layer. The **serous layer** is outermost and consists of the thin visceral peritoneum. The thick muscular layer, or **myometrium**, makes up the bulk of the uterine wall. It is composed of smooth muscle and elastic connective tissue. The vascular layer is called the **endometrium**; it forms the inner lining of the uterus. It consists of epithelium underlaid with a highly vascularized connective tissue, which contains numerous mucous glands. Each month the superficial layers of the endometrium are sloughed off during menstruation.

Vagina

The **vagina**[21] is the organ that receives the penis during intercourse. Often called the *birth canal*, it also provides a passage during childbirth and menstruation. It is a thin-walled tube, about 8 to 10 cm (4 to 5 inches) long, that extends from the cervix of the uterus to the outside (Figs. 19–7 and 19–11). At the end that receives the cervix, its walls curve around to form shallow pockets known as the **fornix**.

The wall of the vagina consists of an outer layer of smooth muscle and an inner mucous membrane.

---

[17]**Corpus luteum:** Latin *corpus* = body + *luteum* = yellow.

[18]**Infundibulum:** Latin *infundibulum* = funnel.

[19]**Fimbriae:** Latin *fimbriae* = fringes.

[20]**Cervix:** Latin *cervix* = neck.

[21]**Vagina:** Latin *vagina* = sheath.

The muscle layer allows the vagina to increase in size, to accommodate the penis during intercourse and passage of a child during childbirth. The mucous membrane contains stratified epithelium to provide protection. It lacks mucous glands, so the vagina relies upon secretions from the uterine tubes, the uterus, or external glands called *vestibular glands* to provide lubrication.

The opening to the outside is called the **vaginal orifice**. In young females the mucosa may extend across this opening, forming a thin barrier called the **hymen** (HĪ-men). The hymen contains blood vessels, so it tends to bleed when it is first penetrated or ruptured in some way. Its rupture may be caused by tampon insertion, a physical activity such as exercise or sports, pelvic examination, or sexual intercourse.

CONCEPTS CHECK

1. What part of the uterine tube collects the ovulated oocyte?

2. What is the narrow portion of the uterus called?

3. What is the inner lining of the uterus called?

External Genitalia

The female reproductive structures that are located external to the vagina are the female external genitalia. They include structures that are collectively called the **vulva**[22] (VUL-va) and are shown in Figure 19–12.

The upper border of the vulva is marked by an elevated, rounded area over the symphysis pubis known as the **mons pubis** (monz PYOO-bis). After puberty, this area is covered with pubic hair. Extending from the mons pubis in a downward direction are two narrow folds of skin that are also covered with hair after puberty, called the **labia majora**[23] (LĀ-bē-a ma-JOR-a). The labia majora are the female counterpart, or *homologue*, to the male scrotum (that is, they are derived from the same embryonic tissue). Medial to the labia majora are two thin, hair-free folds called the **labia minora** (mi-NOR-a). They are covered with mucous membrane that is richly supplied with oil glands.

The labia minora form the outer margins of a region called the **vestibule**,[24] which contains, from the

---

[22]Vulva: Latin *vulva* = a wrapping or covering.
[23]Labia majora: Latin *labia* = lips + *majora* = larger.
[24]Vestibule: *vestibule* = outer chamber.

Figure 19–12
Female external genitalia.
The female vulva contains what structures?

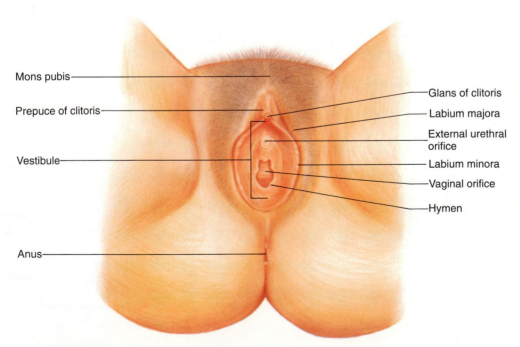

Mons pubis
Prepuce of clitoris
Vestibule
Anus
Glans of clitoris
Labium majora
External urethral orifice
Labium minora
Vaginal orifice
Hymen

upper (anterior) to the lower (posterior) end: the clitoris (kli-TOR-is), the external urethral orifice, and the vaginal orifice. The **clitoris** is the female homologue to the penis. Like the penis, it is composed of erectile tissue and has an anterior end (also called the **glans**) richly supplied with sensory nerve endings that respond to touch. However, there is no corpus spongiosum in the clitoris, and the urethra does not pass through it as it does in the penis. Covering part of the clitoris is a hoodlike fold formed by the union of the labia minora, called the **prepuce**.

Near the vaginal orifice are numerous glands whose mucus secretions help to moisten and lubricate the vestibule and vagina. Known as **vestibular glands**, they are homologues to the bulbourethral glands in the male.

The region between the upper border of the vestibule and the anus is the **perineum** (par'-i-NĒ-um). The soft tissues of the perineum overlie the pelvic outlet. For this reason, the perineum sometimes tears between the vaginal opening and the anus during childbirth. To prevent this, cutting the perineum to widen the birth canal opening is often performed during childbirth, in a procedure known as *episiotomy*.

Mammary Glands

The **mammary glands** are the organs that produce milk for infant nourishment. Located in the breasts, or *mammae*, they consist of modified sweat glands.

The breasts of both males and females contain an external, heavily pigmented **areola** that surrounds a centrally elevated **nipple** (Fig. 19–13). The areola and nipples are very sensitive to touch, and contain smooth muscle fibers that cause them to become erect when stimulated by touch, cold temperature, or sexual arousal. In children, breast structure is similar among males and females, and as in adult males, a rudimentary glandular system is present. Under the influence of increasing levels of estrogen and progesterone, the breasts begin to enlarge by accumulating large amounts of fat. Also, the glandular system develops.

Each adult female mammary gland consists of 15 to 20 **lobes** that radiate around the nipple. The lobes are padded with fat, and ligaments are formed between them that attach the breast to underlying muscle tissue. Within each lobe are smaller chambers called **lobules**, which contain **alveolar glands**. The alveolar glands produce milk when a woman is lactating (the

Figure 19–13
Structure of the mammary glands. (a) Anterior view of the left breast with its right side cut away. (b) Lateral view of the breast, sagittal section, to show its internal anatomy.
What type of tissue occupies most of the breast volume (in an adult female)?

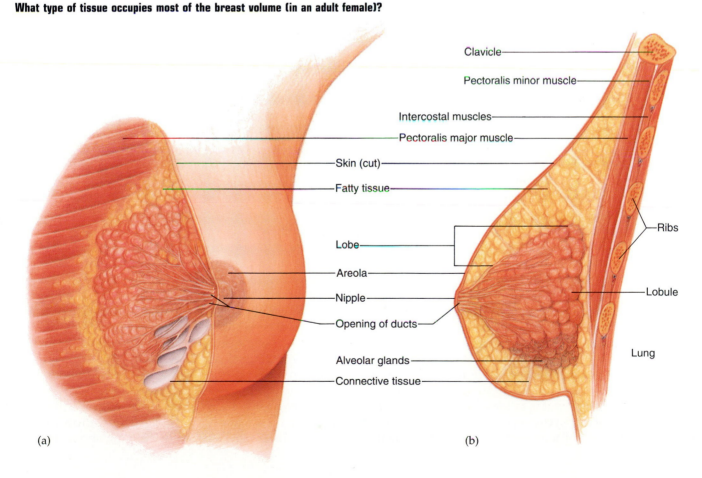

Clavicle
Pectoralis minor muscle
Intercostal muscles
Pectoralis major muscle
Skin (cut)
Fatty tissue
Lobe
Areola
Nipple
Opening of ducts
Alveolar glands
Connective tissue
Ribs
Lobule
Lung

(a)   (b)

process of milk production, which will be discussed further in the next chapter). It is secreted into ducts that open to the outside at the nipple.

PHYSIOLOGY OF FEMALE REPRODUCTION

The female system functions to produce oocytes on a regular, monthly cycle for the purpose of fertilization. This cyclic process is regulated by hormones, while neural mechanisms govern the sexual response.

Like male reproduction, female reproduction is influenced by neural and hormonal mechanisms. Hormones have the greatest influence, for they control reproductive organ development, ovulation, and menstruation and maintain pregnancy. Neural mechanisms govern the physiological responses to sexual stimulation.

Neural Mechanisms

The vulva contains numerous structures that receive an abundant supply of sensory nerves. These nerves respond to touch by sending sensory impulses to the spinal cord. An involuntary reflex arc quickly routes motor impulses along parasympathetic fibers from the cord to the vulva, where erectile tissues soon dilate in response. This is particularly true of the highly sensitive clitoris, which enlarges slightly as its erectile tissues fill with blood when it is stimulated. The vagina responds also, as its walls swell with blood to prepare it for the entry of the penis. The parasympathetic impulses also stimulate secretion of mucus by the vestibular glands for lubrication.

The combined effect of stimulating the sensitive tissues of the vulva and vagina leads to female **orgasm**. Like the male orgasm, it is accompanied by a feeling of intense pleasure that is followed by complete relaxation. A surge of psychological and physical warmth may also be part of the experience. The attainment of

orgasm need not require sexual intercourse, and the ease with which it is experienced varies among individuals. As orgasm begins, a series of reflexes are stimulated that involve peristaltic contractions of muscles associated with the perineum, the walls of the vagina, and the walls of the uterus and uterine tubes. These contractions promote further sexual stimulation, and may also assist the movement of sperm through the female tract.

Hormonal Mechanisms

Hormones influence the female reproductive system by stimulating organ development, and by regulating the cycles that characterize female reproduction. They arise from the hypothalamus, the anterior pituitary gland, and the ovaries.

Development of Reproductive Cycles and Puberty

Throughout childhood, the ovaries grow at a slow rate and continuously secrete small amounts of estrogen. As the child nears puberty, the maturing hypothalamus begins to release gonadotropin-releasing hormone (GnRH). Once released, GnRH diffuses to the anterior pituitary gland and triggers it to secrete follicle-stimulating hormone (FSH) and luteinizing hormone (LH), which target the ovaries. FSH acts upon the primary follicles of the ovary, causing them to begin their process of maturation. As a result of their development, the follicle cells produce increasing levels of estrogen and progesterone. LH promotes their secretion yet further. The increased estrogen and progesterone levels present in the bloodstream for the first time bring about the first episode of menstrual bleeding, or **menarche**, as well as puberty.

Puberty in a young female brings about many body changes, just as it does in a young male. However, the effects of estrogen are quite different from those of testosterone. Estrogen stimulates the enlargement of the vagina, uterus, uterine tubes, and external genitals. It also promotes the accumulation of fat in the breasts, thighs, and buttocks, and stimulates development of the duct system within the mammary glands. In addition, the absorption of calcium and phosphorus into certain bones is affected by estrogen, leading to a wider pelvis and narrow shoulders. These are the female *secondary sexual characteristics*; they are maintained by the presence of estrogen. They are summarized in Table 19–2, along with the effects of other hormones on female reproduction.

The ovaries are also the major source of the second female sex hormone, progesterone. Progesterone works along with estrogen in bringing about changes

Table 19-2   EFFECTS OF HORMONES ON FEMALE REPRODUCTION

Hormone	Primary Source	Effects
GnRH	Hypothalamus	Stimulates anterior pituitary gland to secrete LH and FSH
LH	Anterior pituitary gland	Promotes the development of primary follicles to secondary follicles in the ovaries; high levels induce ovulation
FSH	Anterior pituitary gland	Stimulates the development of primary follicles to secondary follicles in the ovaries
Estrogen	Follicles in the ovaries	Stimulates growth of reproductive organs, regeneration of endometrium lining the uterus, and development of the following secondary sexual characteristics: 1. Accumulation of fat deposits in the breasts, hips, and buttocks 2. Development of bone and muscle to form narrow shoulders and wide hips 3. Development of the ducts and glands in the breasts
Progesterone	Corpus luteum in the ovaries and, during pregnancy, the placenta	Promotes regeneration of endometrium lining the uterus, and during pregnancy helps maintain the uterus and prepares the breasts for lactation

in the uterus during the menstrual cycle. During pregnancy, it is produced by the placenta and assists in maintaining the uterus and preparing the breasts for lactation.

Menstrual Cycle

The **menstrual**[25] (MEN-stroo-al) or **uterine cycle** is a series of cyclic changes that occur in sexually mature females until about the age of 50. It affects the endometrium of the uterus on a monthly basis, and occurs in response to changes in levels of ovarian hormones circulating in the blood. It culminates in menstrual bleeding, or **menses**, a mild hemorrhage caused by the sloughing off of part of the endometrium and its exit from the uterus. A typical menstrual cycle is about 28 days, although the cycle may vary considerably from this average.

As shown in Figure 19–14, the first day of menses marks day 1 of the menstrual cycle. After 3 to 4 days of menses, rising levels of estrogen stimulate the regeneration of the endometrium, which continues for about 14 days. This period of growth (proliferation of the endometrium) is called the **proliferative phase**. Estrogen levels rise during this period because the presence of FSH and LH stimulates the development of follicles in the ovary, which release estrogen as they reach maturity. Estrogen acts on the cells of the uterus, causing them to rapidly divide. Consequently, the endometrium gradually thickens.

As estrogen levels in the blood continue to increase, GnRH secretion by the hypothalamus is also increased by positive feedback. GnRH in turn stimulates more FSH and LH from the pituitary gland, and estrogen secretion increases further in response. At about day 14, LH and FSH levels surge dramatically upward in response to GnRH. The sudden rise in LH stimulates ovulation. Thus, ovulation usually occurs on or about day 14 of the menstrual cycle, or about 10 days after menses stop.

Following ovulation, the follicle cells are converted to a corpus luteum within the ovary, which begins to secrete progesterone and small amounts of estrogen. As a result, progesterone levels begin to rise and estrogen levels decline. Progesterone inhibits continued LH and FSH secretion, so their levels decline also. Progesterone also promotes further development of the endometrium by stimulating cell growth and the secretion of nourishing fluid. For this reason, the period of time between ovulation and the next menses is called the **secretory phase**. About 7 to 8 days after ovulation (about 21 or 22 days of the cycle), the endometrium has become fully prepared to receive the early embryo.

If the oocyte released in ovulation is not fertilized, the corpus luteum begins to degenerate at about day 24 of the cycle. As a result, progesterone and estrogen levels fall rapidly, causing arterioles in the endometrium to constrict. This reduces the supply of blood to the thickened uterine lining, causing damage to capillary walls. Blood leaks from the damaged capillaries and pools beneath the outer layer of the endometrium. Gradually, the outer layer of the en-

---

[25]**Menstrual: Latin** *menstrualis-* = monthly.

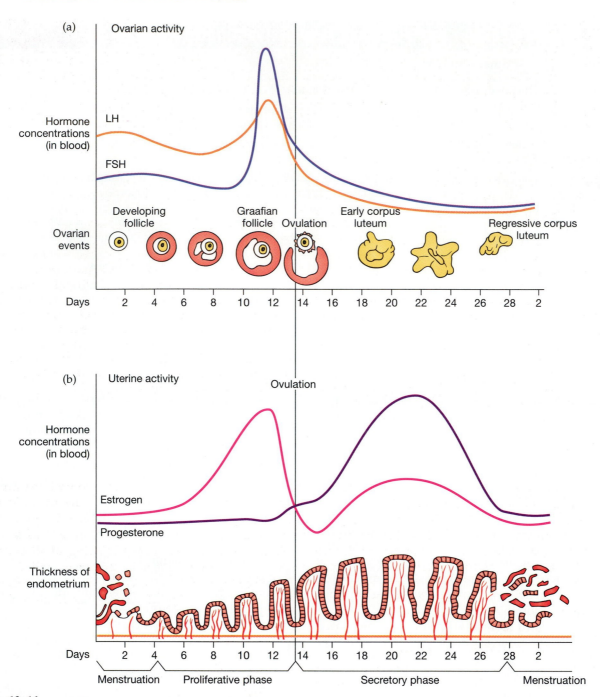

Figure 19–14
The menstrual cycle. Hormonal changes, follicle changes in the ovary, and changes occurring in the endometrium of the uterus are depicted in this illustration relative to each other. (a) Changes occurring in the ovaries. (b) Changes occurring in the uterus.

Approximately how many days from the last day of menstruation does it normally take for ovulation to occur?

Menstrual Cramps and Premenstrual Syndrome

Menstrual cramps, or *dysmenorrhea*, are usually the result of contractions of the uterus that occur before and during menses. The pain that is often experienced during these contractions is the result of an overproduction of prostaglandins. These hormonelike chemicals cause excessive uterine contractions and inflammation in local tissues, which lead to pressure on sensory nerves and its accompanying pain. Prostaglandin action on smooth muscles in other areas of the body can also result in headache, nausea, and vomiting. These powerful chemicals are produced by secretory cells in the uterus during degeneration of the endometrium before its outer layers are sloughed off. Fortunately for many women, the pain and discomfort caused by high levels of prostaglandins can be relieved by taking aspirin and other, similar anti-inflammatory drugs. These chemicals inhibit prostaglandin production. Also, many women experience a reduction in the pain of dysmenorrhea with increasing age, or after pregnancy. However, some women do not gain relief through drugs, increasing age, or pregnancy, for reasons that are not yet understood.

Some women suffer from a condition that has a completely different source from that of menstrual cramps. This is *premenstrual syndrome*, or *PMS*, and is currently thought to be caused by imbalances in levels of estrogen and progesterone during the latter half of the menstrual cycle. Its symptoms include irritability, depression, headache, tiredness, and sore breasts; the condition usually affects women 7 to 10 days before the onset of menses. Treatment includes a proper diet, aspirin, and sedatives, and progesterone therapy is promising.

---

dometrium separates from the uterine wall, creating a flow of blood and degenerating tissue that flows out of the uterus (only about 20 to 200 ml of blood is lost with each menstrual flow). Thus, by about day 28, menses begin and a new menstrual cycle starts all over again.

Menopause

By the age of about 45 or 50 years, a woman's menstrual cycles have usually become less regular and ovulation may not occur each time. This period in life is referred to as the *female climacteric*. About five to seven years later the menstrual cycle will stop completely, at which point the woman is said to have reached **menopause**.

Cessation of the menstrual cycle is caused by changes within the ovary. The follicles that remain are less sensitive to the influences of FSH and LH. Consequently, fewer follicles are stimulated to mature and to secrete estrogen, and fewer corpora lutea are produced to secrete progesterone. The decreased levels of estrogen and progesterone result in a decrease in size of the uterus, a reduction in lubricating fluid secretion by vestibular glands, and a thinning of all epithelial tissues including mucous membranes and the skin. These changes are often accompanied by "hot flashes," which are caused by the sudden vasodilation of blood vessels in the skin.

CONCEPTS CHECK

1. What neural events lead to female orgasm?

2. What is the role of GnRH in developing the first menstrual cycle and puberty?

3. A sudden increase in which hormone stimulates ovulation?

HEALTH CLINIC

Control of Pregnancy: Contraception

The prevention of pregnancy is termed **contraception**. There are many techniques and products available to provide contraception; however, none are completely effective except total abstinence (absence of sex) or sterility. Of the 3 million unwanted pregnancies in the United States each year, about half a million or more occur despite the use of a contraceptive device. The various methods of contraception include barrier methods, prevention of implantation, chemical methods, and surgical sterilization.

Barrier methods include the use of a condom (Fig. a) by the male, or a diaphragm (Fig. b), cervical cap, or vaginal sponge (Fig. c) by the female. The condom is a sheath of animal membrane, rubber, or plastic that is placed over the penis. If it is properly used, a condom may not only prevent unwanted pregnancy, but also protect from sexually transmitted diseases (STDs). The diaphragm is a cup of rubber or plastic that is inserted into the vagina over the external os. A cervical cap is a smaller version of the diaphragm, and a vaginal sponge is a rubber sponge that is inserted into the vagina as well. Each of these three devices should be used in conjunction with spermicidal foams, creams, or jellies, which kill sperm cells on contact (Fig. d). Spermicidal foams, creams, and jellies should not be used exclusively of any other contraceptive method, because they have a high incident rate of failure (about 20%) when used on their own. However, when used as backup to barrier methods, the rate of failure is much lower (less than 10%).

Implantation prevention does not prevent fertilization, as do the barrier methods, but instead keeps the developing embryo from implanting in the uterine wall. This is accomplished by intrauterine devices, or IUDs, which must be surgically inserted into the uterine cavity by a trained professional. Although IUDs are effective in preventing pregnancy (they have a failure rate of 4%), they have caused serious complications such as perforation of the uterus. Their use is becoming less frequent in the United States.

Chemical methods include the use of oral contraceptives (Fig. e), such as synthetic estrogen or progesterone or both. Often called "the pill," these chemicals inhibit LH and FSH secretion from the anterior pituitary gland, thereby preventing the LH surge that causes ovulation. Once ovulation is blocked, fertility is suppressed. Currently, the pill boasts the lowest failure rate of all contraceptive methods other than abstinence or surgery (2.5%). Unfortunately, estrogen and progesterone have bad side effects on some individuals, such as an increased risk of heart attack and stroke in smokers and hypertensive individuals. For this reason, other oral contraceptives are currently being tested for use as alternatives. One such alternative is the so-called minipill, which prevents pregnancy by affecting the uterine lining and cervical mucus. Another potential alternative is the "morning-after" pill, which is a drug introduced by a French pharmaceutical company in the 1980s. Called RU 486, it is used legally in Europe to induce abortion during the first trimester (first three months) of pregnancy. Its use in contraception is due to a recent discovery that, if taken orally the day after sexual intercourse, it will prevent pregnancy, although its mechanism of action is not clearly understood (some researchers have found that RU 486 suppresses ovulation, while others claim that it prevents implantation of a fertilized egg). The use of RU 486 in the United States in any form has not yet been approved by the Federal Drug Administration (FDA).

Surgical methods of contraception are the most reliable, but also the most permanent. They are not recommended for individuals who are considering having children at a later time in life. The male sterilization technique is the *vasectomy*. This procedure is very simple and painless, and does not affect performance of sexual intercourse or other sexual behaviors. As shown in Figure f, it involves cutting the ductus deferens and tying it off, or cauterizing the ends. Female sterilization is a more difficult procedure (Fig. g). The most common method is *tubal ligation*, in which the uterine tubes are cut and tied or cauterized by means of an incision made through the abdominal wall. This procedure closes off the pathway between inserted sperm cells and the oocyte. Another method is a surgical procedure called *abortion*. The most common type of abortion involves the insertion of an instrument through the cervix into the uterine cavity. The instrument scrapes the endometrial surface, and a strong suction pulls the disrupted endometrium and embryo or fetus out. Abortions are usually used in pregnancies that have progressed less than six months. However, controversy shadows over the abortion procedure at present because of the moral opinion of some that it violates the rights of the unborn.

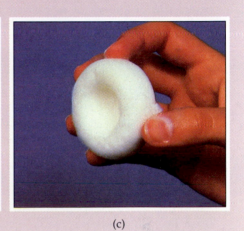

(a)

(b)

(c)

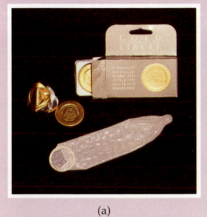

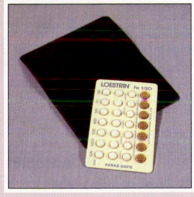

Examples of contraceptive
methods and devices.
(a) Condoms. (b) Diaphragm
with spermicidal jelly.
(c) Vaginal sponge.
(d) Spermicidal foam.
(e) Oral contraceptives.
(f) Vasectomy procedure.
(g) Tubal ligation procedure.

(d)

(e)

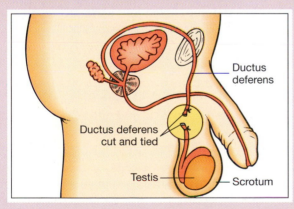

Ductus
deferens

Ductus deferens
cut and tied

Testis

Scrotum

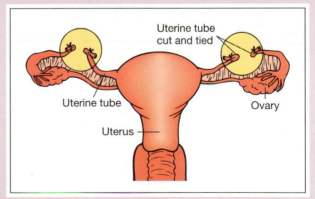

Uterine tube
cut and tied

Uterine tube

Ovary

Uterus

(f)

(g)

CLINICAL TERMS OF THE REPRODUCTIVE SYSTEM

Amenorrhea The absence of menses, or menstrual flow. It is usually caused by a disturbance in hormone levels.

Breast cancer The most common malignant tumor in women. Early diagnosis and treatment improve the prognosis of this dangerous disease. Diagnosis involves finding a lump or discovering the tumor on a mammogram (X-ray of the breast). Treatment includes surgery (often, an operation called a *mastectomy*), examination of axillary lymph nodes for spreading, and chemotherapy and radiation if needed. There appears to be a genetic predisposition, and if it is known that a woman's family has been afflicted, frequent mammograms are encouraged.

Cancer of the prostate A relatively common malignancy, causing enlargement of the prostate gland and leading to painful obstruction of urine flow. The cancer often spreads to pelvic and vertebral bones if not treated early.

Cervical cancer A carcinoma, or cancer arising from epithelial cells. It is highly metastatic, spreading to lymph nodes and on to other tissues quickly. It can arise from an irritation of the cervix due to sexually transmitted disease, insertion of IUDs, and even frequent intercourse with multiple partners.

Cryptorchidism A developmental defect in males characterized by the failure of the testes to descend into the scrotum.

Dilation and curettage (D&C) A surgical procedure in which the cervix is dilated and the endometrium of the uterus is scraped. It may be used in an attempt to regulate menstrual cycles or to remove degenerative tissue.

Endometriosis A condition resulting in severe dysmenorrhea (pain during menstruation), sterility, and painful intercourse, which strikes 1 in 15 women of reproductive age. It is caused by migrations of endometrial tissue that implant into the uterine wall, ovaries, or uterine tubes. This tissue mimics the changes of normal endometrium during the menstrual cycle, and can lead to peritonitis and death. It usually requires surgical removal of the implants and hormone therapy.

Fibroadenoma A common benign tumor that occurs in the breasts of women under the age of 35. It has no association with the development of breast cancer.

Fibrocystic disease Abnormal growth of breast tissue resulting in the formation of cysts, which feel like small pebbles beneath the skin layer. It is associated with breast cancer development.

Gonorrhea A sexually transmitted disease caused by the bacterium *Neisseria gonorrhoeae*. Symptoms include inflammation of the reproductive tract mucous membrane. It is eliminated effectively with antibiotics.

Herpes A sexually transmitted disease caused by a virus known as *herpes simplex genitalis*. It causes ulcers and vesicles on mucous membranes of the reproductive tract and on external genitalia during a periodic eruptive phase. It can be transmitted only during its eruptive phase. In females, it increases risk of cervical cancer.

Hysterectomy Now more commonly known as a uterectomy; the surgical removal of the uterus.

Ovarian benign tumors and cysts Benign growths that occur in the ovaries. They normally occur and regress quickly, providing no complications. Dermoid cysts arise from unfertilized oocytes and often contain hair, teeth, bone, brain, and other types of disorganized tissue.

Pelvic inflammatory disease (PID) An infection that involves the uterine tubes and adjacent tissues. It may be caused by bacterial infection, sexually transmitted diseases, or IUD insertion. It results in scarred tissue—even following successful treatment—which often leads to sterility.

Syphilis A sexually transmitted disease caused by the bacterium *Treponema pallidum*. Symptoms occur in stages, which begin with a sore on the external genitals or mouth and lead into skin rash, lesions on internal organs, and ultimately destruction of the nervous system. Treatments are successful with antibiotics.

Toxic shock syndrome A disorder caused by a toxin produced by a bacterium, *Staphylococcus aureus*. It usually occurs in menstruating women who use high-absorbency tampons, but it can also occur in other women and in men as a result of exposure to the bacterium. Symptoms include fever, vomiting, diarrhea, muscle pain, a drop in blood pressure, and skin rash. Treatment is directed toward relief of symptoms until the effects of the toxin wear off.

Trichomoniasis A sexually transmitted disease caused by the protozoan *Trichomonas vaginalis*. It causes an inflammation of the vaginal mucous membrane, and of the urethra in males. Men may be without symptoms yet still be able to transmit the disease.

Vaginitis Inflammation of the vaginal mucous membrane, usually caused by infection by a fungus (*Candida albicans*), a protozoan parasite (*Trichomonas vaginalis*), or a bacterium (*Gardnerella vaginalis*).

CHAPTER SUMMARY

The overall function of the reproductive system is the production of offspring.

ORGANS OF MALE REPRODUCTION

Testes The primary male organs, which produce sperm and testosterone and are therefore referred to as the male gonads. The two testes are located in the scrotum outside the abdominopelvic cavity.

Structure Each testis contains tightly packed seminiferous tubules. The germinal epithelium of the tubules produces sperm cells by spermatogenesis. Between the tubules are interstitial cells, which secrete testosterone. The tubules unite with the rete testis, and then with efferent ductules, which carry sperm out of the testis.

Spermatogenesis Daughter cells of spermatogonia begin dividing by meiosis at puberty, and continue it throughout life. They first produce primary spermatocytes, which divide again to form spermatids. Spermatids mature into sperm cells within the seminiferous tubules and epididymides.

Ducts The ducts carry sperm from the testes to the urethra.

Epididymis Consists of a single, tightly coiled tube that is shaped like a comma on the posterior side of each testis.

Ductus Deferens Consists of a single, thickened tube that extends from the epididymis through the scrotum, into the pelvic cavity, and around the urinary bladder to empty into the ejaculatory duct. As it travels through the scrotum it passes with other structures to form the spermatic cord.

Urethra Extends from the urinary bladder to the outside, passing through the prostate gland and penis before it emerges.

Accessory Glands Three glands secrete fluids that contribute to semen and serve to support the sperm in a liquid, nutritious broth.

Seminal Vesicles Located at the base of the urinary bladder. They contribute most of the liquid part of semen. This secretion is rich in sugar.

Prostate Gland Encircles the urethra as it exits from the urinary bladder.

Bulbourethral Glands Paired glands located below the prostate gland.

External Genitalia

Scrotum A saclike organ that contains the testes.

Penis A cylindrical organ that is composed of three columns of erectile tissue: two corpora cavernosa and a single corpus spongiosum. The urethra passes through the latter. The glans forms a distal cap over the penis, and the prepuce covers it.

PHYSIOLOGY OF MALE REPRODUCTION

Neural Mechanisms With the presence of sensory stimuli, parasympathetic impulses cause the arteries to the penis to dilate, leading to an increase in blood flow that compresses the veins. This leads to accumulation of blood in the erectile tissues, and the penis becomes erect. Sympathetic impulses soon pass to the ducts, stimulating them to move sperm through by peristalsis. This is called emission. Sensory impulses soon detect the filling of the urethra and motor impulses are routed, causing skeletal muscles at the base of the penis to contract. This results in ejaculation. Orgasm is the pleasurable experience accompanying the last of these events.

Hormonal Mechanisms

Hypothalamic and Pituitary Influences The hypothalamus secretes GnRH as the male approaches puberty. GnRH stimulates the anterior pituitary gland to secrete LH and FSH. LH stimulates development of interstitial cells, and stimulates the release of testosterone from them. FSH prepares the testes to become responsive to testosterone for spermatogenesis.

Testosterone In the presence of FSH, testosterone stimulates spermatogenesis. Testosterone also stimulates the development of secondary sexual characteristics. Testosterone production is regulated by negative feedback control.

ORGANS OF FEMALE REPRODUCTION

Ovaries Paired, oval structures within the abdominopelvic cavity that produce estrogen and oocytes and are, thus, the female gonads.

Structure Each ovary is covered by a layer of visceral peritoneum, a germinal epithelium, and a layer of tough fibrous connective tissue. Internally, it consists of a cortex and a medulla. The cortex contains ovarian follicles in various stages of development.

Oogenesis Primary oocytes are present at birth, and are surrounded by a layer of cells to form the primordial follicles. The first meiotic division does not occur until puberty, at which time the primary oocyte divides into a secondary oocyte and a small first polar body. The secondary oocyte is ready for fertilization; if it occurs, the oocyte undergoes its second meiotic division to produce a zygote and a second polar body.

Follicle Development At puberty, the primordial follicles enlarge to form primary follicles. Under the influence of FSH, about 20 of these enlarge further each month to form secondary follicles, which contain multiple layers of follicle cells and a fluid-filled cavity. As the secondary follicles continue to mature, they attain the ability to secrete estrogen. Only one secondary follicle will reach full maturity each month; it is called a Graafian follicle.

Ovulation The process of expulsion of a Graafian follicle from the ovary into the peritoneal cavity. The follicle cells that remain with the oocyte transform into the corona radiata, and those that remain in the ovary transform into the corpus luteum, which secretes progesterone and estrogen.

Female Accessory Organs

Uterine Tubes Also called fallopian tubes or oviducts. They serve as the site of fertilization and transport the embryo to the uterus. The expanded end near the ovary is the infundibulum; it contains finger-like projections called fimbriae.

Uterus Located near the floor of the pelvic cavity above the urinary bladder. Its cavity houses the developing embryo and fetus in pregnancy, and its inner lining is sloughed off during menstruation. Its main upper portion is the body, and the narrow lower portion is the cervix. The space within the body is the uterine cavity, and that within the cervix is the cervical canal.

Vagina A thin-walled tube between the uterus and the outside. It receives the penis during intercourse. Its muscular walls are lined with mucous membrane.

External Genitalia Also called the *vulva*. There are numerous structures:

Mons Pubis An elevated, rounded area over the symphysis pubis.

Labia Majora Narrow folds of skin covered with hair after puberty.

Labia Minora Narrow folds of mucous membrane extending medial to the labia majora. They border the area called the vestibule, which contains the clitoris, external urethral orifice, vaginal orifice, and vestibular glands.

Clitoris Located at the upper margin of the vestibule, it contains sensitive erectile tissue and is the homologue to the penis. It has a glans and a prepuce.

Vestibular Glands Mucus-secreting glands near the vaginal orifice in the vestibule.

Perineum The region between the lower border of the vestibule and the anus.

Mammary Glands The organs that produce milk. Their development is stimulated by estrogen and progesterone and results in fat deposits and development of the ducts and alveolar glands.

PHYSIOLOGY OF FEMALE REPRODUCTION

Neural Mechanisms Sensory stimuli received by receptors in the external genitalia result in parasympathetic impulses traveling from the spinal cord to the vulva, causing erectile tissues and the vaginal wall to fill with blood. These impulses also stimulate secretion of mucus by the vestibular glands. These events lead to the pleasurable experience of orgasm.

Hormonal Mechanisms

Development of Reproductive Cycles and Puberty As the hypothalamus matures in a growing girl, it begins producing GnRH. GnRH stimulates the production of LH and FSH, which promote the maturation of follicles in the ovaries. As they mature, the follicles begin producing higher levels of estrogen and progesterone, which lead to the first menstruation (menarche) and puberty. The presence of estrogen induces the changes that take place during puberty.

Menstrual Cycle   Averages about 28 days.
  Menses (menstruation)   The first 3 or 4 days of the cycle, caused by the sloughing off of the endometrium.
  Proliferative Phase   Rising levels of estrogen, in response to rising levels of FSH and LH that stimulate development of the follicles, cause regeneration of the endometrium. A peak in LH levels causes ovulation after about 14 days.

Secretory Phase   After ovulation, the corpus luteum develops and secretes progesterone and small amounts of estrogen, and LH and FSH are inhibited. The endometrium thickens further and is prepared for embryo implantation. If fertilization does not occur, the corpus luteum degenerates. As a result, progesterone and estrogen levels fall, stimulating menstruation after about 28 days.

Menopause   After age 45 or 50, the aging of tissues results in a decline of estrogen secretion. Estrogen secretion eventually stops completely about five to seven years later.

KEY TERMS

circumcision (p. 564)
ejaculation (p. 565)
emission (p. 565)
erection (p. 565)
estrogen (p. 570)
fertility (p. 571)

follicle (p. 569)
gamete (p. 557)
gonad (p. 557)
meiosis (p. 558)
menses (p. 575)
menstrual cycle (p. 575)

oocyte (p. 567)
oogenesis (p. 567)
orgasm (p. 565)
ovulation (p. 570)
puberty (p. 566)

semen (p. 562)
sperm (p. 558)
spermatogenesis (p. 558)
sterility (p. 561)
testosterone (p. 558)

QUESTIONS FOR REVIEW

OBJECTIVE QUESTIONS

1. Seminiferous tubules may be found within the:
   a. epididymis
   b. ductus deferens
   c. testes
   d. spermatic cord
2. Testosterone is produced by:
   a. germinal epithelium
   b. interstitial cells
   c. the ovaries
   d. the hypothalamus
3. During spermatogenesis, a primary spermatocyte undergoes_____ to result in four spermatids.
   a. meiosis
   b. fertilization
   c. mitosis
   d. enlargement
4. Spermatids that are newly formed will undergo maturation as they move slowly through the:
   a. seminiferous tubules
   b. rete testis
   c. epididymis
   d. all of the above
5. The spermatic cord consists of the testicular artery and veins, lymph vessels, the cremaster muscle, and:
   a. seminiferous tubules
   b. the epididymis
   c. the ductus deferens
   d. tendons
6. The tube that carries both urine and semen is the:
   a. ductus deferens
   b. ejaculatory duct
   c. urethra
   d. all of the above
7. The male accessory gland located at the base of the urinary bladder that releases much of the liquid part of semen is the:
   a. seminal vesicle
   b. bulbourethral gland
   c. prostate gland
   d. vestibular gland
8. Parasympathetic impulses traveling to the penis in response to sensory stimuli initially cause:
   a. erection
   b. emission
   c. the movement of semen
   d. ejaculation
9. The production of LH by the anterior pituitary gland directly causes:
   a. the production of GnRH
   b. the development of interstitial cells
   c. spermatogenesis
   d. all of the above

10. Testosterone production is regulated by:
    a. negative feedback   c. positive feedback
    b. FSH   d. sensory stimulation
11. Ovarian follicles are located within the:
    a. cortex of the ovaries
    b. uterine wall
    c. medulla of the ovaries
    d. germinal epithelium
12. The first meiotic division in oogenesis occurs:
    a. before birth
    b. each month in a child
    c. at puberty
    d. immediately after fertilization
13. The oocyte present within a Graafian follicle just prior to ovulation is called a:
    a. zygote   c. primary oocyte
    b. secondary oocyte   d. primordial follicle
14. Follicle cells within a secondary follicle secrete:
    a. estrogen   c. progesterone
    b. mucus   d. testosterone
15. Progesterone and small amounts of estrogen are secreted by a mass of modified follicle cells within the ovary, called the:
    a. Graafian follicle   c. corona radiata
    b. corpus luteum   d. placenta
16. Fertilization usually occurs within the:
    a. ovaries   c. uterine tube
    b. uterus   d. vagina
17. The inner lining of the uterus, which is sloughed off and expelled during menstruation, is the:

    a. endometrium   c. myometrium
    b. cervical canal   d. cervix
18. The component of the vulva that contains erectile tissue and is the homologue to the penis is the:
    a. labia majora   c. labia minora
    b. clitoris   d. vestibule
19. Menses, or menstruation, is caused by a decline in:
    a. sexual stimulation
    b. testosterone levels
    c. estrogen and progesterone levels
    d. FSH levels

ESSAY QUESTIONS

1. Describe the changes in semen composition as sperm travels from the testes to the outside. Indicate the ducts it passes through in its journey. What would differ in the case of a man who has undergone a vasectomy?
2. Describe the neural events that lead to orgasm in a male, and those that lead to orgasm in a female.
3. What is the relationship between GnRH and estrogen secretion in a prepubescent (before onset of puberty) female? What is this relationship in a sexually mature woman?
4. What would happen to a woman's menstrual cycle if estrogen production were suddenly stopped, such as would occur if the ovaries were surgically removed?
5. Describe the changes that occur in the female ovaries and mammary glands at puberty.

ANSWERS TO ART LEGEND QUESTIONS

Figure 19–1

What organ contains the testes?   The scrotum contains both testes.

Figure 19-2

What is produced by the interstitial cells (cells of Leydig)?   The male sex hormone testosterone is produced by interstitial cells.

Figure 19-3

What is the function of the sperm cell's flagellum?   The flagellum whips about to provide the sperm cell with mobility.

Figure 19-4

Can you trace the route of sperm travel from the testes to the external urethral orifice?   Sperm travels through the following structures, in order of sequence: seminiferous tubules, rete testis, efferent ductules, epididymis, ductus deferens, urethra, and urethral orifice.

Figure 19-5

What is the function of the penis?   The penis provides a means for inserting sperm cells into the vagina of the female.

Figure 19-6

What part of the brain regulates testosterone production in the testes?   The hypothalamus regulates testosterone production (by negative feedback).

Figure 19-7

Which organs are the female gonads?   The ovaries are the female gonads.

Figure 19-8

What structures within the ovary contain the oocytes?   All ovarian follicles (including primary, secondary, and Graafian follicles) contain oocytes.

Figure 19-9

What are the products of meiosis I in the female?   Female meiosis I, which occurs only in the ovaries, produces a small first polar body and a larger secondary oocyte.

Figure 19-10

After ovulation, the Graafian follicle transforms into what structure?   The Graafian follicle transforms into the corpus luteum after ovulation.

Figure 19–11

What tube transports the oocyte to the uterus after ovulation?   The uterine (or fallopian) tube transports the oocyte to the uterus.

Figure 19-12

The female vulva contains what structures?   The vulva contains the mons pubis, the labia majora, the labia minora, and the vestibule (which contains the clitoris, external urethral orifice, and vaginal orifice).

Figure 19-13

What type of tissue occupies most of the breast volume (in an adult female)?   Fat (adipose) tissue occupies most of the breast volume.

Figure 19-14

Approximately how many days from the last day of menstruation does it normally take for ovulation to occur?   It normally takes about 10 days for ovulation to occur from the last day of menstruation.

HUMAN DEVELOPMENT AND INHERITANCE

LEARNING OBJECTIVES

After studying this chapter, you should be able to:

1. Distinguish between the prenatal and postnatal periods of development.

2. Describe the process of fertilization, and the changes it induces.

3. Describe the formation of the morula and blastocyst.

4. Describe the processes of implantation, germ layer formation, and organogenesis.

5. Explain how the placenta develops and what its roles are in maintaining pregnancy.

6. Identify the main events during the fetal period of development.

7. Describe the events during birth, and the process of lactation.

8. Identify the circulatory changes that occur immediately after birth.

9. Describe the five stages of life.

10. Describe the nature of genetic inheritance and the importance of genetic screening.

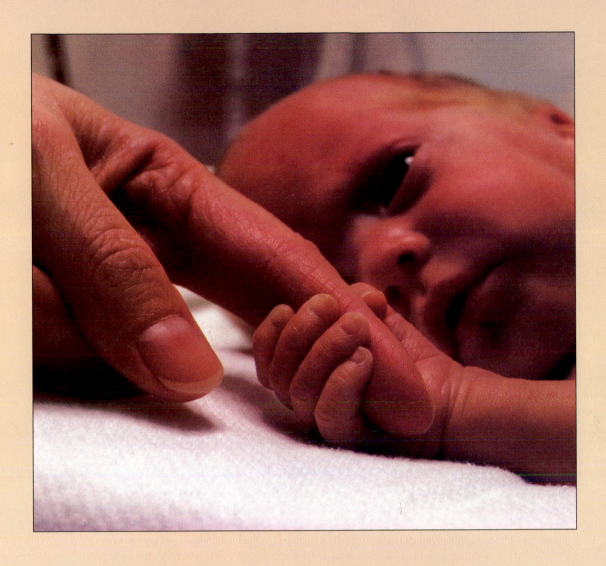

Human development is a continuous process of body change that begins at the moment of fertilization and continues to the death of the individual. It is divided into two periods: **prenatal development** and **postnatal development**. Prenatal development refers to the changes that occur prior to birth. The study of these changes is called *embryology*. Postnatal development refers to the changes that occur after birth, such as body growth, puberty, and senescence. In this chapter, we will focus on the wonders of the prenatal development period by exploring the capacity of cells to group and organize themselves, as they proceed to form a complete human organism from the union of two sex cells. We will also look briefly at some aspects of postnatal development. Following our discussion of development, we will examine the basic concepts of human inheritance.

PRENATAL DEVELOPMENT

Prenatal development involves the differentiation of early embryonic cells into clusters that lead to tissue and organ development. Once these structures have arisen, they undergo rapid growth until birth.

The prenatal period of development begins at the moment of fertilization and continues to birth. It is a dramatic period of rapid changes, during which a single fertilized egg is transformed into a complex individual being. After fertilization, prenatal development is divided into two periods: the embryonic period, which lasts about 8 weeks, and the fetal period, which continues to about 38 weeks, or birth.

Fertilization

Fertilization is the union of an egg cell, or oocyte, with a sperm cell. It begins with the entry of sperm cells into the vagina following ejaculation. Soon after their release, the sperm cells migrate through the uterus and into the uterine tubes by whipping their flagella. Their route is a precarious one, for they must travel against currents of mucus that stream outward and avoid capture by white blood cells in the mucosal linings. Only a small percentage successfully reach the upper one-third of the uterine tube, where the oocyte is usually located shortly after it has been released in ovulation. The journey usually takes about one hour to complete.

The oocyte is capable of being fertilized for up to about 24 hours after ovulation. Although some sperm may remain viable within the female reproductive tract for several days, most of them degenerate after the first 24 hours. Thus, for fertilization to occur, sexual intercourse must take place between 3 days before and 1 day after ovulation.

Although thousands of sperm cells may reach the oocyte in the uterine tube, only one normally participates in actual fertilization. This sperm cell must penetrate the follicle cells of the *corona radiata*, which envelops the oocyte, and continue through the *zona pellucida* before it reaches the oocyte plasma membrane (Fig. 20–1). Penetration is aided by an enzyme (hyaluronidase) that is released from the head of the sperm. Once the zona pellucida has been penetrated, it undergoes biochemical change that prevents further sperm cells from attaching to it.

With the entry of a single sperm cell through the oocyte plasma membrane, the oocyte undergoes its second meiotic division. The nucleus that results quickly migrates to the center of the cell, where it meets the swelled head of the sperm cell and its nucleus. Both of these nuclei are **haploid**; that is, they each contain one-half the chromosomal complement (23 chromo-

somes). Their fusion, which completes the process of fertilization, restores the **diploid** number of chromosomes (46). The resulting fertilized cell is called the **zygote.**[1]

The First Eight Weeks of Life

The first period of prenatal development is the **embryonic period**, which occurs during the first eight weeks following fertilization. This period of rapid change ends when the embryo has developed into a recognizable human being and the main organ systems have begun to form.

---

[1]**Zygote:** Greek *zygotos* = yoked.

Figure 20–1 ▶

Fertilization and the events immediately following. (a) Soon after ovulation, the secondary oocyte drifts into the uterine tube, where it may be encountered by active sperm cells, if intercourse occurs within about 3 days of ovulation (some sperm survive for as long as 72 hours, although most do not). (b) Fertilization begins when a sperm successfully penetrates the corona radiata and zona pellucida and enters the oocyte. The secondary oocyte then completes meiosis II. (c) The sperm head detaches from the midpiece and tail, and the oocyte nucleus and sperm nucleus swell as they approach each other. After making contact, the nuclear membranes begin to disappear and the male and female chromosomes unite, marking the end of fertilization. The fertilized cell is now called a zygote. (d) The united chromosomes attach to the mitotic spindle and undergo a mitotic division. (e) This division is the first cleavage division, which results in two daughter cells. (f) Shortly thereafter, a second mitotic division occurs to form four daughter cells. (g) A third division produces eight daughter cells. (h) A fourth division results in 16 daughter cells, forming a ball of cells known as a *morula*. (i) A color photomicrograph of a secondary oocyte at the moment of fertilization. The sperm has penetrated the oocyte and detached its midpiece and flagellum and is approaching the oocyte nucleus.

Combining of the sperm and oocyte chromosomes results in the formation of what structure?

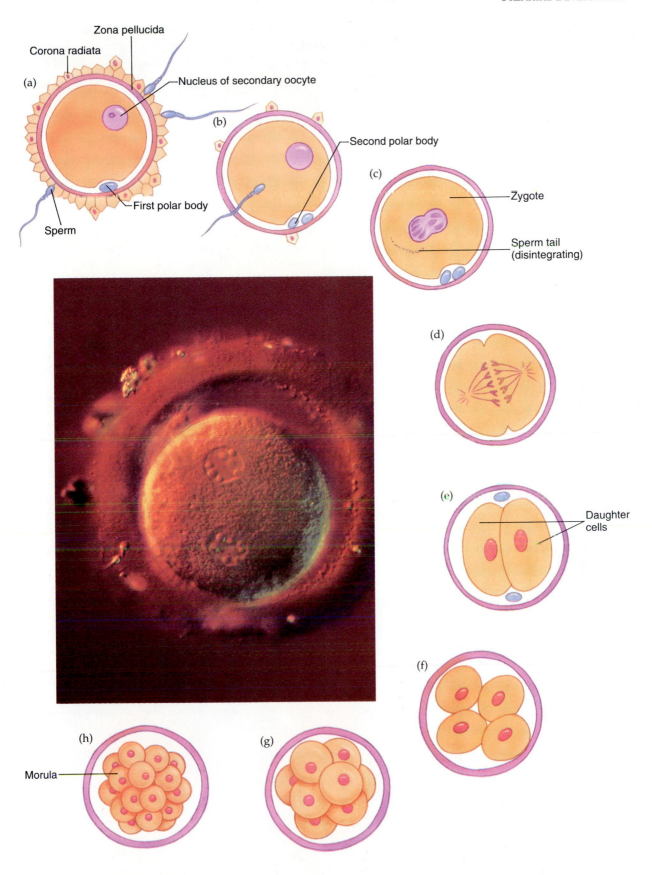

(a)

Corona radiata

Zona pellucida

Nucleus of secondary oocyte

First polar body

Sperm

(b)

Second polar body

(c)

Zygote

Sperm tail
(disintegrating)

(d)

(e)

Daughter
cells

(f)

(g)

(h)

Morula

HEALTH CLINIC

In Vitro Fertilization and Artificial Implantation

For many couples trying to build a new family, creating a child by normal means is not possible. Their condition of **infertility** may be a biological problem that cannot be corrected by current medical procedures. Biological infertility is most frequently caused by blocked uterine tubes in females, and by low sperm counts in males.

To help these people, reproductive biologists have developed a technique whereby a human oocyte may be fertilized outside the body, or *in vitro*, and artificially introduced into the uterus. The fertilization technique is performed in a controlled environment that simulates the conditions within a female's body. Because this environment is artificial and, in fact, is in the form of a glass receptacle, the technique has been called by many the "test tube baby" procedure. Its first success was publicized in 1978, when Louise Joy Brown was born after being "conceived" *in vitro*. Since then, several hundred "test tube" clinics have opened worldwide to provide this alternative to parents in need.

The technique of *in vitro* fertilization involves the removal of an oocyte from a woman, its fertilization in laboratory vessels, and the embryo's brief development within an artificial environment prior to implantation into the mother-to-be's uterus. Prior to oocyte removal, the woman is primed with hormones to promote the development of several oocytes in her ovaries. A specialized syringe, called a *laparoscope*, is then used to aspirate several preovulatory oocytes from Graafian follicles within the woman's ovary. The oocytes are placed in a glass receptacle that contains culture medium and controlled gas mixtures, where they are fertilized with sperm. Fertilization is confirmed by examination under a microscope, and the early development of the zygotes to the blastocyst stage is carefully followed. Once the blastocyst stage is reached, artificial implantation is performed. In this technique, a healthy blastocyst is transferred by laparoscope from its artificial nutrient bath to the uterine wall of the mother-to-be. The blastocysts that are not initially used may be frozen for later use, should the original attempt at implantation fail (these "spares" are often called "Popsicle babies" because of their frozen status).

About 15% to 20% of *in vitro* fertilization and implantation attempts result in pregnancy and the birth of a healthy baby. This percentage is actually quite a success, since natural fertilization results in live birth about 30% of the time. As a result, this technique is increasing in popularity as a viable alternative to overcoming infertility problems.

In vitro fertilization. Once ova have been collected from the woman's ovaries by aspiration, they are transferred to a glass bowl in a nutrient solution. The photograph shows seven ova in solution, prior to addition of the sperm.

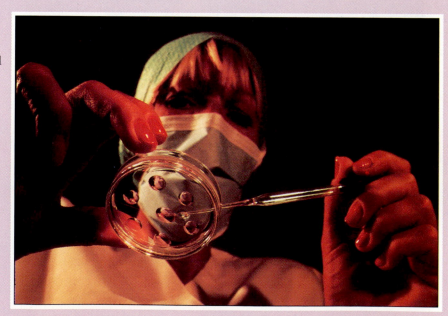

Early Cell Division

Shortly after the zygote has formed, it begins to divide by mitosis. This process is known as **cleavage** and includes a number of successive divisions (Fig. 20–1). Cleavage takes place while the developing zygote is passing down the uterine tube (Fig. 20–2). By the time it reaches the uterus, in about 3 days, it is in the form of a small ball of cells called the **morula**.[2]

Blastocyst Formation

While free in the uterine cavity, the morula undergoes cell division and change, or *differentiation*. Its development includes the movement of fluid into its intercellular spaces, which gradually enlarges the spaces and separates the cells into two portions. The outer layer of cells, called the **trophoblast** (TRŌ-fō-blast), later contributes to the placenta. The inner layer, called the **inner cell mass**, later becomes the **embryo**.[3] Eventually, the intercellular spaces join to form one large cavity called the **blastocyst cavity** (or *blastocele*). At this point the morula has become a **blastocyst**,[4] shown in Figure 20–3a and b.

Implantation

After about 6 days of floating free in the uterine cavity, the trophoblast contacts the inner wall of the uterus, or *endometrium*. This begins the process of **implantation**, which is illustrated in Figure 20–3. Soon the cells of the trophoblast begin to enter the endometrium by the action of enzymes they secrete. Eventually, the blastocyst will become completely buried in the endometrium.

As the blastocyst continues to actively invade the endometrium, the trophoblast differentiates into two layers, the outer of which contains fingerlike processes that penetrate the endometrium. Meanwhile, the embryo continues to differentiate, forming into the shape of a disc. The cells on one side of the disc, the ventral side, form the first of its three **germ layers**.

The three germ layers of the embryo consist of early cells that will eventually differentiate into the tissues and organs of the body. The first layer formed is the **endoderm**.[5] This is followed shortly by the development of cells on the opposite side of the disc, the dorsal side, to form the second germ layer, the **ectoderm**.[6] The middle layer, the **mesoderm**,[7] forms about ten days later. As small spaces begin to appear between the embryo and the trophoblast, the **amniotic cavity**

---

[2]**Morula:** Latin *mortem* = mulberry + *-ula* = little.

[3]**Embryo:** Greek *embryo* = a swelling.

---

[4]**Blastocyst:** Greek *blasto* = sprout + *cyst* = bladder or hollow ball.

[5]**Endoderm:** Greek *endo-* = inner + *-derm* = skin.

[6]**Ectoderm:** Greek *ecto-* = outer + *-derm* = skin.

[7]**Mesoderm:** Greek *meso-* = middle + *-derm* = skin.

Figure 20–2
Site of fertilization and implantation.
Where do fertilization and implantation normally occur?

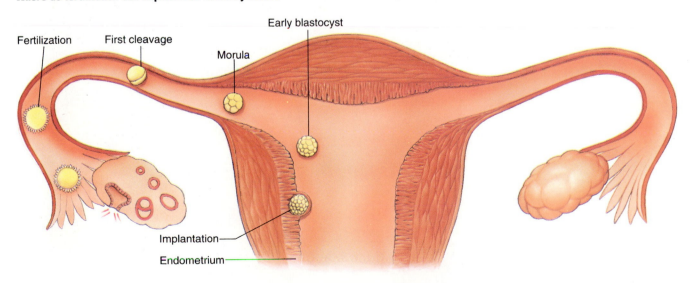

Fertilization   First cleavage   Morula   Early blastocyst

Implantation
Endometrium

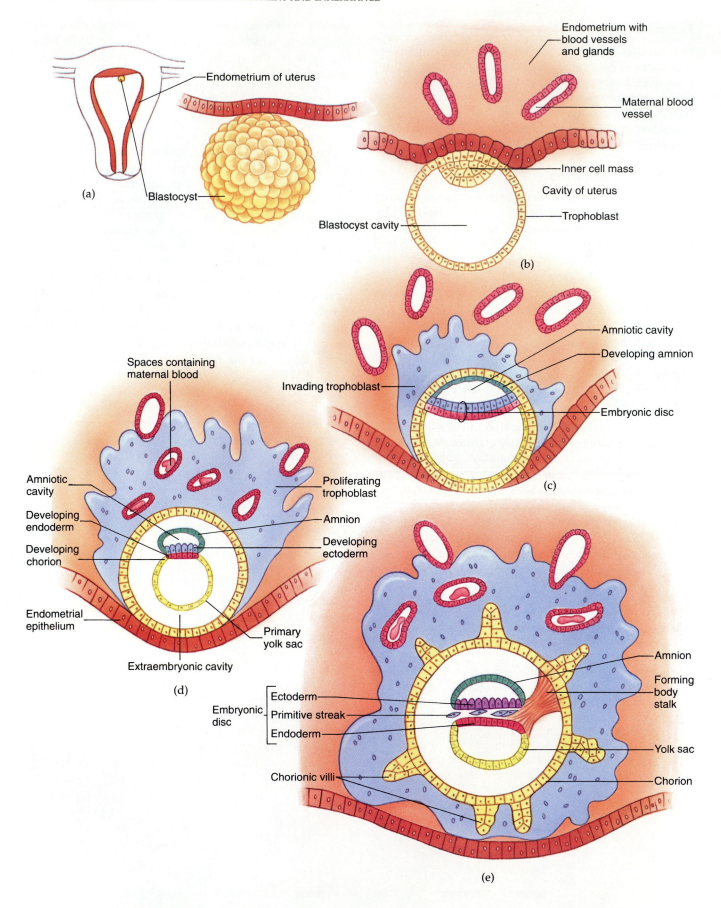

(a)

Endometrium of uterus

Blastocyst

(b)

Endometrium with blood vessels and glands

Maternal blood vessel

Inner cell mass

Cavity of uterus

Trophoblast

Blastocyst cavity

(c)

Invading trophoblast

Amniotic cavity

Developing amnion

Embryonic disc

(d)

Spaces containing maternal blood

Amniotic cavity

Developing endoderm

Developing chorion

Endometrial epithelium

Proliferating trophoblast

Amnion

Developing ectoderm

Primary yolk sac

Extraembryonic cavity

(e)

Embryonic disc
Ectoderm
Primitive streak
Endoderm

Chorionic villi

Amnion

Forming body stalk

Yolk sac

Chorion

◀ *Figure 20–3*
Changes during the second week of life. (a) About six days after fertilization, the blastocyst first contacts the uterine endometrium. (b) Soon after contact, trophoblast cells begin embedding the blastocyst into the endometrium. (c) The blastocyst continues to implant as the embryonic disc first appears and a new amniotic cavity is formed. (d) The primary germ layers become apparent. (e) By the 16th day, two germ layers are present and the embryo is fully embedded within the endometrium.

What are the three primary germ layers?

soon forms. This state of development marks the end of the second week of life. The three primary germ layers and many of the body tissues and organs that differentiate from them are listed in Table 20–1.

Implantation continues into the third week. It becomes complete when the trophoblast differentiates to form the **chorion**,[8] one of the membranes that are later associated with the fetus. The chorion will eventually form the embryonic portion of the **placenta**.[9] A second membrane also begins to develop during the third week. This is the **amnion**,[10] which expands the fluid-filled amniotic cavity until it envelops the embryo. During the latter part of this developmental period, a third membrane arises from the endoderm to form another cavity; this membrane is called the **yolk sac**.

Gastrulation

By the time implantation is completed, the embryo has developed into a flattened disc that is suspended along with its yolk sac and amnion within the chorion (Fig. 20–3e). During the latter part of the third week, the ectoderm layer begins to thicken along the midline of the embryo. This forms the **primitive streak**, which gives rise to the third germ layer, the mesoderm. The mesoderm actually forms when the primitive streak folds inward, or invaginates, resulting in a middle layer of cells. This process is called **gastrulation**, and the embryo is referred to as a **gastrula**[11] during this process (Fig. 20–4a).

Neurulation

During the process of gastrulation, the rodlike **notochord** begins to form as an outgrowth of mesodermal cells from the primitive streak. Once it has formed, portions of the ectoderm invaginate to form the **neural tube** immediately dorsal to it (Fig. 20–4b and c). Other

[8]Chorion: Greek *chorion* = skin or membrane.
[9]Placenta: Greek *placenta* = flat cake.
[10]Amnion: Greek *amnion* = little membrane enveloping the fetus.
[11]Gastrula: *gastr-* = stomach + *-ula* = little.

Table 20–1 EMBRYONIC GERM LAYER DERIVATIVES

Germ Layer	Tissues and Organs
Ectoderm	Brain, spinal cord, peripheral nerves Adrenal medulla Special sensory organs Epidermis, nails, hair, and skin glands Mucous membrane of the mouth, nasal cavity, and anus
Mesoderm	Dense and loose connective tissue Bone, cartilage, and red bone marrow Blood, heart, and blood vessels Dermis Skeletal muscle and most smooth muscle Lymph nodes, spleen, and tonsils Kidneys, ureters, and urinary bladder Reproductive organs
Endoderm	Organs of the alimentary canal Liver and gallbladder Lungs Thyroid, parathyroid, and thymus glands

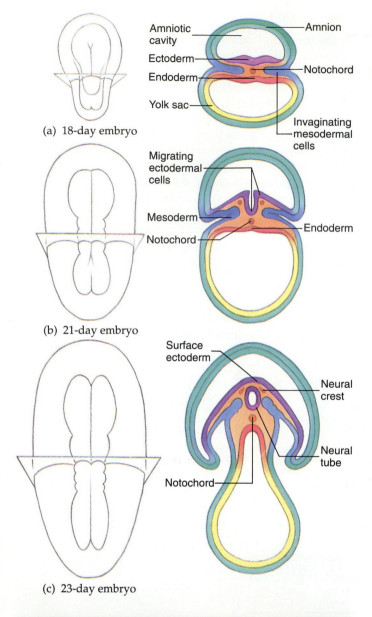

(a) 18-day embryo

Amniotic cavity
Ectoderm
Endoderm
Yolk sac
Amnion
Notochord
Invaginating mesodermal cells

(b) 21-day embryo

Migrating ectodermal cells
Mesoderm
Notochord
Endoderm

(c) 23-day embryo

Surface ectoderm
Neural crest
Neural tube
Notochord

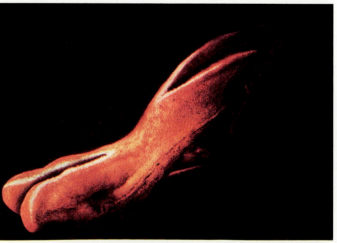

(d)

ectodermal cells develop alongside the neural tube to form the **neural crest**. Still other ectodermal cells continue to develop until they close over the neural tube, eventually forming the body surface layer. This process of ectodermal growth is referred to as **neurulation**, and the embryo during this period is called a **neurula**. The neural tube eventually gives rise to the central nervous system, the neural crest later forms the peripheral nervous system and the medulla of the adrenal gland, and the surface ectodermal cells form the epidermis of the skin.

Development of Body Form

During the third to eighth weeks, the embryo changes dramatically in body form, from the flattened embryonic disc to a form recognizable as human. This is brought about by continued growth into a series of body foldings, which lead to the formation of a head fold and a tail fold (Fig. 20–5). As a result of the foldings, a head cavity called the **foregut** is formed, and a tail cavity called the **hindgut** also forms. A **midgut** formed earlier by the endoderm develops further to become the alimentary canal.

The folding also combines a primitive stalk that connects the embryo to the endometrium with portions of the yolk sac and another membrane, the **allantois** (al-LAN-tō-is). This combination forms the umbilical cord. As the embryo continues to fold, the amnion expands to envelop the embryo. From this point until birth, the embryo is bathed in fluid within the saclike amnion, called **amniotic fluid**.

Many external features of the embryo develop and become evident during this period of folding and differentiation. For example, the head enlarges as the brain develops internally, and facial features appear such as the eyes, ears, and nose. During the fourth and fifth weeks, first arm and then leg buds appear. By the

Figure 20–4
Gastrulation and neurulation. The drawings on the left are orientation guides, showing top views of the embryo at each stage discussed. (a) Cross-sectional view of the embryonic disc showing the three germ layers. Mesodermal cells are in the process of migrating inward at this stage. (b) The migration of cells is complete, resulting in the formation of the mesoderm in the middle. As gastrulation is completed, neurulation begins as ectodermal cells migrate upward and growth continues. (c) Neurulation results in the formation of the neural tube, the neural crest, and a surface layer of cells. (d) Photograph of the human embryo at about 20 days of development. Folding of the ectodermal cells during neurulation is the prominent change during this stage.
The neural tube arises from which primary germ layer?

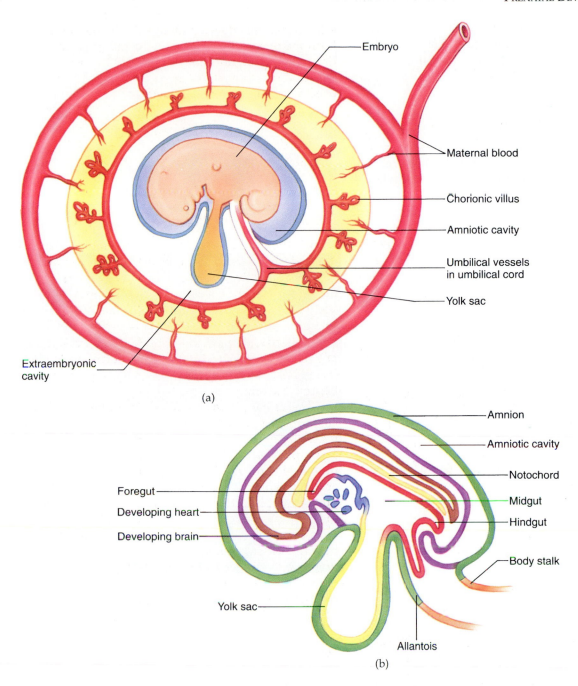

Figure 20–5

Development of body form. (a) As neurulation nears completion, the embryonic body begins to fold. Once folding is complete after about 30 days, the embryo takes on a recognizable form within the amniotic cavity. This lateral view of the embryo at about 30 days of development shows the result of this folding. The maternal support system surrounding the embryo is also shown. (b) A sagittal view through the 30-day embryo and its membranes reveals the early foregut, midgut, and hindgut.

During the first 30 days of development, from what source do you suppose the embryo obtains most of its nourishment?

end of the seventh week, the back of the embryo has straightened, the head is well formed and elevated, the digits are developed, and the tail is regressing.

Development of Organs

The major organs of the body begin to appear and develop during the first eight weeks of life. The process of their development is called **organogenesis** (or'-gan-ō-JEN-e-sis) and is shown in Figure 20–6. Most organs begin to form during the fourth and fifth weeks, when the embryo is in the early stages of folding. Many organs arise as tiny outpocketings of cells originating from the early foregut, midgut, and hindgut. For example, the thymus gland, thyroid gland, lungs, liver, pancreas, and urinary bladder begin as outpocketings. The organs forming the alimentary canal, such as the esophagus, stomach, and small intestine, arise directly from the midgut.

Other organs arise from germ layers that are not associated with the gut formations. For example, the heart is formed from mesoderm that begins as two tubes. By about day 21, the tubes fuse to form a single heart, which almost immediately begins to beat. The kidneys also arise from mesoderm and form an embryonic structure that occupies much of the early body cavity. Most of it degenerates before the eighth week, giving rise to a smaller fetal kidney. The brain arises from the ectoderm, which forms the neural tube before some of its cells begin forming the early brain. By the end of the seventh week, most organs have formed to a point of beginning their early functions.

Development of the Placenta

The **placenta** first appears during the third week of life. It is formed from the chorion of the embryo, which sends fingerlike projections, called **chorionic villi**, into the endometrium (Fig. 20–3e). The endometrium contains pools of maternal blood, known as **lacunae**,[12] which the chorionic villi seek. As development of the embryo continues, the placenta matures until the embryonic blood supply is completely separated from that of the mother by an embryonic capillary wall, a basement membrane, and a thin layer of chorion (Fig. 20–7). This is the *placental blood barrier*, which prevents actual mixing of embryonic and maternal blood. Nutrients, gases, and waste materials are permitted to diffuse across this barrier while blood cells and large molecules are prevented from crossing. In many ways, the placental blood barrier is similar to a capillary network in the tissues of an adult, which also permits only the diffusion of fluids with dissolved solutes through to

[12]Lacunae: Latin *lacunae* = pools or hollows.

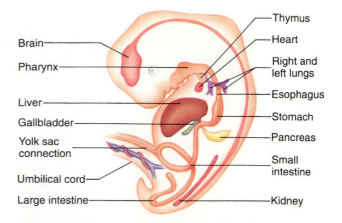

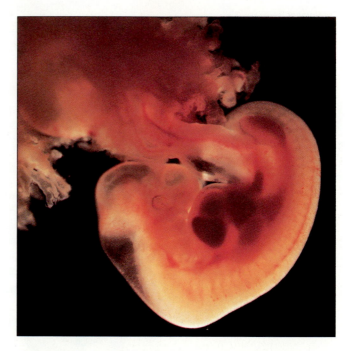

Figure 20–6
The 5-week embryo. (a) By this age many of the organs and glands that arise from the endodermal gut have begun to form. The heart and kidneys have arisen from the mesoderm, and an early brain is forming from the neural tube of the ectoderm. Other organs can also be seen in this lateral view. (b) Photograph of an embryo about 5 weeks old. It measures 1 cm (0.4 inch) in length. Note the hands and feet, which are beginning to bud from the main trunk. **In addition to the brain, name one other organ that is derived from the ectoderm.**

tissue cells. The basic difference lies in the presence of the basement membrane and chorion, which prevent the movement of phagocytes and most proteins.

In the early stages of formation of the placenta, the embryo is connected to it by way of a primitive stalk.

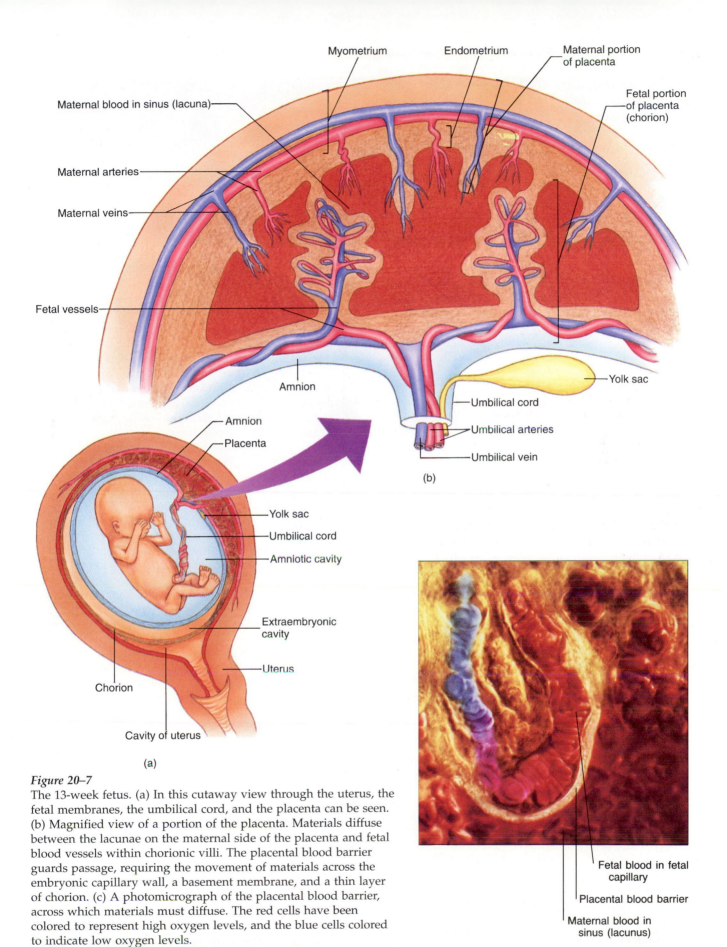

Myometrium   Endometrium   Maternal portion of placenta

Maternal blood in sinus (lacuna)

Maternal arteries

Maternal veins

Fetal portion of placenta (chorion)

Fetal vessels

Amnion

Yolk sac

Umbilical cord

Umbilical arteries

Umbilical vein

(b)

Amnion

Placenta

Yolk sac

Umbilical cord

Amniotic cavity

Extraembryonic cavity

Uterus

Chorion

Cavity of uterus

(a)

Fetal blood in fetal capillary

Placental blood barrier

Maternal blood in sinus (lacunus)

Figure 20–7

The 13-week fetus. (a) In this cutaway view through the uterus, the fetal membranes, the umbilical cord, and the placenta can be seen. (b) Magnified view of a portion of the placenta. Materials diffuse between the lacunae on the maternal side of the placenta and fetal blood vessels within chorionic villi. The placental blood barrier guards passage, requiring the movement of materials across the embryonic capillary wall, a basement membrane, and a thin layer of chorion. (c) A photomicrograph of the placental blood barrier, across which materials must diffuse. The red cells have been colored to represent high oxygen levels, and the blue cells colored to indicate low oxygen levels.

What fetal membrane contributes primarily to the structure of the placenta?

597

The Placental Blood Barrier: Not Always Complete

The movement of most substances from the mother's blood supply to the fetus is prevented by the placental blood barrier. In addition to the small size of pores in the capillary walls, additional protection from migrating substances is provided by a basement membrane of connective tissue and the thin chorion membrane. This barrier is sufficient to prevent blood cells, including wandering phagocytes, and many smaller particles, such as bacteria, fungi, and viruses, from passing from mother to child.

However, some disease-causing substances are known to be transmitted from mother to fetus, often resulting in birth defects or premature death. In most cases, these substances are transmitted only when the placental blood barrier becomes torn. Tearing may occur if the mother receives a traumatic injury during pregnancy, but it frequently occurs during the birth process. The result of a torn barrier is the mixing of blood between the mother and child. If the mother is carrying pathogens or toxins in her bloodstream at the time, such as the virus that causes herpes or the bacteria that cause syphilis, entrance of the harmful particles into the child's bloodstream will result in transmission of the disease. The child may also be in danger if the blood types are not compatible between mother and child and the quantity of blood mixed is significant.

In some cases, a torn placental blood barrier is not a prerequisite for the spread of infection or a toxic substance from mother to child. Evidence is mounting that certain microorganisms and drugs may cross the barrier. For example, it is well established that the virus that causes **rubella** (German measles) is capable of crossing the placental barrier. If the mother is infected during the first trimester, there is great danger that the virus could infect the developing embryo and cause severe birth defects or miscarriage. Another example is the bacterium *Toxoplasma gondii*, the microorganism that causes the infectious disease **toxoplasmosis**. This disease may be contracted by exposure to animal feces, most commonly through house pets such as cats. Although *Toxoplasma* normally has little effect on adults, it can cross the placental barrier quickly to infect the fetus, causing birth defects or miscarriage. Also, a drug administered to pregnant mothers in distress for a brief period during the 1950s, called **thalidomide**, was shown to cause dramatic birth defects in their infants, including the underdevelopment of limbs (arms and legs). Yet another danger is alcohol. If ingested during pregnancy, alcohol has been shown to cause brain dysfunction and growth abnormalities in the newborn (a condition known as **fetal alcohol syndrome**, or **FAS**). It is now suggested that, to insure against this potential disaster, mothers abstain from all sources of alcohol during pregnancy. HIV, the virus that causes **AIDS**, may also be capable of moving from the mother's bloodstream to the child's across the placental barrier. It is well established that HIV does infect the child if the barrier is torn in any way.

From the evidence that continues to accumulate, it is becoming clearer that the developing child may be prone to many hazards, despite the protection of the placental blood barrier. For now, most experts agree that it's best to practice caution during pregnancy: avoid taking medications unless identified as essential by a physician, don't drink alcohol, don't smoke, and if afflicted by an infectious agent, see your physician right away.

As the embryo matures, the stalk thickens and elongates to form the umbilical cord. The umbilical cord carries two umbilical arteries, which originate from the embryo's iliac arteries, and a single umbilical vein. The umbilical arteries carry blood from the embryo to the mother, and the vein carries blood from the mother to the embryo.

In addition to providing a supporting link between the embryo and the mother, the placenta serves as an important endocrine gland. During the second and third months of development, the chorion secretes **human chorionic gonadotropin (HCG)**, which causes the corpus luteum in the mother's ovary to remain functional. As a result, the continued production of progesterone and estrogen by the corpus luteum maintains the endometrium for the first three months of pregnancy. Once the placenta forms by the third month, it takes over this production by secreting estrogen and progesterone on its own. The levels it produces are sufficient to maintain the state of pregnancy, and in fact continually increase through the entire term.

Growth of the Fetus

The embryo becomes a **fetus**[13] about eight weeks after fertilization (Fig. 20–8). The beginning of the fetal period, which continues until birth, is identified by the start of bone ossification. By this time, most of the organs and systems have developed to the point of active functioning. For example, the heart is beating rhythmically, blood is flowing through new vessels, brain neurons are forming synapses, and early kidneys are filtering blood.

The most notable feature of the fetal period is growth. During the next seven months of life, the fetus will have grown from about 2.5 g (0.09 ounce) and 3 cm (over 1 inch) to about 3500 g (7 pounds 11 ounces) and 50 cm (20 inches). Growth is accomplished by the active mitosis of cells in all tissues and organs.

During the fetal period, the skin of the fetus is covered with fine, soft hair called **lanugo**. Also, a waxy coat of sloughed epithelial cells protects the fetus from the harmful effects of waste materials that are present in the amniotic fluid. Under the skin layer, fat is accumulated to insulate and provide for a nutrient reserve.

As the growing fetus nears its 36th week of life, it has developed to the point of being capable of surviving outside the mother. Although birth sometimes occurs earlier, infants usually require life support until 36 weeks or so. By the 38th week, the fetus is said to be *full-term*. It has lost its lanugo covering by this time, although it is still covered with the waxy coat. The bones are mostly ossified, the scalp is often covered with hair, and the fingers and toes contain nails (that often need trimming!). In most cases, the head of the fetus is directed downward toward the cervix.

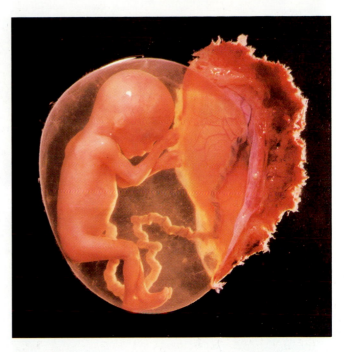

Figure 20–8
Photograph of a developing fetus and the attached placenta at 16 weeks. At this stage the fetus measures about 5 inches in length.
What fetal membrane surrounds the fetus in a bath of fluid?

---

[13]Fetus: Latin *fetus* = offspring.

PARTURITION

Parturition is the process of giving birth to a child. It is regulated by the hypothalamus of the brain.

Pregnancy terminates with the birth process, or **parturition** (Fig. 20–9). For some time prior to this event, placental secretion of estrogen at increasing levels has induced minor contractions of the uterine wall. This is done to prepare the myometrium of the uterus for the powerful, rhythmic contractions necessary for birth. These minor contractions are known as *Braxton-Hicks contractions*; they increase in frequency as full term approaches.

As stretching of the uterine wall due to fetal growth reaches a certain limit, nerve impulses signal the hypothalamus that the time is near. In response, the hypothalamus stimulates the posterior pituitary gland to secrete **oxytocin**. This hormone targets the myometrium, causing it to strengthen its rhythmic contractions. Once the contractions increase in frequency and strength, the process of **labor** begins. Labor refers to the movement of the fetus through the birth canal in response to uterine contractions.

As labor begins, rhythmic contractions of the myometrium start to forcefully push the fetus against the

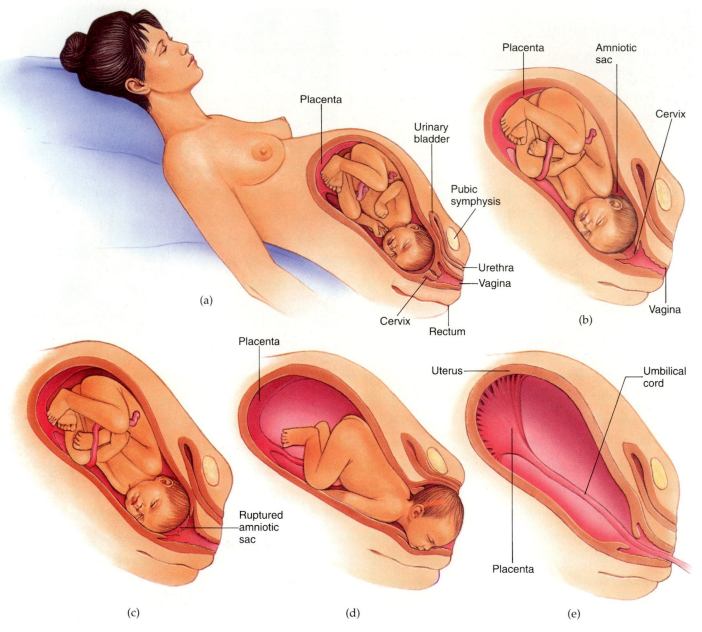

Figure 20–9
Parturition. (a) Location of the fetus in the uterus. (b) The fetus positions toward the birth canal and the cervix begins to dilate. (c) The amniotic sac ruptures. The cervix reaches complete dilation. (d) The fetus passes through the birth canal. (e) The placenta separates from the wall of the uterus and is expelled.

What hormone produced by the pituitary gland stimulates uterine contractions during parturition?

cervix. Early in this process, the amnion (or "water bag") surrounding the fetus ruptures, allowing amniotic fluid to flow through the vagina and to the outside. As contractions continue, the cervical canal dilates in response to the pressure. As a result, the head of the fetus begins to be pushed first into the cervical canal. Dilation of the cervical canal stimulates a positive feedback mechanism, in which uterine contractions become more intense as cervical dilation progresses. Cervical dilation continues until the cervix has expanded to a diameter about the size of the fetus's head (10 cm). This period of labor sometimes lasts about 24 hours, although the time is highly variable among individuals. It is often called the **first stage** of labor.

The **second stage** of labor continues from the time of maximum cervical dilation to the exit of the fetus. During this stage, abdominal muscle contractions are stimulated by the positive feedback mechanism; these contractions further aid labor. The result is a maximum effort that subsides when the fetus finally enters the outside. The second stage may last from 1 minute up to about 1 hour.

Soon after the birth of the fetus, the placenta separates from the uterine wall and is pushed through the birth canal by another wave of contractions. This is the **third stage** of labor, and is usually accompanied by bleeding. The loss of blood is usually restricted as continued uterine contractions compress broken blood vessels.

After parturition, levels of estrogen and progesterone in the blood fall dramatically. This is due to the loss of their temporary source, the placenta. Their levels are restored once the menstrual cycle is resumed by the release of GnRH by the hypothalamus. The first menstrual flow after parturition usually follows about 8 to 12 weeks later.

CONCEPTS CHECK

1. What activity prepares the uterine wall for parturition?

2. What is the role of oxytocin in parturition?

3. What are the three stages of labor?

POSTNATAL DEVELOPMENT

The cycle of life proceeds from birth to death as postnatal development. During this period of life, changes occur in the body in definable stages that are more gradual than those of prenatal development.

The postnatal development period begins immediately after birth and continues until death. It includes the changes that occur during infancy, childhood, adolescence, adulthood, and old age. Soon after birth and throughout the infancy stage, nourishment may be provided by the mother through the natural process of lactation.

Lactation

Lactation[14] is the production of milk by the mammary glands. It usually follows the process of parturition and may continue for two to three years if suckling occurs at a regular, frequent pace.

During pregnancy, the high levels of estrogen and progesterone in the blood that are maintained by the placenta stimulate the development of the duct system and glands in the breasts. They also promote the added accumulation of fat, so that the breasts increase substantially in size. Although **prolactin** is present in the blood during pregnancy, its role in stimulating milk secretion is kept in check by the high levels of estrogen and progesterone.

After parturition, the expulsion of the placenta results in low estrogen and progesterone levels. Also, the rate of prolactin secretion by the anterior pituitary gland is increased. Consequently, milk production is stimulated. During the first several days of milk production, a substance called *colostrum* is released by the mammary glands. Although this type of milk contains small amounts of fat and lactose, it is thought to contain antibodies that protect the nursing child from infections. After several days of production, the milk increases its levels of fat and lactose and the antibodies decline.

The flow of milk is a process that requires nerve stimulation. The mechanical force provided by suckling triggers the production of nerve impulses, which cause the release of **oxytocin** from the posterior pituitary gland. Oxytocin stimulates special contractile cells around the alveolar glands to contract, forcing milk to be ejected through ducts and flow out the nipples. This flow, often called *milk letdown*, may even be stimulated by means other than suckling. For example, the sound of a crying baby or the anticipation of a scheduled feeding may be sufficient to stimulate milk letdown.

As long as nursing continues to be stimulated, prolactin and oxytocin continue to cause the production and release of milk. Once nursing is stopped for more than several days, the ability of the mammary glands to respond to prolactin is lost. As a result, milk production stops until the next pregnancy.

CONCEPTS CHECK

1. In what ways do estrogen levels affect the influence of prolactin?

2. What is the role of oxytocin in lactation?

---

[14]**Lactation:** Latin *lactare* = to give forth milk + *tion* = the act of.

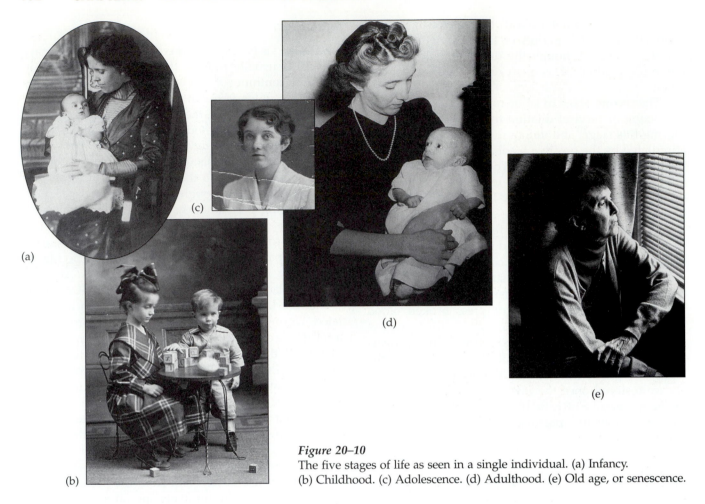

Figure 20–10
The five stages of life as seen in a single individual. (a) Infancy.
(b) Childhood. (c) Adolescence. (d) Adulthood. (e) Old age, or senescence.

Stages of Life

If an individual is fortunate to live the full term of life, currently about 73 years for males and 77 years for females, all five stages of life will have been experienced. Each of these stages is characterized by certain changes that occur in the body, some of which are illustrated in Figure 20–10. The stages are infancy (the first year or two), childhood (from infancy to the onset of puberty at about 8 to 12 years), adolescence (from puberty until 16 to 20 years), adulthood (from the end of adolescence to 65 to 75 years), and old age.

Infancy

The newborn infant, or **neonate**,[15] undergoes many important changes soon after birth. These changes are

[15]**Neonate:** Greek *neo-* = new + Latin *natus* = born.

required to conform the infant from a life of total submersion in a regulated fluid with a constant supply of nutrients and waste removal, to a life of independence in a biologically hostile world. The most dramatic change occurs when the neonate must immediately adjust to living in an air-filled environment. It involves rerouting parts of the circulatory network.

Circulatory Changes

In the moments following birth, the neonate must gasp for air for the first time. This event causes fluids in the lungs to be expelled, and air rushes in to replace them. As the lungs inflate with air during the first breaths, certain changes in the circulatory system take place (Fig. 20–11).

Expansion of the lungs stimulates the flow of blood to them. As a result, the flow of blood from the right ventricle of the heart to the pulmonary circulation increases. This is quickly followed by a decrease in the flow of blood between the right atrium and the

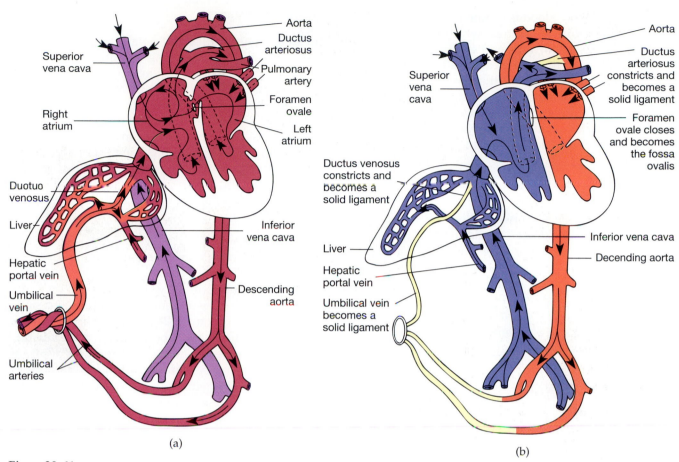

(a)

(b)

Figure 20–11
Circulatory changes. (a) Fetal circulation includes specific structures that are present to support life inside the uterus. They are the foramen ovale, ductus arteriosus, ductus venosus, and umbilical vessels. (b) Changes in fetal circulation occurring at birth.

What major organ of the chest is partially bypassed in the fetal circulatory scheme?

left atrium through the **foramen ovale**,[16] which previously shunted blood away from the lungs in the fetal circulation. At the same time, the increased volume of blood returning to the left atrium from the lungs increases the internal pressure of the left atrium. As a consequence of this increased pressure, blood is pushed against the interatrial septum, forcing a flap of tissue that has developed in that region to close off the foramen ovale. These actions complete the separation of the two atria, and the four-chambered heart can now pump blood efficiently to both the pulmonary and the systemic circulatory routes. Eventually, this flap will fuse with the atrial walls to form the depressed area known as the **fossa ovalis**.[17] Although closure of the

foramen ovale occurs with the first few breaths, complete fusion often takes about a year to complete.

The closing of the foramen ovale is assisted by another change in circulation. In the fetal and neonate circulation, a short artery known as the **ductus arteriosus** connects the pulmonary trunk to the aorta. Before birth, it carries blood from the pulmonary trunk to the aorta in order to bypass the lungs. Soon after birth this artery closes off, allowing blood to flow in one direction as it is pushed through the pulmonary trunk to the lungs.

A third change in circulation occurs when the umbilical cord is cut and tied. In the fetus, fresh blood enters from the placenta by way of one **umbilical vein**, which passes to the liver. The fetal liver manufactures red blood cells, but does not yet function in digestion. Therefore, instead of entering the substance of the liver, blood is shunted to the inferior vena cava, where it enters the fetal circulation by coursing through the **duc-**

---

[16]Foramen ovale: Latin *foramen* = hole or opening; *ovale* = oval-shaped.

[17]Fossa ovalis: Latin *fossa* = depression; *ovalis* = oval-shaped.

tus venosus. Waste blood is drained from the fetus by way of two **umbilical arteries**, which originate from the iliac arteries of the fetus and exit through the umbilical cord. When the umbilical cord is cut, blood no longer flows through the ductus venosus, the umbilical arteries, and the umbilical vein, and they degenerate. The ductus venosus remnant from the fetus becomes a solid ligament, and the umbilical vein becomes part of the falciform ligament associated with the liver.

Other Changes

During the first year or two of life, an infant goes through important changes in addition to the circulatory changes just described. They include rapid brain growth and development, muscle growth and coordination, bone development, and growth of most body tissues and internal organs. By the time the first birthday is reached, the infant can usually stand erect and walk without support, understand many words, and begin the development of language.

Childhood

After the first or second year, the infant has become a child capable of responding to the surrounding world. **Childhood** is characterized as a time of slower body growth. Much of the fat accumulated during infancy is usually lost, and muscles and bones lengthen and straighten. Many of the synapses to be formed in the brain become established, and the ability to learn new information is thought by many to reach an all-time high point. The head growth has slowed more than general body growth, so that body proportions become near to those of adults. Childhood ends as the growing youth approaches puberty.

Adolescence

Puberty usually begins in males between the ages of 10 and 12 years and lasts until 18 years. In females it often begins earlier, between the ages of 8 and 11 years, and lasts only until about 16 years. This stage in life is known as **adolescence**. Many physiological changes occur during this period, many of which affect the emotional behavior of the youth. It is a time during which the sex hormones are produced in high amounts and reproductive organs reach maturity. Secondary sexual characteristics soon appear as a result, often accompanied by a new, uncomfortable self-image. The onset of puberty is usually accompanied by a growth spurt and followed by a slow growth period. The maximum height or stature is usually achieved before 16 or 17 years in females and 18 or 19 years in males.

Adulthood

The attainment of full adult height marks the end of adolescence and the beginning of **adulthood**. This period of life is the longest-lasting, for it extends to old age. It is characterized by a lack of vertical growth, and the slow effects of aging begin to accumulate as the years go by. These effects include increased fat deposition, gradual muscle loss, a decline in metabolic rate, and thinning of the skin. It is during this period of life that factors such as diet, psychological stress, and exercise can affect health in a way that often influences longevity.

Old Age

The last stage in human development is old age, or **senescence**. It usually begins between the ages of 65 and 75 years, and is marked by bone fragility and a consequent loss of posture, muscle atrophy, and slowing of nervous responses. The skin continues to thin and lose its elasticity, and it becomes more heavily pigmented in some areas. The immune response weakens with age, allowing certain opportunistic diseases to take hold, such as pneumonia, various tumors, and infections. All of these changes result from the aging process that is ongoing in body cells beginning with the entry into adulthood and continuing until death. Old age ultimately ends at death, when all body functions stop in response to the failure of one or more vital organs.

CONCEPTS CHECK

1. What two changes in circulation are triggered when the newborn draws its first few breaths?

2. What happens to the newborn when the umbilical cord is cut and tied?

3. What major changes occur in the body of a child as he or she enters adolescence?

GENETIC INHERITANCE

Inheritance is determined by genes and may be predicted through the study of genetics.

The mysteries of who and what we are may be further explored through a study of genetic inheritance. In earlier chapters, we have learned that within the nuclei of body cells are chromosomes, which contain the mol-

A Disease of Old Age: Alzheimer's Disease

Alzheimer's disease is a progressive brain disorder that generally afflicts certain individuals in the older age group (65 or older). It is characterized by a gradual loss of memory and motor skills. Initially, a person's recent memory retrieval abilities are lost. As the disease progresses, long-term memory begins to waver, until the victim becomes unable to remember important events in his or her lifetime, familiar places, and familiar faces—even those of immediate family. With the loss of memory, speech is often impaired, and skills like reading, writing, and sewing become impossible to exercise. In its later stages, victims of Alzheimer's disease are unable to perform even simple tasks, like eating, dressing, and brushing their teeth. Most victims die from this disease within 5 to 15 years after its onset.

Recent evidence arising from research into the cause of Alzheimer's disease indicates that its victims suffer from a selected loss of neurons that synthesize the neurotransmitter acetylcholine. These neurons are located at the base of the brain and are responsible for sensory perception, speech, and language and play a role in learning and memory. The loss of these neurons is thought to cause the gradual deterioration of these functions.

Exactly how these neurons are damaged is a current area of active research. It is known, for example, that about 15% to 20% of Alzheimer's disease patients have inherited the disease. Once investigators identify the actual gene responsible for the genetic form of this disorder, it is expected that the actual cause can be traced to the abnormal production of a protein or an enzyme in neurons. One protein that has already been placed on the suspect list is called beta amyloid, which has been revealed in postmortem studies of Alzheimer's disease patients to form plaques in damaged neurons.

In addition to genetic factors, other factors may also be involved in the development of Alzheimer's disease. For example, aluminum has been found in high concentrations within damaged neurons, pointing to possible environmental poisoning. Also, certain slow-acting viruses have been implicated because they cause other forms of dementia (loss of mental function). Recently, some investigators have suggested that the neuronal damage may be caused by complement, a group of about 20 proteins released during an immune response that help destroy disease-causing microorganisms. This suggests that certain forms of Alzheimer's disease may be caused by the body's own immune response gone awry (*autoimmunity*). However, further research is needed before conclusions can be drawn.

The only treatment currently available for Alzheimer's disease is the use of drugs that increase the production of acetylcholine. They have an effect—albeit extremely limited—in restoring memory that has been lost, at least temporarily. This points to a strong need for focusing research toward a means of prevention, or, possibly, a cure. Since Alzheimer's disease currently afflicts about 6% of persons 65 years of age or older, it is a tragedy that many of us may have to face in the future if a solution is not found soon.

---

ecular "map" of body structure and function, the DNA. Within the structure of DNA are the specific sequences of genes, which code for proteins and enzymes. By coding for proteins and enzymes, the genes determine the nature of our body structure and functions. When the chromosomes of a male gamete (a sperm) combine with the chromosomes of a female gamete (an oocyte), the DNA of both gametes is united and the road to development begins. As we will discover in this section, the combination of gametes during fertilization provides for the variation of traits we see among individuals. The study of how this variation occurs is called **genetics**.

Chromosomes and Genes

Every cell in the human body, with the exception of gametes (and cells without nuclei), contains 23 pairs of chromosomes (46 total chromosomes, the diploid number). One chromosome from each pair comes from the father's sperm, and the other chromosome from the mother's oocyte. The two chromosomes in the pair are called **homologous chromosomes**. Homologous chromosomes carry genes that determine the same trait. For example, if one chromosome carries genes that code for hair color, its homologous chromosome also carries those genes. Genes that carry the same traits—such as hair color, skin color, body height, or shape of the nose—and occupy the same position on homologous chromosomes are called **alleles**.

If the alleles for a certain trait are identical, the person is **homozygous** for that trait. For example, if both chromosomes carry genes that code only for blond hair, the person is homozygous for blond hair. If, on the other hand, the alleles for a certain trait are different, the person is **heterozygous** for that trait. This would occur, for example, if a person had genes on one chromosome that coded for black hair and genes on the other chromosome for blond hair.

Geneticists distinguish between genes and the physical expressions of the traits they determine. The actual genes of an individual that determine a particular trait are called the **genotype**. It is symbolized by a combination of letters that designate respective alleles. For example, albinism is an inherited condition involving the absence or reduction of melanin, the brown pigment responsible for coloration of the skin, hair, and eyes. We may identify the genotype of albinism by the letters aa and the genotype of a person without albinism AA.

The physical expression of the genes, the actual trait, is known as the **phenotype**. For example, the aa genotype produces nonpigmented skin, hair, and eyes; this white appearance is the phenotype. Thus, phenotype refers to visible characteristics, such as height, color, weight, and so on.

Dominant-Recessive Inheritance

The expression of a phenotype is the result of the interactions of the alleles. As we have seen, identical alleles produce homozygous genes, or a homozygous genotype. In our example of albinism, the phenotype of white skin occurs only when the alleles are identical—in other words, when the genotype for albinism is homozygous. This may be designated aa. If the alleles are different, however, the phenotype will be expressed differently—the person will have pigmented skin. This heterozygous genotype may be designated as Aa. In this example, the gene coding for albinism is dominated by the gene coding for pigment. The pigment gene is, therefore, called the **dominant gene**, and the albinism gene is called the **recessive gene**.

The determination of which genes are dominant and which are recessive in a particular combination may be made by the use of a chart called a **Punnett square** (Fig. 20–12). This chart shows the possible combinations of genes for a single trait that would result

Figure 20–12
A Punnett square. (a) A cross between a female homozygous for pigment and a male homozygous for albinism. All offspring in this cross are heterozygous. Their phenotype expresses pigment, since albinism is a recessive trait. (b) A cross between two heterozygous parents. Only one in four offspring will express the albino phenotype.

In a cross between an albino father and a heterozygous mother, what is the probability that an offspring will be albino?

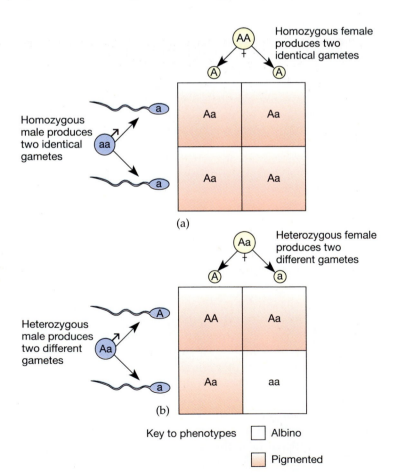

from the combining of female and male gametes. In the first example shown, the male gametes are homozygous for albinism (aa) and the female gametes are homozygous for pigment (AA). The genotypes of their four offspring, all Aa, are indicated in the squares. Because of the dominant nature of the pigment genotype, the phenotypes of the children resulting from this mating would all be pigmented. In the second example, the results of a mating between two of these children are shown. In this case, the possible phenotypes of the offspring would be one albino (aa) and three with pigment (Aa, Aa, and AA). Thus, the chance of producing an albino child from a mating between an albino parent and a parent homozygous for pigment is 0%, but the chance of producing an albino child from a mating between two heterozygous parents is 25% (1 in 4).

Although the Punnett square provides a rapid means of determining the possible outcomes of mating between parents of known genotypes, you should keep in mind that it represents a *probability* of inherited traits. It is entirely possible that all four children from Figure 20–12b would have the heterozygous genotype (Aa) in the real world. However, as the sample size increases (the more offspring are produced),

the probability that the calculated ratio will occur becomes greater. Thus, the Punnett square is a useful technique in producing probabilities of inherited traits, but it cannot predict actual outcome on a small scale.

In addition to our example of albinism, numerous other traits are determined by dominant-recessive genotypes. A partial list is provided in Table 20–2.

Sex-Linked Inheritance

When chromosomes are examined with microscopic techniques, the differences between female and male chromosomes are revealed (Fig. 20–13). These differences appear to be limited to a single pair of chromosomes, known as the **sex chromosomes**. The other 22 pairs of chromosomes are called **autosomes**. The sex chromosomes of a normal female consist of two X-shaped chromosomes (designated XX). In a normal male, the sex chromosomes consist of a single X-shaped chromosome and a smaller Y-shaped chromosome (designated XY). The presence of the Y chromosome makes a person male, while the lack of a Y chromosome makes a person female. This is actually accomplished by one particular gene on the Y chromosome, called the *sex-determining region of the Y (SRY)*

Table 20-2 TRAITS DETERMINED BY DOMINANT-RECESSIVE INHERITANCE	
Dominant Phenotype	**Recessive Phenotype**
Astigmatism	Normal vision
Feet with normal arches	Flattened arches (flat feet)
Normal skin pigmentation	Albinism
Freckles	Absence of freckles
Ability to roll tongue in U shape	Inability to roll tongue
Unattached (free) earlobes	Attached earlobes
Widow's peak hairline	Straight hairline
Polydactyly (extra digits)	Normal number of digits
Syndactyly (webbing between digits)	Normal digits
Huntington's disease (brain deterioration)	Absence of Huntington's disease
Achondroplasia (dwarfism)	Absence of achondroplasia
Absence of cystic fibrosis	Cystic fibrosis
Normal color vision	Total color blindness
Absence of Tay-Sachs disease	Tay-Sachs disease
Absence of progeria (premature aging and death)	Progeria
Normal heart	Congenital heart block
Ability to digest lactose (from milk)	Lactose intolerance
Normal hair growth	Atrichia (absence of hair)

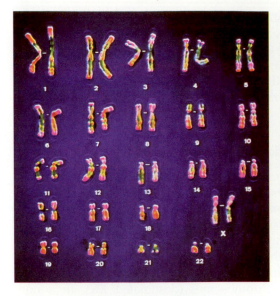

(a) Female chromosomes

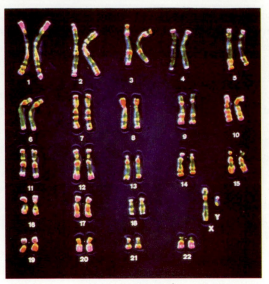

(b) Male chromosomes

Figure 20–13
Human chromosomes. (a) The 23 pairs of female chromosomes. (b) The 23 pairs of male chromosomes.
Which chromosome contains the SRY gene, which determines male development?

gene, which switches on the transformation of embryonic gonadal tissue into testes. Lack of the Y chromosome, and therefore lack of the SRY gene, results in the transformation of the early gonadal tissue into ovaries by other genes. Inherited traits that are determined by genes on the sex chromosomes are called **sex-linked**.

In contrast to autosomes, sex chromosomes are not homologous. The Y chromosome in the male is only about one-third as large as the X chromosome and thereby lacks many of the genes present on the X that code for certain traits. For example, genes that code for the production of certain blood-clotting factors are present on the X chromosome but not on the Y chromosome. Genes found only on the X chromosome are called **X-linked**.

When a female inherits an X-linked trait, expression of the two genes occurs as it would in dominant-recessive inheritance. However, when a male inherits an X-linked gene, the phenotype is always expressed because his Y chromosome has no corresponding gene to mask its effect. For example, the blood-clotting disorder hemophilia ("bleeder's disease") is typically passed from mother to son because hemophilia is an X-linked trait. A female offspring develops hemophilia only on rare occasions, since she must inherit the gene from both parents. Other examples of X-linked traits resulting in disease are listed in Table 20–3.

Genes found only on the Y chromosome are very rare and are called **Y-linked**. One example is the SRY gene already discussed. Another example is hairy ear, which is passed only from father to son. It is not pos-

sible for a female to inherit a Y-linked trait, since females do not have Y chromosomes.

Genetic Screening

Many inherited diseases may be preventable if they are identified early. Early detection includes screening the genetic makeup of both parents before pregnancy, testing the genetics of the unborn child while in the uterus, and screening the newborn infant for disorders.

In cases in which the parents have a family history of inherited disorders, or are members of a population with a high risk of inherited disease, genetic screening before pregnancy may be a method of preventing the birth of a seriously afflicted child. For example, if a member of the mother's family has been diagnosed with cystic fibrosis, it is important for the father to be screened for this defective gene. If he carries it, an offspring will have a high risk of developing this recessive-gene disease. If the parents know this before pregnancy, they may elect to seek alternatives to having children (such as adoption or surrogate pregnancy). Examples of diseases that run a higher risk in certain population groups include sickle cell anemia (African-Americans), Tay-Sachs disease (Jews of Eastern European origin), and Gaucher's disease (also among Jews of Eastern European origin). Screening may involve the analysis of blood by sophisticated techniques to determine the presence of a mutated gene sequence, or a **pedigree analysis**, in which a particular genetic trait is traced by a counselor through several generations.

Table 20-3   X-LINKED TRAITS IN HUMANS

Trait	Description of Disease
Cleft palate	Incomplete fusion of the bony palate.
Deafness	Lack of hearing; includes three types of deafness by X linkage.
Duchenne muscular dystrophy	Degeneration of skeletal and cardiac muscle tissue. Onset is at age 6; disease progresses until death, usually by age 20.
Ectodermal dysplasia	Absence of teeth and sweat glands.
Hemophilia	Lack of certain blood-clotting factors.
Ichthyosis	Rough, scaly skin originating at birth.
Lesch-Nyhan syndrome	Enzyme deficiency resulting in mental retardation, loss of muscular control, and self-mutilating behavior.
Retinitis pigmentosa	Degeneration of the retina of the eye, resulting in blindness.
Testicular feminization syndrome	Males born with female external genitalia, breast development, and blind vagina (XY genotype).

If the parents are at high risk of conceiving a child with a genetic disorder and pregnancy has already occurred, the fetus may be tested for the disease while in the uterus. The procedure in common use for fetal testing is **amniocentesis** (am'-nē-ō-sen-TĒ-sis). Shown in Figure 20–14, amniocentesis involves the insertion of a needle through the mother's abdominal wall and into the amniotic cavity and the withdrawal of a small

Figure 20–14
Amniocentesis. (a) After the fetal position has been determined by ultrasound (shown in the monitor to the rear), a sample of amniotic fluid is withdrawn by inserting a needle attached to a syringe through the abdominal wall. (b) The needle must penetrate the amnion in order to access amniotic fluid.

What risks can you think of that may be associated with amniocentesis?

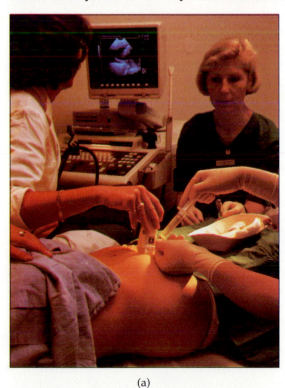

(a)

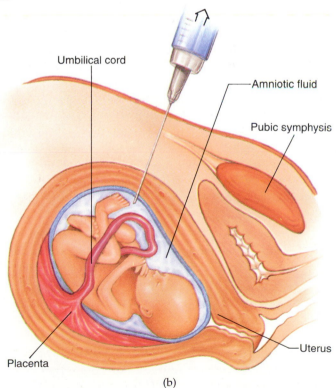

(b)

sample of fluid. Cells floating in the fluid are then subjected to genetic analysis. Amniocentesis may be used to test for many inherited disorders, but it is used mostly in pregnant women past the age of 35 to test for trisomy 21 (Down's syndrome, a form of mental retardation that has a higher incidence in pregnancies in older women).

Although genetic screening is available before and during pregnancy, often the parents are not aware of the risks of inherited diseases, and screening must thereby be performed after their child is born. Newborn infants are routinely screened for a number of common inherited disorders while in the maternity hospital. Anatomical disorders, including congenital hip dysplasia, imperforate anus, microcephaly, and many others may be detected by a physician soon after the child is born. Blood tests are required in most states to test for the metabolic enzyme deficiency phenylketonuria (PKU), several types of retardation, and hemophilia. The anatomical defects that may be detected are treated surgically, and other disorders are treated by counseling the parents on their proper management.

CONCEPTS CHECK

1. What is an allele?

2. How can you inherit a trait that is recessive?

3. How do male and female chromosomes differ?

CLINICAL TERMS OF HUMAN DEVELOPMENT

Abortion The spontaneous or deliberate termination of pregnancy. A spontaneous abortion is usually called a *miscarriage*.

Amniocentesis A diagnostic procedure in which a sample of amniotic fluid is withdrawn from the amniotic cavity during pregnancy.

Cesarean section A surgical procedure that allows an alternative to birth through the birth canal. It involves making an incision through the abdominal wall and uterine wall, and removal of the infant through the incision.

Congenital defects A variety of defective conditions in the newborn that are due to errors made during embryonic or fetal development. They are usually caused by genetic errors, although some may arise from poor nutrition, smoking, or drug abuse by the mother, or infectious diseases originating with the mother.

Ectopic pregnancy The development of an embryo outside of its normal location within the uterus. It may occur in the uterine tube or within the peritoneal cavity and usually results in very serious complications if the pregnancy is not terminated early.

Fetal jaundice The accumulation of bilirubin in the blood of a newborn infant. It is usually caused by the inefficient excretion of bilirubin by the child's liver. It is quite common and usually corrects itself within several days. Treatment is by temporary exposure to high light intensity, which helps degrade the bili rubin.

Hydatid mole A uterine tumor that originates from placental tissue.

Hydramnios An excessive amount of amniotic fluid in the amniotic cavity.

Klinefelter's syndrome A genetic defect that afflicts about 1 out of 500 males, in which at least one extra X chromosome is present (resulting in XXY). It is characterized by reduced testes, long legs, subnormal intelligence, chronic pulmonary disease, breast enlargement, and varicose veins. The severity of the abnormalities increases with greater numbers of X chromosomes, and treatment is available through hormone therapy.

Meconium A fetal discharge that usually includes blood, serous fluid, and fluid wastes from the fetus. It usually originates as amniotic fluid that is inhaled or swallowed by the fetus.

Miscarriage Also called a *spontaneous abortion*; the termination of a pregnancy by natural causes; it occurs in about 70% of all pregnancies. Causes include chromosomal abnormalities in the embryo and improper development of the embryo resulting in its death, among others. Miscarriage during late pregnancy is usually due to partial detachment of the placenta from the uterine wall or obstruction of blood flow through the umbilical cord.

Placenta previa Caused by the attachment of the placenta to the uterine wall near the cervical canal, resulting in its obstruction. It causes bleeding late in pregnancy and is hazardous to both the mother and fetus. Cesarean delivery is required.

Preeclampsia A pregnancy-related syndrome characterized by sudden hypertension, large amounts of protein in the urine, and edema throughout body tissues. It usually disappears after parturition.

Senility A loss of mental or physical ability due to old age.

Septal cardiac defects Failure of the partitions to form between atria and ventricles. It results in a condition called *cyanotic congenital heart disease*, in which the blood in a newborn is getting inadequate amounts of oxygen.

Tubal pregnancy A type of ectopic pregnancy in which the fetus develops within the uterine tube. It causes severe pain and results in rupturing of the tube if the pregnancy is not terminated early.

Turner's syndrome A genetic defect that afflicts females, in which there is an absence of one of the X chromosomes. It occurs about once in every 3000 births and is characterized by underdeveloped ovaries, underdevelopment of the external genitals, underdevelopment of the uterus and vagina, dwarfism, webbing of the skin at the neck, and cardiovascular problems. Treatment includes hormone therapy (mainly estrogens and pituitary growth hormone).

CHAPTER SUMMARY

Prenatal development is the process of change from fertilization to birth, and postnatal development extends from birth until death.

PRENATAL DEVELOPMENT

Divided into two periods: the embryonic period, during the first eight weeks; and the fetal period, which lasts until birth.

Fertilization The union of a sperm cell with an oocyte. Only one sperm cell normally penetrates the corona radiata, zona pellucida, and oocyte membrane. Fertilization results in a diploid zygote.

The First Eight Weeks of Life This time period is the embryonic period.

Early Cell Division The first divisions of the zygote are called cleavage. They result in a small ball of cells, called a morula.

Blastocyst Formation The morula differentiates to form two portions: the outer portion is the trophoblast, and the inner is the blastocyst. The trophoblast contributes to the placenta. The blastocyst becomes the embryo.

Implantation The trophoblast contacts the endometrium after about 6 days. It buries itself and the blastocyst in the endometrium. As implantation proceeds, the cells of the blastocyst differentiate to form the endoderm and ectoderm germ layers. Later during implantation, membranes begin to form around the embryo, including the chorion, amnion, and yolk sac.

Gastrulation The disc-shaped embryo begins to fold inward on itself, forming the middle germ layer, called the mesoderm. Through this process the embryo develops into the gastrula.

Neurulation Portions of the ectoderm invaginate to form the neural tube, while other ectodermal cells form the neural crest and outer surface of cells. The resulting stage of the embryo is called the neurula.

Development of Body Form The disc-shaped embryo folds to form the foregut, midgut, and hindgut, and many external features begin to appear as the embryo begins to take on a more human appearance.

Development of Organs The major organs appear during the first eight weeks, through a process of differentiation called organogenesis.

Development of the Placenta The placenta arises during the third week of life from chorionic villi, which extend into the lacunae of the endometrium. The umbilical cord connects the embryo to the placenta. Before the development of the placenta, HCG secreted by the chorion maintains the corpus luteum in the ovary, keeping progesterone and estrogen levels up. Once it has developed, the placenta secretes large amounts of estrogen and progesterone to help maintain the state of pregnancy.

Growth of the Fetus The fetal period begins at eight weeks of life. This period is marked by rapid growth of all organs and tissues by mitosis. The 38-week old fetus is considered full-term and is ready for birth.

PARTURITION

The process of giving birth.

Braxton-Hicks Contractions The pregnant uterus is prepared in advance for birth by minor contractions of the uterine wall, stimulated by high estrogen levels maintained by the placenta.

Oxytocin Secretion Stretching of the uterine wall signals the hypothalamus to secrete oxytocin, which strengthens contractions of the myometrium. Rising contractions signal the start of labor.

Three Stages of Labor

First Stage Contractions rise in frequency and strength, pushing the fetus against the cervix. In response to the pressure, the cervix dilates, and positive feedback promotes stronger contractions. Once the cervix dilates to 10 cm, the second stage begins.

Second Stage During this shorter stage, the fetus is pushed through the birth canal to the outside.

Third Stage Contractions continue after the child is born in order to push the placenta out as well.

POSTNATAL DEVELOPMENT

The period from birth to death. It includes five main stages of development.

Lactation The process of providing milk for nursing infants. Prolactin is inhibited from stimulating milk secretion until birth by the presence of high blood levels of estrogen. At birth, with the loss of the placenta, estrogen levels drop and prolactin stimulates milk production. Flow of milk, or milk letdown, is stimulated by nerve impulses that trigger the release of oxytocin by the posterior pituitary gland. Oxytocin stimulates contractile cells to contract, squeezing milk out of glands and into ducts.

Stages of Life

Infancy The first year or two of life.

Circulatory Changes Arise with the first breaths. The foramen ovale between the atria closes. The ductus arteriosus between the pulmonary trunk and the aorta closes off, forcing a greater volume of blood to the lungs. Umbilical vessels are severed.

Other Changes Include muscle, bone, and brain development and general growth.

Childhood From infancy to the onset of puberty. It is characterized by body growth and further brain development.

Adolescence From onset of puberty to the attainment of adult height. It is marked by development and

maturity of reproductive organs, completed growth, and secondary sexual characteristics.

Adulthood From the end of adolescence to the beginning of senility. It is a period of no vertical growth and of changes that accompany early signs of aging.

Old Age From about 65 or 75 years of age onward. It is characterized by the accumulating effects of aging on body cells.

GENETIC INHERITANCE

The study of the characteristics we inherit is called *genetics*.

Chromosomes and Genes There are 23 pairs of chromosomes (46 total) in most cells. Genes are located on the chromosomes. A pair of chromosomes consists of one unit inherited from the father's gamete and one unit inherited from the mother's gamete. The pair are called homologous chromosomes. Corresponding genes on homologous chromosomes are called alleles. Homozygous traits occur when alleles are identical; heterozygous traits occur when alleles are not identical. The genes in an individual that determine a particular trait are called the individual's genotype. The physical expression of the genotype is the phenotype.

Dominant-Recessive Inheritance Genotypes may consist of dominant genes, recessive genes, or a combination. Dominant genes are those that express their characteristics when combined with other genes, called recessive genes. The probability that specific genotypes and phenotypes will occur from a combination of gametes can be found by using a Punnett square.

Sex-Linked Inheritance One pair of chromosomes differs between males and females. The pairs are designated XX for female and XY for male. Because the Y (male) chromosome is structurally much smaller than the X, its gene composition differs and is more limited. Therefore, certain inherited defects that would be recessive in a female offspring can be passed directly as phenotypes from mother to son. These defects belong to a group of traits called *X*-linked traits. Y-linked traits are very rare and are passed from father to son.

Genetic Screening Genetic composition must often be determined in order to avoid diseases that may be inherited. Screening for genetic diseases is done by analyzing the genetic composition for mutated genes, or by pedigree analysis. Screening can be done before pregnancy, during pregnancy (usually by amniocentesis), or in the newborn.

QUESTIONS FOR REVIEW

OBJECTIVE QUESTIONS

1. Fusion of the oocyte and sperm cell results in a:
   a. zygote
   b. morula
   c. blastocyst
   d. primary oocyte
2. The trophoblast that forms during differentiation will eventually form the:
   a. embryo
   b. embryonic ectoderm
   c. placenta
   d. blastocyst cavity
3. Implantation normally occurs in the:
   a. chorionic villi
   b. placenta
   c. endometrium
   d. embryo
4. The process that results in the formation of the neural tube and an outer layer of ectoderm is called:
   a. gastrulation
   b. neurulation
   c. implantation
   d. organogenesis
5. The membrane that surrounds the late embryo and fetus like a "water bag" is the:
   a. chorion
   b. yolk sac
   c. amnion
   d. allantois
6. The placenta provides a barrier between the mother and the fetus by preventing large molecules and _____ from passing through.
   a. nutrients
   b. blood cells
   c. gases
   d. all of the above
7. A hormone produced by the placenta during the last 6 months of pregnancy is:
   a. HCG
   b. estrogen
   c. GnRH
   d. prolactin
8. The fetus is considered full-term at:
   a. 3 months
   b. 8 weeks
   c. 38 weeks
   d. birth, regardless of age
9. The direct effect of oxytocin on a pregnant uterus is the stimulation of:
   a. labor
   b. cervical dilation
   c. lactation
   d. all of the above
10. Passage of the fetus through the dilated cervix and the remainder of the birth canal occurs during the:
    a. first stage of labor
    b. third stage of labor
    c. second stage of labor
    d. fourth stage of labor
11. Prolactin stimulates milk production:
    a. after the placenta is lost
    b. only in the presence of oxytocin
    c. throughout pregnancy
    d. in the absence of oxytocin
12. The foramen ovale in the heart closes soon after:
    a. the infant takes the first breaths
    b. blood begins to flow through vessels
    c. the heart begins to beat
    d. childhood begins
13. Blood flow to the lungs increases once the _____ closes.
    a. umbilical vein
    b. aorta
    c. ductus arteriosus
    d. mitral valve
14. The stage of life during which reproductive organs mature and secondary sexual characteristics develop is called:
    a. adolescence
    b. adulthood
    c. childhood
    d. infancy
15. Identical alleles on homologous chromosomes determine traits that are known as:
    a. homozygous
    b. dominant
    c. heterozygous
    d. X-linked
16. An inherited trait that is determined by genes on the X chromosome, but not on the Y chromosome, is called:
    a. autosomal
    b. Y-linked
    c. X-linked
    d. dominant

ESSAY QUESTIONS

1. Describe the events of fertilization, beginning with ovulation and ending with zygote formation. Include a description of where these events usually occur.
2. Describe the sequence of events in their proper order that take place during the first eight weeks of human development.
3. Describe the events associated with parturition. Explain the influences of estrogen and oxytocin.
4. Explain how the placenta develops from early embryonic development. Indicate also its major roles during the last 6 months of pregnancy.
5. Describe the changes that occur with each successive stage of postnatal development.
6. Explain why genetic screening is important for persons in high-risk populations.

ANSWERS TO ART LEGEND QUESTIONS

Figure 20-1
Combining of the sperm and oocyte chromosomes results in the formation of what structure?   A zygote is formed when the sperm and oocyte chromosomes combine.

Figure 20-2
Where do fertilization and implantation normally occur?   Fertilization usually occurs within the upper one-third of the uterine tube, and implantation occurs in the uterine wall.

Figure 20-3
What are the three primary germ layers? The primary germ layers are the endoderm, the mesoderm, and the ectoderm.

Figure 20-4
The neural tube arises from which primary germ layer? The neural tube arises from the ectoderm.

Figure 20-5
During the first 30 days of development, from what source do you suppose the embryo obtains most of its nourishment? The embryo receives nourishment from the yolk sac during its first 30 days.

Figure 20-6
In addition to the brain, name one other organ that is derived from the ectoderm. The spinal cord is also derived from the ectoderm. Other tissues include the epidermis and its derivatives in the skin, the adrenal medulla, and mucous membranes.

Figure 20-7
What fetal membrane contributes primarily to the structure of the placenta? The chorion contributes to the placenta.

Figure 20-8
What fetal membrane surrounds the fetus in a bath of fluid? The amnion surrounds the fetus within a fluid-filled chamber.

Figure 20-9
What hormone produced by the pituitary gland stimulates uterine contractions during parturition? Oxytocin stimulates uterine contractions.

Figure 20-11
What major organ of the chest is partially bypassed in the fetal circulatory scheme? The lungs are partially bypassed.

Figure 20-12
In a cross between an albino father and a heterozygous mother, what is the probability that an offspring will be albino? The probability that an offspring will be albino from such a cross is 25%.

Figure 20-13
Which chromosome contains the SRY gene, which determines male development? The Y chromosome (only) contains the SRY gene.

Figure 20-14
What risks can you think of that may be associated with amniocentesis? Risks connected with amniocentesis include infection to the mother, infection to the fetus, and penetration of the fetus by the needle, leading to hemorrhage and possible miscarriage.

FURTHER READINGS

CHAPTER 1

Clemente, C.D. (Ed.). *Gray's Anatomy of the Human Body, 30th Am. Ed.* Lea & Febiger, Philadelphia. 1985

Crouch, J.E. *Functional Human Anatomy, 4th Ed.* Lea & Febiger, Philadelphia. 1985.

Crowley, L.V. *Introduction to Human Disease.* Wadsworth, Monterey. 1983.

Mulvihill, M.L. *Human Diseases: A Systemic Approach, 3rd Ed.* Appleton & Lange, Norwalk (Connecticut). 1991

Warden-Tamparo, C., and M.A. Lewis. *Diseases of the Human Body.* F.A. Davis Co., Philadelphia. 1989.

CHAPTER 2

Doolittle, R.F. "Proteins." *Scientific American,* 253:88–89. 1985.

Felsenfeld, G. "The Molecules of Life." *Scientific American,* 253:48–57. 1985.

Stryer, L. *Biochemistry, 3rd Ed.* W.H. Freeman & Co., San Francisco. 1987.

CHAPTER 3

Alberts, B., et al. *Molecular Biology of the Cell, 2nd Ed.* Garland Pub., New York. 1991.

Crick, F.H.C. "The Genetic Code." *Scientific American,* 215:55(4). 1966.

Ezzell, C. "A Time to Live, a Time to Die." *Science News,* 142:344 (November). 1992.

Floyd, R.L. "Oxidative Damage to Behavior During Aging." *Science,* 254:1597. 1991.

Hayflick, L. "The Cell Biology of Aging." *Clin. Geriatr. Med.,* 1(1):15–27. 1986.

NIH/CEPH Collaborative Mapping Group. "A Comprehensive Genetic Linkage Map of the Human Genome." *Science,* 258:67. 1992.

Prescott, D.M. *Reproduction of Eukaryotic Cells.* Academic Press, New York. 1976.

Prescott, D.M. *Cells.* Jones & Bartlett Pub., Boston. 1988.

Rusting, R. "Why Do We Age?" *Scientific American,* 267:131. 1992.

Siekevitz, P. "Powerhouse of the Cell." *Scientific American,* 197:131. 1957.

Singer, S.J. and Nicholson, G.L. "The Fluid Mosaic Model of the Structure of Cell Membranes." *Science,* 175:720–731. 1972.

Watson, J.D. *The Double Helix.* New American Library, New York. 1968.

CHAPTER 4

Caplan, A.I. "Cartilage." *Scientific American,* 251(4):84–94. 1984.

di Fiore, M.S. *Atlas of Normal Histology, 6th Ed.* Lea & Febiger, Philadelphia. 1986.

Fawcett, D.W. *A Textbook of Histology, 11th Ed.* W.B. Saunders, Philadelphia. 1986.

Henderson, B.E., R.K. Ross, and M.C. Pike. "Toward the Primary Prevention of Cancer." *Science,* 254: 1131–1137. 1991.

Kessel, R.G., and R.H. Kardon. *Tissues and Organs: A Text-Atlas of Scanning Electron Microscopy, 2nd Ed.* W.H. Freeman & Co., San Francisco. 1989.

Moran, D.T., and J.C. Rowley. *Visual Histology.* Lea & Febiger, Philadelphia. 1988.

CHAPTER 6

Edelson, L.E., and J.M. Fink. "The Immunologic Function of Skin." *Scientific American,* 252:46–53. 1983.

Marples, Mary J. "Life On the Human Skin." *Scientific American,* January. 1969.

Montagna, W., and P.F. Parakkal. *The Structure and Function of the Skin.* Academic Press, New York. 1974.

Pillsbury, D.M., and C.L. Heaton. *A Manual of Dermatology, 2nd Ed.* W.B. Saunders, Philadelphia. 1980.

Skerrett, P.J. "Matrix Algebra Heals Life's Wounds." *Science,* 252:1064–1066. 1991.

CHAPTER 7

Allman, W.F. "The Knee." *Science,* 122–123. 1983.

Basmajian, J.V. *Grant's Method of Anatomy, 10th Ed.* Williams and Wilkins, Baltimore. 1980.

Chase, R.A. *The Bassett Atlas of Human Anatomy.* The Benjamin Cummings Pub. Co., Inc., Redwood City (California). 1989.

Fawcett, D.W. *A Textbook of Histology, 11th Ed.* W.B. Saunders, Philadelphia. 1986.

Yokochi, C., J.W. Rohen, and E.L. Weinreb. *Photographic Atlas of the Human Body, 3rd Ed.* Igaku-Shoin Medical Publishers, Inc., New York. 1989.

CHAPTER 8

Bower, B. "Pumped Up and Strung Out." *Science News,* July, 1991.

Endo, M. "Calcium Release from the Sarcoplasmic Reticulum." *Physiological Reviews,* 57:71. 1971.

Hoyle, G. "How a Muscle is Turned Off and On." *Scientific American,* April. 1970.

Huxley, H.E. "The Mechanism of Muscular Contraction." *Scientific American,* December. 1985.

Rasch, P.J. *Kinesiology and Applied Anatomy, 7th Ed.* Lea & Febiger, Philadelphia. 1989.

Wilson, F.C. *The Musculoskeletal System: Basic Processes and Disorders, 2nd Ed.* J.B. Lippincott, Philadelphia. 1983.

CHAPTER 9

Barr, M.L., and J.A. Kierman. *The Human Nervous System, 4th Ed.* Harper & Row, New York. 1983.

Catterall, W.A. "The Molecular Basis of Neuronal Excitability." *Science,* 223:653. 1984.

Gershon, E.S., and R.O. Rieder. "Major Disorders of Mind and Brain." *Scientific American,* 267:126. 1992.

Goldstein, G.W., and A.L. Betz. "The Blood-Brain Barrier." *Scientific American,* 225:74–83. 1986.

Hoffman, M. "On the Trail of the Errant T Cells of Multiple Sclerosis." *Science,* 254:521. 1991.

Kimura, D. "Sex Differences in the Brain." *Scientific American,* 267:118. 1992.

Marx, J.L. "Regeneration in the Central Nervous System." *Science,* 209:378. 1980.

Shatz, C.J. "The Developing Brain." *Scientific American,* 267:60. 1992.

Travis, J. "Spinal Cord Injuries: New Optimism Blooms for Developing Treatments." *Science,* 258:218. 1992.

CHAPTER 10

Barinaga, M. "Playing 'Telephone' With the Body's Message of Pain." *Science,* 258:1058. 1992.

Cain, W.S. "To Know With the Nose: Keys to Color Identification." *Science,* 203:467. 1979.

Crick, F., and C. Koch. "The Problem of Consciousness." *Scientific American,* 267:152. 1992.

Desimone, R. "The Physiology of Memory: Recordings of Things Past." *Science,* 258:245. 1992.

Glickstein, M. "The Discovery of the Visual Cortex." *Scientific American,* 295:118. 1988.

Kandel, E.R., and R. Hawkins. "The Biological Basis of Learning and Individuality." *Scientific American,* 267:78. 1992.

Koretz, J.F., and G.H. Handelman. "How the Human Eye Focuses." *Scientific American,* 259:92. 1988.

Mischkin, M., and T. Appenzeller. "The Anatomy of Memory." *Scientific American,* 256:80. 1987.

Schnapf, J.L., and D.A. Baylor. "How Photoreceptor Cells Respond to Light." *Scientific American,* 256:40. 1987.

Selkoe, D.J. "Aging Brain, Aging Mind." *Scientific American,* 267:134. 1992.

Squire, L.R., and S. Zola-Morgan. "The Medial Temporal Lobe Memory System." *Science,* 253:1380. 1991.

CHAPTER 11

Allman, W.F. "Steroids in Sports: Do They Work?" *Science,* 4:14. 1983.

Atkinson, M.A., and N. MacLaren. "What Causes Diabetes?" *Scientific American,* 262. 1990.

Crapo, L. *Hormones: Messengers of Life.* W. H. Freeman & Co., San Francisco. 1985.

Martin, C. *Endocrine Physiology.* Oxford University Press, New York. 1985.

McGarry, J.D. "What if Minkowski Had Been Agnostic? An Alternative Angle on Diabetes." *Science,* 258:766. 1992.

Noyori, R., and M. Suzuki. "Organic Synthesis of Prostaglandins: Advanced Biology." *Science,* 259:44. 1993.

Williams, R.H. *Textbook of Endocrinology, 6th Ed.* W.B. Saunders, Philadelphia. 1981.

CHAPTER 12

Bank, A., et al. "Disorders of Human Hemoglobin." *Science,* 207:486. 1980.

Doolittle, R.F. "Fibrinogen and Fibrin." *Scientific American,* 245:126. 1981.

Golde, D.W. "The Stem Cell." *Scientific American,* 265:6. 1991.

Golde, D.W., and J.C. Gasson. "Hormones That Stimulate the Growth of Blood Cells." *Scientific American, 259* (July). 1988.

Perutz, M.F. "Hemoglobin Structure and Respiratory Transport." *Scientific American, 239*:92. 1978.

Pittiglio, D.H., and R.A. Sacher. *Clinical Hematology and Fundamentals of Hemostasis.* F.A. Davis Co., Philadelphia. 1987.

Zucker, M.B. "The Functioning of Blood Platelets." *Scientific American,* 242:86. 1980.

CHAPTER 13

Berne, R.M., and M.N. Levy. *Cardiovascular Physiology, 4th Ed.* C.V. Mosby & Co., St. Louis. 1981.

Cantin, M., and J. Genest. "The Heart as an Endocrine Gland." *Scientific American,* 254:76. 1986.

Cohn, P.F. *Clinical Cardiovascular Physiology.* W.B. Saunders, Philadelphia. 1985.

Johansen, K. "Aneurisms." *Scientific American,* 247:110. 1982.

Lawn, R.M. "Lipoprotein (a) in Heart Disease." *Scientific American,* 266:54. 1992.

Raloff, J. "Hearty Vitamins." *Science News,* 1 August. 1992.

Silberner, J. "Anatomy of Atherosclerosis." *Science News,* 16 March. 1985.

CHAPTER 14

Ada, G.L., and G. Nossal. "The Clonal-Selection Theory." *Scientific American,* 257:62. 1987.

Anderson, R.M., and R.M. May. "Understanding the AIDS Pandemic." *Scientific American,* 266:58. 1992.

Boon, T. "Teaching the Immune System to Fight Cancer." *Scientific American,* 266:82. 1993.

Cohen, Irun R. "The Self, the World, and Autoimmunity." *Scientific American,* 258:52. 1988.

Eisen, H.M. *Immunology.* Harper & Row, Hagerstown (Maryland). 1980.

Gallo, R.C. "The AIDS Virus." *Scientific American,* 256:47. 1987.

Huffer, T.L., D.J. Kanapa, and G.W. Stevenson. *Introduction to Human Immunology.* Jones and Bartlett, Boston. 1986.

Murray, J.E. "Human Organ Transplantation: Background and Consequences." *Science,* 256:1411. 1992.

Salk, J., P.A. Bretscher, P.L. Salk, M. Clerici, and G.M. Shearer. "A Strategy for Prophylactic Vaccination Against HIV." *Science,* 260:1270. 1993.

von Boehmer, H., and P. Kisielow. "How the Immune System Learns About Self." *Scientific American,* 265:4. 1991.

Weiss, R.A. "How Does HIV Cause Aids?" *Science,* 260:1273. 1993.

Yound, J.I., and Z.A. Cohn. "How Killer Cells Kill." *Scientific American,* 258:38. 1988.

CHAPTER 15

Bloom, B.R., and C.J.L. Murray. "Tuberculosis: Commentary on a Reemergent Killer." *Science,* 257:1055. 1992.

Dantzker, D.R. "Physiology and Pathophysiology of Pulmonary Gas Exchange." *Hospital Practice,* January. 1986.

Editors, "The New Tuberculosis Threat." *University of California at Berkeley Wellness Letter,* 8:9 (June). 1992.

Friday, G.A. "The Special Challenges of Managing Asthma in Children." *Journal of Respiratory Diseases,* 12:639. 1991.

Nayle, R.R. "Sudden Infant Death." *Scientific American,* 242 (April). 1980.

Slonim, N.B. *Respiratory Physiology, 4th Ed.* C.V. Mosby & Co., St. Louis. 1981.

Stephenson, J. "Positively Breathtaking." *Harvard Health Letter,* December. 1992.

Stroh, M. "Breathing Lessons." *Science News,* 141:314 (May). 1992.

Weiss, R.A. "TB Trouble: Tuberculosis Is On the Rise Again." *Science News,* 133:92 (Feb.). 1988.

CHAPTER 16

Davenport, H.W. *Physiology of the Digestive Tract, 5th Ed.* Year Book Medical Publishers, Chicago. 1982.

Imrie, C.W., and A.R. Moossa. *Gastrointestinal Emergencies.* Churchill Livingstone, New York. 1987.

Littman, A. "Lactase Deficiency Diagnosis and Management." *Hospital Practice,* January. 1987.

Moog, F. "The Lining of the Small Intestine." *Scientific American,* 245 (November). 1981.

Uvnas-Moberg, K. "The Gastrointestinal Tract in Growth and Reproduction." *Scientific American,* 261 (July). 1989.

Williams, R. *Liver Failure.* Churchill Livingstone, New York. 1986.

CHAPTER 17

Editors. "The Diet/Cancer Connection." *The Johns Hopkins Medical Letter,* 4:4 (May). 1992.

Editors. "The New Vegetarianism." *University of California at Berkeley Wellness Letter,* 9:4 (March). 1993.

Lawn, R.M. "Lipoprotein (a) in Heart Disease." *Scientific American,* 266:54. 1992.

Raloff, J. "Cancer-Fighting Food Additives." *Scientific News,* 141:104 (February). 1992.

CHAPTER 18

Brenner, B., F.L. Coe, and F.C. Rector, Jr. *Renal Physiology in Health and Disease.* W.B. Saunders, Philadelphia. 1987.

Hills, A.G. *Acid-Base Balance.* Williams and Wilkins, Baltimore. 1973.

Murray, J.E. "Human Organ Transplantation: Background and Consequences." *Science,* 256:1411. 1992.

Thomsen, D.E. "Plasma Physics Breaks Stones." *Science News,* 130:157 (September). 1986.

Vander, A.J. *Renal Physiology, 3rd Ed.* McGraw-Hill, New York. 1985.

CHAPTER 19

Bennett, D.D. "Pelvic Inflammatory Disease." *Science News,* 127:263 (April). 1985.

Fackelmann, K.A. "The Neglected Sex Gland." *Science News,* 142:94 (August). 1992.

Grabowski, C.T. *Human Reproduction and Development.* W.B. Saunders, Philadelphia. 1983.

Jones, K.L., et. al. *Dimensions of Human Sexuality.* Wm. C. Brown, Dubuque. 1985.

Marshall, J.C., and R.P. Kelch. "Gonadotropin-releasing Hormone: Role of Pulsatile Secretion in the Regulation of Reproduction." *Science,* 315:1459. 1986.

Wassarman, P.M. "The Biology and Chemistry of Fertilization." *Science,* 235:553. 1987.

CHAPTER 20

Anderson, W.F. "Human Gene Therapy." *Science,* 256:808. 1992.

Bower, B. "Teenage Turning Point." *Science News,* 139:184 (March). 1991.

Collins, F.S. "Cystic Fibrosis: Molecular Biology and Therapeutic Implications." *Science,* 256:774 (May). 1992.

Fackelmann, K.A. "Anatomy of Alzheimer's." *Science News,* 142:394 (December). 1992.

Fischer, K., and A. Lazerson. *Human Development: From Conception Through Adolescence.* W.H. Freeman, San Francisco. 1984.

Lagercrantz, H., and T.A. Slotkin. "The Stress of Being Born." *Scientific American,* 251 (April). 1984.

Moore, K.L. *The Developing Human, 3rd Ed.* W.B. Saunders, Philadelphia. 1982.

Rusting, R. "Why Do We Age?" *Scientific American,* 267:131. 1992.

GLOSSARY

The definitions in this glossary are those that are in most common usage for the terms as they apply in the field of human anatomy and physiology. Most terms you will encounter here are followed by a phonetic guide to pronunciation. Pronunciation is based on standard clinical usage as presented in *Stedman's Medical Dictionary,* which has been widely used as a desk reference in terminology for many years. The pronunciation key follows these rules:

1. The syllable with the strongest accent appears in capital letters. For example,

    science (SĪ-ens) and learning (LER-ning)

2. A secondary accent, in words that have one, is identified by a small accent mark ('). For example,

    homeostasis (hō'-mē-ō-STĀ-sis)

3. Vowels with a line above are pronounced long (the hard sound). For example,

    feet (fēt), base (bās), and site (sīt)

4. Vowels with no line above are short (the soft sound). For example,

    atom (A-tum), duct (dukt), and retina (RE-ti-na)

5. Other phonetic clues to sounds may be found in the following examples:

    oo as in *blue*

    ar as in *fair*

    oy as in *oil*

    ah as in *father* (FAH-ther)

abdomen (AB-dō-men)   The regional area between the diaphragm and pelvis.

abdominal cavity   The membrane-bound space between the diaphragm, abdominal wall, and pelvis.

abdominopelvic (ab-do'-mi-nō-PEL-vik) *cavity*   The membrane-bound space that includes the abdominal cavity and pelvic cavity.

abduction (ab-DUK-shun)   Movement away from the midline.

abortion (a-BOR-shun)   The premature loss or removal of an embryo or fetus.

abscess (AB-ses)   A collection of pus and liquefied tissue in a localized area of the body.

absorption (ab-SORB-shun)   The passage of digested foods from the digestive tract and into the bloodstream.

accommodation (a-kom'-ō-DĀ-shun)   A change in the thickness and curvature of the lens to adjust for vision as an object moves closer to the eye (focusing).

acetabulum (a'-se-TAB-yoo-lum)   A cup-shaped depression on the lateral surface of the coxal (hip) bone. It receives the head of the femur to form the hip joint.

acetylcholine (as-ē-til-KŌ-lēn) **(ACh)**   A neurotransmitter released from motor neurons that stimulates muscle contraction. It is also released by autonomic preganglionic neurons, postganglionic parasympathetic neurons, some postganglionic sympathetic neurons, and some central nervous system neurons.

Achilles (a-KIL-ēs) *tendon*   Also called the *calcaneus tendon;* the common tendon of the calf muscles, which originates from the calcaneus (heel).

acid (AS-id)   Any substance that releases hydrogen ions in a solution. It is characterized by a pH of less than 7.

acidosis (as-i-DŌ-sis)   A disorder characterized by a blood pH less than normal (less than 7.35).

acini (AS-i-nī)   Clusters of cells in the pancreas that secrete digestive enzymes.

acoustic (a-KOOS-tik)   Pertaining to sound or the sense of hearing.

acromegaly (ak'-rō-MEG-a-lē)   A disorder caused by hypersecretion of growth hormone released by the anterior pituitary gland during adulthood. It results in progressive enlargement of the bones of the head, hands, feet, and thorax.

actin (AK-tin)   The contractile protein that makes up the major portion of thin filaments in muscle fibers.

action potential   A rapid change in membrane potential in a neuron or muscle fiber that results in a reversal of electric charges along each side of the plasma membrane. It is also called a *nerve impulse.*

active transport   The movement of materials across the plasma membrane of a cell that requires carrier assistance and energy (in the form of ATP). It results in the movement of materials against a concentration gradient.

acuity (a-KYOO-i-tē)   Sharpness or clarity of a particular sensation.

acute (a-KYOOT)   Describing any disorder that may exhibit a rapid onset and severe symptoms and follow a relatively brief course; not chronic.

adaptation (a-dap-TĀ-shun)   The adjustment of the pupil of the eye to variations in light.

Addison's disease   A disorder caused by the hyposecretion of glucocorticoids by the adrenal gland. It is characterized by hypoglycemia, muscular weakness, anorexia, low blood pressure, and dehydration.

adduction (ad-DUK-shun)   Movement toward the midline.

adenosine triphosphate (a-DEN-e-sēn tri-FOS-fāt) *(ATP)*   A chemical compound composed of a sugar molecule (adenosine) and three phosphate groups. The energy stored in its bonds is used to drive nearly every energy-requiring reaction in the body.

adipose (AD-i-pōs)   A type of loose connective tissue containing adipocytes, or fat cells.

adrenal cortex (a-DRĒ-nal KOR-teks)   The outer region of each adrenal gland, located superior to the kidney. It secretes steroid hormones, including glucocorticoids and mineralocorticoids.

adrenal glands   Two endocrine glands, each situated superior to a kidney. Also called *suprarenal glands.*

adrenal medulla   The inner region of each adrenal gland, which secretes epinephrine and norepinephrine.

adrenaline (a-DREN-a-lin)   A synonym for *epinephrine.*

adrenocorticotropic (a-drē′-nō-kort′-i-kō-TRŌ-pik) *hormone (ACTH)*   A hormone released by the anterior pituitary gland that stimulates the adrenal cortex to produce some of its hormones.

adventitia (ad′-ven-TISH-ya)   The outermost covering of an organ or a tissue.

aerobic (ā-RŌ-bik)   A process requiring molecular oxygen.

afferent (AF-er-ent)   Leading toward a central point. For example, an afferent arteriole of the kidney leads to the glomerulus, and an afferent neuron carries nerve impulses to the central nervous system.

agglutination (a-gloo′-ti-NĀ-shun)   The process by which cells stick together to form clumps.

albumin (al-BYOO-min)   The most abundant and smallest of the plasma proteins. It plays an important role in regulating osmotic pressure in blood.

aldosterone (al-DOS-ter-ōn)   A steroid hormone produced by the adrenal cortex. It facilitates sodium and water reabsorption and potassium excretion in the kidneys.

aldosteronism (al-DOS-ter-ōn-izm)   A disorder caused by hypersecretion of aldosterone by the cortex of the adrenal gland that leads to increased sodium levels in the blood. Its symptoms include muscle paralysis, edema, and high blood pressure.

alimentary canal (al′-i-MEN-ta-rē ca-NAL)   The tube of the digestive system that includes the mouth, esophagus, stomach, small intestine, and large intestine. Also called the *digestive tract* and *gastrointestinal (GI) tract.*

alkaline (AL-ka-līn)   Describing a solution that contains more hydroxyl ions (OH⁻) than hydrogen ions (H⁺). It is characterized by a pH greater than 7.

alkalosis (al-ka-LŌ-sis)   A disorder characterized by a blood pH greater than 7.45.

alleles (a-LĒ-elz)   Genes that control the same trait and are located on the same position in homologous chromosomes.

allergen (AL-ler-jen)   A foreign substance, or antigen, that stimulates a hypersensitivity reaction.

alpha cell   A type of cell in the pancreatic islets of Langerhans that secretes the hormone glucagon.

alveolar (al-VĒ-ō-lar) *duct*   A branch of a respiratory bronchiole within the lungs that leads to alveoli and alveolar sacs.

alveolar sac   Two or more alveoli that share a common opening from an alveolar duct.

alveolus (al-VĒ-ō-lus)   A microscopic air sac within the lungs. Plural is *alveoli.*

Alzheimer's (ALTZ-hī-merz) *disease*   A progressive neurologic disorder characterized by a growing loss of functioning cerebral neurons, resulting in intellectual impairment, memory loss, and personality changes.

amenorrhea (ā-men-ō-RĒ-a)   Absence of menstruation.

amino acid   A class of organic compounds characterized by the presence of a carboxyl group (COOH) and an amino group (NH₂). They are the subunits of proteins and enzymes.

amnesia (am-NĒ-zē-a)   The lack or loss of memory.

amniocentesis (am′-nē-ō-cen-TĒ-sis)   A procedure involving the extraction of fetal amniotic fluid during pregnancy for analysis.

amnion (AM-nē-on)   The innermost fetal membrane, which surrounds the fetus and amniotic fluid. It is also called the "water bag."

amylase (AM-i-lās)   An enzyme produced by the salivary glands and pancreas that cleaves starch, glycogen, and other polysaccharides during chemical digestion.

anabolism (a-NAB-ō-lizm)   Collective term for metabolic processes that synthesize larger molecules from their subunits. It requires energy.

anaerobic (an′-a-RŌ-bik)   Describing a process that does not require molecular oxygen; opposite of *aerobic.*

anal (Ā-nal) *canal*   The terminal 2 or 3 cm of the rectum. It opens to the exterior at the anus.

analgesia (an-al-JĒ-zē-a)   Reduction of pain sensation.

anaphase (AN-a-fāz)   The third stage in mitosis, in which chromatids separate and move to opposite poles of the cell.

anatomic dead space volume   The volume of air that is inhaled and remains in spaces in the upper respiratory tract and does not reach the alveoli to participate in gas exchange. It averages 150 ml in volume.

anatomical position   The reference position, in which the subject is standing erect with the feet facing forward, arms at the sides, and the palms of the hands facing forward (the thumbs to the outside).

anatomy (a-NA-to-mē)   The study of the structure of the body.

androgen (AN-drō-jen)   Any substance that produces or stimulates male characteristics, such as the hormone testosterone.

anemia (a-NĒ-mē-a)   A disorder of the blood in which the number of functional erythrocytes or their hemoglobin content is lower than normal.

anesthesia (an'-es-THĒZ-ē-a)   A loss of sensation to a local body part or more generalized body region.

aneurysm (AN-yoo-rizm)   An abnormal expansion of a portion of a blood vessel wall, caused by a weakening of the wall and/or high blood pressure.

angina pectoris (an-Ī-na PEK-tō-ris)   A pain in the chest that is associated with reduced coronary circulation.

angioplasty (AN-jē-ō-plas'-tē)   A technique involving the insertion of a catheter with a balloon into an occluded blood vessel. Once the device is inserted, air is pumped into the balloon to force the blocked vessel lumen to enlarge.

angiotensin (an'-jē-ō-TEN-sin)   Two forms of a protein (angiotensin I and angiotensin II) involved in the regulation of blood pressure. Angiotensin I is a precursor, and angiotensin II stimulates the secretion of aldosterone from the adrenal cortex.

anion (AN-ī-on)   A negatively charged ion.

anorexia nervosa (an-ō-REKS-ē-a ner-VŌ-sa)   A chronic disorder characterized by a low self-image that leads to a severe, self-induced weight loss.

anoxia (an-OK-sē-a)   Deficiency of oxygen.

antagonist (an-TAG-ō-nist)   A muscle that has an action opposite to that of the agonist (prime mover), such that it relaxes when the agonist contracts and vice versa.

anterior (an-TER-ē-or)   Toward the belly or front of the body. In humans, that direction is also called *ventral*.

anterior horn   A region of the spinal cord gray matter containing the cell bodies of motor neurons. It is also called the *ventral horn*.

anterior root   The structure emerging from the spinal cord on its anterior aspect that contains axons of motor neurons. It is also called the *ventral root*.

anterior pituitary gland   The portion of the pituitary gland located at the base of the brain and composed of glandular epithelium.

antibiotic   A substance produced by a microorganism that inhibits or kills other microorganisms. Literally, it means "against life."

antibody   A protein in the plasma produced by a type of activated white blood cell. Once produced, the antibody binds to an antigen to neutralize it.

anticoagulant (an'-tī-kō-AG-yoo-lant)   A substance that prevents blood clotting or coagulation.

antidiuretic (an'-tī-dī-yoo-RET-ik)   A substance that inhibits urine formation.

antidiuretic hormone (ADH)   A hormone produced by the hypothalamus and released by the posterior pituitary gland that promotes the conservation of water.

antigen (AN-ti-jen)   A substance that stimulates certain white blood cells to produce antibodies.

anuria (a-NOO-rē-a)   A disorder characterized by a daily output of urine of less than 50 ml.

anus (Ā-nus)   The distal end and outlet of the rectum.

aorta (ā-OR-ta)   The main trunk of the systemic circulatory circuit. It originates from the left ventricle.

aortic semilunar valve   One of four heart valves. It consists of three cusps that are attached to the wall of the aorta near its origin from the left ventricle. It is also called the *aortic valve.*

apex (Ā-peks)   The extremity of a conical or pyramidal structure. The apex of the heart is the rounded, inferiormost tip, which points to the left side.

apnea (AP-nē-a)   A temporary cessation of breathing.

apocrine (AP-ō-krin)   One of two types of sudoriferous (sweat) glands. Apocrine glands become active at puberty and respond during periods of emotional stress.

aponeurosis (ap'-ō-noo-RŌ-sis)   A broad, sheetlike tendon joining one muscle with another or with a bone.

appendicitis (a-pen-di-SĪ-tis)   Inflammation of the vermiform appendix.

appendicular (ap'-en-DIK-yoo-lar)   Pertaining to the major appendages, the arms and legs.

aqueous humor (AK-wē-us HYOO-mor)   The watery fluid that fills the anterior compartment of the eye.

arachnoid (a-RAK-noyd)   The middle of three connective tissue coverings (meninges) of the brain and spinal cord.

areolar (a-RĒ-ō-lar)   A type of connective tissue with sparse protein fibers in the matrix (loose connective tissue).

arrector pili (a-REK-tor PĪ-lē)   Smooth muscle fibers attached to hair within the skin.

arrhythmia (a-RITH-mē-a)   Irregular cardiac cycle characterized by a disrupted heart rhythm.

arteriole (ar-TĒ-rē-ōl)   A small branch from a larger artery that delivers blood to a capillary.

artery (AR-ter-ē)   A blood vessel that carries blood away from the heart.

arthritis (ar-THRĪ-tis)   Inflammation of a joint.

articular cartilage (ar-TI-kyoo-lar KAR-ti-lej)   The cartilage that covers the end of a bone where it forms a joint with another bone.

articulation (ar-tik'-yoo-LĀ-shun)   A synonym for a joint, which is a point of contact between two opposing bones.

ascending colon (ā-SEN-ding KŌ-lon)   A segment of the large intestine that extends from the cecum to the transverse colon.

ascites (ā-SĪ-tēz)   An abnormal accumulation of serous fluid within the peritoneal cavity.

asphyxia (as-FIKS-ē-a)   A reduction in oxygen levels in the blood, leading to loss of consciousness.

association area   A region of the brain's cerebral cortex where sensory and motor interconnections occur. It is the primary site of memory, reasoning, and thought processes.

association neuron   A neuron within the central nervous system that interconnects a sensory neuron with a motor neuron. Also called an *interneuron.*

asthma (AS-ma)   A disorder of the respiratory system characterized by excessive production of mucus in the bronchi and bronchiole wall constriction, resulting in shortness of breath.

astigmatism (a-STIG-ma-tizm)   An irregularity of the lens or cornea of the eye, or shape of the eyeball, resulting in an inability to focus clearly on an object.

astrocyte (AS-trō-sīt)   A star-shaped neuroglial cell of the central nervous system.

atelectasis (at'-e-LEK-te-sis)   A collapse of some or all alveoli within the lung, reducing the efficiency of gas exchange.

atherosclerosis (a'-ther-ō-skle-RŌ-sis)   A gradual process of cell damage to the wall of a medium- or large-sized artery, causing a gradual thickening and an accumulation of fat molecules along the wall. This effect reduces the lumen diameter and the flow of blood to an organ.

atmospheric pressure (at'-mos-FER-ik PRESH-ur)   The pressure exerted by air molecules at sea level, which is measured as 760 mm of mercury.

atom (A-tum)   A unit of matter that makes up a chemical element. It consists of a nucleus and electrons.

atomic number   The number of protons in an atom of an element.

atomic weight   A number approximately equal to the number of protons and neutrons in an atom of an element.

atrial natriuretic factor (Ā-trē-al nā'-trē-yoo-RET-ic FAK-tor)   A hormone produced by the atria of the heart in response to atrial stretching. It inhibits aldosterone production, thus lowering blood pressure.

atrioventricular (ā-trē-ō-ven-TRIK-yoo-lar) *(AV) bundle*   A portion of the heart conduction system that serves to carry impulses from the AV node to Purkinje fibers, which distribute the impulses to the walls of the ventricles.

atrioventricular (AV) node   A cluster of specialized cardiac (heart) cells in the right atrial wall that serve as part of the heart conduction system.

atrioventricular (AV) valve   Two of four heart valves, each of which is located between an atrium and a ventricle. The right AV valve is also called the *tricuspid valve,* and the left AV valve is also called the *bicuspid* or *mitral valve.*

atrium (Ā-trē-um)   One of two superior chambers of the heart. Plural form is *atria.*

atrophy (A-trō-fe)   A decrease in size of a tissue or body part, due to lack of use, inadequate nutrition, or disease.

audation (ah-DĀ-shun)   The sense of hearing.

auditory ossicles (AH-di-tor-ē OS-ik-elz)   The three small bones within the middle ear that transmit sound vibrations: the malleus, incus, and stapes.

auditory tube   An epithelium-lined tube that extends between the pharynx and the middle ear. It is also called the *Eustachian tube.*

autoimmune (ah'-tō-i-MYOON) *disease*   A disorder of the immune response in which a person's own components of immunity attack functional cells.

autonomic ganglion (ah'-tō-NOM-ik GĀNG-lē-on)   A cluster of neuron cell bodies that lie outside the central nervous system, serving either the sympathetic or the parasympathetic division.

autonomic nervous system (ANS)   A division of the peripheral nervous system that serves body functions not requiring conscious control. It is divided into sympathetic and parasympathetic subdivisions.

autosome (AH-tō-sōm)   Any chromosome other than a sex chromosome.

avitaminosis (ā-vīt-a-min-Ō-sis)   A deficiency of a vitamin in the diet.

axial (AK-sē-al)   The body region along the midline, which includes the head, neck, and trunk.

axilla (ak-SIL-a)   The small depression beneath the arm where it joins the trunk of the body. It is also called the *armpit.*

axon (AK-son)   A long process of a neuron that carries a nerve impulse away from the cell body.

B cell   A type of lymphocyte that reacts against antigens by changing into a plasma cell and producing antibodies that inactivate the antigens.

baroreceptor (bar'-ō-rē-SEP-tor)   Sensory nerve ending that detects the degree of stretch in the walls of the heart and arteries. It is also called a *pressoreceptor.*

basal ganglia (BĀ-sal GĀNG-lē-a)   Clusters of neuron cell bodies that make up the central gray matter in each of the cerebral hemispheres. They are also called *cerebral nuclei.*

basal metabolic rate (BMR)   The rate of body metabolism measured under ideal conditions of minimal activity.

base   A hydrogen-ion (proton) acceptor, or nonacid substance. It is characterized by an excess of hydroxide ions (OH^-) in solution and a pH greater than 7.

basement membrane   A thin layer of extracellular material that underlies epithelium.

basilar (BĀS-i-lar) *membrane*   One of two membranes in the inner ear that form the cochlear duct. It supports the organ of Corti.

basophil (BĀS-ō-fil)   A type of white blood cell that is characterized by large cytoplasmic granules that stain blue with basic dyes.

belly   The abdomen. Also the large, fleshy portion of a muscle between the tendons of insertion and origin.

benign (bē-NĪN)   Not threatening. When used to describe one of two forms of cancer, it is the type that is noninvasive and nonmetastasizing (does not spread to other areas of the body).

beta (BĀ-ta) *cell*   A cell in the pancreatic islets of Langerhans that secretes the hormone insulin.

bicuspid (bī-KUS-pid) *valve*   The left atrioventricular (AV) valve of the heart. It is also called the *mitral valve.*

bile   A fluid secreted by the liver and stored in the gallbladder. It assists in the digestion of fats within the small intestine.

bilirubin (bil-ē-ROO-bin)   An orange or red pigment resulting from the breakdown of hemoglobin in the liver, and excreted as a waste material in the bile.

biliverdin (bil-ē-VER-din)   A green pigment resulting from the breakdown of hemoglobin in the liver, and excreted as a waste material in the bile.

blastocyst (BLAS-tō-sist)   A structure that develops during early embryological development; a hollow ball of cells with an inner cell mass and an outer trophoblast layer.

blind spot   A region of the retina where no photoreceptive cells are present, because it is the exit point of the optic nerve.

blood pressure (BP)   The amount of pressure exerted by the force of blood pressing against the walls of a blood vessel. This force is initiated mainly by the contraction of the left ventricle of the heart.

body cavity   A space in the body that is internally lined by a membrane and contains structures including organs.

bolus (BŌ-lus)   A soft, rounded mass of food that is pushed from the mouth to the stomach by swallowing and peristalsis.

bony labyrinth (BŌ-nē LAB-i-rinth)   The portion of the inner ear that consists of cavities within the temporal bone, forming the vestibule, the cochlea, and the semicircular canals.

Bowman's capsule (BŌ-manz CAP-sool)   The part of a kidney nephron that surrounds the glomerulus.

brachial (BRĀ-kē-al)   Pertaining to the upper arm.

brachial plexus (BRĀ-kē-al PLEK-sus)   A network of nerves from the anterior rami of spinal nerves C5, C6, C7, C8, and T1. The nerves of the brachial plexus supply the upper appendages.

bradycardia (brād′-i-KAR-dē-a)   A reduced heartbeat or pulse rate.

brain stem   The inferior portion of the brain that consists of the midbrain, pons, and medulla oblongata.

Broca's (BRŌ-kaz) *area*   The motor area within the frontal lobe of the brain that coordinates motor activities required for speech.

bronchial tree (BRONG-kē-al)   A portion of the respiratory conduction zone that consists of a series of tubes that branch from the trachea (the primary bronchi), and continue to branch after entering each lung (into secondary and tertiary bronchi).

bronchiole (BRONG-kē-ol)   A series of small tubes that arise as branches from tertiary bronchi within each lung.

bronchitis (brong-KĪ-tis)   A disorder of the respiratory tract, characterized by inflammation of the bronchial walls and excessive mucus production, resulting in repeated coughing.

bronchus (BRONG-kus)   Any one of the air passageways that carry air between the trachea and the bronchioles. Plural form is *bronchi.*

buccal (BUK-al)   Pertaining to the mouth or the cheeks.

buffer (BUF-er)   A chemical that reacts with an acid or a base to form a weaker acid or base, thereby minimizing a change in pH.

bulbourethral (bul′-bō-yoo-RĒ-thral) *glands*   A pair of small glands in the male reproductive system located inferior to the prostate gland. Their secretions contribute to semen. Also called *Cowper's glands.*

bulimia (boo-LIM-ē-a)   A disorder characterized by overeating followed by self-induced purging (such as vomiting, fasting, vigorous exercise, or use of laxatives).

burn   An injury that causes damage to proteins as a result of exposure to heat, acid, ultraviolet light, or other agents.

bursa (BUR-sa)   A sac of synovial fluid located at major friction points of joints.

bursitis (bur-SĪ-tis)   Inflammation of a bursa.

calcaneus (kal-KĀ-nē-us)   The largest tarsal bone (of the foot). It forms the heel.

calcitonin (kal-si-TŌ-nin)   A hormone secreted by the thyroid gland that lowers calcium (and phosphorus) levels in the blood.

callus (KAL-us)   A growth of new bone tissue in and around a fractured region of bone.

calorie (KAL-ō-rē)   A unit of energy in the form of heat. A calorie is the amount of heat required to raise 1 gram of water 1°C from 14° to 15°C.

calyx (KAL-iks)   A cuplike extension of the renal pelvis (pelvis of the kidney). Plural form is *calyces.*

canaliculus (kan′-a-LIK-yoo-lus)   A small channel within compact bone tissue that connects lacunae. Plural form is *canaliculi.*

cancer (KAN-ser)   A disorder of cells in which cell division and growth occur without restraint, producing cells that are not functional and are capable of spreading throughout the body. A malignant tumor.

capillary (KAP-i-lar′-ē)   A microscopic blood vessel that interconnects arterioles with venules. The capillary wall is a single cell layer in thickness and is the only site of nutrient diffusion between the bloodstream and body cells.

carbohydrate (kar′-bō-HĪ-drat)   An organic molecule made up of one or more monosaccharides bound together. It includes sugars, starches, and glycogen.

carbonic anhydrase (kar-BON-ic an-HĪ-drās)   An enzyme in red blood cells that promotes the reaction between carbon dioxide and water to form carbonic acid, and vice versa.

carcinoma (kar-si-NŌ-ma)   A cancer (malignant tumor) that arises from epithelial tissue.

cardiac (KAR-dē-ak) *arrest*   The cessation of functional heart activity. The heart may stop completely, or may enter ventricular fibrillation.

cardiac cycle   One complete series of heart contractions (one systole and one diastole), which constitutes a single heartbeat.

cardiac muscle   One of three types of muscle tissue. It is characterized by striations and involuntary contractions and makes up the bulk of the heart wall.

cardiac output   The volume of blood pumped from the left ventricle per minute. It is calculated by multiplying stroke volume and heart rate and averages about 5.2 liters per minute under resting conditions.

carpals (KAR-palz)   The eight individual bones of the wrist. As a group, the bones of the wrist are referred to as the *carpus.*

cartilage (KAR-ti-lej)   A type of connective tissue characterized by the presence of a matrix containing a dense distribution of proteins and a thickened ground substance. The matrix is secreted by chondroblasts.

cartilaginous (kar-ti-LA-jin-us) *joint*   One of three types of joints. It is characterized by the presence of cartilage that connects opposing bones.

castration (kas-TRA-shun)   The removal of the testes.

catabolism (ka-TAB-ō-lizm)   A type of metabolism in which molecules are broken down into smaller subunits for the release of energy stored within their chemical bonds.

catalyst (KA-ta-list)   A substance that accelerates the rate of a chemical reaction. An example of a catalyst is an enzyme.

cataract (KAT-a-rakt)   The loss of transparency to the lens, reducing visual acuity.

catheter (KATH-i-ter)   A plastic tube that is inserted into a body cavity. It is used to remove body fluids, such as blood or urine, or to introduce fluids, such as saline or glucose and water.

cation (CAT-ī-on)   A positively charged ion.

cauda equina (KAH-da e-KWĪ-na)   The group of spinal nerves and roots at the inferior end of the spinal cord.

cecum (SĒ-kum)   The proximal end of the large intestine, which receives the ileum of the small intestine.

cell   The basic living unit of multicellular organisms.

cell body   The portion of a neuron that contains the nucleus and much of the cytoplasm. Also called the *soma*.

cell-mediated immunity   Mechanisms of resistance to infection that involve the activity of T cells. Also called *cellular immunity*.

cellular respiration   The cellular process of catabolism, in which molecules are broken down into their subunits for the release of energy contained in their chemical bonds.

cementum (sē-MEN-tum)   Calcified tissue covering the root of a tooth.

central canal   A small channel extending the length of the spinal cord that is continuous with the ventricles of the brain. It contains cerebrospinal fluid.

central fovea (FŌ-vē-a)   A small depression in the center of the macula lutea of the retina. It contains cone cells (only) and is the area of optimal visual acuity (clearest vision).

central nervous system (CNS)   A major division of the nervous system that contains the brain and spinal cord.

centriole (SEN-trē-ōl)   Cylindrical structures within the cytoplasm of a cell, consisting of microtubules, which play a role in cell division.

cephalic (se-FAL-ik)   Pertaining to the head region.

cerebellar peduncle (ser'-e-BEL-ar pe-DUNG-kul)   A bundle of nerve fibers connecting the cerebellum and the brain stem.

cerebellum (ser'-e-BEL-um)   A functional region of the hindbrain located inferior to the cerebrum. It coordinates muscle movement.

cerebral aqueduct (SER-e-bral AK-we-dukt)   A channel through the midbrain containing cerebrospinal fluid. It connects the third and fourth ventricles of the brain.

cerebral cortex   The outer layer of the cerebrum, which is composed of gray matter.

cerebrospinal fluid (ser'-ē-brō-SPĪ-nal FLOO-id) *(CSF)*   A clear fluid produced as a filtrate from blood in the choroid plexuses of the brain. It circulates through the ventricles of the brain, the central canal of the spinal cord, and the subarachnoid space of the brain and spinal cord.

cerebrovascular (ser'-ē-brō-VAS-kyoo-lar) *accident (CVA)*   Destruction of brain tissue caused by a damaged blood vessel supplying the brain. Also called *stroke*.

cerebrum (SER-ē-brum)   The largest functional region of the brain; the convoluted mass that lies superior to all other parts of the brain. It is the main site of integration of sensory and motor impulses.

cerumen (sē-ROO-men)   A type of sebum (oil secretion) produced in the external auditory canal. Also called *earwax*.

cervical (SER-vi-kal)   Pertaining to the neck region.

cervical plexus (PLEK-sus)   A branching network of nerves originating from the anterior rami of the first four cervical nerves.

cervix (SER-viks)   The narrow, constricted part of the uterus that lies between the vagina and the body of the uterus.

chemical formula (KEM-i-kal FOR-myoo-la)   A group of atomic symbols used to indicate the chemical composition of a molecule.

chemoreceptor (kē'-mō-rē-CEP-tor)   A sensory receptor that is stimulated by the presence of certain chemicals, such as carbon dioxide or oxygen.

chemotherapy (kē'-mō-THER-a-pē)   The clinical treatment of disease by the use of chemicals, including compounds that kill cells (called *cytotoxic agents*).

cholecystectomy (kō'-lē-sis-TEK-tō-mē)   The surgical removal of the gallbladder.

cholesterol (kō'-LES-ter-al)   A type of lipid classified as a steroid. It is produced by body cells and is an important component in plasma membranes and is used in the synthesis of other steroid compounds.

chondrocyte (KON-drō-sīt)   A mature cartilage cell.

chondroitin sulfate (kon-DRO-i-tin SUL-fāt)   A semisolid material that forms part of the matrix in certain connective tissues.

chordae tendineae (KOR-dē ten-DIN-ē-ē)   Strands of connective tissue in the heart that anchor atrioventricular valves to papillary muscles.

chorion (KOR-ē-on)   The outermost fetal membrane. Its fingerlike projections (called *chorionic villi*) form part of the placenta, serving a protective and nourishing function to the fetus.

choroid (KŌ-royd)   Part of the vascular tunic of the eyeball. It lines most of the internal surface of the sclera, thereby forming the middle layer of the wall of the eye.

choroid plexus (KŌ-royd PLEK-sus) A mass of specialized capillaries in the ventricles of the brain, from which cerebrospinal fluid is produced.

chromatin (KRŌ-ma-tin) The bundled mass of genetic material in the nucleus of a cell, consisting mostly of DNA. It is visible only during interphase.

chromosome (KRŌ-mō-sōm) One of the structures (46 in human cells) within the cell nucleus that contain genetic material. Chromosomes become visible during cell division.

chronic (KRON-ik) A long-term or frequently recurring disease; not acute.

chyme (kīm) Semifluid, partially digested food material found in the stomach and small intestine.

cilia (SIL-ē-a) Hairlike processes associated with a cell that are a modification of the plasma membrane. Their movement generates a flow of fluid (usually mucus) in the extracellular environment. Singular form is *cilium.*

ciliary body (SIL-ē-ar-ē BOD-ē) Part of the vascular tunic of the eyeball, along with the choroid and iris. It suspends the lens and consists of the ciliary muscle and ligaments.

circadian rhythm (ser'-KĀ-dē-an RI-thum) A pattern of repeated activity associated with 24-hour cycles.

circumcision (ser'-kum-SI-shun) The surgical removal of the prepuce (foreskin), which is a fold of skin over the glans of the penis.

circumduction (ser'-kum-DUK-shun) A movement at a synovial joint such that the distal end of the bone draws a circular path.

cirrhosis (si-RŌ-sis) A generalized disorder of the liver, in which functional liver cells die and are replaced by nonfunctional connective tissue and fat.

cisterna chyli (sis-TER-na KĪ-lē) The expanded origin of the thoracic duct, located at its inferior end.

clavicle (KLA-vik-el) One of two bones connecting the sternum and the upper appendage. Also called the *collarbone.*

cleavage (KLĒ-vij) The initial mitotic divisions that follow fertilization of the secondary oocyte.

cleft palate (kleft PAL-at) A congenital disorder in which a child is born with an incomplete fusion of the hard palate (the palatine processes of the maxillary bones) at the midline. It is often associated with cleft lip, in which the lip is split.

clitoris (kli-TOR-is) The female erectile organ located at the anterior junction of the labia minora.

coagulation (ko-ag-yoo-LĀ-shun) The process by which a blood clot is formed.

coccyx (KOK-siks) The fused bones at the end of the vertebral column.

cochlea (KŌK-lē-a) The portion of the inner ear that contains the receptors of hearing (the organ of Corti).

codon (KŌ-don) A set of three nucleotides in a messenger RNA molecule that is complementary to a set of three nucleotides in a DNA molecule. A single codon provides a message specific for a single amino acid during protein synthesis.

coenzyme (ko-EN-zīm) A nonprotein organic molecule that activates an enzyme.

coitus (KO-i-tus) Sexual intercourse, or copulation.

colitis (kō-LĪ-tis) Chronic inflammation of the large intestine, characterized by periodic bouts of intestinal cramping, excessive mucus production, and diarrhea.

collagen (KOL-a-jen) A protein that is an abundant component of connective tissue.

colliculi (kō-LIK-yoo-lī) Small elevations on the midbrain concerned with visual and auditory reflexes. Also called *corpora quadrigemina.*

colon (KŌ-lon) The division of the large intestine containing the ascending, transverse, descending, and sigmoid sections.

colostrum (kō-LOS-trum) A cloudy white fluid secreted by the mammary glands several days before or after birth, which precedes production of true milk.

coma (KŌ-ma) A mental state of unconsciousness, in which there is a complete unresponsiveness to external stimuli.

common bile duct A tube extending from the union of the common hepatic duct and the cystic duct to the duodenum (of the small intestine) that transports bile.

compact bone One of two types of bone tissue. It is characterized by a dense matrix filled with mineral salts and collagen arranged in lamellae that surround a central osteonic (haversian) canal. Also called *dense bone.*

complement (KOM-ple-ment) A group of proteins in the blood that assist in resistance against infection by destroying foreign or diseased cells.

compound A molecule composed of two or more different types of atoms.

conchae (KONG-kē) Scroll-like bones in the facial skeleton forming the superior, middle, and inferior shelves and meati of the nasal cavity.

concussion (kon-KUSH-un) A condition of the brain caused by traumatic injury that often results in an abrupt, temporary loss of consciousness followed by nausea and headache.

conductivity (kon'-duk-TIV-i-tē) A physiological property in which a nerve impulse is carried from one part of a cell to another.

condyle (KON-dīl) A rounded process of a bone.

cone A photoreceptor in the retina of the eye that is involved in color vision and high visual acuity.

congenital (kon-JEN-i-tal) Present before or at the time of birth.

congestive heart failure A failure of the heart to meet the oxygen demands of the body, leading to a loss of function to vital organs if not corrected.

conjunctiva (kon'-junk-TĒ-va) A thin, transparent membrane lining the outer surface of the cornea of the eye and inner surface of the eyelids.

conjunctivitis (kon'-junk-ti-VĪ-tis) Inflammation of the conjunctiva. Also called *pinkeye.*

connective tissue One of the four basic types of tissue in the body. It is characterized by an abundance of extracellular material with relatively few cells, and functions in the support and binding of body structures.

contraception (kon'-tra-SEP-shun) The prevention of conception or pregnancy.

contractility (kon'-trak-TIL-i-te) A physiological property in which a cell or its parts can shorten or otherwise change form for the purpose of movement.

contusion (kon'-TOO-shun) A condition in which the tissue deep to the skin has been damaged by trauma but the skin remains unbroken. Also called a *bruise*.

conus medullaris (KO-nus med'-yoo-LAR-is) The tapered terminal end of the spinal cord.

convulsion (kon-VUL-shun) A series of involuntary, tetanic contractions of a group of muscles.

cornea (KOR-ne-a) The transparent anterior portion of the fibrous tunic covering the eye.

corona radiata (ko-RO-na ra-de-A-ta) Several layers of follicle cells that form a protective mantle around the secondary oocyte.

coronal (ko-RO-nal) *plane* A plane that extends vertically to divide the body into anterior and posterior portions. Also called a *frontal plane*.

coronary circulation (KO-ro-nar-e ser'-kyoo-LA-shun) The circulatory pathway of blood to the heart wall from the aorta (by way of coronary arteries) and its return to the right atrium (by way of coronary veins).

coronary sinus An expanded venous channel on the posterior surface of the heart into which coronary veins empty.

corpus callosum (KOR-pus ka-LO-sum) A bundle of nerve fibers forming a band of white matter that interconnects the two cerebral hemispheres of the brain.

corpus luteum (LOO-te-um) A structure within the ovary that forms from a ruptured Graafian follicle and functions as an endocrine gland by secreting female hormones.

cortex (KOR-teks) The outer portion of an organ.

cortisol (KOR-ti-sol) A glucocorticoid hormone secreted by the adrenal cortex that inhibits inflammation.

costal cartilage (KOS-tel CAR-ti-lej) A band of hyaline cartilage that connects a true rib with the sternum.

covalent bond (ko-VA-lent bond) A chemical bond that results when two atoms share a pair of electrons.

coxal bones (KOK-sal bonz) The two bones that form the pelvis. Also called *os coxae* or *innominate bones*.

cranial cavity (KRA-ne-al CAV-i-te) The cavity within the skull that contains the brain.

cranial nerve One of 12 pairs of nerves that originate from the brain.

cranium (KRA-ne-um) The skeletal portion of the skull that forms the cranial cavity.

crenation (kre-NA-shun) The shrinkage of a cell caused by contact with a hypotonic solution.

cretinism (KRE-tin-izm) A congenital disorder resulting in hypothyroidism, which causes physical and mental retardation.

cristae (KRIS-te) Regions within the ampulla of each semicircular canal (of the inner ear) that serve as receptors for dynamic equilibrium.

crown The part of a tooth that is exposed and covered with enamel.

cryptorchidism (krip-TOR-ki-dizm) A condition in which the testes fail to descend into the scrotum.

cupula (KUP-yoo-la) A gelatinous mass covering the hair cells within a crista of the semicircular canals (of the inner ear).

Cushing's syndrome A disorder caused by hypersecretion of glucocorticoids by the adrenal cortex, resulting in the development of thin legs, a rounded ("moon") face, a large, fatty abdomen, and slow wound healing.

cutaneous (kyoo-TA-ne-us) *membrane* One of the three types of epithelial membranes found in the body; also known as the *skin*.

cyanosis (si-a-NO-sis) A condition characterized by insufficient oxygenation of blood resulting in blue or dark-purple discoloration in the lips and nail beds.

cyst (sist) An abnormal sac of fluid surrounded by a connective tissue wall.

cystic duct (SIS-tik dukt) A tube that transports bile from the gallbladder to the common bile duct.

cystitis (sis-TI-tis) Inflammation of the urinary bladder.

cytoplasm (SI-to-plazm) The material of a cell located within the plasma membrane and outside the nuclear membrane, and containing the cellular organelles.

cytoskeleton (ci'-to-SKEL-e-ton) A complex supportive network of microtubules and microfilaments within the cytoplasm of a cell.

cytosol (SI-to-sol) The thickened fluid of the cytoplasm that lies outside the cellular organelle membranes.

deciduous (de-SID-yoo-us) *teeth* The temporary set of teeth that are shed and replaced by permanent teeth. Also called *primary teeth*.

decubitus ulcer (de-KYOO-bi-tus UL-ser) The destruction of tissue in a local region due to the persistent reduction of blood supply caused by constant pressure, such as from a cast or from lying in bed for long periods of time. Also called a *bed sore*.

deep A directional term meaning away from the surface of the body.

deep fascia (FASH-e-a) A sheet of connective tissue covering the external surface of a muscle. Also called the *epimysium*.

defecation (def'-e-KA-shun) The discharge of feces from the rectum through the anus. Also called *elimination*.

deglutition (de-gloo-TI-shun) The process of swallowing.

dementia (de-MEN-she-a) A mental disorder that results in the permanent loss of intellectual functions, including memory, behavior, and judgment.

denaturation (de-na'-tyoo-RA-shun) A change in the structure of a protein caused by breaking its stabilizing bonds, causing it to lose its structural or functional role in the body.

dendrite (DEN-drit) A cytoplasmic extension from the cell body of a neuron that carries a nerve impulse toward the cell body.

dental caries (DEN-tal CAR-ez) A disorder of teeth resulting in the gradual demineralization of the enamel and dentin that may invade the pulp if not repaired. Also called *cavities*.

dentin (DEN-tin) The bonelike material forming the bulk of a tooth.

deoxyribonucleic acid (dē-ok′-sē-ri-bō-nyoo-KLĒ-ik A-sid) **(DNA)** A nucleic acid in the shape of a double helix that contains the genetic information necessary for protein synthesis.

depolarization (dē′-pō-lar-i-ZĀ-shun) A reduction in voltage across a plasma membrane, such that the charge becomes less negative on the inside of the membrane relative to the outside.

depression (dē-PRE-shun) Movement of a body part downward.

dermis (DER-mis) The layer of the skin lying deep to the epidermis and composed of dense irregular connective tissue.

descending colon (dē-SEN-ding KŌ-lon) The segment of the large intestine between the transverse colon and the sigmoid colon.

diabetes insipidus (dī-a-BĒ-tez in-SIP-i-dus) A disorder caused by the hyposecretion of antidiuretic hormone (ADH) by the posterior pituitary gland, resulting in the excretion of large amounts of urine and excessive thirst.

diabetes mellitus (MEL-i-tus) A disorder caused by the hyposecretion of insulin by the pancreatic islets of Langerhans, resulting in hyperglycemia and its associated symptoms.

diagnosis (dī′-ag-NŌ-sis) The determination of a disease based on signs and symptoms.

diapedesis (dī′-a-pe-DĒ-sis) The process by which white blood cells move through the walls of blood vessels.

diaphragm (DĪ-A-fram) An internal, circular muscle dividing the thoracic cavity from the abdominopelvic cavity.

diaphysis (dī-A-fi-sis) The shaft of a long bone.

diarrhea (dī-a-RĒ-a) Passage of a liquefied feces caused by increased motility of the large intestine.

diastole (dī-AS-tō-lē) A part of the cardiac cycle characterized by relaxation of the heart chambers, during which they fill with blood.

diastolic (dī-as-TOL-ik) *pressure* The force of blood pressure exerted on the walls of arteries during ventricular relaxation.

diencephalon (dī-en-CEF-a-lon) A region of the brain located inferior to the cerebrum. It is part of the forebrain and contains the thalamus, hypothalamus, and pineal gland.

diffusion (di-FYOO-shun) The passive movement of molecules or ions from a region of higher concentration to a region of lower concentration.

digestion (dī-JES-chun) The breakdown of food particles into units small enough to be absorbed.

digestive tract See *alimentary canal.*

dilate (DĪ-lat) To enlarge in size or expand.

diploid (DI-ployd) The complete set of chromosomes within a cell, designated as 2*N*. The diploid number of chromosomes in human cells is 46.

disaccharide (dī-SAK-a-rīd) A carbohydrate molecule containing two monosaccharides.

disease (di-ZĒZ) Any state of the body other than that of optimal health.

dislocation (dis-lō-KĀ-shun) A disorder in which the bones forming a joint become separated from their normal position. Also called *luxation.*

distal (DIS-tal) A directional term identifying a body part located farther from the origin or point of attachment to the trunk relative to another.

distal convoluted tubule A segment of the renal tubule (of the kidney nephron) that extends from the loop of Henle to the collecting duct.

diuretic (dī-yoo-RET-ik) A chemical that inhibits sodium reabsorption by the kidneys, resulting in increased urine volume.

diverticulitis (di-ver-ti′-kyoo-LI-tis) Inflammation of a sac or pouch that has formed in the wall of the large intestine.

dominant (DO-mi-nant) In genetics, describing the inherited gene that is expressed as a trait when combined with a gene for an opposing trait, or *recessive* gene.

dorsal (DOR-sal) A directional term indicating toward the back side, or posterior.

dorsal cavity A major body cavity containing the cranial cavity and the vertebral canal.

dorsiflexion (dor′-si-FLEK-shun) Movement of the foot toward the dorsal side (upward).

Down's syndrome A congenital disease caused by an extra chromosome (chromosome 21 is repeated). Symptoms include mental retardation, enlarged tongue, shortened fingers, and a skull reduced in size and flattened back to front. Also called *trisomy 21.*

ductus deferens (DUK-tus DEF-er-enz) The tube that conducts sperm from the epididymis in the testis to the ejaculatory duct. Also called *vas deferens* and *seminal duct.*

duodenum (doo-ō-DĒ-num) The first segment of the small intestine, which extends from the pyloric valve to the jejunum.

dura mater (DYOO-ra MĀ-ter) The outer of the three meninges that surround the brain and spinal cord.

dynamic equilibrium The maintenance of balance in response to sudden movements of the head.

dysmenorrhea (dis′-men-ō-RĒ-a) Painful menstruation.

dyspnea (DISP-nē-a) Shortness of breath.

eccrine (EK-rin) A type of sweat gland that functions in the maintenance of body temperature.

ectoderm (EK-tō-derm) One of three primary germ layers in the developing embryo. It gives rise to the nervous system and to the epidermis and its derivatives.

ectopic (ek-TOP-ik) Not in the normal location; for example, ectopic pregnancy.

edema (e-DĒ-ma) An abnormal accumulation of interstitial fluid in the extracellular environment.

effector (e-FEK-tor) Any muscle or gland that responds to a stimulus from a motor neuron.

efferent arteriole (EF-er-ent ar-TĒ-rē-ol) An arteriole that transports blood away from the glomerulus of a nephron (in the kidney).

efferent ductules (DUK-tyoo-elz) Small, coiled tubes that transport sperm from the rete testis to the epididymis.

efferent neuron (NYOO-ron) A neuron that carries impulses away from the central nervous system. Also called *motor neuron.*

ejaculation (ē-jak′-yoo-LĀ-shun)   The expulsion of semen from the penis, which is accomplished by reflexive muscle contractions.

ejaculatory duct (ē-JAK-yoo-la-tō-rē dukt)   A short tube that carries sperm from the ductus deferens to the urethra.

elasticity (ē-las-TI-si-tē)   The physiological property whereby tissue tends to return to its original shape after extension or contraction.

electrocardiogram (ē-lek′-trō-KAR-dē-ō-gram) **(ECG or EKG)**   A recording of the electrical changes that occur in the heart during a cardiac cycle.

electroencephalogram (ē-lek′-trō-en-SEF-a-lō-gram) **(EEG)**   A recording of the electrical events of the brain.

electrolyte (ē-LEK-trō-līt)   A chemical that can separate into ions and conduct an electric current when in a water solution.

electron (ē-LEK-tron)   A negatively charged atomic particle found in the orbitals surrounding the atom's nucleus.

electron transport chain   A chain of energy-relay molecules attached to the inner mitochondrial membrane of a cell. The chain transports energy from one molecule to the next during cellular respiration, generating molecules of ATP in the process.

element (EL-e-ment)   A basic chemical substance composed of one type of atom.

elevation (el-e-VĀ-shun)   Movement of a body part upward.

embolism (EM-bō-lizm)   An obstruction or closure of a blood vessel by an embolus (a substance transported by the blood, such as a blood clot, air bubble, or fat).

embryo (EM-brē-ō)   In the human, the developing organism during its first eight weeks of life following fertilization.

emesis (EM-e-sis)   A clinical synonym for vomiting.

emulsification (ē-mul-si-fi-CĀ-shun)   The partial digestion of large fat globules into smaller particles of fat.

enamel (ē-NAM-el)   The hardened outer covering on the crown of a tooth.

endocardium (en-dō-KAR-dē-um)   The inner layer of the heart wall, which forms a thin, smooth lining covering the chambers and valves.

endochondral ossification (en-dō-KON-dral os′-i-fi-KĀ-shun)   The development of bone such that bone tissue forms in replacement of hyaline cartilage.

endocrine gland (EN-dō-krin gland)   One of two main categories of glands, in which the products are secreted into the extracellular space and transported by the bloodstream. Also called *ductless gland.*

endocytosis (en′-dō-sī-TŌ-sis)   The active process of bulk transport of material into a cell. It includes phagocytosis and pinocytosis.

endoderm (EN-dō-derm)   One of the three primary germ layers in an embryo. It begins as the inner layer and later forms the organs of the alimentary canal and the respiratory tract.

endolymph (EN-dō-lymf)   The fluid within the membranous labyrinth of the inner ear.

endometriosis (en′-dō-mē-trē-Ō-sis)   The proliferation of endometrial tissue (from the uterine lining) outside the uterus.

endometrium (en′-dō-MĒ-trē-um)   The inner, vascular layer of the uterus.

endomysium (en′-dō-MĪ-sē-um)   The deepest layer of connective tissue associated with a muscle. It surrounds individual muscle fibers.

endoneurium (en′-dō-NYOO-rē-um)   The deepest layer of connective tissue associated with a nerve. It surrounds individual nerve fibers (myelinated axons of neurons).

endoplasmic reticulum (en′-dō-PLAZ-mik re-TlK-yoo-lum) **(ER)**   A cytoplasmic organelle that consists of a series of tubules with a hollow center. Smooth ER functions in the transport of cellular products, and rough ER (which has ribosomes attached) serves as a site for protein synthesis.

endoscope (EN-dō-skōp)   A clinical device consisting of a hollow tube with an illuminated end that can be inserted into small openings in the body during examination or surgery.

endosteum (en-DŌ-stē-um)   A membrane lining the medullary cavity within a bone and containing osteoblasts and osteoclasts.

endothelium (en′-dō-THĒ-lē-um)   A layer of simple squamous epithelium lining the inside of blood vessels and the heart chambers.

energy (EN-er-gē)   The capacity to do work.

enuresis (en′-yoo-RĒ-sis)   The involuntary, uncontrolled release of urine after the age of 3 years.

enzyme (EN-zīm)   A protein that performs the role of catalyst in a chemical reaction.

eosinophil (ē′-ō-SIN-ō-fil)   A type of granulated white blood cell characterized by a cytoplasm that accepts the stain eosin.

ependymal (e-pen-DĪ-mal) *cells*   Those neuroglial cells in the brain that line the ventricles. Also called *ependymocytes.*

epicardium (ep′-i-KAR-dē-um)   The thin outer layer of the heart wall. Also called the *visceral pericardium.*

epidemiology (ep′-i-dē-mē-OL-ō-gē)   A branch of medicine that uses statistical data to determine the occurrence and distribution of disease.

epidermis (ep′-i-DERM-is)   The superficial layer of skin, composed of stratified squamous epithelium.

epididymis (ep′-i-DlD-i-mis)   An organ in the male reproductive system that consists of a coiled tube located within the scrotum.

epiglottis (ep′-i-GLOT-is)   A part of the larynx that consists of a leaf-shaped piece of hyaline cartilage that forms a movable lid over the opening into the trachea, called the *glottis.*

epilepsy (EP-i-lep-sē)   A brain disorder characterized by periods of uncontrollable, disturbed brain activity resulting in minor or major convulsions.

epimysium (ep′-i-MĪ-zē-um)   The outer layer of connective tissue associated with muscle; it surrounds the whole muscle. Also called *deep fascia.*

epinephrine (ep′-i-NEF-rin)   A hormone secreted by the adrenal medulla that stimulates a response similar to those resulting from sympathetic stimulation. Also called *adrenaline*.

epineurium (ep′-i-NYOO-rē-um)   The outermost layer of connective tissue associated with a nerve. It surrounds the whole nerve.

epiphyseal (ep′-i-FIZ-ē-al) *line*   A line of calcified bone visible in a section through bone that is the remnant of the epiphyseal plate.

epiphyseal plate   A region of cartilage between the epiphysis and diaphysis that produces lengthwise growth of a bone.

epiphysis (e-PIF-i-sis)   The end of a long bone that contains spongy bone tissue.

epithelial (ep′-i-THĒ-lē-al) *tissue*   One of four primary tissue types. It is characterized by a close arrangement of cells with little intercellular material. Also called *epithelium*.

eponychium (ep′-ō-NIK-ē-um)   A narrow region of stratum corneum at the proximal end of a nail. Also called *cuticle*.

equilibrium (ē′-kwi-LI-brē-um)   A state of balance between two opposing processes.

erection (ē-REK-shun)   The stiffened, enlarged state of the penis or clitoris caused by a temporary increase in blood flow during sexual stimulation.

erythrocyte (ē-RITH-rō-sīt)   A synonym for *red blood cell*.

erythropoiesis (ē-rith′-rō-poy-Ē-sis)   The process by which erythrocytes are formed.

erythropoietin (ē-rith′-rō-POY-ē-tin)   A hormone secreted by kidney cells that stimulates the production of erythrocytes.

esophagus (e-SOF-a-gus)   A tubular segment of the alimentary canal between the pharynx and the stomach.

essential nutrients   Nutrients that are required for maintaining good health but are not manufactured by body cells and must be obtained from the diet.

estrogen (ES-trō-jen)   The primary female sex hormone secreted by the ovaries. It refers to any one of several similar compounds, including beta estradiol, estrone, and estriol, all of which produce the secondary sex characteristics and development of reproductive organs.

etiology (e′-tē-OL-ō-jē)   The study of the causes of diseases.

eustachian (yoo-STĀ-shē-an) *tube*   See *auditory tube*.

eversion (ē-VER-zhun)   Movement of the sole of the foot in an outward direction.

excitability (ek-sīt′-a-BIL-i-tē)   A physiological property in which a cell is capable of receiving and responding to stimuli.

excretion (ek-SKRĒ-shun)   The process by which metabolic waste materials are removed from a cell, a tissue, or an entire body.

exocrine (EK-sō-krin) *gland*   One of two main categories of glands, in which the products are released into ducts that transport them to the body surface or into body cavities.

exocytosis (ek′-sō-sī-TŌ-sis)   The active cellular process by which materials are transported out of a cell and into the extracellular environment.

expiration (ek′-spi-RĀ-shun)   The process of expelling air from the lungs to the external environment, or breathing out. Also called *exhalation*.

extensibility (ek′-sten-si-BIL-i-tē)   A physiological property in which a tissue can stretch in response to being pulled.

extension (ek-STEN-shun)   Movement of a body part such that the angle between opposing bones is increased; returning to original position after flexion.

external auditory canal   The epidermis-lined tube of the external ear extending from the auricle to the tympanic membrane. It passes through the hole in the temporal bone called the *external auditory meatus*.

external ear   The outer part of the ear, which consists of the appendage known as the *auricle*, the external auditory canal, and the tympanic membrane.

external nares (NAR-ēz)   The openings of the nose between the external environment and the nasal cavity. Also called *nostrils*.

external respiration   The exchange of respiratory gases between the lungs and the bloodstream.

extracellular environment (ek′-stra-CEL-yoo-lar en-VĪ-ron-ment)   The body space outside the plasma membrane of cells.

extracellular fluid (ECF)   The fluid outside the plasma membrane of cells, including interstitial fluid and blood plasma.

extrinsic (ek-STRIN-sik)   Pertaining to an external origin.

facet (FA-set)   A smooth articular surface on a bone.

facilitated diffusion (fa-SIL-i-tā-ted di-FYOO-shun)   The passive movement of molecules across a plasma membrane that requires the assistance of membrane carrier proteins.

facilitation (fa-sil-i-TĀ-shun)   The process by which a neuron becomes more excitable as a result of an accumulation of incoming subthreshold impulses.

falciform ligament (FAL-si-form LIG-a-ment)   A part of the parietal peritoneum that is located between the right and left lobes of the liver.

fallopian tube   See *uterine tube*.

fascia (FASH-ē-a)   A sheet or band of dense connective tissue that structurally supports organs and tissues. Deep fascia surrounds muscle, and superficial fascia separates the skin and muscle layers.

fascicle (FAS-i-kul)   A bundle of skeletal muscle fibers (cells) that forms a part of a muscle.

fat   A lipid compound formed from one molecule of glycerol and three molecules of fatty acids. It is the body's most concentrated form of energy and also serves to insulate the body from external temperature changes. It is stored within cells making up adipose tissue.

fatigue (fa-TĒG)   The inability of a muscle to maintain its strength of contraction or tension, which may be due to insufficient oxygen, lack of ATP, or lactic acid accumulation.

fauces (FAW-sēs)   The opening into the pharynx from the oral cavity (mouth).

feces (FĒ-sēz)   Waste material discharged from the large intestine during defecation.

fertilization (fer'-ti-li-ZĀ-shun)   The union of a sperm cell with a secondary oocyte.

fetal alcohol syndrome (FAS)   A disorder of newborn infants that is caused by alcohol consumption by the mother and its exposure to the infant in the uterus. It is characterized by neurological retardation, slow growth, and defective organs.

fetus (FĒ-tus)   The early developmental stage from eight weeks after fertilization to the time of birth.

fever (FĒ-ver)   An elevation in body temperature above its normal value (of 37°C).

fibrin (FĪ-brin)   An insoluble protein in the blood that is formed from fibrinogen and is required for blood clotting.

fibrinogen (fī-BRIN-ō-jen)   A large plasma protein that plays an important role in blood clotting. It is converted to fibrin by thrombin.

fibroblast (FĪ-brō-blast)   A large cell in connective tissue that manufactures much of the intercellular material.

fibrosis (fī-BRŌ-sis)   The abnormal production of fibrous (dense) connective tissue.

fibrous (FĪ-brus) *joint*   One of three general types of joints in the body. Fibrous joints are characterized by the presence of dense connective tissue between opposing bones. They allow little or no movement between bones.

fibrous tunic   The outer wall of the eyeball, which is composed of dense connective tissue. It contains the sclera and the cornea.

filtration (fil-TRĀ-shun)   The process by which small molecules are forced across a permeable membrane by a pressure gradient, leaving larger molecules behind. The molecules that penetrate the membrane form the filtrate.

filum terminale (FĪ-lum ter-mi-NAL-ē)   Connective tissue that extends beyond the conus medullaris of the spinal cord inferiorly into the coccyx.

fimbriae (FIM-brē-ē)   The fingerlike extensions of the uterine tube at its proximal end (near the ovary).

fissure (FISH-er)   A cleft or groove separating two parts, such as the cerebral hemispheres of the brain; a deep sulcus of the brain.

flagellum (fla-JEL-um)   A single, long extension of a cell composed of protein filaments to provide mobility. Among human cells, it is found only in sperm cells.

flexion (FLEK-shun)   Movement of a body part such that the angle between bones is decreased.

follicle-stimulating hormone (FSH)   A hormone secreted by the anterior pituitary gland that stimulates the development of ova and stimulates the ovaries to secrete estrogen in the female, and stimulates sperm production in the male.

fontanel (fon'-ta-NEL)   An area in the skull of a newborn that is covered by a fibrous membrane instead of cranium (bone).

foramen (fō-RĀ-men)   An opening or passage through bone. Plural is *foramina.*

foramen magnum (MAG-num)   The large opening at the base of the skull through which the spinal cord passes.

fossa (FOS-a)   A shallow depression or groove in a bone.

fourth ventricle (forth VEN-tri-kul)   The cavity in the brain located between the cerebellum and the midbrain and pons.

fovea centralis (FŌ-vē-a cen-TRAL-is)   The region of the retina that consists of cone cells but no rod cells. It is the area of highest visual acuity (sharpness of vision).

fracture (FRAK-shur)   A break in a bone.

frontal plane   A plane that extends in a vertical direction dividing the human body into front (anterior) and back (posterior) portions. Also called a *coronal plane.*

furuncle (FYOOR-ung-kul)   A painful swelling below the skin surface caused by bacterial infection and inflammation of a hair follicle or sebaceous gland. Also called a *boil.*

gallbladder (GAWL-blad-er)   A small, saclike organ located beneath the liver that stores bile.

gallstone (GAWL-stōn)   A concretion of bile salts and cholesterol that may form within any of the biliary tubes between the liver and the duodenum, causing a blockage of bile flow. Also called *biliary calculus.*

gamete (GAM-ēt)   A sex cell. It may be male (sperm cell) or female (oocyte).

ganglion (GANG-lē-on)   A cluster of neuron cell bodies located outside the central nervous system.

gastric (GAS-trik) *gland*   Any one of several types of glands in the stomach mucosa that contribute to the gastric juice.

gastrin (GAS-trin)   A hormone secreted by cells in the stomach wall that promotes the secretion of gastric juice and stimulates peristalsis.

gastrula (GAS-troo-la)   A stage in the development of the embryo characterized by an early invagination of groups of cells that results in the establishment of the primary germ layers. The process of cell movement during this phase is called *gastrulation.*

gene   A segment of a DNA molecule that contains the information needed to synthesize one complete polypeptide chain.

genitalia (jen'-i-TĀL-ya)   The reproductive organs.

genome (JĒ-nom)   The complete genetic makeup of an organism.

genotype (JĒ-nō-tīp)   The inherited makeup of an individual.

germinal (JER-mi-nal) *epithelium*   A layer of epithelial cells covering the ovaries.

gestation (jes-TĀ-shun)   The period of development prior to birth.

giantism (JĪ-ant-izm)   Abnormal, excessive growth during childhood caused by hypersecretion of growth hormone (GH) by the anterior pituitary gland.

gingivae (JIN-ji-vē)   The mucous membrane covering the alveolar processes of the maxillary bones and mandible. Also called the *gums.*

gingivitis (jin'-ji-VĪ-tis)   Inflammation of the gingivae (gums).

gland A specialization of epithelial tissue that secretes substances. It may consist of a single cell or a multicellular arrangement.

glans penis (glanz PĒ-nis) The slightly enlarged, distal end of the penis.

glaucoma (glaw-KŌ-ma) A disorder of the eye characterized by an increased intraocular pressure that is usually caused by an excess of aqueous humor.

glomerulus (glō-MER-yoo-lus) One of many specialized capillary networks located in the kidney cortex, each of which is encapsulated by a Bowman's capsule. It is part of the kidney nephron and is the site of kidney filtration.

glottis (GLOT-is) The opening into the larynx from the pharynx.

glucagon (GLOO-ka-gon) A hormone secreted by the alpha cells of the pancreas that increases glucose levels in the blood.

glucocorticoids (gloo'-kō-KORT-i-koydz) A group of hormones secreted by the adrenal cortex that influence metabolism.

glucose (GLOO-kōs) A monosaccharide that serves as the preferred energy source for the body.

glycerol (GLI-ser-ol) An organic compound that is a subunit of fat.

glycogen (GLĪ-kō-jen) A polysaccharide composed of glucose subunits that is manufactured by the liver to serve as a storage form of energy.

glycolysis (glī-KOL-i-sis) A series of chemical reactions during cellular respiration that results in the breakdown of glucose into pyruvic acid.

glycoprotein (glī-kō-PRŌ-ten) An organic molecule composed of carbohydrate and protein subunits.

glycosuria (glī-kō-SOO-rē-a) The presence of glucose in the urine. It may be temporary following a meal rich in sugar, or it may be long-lasting and frequent and be an indication of a disorder.

goblet cell A unicellular gland often in the shape of a goblet that secretes mucus. Also called a *mucous cell.*

goiter (GOY-ter) An enlargement of the thyroid gland caused by iodine deficiency.

Golgi apparatus (GŌL-jē ap'-a-RAT-us) A cellular organelle characterized by a series of flattened, hollow cisternae. It serves as a site of anabolic activities.

gonad (GŌ-nad) An organ that produces gametes and sex hormones. In the male the gonads are the testes, and in the female, the ovaries.

gonadotropic (gō'-nad-ō-TRŌ-pik) *hormone* A hormone that regulates the activities of the gonads.

gonorrhea (gon'-ō-RĒ-a) A sexually transmittable disease (STD) caused by the bacterium *Neisseria gonorrhoeae.*

gout (gowt) A disorder caused by excessive uric acid in the blood. The uric acid crystallizes and deposits in joints, kidneys, and soft tissue, resulting in localized pain.

gray matter Nerve tissue in the brain and spinal cord that contains neuron cell bodies, dendrites, and non-myelinated axons and therefore appears gray or non-white in color.

greater omentum (GRĀ-ter ō-MEN-tum) A large fold of the serosa of the stomach (or visceral peritoneum) that covers over the abdominal cavity.

groin (groyn) The region of the body located between the thigh and the trunk.

growth hormone (GH) A hormone secreted by the anterior lobe of the pituitary gland that stimulates growth of body tissues. Also called *somatotropin.*

gustation (gus-TĀ-shun) The sense of taste.

gyrus (JĪ-rus) An upfolding convolution on the cerebral surface of the brain. Plural is *gyri.*

hair A threadlike outgrowth of the skin that is composed of columns of keratinized cells.

hair follicle A region of epithelial tissue surrounding the root of a hair where the hair originates.

hamstring (HAM-streng) The posterior muscles of the thigh that flex the leg at the knee.

haploid (HAP-loyd) Having one-half the normal number of chromosomes in a cell, designated as 1N. Gametes are the only haploid cells in the body.

hard palate (PAL-at) The anterior portion of the roof of the mouth. It is formed by the maxillary and palatine bones and is lined with mucous membrane.

haustra (HAWS-tra) The pouches in the large intestine that form when the taenia coli muscle contracts.

haversian system See osteon.

head The region of the body superior to the neck. The rounded, proximal end of a long bone. The proximal attachment of a muscle to a bone.

heart The hollow muscular organ within the thoracic cavity that propels blood through the circulatory network.

heart rate The number of heartbeats (complete cardiac cycles) per minute.

hematocrit (hē-MAT-ō-krit) Procedure to measure the percentage of red blood cells in a sample of blood, which involves centrifuging the sample and measuring the red blood cell volume relative to other blood components. Also, the percentage obtained by means of this test.

hematoma (hē'-ma-TŌ-ma) The presence of blood within a body cavity or space.

hematopoiesis (hēm'-a-tō-poy-e-sis) The production of blood cells in the red bone marrow. Also called *hemopoiesis.*

hematuria (hē'-ma-TOOR-ē-a) The condition of blood in the urine.

hemoglobin (hē'-mō-GLŌ-bin) A complex protein in red blood cells involved in the transport of oxygen and carbon dioxide.

hemolysis (hē-MOL-i-sis) The bursting of a red blood cell resulting from disruption of the plasma membrane by toxins, freezing or thawing, or exposure to a hypotonic solution.

hemophilia (hē'-mō-FĒ-lē-a) An inherited disorder in which the blood-clotting mechanism is incomplete, resulting in a partial or complete inability to stop blood leakage following an injury.

hemorrhage (HEM-or-rij) The loss of blood from the circulatory system, or bleeding.

hemorrhoids (HEM-ō-roydz)   Varicosed (dilated) blood vessels in the anal region. Also called *piles*.

hemostasis (hē'-mō-STĀ-sis)   The stoppage of bleeding.

hepatic (he-PAT-ic)   Pertaining to the liver.

hepatic portal circulation   A circulatory network within the systemic system that involves the transport of blood from the alimentary canal to the liver. Its main vessel is the hepatic portal vein.

hepatitis (hep'-a-TĪ-tis)   Inflammation of the liver.

hernia (HER-nē-a)   The protrusion of a body part through a body cavity wall or membrane.

heterozygous (het'-er-ō-ZĪ-gus)   Having two different genes for the same trait.

hilum (HĪ-lum)   An area of an organ, usually a depression, where blood vessels and nerves enter or exit. Also called a *hilus*.

hirsutism (HER-soot-izm)   The excessive growth of male-pattern hair in females (and children) due to high levels of androgens.

histamine (HlS-ta-mēn)   A chemical released by certain blood cells (mast cells, basophils, and platelets) when they are injured. It causes vasodilation, increased permeability of blood vessels, and bronchiole constriction.

histology (his-TO-lō-jē)   The microscopic study of tissues.

homeostasis (hō'-mē-ō-STĀ-sis)   A condition of equilibrium, or physiological stability, of body systems, in which the internal environment of the body remains relatively constant.

homologous (hō-MOL-ō-gus)   Describing two organs that correspond in structure, position, and origin.

homozygous (hō-mō-ZĪ-gus)   Having two identical genes for a particular trait.

horizontal plane   A plane that extends perpendicular to the length of the body, dividing it into superior and inferior portions. Also called *transverse plane*.

hormone (HOR-mōn)   A substance secreted by endocrine tissue that changes the physiological activity of the target cell.

humoral immunity (HYOO-mor-al im-MYOO-ni-tē)   Specific resistance to circulating pathogens by the production of antibodies.

hydrocephalus (hī-drō-SEF-a-lus)   Abnormal accumulation of cerebrospinal fluid in the brain ventricles of a child whose cranial bones have not yet fused, resulting in cranial enlargement.

hydrostatic pressure   The pressure exerted by a fluid.

hymen (HĪ-men)   A thin fold of mucous membrane rich in blood vessels at the vaginal orifice.

hyperplasia (hī'-per-PLĀ-zē-a)   An increased production and growth of cells beyond normal limits.

hyperpolarization (hī'-per-pōl-a-ri-ZĀ-shun)   The increase of negativity on the inside of a cell's plasma membrane, resulting in an increase in voltage difference.

hypersecretion (hī'-per-sē-KRĒ-shun)   The increased secretion of a product by a gland beyond its normal levels.

hypertension (hī'-per-TEN-shun)   The condition of high blood pressure.

hyperthermia (hī'-per-THERM-ē-a)   An elevation in body temperature above normal levels.

hypertonic (hī'-per-TON-ik)   The state of a solution having a greater concentration of dissolved particles than the solution it is compared with.

hypertrophy (hī'-PER-trō-fē)   The abnormal enlargement or growth of a cell, a tissue, or an organ.

hyperventilation (hī'-per-vent-i-LĀ-shun)   A respiration rate that is higher than normal.

hypodermis (hī'-pō-DERM-is)   The area of the body between the dermis of the skin and skeletal muscle.

hyposecretion (hī'-pō-se-KRĒ-shun)   The diminished secretion of a product by a gland.

hypothalamus (hī'-pō-THAL-a-mus)   The small interior portion of the diencephalon in the brain. It functions mainly in the control of involuntary activities, including endocrine gland regulation, sleep, thirst, and hunger.

hypothermia (hī'-pō-THERM-ē-a)   A depression of body temperature below normal levels.

hypotonic (hī'-pō-TON-ik)   The state of a solution having a lower concentration of dissolved particles than the solution it is compared with.

ileocecal valve (il'-ē-ō-SĒ-kal valv)   A fold of mucous membrane between the ileum (of the small intestine) and the cecum (of the large intestine). Also called the *ileocecal sphincter*.

ileum (IL-ē-um)   The distal segment of the small intestine.

immunity (i-MYOON-i-tē)   The state of resistance to infection.

immunoglobulin (i-myoon'-ō-GLOB-yoo-lin) *(Ig)*   A protein antibody produced by plasma cells during an immune response to infection. There are five types of immunoglobulins based on the nature of their protein structure.

immunosuppression (i-myoo'-nō-su-PRE-shun)   Inhibition of the immune response.

implantation (im-plan-TĀ-shun)   The attachment of an embryonic blastocyst to the lining of the uterus. It normally occurs about 8 days following fertilization.

impotence (IM-pō-tens)   The inability to maintain an erection long enough for sexual intercourse.

in utero (in YOO-ter-ō)   Within the uterus.

in vitro (in VĒ-trō)   Outside the body, such as in a culture bottle.

in vivo (in VĒ-vō)   Inside the living body.

incontinence (in-KON-tin-ens)   The inability to retain urine, semen, or feces due to a lack of sphincter tone or nervous control.

infarction (in-FARK-shun)   The death of tissue caused by a reduction in blood supply.

infection (in-FEK-shun)   A challenge to body health by the entry and multiplication of microorganisms or foreign particles.

inferior (in-FĒR-ē-or)   A directional term describing a location farther from the head than something else.

infertility (in-fer-TIL-i-tē)   An inability to fertilize or conceive. Also called *sterility*.

inflammation (in'-fla-MĀ-shun)   A localized response to tissue injury to promote the migration of white blood cells to the injury site. It is characterized by the symptoms of redness, pain, heat, and swelling.

infundibulum (in′-fun-DIB-yoo-lum)   The narrow connection between the hypothalamus of the brain and the pituitary gland. Also, the funnel-shaped distal end of a uterine tube that opens near an ovary.

ingestion (in-GES-chun)   The intake of food or liquid by the mouth.

inguinal (IN-gwi-nal)   Pertaining to the groin region (between the hip and thigh).

inorganic (in′-or-GAN-ik)   Describing a chemical substance whose molecular structure is not based on a chain of carbon atoms.

insertion (in-SER-shun)   The attachment of a muscle by its tendon to a movable bone.

inspiration (in-spi-RĀ-shun)   The act of drawing air into the lungs. Also called *inhalation*.

insulin (IN-soo-lin)   A hormone produced by the beta cells of the pancreas. It decreases glucose levels in the blood by stimulating glucose uptake by cells.

integumentary (in-teg′-yoo-MEN-tar-ē)   Pertaining to the skin and its accessory organs.

intercalated disc (in-ter′-ka-LĀ-ted disk)   A transverse thickening of a cardiac muscle cell's sarcolemma at its boundary with an adjacent cell. It aids in the conduction of an impulse from one cardiac cell to another.

intercellular (in′-ter-SEL-yoo-lar) *environment*   The area between cells.

interferon (in′-ter-FER-on)   A protein produced by virus-infected cells that stimulates noninfected cells to produce antiviral proteins, which inhibit viral replication.

internal (in′-TER-nal)   A directional term describing a location deep to the surface of the skin relative to something else.

internal nares (NAR-ēz)   The paired openings between the nasal cavity and the nasopharynx through which air passes. Also called *choanae*.

internal respiration   The exchange of respiratory gases between the blood and body cells.

interphase (IN-ter-fāz)   The period in the life cycle of a cell between cell divisions during which growth and DNA replication occur.

interstitial (in′-ter-STI-shul) *cells*   Cells in the testes located between seminiferous tubules that secrete testosterone. Also called *cells of Leydig*.

interstitial fluid (in′-ter-STI-shul FLOO-id)   The portion of extracellular fluid that fills the tissue spaces between cells. Also called *tissue fluid* and *intercellular fluid*.

intervertebral disc (in′-ter-VER-te-bral disk)   A cartilaginous joint that consists of a pad of fibrocartilage located between two adjacent vertebrae.

intestinal gland (in-TES-tin-al gland)   A tubular gland in the mucosa of the small intestine that secretes digestive enzymes. Also called *crypt of Lieberkühn*.

intracellular (in′-tra-SEL-yoo-lar) *environment*   The space within a cell.

intracellular fluid (ICF)   The fluid within cells.

intramembranous ossification (in′-tra-MEM-bra-nus os′-i-fi-KĀ-shun)   The development of bone from fetal connective tissue membranes.

intrauterine device (in′-tra-YOO-ter-in de-VĪS) *(IUD)*   A small artificial object inserted into the uterus to prevent pregnancy.

intrinsic factor (in-TRIN-sik FAK-tor)   A substance secreted by parietal cells of the stomach mucosa that facilitates vitamin B_{12} absorption.

invagination (in-vaj′-in-Ā-shun)   A folding inward of a body cavity wall into the body cavity.

inversion (in-VER-zhun)   Movement of the foot inward so that the sole of the foot faces medially.

involuntary (in-VOL-un-tar-ē)   Describing an activity that does not require conscious control.

ion (Ī-on)   An atom or a group of atoms with an electrical charge.

ionic bond (ī-ON-ik bond)   A chemical bond that is formed between atoms when one atom loses an electron and the other atom gains an electron.

ionization (ī′-on-i-ZĀ-shun)   The chemical process by which molecules dissociate in solution to form ions.

iris (Ī-ris)   A part of the vascular tunic of the eye. It is located on the anterior side of the eyeball and is composed of smooth muscle fibers that regulate the amount of light entering the eye. The iris is the colored part of the eye surrounding the pupil.

irritability (ir′-i-ta-BIL-i-tē)   The physiological process that permits an organism to react to changes in the environment.

ischemia (is-KĒ-mē-a)   A reduced, insufficient blood supply to an organ caused by an obstruction.

islet of Langerhans (ī-let of LANG-er-hanz)   One of numerous clusters of endocrine cells within the pancreas.

isometric (ī′-so-MET-rik) *contraction*   A type of muscle contraction in which tension is produced although little or no movement occurs.

isotonic (ī′-so-TON-ik) *contraction*   A type of muscle contraction in which the tension is held constant to produce movement.

isotonic solution   A solution that contains an equal amount of solutes relative to another solution.

isotope (Ī-so-tōp)   A chemical element that has the same number of protons as another but a different number of neutrons. A radioactive isotope, or radioisotope, changes into other elements as it decays and emits radiation.

jaundice (JAWN-dis)   A disorder characterized by yellowing of the skin and sclera (of the eyes) that is caused by the accumulation of breakdown products of red blood cells in the blood.

jejunum (je-JOO-num)   The middle segment of the small intestine.

joint (joynt)   A point of contact between two opposing bones. Also called *articulation*.

juxtaglomerular apparatus (juks′-ta-glō-MER-yoo-lar ap′-a-RAT-us)   A structure located in a kidney nephron that is composed of cells from the distal convoluted tubule and the afferent arteriole. It secretes renin in response to a decrease in blood pressure.

keratin (KER-a-tin)   A waterproofing protein present in the epidermis, nails, and hair.

ketone bodies (KĒ-tōn BOD-ēz) Compounds produced during excessive fat utilization (fat catabolism).

ketosis (kē-TŌ-sis) A metabolic disorder characterized by the excessive production of ketone bodies.

kidney stones A concretion of substances, including uric acid and calcium salts, that may form in the urinary tract to cause blockage. Also called *renal calculi*.

Korotkoff (kō-ROT-kof) *sounds* The audible sounds heard over an artery when blood pressure is measured with a sphygmomanometer.

kyphosis (kī-FŌ-sis) A disorder in which the thoracic curvature of the vertebral column is exaggerated, resulting in a round-shouldered or "hunchback" appearance.

labium (LĀ-bē-um) A synonym for lip. Plural is *labia*.

labor (LĀ-bor) The process of childbirth.

labyrinth (LAB-i-rinth) The system of interconnecting tubes of the inner ear.

laceration (las'-er-Ā-shun) An irregular wound in the skin.

lacrimal (LAK-ri-mal) Pertaining to the production or release of tears.

lactation (lak-TĀ-shun) The production of milk by the mammary glands.

lacteal (lak-TĒ-al) A small lymphatic vessel located within a villus of the small intestine that transports fat.

lactic acid (LAK-tik A-sid) A three-carbon molecule that is formed from pyruvic acid during anaerobic respiration.

lacuna (la-KOO-na) A chamber within bone or cartilage matrix that houses a cell (an osteocyte or a chondrocyte). Plural is *lacunae*.

lamellae (la-MEL-ē) Concentric rings of hardened bone matrix found in compact bone.

lanugo (la-NOO-gō) Fine hairs covering the skin of a fetus and newborn child.

large intestine The final segment of the alimentary canal, consisting of a large tube that forms the feces, which are expelled by the process of defecation.

laryngopharynx (la-ring'-ō-FAR-inks) The inferior part of the pharynx, which opens to the esophagus (posteriorly) and to the larynx (anteriorly).

larynx (LAR-inks) A boxlike cartilaginous organ in the respiratory tract located between the pharynx and the trachea.

lateral (LA-ter-al) A directional term describing a structure that is located farther from the vertical midline of the body relative to another structure.

lens An oval, transparent structure located between the posterior iris and the vitreous humor of the eyeball. It is connected to the vascular tunic by suspensory ligaments.

lesion (LĒ-zhun) Any structural change in a body part or region caused by disease.

lesser omentum (LES-er ō-MEN-tum) A fold of the peritoneum that extends between the liver and the medial margin of the stomach.

lethargy (LETH-ar-jē) A condition of drowsiness or lack of motivational energy.

leukemia (loo-KĒ-mē-a) A malignant disease of blood-forming tissue characterized by the abnormal production of immature or mutated white blood cells.

leukocyte (LOO-kō-sīt) A white blood cell. Also spelled *leucocyte*.

ligament (LlG-a-ment) A band or cord of dense connective tissue that extends from one bone to another to provide a joint with structural stability.

limbic system A functional region of the brain consisting of a group of interconnected parts that play a central role in emotions and memory.

lingual (LIN-gwal) Pertaining to the tongue; for example, the lingual frenulum, which connects the tongue to the floor of the mouth.

lipase (LĪ-pās) An enzyme that breaks down lipids.

lipid (LI-pid) An organic compound that is usually insoluble in water but soluble in alcohol, ether, and chloroform. Lipids include fats, phospholipids, and steroids.

lipoprotein (li'-pō-PRŌ-tēn) A protein-lipid complex produced by the liver that transports cholesterol and triglycerides through the bloodstream. Low-density lipoproteins (LDLs) are associated with an increased risk of atherosclerosis, whereas high-density lipoproteins (HDLs) are associated with a reduced risk.

liver A large digestive organ in the superior right corner of the abdominopelvic cavity that functions mainly in the interconversion of energy-storage molecules, detoxification of blood, and production of bile.

lordosis (lor-DŌ-sis) A disorder in which the lumbar curvature of the vertebral column is exaggerated.

lumbar (LUM-bar) Pertaining to the region of the back between the ribs and pelvis, or the loins.

lumen (LOO-men) The potential space within a tubular structure.

lung One of two large organs in the thoracic cavity that function in the exchange of respiratory gases.

lunula (LOO-nyoo-la) The white, crescent-shaped area at the proximal end of a nail.

luteinizing (LOO-tē-in-ī-z-ing) *hormone (LH)* A hormone secreted by the anterior lobe of the pituitary gland that in females stimulates ovulation, stimulates progesterone secretion by the corpus luteum, and prepares the mammary glands for milk secretion, and in males stimulates testosterone production by the testes.

lymph (limf) The slow-moving fluid within lymphatic vessels of the lymphatic system.

lymph node A small, oval organ located within the lymphatic vessel network.

lymphatic vessel A hollow, tubular structure similar to a vein that transports lymph in a direction leading toward the heart.

lymphocyte (LlM-fō-cīt) A type of white blood cell lacking large granules in the cytoplasm that plays a central role in immunity.

lymphoid tissue A specialized type of connective tissue that contains an abundance of lymphocytes. Also called *lymphatic tissue*.

lymphokines (LlM-fō-kīnz) A class of proteins produced by T lymphocytes that serve multiple functions in the immune response.

lysosome (LĪ-sō-sōm)   A cellular organelle that contains digestive enzymes.

macrophage (MAK-rō-fāj)   A large phagocytic cell that originated from a monocyte.

macula (MAK-yoo-la)   One of the sensory structures in the vestibule of the inner ear. It serves as a receptor for static equilibrium.

macula lutea (MAK-yoo-la LOO-tē-a)   A yellow-colored depression in the retina of the eye.

malignant (ma-LIG-nant)   Referring to diseases that tend to worsen and cause death; a malignant tumor is a cancer.

malleus (MAL-e-us)   The lateral ear bone that contacts the tympanic membrane; the hammer.

malnutrition (mal′-noo-TRI-shun)   The state of insufficient or poor nutrition.

mammary (MAM-a-rē) *gland*   A modified sweat gland in the breast that serves as the gland of milk secretion for nourishment of the young.

mammogram (MAM-ō-gram)   An X-ray photograph of the breast for evaluation of potential disease. The procedure for obtaining a mammogram is called *mammography.*

marrow (MAR-ō)   The soft, highly vascularized tissue within bone. It includes yellow marrow, consisting of adipose tissue; and red marrow, which consists of blood-forming tissue.

mast cell   A basophil that has migrated into loose connective tissue and secretes heparin (an anticoagulant) and serotonin (which promotes inflammation).

mastectomy (mas-TEK-tō-mē)   Surgical removal of a breast and surrounding tissues.

mastication (mas′-ti-KĀ-shun)   A synonym for the muscular act of chewing.

matrix (MĀ-triks)   The intercellular material in connective tissue.

matter (MA-ter)   Any substance that has weight and occupies space.

mechanoreceptor (me-KAN-ō-rē-sep′-tor)   A receptor that responds to a mechanical change of the receptor itself.

medial (MĒ-dē-al)   A directional term describing a part lying nearer to the vertical midline of the body relative to another part.

mediastinum (mē′-dē-as-TĪ-num)   A partition between the two peural cavities in the chest that consists of the heart, part of the esophagus, part of the trachea, and the major vessels of the heart.

medulla (me-DUL-a)   An inner, or deeper, part of an organ; for example, the medulla of the kidneys and the medulla of the adrenal gland.

medulla oblongata (me-DUL-a ob′-long-GA-ta)   The inferior part of the brain stem.

medullary cavity (med-YOO-lar-ē KAV-i-tē)   The potential space within the shaft of a long bone that contains yellow marrow.

meiosis (mī-Ō-sis)   The process of cell division resulting in the formation of haploid gametes.

melanin (MEL-a-nin)   A dark pigment released into some parts of the body, such as the skin.

melanocyte (MEL-an-ō-sīt′)   A cell normally located deep to the epidermis in the skin that secretes melanin.

melanocyte-stimulating hormone (MSH)   A hormone secreted by the anterior pituitary gland that may stimulate melanin production.

melanoma (mel-a-NŌ-ma)   A highly metastatic malignancy arising from melanocytes in the skin. Also called *malignant melanoma.*

melatonin (mel′-a-TŌ-nin)   A hormone secreted by the pineal gland that may play a role in circadian rhythms.

membrane (MEM-brān)   A thin sheet of tissue that lines or covers body structures. It may contain a thin layer of epithelial tissue and connective tissue, or only connective tissue.

membranous labyrinth (MEM-bra-nus LAB-i-rinth)   The portion of the inner ear located inside the bony labyrinth that contains perilymph fluid. It consists of the membranous semicircular canals, the saccule and utricle, and the cochlear duct.

memory (MEM-or-ē)   A function of the brain that consists of conscious recognition of past events.

menarche (me-NAR-kē)   The onset of menstruation.

meninges (me-NIN-jēz)   The three membranes covering the brain and spinal cord. Singular is *meninx.*

meningitis (me-in-JĪ-tis)   Inflammation of the meninges.

menopause (MEN-ō-pawz)   Termination of menstrual cycles.

menstrual cycle (MEN-stroo-al SĪ-kel)   The series of changes that occur on a regular, approximately 28-day cycle in the nonpregnant uterine lining.

menstruation (men′-stroo-Ā-shun)   The periodic discharge of the inner uterine lining characterized by a flow of blood, mucus, and epithelial cells that usually lasts for 3 to 5 days. Also called *menses.*

mesentery (MES-en-ter′-ē)   A fold of the peritoneum that attaches the small intestine to the posterior abdominal wall.

mesoderm (MEZ-ō-derm)   The middle of the three primary germ layers in a developing embryo that forms the muscles, the heart and blood vessels, and the connective tissues.

metabolic (MET-a-bol-ik) *rate*   The rate at which chemical reactions occur in the body.

metabolism (me-TAB-ō-lizm)   The sum of all chemical reactions occurring in the body, including anabolic (synthetic) and catabolic (decomposition) reactions.

metacarpus (met′-a-KAR-pus)   A collective term for the five bones (each of which is called a *metacarpal*) of the palm of the hand.

metaphase (MET-a-fāz)   A stage in mitosis characterized by alignment of the chromosomes in the center of the cell.

metastasis (me-TAS-ta-sis)   The spread of a primary malignant tumor into other tissues by its invasive growth into surrounding tissues and by the spread of tumor cells by way of lymphatic and circulatory channels into distant tissues.

metatarsus (met′-a-TAR-sus)   A collective term for the five bones (each of which is called a *metatarsal)* of the foot.

microfilament (mī′-krō-FIL-a-ment)   A rod-shaped component of cytoplasm composed of protein that provides a means of mobility for the cell.

microglia (mī′-krō-GLĒ-a)   A type of neuroglia in the brain characterized by its small size and phagocytic function.

microtubule (mī′-krō-TOOB-yool)   A tube-shaped component of cytoplasm composed of protein that provides support and shape for the cell.

microvilli (mī′-krō-VIL-ī)   Microscopic extensions of the cell membrane filled with cytoplasm that serve to increase the absorptive surface area of the cell.

micturition (mik′-tyoo-RISH-un)   The act of discharging urine from the urinary bladder to the exterior. Also called *urination.*

midbrain (MĪD-brān)   The superior part of the brain stem, located between the diencephalon and the pons. It serves as a relay center for impulses. Also called the *mesencephalon.*

middle ear   The area of the ear between the tympanic membrane of the outer ear and the bony labyrinth of the inner ear. It is an epithelium-lined space that houses the three ear ossicles. Also called the *tympanic cavity.*

midsagittal (MID-saj-i-tal)   A plane that extends vertically through the body, dividing it into equal right and left portions.

mineral (MIN-er-al)   An inorganic nutrient necessary for normal metabolic activities.

mineralocorticoid (min′-er-al-ō-KOR-ti-koyd)   Any one of a group of hormones secreted by the adrenal cortex that influence the concentrations of electrolytes in the body.

mitochondrion (mīt′-ō-KON-drē-on)   A cellular organelle that consists of a double layer of plasma membrane where many of the catabolic activities of the cell take place.

mitosis (mi-TŌ-sis)   The division of a cell's nucleus into two daughter nuclei, each of which contains the same genetic composition as the original parent. When mitosis is followed by cytokinesis, equal division of the whole cell results.

mitral valve (MĪ-tral valv)   A synonym for the left atrioventricular valve. Also called the *bicuspid valve.*

mixed nerve   A nerve that contains axons from sensory and motor neurons.

molecule (MOL-e-kyool)   A combination of two or more atoms chemically bound.

monocyte (MON-ō-sit)   A large, agranular white blood cell that is phagocytic.

monosaccharide (mon′-ō-SAK-a-rīd)   The basic building block of any carbohydrate. It is a simple sugar like glucose or fructose.

mons pubis (monz PYOO-bis)   The elevated, hair-covered body surface area over the symphysis pubis in females.

morula (MOR-yoo-la)   A solid mass of cells resulting from cleavage of the zygote during early embryonic development.

motor end plate   The portion of the sarcolemma of a muscle fiber that is in close association with a motor neuron.

motor nerve   A nerve that contains axons from motor neurons and thereby transmits impulses away from the central nervous system.

mucosa (myoo-KŌ-sa)   An epithelial membrane that lines a body cavity or organ and contains cells that secrete mucus. Also called *mucous membrane.*

mucus (MYOO-kus)   A thick fluid secretion from mucous cells.

multiple sclerosis (MUL-ti-pul skler-Ō-sis)   An inherited disorder of the brain in which oligodendrocytes surrounding certain axons degenerate, leading to lesions that cause a gradual, progressive loss of function.

mumps   Inflammation of the parotid (salivary) glands, causing swelling, pain, and fever.

murmur (MER-mer)   An abnormal sound produced during a cardiac cycle that is an indication of faulty heart valve operation.

muscle   An organ composed of skeletal muscle tissue and its associated connective tissue that functions mainly in the production of movement of the skeleton.

muscle fiber   A synonym for *muscle cell.*

muscle tissue   One of the four primary types of tissue in the body, characterized by its specialization to contract.

muscle tone   A sustained, partial contraction of a muscle.

muscular dystrophy   An inherited disease characterized by the progressive degeneration of muscle.

muscularis (mus′-kyoo-LAR-is)   A layer of smooth muscle tissue within the wall of an organ.

mutagenesis (myoo-tā-JEN-e-sis)   The development of a mutation by some factor.

mutation (myoo-TĀ-shun)   A change in the sequence of bases in the DNA molecule resulting in an alteration of the genetic code.

myasthenia gravis (mī-as-THĒ-nē-a GRAV-is)   A disorder of muscle resulting in a loss of strength caused by destruction of acetylcholine receptors by the body's own antibodies, which inhibits muscle contraction.

myelin sheath (MĪ-e-lin shēth)   A white, segmented insulative cover over the axons of many peripheral neurons that is produced by Schwann cells. A neuron axon that is covered by the myelin sheath is said to be *myelinated.*

myocardial infarction (mī′-ō-KAR-dē-al in-FARK-shun) **(MI)**   The death of cardiac muscle tissue caused by an interruption in blood supply. It leads to a *heart attack.*

myocardium (mī′-ō-KAR-dē-um)   The primary layer of the heart wall, which is composed of cardiac muscle tissue.

myofibril (mī′-ō-FĪ-bril)   A rod-shaped component of a muscle fiber, which extends the length of the fiber and is composed of thin and thick filaments of protein.

myogram (MĪ-ō-gram)   A recording of muscle contraction by an apparatus known as a *myograph.*

myometrium (mī'-ō-MĒ-trē-um)   The smooth muscle layer in the wall of the uterus.

myopia (mī-Ō-pē-a)   A defect in vision resulting in difficulty focusing on distant objects. Also called *nearsightedness*.

myosin (MĪ-ō-sin)   A contractile protein that provides cells with the ability to contract, particularly evident in muscle fibers, where it makes up thick filaments.

myxedema (miks-e-DĒ-ma)   A disorder caused by hypothyroidism in adults, which is characterized by a low metabolic rate and swelling of the face (producing a "moon face" appearance).

nail   A thin, hard plate mostly of keratin that is derived from the epidermis and develops at the distal end of the fingers and toes.

nasal cavity   The space within the nose that is lined with mucous membrane and divided by the nasal septum into right and left chambers.

nasal septum   A vertical partition dividing the nasal cavity into right and left chambers that is composed of bone and cartilage covered with mucous membrane.

nasopharynx (nā'-sō-FAR-inks)   The superior portion of the pharynx, which transports air between the internal nares and the oropharynx.

nausea (NAW-sē-a)   A sensation of discomfort that may include loss of appetite, dizziness, and stomach upset.

necrosis (ne-KRŌ-sis)   Death of a cell, a group of cells, or a tissue due to disease.

negative feedback   A mechanism in which a stimulus causes a response that reverses or reduces the stimulus, returning the body function to a more stable state or homeostasis.

neonatal (nē'-ō-NĀ-tal)   Pertaining to the period of development within the first four weeks after birth.

neoplasm (NĒ-ō-plazm)   A tumor, or abnormal growth of cells, that may be benign or malignant.

nephron (NE-fron)   One of many microscopic, tubular structures within each kidney where the functions of filtration, reabsorption, and secretion occur.

nerve   An organ of the nervous system composed of a bundle of neuron axons invested and surrounded by connective tissue and blood vessels, which functions in the conduction of an impulse from one area of the body to another.

neuralgia (nyoo-RAL-jē-a)   A sensation of pain that can be felt along a peripheral nerve.

neurilemma (nyoo'-ri-LEM-a)   The outer layer of a myelin sheath associated with a nerve fiber that contains the nucleus and much of the cytoplasm of a Schwann cell.

neuritis (nyoo-RĪ-tis)   Inflammation of a nerve or group of nerves.

neuroglia (nyoo'-RŌG-lē-a)   Supportive cells of the nervous system that are most prevalent in the brain and spinal cord.

neuromuscular junction (nyoo'-rō-MUS-kyoo-lar JUNK-shun)   The area of contact between the terminal end of a motor neuron and the sarcolemma of a skeletal muscle fiber.

neuron (NYOO-ron)   A cell of nerve tissue characterized by its specialization to conduct impulses (conductivity).

neurosecretory (nyoo'-rō-sē-KRĒ-tor-ē) *cells*   Neurons that extend from the hypothalamus to the posterior lobe of the pituitary gland and secrete the hormones oxytocin (OT) and antidiuretic hormone (ADH).

neurotransmitter (nyoo'-rō-TRANS-mit-er)   A molecule that transmits or inhibits the transmission of a nerve impulse from one neuron to another across a synapse.

neutrophil (NOO-trō-fil)   A type of granular, phagocytic white blood cell characterized by a cytoplasm that stains pink in a neutral stain.

node of Ranvier (ran'-vē-Ā)   A gap in the myelin sheath covering a nerve fiber, which accelerates the rate of impulse conduction.

norepinephrine (nor'-ep-i-NEF-rin) *(NE)*   A hormone secreted by the adrenal medulla that stimulates a response similar to that produced by sympathetic stimulation (the fight-or-flight response).

nucleic acid (noo-KLĀ-ik A-sid)   An organic compound that is composed of numerous subunits called *nucleotides*. DNA and RNA are examples of nucleic acids.

nucleolus (noo-KLĒ-ō-lus)   A spherical body within the nucleus of a cell that is not bound by a plasma membrane, which functions in the storage of ribosomal RNA.

nucleotide (NOO-klē-ō-tīd')   An organic compound consisting of a simple sugar, a nitrogenous base, and a phosphate group.

nucleus (NOO-klē-us)   The largest structure in a cell. It contains the genetic material to determine protein structure and function—the DNA—and is enveloped by a double-layered plasma membrane. Also, the dense core of an atom that contains protons and neutrons.

nutrient (NOO-trē-ent)   Any chemical substance that provides the body with the capability to form new compounds, generate and use energy, or perform body activities.

nutrition (noo-TRI-shun)   The process by which nutrients are taken into the body and utilized by body cells.

obesity (ō-BĒS-i-tē)   A disorder characterized by a body weight that is 10% to 20% over a desirable standard weight as a result of the accumulation of fat.

occipital (ok-SIP-i-tal)   Pertaining to the lower back portion of the head.

occlusion (o-KLOO-zhun)   A blockage or closure.

olfactory (ōl-FAK-tor-ē)   Pertaining to the sense of smell.

oligodendrocyte (ō-lig-ō-DEN-drō-sīt)   A neuroglial cell found in the brain and spinal cord that produces a supportive myelin sheath around CNS axons.

oncogene (ON-kō-jēn)   A gene that can transform the cell it is a part of into a cancerous cell.

oncology (on-KOL-ō-jē)   The study or science of tumors.

oocyte (Ō-ō-sīt)   A gamete produced within an ovary. Also called *ovum* or *egg*.

oogenesis (ō'-ō-JEN-e-sis)   The meiotic formation and development of an oocyte.

optic (OP-tik)   Pertaining to the sense of vision or to the eye.

optic chiasma (OP-tik kī-AZ-ma)   The point at which the two optic nerves cross on the ventral aspect of the brain.

optic disc   The area on the retina where the optic nerve exits the eye; it contains no rod or cone cells. Also called the *blind spot.*

orbit (OR-bit)   One of two large depressions in the skull that are each bordered by seven bones and house the eyeball and associated structures. Also called *eye socket.*

organ (OR-gan)   An organized combination of two or more different types of tissues that performs a general function.

organ of Corti (KOR-tī)   The structure within the inner ear that contains receptor cells sensitive to sound vibrations.

organelle (or-gan-EL)   A component of a cell that has a consistent, similar structure in other cells and performs a particular function.

organic (or-GAN-ik)   Describing a chemical substance whose molecular structure is based on a carbon skeleton.

organism (OR-gan-izm)   A complete living being; a whole individual.

organogenesis (or'-gan-ō-JEN-e-sis)   The formation of organs during embryonic development.

orgasm (OR-gazm)   Nervous events (sensory and motor) associated with sexual climax.

orifice (OR-i-fis)   An opening into the body or into a structure.

origin (OR-i-jin)   The point of attachment of a muscle's tendon to a stationary bone.

oropharynx (or'-ō-FAR-inks)   The middle portion of the pharynx, located between the nasopharynx and the laryngopharynx. The oral cavity opens into it (by way of the fauces).

osmosis (oz-MŌ-sis)   The net movement of water molecules across a selectively permeable membrane from a region of high water concentration (low solutes) to a region of low water concentration (high solutes).

osmotic pressure (oz-MO-tik PRE-shur)   The pressure required to prevent the movement of water across a selectively permeable membrane.

osseous (OS-ē-us)   Pertaining to bone.

ossification (os'-i-fi-KĀ-shun)   Bone formation. Also called *osteogenesis.*

osteoblast (ŌS-tē-ō-blast)   A type of bone cell characterized by its mobility and by its ability to produce bone matrix.

osteoclast (ŌS-tē-ō-klast)   A type of bone cell characterized by its ability to dissolve bone matrix.

osteocyte (ŌS-tē-ō-sīt)   A type of bone cell characterized by its immobile location within a lacuna and by a reduced ability to produce bone matrix.

osteomalacia (ōs'-tē-ō-ma-LĀ-shē-a)   A disorder characterized by the softening of bones that is caused by vitamin D deficiency in the diet.

osteomyelitis (ōs'-tē-ō-mī-i-LĪ-tis)   Inflammation of the soft tissue in bone (bone marrow).

osteon (ŌS-tē-on)   An organized arrangement of bone tissue in adult compact bone such that the bone matrix concentrically surrounds a central canal containing a blood vessel. Also called a *Haversian system.*

osteonic canal   The central canal in an osteon that contains a blood vessel. Also called *haversian canal.*

osteoporosis (ōs'-tē-ō-por-Ō-sis)   A disorder common in old age characterized by a decreased bone mass and a subsequent increased tendency to fracture caused by decreased levels of estrogen and/or calcium.

otic (O-tik)   Pertaining to the ear.

otitis media (ō-TĪ-tis MĒ-dē-a)   An acute infection of the middle ear that causes symptoms of pain and local fever.

otolith (Ō-tō-lith)   A small crystal of calcium carbonate located within the macula of the inner ear that participates in the sensation of static equilibrium.

oval window   The membrane-covered opening between the stapes and the inner ear.

ovary (Ō-var-ē)   The female gonad, or primary reproductive organ, which produces gametes and female sex hormones.

ovulation (ov-yoo-LĀ-shun)   The release of a secondary oocyte from a Graafian follicle in an ovary.

oxidation (ok-si-DĀ-shun)   A chemical event in which electrons and hydrogen ions are removed from a molecule, or oxygen is added to a molecule.

oxygen debt (OK-si-jen det)   The amount of oxygen necessary to eliminate (by oxidation) lactic acid that has accumulated in muscle as a result of muscle activity.

oxytocin (ok'-sē-TŌ-sin)   *(OT)*   A hormone secreted by neurosecretory cells in the hypothalamus and released by the posterior lobe of the pituitary gland that stimulates contraction of the pregnant uterus and the ducts of mammary glands.

pacemaker (PĀS-mā-ker)   An electronic device surgically inserted under the skin in the left side of the chest that is attached to the sinoatrial node of the heart by way of an electrode. It relays a signal to the heart to stimulate each cardiac cycle in patients whose SA node is not functional.

Pacinian corpuscle (pa-SIN-ē-an KOR-pus-el)   A receptor located in the dermis that responds to touch (pressure).

palate (PAL-at)   The mucous membrane–lined structure forming the roof of the mouth. The anterior end is the hard palate, and the posterior end is the soft palate.

palpate (PAL-pāt)   To examine by touch.

pancreas (PAN-krē-as)   A soft, oblong organ located posterior to the stomach in the abdominal cavity. The pancreas secretes digestive enzymes, and hormones that regulate blood sugar.

Pap smear   A diagnostic test involving the removal of cells from the uterine lining or other female reproductive organs and their microscopic examination for malignancies. Also called the *Papanicolaou test.*

papilla (pa-PIL-a)   A small, finger-shaped projection.

paralysis (pa-RAL-i-sis)   A loss of motor function due to a lesion of nervous or muscle tissue.

parasagittal (par'-a-SAJ-i-tal) *plane*   A plane that extends vertically through the body, dividing it into unequal right and left portions.

parasympathetic division (par'-a-simp-a-THE-tik di-VI-zhun)   The component of the autonomic nervous system that stimulates activities that conserve body energy.

parathyroid (par'-a-THĪ-royd) *gland*   One of four or five pea-shaped glands located embedded in the posterior side of the thyroid gland.

parathyroid hormone (PTH)   A hormone secreted by the parathyroid glands that increases calcium levels in the blood by stimulating osteoclast activity.

parietal (pa-RĪ-e-tal)   Pertaining to the outer wall of a cavity or an organ.

parietal cell   A cell in the stomach mucosa that secretes hydrochloric acid and intrinsic factor.

parietal pericardium (par'-i-KAR-dē-um)   The outer serous membrane covering the heart. Also called the *pericardial sac.*

parietal pleura (PLOO-ra)   The outer serous membrane associated with each lung. It is attached to the inner thoracic wall.

Parkinson's disease   A progressive disease that leads to the degeneration of the basal ganglia and substantia nigra of the cerebrum and is characterized by muscle tremors and the gradual loss of motor function. Also called *Parkinsonism.*

parotid (pa-ROT-id) *glands*   A pair of salivary glands, each of which is located between the skin of the cheek and the masseter muscle on a side of the face.

partial pressure (PAR-shal PRESH-er) The pressure caused by one gas in a mixture of gases.

parturition (par'-too-RISH-un)   The process of childbearing, or giving birth.

pathogen (PATH-ō-jen)   A disease-causing organism.

pathology (path-O-lō-jē)   The study or science of diseases.

pectoral (PEK-tor-al)   Pertaining to the upper trunk or chest region.

pelvic (PEL-vik)   Pertaining to the base of the trunk region or pelvis.

pelvic cavity   The inferior portion of the abdominopelvic cavity bordered by the pelvis.

pelvic inflammatory disease (PID)   A general term for a bacterial infection of any of the organs within the pelvic cavity.

pelvis   The bowl-like base of the axial skeleton formed by the two pelvic (innominate) bones and the sacrum.

penis (PĒ-nis)   The external reproductive organ of the male through which most of the urethra extends.

pepsin (PEP-sin)   An enzyme initially secreted by zymogenic (chief) cells in the stomach mucosa in the inactive form of pepsinogen that, when activated by the presence of hydrochloric acid into pepsin, can digest protein.

pericardial (par'-i-KAR-dē-al) *cavity*   A narrow space between the outer wall of the heart (the visceral peri-cardium) and the parietal pericardium that contains pericardial fluid.

pericardium (par'-i-KAR-dē-um)   The serous membrane associated with the heart that is composed of two layers, the inner, visceral pericardium and the outer, parietal pericardium.

perichondrium (par'-i-KON-dre-um)   A layer of dense connective tissue that envelops cartilage.

perilymph (PAR-i-limf)   The fluid within the membranous labyrinth of the inner ear.

perimysium (par'-i-MĪ-sē-um)   An extension of the epimysium of muscle that invaginates inward to divide a muscle into bundles.

perineum (par'-i-NĒ-um)   The area between the anus and the posterior border of the external genitalia.

perineurium (par'-i-NYOO-rē-um)   An extension of the epineurium of a nerve that invaginates inward to wrap around bundles of nerve fibers.

periodontal ligament (par'-ē-ō-DON-tal LIG-a-ment)   A connective tissue membrane that surrounds a tooth and connects it to the bone of the jaw.

periosteum (par'-ē-OS-tē-um)   The connective tissue covering around a bone that is important in bone growth, nutrition, and repair.

peripheral (per-I-fer-al) *nervous system (PNS)*   The division of the nervous system consisting of nerves and ganglia located between the central nervous system and the body surfaces.

peripheral resistance   The slowing of blood flow through vessels caused by contact between blood and vessel walls, resulting in friction.

peristalsis (par'-i-STAL-sis)   The sequential contraction of muscles surrounding a tubular structure that forces material through its lumen.

peritoneal (par'-i-tō-NĒ-al) *cavity*   The space between the parietal peritoneum and the visceral peritoneum that contains a small amount of fluid.

peritoneum (par'-i-tō-NĒ-um)   The extensive serous membrane associated with the abdominopelvic cavity.

peritonitis (par'-i-tō-NĪ-tis)   Inflammation of the peritoneum, usually caused by bacterial infection.

peroxisomes (per'-OX-i-sōmz)   Small, spherical cellular organelles similar to lysosomes that play catabolic roles in the cell.

Peyer's patches   Clusters of lymphatic tissue containing numerous white blood cells, located in the wall of the small intestine.

pH   A measure of the concentration of hydrogen ions in a solution, according to a graded scale of decreasing ions from 1.0 to 14.0.

phagocytosis (fag'-ō-sī-TŌ-sis)   A type of cytosis in which bulk solid materials may be transported into a cell. It is performed by white blood cells for the removal of harmful particles and cells from the body.

phalanx (FĀ-lanks)   A bone of a finger or toe. Plural is *phalanges.*

pharynx (FAR-inks)   A tube that extends from the level of the internal nares to its union with the larynx and transports air, food, and liquid.

phenotype (FĒN-ō-tīp)   A physical characteristic that is an expression of the genetic makeup of an individual.

phenylketonuria (fen'-il-kē'-tō-NYOOR-ē-a) *(PKU)*   A metabolic disorder characterized by elevated levels of the amino acid phenylalanine in the bloodstream.

phospholipid (fos'-fō-LI-pid)   A lipid molecule that contains a phosphate group associated with two fatty acid molecules and a glycerol molecule and is a major component of plasma membranes.

photoreceptor (fō'-tō-rē-SEP-tor)   A sensory receptor that responds to changes in light by initiating a nerve impulse.

physiology (fiz'-ē-O-lō-jē)   The study or science of the functions associated with an organism's survival.

pia mater (PĒ-a MĀ-ter)   The innermost of the three meninges surrounding the brain and spinal cord.

pineal (pī-NĒ-al) *gland*   A small endocrine gland located at the posterior end of the diencephalon, forming a part of the roof of the third ventricle. Also called the *epithalamus.*

pinocytosis (pin'-ō-sī-TŌ-sis)   A type of exocytosis in which bulk amounts of fluid are transported into the cell.

pituitary (pi-TOO-i-tar-ē) *gland*   A small, functionally important endocrine gland located inferior to the hypothalamus and attached to it by way of a short stalk. Also called the *hypophysis.*

placenta (pla-SEN-ta)   A structure whose origin is shared by embryonic cells and the uterine lining that provides a means of material transport between the mother and the developing unborn child.

plasma (PLAZ-ma)   The extracellular fluid that forms a portion of blood.

plasma cell   A differentiated white blood cell that secretes antibodies.

plasma membrane   A microscopic barrier associated with cells that is composed mainly of a phospholipid bilayer and protein. The outer plasma membrane of a cell is also called the *cell membrane.*

platelet (PLĀT-let)   A formed element of blood that is active in blood clot formation.

pleura (PLOOR-a)   The serous membrane associated with the lungs. It consists of the inner, visceral pleura and the outer, parietal pleura. Plural is *pleurae.*

pleural cavity   A narrow space between the visceral and parietal pleurae that contains pleural fluid.

plexus (PLEKS-us)   A network of interconnecting nerves, veins, or lymphatic vessels.

pneumonia (noo-MŌ-nē-a)   Acute infection of the lung alveoli that triggers an inflammatory response in lung tissue.

podocyte (PŌ-dō-sīt)   A fenestrated cell forming the visceral layer of a Bowman's capsule in the kidneys.

polar body (PŌ-lar BO-dē)   A nonfunctional cell of reduced size produced during oogenesis.

polarization (pō'-lar-i-ZĀ-shun)   A state in which opposite charges occur on the same structure. It occurs along a cell's plasma membrane when the inner surface carries a negative charge with respect to the outer surface, which is positively charged.

poliomyelitis (pō'-lē-ō-mī-e-LĪ-tis)   An acute viral infection that causes fever, headache, backache, and deep muscle pain. A severe form, called *bulbar polio,* can cause paralysis or death.

polycythemia (pol'-ē-sī-THĒ-mē-a)   A disorder characterized by a hematocrit above normal levels.

polypeptide (pol'-ē-PEP-tīd)   A protein that consists of more than ten amino acid subunits bonded together.

polysaccharide (pol'-ē-SAK-a-rīd)   A carbohydrate that consists of a long chain of simple sugars bonded together.

pons (ponz)   A part of the brain located between the midbrain and the medulla oblongata.

popliteal (pop'-li-Tē-al)   Pertaining to the area posterior to the knee joint.

posterior (pō-STĒR-ē-or)   A directional term describing the location of a part toward the back or rear side relative to another part. In humans it is a synonym for *dorsal.*

posterior horn   A region of the spinal cord gray matter containing sensory neuron cell bodies. Also called the *dorsal horn.*

posterior pituitary gland   The part of the pituitary gland at the base of the brain that consists of the axons of neurons originating in the hypothalamus and supporting tissue.

posterior root   The structure merging with the spinal cord on its posterior aspect that contains sensory nerves. Also called the *dorsal root.*

postganglionic neuron (pōst'-gang-lē-ON-ik NYOO-ron)   A neuron of the autonomic nervous system that conducts impulses from a ganglion to an effector, such as smooth muscle, cardiac muscle, or a gland.

postnatal (pōst'-NĀ-tal)   Pertaining to any time period after childbirth.

preganglionic neuron (prē'-gang-lē-ON-ik NYOO-ron)   A neuron of the autonomic nervous system that conducts impulses from the central nervous system to a ganglion.

pregnancy (PREG-nan-sē)   A state in which a new organism develops and is nourished within a female's uterus.

premenstrual syndrome (prē-MEN-stroo-al SĪN-drōm) *(PMS)*   A disorder in females that is characterized by abnormal levels of pain and discomfort, leading to increased irritability and stress, and is associated with hormonal cycling during the menstrual cycle.

prepuce (PRĒ-pyoos)   The skin that partially covers the glans of the penis (in the male) or clitoris (in the female).

primary germ layer   One of three layers of cells that differentiate during the embryonic stage to give rise to all tissues in the body. They are the endoderm, mesoderm, and ectoderm.

process (PRO-ses)   A prominent projection on a bone.

progesterone (prō-JES-ter-ōn)   A female sex hormone secreted by the corpus luteum in the ovary and, during pregnancy, by the placenta.

prognosis (prog-NŌ-sis)   A forecast of the likely effects of a disease.

prolactin (prō-LAK-tin) *(PRL)*   A hormone secreted by the anterior pituitary gland that stimulates and maintains milk secretion by the mammary glands.

pronation (prō-NĀ-shun)   Movement of the hand such that the palm is turned downward (inferiorly) or backward (posteriorly).

prophase (PRŌ-fāz)   The first stage of mitosis, during which the chromosomes become visible as chromatid pairs joined at the centromere.

proprioception (PRŌ-prē-ō-sep′-shun)   The sense of body position and direction as perceived by receptors, called *proprioceptors,* located in the muscles, joints, and the inner ear. Also called *kinesthesia.*

prostaglandin (prō′-sta-GLAN-din)   A hormonelike substance synthesized in small quantities that triggers responses in nearby organs.

prostate (PRŌ-stāt) *gland*   A walnut-shaped gland surrounding the urethra as it emerges from the urinary bladder in males. Its secretions contribute to semen.

prosthesis (prōs-THĒ-sis)   An artificial device used to assist in mechanical movements of the body in place of missing or nonfunctional body parts.

protein (PRŌ-tēn)   An organic compound composed of amino acid subunits.

prothrombin (prō-THROM-bin)   An inactive protein produced by the liver that is converted to the active enzyme thrombin during the process of blood clot formation.

proton (PRŌ-ton)   A subatomic particle located in the nucleus of an atom that carries a positive charge.

proximal (PROKS-i-mal)   A directional term indicating a body part that is located nearer to the origin or point of attachment to the trunk than another; opposite of *distal.*

pseudopod (SOO-dō-pod)   A streaming extension of plasma membrane–bound cytoplasm from a mobile cell.

psychosomatic (sī′-kō-sō-MAT-ik)   Pertaining to a condition in which a sensation or actual lesion is the result of emotional stress.

puberty (PYOO-ber-tē)   The stage of body development during which reproductive organs become functional.

pulmonary (PUL-mo-nar-ē)   Pertaining to the lungs and their function of gas exchange.

pulmonary circulation   The circuit of blood flow to the lungs, through lung capillaries, and back to the left atrium of the heart.

pulp cavity   The space within a tooth that is filled with pulp, which consists of connective tissue containing blood vessels and nerves.

pulse   The rhythmic flow of blood through arteries that can be felt by pressing against arteries near the skin. The pulse rate is equal to the heart rate.

pupil (PYOO-pil)   The small hole through the center of the iris in the eye through which light passes.

Purkinje fibers (per-KIN-jē FĪ-berz)   Specialized cardiac muscle cells that conduct impulses from the ventricular bundle (His bundle) of the heart conduction system to the ventricles.

pus   A liquid containing blood plasma, dead or dying leukocytes, and other cell debris that forms during inflammation.

pyloric sphincter (pī-LOR-ik SFENK-ter)   A circular band of smooth muscle at the union of the stomach and small intestine that controls the movement of material between them. Also called the *pyloric valve.*

pyorrhea (pī-ō-RĒ-a)   A discharge of pus, especially from the gums, as a result of bacterial infection.

pyruvic acid (pī-ROO-vik A-sid)   A three-carbon molecule that is formed as the product of glycolysis (the breakdown of glucose).

radioactive (rā′-dē-ō-AK-tiv)   A property of an unstable isotope whereby it releases energy in the form of alpha, beta, or gamma particles as it decays.

reactivity (rē′-ak-TIV-i-tē)   The readiness with which an atom forms or breaks a chemical bond with another atom.

receptor (rē-SEP-tor)   A structure that is capable of responding to a stimulus by initiating a nerve impulse.

recessive (rē-SES-iv)   In genetics, pertaining to a gene whose trait is not expressed when it is combined with a gene for an opposing trait, the dominant gene.

recombinant DNA   DNA that is synthesized by genetic engineering, in which DNA material from one source may be combined with DNA from a different source.

recruitment (rē-KROOT-ment)   The process of increasing the number of motor units to increase strength of a muscle contraction.

rectum (REK-tum)   The distal portion of the large intestine.

reduction (rē-DUK-shun)   A chemical reaction in which electrons or hydrogen ions are added to a molecule, or oxygen is removed.

referred pain   Pain that is perceived from an area of the body other than that which originated it.

reflex (RĒ-fleks)   A rapid response to a stimulus in order to restore homeostasis.

reflex arc   The nerve pathway that results in a reflex. It includes a receptor, a sensory neuron, a center (association neuron) in the central nervous system, a motor neuron, and an effector.

refraction (rē-FRAK-shun)   The bending of light as it passes from one medium to another of a different density.

regeneration (rē-jen-er-Ā-shun)   The repair of tissue by the replacement of damaged or dead cells by new cells with the same characteristics.

regulating factor   A substance released by the hypothalamus that may stimulate or inhibit activity of the anterior pituitary gland.

regurgitation (rē-ger′-ji-TĀ-shun)   The reflux of stomach contents. Also called vomiting.

relapse (RĒ-laps)   The return of a disease after its apparent elimination.

remodeling (rē-MOD-el-ing)   The replacement of old bone tissue with new.

renal (RĒ-nal)   Pertaining to the kidneys.

renal corpuscle (KOR-pus-el)   The portion of a kidney nephron consisting of the Bowman's capsule and glomerulus.

renal pelvis (PEL-vis)   A membrane-lined basin within the renal sinus of each kidney.

renal pyramid (PlR-a-mid)   One of about eight to ten cone-shaped structures in each kidney extending from the medulla to the cortex and containing the renal tubules.

renal sinus (SĪ-nus)   A potential space within each kidney extending from the hilum to the medulla and containing the renal pelvis.

renal tubule (TOO-byool)   A part of a nephron of the kidney consisting of a microscopic tube extending from the Bowman's capsule to a collecting duct. The functions of reabsorption and secretion occur across its walls.

renin (RE-nin)   An enzyme secreted by the kidney that converts a plasma protein (angiotensinogen) to angiotensin I, thereby performing a precursory role in the regulation of blood pressure.

reproduction (rē′-prō-DUK-shun)   The process by which offspring are produced.

residual volume (re-ZlD-yoo-al VOL-yoom)   The amount of air remaining in the lungs after a forced expiration.

respiratory center (RES-pir-a-tor-ē SEN-ter)   A cluster of neurons in the pons and medulla oblongata that control inspiration and expiration.

respiratory membrane   The barrier associated with each alveolus in the lungs that must be crossed by gas molecules during gas exchange. It consists of the alveolar epithelium, a basement membrane, and the endothelium of a capillary.

resting potential   The difference in charge across the plasma membrane of a cell that has not been stimulated by an action potential.

reticular formation (re-TlK-yoo-lar for-MĀ-shun)   A network of scattered neurons throughout the brain stem that play a role in daily sleep-wake cycles.

retina (RET-i-na)   The light-sensitive inner layer of the eye that contains rod and cone cells.

retroperitoneal (re′-trō-par′-i-tō-NĒ-al)   Pertaining to a structure lying external to the parietal peritoneum.

Rh factor   An inherited antigen on the surface of red blood cells.

rhodopsin (rō-DOP-sin)   A pigment in rod cells of the eye that plays a role in photoreception. Also called *visual purple.*

ribonucleic acid (rī′-bō-nyoo-KLĒ-ik A-sid) *(RNA)*   A single-stranded nucleic acid molecule composed of nucleotides with one of four possible bases (adenine, cytosine, guanine, or uracil), a ribose sugar, and a phosphate group.

ribosome (RĪ-bō-sōm)   A microscopic, spherical structure within the cytoplasm of a cell composed of RNA and protein that serves as an attachment site for messenger RNA during protein synthesis.

rickets (RIK-ets)   A disorder of bone in children caused by insufficient calcium absorption due to a vitamin D deficiency and characterized by soft, often bowed long bones.

rigor mortis (RI-gor MOR-tis)   A state of partial contraction of muscles after death due to the unavailability of ATP.

rod cell   A photoreceptor cell in the retina of the eye that detects very low levels of light.

rotation (rō-TĀ-shun)   The movement of a bone around its own (longitudinal) axis.

rotator cuff   A group of four muscles that attach the humerus to the scapula.

round window   The membrane-covered opening between the middle ear and the inner ear that is not in contact with the auditory ossicles.

rugae (ROO-jē)   Folds or ridges in the mucosa of an organ with a large lumen, such as the stomach or vagina.

saccule (SAK-yool)   One of two sacs within the vestibule of each inner ear that house the receptors of static equilibrium.

sagittal (SAJ-i-tal)   A vertical plane that divides the body into right and left portions. The midsagittal plane divides the body into equal halves, and a parasagittal plane divides it into unequal portions.

saliva (sa-LĪ-va)   A fluid secretion by the salivary glands deposited into the mouth to lubricate and begin digestion of food before swallowing.

salivary (SAL-i-var-ē) *gland*   One of several exocrine glands in the facial region that secrete saliva into the mouth to initiate the digestive process.

salt   A molecule that consists of a positively charged ion (other than hydrogen ion) and a negatively charged ion (other than hydroxyl ion).

sarcolemma (sar′-kō-LEM-a)   The plasma membrane covering the outer surface of a muscle fiber.

sarcoma (sar-KŌ-ma)   A tumor that originates from connective tissue.

sarcomere (SAR-kō-mēr)   A contractile microscopic subunit of striated muscle (skeletal and cardiac muscle).

sarcoplasm (SAR-kō-plazm)   The cytoplasm of a muscle fiber.

sarcoplasmic reticulum (sar′-kō-PLAZ-mik re-TlK-yoo-lum)   A cellular organelle found only in muscle cells that stores calcium ions and is similar in structure to the endoplasmic reticulum found in other cells.

satiety (sa-TĪ-e-tē)   The sensation of fullness or satisfaction, as opposed to hunger or thirst.

saturated fat   A fat that contains only single bonds between its carbon atoms and the maximum number possible of bonded hydrogen atoms. It is found in animal sources of foods.

Schwann cell   A type of neuroglial cell that forms myelin sheaths around axons of peripheral nerves.

sclera (SKLĒ-ra)   The posterior part of the outer, fibrous tunic covering the eyeball; the white of the eye.

sclerosis (skler-Ō-sis)   A hardening and loss of elasticity of a body part associated with disease.

scoliosis (skō′-lē-Ō-sis)   An abnormal lateral curvature of the vertebral column.

scrotum (SKRŌ-tum)   An external genital organ of the male consisting of a skin-covered sac that contains the testes.

sebaceous (se-BĀ-shus) *gland*   An exocrine gland located in the dermis that secretes an oily substance called *sebum*. It is usually associated with a hair follicle.

sebum (SĒ-bum)   An oily secretion of a sebaceous gland.

secretion (se-KRĒ-shun)   A substance produced and released by a cell that serves a useful benefit.

semen (SĒ-men)   A reproductive fluid discharged by a male during ejaculation that contains sperm cells and secretions from the seminal vesicles, prostate gland, and bulbourethral glands.

semicircular canal (se'-mi-SER-kyoo-lar ca-NAL)   One of three looping canals in each temporal bone that form a part of the inner ear. It contains perilymph fluid and the receptors for equilibrium.

semilunar (SL) valve (sem-i-LOO-nar valv)   One of two heart valves located between a ventricle and a major artery. The aortic valve is located between the left ventricle and aorta, and the pulmonary valve is located between the right ventricle and pulmonary trunk.

seminal vesicle (SEM-i-nal VES-i-kel)   One of a pair of convoluted glands of the male reproductive system located posterior to the urinary bladder that secrete part of the semen.

seminiferous tubule (sem'-i-NIF-er-us TOO-byool)   A microscopic, tightly packed tube within each testis where sperm cells develop and from which they exit.

senescence (se-NES-ens)   The developmental stage of life beginning between 65 and 75 years of age. Also known as *old age*.

senility (se-NIL-i-tē)   A loss of mental or physical functions due to increased age.

sensation (sen-SĀ-shun)   A state of awareness of an internal or external environmental change.

septicemia (sep'-ti-SĒ-mē-a)   An infection of the blood caused by a bacterial pathogen. Also called *blood poisoning*.

septum (SEP-tum)   A barrier between two spaces, such as the interventricular septum of the heart and the nasal septum.

serosa (ser-Ō-sa)   Any serous membrane. Also, the outer, serous membrane layer of a visceral organ.

serous membrane (SER-us MEM-brān)   An epithelial membrane that lines a body cavity or covers an organ and secretes small amounts of fluid.

serum (SER-um)   A fluid used in clinical situations that is composed of plasma less its blood-lotting factors.

sesamoid (SES-a-moyd) *bones*   Small bones formed and located within major tendons or ligaments. For example, the patella (kneecap) is a sesamoid bone.

sex chromosomes   The chromosomes that contain the sex-determining genes. In humans, they are the 23rd pair, which are designated XX (female) and XY (male).

sexually transmitted disease (STD)   A disease that is contracted as a result of physical contact during sexual activity. Also called *venereal disease*.

shingles   An acute viral infection that affects major nerve tracts along the back, resulting in fever and the formation of skin pustules.

shock   An acute condition in which the cardiovascular system is unable to deliver oxygenated blood to vital organs and tissues, resulting in acidosis, reduced mental functions, a weak, rapid pulse, and accelerated heart rate.

sigmoid colon (SIG-moyd KŌ-lon)   The distal segment of the colon, located between the descending colon and the rectum.

sign (sīn)   An evidence of disease that can be observed or measured, such as fever, inflammation, or a lesion.

sinoatrial node (sin'-ō-Ā-trē-al nōd) *(SA node)*   A cluster of specialized cardiac muscle cells in the wall of the right atrium that initiate each cardiac cycle. Also called the *sinuatrial node* or *pacemaker*.

sinus (SĪ-nus)   A space within a bone lined with mucous membrane, such as the frontal and maxillary sinuses in the head. Also, a modified vein with an enlarged lumen for blood storage.

sinusitis (sī'-nyoo-SĪ-tis)   Inflammation of the mucous membranes lining all or any one of the sinuses in the head.

skeletal muscle tissue   One of three types of muscle tissue in the body, characterized by the presence of visible striations and conscious control over its contraction. It attaches to bones to form the muscles of the body.

skull   The group of bones that make up the supporting framework and body of the head.

sliding-filament mechanism   The mechanism of muscle contraction in which protein filaments slide across one another to cause muscle subunits called *sarcomeres* to shorten in length. When all sarcomeres of a muscle shorten in unison, the muscle contracts.

small intestine   The organ of the alimentary canal located between the stomach and the large intestine that functions in the final digestion and absorption of nutrients.

smooth (visceral) muscle   One of three types of muscle tissue in the body, characterized by the lack of visible striations and by unconscious control over its contraction. It forms part of the walls of hollow organs and blood vessels.

soft palate   The posterior portion of the bridge forming the roof of the mouth, consisting of skeletal muscle covered with mucous membrane.

solute (SOL-yoot)   A dissolved substance in a solution.

solution (sō-LOO-shun)   A homogeneous mixture of a dissolved substance (solute) in a liquid medium (solvent).

solvent (SOL-vent)   A liquid medium in which a substance may dissolve to form a solution.

somatic (sō-MA-tik)   Pertaining to the body. For example, a somatic cell is any body cell other than a sex cell.

somatic nervous system   The component of the peripheral nervous system that conveys impulses associated with conscious sensory and motor activities.

somesthetic (sō'-mes-THE-tik)   Pertaining to consciously perceived sensations.

spasm   An involuntary contraction of a large group of muscles, causing a sudden response.

spermatic cord (sper-MA-tik kord)   A narrow bundle of tissue in the male reproductive system extending from the epididymis to the inguinal canal, consisting of the ductus deferens, cremaster muscle, blood vessels, lymphatic vessels, nerves, and connective tissue.

spermatogenesis (sper-ma'-tō-GEN-e-sis)   The production of sperm cells by meiosis within the seminiferous tubules of the testes.

spermatozoa (sper-ma'-tō-ZŌ-a)   The male gametes, or reproductive cells. Also called *sperm cells*. Singular form is *spermatozoon*.

sphincter (SFĒNK-ter)   A circular band of smooth muscle surrounding an opening, which serves to control the movement of materials through.

sphygmomanometer (sfig'-mō-ma-NO-me-ter)   An instrument that measures arterial blood pressure.

spina bifida (SPĪ-na BlF-i-da)   A congenital defect in the development of the vertebral column in which the two halves of the vertebral arch of one or more vertebrae fail to fuse along the midline.

spinal cord   A long, narrow organ of the central nervous system that extends through the vertebral canal and connects the peripheral nervous system with the brain.

spinal nerve   One of 31 pairs of nerves that extend between the spinal cord and another part of the body.

spleen   A soft, glandular organ that is part of the lymphatic system and is located in the upper left region of the abdomen behind the stomach.

spongy bone   One of two types of bone tissue, characterized by the presence of spaces filled with red marrow between thin bone spicules called *trabeculae*.

sprain   Injury to a joint as a result of twisting or wrenching without complete dislocation, resulting in local inflammation and pain.

squamous (SKWĀ-mus)   Flat, scalelike.

starvation (star-VĀ-shun)   A diseased state that is caused by the loss of stored energy due to a lack of nutritional intake and in which functional tissue is metabolized to compensate.

static equilibrium (STA-tik ē'-kwi-LIB-rē'-um)   The maintenance of balance based on sensory information about the position of the head.

stenosis (ste-NŌ-sis)   An abnormal constriction or narrowing of a body opening.

sterile (STE-ril)   In the absence of living organisms. Also, the inability to produce offspring.

stimulus (STI-myoo-lus)   A change in the environment sufficient to cause a plasma membrane to depolarize.

stomach (STO-muk)   A large, hollow organ in the alimentary canal located between the esophagus and small intestine that plays a prominent role in digestion.

stretch receptor   A receptor in the walls of the bronchi and lungs that stimulates the respiratory center in the brain stem to prevent overinflation of the lungs.

stroke volume   The volume of blood forced out of one ventricle during a single systole of a cardiac cycle.

subarachnoid space (sub-a-RAK-noyd spās)   The narrow space between the arachnoid and pia mater surrounding the brain and spinal cord, which contains circulating cerebrospinal fluid.

subcutaneous layer (sub'-kyoo-TĀ-nē-us LĀ-yer)   The layer of loose connective tissue and adipose tissue deep to the dermis of the skin. Also called *hypodermis* and *superficial fascia*.

sublingual (sub'-LlNG-wal) *glands*   A pair of salivary glands located in the floor of the mouth deep to the mucous membrane.

submandibular (sub'-man-DlB-yoo-lar) *glands*   A pair of salivary glands located along the inner surface of the jaw in the floor of the mouth. Also called *submaxillary glands*.

submucosa (sub'-myoo-KŌ-sa)   A layer of connective tissue located external to a mucous membrane.

substrate (SUB-strāt)   A substance that reacts with an enzyme.

subthreshold stimulus (sub-THRESH-hold STI-myoo-lus)   A change in environment that is less than sufficient to cause a depolarization in a plasma membrane.

sudoriferous (soo'-dor-I-fer-us) *gland*   An exocrine gland located in the skin that secretes sweat. Also called *sweat gland*.

sulcus (SUL-kus)   A shallow groove or depression.

summation (sum-MĀ-shun)   Process whereby a muscle contraction of increased strength results from a rapid succession of stimuli.

superficial (soo'-per-FlSH-al)   A directional term indicating the location of a part that is toward or nearer to the body surface relative to another part.

superior (soo'-PĒR-ē-or)   A directional term indicating the location of a part that is nearer to the head region than another part. Also called *craniad* or *cephalad*.

supination (soo'-pi-NĀ-shun)   Rotation of the forearm such that the palm of the hand is turned anteriorly or superiorly.

surfactant (ser-FAK-tant)   A lipid secretion produced by the lungs that reduces surface tension within alveoli.

suture (SOO-cher)   A type of tight-fitting fibrous joint that permits little or no movement between opposing bones.

sweat gland   An exocrine gland located in the skin that secretes sweat. Also called *sudoriferous gland*.

sympathetic division (simp'-a-THE-tik di-VI-zhun)   A division of the autonomic nervous system that functions mainly in stimulating emergency responses (fight-or-flight).

symptom (SlMP-tom)   A sensation that can be correlated with an origin from disease.

synapse (sin-APS)   The junction between the axon of one neuron and the dendrite or cell body of another neuron.

synapsis (sin-AP-sis)   The pairing of homologous chromosomes during prophase of meiosis I.

synaptic cleft (sin-AP-tik cleft)   A part of the synapse that consists of the space between neurons or neuron and muscle fiber across which the neurotransmitter must diffuse.

synaptic end bulb   The expanded distal end of a neuron's axon that contains numerous synaptic vesicles and mitochondria.

synaptic vesicles   Microscopic sacs within synaptic end bulbs of axons that store neurotransmitter.

syndrome (SIN-drōm)   A collection of signs and symptoms that characterize a particular disease.

synergist (SIN-er-jist)   A muscle in a group action that assists the prime mover by keeping other structures stable.

synovial fluid   The liquid secretion of epithelial cells in the synovial membrane lining a synovial joint, which serves as a lubricant and shock absorber.

synovial joint (sin-Ō-vē-al joynt)   A type of joint characterized by the presence of a membrane-lined cavity, called the *synovial cavity*, between opposing bones.

synthesis (SIN-the-sis)   A chemical reaction resulting in the production of a compound by combining atoms, ions, or molecules.

syphilis (SIF-i-lis)   A sexually transmitted disease caused by the bacterium *Treponema pallidum.*

system (SIS-tem)   An organized combination of organs and associated structures that share a common function.

systemic circulation   The major circulatory network of the body, which carries oxygenated blood from the left ventricle throughout the body (except the lungs) and returns deoxygenated blood to the right atrium.

systemic lupus erythematosus (LOO-pus ē-rith'-e-ma-TŌ-sus) *(SLE)*   A progressive, autoimmune collagen disease that causes widespread inflammation in tissues.

systole (SIS-tō-lē)   A stage in the cardiac cycle during which the heart chambers (especially the ventricles) contract.

systolic blood pressure   The force exerted by blood against arterial walls during ventricular contraction.

T cell   A type of lymphocyte whose stem cells arise from the thymus gland and become sensitized into several forms, including helper T cells, suppressor T cells, killer T cells, and memory T cells, all of which play a particular role in the immune response.

tachycardia (tak'-ē-KAR-dē-a)   A heart rate and pulse rate that is higher than normal.

taenia coli (TĒ-nē-a KŌ-lī)   Three flat bands of smooth muscle that extend longitudinally along the length of the large intestine.

target cell   A cell that is affected by a hormone in such a way that its functions become altered.

tarsus (TAR-sus)   The seven bones of the ankle as a collective unit.

Tay-Sachs (tā-saks) *disease*   An inherited, progressive disease that causes a fatal deterioration of the central nervous system due to a deficient lysosomal enzyme.

tectorial membrane (tek-TOR-ē-al MEM-brān)   A thin membrane in the inner ear that projects over the receptor hair cells of the organ of Corti.

telophase (TEL-ō-fāz)   The final stage in mitosis, which includes the formation of two daughter nuclei and the division of the cytoplasm to form two separate, but genetically identical, daughter cells.

tendon (TEN-don)   A band of dense connective tissue that extends from the muscle to attach to a bone.

testis (TES-tis)   One of the pair of male gonads (sex glands), located within the scrotum, which produce sperm cells and testosterone. Plural is *testes.*

testosterone (tes-TOS-ter-ōn)   The male sex hormone secreted by the interstitial cells (cells of Leydig) within the testes.

tetanus (TET-an-us)   A smooth, sustained contraction of muscle produced by a rapid sequence of stimuli. Also, an infectious disease resulting in severe muscle spasms, lockjaw, and exaggerated reflexes that is caused by toxins released from the bacterium *Clostridium tetani.*

thalamus (THAL-a-mus)   A bilobed endocrine gland located in the anterior neck region that produces hormones influencing growth and metabolism and calcium levels in the blood.

thigh (thī)   The proximal part of the leg, extending from the hip to the knee.

thoracic cavity (thō-RAS-ik CAV-i-tē)   The part of the anterior (ventral) body cavity located superior to the diaphragm, which contains the heart, lungs, and mediastinum.

thoracic duct   The main collecting trunk of the lymphatic circulation, which extends along the back of the chest to the right subclavian vein. It drains lymph from all areas of the body but the right side of the head, neck, and chest and the right arm.

thorax (THOR-aks)   The region of the trunk located superior to the diaphragm. Also called the *chest.*

threshold stimulus (THRESH-hōld STI-myoo-lus)   A change in environment great enough to cause an action potential (nerve impulse).

thrombin (THROM-bin)   An activated protein involved in the formation of a blood clot by converting fibrinogen to fibrin.

thrombocyte (THROM-bō-sīt)   The formed elements in blood that play a prominent role in blood clotting. Also called *platelets.*

thrombosis (throm-BŌ-sis)   A condition in which a blood clot has formed in a vein or, in some cases, an artery.

thrombus (THROM-bus)   A blood clot that has formed in a vein or an artery.

thymus (THĪ-mus) *gland*   A glandular lymphatic organ located superior to the heart that produces T lymphocytes during early childhood and degenerates by adulthood.

thyroid cartilage (THĪ-royd CAR-ti-lij)   The largest piece of hyaline cartilage of the larynx. Also called *Adam's apple.*

thyroid gland   An endocrine gland located on the anterior side of the neck that secretes hormones involved in growth and metabolism and in maintaining calcium levels in the blood.

thyroid-stimulating hormone (TSH)   A hormone secreted by the anterior pituitary gland that stimulates the production and secretion of hormones released by the thyroid gland.

thyroxine (thī-ROK-sin)   A hormone secreted by the thyroid gland that regulates growth and metabolism.

tidal volume   The volume of air inhaled and exhaled during a normal, relaxed breathing cycle.

tissue (TI-shoo)   A group of similar cells that combine to perform a common function.

tongue (tung)   The muscular organ of the digestive system that is anchored to the floor of the mouth and wall of the pharynx, and that plays a role in swallowing and speech formation.

tonsil (TON-sil)   A small organ of the lymphatic system that consists of an aggregation of fixed lymphocytes and connective tissue embedded in a mucous membrane. There are three pairs (pharyngeal, palatine, and lingual), all of which play a role in the immune response.

topical (TOP-i-kal)   Term describing application of a medication directly onto the surface of the skin.

total lung capacity   The maximum air capacity of the lungs, which is the sum of the inspiratory reserve volume, expiratory reserve volume, tidal volume, and reserve volume.

toxic (TOK-sik)   Poisonous, or causing illness.

toxic shock syndrome   A collection of symptoms in females, including high fever, headache, sore throat, fatigue, and abdominal pain, that is caused by the bacterium *Staphylococcus aureus* and can usually be traced to the use of tampons during menstruation.

trabecula (tra-BEK-yoo-la)   A thin plate of bone within spongy bone tissue. Also, a band of supportive connective tissue extending to the interior of an organ from its outer wall.

trachea (TRĀ-kē-a)   An organ of the respiratory system that consists of a long tube supported by rings of cartilage extending from the pharynx to the bronchi.

transcription (trans-KRIP-shun)   A stage in protein synthesis during which a molecule of RNA is synthesized from a template consisting of a single strand of DNA.

transfusion (trans-FYOO-zhun)   The clinical transfer of blood or blood products directly into the bloodstream.

translation (trans-LĀ-shun)   A stage in protein sythesis during which a polypeptide is constructed on a ribosome by the addition of amino acids in a sequence determined by messenger RNA.

transplantation (trans'-plan-TĀ-shun)   The surgical replacement of nonfunctional tissues or organs with functioning ones.

transverse colon (TRANS-vers KŌ-lon)   The segment of the colon that extends from its union with the ascending colon to its union with the descending colon.

trauma (TRAW-ma)   An injury resulting in a disturbance of homeostasis, which may affect either the physical or the mental state.

treppe (TREP-ē)   A response of muscle characterized by a gradual, steplike increase in contraction that is caused by a series of rapid stimuli of equal strength.

tricuspid valve (trī-KUS-pid valv)   The heart valve located between the right atrium and right ventricle. Also called the *right atrioventricular (AV) valve.*

triglyceride (trī-GLIS-er-īd)   A lipid composed of three fatty acids and one glycerol. Also called *fat.*

trigone (TRĪ-gōn)   A triangular region on the inner wall of the urinary bladder bordered by the entrance of the two ureters and the exit of the urethra.

triiodothyronine (trī-ī-ō-dō-THĪ-rō-nin)   A hormone secreted by the thyroid gland that regulates growth and metabolism.

tropic (TRŌ-pik) *hormone*   A hormone whose target cells are within another endocrine gland.

trunk   The region of the body to which the appendages are attached; includes the chest, abdomen, and back.

tubercle (TOO-ber-kul)   A small, rounded process on the surface of a bone.

tuberculosis (too-ber'-kyoo-LŌ-sis)   A disease caused by the bacterium *Mycobacterium tuberculosis* that involves an attack on the functional tissue in soft organs, especially the lungs, converting it to fibrous plaques of connective tissue.

tumor (TOO-mer)   An excess of tissue produced by unregulated growth of cells. Also called a *neoplasm.*

twitch   A brief, rapid contraction of a muscle that is brought on by a single stimulus.

tympanic membrane (tim-PAN-ik MEM-brān)   A thin membrane between the external auditory canal and the tympanic cavity, separating the external ear from the middle ear. Also called the *eardrum.*

ulcer (UL-ser)   A lesion on the skin or mucous membrane with evidence of necrosis that opens to the exterior.

umbilical cord (um-BIL-i-kal kord)   The ropelike structure containing the umbilical arteries and umbilical vein that connects a fetus with the placenta.

uremia (yoo'-RĒ-mē-a)   The toxic accumulation of urea and uric acid in the blood, usually as a result of kidney disease.

ureter (YOO-re-ter)   A long, narrow tube that extends from a kidney to the urinary bladder and transports urine.

urethra (yoo-RĒ-thra)   A tube extending from the urinary bladder to the exterior that carries urine in females and urine and semen in males.

urinalysis (yoo'-rin-AL-i-sis)   The clinical analysis of a urine sample to determine its chemical composition as a diagnostic tool.

urinary bladder (YOO-ri-nar'-ē BLAD-der)   A hollow muscular organ located at the floor of the pelvic cavity that temporarily stores urine.

urine (YOO-rin)   The fluid produced by the kidneys that is expelled out the urethra; it contains water, metabolic waste materials, and excess salts.

uterine tube (YOO-ter-in toob)   One of two tubes that transport ova from the ovaries to the uterus in the female reproductive system. Also called *fallopian tubes* or *oviducts.*

uterus (YOO-ter-us)   A hollow muscular organ in the female reproductive system that serves as a site of embryo implantation and development, and of menstruation.

utricle (YOO-tri-kul)   One of two divisions of the membranous labyrinth within the vestibule of each inner ear that contain receptor organs for static equilibrium.

uvula (YOO-vyoo-la)   A fingerlike projection of skeletal muscle covered with mucous membrane at the posterior end of the soft palate.

vaccine (vak-SĒN)   A preparation composed of killed or altered microorganisms or their parts that is introduced into the body by inoculation or ingestion in order to stimulate the immune response.

vagina (va-JĪ-na)   A tubular, muscular organ of the female reproductive system extending between the vulva and the uterus.

varicosity (var-i-KOS-i-tē)   An abnormal condition in which a superficial vein loses its wall elasticity and valve durability, and becomes swollen.

vascular (VAS-kyoo-lar)   Pertaining to or containing blood vessels.

vasectomy (va-SEK-to-mē)   A surgical procedure involving the sterilization of a male by cutting through the ductus deferens and closing the exposed ends.

vasoconstriction (vā′-sō-con-STRIK-shun)   The restriction of blood flow through a blood vessel by the contraction of smooth muscles in its wall, which narrows its lumen.

vasodilation (vā′-sō-dī-LĀ-shun)   An increase in blood flow through a blood vessel by the relaxation of smooth muscles in its wall, which expands its lumen.

vein (vān)   A blood vessel that transports blood from body tissues to the heart.

ventral (VEN-tral)   A directional term describing the location of a part nearer to the anterior or front side of the body relative to another. In humans, also called *anterior*.

ventral cavity   The body cavity located on the anterior side of the trunk containing the thoracic and abdominopelvic cavities.

ventricle (VEN-tri-kul)   One of the two inferior, highly muscular chambers of the heart that push blood into major arteries during their contraction.

ventricular fibrillation (ven-TRI-kyoo-lar fib-ri-LĀ-shun)   Abnormal, unsynchronized contraction of the heart chambers that leads to heart failure.

venule (VEN-yool)   A small vein that collects deoxygenated blood from a capillary network and conveys it to a larger vein.

vermiform appendix (VER-mi-form a-PEN-diks)   A small, closed-ended tube extending from the cecum of the large intestine.

vermis (VER-mis)   The central, constricted part of the cerebellum that separates the two cerebellar hemispheres.

vertebral canal (VER-te-bral ka-NAL)   A cavity extending through the vertebral column that is formed by the vertebral foramina of all the vertebrae, through which the spinal cord extends.

vertebral column   The skeleton of the back, which is composed of 26 vertebrae and associated tissues. Also called the *backbone, spine,* or *spinal column.*

vesicle (VES-i-kul)   A small sac containing a fluid. In the cell, it is a membranous sac within the cytoplasm that contains cellular products or waste materials.

vestibular membrane (ves-TIB-yoo-lar MEM-brān)   A thin membrane inside the membranous labyrinth of the inner ear.

vestibule (VES-ti-byool)   A small space that opens into a larger cavity or canal. A vestibule is found in the inner ear, mouth, nose, and vagina.

villus (VIL-lus)   A small, fingerlike projection of the small intestinal wall that contains connective tissue, blood vessels, and a lymphatic vessel, and which functions in the absorption of nutrients. Plural is *villi.*

visceral (VIS-er-al)   Pertaining to the internal components (mainly the organs) of a body cavity; pertaining to the outer surface of an internal organ.

visceral effector (e-FEK-tor)   A body part that receives and responds to autonomic innervation. Visceral effectors include smooth muscle, cardiac muscle, and glands.

visceral peritoneum (pār′-i-tō-NĒ-um)   A serous membrane that covers the surfaces of abdominal organs.

visceral pleura (PLOO-ra)   A serous membrane that covers the outer surface of each lung.

viscosity (vis-KOS-i-tē)   The tendency of a fluid to resist flow because of the density of its molecules.

vital capacity   The total amount of exchangeable air through the lungs, which is obtained by adding inspiratory reserve volume, tidal volume, and expiratory reserve volume.

vitamin (VĪ-ta-min)   An organic molecule that is required in small amounts to maintain normal metabolism.

vitreous humor (VI-trē-us HYOO-mer)   A mass of gelatinous material located within the eyeball in the posterior cavity located between the lens and the retina. Also called *vitreous body.*

vocal cords (VŌ-kal kordz)   Folds of mucous membrane within the larynx that produce sound when they vibrate.

vulva (VUL-va)   The external genitalia of the female reproductive system. Also called *pudendum.*

white matter   A type of nerve tissue composed mainly of the myelinated axons of neurons.

yellow marrow   A collection of fat storage (adipose) and other tissues found within the medullary cavities of bones.

yolk sac   A fetal membrane surrounding a temporary food storage during the early embryonic stage, which later helps to form the umbilical cord in the fetus.

zygote (ZĪ-gōt)   The single diploid cell resulting from the union of an oocyte with a sperm cell by fertilization.

zymogenic (zī′-mō-GEN-ik) *cell*   A cell within a gastric gland of the stomach mucosa that secretes a precursor protein, pepsinogen, of the digestive enzyme pepsin.

PHOTO CREDITS

UNIT OPENERS

Part I: David Phillips/Visuals Unlimited, p. 1; Part II: courtesy of Dr. Patrick McDonough, San Diego State University, p. 133; Part III: © Secchi-Lecaque/Roussel-UCLAF/CNRI/Science Photo Library/Photo Researchers, p. 237; Part IV: Fred Hossler/Visuals Unlimited, p. 343; Part V: © G. Shih-R. Kessel/Visuals Unlimited, p. 441; Part VI: © Cabisco/Visuals Unlimited, p. 555.

CHAPTER OPENERS

Chapters 1, 2, 5, 6, 7, 9, 11, 14, 15, 17, 19, © David Madison; Chapter 3, © Custom Medical Stock Photo; Chapter 4, © Will & Deni McIntyre/ Photo Researchers; Chapter 8, © Renee Lynn/David Madison; Chapter 10, courtesy of Bruce Wingerd; Chapter 12, Photo Researchers; Chapter 13, © Vladimir Lange, M.D./The Image Bank; Chapter 16, © Allen Birnbach/ Westlight; Chapter 18, © Kay Chernush/The Image Bank; Chapter 20, © Henley & Savage/Tony Stone Worldwide.

FIGURES

Chapter 1: *Fig. 1.1, 1.2, 1.7(a), 1.7(b), Robert Goynes, Sr. & Victor Smith; 1.4(b), © John D. Cunningham/Visuals Unlimited; 1.4(c), © Robert Brenner/Photo Edit; un1-(a)1, p. 15, © Columbus Hospital/ Custom Medical Stock Photo; un1-(a)2, un1-(a)3, p. 15, Dan McCoy/ Rainbow; un1-(a)4, Lutheran Hospital/Peter Arnold, Inc.; un1-(a)5, p. 15, © CNRI/SPL/Photo Researchers; un1-(a)6, p. 15, courtesy of Erik Courcesne & Associates/San Diego Children's Hospital and UCSD; 1.10(b), Peter Arnold, Inc.; un1-(b)1, p. 19, © Dr. Gopal Murti/CNRI/ Phototake, NYC; un1-(b)2, p. 19, CNRI/Phototake, NYC; un1-(b)3, p. 19, © Manfred Kage/Peter Arnold, Inc.; un1-(b)4, p. 19, CNRI/Science Photo Library/Photo Researchers.*

Chapter 2: *Fig. 2.1, Peter Arnold, Inc.; un2-(a)1, p. 33, courtesy of Bruce Wingerd; un2-(a)2, p. 33, © James Stevenson/Science Photo Library/Photo Researchers; un2-(a)3, p. 33, Kings College School of Medicine, Department of Surgery/Science Photo Library/Photo Researchers; 2.17(b), © Martin M. Rotker/Science Source/Photo Researchers, Inc.; 2.23(b), Dr. A. Lesk, Laboratory of Molecular Biology/Science Photo Library/Photo Researchers.*

Chapter 3: *Fig. 3.2(a), Photo Researchers, Inc.; 3.2(b), © Secchi-Lecaque/ Roussel-UCLAF/CNRI/Science Photo Library/Photo Researchers; 3.2(c), Dr. David Scott/CNRI/Phototake, NYC; 3.3(b), © Secchi-Lecaque/Roussel-UCLAF/CNRI/Science Photo Library/Photo Researchers; 3.6, from R.G. Kessel and R.H. Kardon,* Tissues and Organs: a text-atlas of scanning electron microscopy. *Micrographs courtesy of Professor R.G. Kessel; 3.8(a), © Stanley Fiegler/Visuals Unlimited; 3.8(b), 3.8(c), © David M.*

Phillips/Visuals Unlimited; 3.10(d), © Biological Photo Service; 3.12(b), © Don Fawcett/Photo Researchers; 3.13(b), © Biophoto Assoc/Science Source/Photo Researchers; 3.14(b), © Keith Porter/Photo Researchers; 3.16(b), © Dr. Don Fawcett/Photo Researchers; 3.19(a-d), © John Cunningham/Visuals Unlimited.

Chapter 4: *Fig. 4.1(b), 4.2(b), 4.3(b), 4.4(b), 4.6(b), 4.8(a-c), 4.9(a), 4.10(b), 4.11(b), 4.12(a), 4.12(b), 4.14(b), 4.16(a), 4.16(b), 4.17(b), courtesy of Bruce Wingerd; 4.5(b), © Ed Reschke/Peter Arnold, Inc.; 4.15(a), © Manfred Kage/Peter Arnold, Inc.; un4-(a)1, p. 104, © Boehringer Ingelheim International GmbH. Photo by Lennart Nilsson, from* The Incredible Machine, *published by The National Geographic Society; un4-(a)2, p. 104, © Boehringer Ingelheim International GmbH. Photo by Lennart Nilsson, from* The Incredible Machine, *published by The National Geographic Society.*

Chapter 5: *Fig. 5.2(b), courtesy of Bruce Wingerd; 5.3(b), Ron Mensching/Phototake, NYC.*

Chapter 6: *Fig. 6.2(b); courtesy of Bruce Wingerd; un6-a, p. 139, Custom Medical Stock Photo; un6-b, p. 139, Biophoto Assoc./Science Source; un6-c, p. 139, Custom Medical Stock Photo.*

Chapter 7: *Fig. 7.1(a), James Stevenson/Science Photo Library/Photo Researchers; 7.1(b), © Ron Mensching/Phototake, NYC; 7.1(c), Photo Researchers; 7.1(d), CNRI/Phototake, NYC; 7.3, from R.G. Kessel and R.H. Kardon,* Tissues and Organs: a text-atlas of scanning electron microscopy. *Micrographs courtesy of Professor R.G. Kessel; 7.4(c), courtesy of Bruce Wingerd; un7-(a)1, p. 160, © Scott Camazine/Photo Researchers; un7-(a)2, p. 160, © SIU/Photo Researchers; un7-(a)3, p. 160, © J & L Weber/Peter Arnold, Inc.; 7.14(c), © C. Golvaux Communications/ Phototake, NYC; 7.16(b), from* Color Atlas of Human Anatomy, *2nd Edition by R.M.H. McMinn & R.T. Hutchings, © 1988 Year Book Medical Publishers, Inc.; 7.17, © Michael Newman/PhotoEdit; 7.22, © James Stevenson/Science Photo Library/Photo Researchers; 7.29(b), © Michael Gabridge/Visuals Unlimited; 7.30(b), from* Color Atlas of Human Anatomy, *2nd Edition by R.M.H. McMinn & R.T. Hutchings, © 1988 by Year Book Medical Publishers, Inc.; un7-(b)1, p. 183, © Martin Rotker/Phototake, NYC; 7.35(all), Robert Goynes, Sr. & Victor Smith.*

Chapter 8: *Fig. 8.1(a), © M.I. Walker/Photo Researchers; 8.1(b), 8.1(c), © Biophoto Associates/Photo Researchers; 8.4(b), 8.5(b), courtesy of Bruce Wingerd.*

Chapter 9: *Fig. 9.10(d), 9.18(c); Custom Medical Stock Photo; 9.19(b), © Manfred Kage/Peter Arnold, Inc.; 9.20, from R.G. Kessel and R.H. Kardon,* Tissues and Organs: a text-atlas of scanning electron microscopy. *Micrographs courtesy of Professor R.G. Kessel.*

Chapter 10: *Fig. 10.5(b), 10.12, courtesy of Bruce Wingerd.*

Chapter 11: *Fig. 11.11(b), © Martin Rotker/Phototake, NYC; 11.14(c), © CNRI/Phototake, NYC; 11.15(b), un11-a, p. 338, © Biophoto Associates/ Photo Researchers; un11-b(both), p. 338, © Science VU/Visuals Unlimited.*

Chapter 12: *Fig. 12.1, © Simon Fraser/Science Photo Library/Photo Researchers; 12.2(a), 12.2(b), 12.2(c), © Whiteley Illustration Services; 12.2(e), © James Hayden/Phototake; 12.3, © Fred Hossler/Visuals Unlimited; un12-a, p. 352, © SUI/Photo Researchers; 12.5, © David M. Phillips/Visuals Unlimited; 12.6, © Boehringer Ingelheim International GmbH. Photo by Lennart Nilsson, from* The Incredible Machine, *published by The National Geographic Society; 12.12(a), © Dennis Kunkel/Phototake.*

Chapter 13: *Fig. 13.4(b), © Cabisco/Visuals Unlimited; 13.11(b), Visuals Unlimited; 13.12(a), 13.12(b), © SIU/Visuals Unlimited; 13.12(c), © F.C.F. Earney/Visuals Unlimited; 13.12(d), © Fernando Guttierrez/ Phototake, NYC; 13.13(b), © Cabisco/Visuals Unlimited; un13-a, p. 392, © SIU/Peter Arnold, Inc., un13-b, p. 393, © W. Ober/Visuals Unlimited; un13-c, p. 393, © Sloop-Ober/Visuals Unlimited; un13-f, p. 393, Peter Arnold, Inc.*

Chapter 14: *Fig. 14.3(b), © John D. Cunningham/Visuals Unlimited; 14.5(a-c), © Boehringer Ingelheim International GmbH. Photo by Lennart Nilsson, from* The Incredible Machine, *published by The National Geographic Society; 14.12, © Secchi-Lecaque/Roussel-UCLAF/CNRI/ Science Photo Library/Photo Researchers; 14.13, © Baylor College of Medicine/Peter Arnold, Inc.; 14.15, © Boehringer Ingelheim International GmbH. Photo by Lennart Nilsson.*

Chapter 15: *Fig. 15.7(c), Science Photo Library/Photo Researchers; 15.8(a), © David M. Phillips/Visuals Unlimited; 15.12, © Paul Fry/Peter Arnold, Inc.; un15-a1, p. 450, © Martin Rotker/Phototake, NYC; un15-a2, p. 450, © Phototake, NYC; un15-b2, p. 462, courtesy of C. Petree & J. Holtmeier.*

Chapter 16: *Fig. 16.8(b-e), © Dr. Stanley L. Gibbs/Peter Arnold, Inc.; 16.14(b), © Fred Hossler/Visuals Unlimited; 16.20(c), © G. Shih-R. Kessel/Visuals Unlimited; 16.20(e), © VU/SIU/Visuals Unlimited; un16-a1, © SIU/Biomed Comm./Custom Medical Stock Photo; un16-a2, Custom Medical Stock Photo.*

Chapter 18: *Fig. 18.4(d), © Biophoto Associates/Photo Researchers; 18.5(d), © David M. Phillips/Visuals Unlimited; un18-a2, Peter Arnold, Inc.*

Chapter 19: *Fig. 19.2(c), 19.2(d), © Fred Hossler/Visuals Unlimited; 19.3(c), © David M. Phillips/Visuals Unlimited; un19-a1a, p. 579, © M. Long/Visuals Unlimited; un19-a1b, p. 579, © SIU/Visuals Unlimited; un19-a1c, p. 579, © SIU/Visuals Unlimited; un19-a1d, p. 579, © M. Long/Visuals Unlimited; un19-a1e, p. 579, © Stanley L. Fiegler/Visuals Unlimited.*

Chapter 20: *Fig. 20.1(c), 20.4(d), 20.6(b), 20.7(c), 20.8, © Lennart Nilsson, from the book* A Child Is Born, *Dell Publishing Company; 20.10(a-e), courtesy of Whiteley Illustrations Services; 20.13(a), 20.13(b), © CNRI/SPL/Science Source; 20.14(a), © Lennart Nilsson, from the book* A Child Is Born, *Dell Publishing Company; un20-a1, © Lennart Nilsson, from the book* A Child Is Born, *Dell Publishing Company.*

Boldface numbers indicate pages containing figures.